Student Solutions Manual

Laurel Technical Services

Intermediate Algebra

THIRD EDITION

K. Elayn Martin-Gay

Prentice Hall

Upper Saddle River, NJ 07458

Executive Editor: Karin E. Wagner
Project Manager: Mary Beckwith
Special Projects Manager: Barbara A. Murray
Production Editor: Wendy A. Perez
Supplement Cover Manager: Paul Gourhan
Supplement Cover Designer: PM Workshop Inc.
Manufacturing Buyer: Lisa McDowell

Printed in the United States of America

10 9 8 7 6

ISBN 0-13-017329-0

Prentice-Hall International (UK) Limited, London
Prentice-Hall of Australia Pty. Limited, Sydney
Prentice-Hall Canada, Inc., Toronto
Prentice-Hall Hispanoamericana, S.A., Mexico
Prentice-Hall of India Private Limited, New Delhi
Pearson Education Asia Pte. Ltd., Singapore
Prentice-Hall of Japan, Inc., Tokyo
Editora Prentice-Hall do Brazil, Ltda., Rio de Janeiro

Table of Contents

Chapter 1

Exercise Set 1.2

1. $5x = (5)(7) = 35$

3. $9.8z = (9.8)(3.1) = 30.38$

5. $ab = \left(\frac{1}{2}\right)\left(\frac{3}{4}\right) = \frac{3}{8}$

7. $3x + y = (3)(6) + 4 = 18 + 4 = 22$

9. $400t = (400)(5) = 2000$ miles

11. $lw = (5.1)(4) = 20.4$
The display needs 20.4 square feet of floor space.

13. $7098t = (7098)(5.2) = \$36{,}909.60$

15. $\{1, 2, 3, 4, 5\}$

17. $\{11, 12, 13, 14, 15, 16\}$

19. $\{0\}$

21. $\{0, 2, 4, 6, 8\}$

23. 

25. 

27. 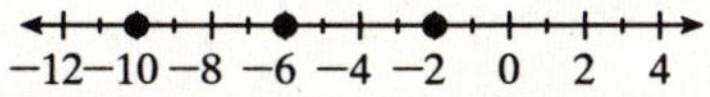

29. Answers may vary.

31. $\sqrt{36} = 6$
$\{3, 0, \sqrt{36}\}$

33. $\sqrt{36} = 6$
$\{3, \sqrt{36}\}$

35. $\{\sqrt{7}\}$

37. $-11 \in \{x | x \text{ is an integer}\}$

39. $0 \notin \{x | x \text{ is a positive integer}\}$

41. $12 \notin \{1, 3, 5, \ldots\}$

43. $0 \notin \{1, 2, 3, \ldots\}$

45. True; every integer is a real number.

47. True; −1 is an integer.

49. False; 0 is not a natural number.

51. False; $\sqrt{5}$ is an irrational number.

53. True; every natural number is an integer.

55. False; the number $\sqrt{7}$, for example, is a real number, but it is not a rational number.

57. Answers may vary.

59. $-|2| = -2$ since $|2| = 2$.

61. $|-4| = 4$ since −4 is located 4 units from 0 on the number line.

63. $|0| = 0$ since 0 is located 0 units from 0 on the number line.

65. $-|-3| = -3$ since $|-3| = 3$.

67. Answers may vary.

69. The opposite of −6.2 is $-(-6.2) = 6.2$.

71. The opposite of $\frac{4}{7}$ is $-\frac{4}{7}$.

73. The opposite of $-\frac{2}{3}$ is $-\left(-\frac{2}{3}\right) = \frac{2}{3}$.

75. The opposite of 0 is 0.

77. $2x$

79. $2x + 5$

81. $x - 10$

83. $x + 2$

85. $\frac{x}{11}$

87. $3x + 12$

89. $x - 17$

91. $2(x + 3)$

93. $\frac{5}{4 - x}$

95. China = 137
USA = 102
France = 93
Spain = 71
Hong Kong = 59

Exercise Set 1.3

1. $-3 + 8 = 5$

3. $-14 + (-10) = -24$

5. $-4.3 - 6.7 = -11$

7. $13 - 17 = -4$

9. $\frac{11}{15} - \left(-\frac{3}{5}\right) = \frac{11}{15} + \frac{9}{15} = \frac{20}{15} = \frac{4}{3}$

11. $19 - 10 - 11 = 9 - 11 = -2$

13. Since the signs of the two numbers are different, the product is negative. Thus $(-5)(12) = -60$.

15. Since the signs of the two numbers are the same, the product is positive. Thus $(-8)(-10) = 80$.

17. Since the signs of the two numbers are the same, the quotient is positive. Thus $\frac{-12}{-4} = 3$.

19. $\frac{0}{-2} = 0$

21. $(-4)(-2)(-1) = 8(-1) = -8$

23. $\frac{-6}{7} \div 2 = \frac{-6}{7} \cdot \frac{1}{2} = \frac{-3}{7} = -\frac{3}{7}$

25. $\left(-\frac{2}{7}\right)\left(-\frac{1}{6}\right) = \frac{2}{42} = \frac{1}{21}$

27. $-7^2 = -(7 \cdot 7) = -49$

29. $(-6)^2 = (-6)(-6) = 36$

31. $(-2)^3 = (-2)(-2)(-2) = 4(-2) = -8$

33. Answers may vary.

35. $\sqrt{49} = 7$ since 7 is positive and $7^2 = 49$.

37. $\sqrt{\frac{1}{9}} = \frac{1}{3}$ since $\frac{1}{3}$ is positive and $\left(\frac{1}{3}\right)^2 = \frac{1}{9}$.

39. $\sqrt[3]{64} = 4$ since $4^3 = 64$.

41. $\sqrt[4]{81} = 3$ since $3^4 = 81$.

43. $3(5 - 7)^4 = 3(-2)^4 = 3(16) = 48$

45. $-3^2 + 2^3 = -9 + 8 = -1$

47. $\frac{3-(-12)}{-5} = \frac{3+12}{-5} = \frac{15}{-5} = -3$

49. $|3.6 - 7.2| + |3.6 + 7.2| = |-3.6| + |10.8|$
$= 3.6 + 10.8$
$= 14.4$

51. $\frac{(3-\sqrt{9})-(-5-1.3)}{-3}=\frac{(3-3)-(-6.3)}{-3}$
$=\frac{0+6.3}{-3}$
$=\frac{6.3}{-3}$
$=-2.1$

53. $\frac{|3-9|-|-5|}{-3}=\frac{6-5}{-3}=\frac{1}{-3}=-\frac{1}{3}$

55. $(-3)^2+2^3=9+8=17$

57. $\frac{3(-2+1)}{5}-\frac{-7(2-4)}{1-(-2)}=\frac{3(-1)}{5}-\frac{-7(-2)}{1+2}$
$=\frac{-3}{5}-\frac{14}{3}$
$=\frac{-9}{15}-\frac{70}{15}$
$=-\frac{79}{15}$

59. $\frac{\left(\frac{-3}{10}\right)}{\left(\frac{42}{50}\right)}=-\frac{3}{10}\cdot\frac{50}{42}=-\frac{5}{14}$

61. $\frac{-1.682-17.895}{(-7.102)(-4.691)}=\frac{-19.577}{33.3155}=-0.5876$

63. $x^2+z^2=(-2)^2+3^2=4+9=13$

65. $-5(-x+3y)=-5(-(-2)+3(-5))$
$=-5(2-15)$
$=-5(-13)$
$=65$

67. $\frac{3z-y}{2x-z}=\frac{3(3)-(-5)}{2(-2)-3}$
$=\frac{9+5}{-4-3}$
$=\frac{14}{-7}$
$=-2$

69. $\frac{y_2-y_1}{x_2-x_1}=\frac{2-(-3)}{4-2}=\frac{2+3}{2}=\frac{5}{2}$

71. **a.**

y	5	7	10	100
$8+2y$	18	22	28	208

b. Increase

73. **a.**

x	10	100	1000
$\frac{100x+5000}{x}$	600	150	105

b. Decrease

75. $1-\frac{1}{5}-\frac{3}{7}=\frac{35}{35}-\frac{7}{35}-\frac{15}{35}=\frac{13}{35}$

77. 10,203 – 5998 = 4205 meters

79. b

81. d

83. Yes. Two players have 6 points each (third player has 0 points), or two players have 5 points each (third has 2 points).

85. $\sqrt{273}=16.5227$

87. $\sqrt{19.6}=4.4272$

89. 13.2%

91. 12.9% – 2.1% = 10.8%

Exercise Set 1.4

1. $4c=7$

3. $3(x+1)=7$

5. $\frac{n}{5}=4n$

7. $z-2=2z$

9. $0>-2$

11. $\frac{12}{3}=\frac{8}{2}$

13. $-7.9<-7.09$

15. $7x\le-21$

17. $-2 + x \neq 10$

19. $2(x-6) > \frac{1}{11}$

21. $y - 7 = 6$

23. $2(x - 6) = -27$

25. $-5; \frac{1}{5}$

27. $8; -\frac{1}{8}$

29. $\frac{1}{4}; -4$

31. 0; undefined

33. $-\frac{7}{8}; \frac{8}{7}$

35. Zero. For every real number x, $0 \cdot x \neq 1$, so 0 has no reciprocal. It is the only real number that has no reciprocal because if $x \neq 0$, then $x \cdot \frac{1}{x} = 1$ by definition.

37. $7x + y = y + 7x$

39. $z \cdot w = w \cdot z$

41. $\frac{1}{3} \cdot \frac{x}{5} = \frac{x}{5} \cdot \frac{1}{3}$

43. No, subtraction is not commutative. For example, $7 - 4 \neq 4 - 7$.

45. $5 \cdot (7x) = (5 \cdot 7)x$

47. $(x + 1.2) + y = x + (1.2 + y)$

49. $(14z) \cdot y = 14(z \cdot y)$

51. $12 - (5 - 3) = 10$; $(12 - 5) - 3 = 4$; Subtraction is not associative.

53. $3(x + 5) = 3 \cdot x + 3 \cdot 5 = 3x + 15$

55. $-(2a + b) = -1(2a + b) = -2a - b$

57. $2(6x + 5y + 2z) = 12x + 10y + 4z$

59. $2y - 6 = \frac{1}{8}$

61. $\frac{n+5}{2} > 2n$

63. $6 + 3x$

65. 0

67. 7

69. $(10 \cdot 2)y$

71. $a(b + c) = ab + ac$

73. In words:

Value of a dime	·	Number of dimes

Translate: $0.1 \cdot d$ or $0.1d$

75. If two numbers have a sum of 112 and one number is x, then the other number is the "rest of 112." So, in other words, we have

One hundred twelve	–	x

Translate: $112 - x$

77. In words:

Ninety	–	$5x$

Translate: $90 - 5x$

79. In words:

Cost of a book	·	Number of books

Translate: 35.61y$

81. The next even integer would be 2 more than the given even integer. In words:

Even integer	+	Two

Translate: $2x + 2$

83.
$$\begin{aligned} 5y - 14 + 7y - 20y &= 5y + 7y - 20y - 14 \\ &= (5 + 7 - 20)y - 14 \\ &= -8y - 14 \end{aligned}$$

85. $-11c-(4-2c)=-11c-4+2c$
$=-11c+2c-4$
$=(-11+2)c-4$
$=-9c-4$

87. $(8-5y)-(4+3y)=8-5y-4-3y$
$=-5y-3y+8-4$
$=(-5-3)y+4$
$=-8y+4$
$=4-8y$

89. $-4(y+3)-7y+1=-4y-12-7y+1$
$=-4y-7y-12+1$
$=(-4-7)y-11$
$=-11y-11$

91. $-(8-t)+(2t-6)=-8+t+2t-6$
$=t+2t-8-6$
$=(1+2)t-14$
$=3t-14$

93. $5(2z-6)+10(3-z)=10z-30+30-10z$
$=10z-10z-30+30$
$=(10-10)z+0$
$=0$

95. $7n+3(2n-6)-2=7n+6n-18-2$
$=(7+6)n-20$
$=13n-20$

97. $5.8(-9.6-31.2y)-18.65$
$=-55.68-180.96y-18.65$
$=-180.96y-55.68-18.65$
$=-180.96y-74.33$

99. $6.5y-4.4(1.8x-3.3)+10.95$
$=6.5y-7.92x+14.52+10.95$
$=6.5y-7.92x+25.47$

101. It is not the case that two rectangles with the same perimeter will necessarily have the same area. Take a rectangle that is 5 in. by 5 in. and another that is 8 in. by 2 in. Both have the same perimeter, but the areas are 25 square inches and 16 square inches respectively.

103. 70 million

105. 35 million

107. $(4.1\%)\cdot 3=12.3\%$

Chapter 1 Review Exercises

1. $7x=7(3)=21$

2. $st=(1.6)(5)=8$

3. One hour is 60(60) = 3600 seconds.
$90t=90(3600)=324{,}000$
324,000 wing beats per hour

4. $\{-1, 1, 3\}$

5. $\{-2, 0, 2, 4, 6\}$

6. $\varnothing$

7. $\varnothing$

8. $\{6, 7, 8, \ldots\}$

9. $\{\ldots, -1, 0, 1, 2\}$

10. True

11. False

12. True, since $\sqrt{169}=13$.

13. True, since zero is not an element of the empty set.

14. False, since π is irrational.

15. True, since π is a real number.

16. False, since $\sqrt{4}=2$.

17. True, since -9 is a rational number.

18. True

19. True, since C is not a subset of B.

20. False, since all integers are rational numbers.

21. True, since the empty set is a subset of all sets.

22. True, since every set is a subset of itself.

23. True, since every element of D is also an element of C.

24. True, since every integer is a real number.

25. True, since every irrational number is also a real number.

26. False, since B does not contain the set $\{5\}$.

27. True, since $\{5\}$ is a subset of B.

28. $\left\{5, \frac{8}{2}, \sqrt{9}\right\}$

29. $\left\{5, \frac{8}{2}, \sqrt{9}\right\}$

30. $\left\{5, -\frac{2}{3}, \frac{8}{2}, \sqrt{9}, 0.3, 1\frac{5}{8}, -1\right\}$

31. $\left\{\sqrt{7}, \pi\right\}$

32. $\left\{5, -\frac{2}{3}, \frac{8}{2}, \sqrt{9}, 0.3, \sqrt{7}, 1\frac{5}{8}, -1, \pi\right\}$

33. $\left\{5, \frac{8}{2}, \sqrt{9}, -1\right\}$

34. The opposite of $-\frac{3}{4}$ is $-\left(-\frac{3}{4}\right) = \frac{3}{4}$.

35. The opposite of 0.6 is –0.6.

36. The opposite of 0 is –0 = 0.

37. The opposite of 1 is –1.

38. The reciprocal of $-\frac{3}{4}$ is $\frac{1}{\left(-\frac{3}{4}\right)} = -\frac{4}{3}$.

39. The reciprocal of 0.6 is $\frac{1}{0.6}$.

40. The reciprocal of 0 is $\frac{1}{0}$ which is undefined.

41. The reciprocal of 1 is $\frac{1}{1} = 1$.

42. $-7 + 3 = -4$

43. $-10 + (-25) = -35$

44. $5(-0.4) = -2$

45. $(-3.1)(-0.1) = 0.31$

46. $-7 - (-15) = -7 + 15 = 8$

47. $9 - (-4.3) = 9 + 4.3 = 13.3$

48. $(-6)(-4)(0)(-3) = 0$

49. $(-12)(0)(-1)(-5) = 0$

50. $(-24) \div 0$ is undefined.

51. $0 \div (-45) = 0$

52. $(-36) \div (-9) = 4$

53. $(60) \div (-12) = -5$

54. $\left(-\frac{4}{5}\right) - \left(-\frac{2}{3}\right) = -\frac{12}{15} + \frac{10}{15} = -\frac{2}{15}$

55. $\left(\frac{5}{4}\right) - \left(-2\frac{3}{4}\right) = \frac{5}{4} + \frac{11}{4} = \frac{16}{4} = 4$

56. $1 - \frac{1}{4} - \frac{1}{3} = \frac{12}{12} - \frac{3}{12} - \frac{4}{12} = \frac{5}{12}$

57.
$$\begin{aligned} -5 + 7 - 3 - (-10) &= 2 - 3 + 10 \\ &= -1 + 10 \\ &= 9 \end{aligned}$$

58.
$$\begin{aligned} 8 - (-3) + (-4) + 6 &= 8 + 3 - 4 + 6 \\ &= 11 - 4 + 6 \\ &= 7 + 6 \\ &= 13 \end{aligned}$$

59. $3(4-5)^4 = 3(-1)^4 = 3(1) = 3$

60. $6(7-10)^2 = 6(-3)^2 = 6(9) = 54$

61. $\left(-\frac{8}{15}\right)\cdot\left(-\frac{2}{3}\right)^2 = -\frac{8}{15}\cdot\frac{4}{9} = -\frac{32}{135}$

62. $\left(-\frac{3}{4}\right)^2\cdot\left(-\frac{10}{21}\right) = \left(\frac{9}{16}\right)\left(-\frac{10}{21}\right) = -\frac{15}{56}$

63. $-\frac{6}{15}\div\frac{8}{25} = -\frac{6}{15}\cdot\frac{25}{8} = -\frac{150}{120} = -\frac{5}{4}$

64. $\frac{4}{9}\div-\frac{8}{45} = \frac{4}{9}\cdot\left(-\frac{45}{8}\right) = -\frac{180}{72} = -\frac{5}{2}$

65. $$\begin{aligned} -\frac{3}{8}+3(2)\div 6 &= -\frac{3}{8}+6\div 6 \\ &= -\frac{3}{8}+1 \\ &= -\frac{3}{8}+\frac{8}{8} \\ &= \frac{5}{8} \end{aligned}$$

66. $$\begin{aligned} 5(-2)-(-3)-\frac{1}{6}+\frac{2}{3} &= -10+3-\frac{1}{6}+\frac{2}{3} \\ &= -7-\frac{1}{6}+\frac{2}{3} \\ &= -\frac{42}{6}-\frac{1}{6}+\frac{4}{6} \\ &= -\frac{39}{6} \\ &= -6\frac{1}{2} \end{aligned}$$

67. $$\begin{aligned} \left|2^3-3^2\right|-|5-7| &= |8-9|-|-2| \\ &= |-1|-2 \\ &= 1-2 \\ &= -1 \end{aligned}$$

68. $$\begin{aligned} \left|5^2-2^2\right|+|9\div(-3)| &= |25-4|+|-3| \\ &= |21|+3 \\ &= 21+3 \\ &= 24 \end{aligned}$$

69. $$\begin{aligned} (2^3-3^2)-(5-7) &= (8-9)-(-2) \\ &= -1+2 \\ &= 1 \end{aligned}$$

70. $$\begin{aligned} (5^2-2^4)+[9\div(-3)] &= (25-16)+(-3) \\ &= 9+(-3) \\ &= 6 \end{aligned}$$

71. $$\begin{aligned} \frac{(8-10)^3-(-4)^2}{2+8(2)\div 4} &= \frac{(-2)^3-16}{2+16\div 4} \\ &= \frac{-8-16}{2+4} \\ &= \frac{-24}{6} \\ &= -4 \end{aligned}$$

72. $$\begin{aligned} \frac{(2+4)^2+(-1)^5}{12\div 2\cdot 3-3} &= \frac{6^2+(-1)}{6\cdot 3-3} \\ &= \frac{36-1}{18-3} \\ &= \frac{35}{15} \\ &= \frac{7}{3} \end{aligned}$$

73. $$\begin{aligned} \frac{(4-9)+4-9}{10-12\div 4\cdot 8} &= \frac{-5+4-9}{10-3\cdot 8} \\ &= -\frac{-1-9}{10-24} \\ &= \frac{-10}{-14} \\ &= \frac{5}{7} \end{aligned}$$

74. $$\begin{aligned} \frac{3-7-(7-3)}{15+30\div 6\cdot 2} &= \frac{-4-(4)}{15+5\cdot 2} \\ &= \frac{-8}{15+10} \\ &= -\frac{8}{25} \end{aligned}$$

75. $\frac{\sqrt{25}}{4+3\cdot 7} = \frac{5}{4+21} = \frac{5}{25} = \frac{1}{5}$

76. $\frac{\sqrt{64}}{24-8\cdot 2} = \frac{8}{24-16} = \frac{8}{8} = 1$

77. $x^2 - y^2 + z^2 = 0^2 - 3^2 + (-2)^2$
$= 0 - 9 + 4$
$= -5$

78. $\frac{5x+z}{2y} = \frac{5(0)+(-2)}{2(3)} = \frac{0-2}{6} = \frac{-2}{6} = -\frac{1}{3}$

79. $\frac{-7y-3z}{-3} = \frac{-7(3)-3(-2)}{-3}$
$= \frac{-21-(-6)}{-3}$
$= \frac{-21+6}{-3}$
$= \frac{-15}{-3}$
$= 5$

80. $(x - y + z)^2 = (0 - 3 + (-2))^2$
$= (-3-2)^2$
$= (-5)^2$
$= 25$

81. **a.** When $r = 1$,
$2\pi(1) \approx 2(3.14) = 6.28$.
When $r = 10$,
$2\pi r = 2\pi(10) \approx 20(3.14) = 62.8$.
When $r = 100$,
$2\pi r = 2\pi(100) \approx 200(3.14) = 628$.

r	1	10	100
$2\pi r$	6.28	62.8	628

b. As the radius increases, the circumference increases.

82. $5xy - 7xy + 3 - 2 + xy$
$= 5xy - 7xy + xy + 3 - 2$
$= (5 - 7 + 1)xy + (3 - 2)$
$= (-1)xy + 1$
$= -xy + 1$

83. $4x + 10x - 19x + 10 - 19$
$= (4 + 10 - 19)x + (10 - 19)$
$= -5x + (-9)$
$= -5x - 9$

84. $6x^2 + 2 - 4(x^2 + 1) = 6x^2 + 2 - 4x^2 - 4$
$= 6x^2 - 4x^2 + 2 - 4$
$= (6-4)x^2 + (2-4)$
$= 2x^2 - 2$

85. $-7(2x^2 - 1) - x^2 - 1$
$= -14x^2 + 7 - x^2 - 1$
$= -14x^2 - x^2 + 7 - 1$
$= (-14-1)x^2 + (7-1)$
$= -15x^2 + 6$

86. $(3.2x - 1.5) - (4.3x - 1.2)$
$= 3.2x - 1.5 - 4.3x + 1.2$
$= 3.2x - 4.3x - 1.5 + 1.2$
$= (3.2 - 4.3)x - 0.3$
$= -1.1x - 0.3$

87. $(7.6x + 4.7) - (1.9x + 3.6)$
$= 7.6x + 4.7 - 1.9x - 3.6$
$= 7.6x - 1.9x + 4.7 - 3.6$
$= (7.6 - 1.9)x + 1.1$
$= 5.7x + 1.1$

88. $12 = -4x$

89. $n + 2n = -15$

90. $4(y + 3) = -1$

91. $6(t - 5) = 4$

92. $z - 7 = 6$

93. $9x - 10 = 5$

94. $x - 5 \geq 12$

95. $-4 < 7y$

96. $\frac{2}{3} \neq 2\left(n + \frac{1}{4}\right)$

97. $t + 6 \leq -12$

98. Associative Property of Addition

99. Distributive Property

100. Additive Inverse Property

101. Commutative Property of Addition

102. Associative and Commutative Properties of Multiplication
To see this:
$(XY)Z = X(YZ) = (YZ)X$

103. Multiplicative Inverse Property

104. Multiplication Property of Zero

105. Associative Property of Multiplication

106. Additive Identity Property

107. Multiplicative Identity Property

108. $5x - 15z = 5(x - 3z)$

109. $(7 + y) + (3 + x) = (3 + x) + (7 + y)$

110. $0 = 2 + (-2)$, for example

111. $1 = 2 \cdot \frac{1}{2}$, for example

112. $[(3.4)(0.7)]5 = (3.4)[(0.7)5]$

113. $7 = 7 + 0$

114. $-9 > -12$

115. $0 > -6$

116. $-3 < -1$

117. $7 = |-7|$

118. $-5 < -(-5)$, since $-(-5) = 5$

119. $-(-2) > -2$, since $-(-2) = 2$

Chapter 1 Test

1. True; -2.3 lies to the right of -2.33 on the number line.

2. False; $-6^2 = -36$, while $(-6)^2 = 36$.

3. False; $-5 - 8 = -13$, while $-(5 - 8) = -(-3) = 3$.

4. False; $(-2)(-3)(0) = 0$, while $\frac{(-4)}{0}$ is undefined.

5. True; natural numbers are positive integers.

6. False; for example, $\frac{1}{2}$ is a rational number that is not an integer.

7. $5 - 12 \div 3(2) = 5 - 4 \cdot 2 = 5 - 8 = -3$

8. $$\begin{aligned} |4-6|^3 - (1-6^2) &= |-2|^3 - (1-36) \\ &= 2^3 - (-35) \\ &= 8 + 35 \\ &= 43 \end{aligned}$$

9. $$\begin{aligned} (4-9)^3 - |-4-6|^2 &= (-5)^3 - |-10|^2 \\ &= -125 - 10^2 \\ &= -125 - 100 \\ &= -225 \end{aligned}$$

10. $$\begin{aligned} &\left[3|4-5|^5 - (-9)\right] \div (-6) \\ &= \left[3|-1|^5 + 9\right] \div (-6) \\ &= (3 \cdot 1^5 + 9) \div (-6) \\ &= (3 + 9) \div (-6) \\ &= 12 \div (-6) \\ &= -2 \end{aligned}$$

11. $$\begin{aligned} \frac{6(7-9)^3 + (-2)}{(-2)(-5)(-5)} &= \frac{6(-2)^3 - 2}{10(-5)} \\ &= \frac{6(-8) - 2}{-50} \\ &= \frac{-48 - 2}{-50} \\ &= \frac{-50}{-50} \\ &= 1 \end{aligned}$$

12. $q^2 - r^2 = 4^2 - (-2)^2 = 16 - 4 = 12$

13. $\frac{5t - 3q}{3r - 1} = \frac{5(1) - 3(4)}{3(-2) - 1} = \frac{5 - 12}{-6 - 1} = \frac{-7}{-7} = 1$

14. a. When $x = 1$, $5.75x = 5.75(1) = 5.75$.
When $x = 3$, $5.75x = 5.75(3) = 17.25$.
When $x = 10$,
$5.75x = 5.75(10) = 57.50$.
When $x = 20$,
$5.75x = 5.75(20) = 115.00$.

x	1	3	10	20
$5.75x$	5.75	17.25	57.50	115.00

b. As the number of adults increases the total cost increases.

15. $2|x+5| = 30$

16. $\frac{(6-y)^2}{7} < -2$

17. $\frac{9z}{|-12|} \neq 10$

18. $3\left(\frac{n}{5}\right) = -n$

19. $20 = 2x - 6$

20. $-2 = \frac{x}{x+5}$

21. Distributive Property

22. Associative Property of Addition

23. Additive Inverse Property

24. Multiplication Property of Zero

25. $0.05n + 0.1d$

26. $4y^2 + 10 - 2(y^2 + 10)$
$= 4y^2 + 10 - 2y^2 - 20$
$= 4y^2 - 2y^2 + 10 - 20$
$= (4-2)y^2 - 10$
$= 2y^2 - 10$

27. $(8.3x - 2.9) - (9.6x - 4.8)$
$= 8.3x - 2.9 - 9.6x + 4.8$
$= 8.3x - 9.6x - 2.9 + 4.8$
$= (8.3 - 9.6)x + 1.9$
$= -1.3x + 1.9$

Chapter 2

Section 2.1

Mental Math

1. $3x+5x+6+15=8x+21$

2. $8y+3y+7+11=11y+18$

3. $5n+n+3-10=6n-7$

4. $m+2m+4-8=3m-4$

5. $8x-12x+5-6=-4x-1$

6. $4x-10x+13-16=-6x-3$

Exercise Set 2.1

1. $-3x=36$

$\frac{-3x}{-3}=\frac{36}{-3}$

$x=-12$

Check:

$-3(-12) \stackrel{?}{=} 36$

$36=36$ True

The solution is -12.

3. $x+2.8=1.9$

$x+2.8-2.8=1.9-2.8$

$x=-0.9$

Check:

$-0.9+2.8 \stackrel{?}{=} 1.9$

$1.9=1.9$ True

The solution is -0.9.

5. $5x-4=26$

$5x-4+4=26+4$

$5x=30$

$\frac{5x}{5}=\frac{30}{5}$

$x=6$

Check:

$5(6)-4 \stackrel{?}{=} 26$

$30-4 \stackrel{?}{=} 26$

$26=26$ True

The solution is 6.

7. $-4=3x+11$

$-4-11=3x+11-11$

$-15=3x$

$\frac{-15}{3}=\frac{3x}{3}$

$-5=x$

Check:

$-4 \stackrel{?}{=} 3(-5)+11$

$-4 \stackrel{?}{=} -15+11$

$-4=-4$ True

The solution is -5.

9. $-4.1-7z=3.6$

$-4.1-7z+4.1=3.6+4.1$

$-7z=7.7$

$\frac{-7z}{-7}=\frac{7.7}{-7}$

$z=-1.1$

Check:

$-4.1-7(-1.1) \stackrel{?}{=} 3.6$

$-4.1+7.7 \stackrel{?}{=} 3.6$

$3.6=3.6$ True

The solution is -1.1.

11. $5y+12=2y-3$

$5y+12-12=2y-3-12$

$5y=2y-15$

$5y-2y=2y-15-2y$

$3y=-15$

$\frac{3y}{3}=\frac{-15}{3}$

$y=-5$

Check:

$5(-5)+12 \stackrel{?}{=} 2(-5)-3$

$-25+12 \stackrel{?}{=} -10-3$

$-13=-13$ True

The solution is -5.

13. $8x-5x+3=x-7+10$
$3x+3=x+3$
$2x=0$
$x=0$

Check:
$8(0)-5(0)+3 \stackrel{?}{=} 0-7+10$
$0-0+3 \stackrel{?}{=} 3$
$3=3$ True

The solution is 0.

15. $5x+12=2(2x+7)$
$5x+12=4x+14$
$x=2$

Check:
$5(2)+12 \stackrel{?}{=} 2[2(2)+7]$
$10+12 \stackrel{?}{=} 2(4+7)$
$22 \stackrel{?}{=} 2(11)$
$22=22$ True

The solution is 2.

17. $3(x-6)=5x$
$3x-18=5x$
$-18=2x$
$-9=x$

Check:
$3(-9-6) \stackrel{?}{=} 5(-9)$
$3(-15) \stackrel{?}{=} -45$
$-45=-45$ True

The solution is –9.

19. $-2(5y-1)-y=-4(y-3)$
$-10y+2-y=-4y+12$
$-11y+2=-4y+12$
$-7y=10$
$y=-\frac{10}{7}$

Check:
$$-2\left[5\left(-\frac{10}{7}\right)-1\right]-\left(-\frac{10}{7}\right) \stackrel{?}{=} -4\left(-\frac{10}{7}-3\right)$$
$$-2\left(-\frac{50}{7}-1\right)+\frac{10}{7} \stackrel{?}{=} -4\left(-\frac{31}{7}\right)$$
$$-2\left(-\frac{57}{7}\right)+\frac{10}{7} \stackrel{?}{=} \frac{124}{7}$$
$$\frac{114}{7}+\frac{10}{7} \stackrel{?}{=} \frac{124}{7}$$
$$\frac{124}{7}=\frac{124}{7} \quad \text{True}$$

The solution is $-\frac{10}{7}$.

21. a. $4(x+1)+1=4x+4+1=4x+5$

b. $4(x+1)+1=-7$
$4x+5=-7$
$4x=-12$
$x=-3$

The solution is –3.

c. Answers may vary.

23.
$$\frac{x}{2}+\frac{2}{3}=\frac{3}{4}$$
$$12\left(\frac{x}{2}+\frac{2}{3}\right)=12\left(\frac{3}{4}\right)$$
$$6x+8=9$$
$$6x=1$$
$$x=\frac{1}{6}$$

Check:
$$\frac{\frac{1}{6}}{2}+\frac{2}{3} \stackrel{?}{=} \frac{3}{4}$$
$$\frac{1}{12}+\frac{2}{3} \stackrel{?}{=} \frac{3}{4}$$
$$\frac{1}{12}+\frac{8}{12} \stackrel{?}{=} \frac{3}{4}$$
$$\frac{9}{12} \stackrel{?}{=} \frac{3}{4}$$
$$\frac{3}{4}=\frac{3}{4} \quad \text{True}$$

The solution is $\frac{1}{6}$.

25. $\frac{3t}{4} - \frac{t}{2} = 1$

$4\left(\frac{3t}{4} - \frac{t}{2}\right) = 4(1)$

$3t - 2t = 4$

$t = 4$

Check:

$\frac{3(4)}{4} - \frac{4}{2} \stackrel{?}{=} 1$

$3 - 2 \stackrel{?}{=} 1$

$1 = 1$ True

The solution is 4.

27. $\frac{n-3}{4} + \frac{n+5}{7} = \frac{5}{14}$

$28\left(\frac{n-3}{4}\right) + 28\left(\frac{n+5}{7}\right) = 28\left(\frac{5}{14}\right)$

$7(n-3) + 4(n+5) = 2(5)$

$7n - 21 + 4n + 20 = 10$

$11n - 1 = 10$

$11n = 11$

$n = 1$

Check:

$\frac{1-3}{4} + \frac{1+5}{7} \stackrel{?}{=} \frac{5}{14}$

$\frac{-2}{4} + \frac{6}{7} \stackrel{?}{=} \frac{5}{14}$

$-\frac{1}{2} + \frac{6}{7} \stackrel{?}{=} \frac{5}{14}$

$-\frac{7}{14} + \frac{12}{14} \stackrel{?}{=} \frac{5}{14}$

$\frac{5}{14} = \frac{5}{14}$ True

The solution is 1.

29. $0.6x - 10 = 1.4x - 14$

$4 = 0.8x$

$5 = x$

Check:

$0.6(5) - 10 \stackrel{?}{=} 1.4(5) - 14$

$3 - 10 \stackrel{?}{=} 7 - 14$

$-7 = -7$ True

The solution is 5.

31. $4(n+3) = 2(6+2n)$

$4n + 12 = 12 + 4n$

$4n + 12 = 4n + 12$

This is true for all n.

Therefore, all real numbers are solutions.

33. $3(x-1) + 5 = 3x + 7$

$3x - 3 + 5 = 3x + 7$

$3x + 2 = 3x + 7$

$3x + 2 - 3x = 3x + 7 - 3x$

$2 = 7$

This is false for any x. Therefore, the solution set is $\varnothing$.

35. Answers may vary.

37. $-9x = -72$

$x = 8$

The solution is 8.

39. $x - 1.7 = -7.6$

$x = -5.9$

The solution is –5.9.

41. $6x + 9 = 51$

$6x = 42$

$x = 7$

The solution is 7.

43. $-5x + 1.5 = -19.5$

$-5x = -21$

$x = 4.2$

The solution is 4.2.

45. $x - 10 = -6x + 4$

$7x = 14$

$x = 2$

The solution is 2.

47. $3x - 4 - 5x = x + 4 + x$

$-2x - 4 = 2x + 4$

$-4x = 8$

$x = -2$

The solution is –2.

49. $5(y+4) = 4(y+5)$

$5y + 20 = 4y + 20$

$y = 0$

The solution is 0.

51. $$0.6x-10=1.4x-14$$
$$4=0.8x$$
$$x=5$$
The solution is 5.

53. $$6x-2(x-3)=4(x+1)+4$$
$$6x-2x+6=4x+4+4$$
$$4x+6=4x+8$$
$$6=8$$
Therefore, no solution exists.

55. $$\frac{3}{8}+\frac{b}{3}=\frac{5}{12}$$
$$24\left(\frac{3}{8}+\frac{b}{3}\right)=24\left(\frac{5}{12}\right)$$
$$9+8b=10$$
$$8b=1$$
$$b=\frac{1}{8}$$
The solution is $\frac{1}{8}$.

57. $$z+3(2+4z)=6(z+1)+5z$$
$$z+6+12z=6z+6+5z$$
$$13z+6=11z+6$$
$$2z=0$$
$$z=0$$
The solution is 0.

59. $$\frac{3t+1}{8}=\frac{5+2t}{7}+2$$
$$56\left(\frac{3t+1}{8}\right)=56\left(\frac{5+2t}{7}+2\right)$$
$$7(3t+1)=8(5+2t)+112$$
$$21t+7=40+16t+112$$
$$21t+7=16t+152$$
$$5t=145$$
$$t=29$$
The solution is 29.

61. $$\frac{m-4}{3}-\frac{3m-1}{5}=1$$
$$15\left(\frac{m-4}{3}-\frac{3m-1}{5}\right)=15(1)$$
$$5(m-4)-3(3m-1)=15$$
$$5m-20-9m+3=15$$
$$-4m-17=15$$
$$-4m=32$$
$$m=-8$$
The solution is –8.

63. $$5(x-2)+2x=7(x+4)-38$$
$$5x-10+2x=7x+28-38$$
$$7x-10=7x-10$$
$$0=0$$
This is true for all x.
Therefore, all real numbers are solutions.

65. $$y+0.2=0.6(y+3)$$
$$y+0.2=0.6y+1.8$$
$$0.4y=1.6$$
$$y=4$$
The solution is 4.

67. $$2y+5(y-4)=4y-2(y-10)$$
$$2y+5y-20=4y-2y+20$$
$$7y-20=2y+20$$
$$5y=40$$
$$y=8$$
The solution is 8.

69. $$2(x-8)+x=3(x-6)+2$$
$$2x-16+x=3x-18+2$$
$$3x-16=3x-16$$
This is true for all x.
Therefore, all real numbers are solutions.

71. $$\frac{3x-1}{9}+x=\frac{3x+1}{3}+4$$
$$9\left(\frac{3x-1}{9}+x\right)=9\left(\frac{3x+1}{3}+4\right)$$
$$(3x-1)+9x=3(3x+1)+36$$
$$3x-1+9x=9x+3+36$$
$$12x-1=9x+39$$
$$3x=40$$
$$x=\frac{40}{3}$$
The solution is $\frac{40}{3}$.

73.
$$1.5(4-x)=1.3(2-x)$$
$$10[1.5(4-x)]=10[1.3(2-x)]$$
$$15(4-x)=13(2-x)$$
$$60-15x=26-13x$$
$$-2x=-34$$
$$x=17$$
The solution is 17.

75.
$$-2(b-4)-(3b-1)=5b+3$$
$$-2b+8-3b+1=5b+3$$
$$-5b+9=5b+3$$
$$-10b=-6$$
$$b=\frac{6}{10}=\frac{3}{5}$$
The solution is $\frac{3}{5}$.

77.
$$\frac{1}{4}(a+2)=\frac{1}{6}(5-a)$$
$$12\cdot\frac{1}{4}(a+2)=12\cdot\frac{1}{6}(5-a)$$
$$3(a+2)=2(5-a)$$
$$3a+6=10-2a$$
$$5a=4$$
$$a=\frac{4}{5}$$
The solution is $\frac{4}{5}$.

79.
$$3.2x+4=5.4x-7$$
$$3.2x+4-4=5.4x-7-4$$
$$3.2x=5.4x-11$$
From this we see that $K=-11$.

81.
$$\frac{x}{6}+4=\frac{x}{3}$$
$$6\left(\frac{x}{6}+4\right)=6\left(\frac{x}{3}\right)$$
$$x+24=2x$$
From this we see that $K=24$.

83.
$$2.569x=-12.48534$$
$$\frac{2.569x}{2.569}=\frac{-12.48534}{2.569}$$
$$x=-4.86$$
The solution is –4.86.

85.
$$2.86z-8.1258=-3.75$$
$$2.86z=4.3758$$
$$\frac{2.86z}{2.86}=\frac{4.3758}{2.86}$$
$$z=1.53$$
The solution is 1.53.

87. Not a fair game

89. $8+x$

91. $8-x$

93. $2x-5$

95.
$$7x^2+2x-3=6x(x+4)+x^2$$
$$7x^2+2x-3=6x^2+24x+x^2$$
$$7x^2+2x-3=7x^2+24x$$
$$2x-3=24x$$
$$-3=22x$$
$$-\frac{3}{22}=x$$
The solution is $-\frac{3}{22}$.

97.
$$x(x+1)+16=x(x+5)$$
$$x^2+x+16=x^2+5x$$
$$x+16=5x$$
$$16=4x$$
$$4=x$$
The solution is 4.

Exercise Set 2.2

1. $y+y+y+y=4y$

3. $z+(z+1)+(z+2)=3z+3$

5. $5x+10(x+3)=5x+10x+30$
$=(15x+30)$ cents

7. $4x+3(2x+1)=4x+6x+3=10x+3$

9.
$$4(x-2)=2+6x$$
$$4x-8=2+6x$$
$$-10=2x$$
$$-5=x$$
The number is –5.

11. $x+5x=270$
$6x=270$
$x=45$
$5x=5(45)=225$
The numbers are 45 and 225.

13. $30\% \cdot 260 = 0.30 \cdot 260 = 78$

15. $12\% \cdot 16 = 0.12 \cdot 16 = 1.92$

17. $29\% \cdot 2271 = 0.29 \cdot 2271 = 658.59$
Approximately 658.59 million acres are federally owned.

19. $47\% \cdot 110{,}000 = 0.47 \cdot 110{,}000 = 51{,}700$
You would expect 51,700 homes to have computers.

21. $100 - 12 - 39 - 8 - 21 = 20$
20% of credit union loans are for credit cards and other unsecured loans.

23. $39\% \cdot 300 = 0.39 \cdot 300 = 117$
We would expect 117 of the loans to be automobile loans.

25. $3x+17.5=199$
$3x=181.5$
$x=60.5$
$x+7.7=60.5+7.7=68.2$
$x+9.8=60.5+9.8=70.3$
The Dallas/Ft. Worth airport has 60.5 million annual arrivals and departures. The Atlanta airport has 68.2 million annual arrivals and departures. The Chicago airport has 70.3 million annual arrivals and departures.

27. Let x = seats in B737-200;
then $x + 104$ = seats in B767-300ER.
$x+x+104=328$
$2x+104=328$
$2x=224$
$x=112$
$x+104=112+104=216$
The B737-200 aircraft has 112 seats.
The B767-300ER aircraft has 216 seats.

29. Let x = price before taxes.
$x+0.08x=464.40$
$1.08x=464.40$
$x=430$
The price was $430 before taxes.

31. a. Let x = number of telephone company operators in 1996
$x-0.47x=26{,}000$
$0.53x=26{,}000$
$x\approx 49{,}057$
There were approximately 49,057 operators in 1996.

b. Answers may vary.

33. $20-75\%\cdot 20 = 20-0.75\cdot 20$
$=20-15$
$=5$
The lifespan is about 5 years.

35. Let x = number in millions of returns electronically filed in 1998; then $x+.24x$ = number in 1999.
$x+.24x=21.1$
$1.24x=21.1$
$x\approx 17$
There were approximately 17 million electronically filed returns in 1998.

37. Let x = length of a side of the square; then $x+6$ = length of a side of the triangle.
$4x=3(x+6)$
$4x=3x+18$
$x=18$
$x+6=24$
The square's sides are 18 centimeters.
The triangle's sides are 24 centimeters.

39. Let x = width of room;
then $2x + 2$ = length of room.
$2x+2(2x+2)=40$
$2x+4x+4=40$
$6x+4=40$
$6x=36$
$x=6$
$2x+2=14$
The width is 6 centimeters and the length is 14 centimeters.

41. Let x = width; then $5(x + 1)$ = height.
$$x+5(x+1)=55.4$$
$$x+5x+5=55.4$$
$$6x+5=55.4$$
$$6x=50.4$$
$$x=8.4$$
$5(x + 1) = 47$
The width is 8.4 meters and the height is 47 meters.

43. Let x = measure of second angle; then $2x$ = measure of first angle, and $3x - 12$ = measure of third angle.
$$x+2x+3x-12=180$$
$$6x-12=180$$
$$6x=192$$
$$x=32$$
$2x = 2 \cdot 32 = 64$
$3x - 12 = 3 \cdot 32 - 12 = 84$
The angles measure 64°, 32°, and 84°.

45.
$$x+20+x=180$$
$$2x+20=180$$
$$2x=160$$
$$x=80$$
$x + 20 = 80 + 20 = 100$
The angles measure 80° and 100°.

47.
$$x+5x=90$$
$$6x=90$$
$$x=15$$
$5x = 5 \cdot 15 = 75$
The angles measure 15° and 75°.

49. Let x = measure of the angle; then $180 - x$ = measure of its supplement.
$$x=3(180-x)+20$$
$$x=540-3x+20$$
$$4x=560$$
$$x=140$$
$180 - x = 180 - 140 = 40$
The angles measure 140° and 40°.

51. Let x = first integer; then $(x + 1)$ = next consecutive integer, and $(x + 2)$ = third consecutive integer.
$$x+(x+1)+(x+2)=228$$
$$3x+3=228$$
$$3x=225$$
$$x=75$$
$x + 1 = 75 + 1 = 76$
$x + 2 = 75 + 2 = 77$
The integers are 75, 76 and 77.

53. Let x = 1st integer; then $x + 2$ = 2nd integer, and $x + 4$ = 3rd integer.
$$2x+x+4=268,222$$
$$3x+4=268,222$$
$$3x=268,218$$
$$x=89,406$$
$x + 2 = 89,406 + 2 = 89,408$
$x + 4 = 89,406 + 4 = 89,410$
Fallon's zip code is 89406, Fernley's is 89408, and Gardnerville's is 89410.

55. Let x = 1st integer; then $x + 1$ = 2nd integer, and $x + 2$ = 3rd integer.
$$x+x+1+x+2=3(x+1)$$
$$3x+3=3x+3$$
Since this is an identity, any three consecutive integers will work.

57.
$$24x=100+20x$$
$$4x=100$$
$$x=25$$
25 skateboards

59.
$$7.50x=4.50x+2400$$
$$3x=2400$$
$$x=800$$
800 books

61. It loses money.

63. Let x = number (in millions) of trees' worth of newsprint recycled; then $x + 30$ = number (in millions) of trees' worth of newsprint either recycled or discarded.

$$x = 0.27(x+30)$$
$$x = 0.27x + 8.1$$
$$0.73x = 8.1$$
$$x \approx 11$$

About 11 million trees' worth of newsprint is recycled.

65. $2a+b-c = 2(5)+(-1)-3$
$= 10-1-3$
$= 6$

67. $4ab-3bc = 4(-5)(-8)-3(-8)(2)$
$= 160+48$
$= 208$

69. $n^2-m^2 = (-3)^2-(-8)^2 = 9-64 = -55$

71. $P + Prt = 3000 + 3000(0.0325)(2) = 3195$

73. $\frac{1}{3}Bh = \frac{1}{3}(53.04)(6.89) = 121.8152$

Section 2.3

Mental Math

1. $2x+y=5$
$y=5-2x$

2. $7x-y=3$
$7x-3=y$ or $y=7x-3$

3. $a-5b=8$
$a=5b+8$

4. $7r+s=10$
$s=10-7r$

5. $5j+k-h=6$
$k=h-5j+6$

6. $w-4y+z=0$
$z=4y-w$

Exercise Set 2.3

1. $D=rt$
$\frac{D}{r}=\frac{rt}{r}$
$\frac{D}{r}=t$
$t=\frac{D}{r}$

3. $I=PRT$
$\frac{I}{PT}=\frac{PRT}{PT}$
$\frac{I}{PT}=R$
$R=\frac{I}{PT}$

5. $9x-4y=16$
$9x-4y-9x=16-9x$
$-4y=16-9x$
$\frac{-4y}{-4}=\frac{16-9x}{-4}$
$y=\frac{9x-16}{4}$

7. $P=2L+2W$
$P-2L=2W$
$\frac{P-2L}{2}=\frac{2W}{2}$
$\frac{P-2L}{2}=W$
$W=\frac{P-2L}{2}$

9. $J=AC-3$
$J+3=AC$
$\frac{J+3}{C}=\frac{AC}{C}$
$\frac{J+3}{C}=A$
$A=\frac{J+3}{C}$

11.
$$\begin{aligned} W &= gh - 3gt^2 \\ W &= g(h - 3t^2) \\ \frac{W}{h-3t^2} &= \frac{g(h-3t^2)}{h-3t^2} \\ \frac{W}{h-3t^2} &= g \\ g &= \frac{W}{h-3t^2} \end{aligned}$$

13.
$$\begin{aligned} T &= C(2 + AB) \\ T &= 2C + ABC \\ T - 2C &= 2C + ABC - 2C \\ T - 2C &= ABC \\ \frac{T-2C}{AC} &= \frac{ABC}{AC} \\ \frac{T-2C}{AC} &= B \\ B &= \frac{T-2C}{AC} \end{aligned}$$

15.
$$\begin{aligned} C &= 2\pi r \\ \frac{C}{2\pi} &= \frac{2\pi r}{2\pi} \\ \frac{C}{2\pi} &= r \\ r &= \frac{C}{2\pi} \end{aligned}$$

17.
$$\begin{aligned} E &= I(r + R) \\ E &= Ir + IR \\ E - IR &= Ir + IR - IR \\ E - IR &= Ir \\ \frac{E-IR}{I} &= \frac{Ir}{I} \\ r &= \frac{E-IR}{I} \end{aligned}$$

19.
$$\begin{aligned} s &= \frac{n}{2}(a + L) \\ 2 \cdot s &= 2 \cdot \frac{n}{2}(a + L) \\ 2s &= n(a + L) \\ 2s &= an + Ln \\ 2s - an &= an + Ln - an \\ 2s - an &= Ln \\ \frac{2s-an}{n} &= \frac{Ln}{n} \\ L &= \frac{2s-na}{n} \end{aligned}$$

21.
$$\begin{aligned} N &= 3st^4 - 5sv \\ N - 3st^4 &= 3st^4 - 5sv - 3st^4 \\ N - 3st^4 &= -5sv \\ \frac{N-3st^4}{-5s} &= \frac{-5sv}{-5s} \\ \frac{3st^4-N}{5s} &= v \\ v &= \frac{3st^4-N}{5s} \end{aligned}$$

23.
$$\begin{aligned} S &= 2LW + 2LH + 2WH \\ S - 2LW &= 2LW + 2LH + 2WH - 2LW \\ S - 2LW &= 2LH + 2WH \\ S - 2LW &= H(2L + 2W) \\ \frac{S-2LW}{2L+2W} &= \frac{H(2L+2W)}{2L+2W} \\ \frac{S-2LW}{2L+2W} &= H \\ H &= \frac{S-2LW}{2L+2W} \end{aligned}$$

25. $A = P\left(1+\frac{r}{n}\right)^{nt} = 3500\left(1+\frac{0.03}{n}\right)^{10n}$

n	1	2	4	12	365
A	$4703.71	$4713.99	$4719.22	$4722.74	$4724.45

27. 12 times a year; you earn more interest

29. $A = P\left(1+\frac{r}{n}\right)^{nt} = 6000\left(1+\frac{0.04}{n}\right)^{5n}$

a. $n = 2$

$A = 6000\left(1+\frac{0.04}{2}\right)^{5\cdot 2} \approx 7313.97$

$7313.97

b. $n = 4$

$A = 6000\left(1+\frac{0.04}{4}\right)^{5\cdot 4} \approx 7321.14$

$7321.14

c. $n = 12$

$A = 6000\left(1+\frac{0.04}{12}\right)^{5\cdot 12} \approx 7325.98$

$7325.98

31. $C = \frac{5}{9}(F-32)$

$C = \frac{5}{9}(104-32)$

$C = \frac{5}{9}(72)$

$C = 40°$

The day's high temperature was 40°C.

33. $d = rt$ or $t = \frac{d}{r}$ where $d = 2(90) = 180$ mi and $r = 50$ mi/hr, so $t = \frac{180}{50} = 3.6$ hr

Thus, she takes 3.6 hours or 3 hours, 36 minutes to make the round trip.

35. $A = s^2$ where $s = 64$ ft, so $A = 64^2 = 4096$ ft^2

Thus, $\frac{4096}{24} \approx 171$ packages of tiles should be bought.

37. $V = \pi r^2 h$
1 mile = 5280 feet
1.3 miles = 6864 feet
$3800 = \pi r^2(6864)$
$0.42 = r$
The radius of the hole is 0.42 feet.

39. $\frac{168 \text{ mi}}{1 \text{ hr}} \cdot \frac{5280 \text{ ft}}{1 \text{ mi}} \cdot \frac{1 \text{ hr}}{60 \text{ min}} \cdot \frac{1 \text{ min}}{60 \text{ sec}}$
$= 246.4$ ft/sec
$d = rt$
$60.5 \text{ feet} = 246.4 \text{ ft / sec} \cdot t$
$.25 \text{ sec} = t$
The ball would reach the plate in approximately 0.25 seconds.

41. $C = \pi d$
$= \pi \cdot 41.125$
$= 41.125\pi$ ft
≈ 129.1325 ft
The circumference of Eartha is $41.125\pi \approx 129.1325$ feet.

43. cost per person $= \frac{\$1.7 \text{ billion}}{250 \text{ million people}}$
$= \$6.80$ per person

45. $A = P\left(1 + \frac{r}{n}\right)^{nt}$

$A = 10,000\left(1 + \frac{0.085}{4}\right)^{4 \cdot 2}$
$= 10,000(1 + 0.02125)^8$
$= 10,000(1.183195)$
$= \$11,831.96$
$11,831.96 – $10,000 = $1831.96

47. The area of one pair of walls is $2 \cdot 14 \cdot 8 = 224 \text{ ft}^2$ and the area of the other walls is $2 \cdot 16 \cdot 8 = 256 \text{ ft}^2$ for a total of 480 ft^2. Multiplying by 2, the number of coats, yields 960 ft^2. Dividing this by 500 yields 1.92. Thus, 2 gallons should be purchased.

49. a. $V = \pi r^2 h$
$V = \pi(4.2)^2(21.2)$
$V = 1174.86$
The volume of the cylinder is 1174.86 cubic meters.

b. $V = \frac{4}{3}\pi r^3$
$V = \frac{4}{3}\pi(4.2)^3$
$V = 310.34$
The volume of the sphere is 310.34 cubic meters.

c. $V = 1174.86 + 310.34 = 1485.20$
The volume of the tank is 1485.20 cubic meters.

51. $d = rt$ or $t = \frac{d}{r}$ where d = 135 mi and r = 60 mi/hr so $t = \frac{135}{60} = 2.25$ hr
Thus, it takes him 2.25 hours or 2 hours and 15 minutes.

53. $80.5 \text{ months} \cdot \frac{30 \text{ days}}{1 \text{ month}} \cdot \frac{24 \text{ hours}}{1 \text{ day}}$
$= 57,960$ hours
$d = rt$ or $r = \frac{d}{t}$
$r = \frac{2 \text{ billion}}{57,960} \approx 34,507$ miles per hour

55.

Planet	AU from Sun
Mercury	0.388
Venus	0.723
Earth	1.000
Mars	1.523
Jupiter	5.202
Saturn	9.538
Uranus	19.193
Neptune	30.065
Pluto	39.505

57. $\frac{1}{4}$

59. $\frac{3}{8}$

61. $\frac{3}{8}$

63. $\frac{3}{4}$

65. 1

67. 1

69. $\{2, 3\}$

71. $x - 3 \geq -7$
$x \geq -4$
$\{-3, -2, -1, 0, 1, 2, 3\}$

73. Answers may vary.

Section 2.4

Mental Math

1. $x - 2 < 4$
$x < 6$
$\{x|x < 6\}$

2. $x - 1 > 6$
$x > 7$
$\{x|x > 7\}$

3. $x + 5 \geq 15$
$x \geq 10$
$\{x|x \geq 10\}$

4. $x + 1 \leq 8$
$x \leq 7$
$\{x|x \geq 10\}$

5. $3x > 12$
$x > 4$
$\{x|x > 4\}$

6. $5x < 20$
$x < 4$
$\{x|x < 4\}$

7. $\frac{x}{2} \leq 1$
$x \leq 2$
$\{x|x \leq 2\}$

8. $\frac{x}{4} \geq 2$
$x \geq 8$
$\{x|x \geq 8\}$

Exercise Set 2.4

1. $(-\infty, -3)$

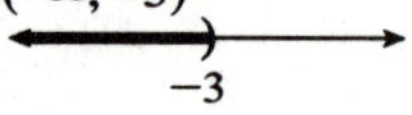

3. $[0.3, \infty)$

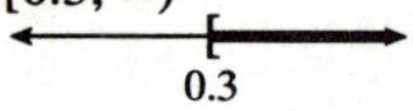

5. $(5, \infty)$

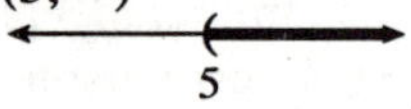

7. $(-2, 5)$

9. $(-1, 5)$

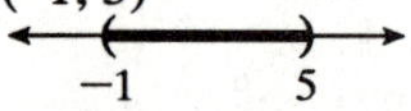

11. Answers may vary.

13. The only solution is 100.

15. The solutions are –5 and –1,000,000.

17. $x - 7 \geq -9$
$x \geq -2$
$[-2, \infty)$

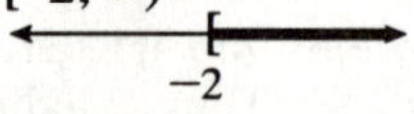

19. $7x < 6x + 1$
$x < 1$
$(-\infty, 1)$

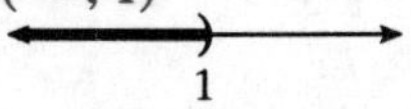

21. $8x - 7 \le 7x - 5$
$x - 7 \le -5$
$x \le 2$
$(-\infty, 2]$

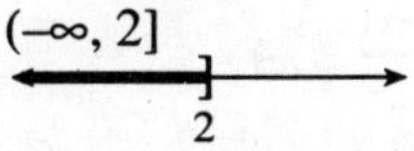

23. $2 + 4x > 5x + 6$
$-4 > x$
$(-\infty, -4)$

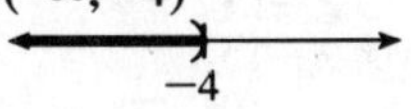

25. $\frac{3}{4}x \ge 2$
$x \ge \frac{8}{3}$
$\left[\frac{8}{3}, \infty\right)$

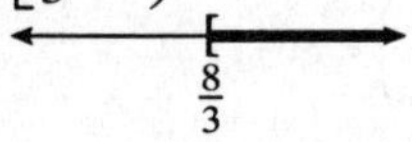

27. $5x < -23.5$
$x < -4.7$
$(-\infty, -4.7)$

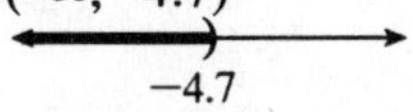

29. $-3x \ge 9$
$x \le -3$
$(-\infty, -3]$

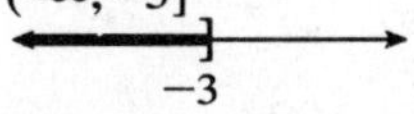

31. $-x < -4$
$x > 4$
$(4, \infty)$

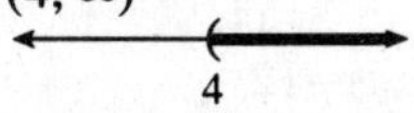

33. $-2x + 7 \ge 9$
$-2x \ge 2$
$x \le -1$
$(-\infty, -1]$

35. $15 + 2x \ge 4x - 7$
$15 \ge 2x - 7$
$22 \ge 2x$
$11 \ge x$
$(-\infty, 11]$

37. $3(x - 5) < 2(2x - 1)$
$3x - 15 < 4x - 2$
$-15 < x - 2$
$-13 < x$
$(-13, \infty)$

39. $\frac{1}{2} + \frac{2}{3} \ge \frac{x}{6}$
$6\left(\frac{1}{2} + \frac{2}{3}\right) \ge 6\left(\frac{x}{6}\right)$
$3 + 4 \ge x$
$7 \ge x$
$(-\infty, 7]$

41. $4(x - 1) \ge 4x - 8$
$4x - 4 \ge 4x - 8$
$-4 \ge -8$
All real numbers
$(-\infty, \infty)$

43. $7x < 7(x - 2)$
$7x < 7x - 14$
$0 < -14$
No solution
$\varnothing$

45. $4(2x + 1) > 4$
$2x + 1 > 1$
$2x > 0$
$x > 0$
$(0, \infty)$

47. $\frac{x + 7}{5} > 1$
$x + 7 > 5$
$x > -2$
$(-2, \infty)$

49. $\frac{-5x+11}{2} \le 7$
$-5x+11 \le 14$
$-5x \le 3$
$x \ge -\frac{3}{5}$
$\left[-\frac{3}{5}, \infty\right)$

51. $8x-16.4 \le 10x+2.8$
$-19.2 \le 2x$
$-9.6 \le x$
$x \ge -9.6$
$[-9.6, \infty)$

53. $2(x-3) > 70$
$x-3 > 35$
$x > 38$
$(38, \infty)$

55. Answers may vary.

57. $-5x+4 \le -4(x-1)$
$-5x+4 \le -4x+4$
$0 \le x$
$x \ge 0$
$[0, \infty)$

59. $\frac{1}{4}(x-7) \ge x+2$
$x-7 \ge 4x+8$
$-15 \ge 3x$
$-5 \ge x$
$x \le -5$
$(-\infty, -5]$

61.
$\frac{2}{3}(x+2) < \frac{1}{5}(2x+7)$
$15\left(\frac{2}{3}(x+2)\right) < 15\left(\frac{1}{5}(2x+7)\right)$
$10(x+2) < 3(2x+7)$
$10x+20 < 6x+21$
$4x < 1$
$x < \frac{1}{4}$
$\left(-\infty, \frac{1}{4}\right)$

63. $4(x-6)+2x-4 \ge 3(x-7)+10x$
$4x-24+2x-4 \ge 3x-21+10x$
$6x-28 \ge 13x-21$
$-7 \ge 7x$
$-1 \ge x$
$x \le -1$
$(-\infty, -1]$

65. $\frac{5x+1}{7} - \frac{2x-6}{4} \ge -4$
$28\left(\frac{5x+1}{7} - \frac{2x-6}{4}\right) \ge 28(-4)$
$4(5x+1)-7(2x-6) \ge -112$
$20x+4-14x+42 \ge -112$
$6x+46 \ge -112$
$6x \ge -158$
$x \ge -\frac{79}{3}$
$\left[-\frac{79}{3}, \infty\right)$

67. $\frac{-x+2}{2} - \frac{1-5x}{8} < -1$
$8\left(\frac{-x+2}{2} - \frac{1-5x}{8}\right) < 8(-1)$
$4(-x+2)-(1-5x) < -8$
$-4x+8-1+5x < -8$
$x+7 < -8$
$x < -15$
$(-\infty, -15)$

69. $0.8x+0.6x \ge 4.2$
$1.4x \ge 4.2$
$x \ge 3$
$[3, \infty)$

71. $\frac{x+5}{5} - \frac{3+x}{8} \ge -\frac{3}{10}$
$40\left(\frac{x+5}{5} - \frac{3+x}{8}\right) \ge 40\left(-\frac{3}{10}\right)$
$8(x+5)-5(3+x) \ge 4(-3)$
$8x+40-15-5x \ge -12$
$3x+25 \ge -12$
$3x \ge -37$
$x \ge -\frac{37}{3}$
$\left[-\frac{37}{3}, \infty\right)$

73. $\frac{x+3}{12}+\frac{x-5}{15}<\frac{2}{3}$

$60\left(\frac{x+3}{12}+\frac{x-5}{15}\right)<60\left(\frac{2}{3}\right)$

$5(x+3)+4(x-5)<20(2)$

$5x+15+4x-20<40$

$9x-5<40$

$9x<45$

$x<5$

$(-\infty, 5)$

75. Let x = her score on the final. Then

$\frac{72+67+82+79+2x}{6}\geq 60$

$300+2x\geq 360$

$2x\geq 60$

$x\geq 30$

Therefore, she must score at least 30 on the final exam.

77. Let x = the weight of the luggage and cargo. Then

$6(160)+x\leq 2000$

$960+x\leq 2000$

$x\leq 1040$

The plane can carry a maximum of 1040 pounds of luggage and cargo.

79. $0.33+0.22(x-1)\leq 4.00$

$100[0.33+0.22(x-1)]\leq 100(4.00)$

$33+22(x-1)\leq 400$

$33+22x-22\leq 400$

$22x+11\leq 400$

$22x\leq 389$

$x\leq\frac{389}{22}$

$x\leq 17.69$

Thus, at most 17 ounces can be mailed for $4.00.

81. Let n = the number of calls made in a given month. Then solve

$25<13+0.06n$

$12<0.06n$

$200<n$

$n>200$

Therefore, Plan 1 is more economical than Plan 2 when more than 200 calls are made.

83. $F\geq\frac{9}{5}C+32$

$F\geq\frac{9}{5}(500)+32$

$F\geq 932°$

Glass is a liquid at temperatures of 932°F or higher.

85. $s=2806.6t+32{,}558$

a. $50{,}000<2806.6t+32{,}558$

$17{,}442<2806.6t$

$6.21<t$

Beginning salaries will be greater than $50,000 in 1995 + 7 = 2002.

b. Answers may vary.

87. The consumption of whole milk is decreasing. The graph of the line is going down over time.

89. $t=2005-1994=11$

$w=-0.18t+8.72$

$w=-0.18(11)+8.72$

$w=6.74$

The consumption of whole milk will be about 6.74 gallons per person per year.

91. $-0.18t+8.72\leq 7$

$-0.18t\leq -1.72$

$t\geq 9.6$

$1994+10=2004$

Consumption of whole milk will be less than 7 gallons per person per year beginning in 2004.

93. Answers may vary.

95. $-0.18t+8.72=0.26t+5.86$

$8.72=0.44t+5.86$

$2.86=0.44t$

$6.5=t$

$t\approx 7$

$1994+7=2001$

The consumption of whole milk will be the same as the consumption of skim milk in 2001.

97. $x\geq 0$ and $x\leq 7$

The integers are 0, 1, 2, 3, 4, 5, 6, 7 or $\{0, 1, 2, 3, 4, 5, 6, 7\}$.

99. $x < 6$ and $x < -5$
$\{\ldots, -9, -8, -7, -6\}$

101. $\{x \mid -7 < x \le 1\}$
$(-7, 1]$

103. $\{x \mid -2.5 \le x < 5.3\}$
$[-2.5, 5.3)$
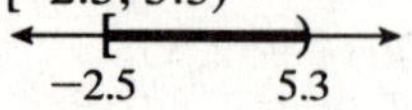

Exercise Set 2.5

1. $C \cup D = \{2, 3, 4, 5, 6, 7\}$

3. $A \cap D = \{4, 6\}$

5. $A \cup B = \{\ldots, -2, -1, 0, 1, \ldots\}$

7. $B \cap D = \{5, 7\}$

9. $B \cup C = \{x \mid x \text{ is an odd integer or } x = 2 \text{ or } x = 4\}$

11. $A \cap C = \{2, 4\}$

13. $x < 5$ and $x > -2$
$-2 < x < 5$
$(-2, 5)$

15. $x + 1 \ge 7$ and $3x - 1 \ge 5$
$x \ge 6$ and $3x \ge 6$
$x \ge 2$

$x \ge 6$
$[6, \infty)$
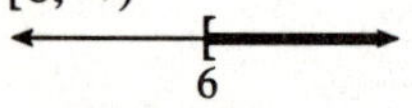

17. $4x + 2 \le -10$ and $2x \le 0$
$4x \le -12$ and $x \le 0$
$x \le -3$
$(-\infty, -3]$
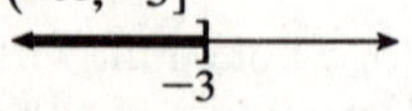

19. $5 < x - 6 < 11$
$11 < x < 17$
$(11, 17)$

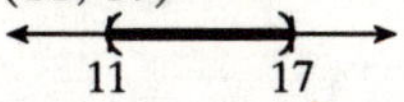

21. $-2 \le 3x - 5 \le 7$
$3 \le 3x \le 12$
$1 \le x \le 4$
$[1, 4]$

23. $1 \le \frac{2}{3}x + 3 \le 4$
$-2 \le \frac{2}{3}x \le 1$
$-3 \le x \le \frac{3}{2}$
$\left[-3, \frac{3}{2}\right]$

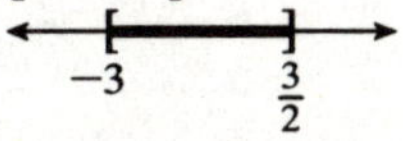

25. $-5 \le \frac{x+1}{4} \le -2$
$-20 \le x + 1 \le -8$
$-21 \le x \le -9$
$[-21, -9]$

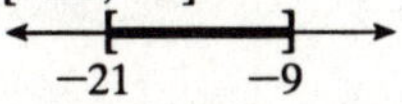

27. $x < -1$ or $x > 0$
$(-\infty, -1) \cup (0, \infty)$

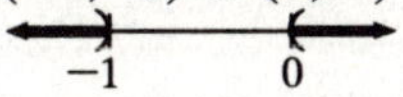

29. $-2x \le -4$ or $5x - 20 \ge 5$
$x \ge 2$ or $5x \ge 25$
$x \ge 5$

$x \ge 2$
$[2, \infty)$

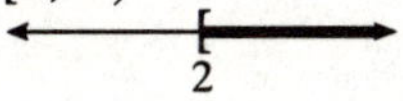

31. $3(x-1) < 12$ or $x + 7 > 10$
$x - 1 < 4$ or $x > 3$
$x < 5$
All real numbers.
$(-\infty, \infty)$

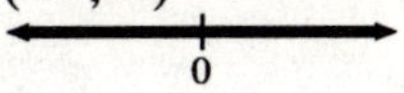

33. Answers may vary.

35. $x < 2$ and $x > -1$
$-1 < x < 2$
$(-1, 2)$

37. $x < 2$ or $x > -1$
All real numbers
$(-\infty, \infty)$

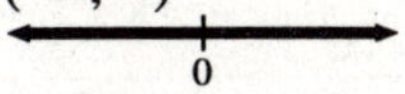

39. $x \ge -5$ and $x \ge -1$
$x \ge -1$
$[-1, \infty)$

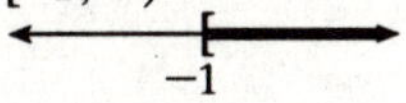

41. $x \ge -5$ or $x \ge -1$
$x \ge -5$
$[-5, \infty)$

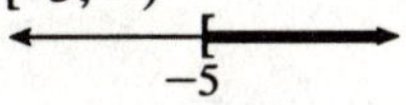

43. $0 \le 2x - 3 \le 9$
$3 \le 2x \le 12$
$\frac{3}{2} \le x \le 6$
$\left[\frac{3}{2}, 6\right]$

45. $\frac{1}{2} < x - \frac{3}{4} < 2$
$\frac{5}{4} < x < \frac{11}{4}$
$\left(\frac{5}{4}, \frac{11}{4}\right)$

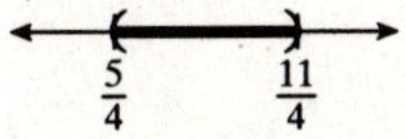

47. $x+3\geq 3$ and $x+3\leq 2$
$x\geq 0$ and $x\leq -1$
No solutions exist.
$\varnothing$

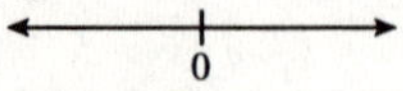

49. $3x\geq 5$ or $-x-6<1$
$x\geq \frac{5}{3}$ or $-x<7$
$x>-7$
$(-7, \infty)$

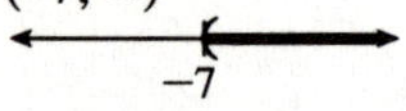

51. $0<\frac{5-2x}{3}<5$
$0<5-2x<15$
$-5<-2x<10$
$\frac{-5}{-2}>\frac{-2x}{-2}>\frac{10}{-2}$
$-5<x<\frac{5}{2}$
$\left(-5, \frac{5}{2}\right)$

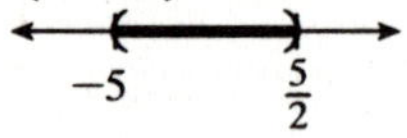

53. $-6<3(x-2)\leq 8$
$-2<x-2\leq \frac{8}{3}$
$0<x\leq \frac{14}{3}$
$\left(0, \frac{14}{3}\right]$

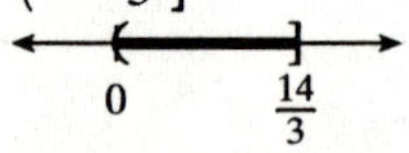

55. $-x+5>6$ and $1+2x\leq -5$
$-x>1$ and $2x\leq -6$
$x<-1$ and $x\leq -3$
$x\leq -3$
$(-\infty, -3]$

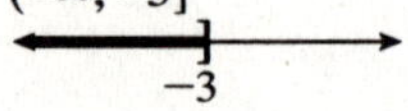

57. $3x+2\leq 5$ or $7x>29$
$3x\leq 3$ or $x>\frac{29}{7}$
$x\leq 1$
$(-\infty, 1]\cup\left(\frac{29}{7}, \infty\right)$

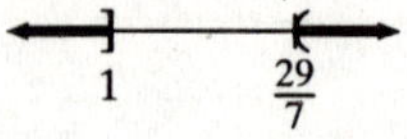

59. $5-x>7$ and $2x+3\geq 13$
$-x>2$ and $2x\geq 10$
$x<-2$ and $x\geq 5$
$\varnothing$

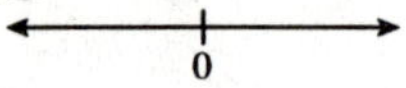

61. $-\frac{1}{2}\leq \frac{4x-1}{6}<\frac{5}{6}$
$-3\leq 4x-1<5$
$-2\leq 4x<6$
$-\frac{1}{2}\leq x<\frac{3}{2}$
$\left[-\frac{1}{2}, \frac{3}{2}\right)$

63. $\frac{1}{15}<\frac{8-3x}{15}<\frac{4}{5}$
$1<8-3x<12$
$-7<-3x<4$
$-\frac{4}{3}<x<\frac{7}{3}$
$\left(-\frac{4}{3}, \frac{7}{3}\right)$

65. $0.3<0.2x-0.9<1.5$
$1.2<0.2x<2.4$
$6<x<12$
$(6, 12)$

67. $-29 \le C \le 35$
$$-29 \le \frac{5}{9}(F-32) \le 35$$
$$-52.2 \le F-32 \le 63$$
$$-20.2 \le F \le 95$$
$$-20.2^\circ \le F \le 95^\circ$$

69. $70 \le \frac{68+65+75+78+2x}{6} \le 79$
$$420 < 286+2x \le 474$$
$$134 \le 2x \le 188$$
$$67 \le x \le 94$$
If Christian scores between 67 and 94 inclusive on his final exam, he will receive a C in the course.

71. The years that the consumption of pork was greater than 48 pounds per person were 1992, 1993, 1994, and 1995.
The years that the consumption of chicken was greater than 48 pounds per person were 1993, 1994, 1995, and 1996.
The years in common are 1993, 1994, and 1995.

73. $|-7|-|19| = 7-19 = -12$

75. $-(-6)-|-10| = 6-10 = -4$

77. $|x| = 7$
$x = -7, 7$

79. $|x| = 0$
$x = 0$

81. $2x-3 < 3x+1 < 4x-5$
$2x-3 < 3x+1$ and $3x+1 < 4x-5$
$-4 < x$ and $6 < x$
$x > 6$
$(6, \infty)$

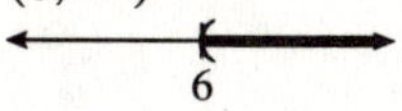

83. $-3(x-2) \le 3-2x \le 10-3x$
$-3x+6 \le 3-2x$ and $3-2x \le 10-3x$
$3 \le x$ and $x \le 7$
$3 \le x \le 7$
$[3, 7]$

85. $5x-8 < 2(2+x) < -2(1+2x)$
$5x-8 < 4+2x$ and $4+2x < -2-4x$
$3x < 12$ and $6 < -6x$
$x < 4$ and $-1 > x$
$x < -1$
$(-\infty, -1)$

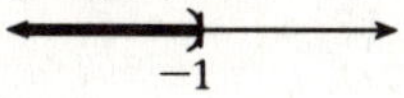

Section 2.6

Mental Math

1. $|-7| = 7$

2. $|-8| = 8$

3. $-|5| = -5$

4. $-|10| = -10$

5. $-|-6| = -6$

6. $-|-3| = -3$

7. $|-3|+|-2|+|-7| = 3+2+7 = 12$

8. $|-1|+|-6|+|-8| = 1+6+8 = 15$

Exercise Set 2.6

1. $|x| = 7$
$x = 7$ or $x = -7$
The solution set is $\{7, -7\}$.

3. $|3x| = 12.6$
$3x = 12.6$ or $3x = -12.6$
$x = 4.2$ or $x = -4.2$
The solution set is $\{4.2, -4.2\}$.

5. $|2x-5|=9$
$2x-5=9$ or $2x-5=-9$
$2x=14$ or $2x=-4$
$x=7$ or $x=-2$
The solution set is $\{7, -2\}$.

7. $\left|\frac{x}{2}-3\right|=1$
$\frac{x}{2}-3=1$ or $\frac{x}{2}-3=-1$
$\frac{x}{2}=4$ or $\frac{x}{2}=2$
$x=8$ or $x=4$
The solution set is $\{8, 4\}$.

9. $|z|+4=9$
$|z|=5$
$z=5$ or $z=-5$
The solution set is $\{5, -5\}$.

11. $|3x|+5=14$
$|3x|=9$
$3x=9$ or $3x=-9$
$x=3$ or $x=-3$
The solution set is $\{3, -3\}$.

13. $|2x|=0$
$2x=0$
$x=0$
The solution set is $\{0\}$.

15. $|4n+1|+10=4$
$|4n+1|=-6$ which is impossible.
The solution set is $\varnothing$.

17. $|5x-1|=0$
$5x-1=0$
$5x=1$
$x=\frac{1}{5}$
The solution set is $\left\{\frac{1}{5}\right\}$.

19. $|x|=5$

21. $|5x-7|=|3x+11|$
$5x-7=3x+11$ or $5x-7=-(3x+11)$
$2x=18$ or $5x-7=-3x-11$
$x=9$ or $8x=-4$
$x=9$ or $x=-\frac{1}{2}$
The solution set is $\left\{9, -\frac{1}{2}\right\}$.

23. $|z+8|=|z-3|$
$z+8=z-3$ or $z+8=-(z-3)$
$8=-3$ or $z+8=-z+3$
no solution or $2z=-5$
$z=-\frac{5}{2}$
The solution set is $\left\{-\frac{5}{2}\right\}$.

25. Answers may vary.

27. $|x|=4$
$x=4$ or $x=-4$
The solution set is $\{4, -4\}$.

29. $|y|=0$
$y=0$
The solution set is $\{0\}$.

31. $|z|=-2$ is impossible.
The solution set is $\varnothing$.

33. $|7-3x|=7$
$7-3x=7$ or $7-3x=-7$
$-3x=0$ or $-3x=-14$
$x=0$ or $x=\frac{14}{3}$
The solution set is $\left\{0, \frac{14}{3}\right\}$.

35. $|6x|-1=11$
$|6x|=12$
$6x=12$ or $6x=-12$
$x=2$ or $x=-2$
The solution set is $\{2, -2\}$.

37. $|4p| = -8$ is impossible.
The solution set is $\varnothing$.

39. $|x-3|+3=7$
$|x-3|=4$
$x-3=4$ or $x-3=-4$
$x=7$ or $x=-1$
The solution set is $\{7, -1\}$.

41. $\left|\frac{z}{4}+5\right| = -7$ is impossible.
The solution set is $\varnothing$.

43. $|9v-3| = -8$ is impossible.
The solution set is $\varnothing$.

45. $|8n+1| = 0$
$8n+1=0$
$8n=-1$
$n=-\frac{1}{8}$
The solution set is $\left\{-\frac{1}{8}\right\}$.

47. $|1+6c|-7=-3$
$|1+6c|=4$
$1+6c=4$ or $1+6c=-4$
$6c=3$ or $6c=-5$
$c=\frac{1}{2}$ or $c=-\frac{5}{6}$
The solution set is $\left\{\frac{1}{2}, -\frac{5}{6}\right\}$.

49. $|5x+1|=11$
$5x+1=11$ or $5x+1=-11$
$5x=10$ or $5x=-12$
$x=2$ or $x=-\frac{12}{5}$
The solution set is $\left\{2, -\frac{12}{5}\right\}$.

51. $|4x-2|=|-10|$ or $|4x-2|=10$
$4x-2=10$ or $4x-2=-10$
$4x=12$ or $4x=-8$
$x=3$ or $x=-2$
The solution set is $\{3, -2\}$.

53. $|5x+1|=|4x-7|$
$5x+1=4x-7$ or $5x+1=-(4x-7)$
$x=-8$ or $5x+1=-4x+7$
$x=-8$ or $9x=6$
$x=-8$ or $x=\frac{2}{3}$
The solution set is $\left\{-8, \frac{2}{3}\right\}$.

55. $|6+2x|=-|-7|$ or $|6+2x|=-7$, which is impossible.
The solution set is $\varnothing$.

57. $|2x-6|=|10-2x|$
$2x-6=10-2x$ or $2x-6=-(10-2x)$
$4x=16$ or $2x-6=-10+2x$
$x=4$ or $-6=-10$
$-6=-10$ is impossible.
The solution set is $\{4\}$.

59. $\left|\frac{2x-5}{3}\right|=7$
$\frac{2x-5}{3}=7$ or $\frac{2x-5}{3}=-7$
$2x-5=21$ or $2x-5=-21$
$2x=26$ or $2x=-16$
$x=13$ or $x=-8$
The solution set is $\{13, -8\}$.

61. $2+|5n|=17$
$|5n|=15$
$5n=15$ or $5n=-15$
$n=3$ or $n=-3$
The solution set is $\{3, -3\}$.

63. $\left|\frac{2x-1}{3}\right|=|-5|$
$\left|\frac{2x-1}{3}\right|=5$
$\frac{2x-1}{3}=5$ or $\frac{2x-1}{3}=-5$
$2x-1=15$ or $2x-1=-15$
$2x=16$ or $2x=-14$
$x=8$ or $x=-7$
The solution set is $\{8, -7\}$.

65. $|2y-3| = |9-4y|$

$2y-3 = 9-4y$ or $2y-3 = -(9-4y)$

$6y = 12$ or $2y-3 = -9+4y$

$y = 2$ or $6 = 2y$

$y = 2$ or $3 = y$

The solution set is {2, 3}.

67. $\left|\frac{3n+2}{8}\right| = |-1|$ or $\left|\frac{3n+2}{8}\right| = 1$

$\frac{3n+2}{8} = 1$ or $\frac{3n+2}{8} = -1$

$3n+2 = 8$ or $3n+2 = -8$

$3n = 6$ or $3n = -10$

$n = 2$ or $n = -\frac{10}{3}$

The solution set is $\left\{2, -\frac{10}{3}\right\}$.

69. $|x+4| = |7-x|$

$x+4 = 7-x$ or $x+4 = -(7-x)$

$2x = 3$ or $x+4 = -7+x$

$x = \frac{3}{2}$ or $4 = -7$

Since $4 = -7$ is false, the only solution is $x = \frac{3}{2}$.

The solution set is $\left\{\frac{3}{2}\right\}$.

71. $\left|\frac{8c-7}{3}\right| = -|-5|$ or $\left|\frac{8c-7}{3}\right| = -5$ which is impossible.
The solution set is $\varnothing$.

73. Answers may vary.

75. 13% of Disney's operating income came from consumer products.

77. $32\%(3.4 \text{ billion}) = 0.32(3.4 \text{ billion}) = 1.088 \text{ billion}$

\$1.088 billion is expected from the media networks segment.

79. $|x| \geq -2$

$-2, -1, 0, 1, 2$, for example

81. $|y| < 0$

No solution

Exercise Set 2.7

1. $|x| \leq 4$

$-4 \leq x \leq 4$

[−4, 4]

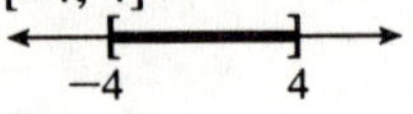

3. $|x-3| < 2$

$-2 < x-3 < 2$

$1 < x < 5$

(1, 5)

5. $|x+3| < 2$

$-2 < x+3 < 2$

$-5 < x < -1$

(−5, −1)

7. $|2x+7| \leq 13$

$-13 \leq 2x+7 \leq 13$

$-20 \leq 2x \leq 6$

$-10 \leq x \leq 3$

[−10, 3]

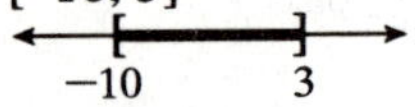

9. $|x|+7 \leq 12$

$|x| \leq 5$

$-5 \leq x \leq 5$

[−5, 5]

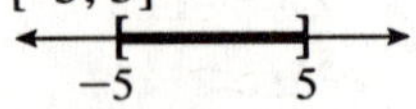

11. $|3x-1| < -5$

No real solutions.

$\varnothing$

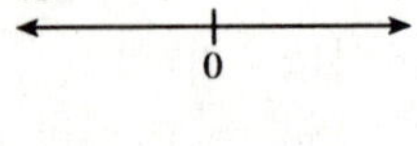

13. $|x-6|-7 \le -1$
$|x-6| \le 6$
$-6 \le x-6 \le 6$
$0 \le x \le 12$
$[0, 12]$

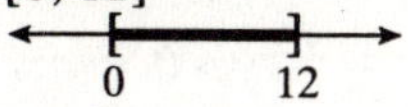

15. $|x| > 3$
$x < -3$ or $x > 3$
$(-\infty, -3) \cup (3, \infty)$

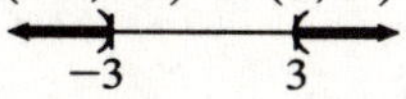

17. $|x+10| \ge 14$
$x+10 \le -14$ or $x+10 \ge 14$
$x \le -24$ or $x \ge 4$
$(-\infty, -24] \cup [4, \infty)$

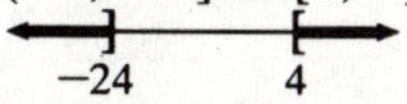

19. $|x|+2 > 6$
$|x| > 4$
$x < -4$ or $x > 4$
$(-\infty, -4) \cup (4, \infty)$

21. $|5x| > -4$
All real numbers. $(-\infty, \infty)$

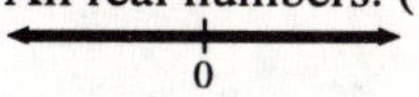

23. $|6x-8|+3 > 7$
$|6x-8| > 4$
$6x-8 < -4$ or $6x-8 > 4$
$6x < 4$ or $6x > 12$
$x < \frac{2}{3}$ or $x > 2$
$\left(-\infty, \frac{2}{3}\right) \cup (2, \infty)$

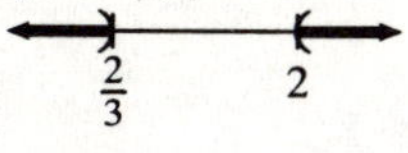

25. $|x| \le 0$
$|x| = 0$
$x = 0$
$\{0\}$

27. $|8x+3| > 0$ only excludes $|8x+3| = 0$
$8x+3 = 0$
$8x = -3$
$x = -\frac{3}{8}$
All real numbers except $-\frac{3}{8}$.
$\left(\infty, -\frac{3}{8}\right) \cup \left(-\frac{3}{8}, \infty\right)$

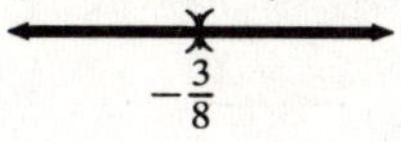

29. $|x| < 7$

31. $|x| \le 5$

33. $|x| \le 2$
$-2 \le x \le 2$
$[-2, 2]$

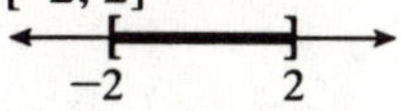

35. $|y| > 1$
$y < -1$ or $y > 1$
$(-\infty, -1) \cup (1, \infty)$

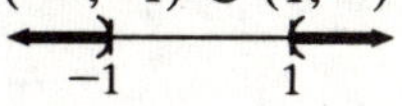

37. $|x-3| < 8$
$-8 < x-3 < 8$
$-5 < x < 11$
$(-5, 11)$

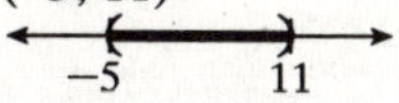

39. $|0.6x-3| > 0.6$
$0.6x-3 < -0.6$ or $0.6x-3 > 0.6$
$0.6x < 2.4$ or $0.6x > 3.6$
$x < 4$ or $x > 6$
$(-\infty, 4) \cup (6, \infty)$

41. $5+|x| \le 2$
$|x| \le -3$
No real solution.
$\varnothing$

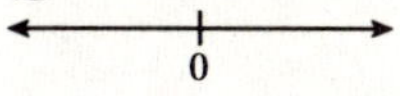

43. $|x| > -4$

All real numbers.

$(-\infty, \infty)$

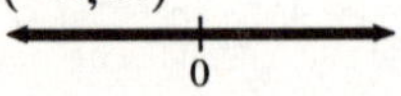

45. $|2x - 7| \le 11$

$-11 \le 2x - 7 \le 11$

$-4 \le 2x \le 18$

$-2 \le x \le 9$

$[-2, 9]$

47. $|x + 5| + 2 \ge 8$

$|x + 5| \ge 6$

$x + 5 \le -6$ or $x + 5 \ge 6$

$x \le -11$ or $x \ge 1$

$(-\infty, -11] \cup [1, \infty)$

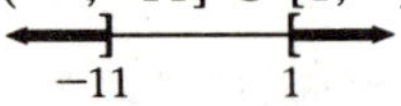

49. $|x| > 0$ excludes only $|x| = 0$, or $x = 0$

All reals except $x = 0$

$(-\infty, 0) \cup (0, \infty)$

51. $9 + |x| > 7$

$|x| > -2$

All real numbers.

$(-\infty, \infty)$

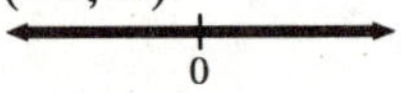

53. $6 + |4x - 1| \le 9$

$|4x - 1| \le 3$

$-3 \le 4x - 1 \le 3$

$-2 \le 4x \le 4$

$-\frac{1}{2} \le x \le 1$

$\left[-\frac{1}{2}, 1\right]$

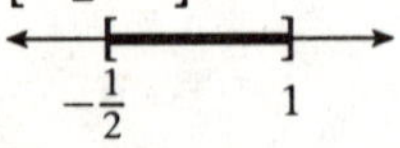

55. $\left|\frac{2}{3}x + 1\right| > 1$

$\frac{2}{3}x + 1 < -1$ or $\frac{2}{3}x + 1 > 1$

$\frac{2}{3}x < -2$ or $\frac{2}{3}x > 0$

$x < -3$ or $x > 0$

$(-\infty, -3) \cup (0, \infty)$

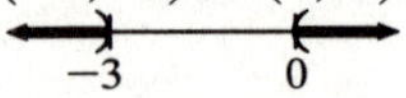

57. $|5x + 3| < -6$

No real numbers.

$\varnothing$

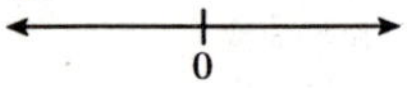

59. $|8x + 3| \ge 0$

All real numbers.

$(-\infty, \infty)$

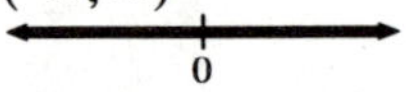

61. $|1 + 3x| + 4 < 5$

$|1 + 3x| < 1$

$-1 < 1 + 3x < 1$

$-2 < 3x < 0$

$-\frac{2}{3} < x < 0$

$\left(-\frac{2}{3}, 0\right)$

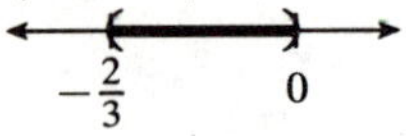

63. $|x| - 3 \ge -3$

$|x| \ge 0$

All real numbers.

$(-\infty, \infty)$

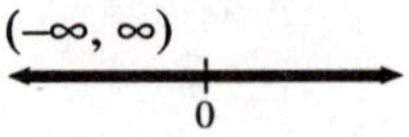

65. $|8x| - 10 > -2$

$|8x| > 8$

$8x < -8$ or $8x > 8$

$x < -1$ or $x > 1$

$(-\infty, -1) \cup (1, \infty)$

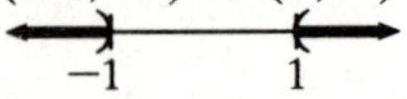

67. $\left|\frac{x+6}{3}\right| > 2$

$\frac{x+6}{3} < -2$ or $\frac{x+6}{3} > 2$

$x+6 < -6$ or $x+6 > 6$

$x < -12$ or $x > 0$

$(-\infty, -12) \cup (0, \infty)$

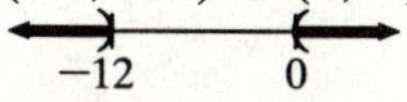

69. $|2(3+x)| > 6$

$2|3+x| > 6$

$|3+x| > 3$

$3+x < -3$ or $3+x > 3$

$x < -6$ or $x > 0$

$(-\infty, -6) \cup (0, \infty)$

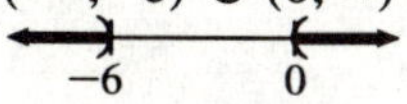

71. $\left|\frac{5(x+2)}{3}\right| < 7$

$\frac{5}{3}|x+2| < 7$

$|x+2| < \frac{21}{5}$

$-\frac{21}{5} < x+2 < \frac{21}{5}$

$-\frac{31}{5} < x < \frac{11}{5}$

$\left(-\frac{31}{5}, \frac{11}{5}\right)$

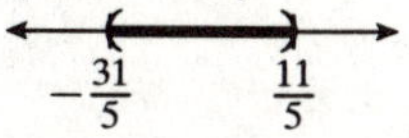

73. $-15+|2x-7| \le -6$

$|2x-7| \le 9$

$-9 \le 2x-7 \le 9$

$-2 \le 2x \le 16$

$-1 \le x \le 8$

$[-1, 8]$

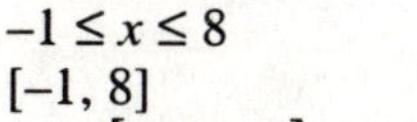

75. $\left|2x+\frac{3}{4}\right| - 7 \le -2$

$\left|2x+\frac{3}{4}\right| \le 5$

$-5 \le 2x+\frac{3}{4} \le 5$

$-\frac{23}{4} \le 2x \le \frac{17}{4}$

$-\frac{23}{8} \le x \le \frac{17}{8}$

$\left[-\frac{23}{8}, \frac{17}{8}\right]$

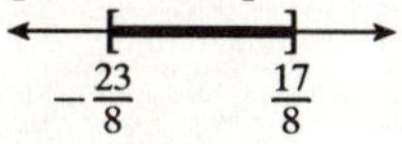

77. $|2x-3| < 7$

$-7 < 2x-3 < 7$

$-4 < 2x < 10$

$-2 < x < 5$

The solution set is $(-2, 5)$.

79. $|2x-3| = 7$

$2x-3 = 7$ or $2x-3 = -7$

$2x = 10$ or $2x = -4$

$x = 5$ or $x = -2$

The solution set is $\{5, -2\}$.

81. $|x-5| \ge 12$

$x-5 \le -12$ or $x-5 \ge 12$

$x \le -7$ or $x \ge 17$

The solution set is $(-\infty, -7] \cup [17, \infty)$.

83. $|9+4x| = 0$

$9+4x = 0$

$4x = -9$

$x = -\frac{9}{4}$

The solution set is $\left\{-\frac{9}{4}\right\}$.

85. $|2x+1|+4 < 7$

$|2x+1| < 3$

$-3 < 2x+1 < 3$

$-4 < 2x < 2$

$-2 < x < 1$

The solution set is $(-2, 1)$.

87. $|3x-5|+4=5$
$|3x-5|=1$
$3x-5=-1$ or $3x-5=1$
$3x=4$ or $3x=6$
$x=\frac{4}{3}$ or $x=2$
The solution set is $\left\{\frac{4}{3}, 2\right\}$.

89. $|x+11|=-1$
The solution set is $\varnothing$.

91. $\left|\frac{2x-1}{3}\right|=6$
$\frac{2x-1}{3}=-6$ or $\frac{2x-1}{3}=6$
$2x-1=-18$ or $2x-1=18$
$2x=-17$ or $2x=19$
$x=-\frac{17}{2}$ or $x=\frac{19}{2}$
The solution set is $\left\{-\frac{17}{2}, \frac{19}{2}\right\}$.

93. $\left|\frac{3x-5}{6}\right|>5$
$\frac{3x-5}{6}<-5$ or $\frac{3x-5}{6}>5$
$3x-5<-30$ or $3x-5>30$
$3x<-25$ or $3x>35$
$x<-\frac{25}{3}$ or $x>\frac{35}{3}$
The solution set is $\left(-\infty, -\frac{25}{3}\right)\cup\left(\frac{35}{3}, \infty\right)$.

95. Answers may vary.

97. $|3.5-x|<0.05$
$-0.05<3.5-x<0.05$
$-3.55<-x<-3.45$
$3.55>x>3.45$
$3.45<x<3.55$

99. P(rolling a 2) = $\frac{1}{6}$

101. P(rolling a 7) = 0

103. P(rolling a 1 or 3) = $\frac{1}{6}+\frac{1}{6}=\frac{2}{6}=\frac{1}{3}$

105. $3x-4y=12$
$3(2)-4y=12$
$6-4y=12$
$-4y=6$
$y=-\frac{6}{4}$
$y=-\frac{3}{2}=-1.5$

107. $3x-4y=12$
$3x-4(-3)=12$
$3x+12=12$
$3x=0$
$x=0$

Chapter 2 Review Exercises

1. $4(x-5)=2x-14$
$4x-20=2x-14$
$2x=6$
$x=3$

2. $x+7=-2(x+8)$
$x+7=-2x-16$
$3x=-23$
$x=-\frac{23}{3}$

3. $3(2y-1)=-8(6+y)$
$6y-3=-48-8y$
$14y=-45$
$y=-\frac{45}{14}$

4. $-(z+12)=5(2z-1)$
$-z-12=10z-5$
$-7=11z$
$-\frac{7}{11}=z$

5. $n-(8+4n)=2(3n-4)$
$n-8-4n=6n-8$
$-3n=6n$
$0=9n$
$n=0$

6. $4(9v+2)=6(1+6v)-10$
$36v+8=6+36v-10$
$36v+8=36v-4$
$8=-4$
No solution

7. $0.3(x-2)=1.2$
$10[0.3(x-2)]=10(1.2)$
$3(x-2)=12$
$3x-6=12$
$3x=18$
$x=6$

8. $1.5=0.2(c-0.3)$
$1.5=0.2c-0.06$
$100(1.5)=100(0.2c-0.06)$
$150=20c-6$
$156=20c$
$7.8=c$

9. $-4(2-3h)=2(3h-4)+6h$
$-8+12h=6h-8+6h$
$-8+12h=12h-8$
$-8=-8$
All real numbers

10. $6(m-1)+3(2-m)=0$
$6m-6+6-3m=0$
$3m=0$
$m=0$

11. $6-3(2g+4)-4g=5(1-2g)$
$6-6g-12-4g=5-10g$
$-6-10g=5-10g$
$-6=5$
No solution

12. $20-5(p+1)+3p=-(2p-15)$
$20-5p-5+3p=-2p+15$
$15-2p=-2p+15$
$15=15$
All real numbers

13. $\frac{x}{3}-4=x-2$
$3\left(\frac{x}{3}-4\right)=3(x-2)$
$x-12=3x-6$
$-6=2x$
$-3=x$

14. $\frac{9}{4}y=\frac{2}{3}y$
$12\left(\frac{9}{4}y\right)=12\left(\frac{2}{3}y\right)$
$3(9y)=4(2y)$
$27y=8y$
$19y=0$
$y=0$

15. $\frac{3n}{8}-1=3+\frac{n}{6}$
$24\left(\frac{3n}{8}-1\right)=24\left(3+\frac{n}{6}\right)$
$9n-24=72+4n$
$5n=96$
$n=\frac{96}{5}$

16. $\frac{z}{6}+1=\frac{z}{2}+2$
$6\left(\frac{z}{6}+1\right)=6\left(\frac{z}{2}+2\right)$
$z+6=3z+12$
$-6=2z$
$-3=z$

17. $\frac{y}{4}-\frac{y}{2}=-8$
$4\left(\frac{y}{4}-\frac{y}{2}\right)=4(-8)$
$y-2y=-32$
$-y=-32$
$y=32$

18. $\frac{2x}{3}-\frac{8}{3}=x$
$2x-8=3x$
$-8=x$

19. $\frac{b-2}{3}=\frac{b+2}{5}$

$15\left(\frac{b-2}{3}\right)=15\left(\frac{b+2}{5}\right)$

$5b-10=3b+6$

$2b=16$

$b=8$

20. $\frac{2t-1}{3}=\frac{3t+2}{15}$

$15\left(\frac{2t-1}{3}\right)=15\left(\frac{3t+2}{15}\right)$

$5(2t-1)=3t+2$

$10t-5=3t+2$

$7t=7$

$t=1$

21. $\frac{2(t+1)}{3}=\frac{2(t-1)}{3}$

$3\left[\frac{2(t+1)}{3}\right]=3\left[\frac{2(t-1)}{3}\right]$

$2t+2=2t-2$

$2=-2$

No solution

22. $\frac{3a-3}{6}=\frac{4a+1}{15}+2$

$30\left(\frac{3a-3}{6}\right)=30\left(\frac{4a+1}{15}+2\right)$

$5(3a-3)=2(4a+1)+30\cdot 2$

$15a-15=8a+2+60$

$15a-15=8a+62$

$7a=77$

$a=11$

23. $\frac{x-2}{5}+\frac{x+2}{2}=\frac{x+4}{3}$

$30\left(\frac{x-2}{5}+\frac{x+2}{2}\right)=30\left(\frac{x+4}{3}\right)$

$6(x-2)+15(x+2)=10(x+4)$

$6x-12+15x+30=10x+40$

$21x+18=10x+40$

$11x=22$

$x=2$

24. $\frac{2z-3}{4}-\frac{4-z}{2}=\frac{z+1}{3}$

$12\left(\frac{2z-3}{4}-\frac{4-z}{2}\right)=12\left(\frac{z+1}{3}\right)$

$3(2z-3)-6(4-z)=4(z+1)$

$6z-9-24+6z=4z+4$

$12z-33=4z+4$

$8z=37$

$z=\frac{37}{8}$

25. $2(x-3)=3x+1$

$2x-6=3x+1$

$-7=x$

The number is -7.

26. Let x = smaller number,
then $x + 5$ = larger number.

$x+x+5=285$

$2x=280$

$x=140$

$x+5=145$

The numbers are 140 and 145.

27. $40\% \cdot 130=0.40\cdot 130=52$

28. $1.5\% \cdot 8=0.015\cdot 8=0.12$

29. Let x = 1998 earnings for a high school graduate.

$x+0.3047x=29{,}872$

$1.3047x=29{,}872$

$x\approx 22{,}896$

Rounded to the nearest dollar, the average annual salary for a high school graduate in 1998 was $22,896.

30. Let n = the first integer, then
$n + 1$ = the second integer,
$n + 2$ = the third integer, and
$n + 4$ = the fourth integer.

$(n+1)+(n+2)+(n+3)-2n=16$

$n+6=16$

$n=10$

Therefore, the integers are 10, 11, 12, and 13.

31. Let x = smaller odd integer,
then $x + 2$ = larger odd integer.

$$5x = 3(x+2)+54$$
$$5x = 3x+6+54$$
$$2x = 60$$
$$x = 30$$

Since this is not odd, no such consecutive odd integers exist.

32. Let w = the width of the playing field then $2w - 5$ = the length of the playing field.

$$2w+2(2w-5) = 230$$
$$2w+4w-10 = 230$$
$$6w = 240$$
$$w = 40$$

Then $2w-5 = 2(40)-5$
$= 80-5$
$= 75$

Therefore, the field is 75 meters long and 40 meters wide.

33. Let n = the number of miles driven.

$$2(29.95)+0.15(n-200) = 83.6$$
$$59.9+0.15n-30 = 83.6$$
$$29.9+0.15n = 83.6$$
$$0.15n = 53.7$$
$$n = 358$$

He drove 358 miles.

34. Solve $R = C$.

$$16.50x = 4.50x+3000$$
$$12x = 3000$$
$$x = 250$$

Thus, 250 calculators must be produced and sold in order to break even.

35. Solve $R = C$.

$$40x = 20x+100$$
$$20x = 100$$
$$x = 5$$
$$R = 40\cdot 5 = 200$$

She will break even if she sells 5 plants. The revenue will be $200.

36. $V = lwh$

$$w = \frac{V}{lh}$$

37. $C = 2\pi r$

$$\frac{C}{2\pi} = r$$

38. $5x-4y = -12$

$$5x+12 = 4y$$
$$y = \frac{5x+12}{4}$$

39. $5x-4y = -12$

$$5x = 4y-12$$
$$x = \frac{4y-12}{5}$$

40. $y-y_1 = m(x-x_1)$

$$m = \frac{y-y_1}{x-x_1}$$

41.

$$y-y_1 = m(x-x_1)$$
$$y-y_1 = mx-mx_1$$
$$y-y_1+mx_1 = mx$$
$$\frac{y-y_1+mx_1}{m} = x$$

42.

$$E = I(R+r)$$
$$E = IR+Ir$$
$$E-IR = Ir$$
$$\frac{E-IR}{I} = r$$

43.

$$S = vt+gt^2$$
$$S-vt = gt^2$$
$$\frac{S-vt}{t^2} = g$$

44. $T = gr+gvt$

$$T = g(r+vt)$$
$$g = \frac{T}{r+vt}$$

45.

$$I = Prt+P$$
$$I = P(rt+1)$$
$$\frac{I}{rt+1} = P$$

46.
$$A = \frac{h}{2}(B+b)$$
$$2A = hB + hb$$
$$2A - hb = hB$$
$$\frac{2A - hb}{h} = B$$

47.
$$V = \frac{1}{3}\pi r^2 h$$
$$3V = \pi r^2 h$$
$$\frac{3V}{\pi r^2} = h$$

48.
$$R = \frac{r_1 + r_2}{2}$$
$$2R = r_1 + r_2$$
$$r_1 = 2R - r_2$$

49.
$$\frac{V_1}{T_1} = \frac{V_2}{T_2}$$
$$T_2 V_1 = T_1 V_2$$
$$T_2 = \frac{T_1 V_2}{V_1}$$

50. $A = P\left(1+\frac{r}{n}\right)^{nt} = 3000\left(1+\frac{0.03}{n}\right)^{7n}$

a. $A = 3000\left(1+\frac{0.03}{2}\right)^{14} \approx \$3695.27$

b. $A = 3000\left(1+\frac{0.03}{52}\right)^{364} \approx \$3700.81$

51.
$$C = \frac{5}{9}(F-32)$$
$$C = \frac{5}{9}(90-32)$$
$$C = \frac{5}{9}(58)$$
$$C = \frac{290}{9} \approx 32.2$$

90°F is $\frac{290°}{9}$C ≈ 32.2°C.

52. Let x = original width,
then $x + 2$ = original length.
$$(x+4)(x+2+4) = x(x+2) + 88$$
$$(x+4)(x+6) = x^2 + 2x + 88$$
$$x^2 + 10x + 24 = x^2 + 2x + 88$$
$$8x = 64$$
$$x = 8$$
$x + 2 = 10$
Original width is 8 in. and the original length is 10 in.

53. Area = $18 \times 21 = 378$ ft^2
Packages = $\frac{378}{24} = 15.75$
16 packages are needed.

54. $V_{box} = lwh = 8 \cdot 5 \cdot 3 = 120$ in^3, while
$V_{cyl} = \pi r^2 h = \pi \cdot 3^2 \cdot 6 = 54\pi \approx 170$ in^3
Therefore, the cylinder holds more ice cream.

55. $d = rt$ or $r = \frac{d}{t}$
11:00 A.M. to 1:15 P.M. is 2.25 hours.
$r = \frac{130}{2.25} \approx 58$
His average speed was 58 mph.

56.
$$3(x-5) > -(x+3)$$
$$3x - 15 > -x - 3$$
$$4x > 12$$
$$x > 3$$
The solution set is $(3, \infty)$.

57.
$$-2(x+7) \ge 3(x+2)$$
$$-2x - 14 \ge 3x + 6$$
$$-20 \ge 5x$$
$$-4 \ge x$$
The solution set is $(-\infty, -4]$.

58.
$$4x - (5 + 2x) < 3x - 1$$
$$4x - 5 - 2x < 3x - 1$$
$$2x - 5 < 3x - 1$$
$$-4 < x$$
The solution set is $(-4, \infty)$.

59. $3(x-8) < 7x + 2(5-x)$
$3x - 24 < 7x + 10 - 2x$
$3x - 24 < 5x + 10$
$-34 < 2x$
$-17 < x$
The solution set is $(-17, \infty)$.

60. $24 \geq 6x - 2(3x-5) + 2x$
$24 \geq 6x - 6x + 10 + 2x$
$24 \geq 10 + 2x$
$14 \geq 2x$
$7 \geq x$
The solution set is $(-\infty, 7]$.

61. $48 + x \geq 5(2x+4) - 2x$
$48 + x \geq 10x + 20 - 2x$
$48 + x \geq 8x + 20$
$28 \geq 7x$
$4 \geq x$
The solution set is $(-\infty, 4]$.

62. $\frac{x}{3} + \frac{1}{2} > \frac{2}{3}$
$6\left(\frac{x}{3} + \frac{1}{2}\right) > 6\left(\frac{2}{3}\right)$
$2x + 3 > 4$
$2x > 1$
$x > \frac{1}{2}$

The solution set is $\left(\frac{1}{2}, \infty\right)$.

63. $x + \frac{3}{4} < -\frac{x}{2} + \frac{9}{4}$
$4\left(x + \frac{3}{4}\right) < 4\left(-\frac{x}{2} + \frac{9}{4}\right)$
$4x + 3 < -2x + 9$
$6x < 6$
$x < 1$
The solution set is $(-\infty, 1)$.

64. $\frac{x-5}{2} \leq \frac{3}{8}(2x+6)$
$8\left(\frac{x-5}{2}\right) \leq 8\left[\frac{3}{8}(2x+6)\right]$
$4(x-5) \leq 3(2x+6)$
$4x - 20 \leq 6x + 18$
$-38 \leq 2x$
$-19 \leq x$
The solution set is $[-19, \infty)$.

65. $\frac{3(x-2)}{5} > \frac{-5(x-2)}{3}$
$15\left[\frac{3(x-2)}{5}\right] > 15\left[\frac{-5(x-2)}{3}\right]$
$9(x-2) > -25(x-2)$
$9x - 18 > -25x + 50$
$34x > 68$
$x > 2$
The solution set is $(2, \infty)$.

66. Let n = the number of pounds of laundry.
$25 < 0.9(10) + 0.8(n-10)$
$25 < 9 + 0.8n - 8$
$25 < 1 + 0.8n$
$24 < 0.8n$
$30 < n$
Therefore, it is more economical to use the housekeeper for more than 30 pounds of laundry per week.

67. $500 \leq F \leq 1000$
$500 \leq \frac{9}{5}C + 32 \leq 1000$
$468 \leq \frac{9}{5}C \leq 968$
$260 \leq C \leq 538$
So rounded to the nearest degree, firing temperatures range from 260°C to 538°C.

68. Let x = the minimum score that the last judge can give.

$$\frac{9.5+9.7+9.9+9.7+9.7+9.6+9.5+x}{8} \geq 9.65$$
$$67.6+x \geq 77.2$$
$$x \geq 9.6$$

Therefore, the last judge must give Nana at least a 9.6 for her to win a silver medal.

69. Let x = the amount she saves each summer.

$4000 \leq 2x+500 \leq 8000$
$3500 \leq 2x \leq 7500$
$1750 \leq x \leq 3750$

She must save between \$1750 and \$3750 each summer.

70. $1 \leq 4x-7 \leq 3$

$8 \leq 4x \leq 10$

$2 \leq x \leq \frac{5}{2}$

The solution set is $\left[2, \frac{5}{2}\right]$.

71. $-2 \leq 8+5x < -1$

$-10 \leq 5x < -9$

$-2 \leq x < -\frac{9}{5}$

The solution set is $\left[-2, -\frac{9}{5}\right)$.

72. $-3 < 4(2x-1) < 12$

$-\frac{3}{4} < 2x-1 < 3$

$\frac{1}{4} < 2x < 4$

$\frac{1}{8} < x < 2$

The solution set is $\left(\frac{1}{8}, 2\right)$.

73. $-6 < x-(3-4x) < -3$

$-6 < x-3+4x < -3$

$-6 < 5x-3 < -3$

$-3 < 5x < 0$

$-\frac{3}{5} < x < 0$

The solution set is $\left(-\frac{3}{5}, 0\right)$.

74. $\frac{1}{6} < \frac{4x-3}{3} \le \frac{4}{5}$

$30\left(\frac{1}{6}\right) < 30\left(\frac{4x-3}{3}\right) \le 30\left(\frac{4}{5}\right)$

$5 < 10(4x-3) \le 24$

$5 < 40x - 30 \le 24$

$35 < 40x \le 54$

$\frac{7}{8} < x \le \frac{27}{20}$

The solution set is $\left(\frac{7}{8}, \frac{27}{20}\right]$.

75. $0 \le \frac{2(3x+4)}{5} \le 3$

$5(0) \le 5\left[\frac{2(3x+4)}{5}\right] \le 5(3)$

$0 \le 6x + 8 \le 15$

$-8 \le 6x \le 7$

$-\frac{4}{3} \le x \le \frac{7}{6}$

The solution set is $\left[-\frac{4}{3}, \frac{7}{6}\right]$.

76. $x \le 2$ and $x > -5$

$-5 < x \le 2$

The solution set is $(-5, 2]$.

77. $x \le 2$ or $x > -5$

The solution set is $(-\infty, \infty)$.

78. $3x - 5 > 6$ or $-x < -5$

$3x > 11$ or $x > 5$

$x > \frac{11}{3}$ or $x > 5$

The solution set is $\left(\frac{11}{3}, \infty\right)$.

79. $-2x \le 6$ and $-2x + 3 < -7$

$x \ge -3$ and $-2x < -10$

$x \ge -3$ and $x > 5$

The solution set is $(5, \infty)$.

80. $|x - 7| = 9$

$x - 7 = 9$ or $x - 7 = -9$

$x = 16$ or $x = -2$

The solution set is $\{16, -2\}$.

81. $|8 - x| = 3$

$8 - x = 3$ or $8 - x = -3$

$5 = x$ or $11 = x$

The solution set is $\{5, 11\}$.

82. $|2x + 9| = 9$

$2x + 9 = 9$ or $2x + 9 = -9$

$2x = 0$ or $2x = -18$

$x = 0$ or $x = -9$

The solution set is $\{0, -9\}$.

83. $|-3x + 4| = 7$

$-3x + 4 = 7$ or $-3x + 4 = -7$

$-3x = 3$ or $-3x = -11$

$x = -1$ or $x = \frac{11}{3}$

The solution set is $\left\{-1, \frac{11}{3}\right\}$.

84. $|3x - 2| + 6 = 10$

$|3x - 2| = 4$

$3x - 2 = 4$ or $3x - 2 = -4$

$3x = 6$ or $3x = -2$

$x = 2$ or $x = -\frac{2}{3}$

The solution set is $\left\{2, -\frac{2}{3}\right\}$.

85. $5 + |6x + 1| = 5$

$|6x + 1| = 0$

$6x + 1 = 0$

$6x = -1$

$x = -\frac{1}{6}$

The solution set is $\left\{-\frac{1}{6}\right\}$.

86. $-5 = |4x - 3|$

The solution set is $\varnothing$.

87. $|5 - 6x| + 8 = 3$

$|5 - 6x| = -5$

The solution set is $\varnothing$.

88. $|7x| - 26 = -5$
$|7x| = 21$
$7x = 21$ or $7x = -21$
$x = 3$ or $x = -3$
The solution set is $\{3, -3\}$.

89. $-8 = |x-3| - 10$
$2 = |x-3|$
$x - 3 = 2$ or $x - 3 = -2$
$x = 5$ or $x = 1$
The solution set is $\{1, 5\}$.

90. $\left|\frac{3x-7}{4}\right| = 2$
$\frac{3x-7}{4} = 2$ or $\frac{3x-7}{4} = -2$
$3x - 7 = 8$ or $3x - 7 = -8$
$3x = 15$ or $3x = -1$
$x = 5$ or $x = -\frac{1}{3}$
The solution set is $\left\{5, -\frac{1}{3}\right\}$.

91. $\left|\frac{9-2x}{5}\right| = -3$
The solution set is $\varnothing$.

92. $|6x+1| = |15+4x|$
$6x + 1 = 15 + 4x$ or $6x + 1 = -(15 + 4x)$
$2x = 14$ or $6x + 1 = -15 - 4x$
$x = 7$ or $10x = -16$
$x = 7$ or $x = -\frac{16}{10} = -\frac{8}{5}$
The solution set is $\left\{7, -\frac{8}{5}\right\}$.

93. $|x-3| = |7+2x|$
$x - 3 = 7 + 2x$ or $x - 3 = -(7 + 2x)$
$-10 = x$ or $x - 3 = -7 - 2x$
$x = -10$ or $3x = -4$
$x = -10$ or $x = -\frac{4}{3}$
The solution set is $\left\{-10, -\frac{4}{3}\right\}$.

94. $|5x-1| < 9$
$-9 < 5x - 1 < 9$
$-8 < 5x < 10$
$-\frac{8}{5} < x < 2$
The solution set is $\left(-\frac{8}{5}, 2\right)$.

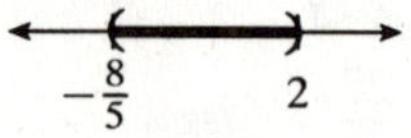

95. $|6+4x| \ge 10$
$6 + 4x \le -10$ or $6 + 4x \ge 10$
$4x \le -16$ or $4x \ge 4$
$x \le -4$ or $x \ge 1$
The solution set is $(-\infty, -4] \cup [1, \infty)$.

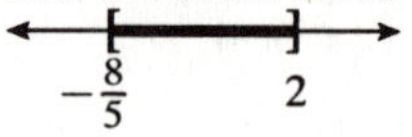

96. $|3x| - 8 > 1$
$|3x| > 9$
$3x < -9$ or $3x > 9$
$x < -3$ or $x > 3$
The solution set is $(-\infty, -3) \cup (3, \infty)$.

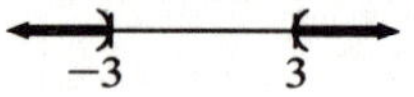

97. $9 + |5x| < 24$
$|5x| < 15$
$-15 < 5x < 15$
$-3 < x < 3$
The solution set is $(-3, 3)$.

98. $|6x-5| \le -1$
The solution set is $\varnothing$.

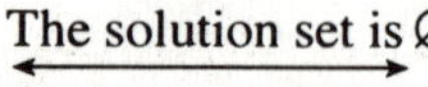

99. $|6x-5| \ge -1$
Since $|6x-5|$ is nonnegative for all real numbers x, the solution set is $(-\infty, \infty)$.

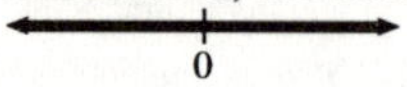

100. $\left|3x+\frac{2}{5}\right|\ge 4$

$3x+\frac{2}{5}\le -4$ or $3x+\frac{2}{5}\ge 4$

$3x\le -\frac{22}{5}$ or $3x\ge \frac{18}{5}$

$x\le -\frac{22}{15}$ or $x\ge \frac{6}{5}$

The solution set is $\left(-\infty, -\frac{22}{15}\right]\cup\left[\frac{6}{5}, \infty\right)$.

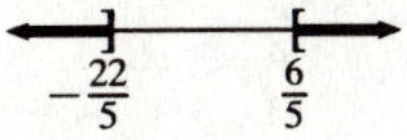

101. $\left|\frac{4x-3}{5}\right|<1$

$-1<\frac{4x-3}{5}<1$

$-5<4x-3<5$

$-2<4x<8$

$-\frac{1}{2}<x<2$

The solution set is $\left(-\frac{1}{2}, 2\right)$.

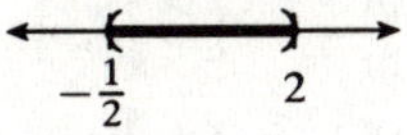

102. $\left|\frac{x}{3}+6\right|-8>-5$

$\left|\frac{x}{3}+6\right|>3$

$\frac{x}{3}+6<-3$ or $\frac{x}{3}+6>3$

$\frac{x}{3}<-9$ or $\frac{x}{3}>-3$

$x<-27$ or $x>-9$

The solution set is $(-\infty, -27)\cup(-9, \infty)$.

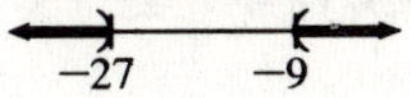

103. $\left|\frac{4(x-1)}{7}\right|+10<2$

$\left|\frac{4(x-1)}{7}\right|<-8$

The solution set is $\varnothing$.

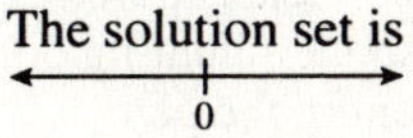

Chapter 2 Test

1. $8x+14=5x+44$

$3x=30$

$x=10$

2. $3(x+2)=11-2(2-x)$

$3x+6=11-4+2x$

$3x+6=7+2x$

$x=1$

3. $3(y-4)+y=2(6+2y)$

$3y-12+y=12+4y$

$4y-12=12+4y$

$-12=12$

No solution

4. $7n-6+n=2(4n-3)$

$8n-6=8n-6$

$-6=-6$

All real numbers

5. $\frac{z}{2}+\frac{z}{3}=10$

$6\left(\frac{z}{2}+\frac{z}{3}\right)=6(10)$

$3z+2z=60$

$5z=60$

$z=12$

6. $\frac{7w}{4}+5=\frac{3w}{10}+1$

$20\left(\frac{7w}{4}+5\right)=20\left(\frac{3w}{10}+1\right)$

$35w+100=6w+20$

$29w=-80$

$w=-\frac{80}{29}$

7. $|6x-5|=1$

$6x-5=-1$ or $6x-5=1$

$6x=4$ or $6x=6$

$x=\frac{2}{3}$ or $x=1$

8. $|8-2t|=-6$

No solution

9. $3x - 4y = 8$
$3x - 8 = 4y$
$y = \frac{3x-8}{4}$

10. $4(2n - 3m) - 3(5n - 7m) = 0$
$8n - 12m - 15n + 21m = 0$
$9m - 7n = 0$
$9m = 7n$
$\frac{9m}{7} = n$

11. $S = gt^2 + gvt$
$S = g(t^2 + vt)$
$g = \frac{S}{t^2 + vt}$

12. $F = \frac{9}{5}C + 32$
$F - 32 = \frac{9}{5}C$
$C = \frac{5}{9}(F - 32)$

13. $3(2x - 7) - 4x > -(x + 6)$
$6x - 21 - 4x > -x - 6$
$2x - 21 > -x - 6$
$3x > 15$
$x > 5$
The solution set is $(5, \infty)$.

14. $8 - \frac{x}{2} \le 7$
$2\left(8 - \frac{x}{2}\right) \le 2(7)$
$16 - x \le 14$
$2 \le x$
The solution set is $[2, \infty)$.

15. $-3 < 2(x - 3) \le 4$
$-3 < 2x - 6 \le 4$
$3 < 2x \le 10$
$\frac{3}{2} < x \le 5$
The solution set is $\left(\frac{3}{2}, 5\right]$.

16. $|3x + 1| > 5$
$3x + 1 < -5$ or $3x + 1 > 5$
$3x < -6$ or $3x > 4$
$x < -2$ or $x > \frac{4}{3}$
The solution set is $(-\infty, -2) \cup \left(\frac{4}{3}, \infty\right)$.

17. $x \ge 5$ and $x \ge 4$
$x \ge 5$
The solution set is $[5, \infty)$.

18. $x \ge 5$ or $x \ge 4$
$x \ge 4$
The solution set is $[4, \infty)$.

19. $-x > 1$ and $3x + 3 \ge x - 3$
$x < -1$ and $2x \ge -6$
$x < -1$ and $x \ge -3$
$-3 \le x < -1$
The solution set is $[-3, -1)$.

20. $6x + 1 > 5x + 4$ or $1 - x > -4$
$x > 3$ or $5 > x$
$x > 3$ or $x < 5$
The solution set is $(-\infty, \infty)$.

21. $12\% \cdot 80 = 0.12 \cdot 80 = 9.6$

22. Let x = employees in 1996.
$x + 1.18x = 461{,}000$
$2.18x = 461{,}000$
$x \approx 211{,}468$
The number of people employed in these occupations in 1996 was 211,468.

23. Recall that $C = 2\pi r$. Here $C = 78.5$.
$78.5 = 2\pi r$
$r = \frac{78.5}{2\pi} = \frac{39.25}{\pi}$
Also, recall that $A = \pi r^2$.
$A = \pi\left(\frac{39.25}{\pi}\right)^2 \approx \frac{39.25^2}{3.14} \approx 490.63$
Dividing this by 60 yields approximately 8.18. Therefore, about 8 hunting dogs could safely be kept in the pen.

24. Solve $R > C$.
$7.4x > 3910 + 2.8x$
$4.6x > 3910$
$x > 850$
Therefore, more than 850 sunglasses must be produced and sold in order for them to yield a profit.

25. $A = P\left(1+\frac{r}{n}\right)^{nt}$
$= 2500\left(1+\frac{0.035}{4}\right)^{4\cdot 10} \approx \$3542.27$

26. Let x = average fresh market price for broccoli in 1995, then $x + .27x$ = average fresh market price for broccoli in 1998.
$x + .27x = 1.10$
$1.27x = 1.10$
$x \approx 0.87$
The average frexh market price for broccoli in 1995 was \$0.87 per pound, to the nearest cent.

27. $P = 0.5rAV^3$
$= 0.5(1.225)(3.14)(9)^3$
$= 1402$ watts (to the nearest whole watt).

Chapter 3

Section 3.1

Graphing Calculator Explorations

1.

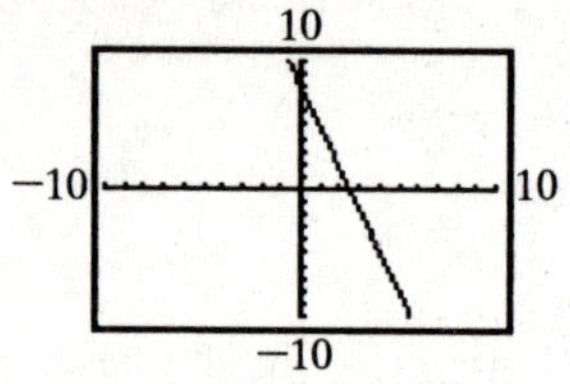

3.

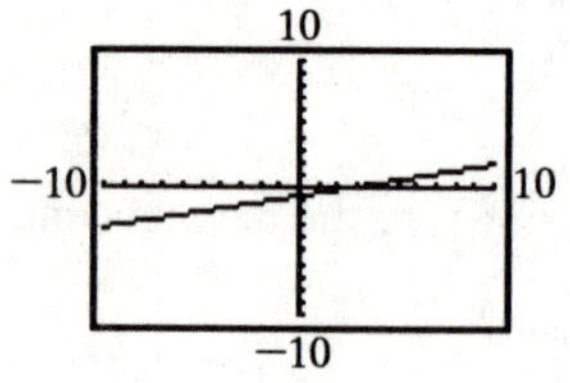

5.

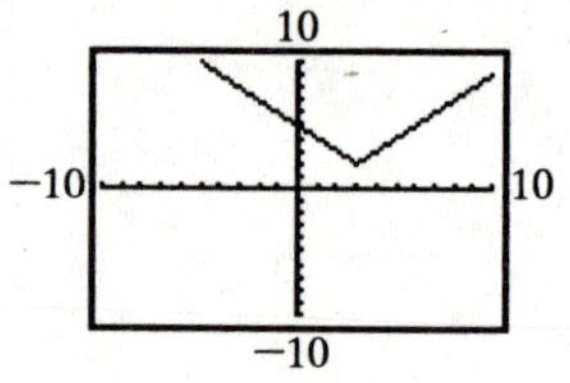

7.

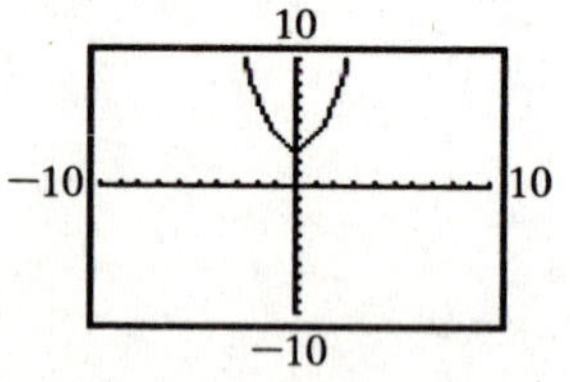

Mental Math

1. Point A is (5, 2).
2. Point B is (2, 5).
3. Point C is (3, –1).
4. Point D is (–1, 3).
5. Point E is (–5, –2).
6. Point F is (–3, 5).
7. Point G is (–1, 0).
8. Point H is (0, –3).

Exercise Set 3.1

1. (3, 2) in quadrant I

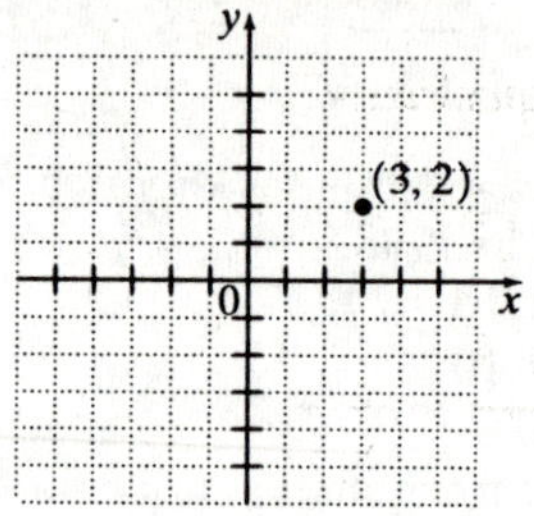

3. (–5, 3) in quadrant II

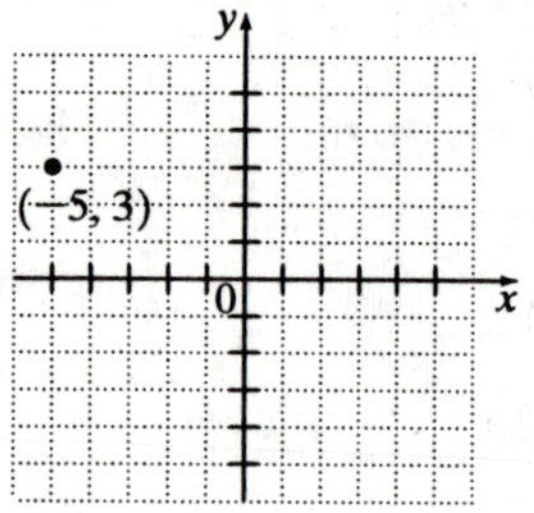

5. $\left(5\frac{1}{2}, -4\right)$ in quadrant IV

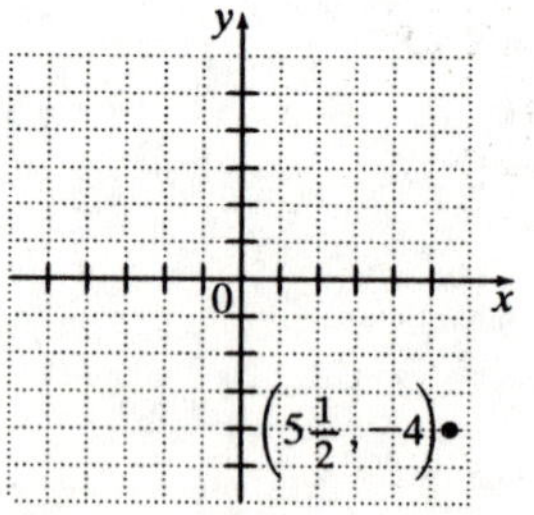

7. (0, 3.5) on y-axis

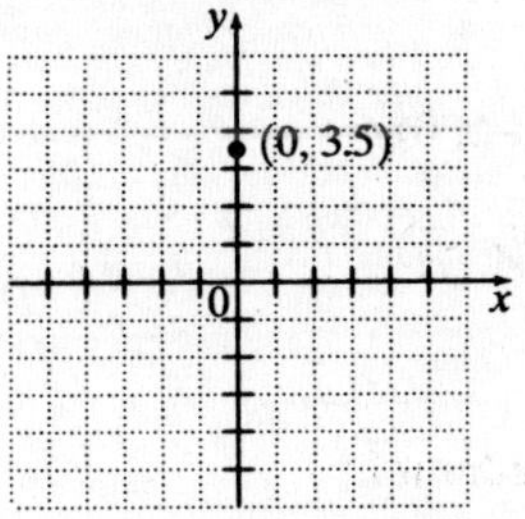

9. (–2, –4) in quadrant III

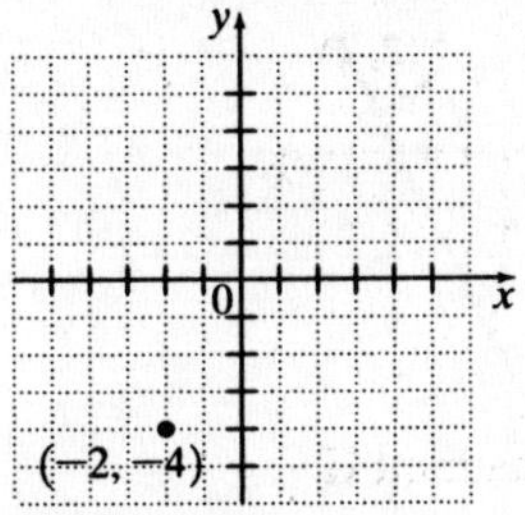

11. quadrant IV

13. x-axis

15. quadrant III

17. Let $x = 0, y = 5$

$y = 3x - 5$
$5 = 3 \cdot 0 - 5$
$5 = -5$
False; No

Let $x = -1, y = -8$

$y = 3x - 5$
$-8 = 3(-1) - 5$
$-8 = -3 - 5$
$-8 = -8$
True; Yes

19. Let $x = 1, y = 0$

$-6x + 5y = -6$
$-6(1) + 5(0) = -6$
$-6 + 0 = -6$
$-6 = -6$
True; Yes

Let $x = 2, y = \frac{6}{5}$

$-6x + 5y = -6$
$-6(2) + 5\left(\frac{6}{5}\right) = -6$
$-12 + 6 = -6$
$-6 = -6$
True; Yes

21. Let $x = 1, y = 2$

$y = 2x^2$
$2 = 2(1)^2$
$2 = 2$
True; Yes

Let $x = 3, y = 18$

$y = 2x^2$
$18 = 2(3)^2$
$18 = 2(9)$
$18 = 18$
True; Yes

23. Let $x = 2, y = 8$

$y = x^3$
$8 = 2^3$
$8 = 8$
True; Yes

Let $x = 3, y = 9$

$y = x^3$
$9 = 3^3$
$9 = 27$
False; No

25. Let $x = 1$, $y = 3$

$y = \sqrt{x} + 2$

$3 = \sqrt{1} + 2$

$3 = 1 + 2$

$3 = 3$

True; Yes

Let $x = 4$, $y = 4$

$y = \sqrt{x} + 2$

$4 = \sqrt{4} + 2$

$4 = 2 + 2$

$4 = 4$

True; Yes

27. Linear

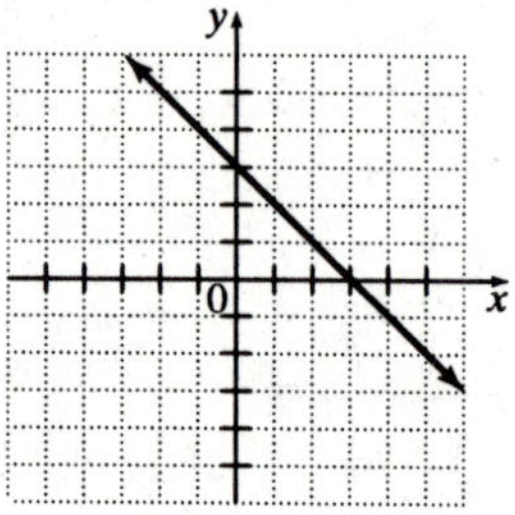

29. Linear

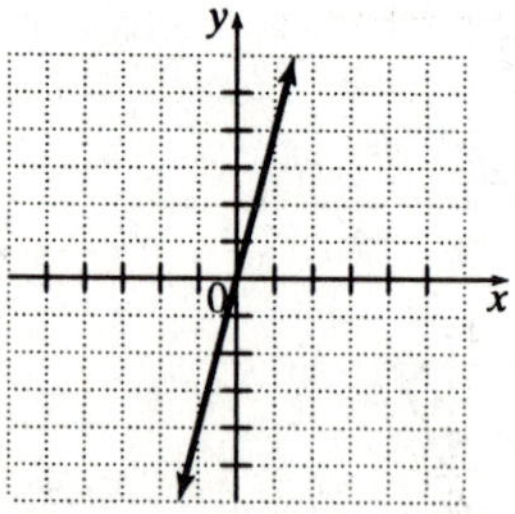

31. Linear

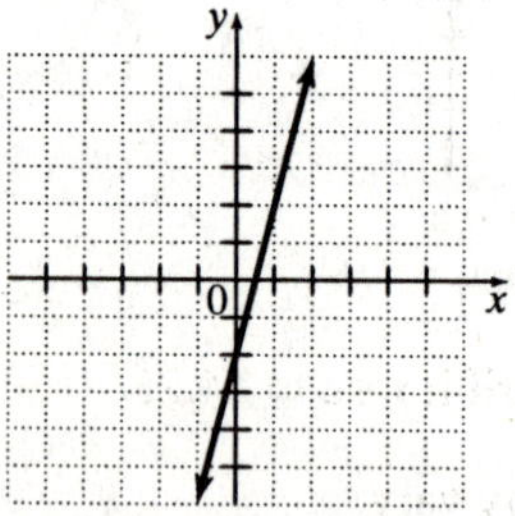

33. Not linear

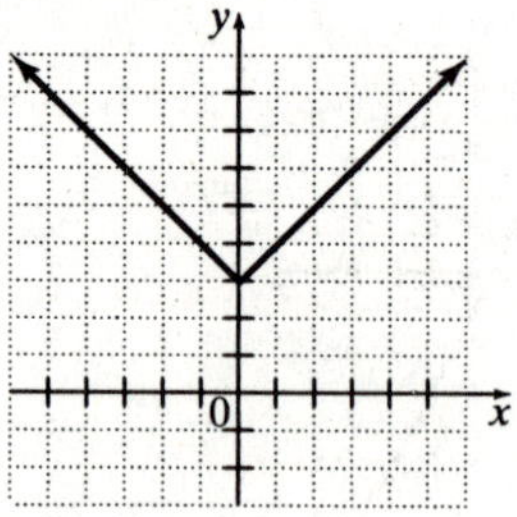

35. Linear

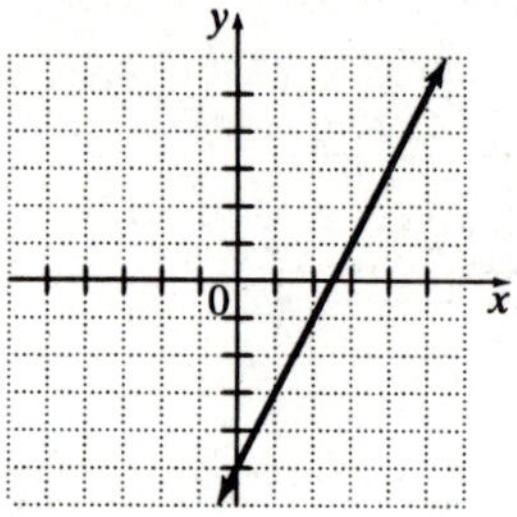

37. Not linear

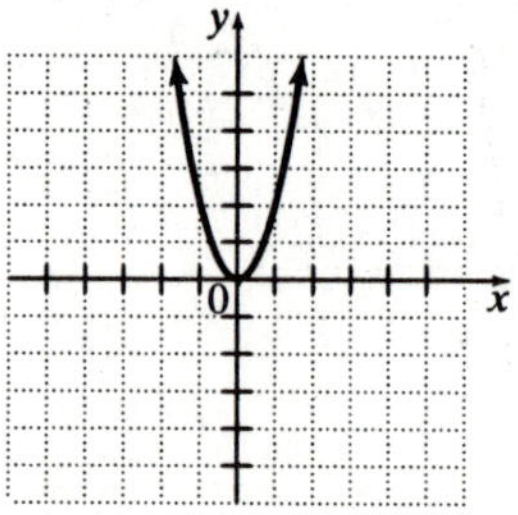

39. Not linear

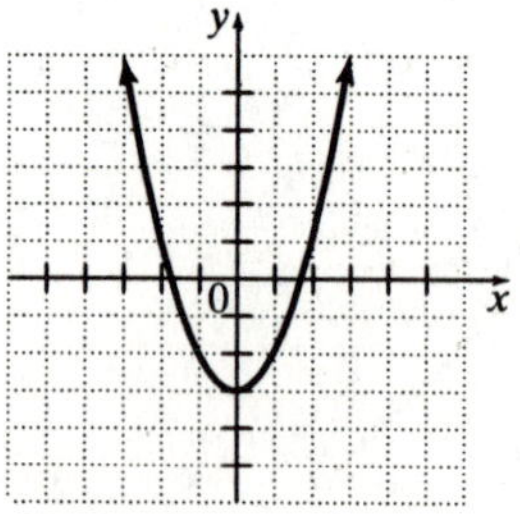

41. Linear

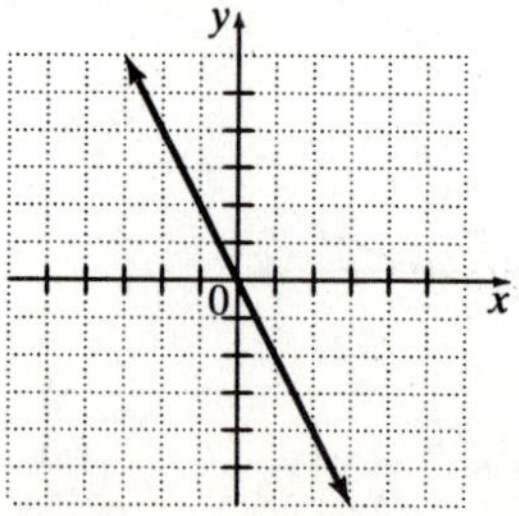

43. Linear

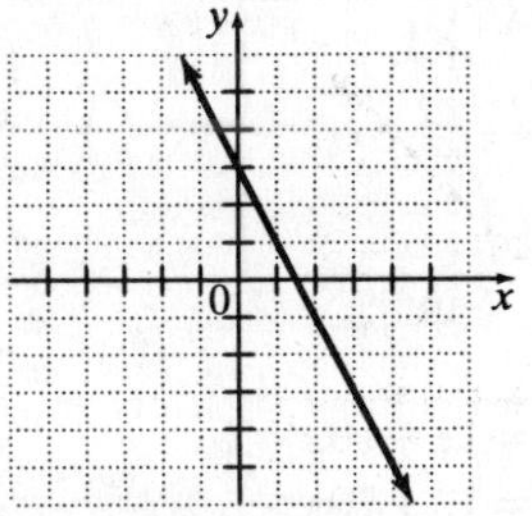

45. Not linear

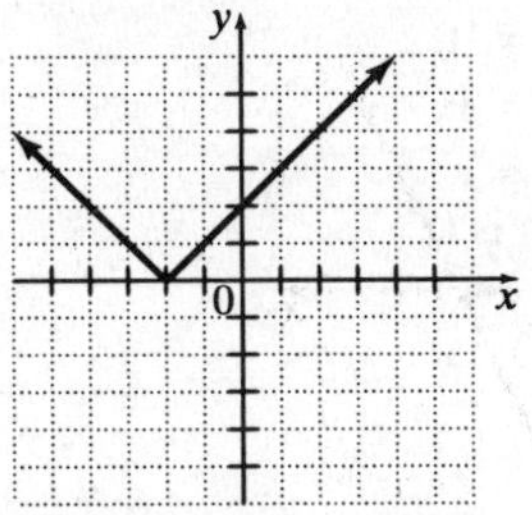

47. Not linear

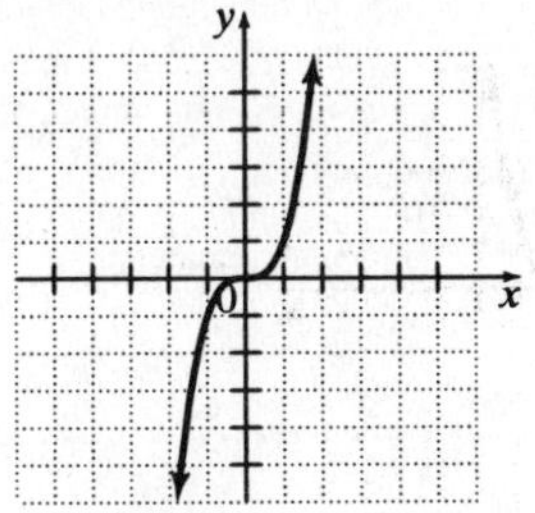

49. Not linear

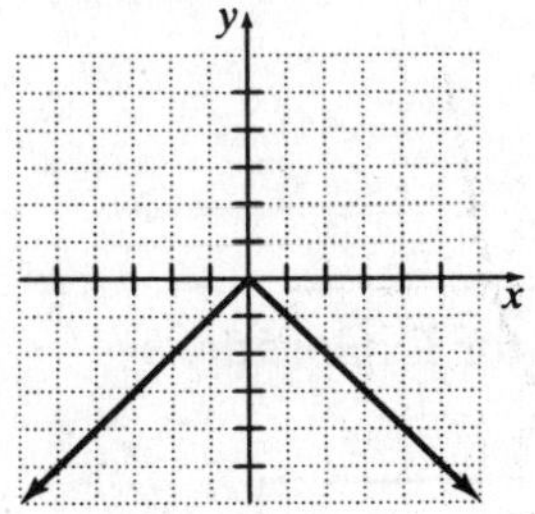

51. Linear

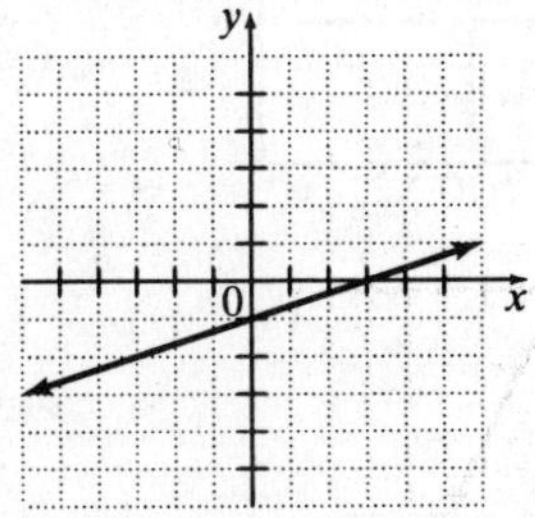

53. Linear

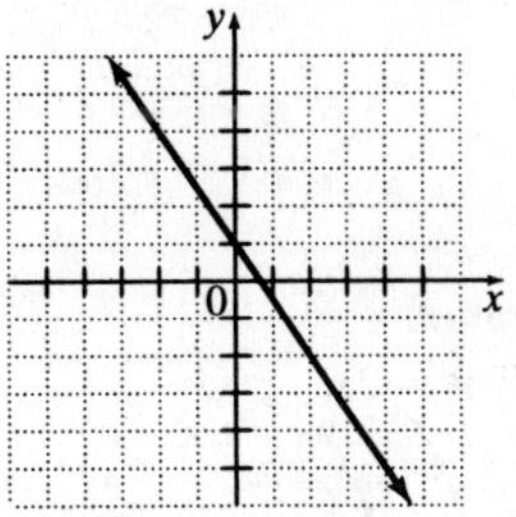

55. $y = x^2 - 4x + 7$

x	y
0	7
1	4
2	3
3	4
4	7

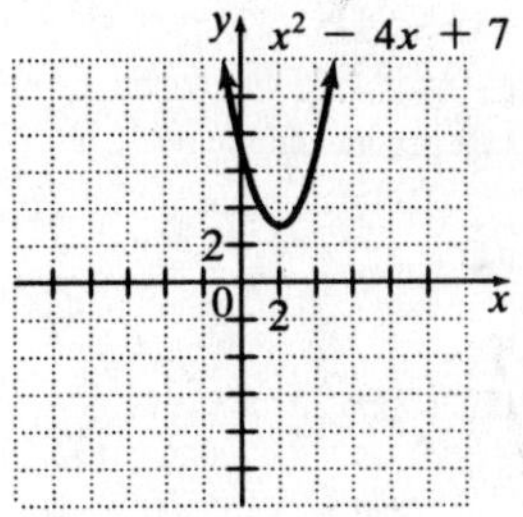

57. a.

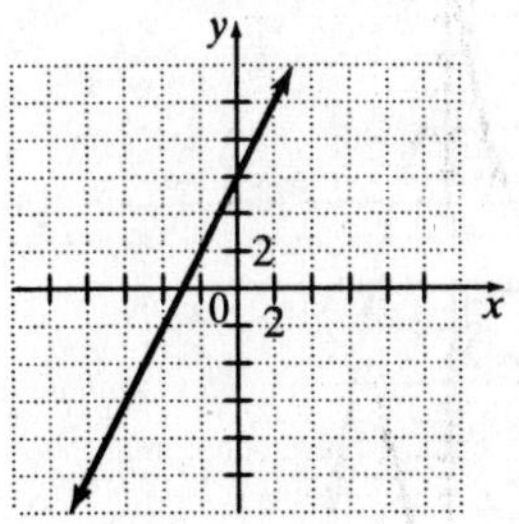

b. 14 inches

59. 1991: in April it rose to \$4.25.

61. Answers may vary.

63. \$7000

65. 7000 – 6500 = \$500

67. Answers may vary.

69. B

71. C

73. They are parallel.

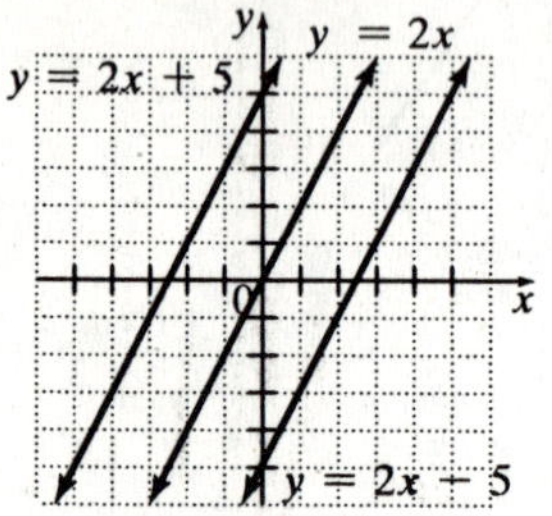

75. Answers may vary.

77. $y = -3 - 2x$

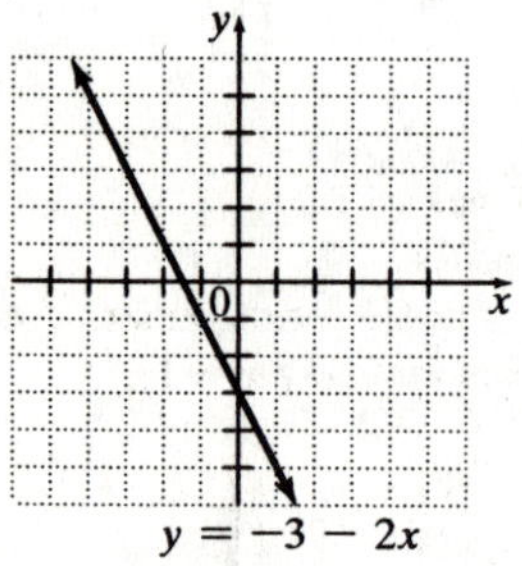

79. $y = 5 - x^2$

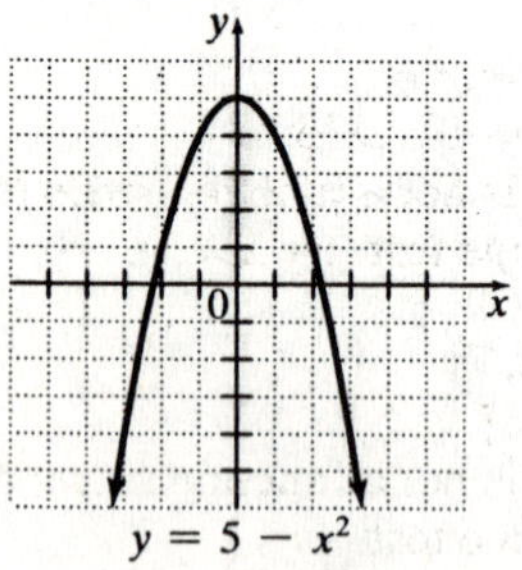

81.

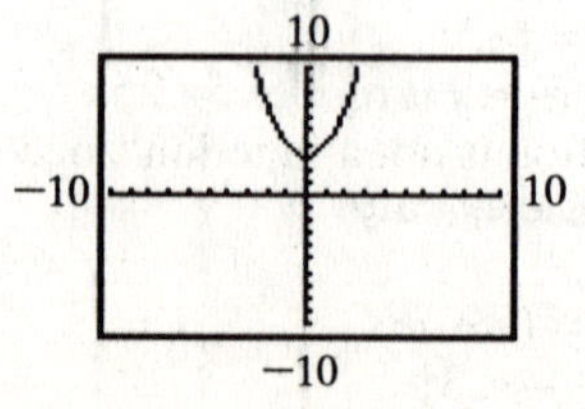

83.

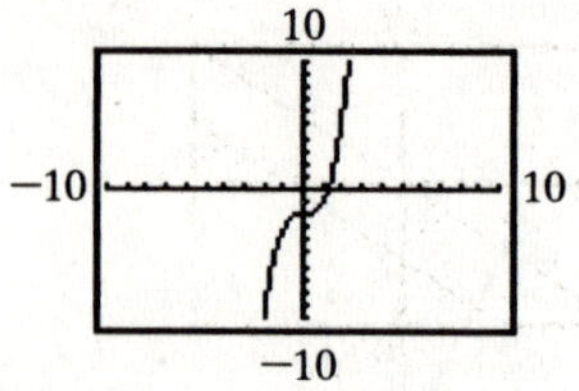

85. $5+7(x+1)=12+10x$
$5+7x+7=12+10x$
$7x+12=12+10x$
$7x=10x$
$0=3x$
$0=x$
The solution is 0.

87. $\frac{1}{6}+2x=\frac{2}{3}$
$1+12x=4$
$12x=3$
$x=\frac{1}{4}$

The solution is $\frac{1}{4}$.

89. $-3x > 18$
$x < -6$
$(-\infty, -6)$

91. $9x + 8 \le 6x - 4$
$3x \le -12$
$x \le -4$
$(-\infty, -4]$

Section 3.2

Graphing Calculator Explorations

1.

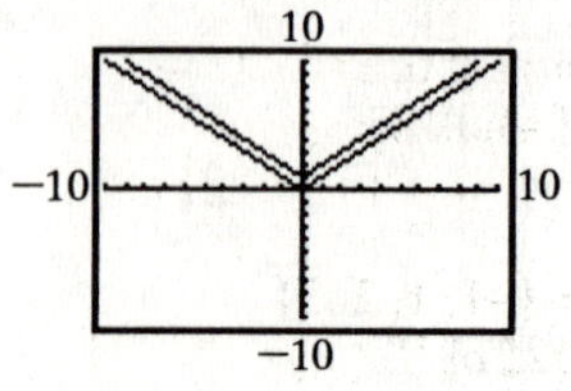

3.

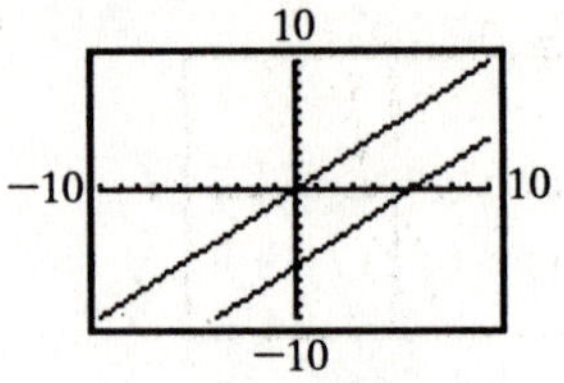

5.

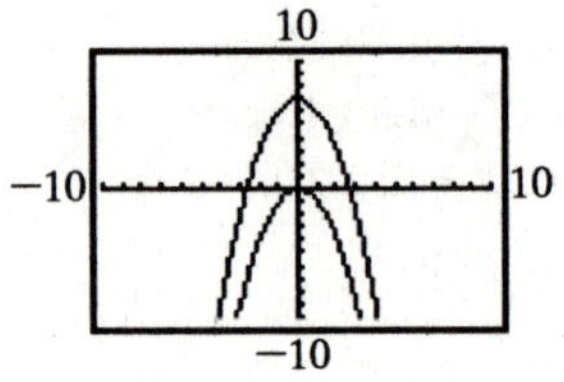

Exercise Set 3.2

1. Domain = {–1, 0, –2, 5}
Range = {7, 6, 2}
Function

3. Domain = {–2, 6, –7}
Range = {4, –3, –8}
The relation is not a function since –2 is paired with both 4 and –3.

5. Domain = {1}
Range = {1, 2, 3, 4}
The relation is not a function since 1 is paired with both 1 and 2 for example.

7. Domain = $\left\{\frac{3}{2}, 0\right\}$

Range = $\left\{\frac{1}{2}, -7, \frac{4}{5}\right\}$

The relation is not a function since $\frac{3}{2}$ is paired with both $\frac{1}{2}$ and –7.

9. Domain = {–3, 0, 3}
Range = {–3, 0, 3}
Function

11. Domain = {–1, 1, 2, 3}
Range = {2, 1}
Function

13. Domain = {Colorado, Alaska, Delaware, Illinois, Connecticut, Texas}
Range = {6, 1, 20, 30}
Function

15. Domain = {32°, 104°, 212°, 50°}
Range = {0°, 40°, 10°, 100°}
Function

17. Domain = {2, –1, 5, 100}
Range = {0}
Function

19. Function

21. Answers may vary.

23. Function

25. Not a function

27. Function

29. Domain = [0, ∞)
Range = All reals
The relation is not a function since it fails the vertical line test (try $x = 1$).

31. Domain = [–1, 1]
Range = All reals
The relation is not a function since it fails the vertical line test (try $x = 0$).

33. Domain = (–∞, ∞)
Range = (–∞, –3] ∪ [3, ∞)
The relation is not a function since it fails the vertical line test (try $x = 3$).

35. Domain = [2, 7]
Range = [1, 6]
The relation is not a function since it fails the vertical line test.

37. Domain = {–2}
Range = (–∞, ∞)
The relation is not a function since it fails the vertical line test.

39. Domain = (–∞, ∞)
Range = (–∞, 3]
Function

41. Answers may vary.

43. Yes

45. No

47. Yes

49. Yes

51. Yes

53. No

55. $f(x) = 3x + 3$
$f(4) = 3(4) + 3 = 12 + 3 = 15$

57. $h(x) = 5x^2 - 7$
$$\begin{aligned} h(-3) &= 5(-3)^2 - 7 \\ &= 5(9) - 7 \\ &= 45 - 7 \\ &= 38 \end{aligned}$$

59. $g(x) = 4x^2 - 6x + 3$
$$\begin{aligned} g(2) &= 4(2)^2 - 6(2) + 3 \\ &= 4(4) - 12 + 3 \\ &= 16 - 12 + 3 \\ &= 7 \end{aligned}$$

61. $g(x) = 4x^2 - 6x + 3$
$$\begin{aligned} g(0) &= 4(0)^2 - 6(0) + 3 \\ &= 4(0) - 0 + 3 \\ &= 0 - 0 + 3 \\ &= 3 \end{aligned}$$

63. $f(x) = \frac{1}{2}x$

a. $f(0) = \frac{1}{2}(0) = 0$

b. $f(2) = \frac{1}{2}(2) = 1$

c. $f(-2) = \frac{1}{2}(-2) = -1$

65. $g(x) = 2x^2 + 4$

a. $$\begin{aligned} g(-11) &= 2(-11)^2 + 4 \\ &= 2(121) + 4 \\ &= 242 + 4 \\ &= 246 \end{aligned}$$

b. $$\begin{aligned} g(-1) &= 2(-1)^2 + 4 \\ &= 2(1) + 4 \\ &= 2 + 4 \\ &= 6 \end{aligned}$$

c. $$\begin{aligned} g\left(\frac{1}{2}\right) &= 2\left(\frac{1}{2}\right)^2 + 4 \\ &= 2\left(\frac{1}{4}\right) + 4 \\ &= \frac{1}{2} + \frac{8}{2} \\ &= \frac{9}{2} \end{aligned}$$

67. $f(x) = -5$

a. $f(2) = -5$

b. $f(0) = -5$

c. $f(606) = -5$

69. $f(x) = 1.3x^2 - 2.6x + 5.1$

a. $$\begin{aligned} f(2) &= 1.3(2)^2 - 2.6(2) + 5.1 \\ &= 1.3(4) - 5.2 + 5.1 \\ &= 5.2 - 5.2 + 5.1 \\ &= 5.1 \end{aligned}$$

b. $$\begin{aligned} f(-2) &= 1.3(-2)^2 - 2.6(-2) + 5.1 \\ &= 1.3(4) + 5.2 + 5.1 \\ &= 5.2 + 5.2 + 5.1 \\ &= 15.5 \end{aligned}$$

c. $$\begin{aligned} f(3.1) &= 1.3(3.1)^2 - 2.6(3.1) + 5.1 \\ &= 1.3(9.61) - 8.06 + 5.1 \\ &= 12.493 - 8.06 + 5.1 \\ &= 9.533 \end{aligned}$$

71. (1, −10)

73. −2

75. − 4, 0

77. Infinite number
The reason is that a graph is a function as long as it passes the vertical line test. So it does not matter if the equation of the graph takes on the value 0 many times.

79. a. \$13.4 billion

b. \$13.672 billion

81. \$34.374 billion

83. $f(x) = x + 7$

85. 25π square centimeters

87. 2744 cubic inches

89. 166.38 centimeters

91. 163.2 milligrams

93. a. $c(x) = 1.7x + 88$
$c(2) = 1.7(2) + 88 = 3.4 + 88 = 91.4$
The per capita consumption of poultry was 91.4 pounds in 1997.

b. 106.7 pounds

95. $x - y = -5$

x	0	−5	1
y	5	0	6

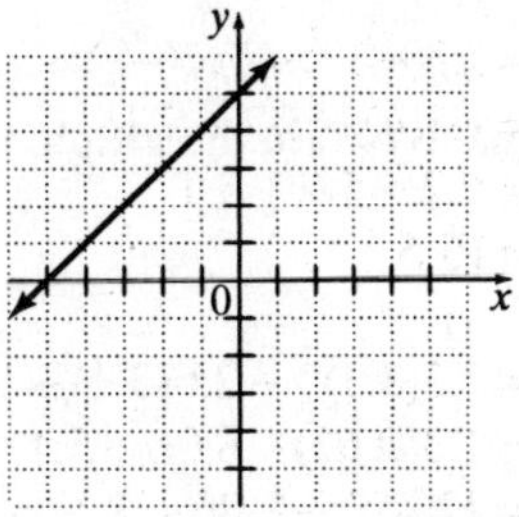

97. $7x + 4y = 8$

x	0	$\frac{8}{7}$	$\frac{12}{7}$
y	2	0	−1

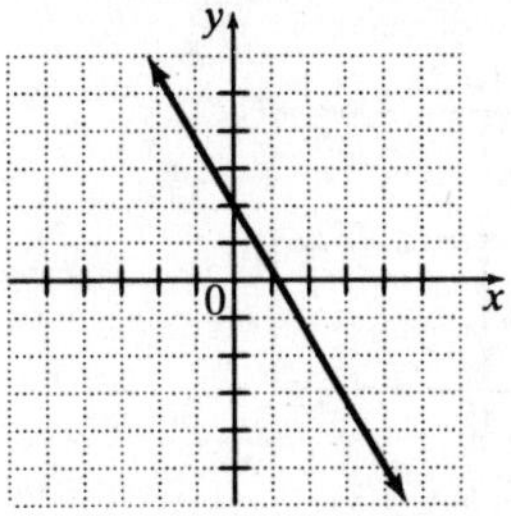

99. $y = 6x$

x	0	0	−1
y	0	0	−6

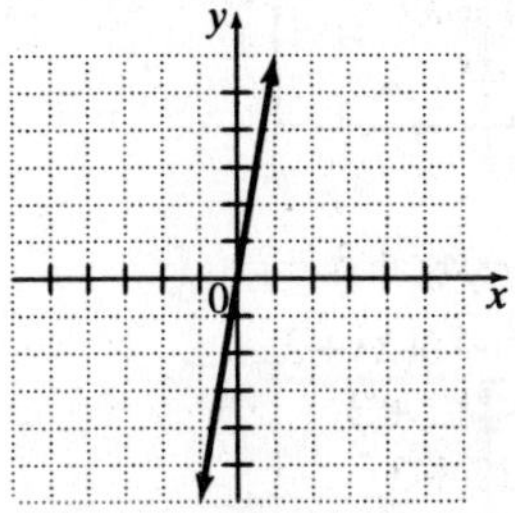

101. Yes; 170 meters

103. $g(x) = -3x + 12$
$g(s) = -3s + 12$
$g(r) = -3r + 12$

105. $f(x) = x^2 - 12$
$f(12) = 12^2 - 12 = 132$
$f(a) = a^2 - 12$

Section 3.3

Graphing Calculator Explorations

1. $y = \dfrac{x}{3.5}$

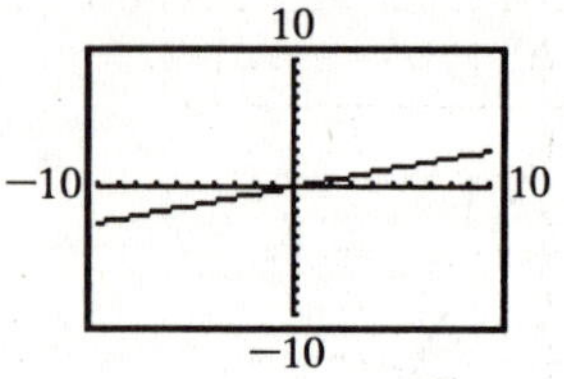

3. $y = -\dfrac{5.78}{2.31}x + \dfrac{10.98}{2.31}$

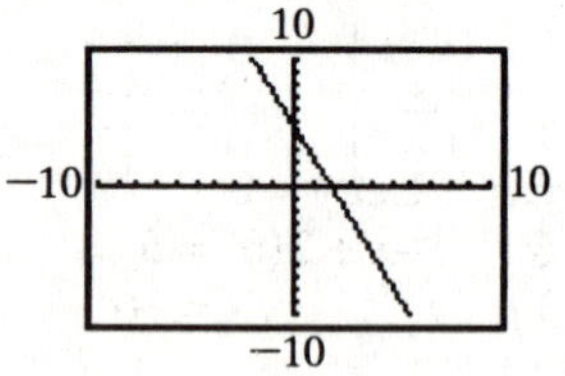

5. $y = |x| + 3.78$

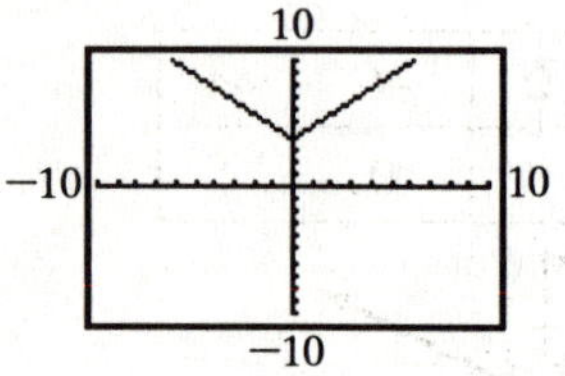

7. $y = 5.6x^2 + 7.7x + 1.5$

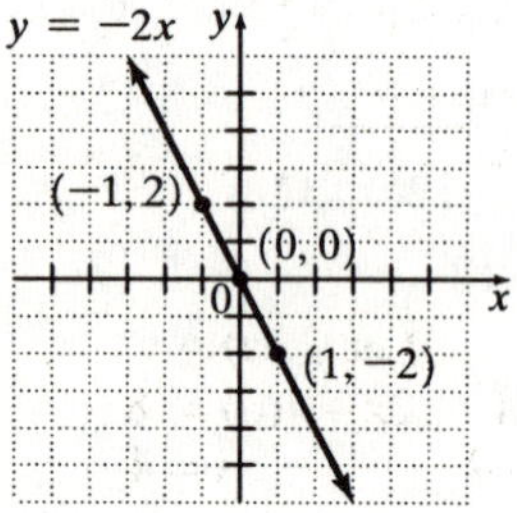

Exercise Set 3.3

1. $f(x) = -2x$
Plot the points to obtain the graph.

x	y
0	0
−1	2
1	−2

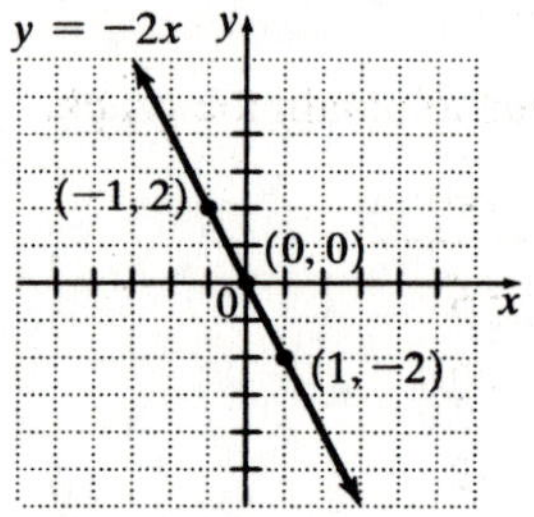

3. $f(x) = -2x + 3$
Plot the points to obtain the graph.

x	y
0	3
1	1
−1	5

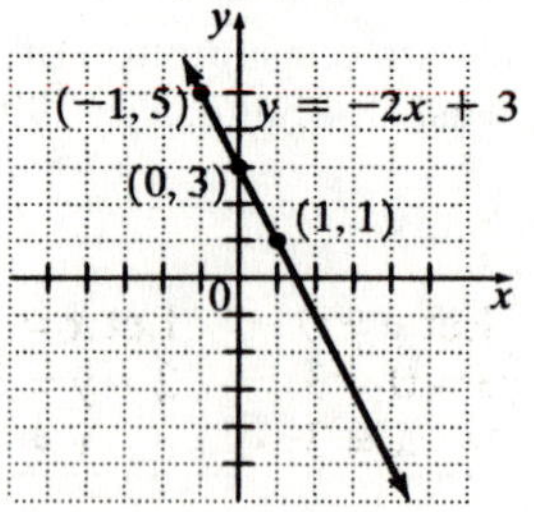

5. $f(x) = \dfrac{1}{2}x$
Plot the points to obtain the graph.

x	y
0	0
2	1
−2	−1

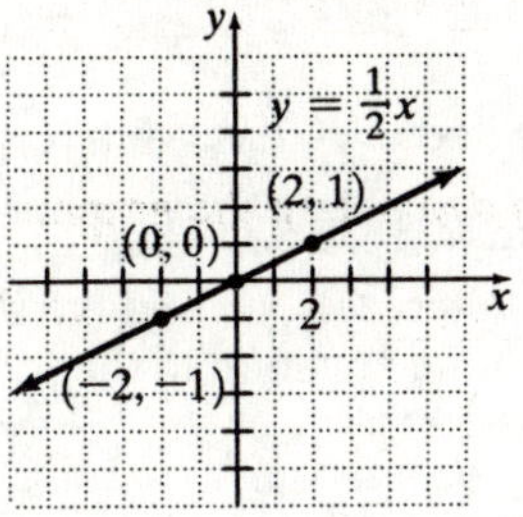

7. $f(x) = \frac{1}{2}x - 4$

Plot the points to obtain the graph.

x	y
0	−4
2	−3
4	−2

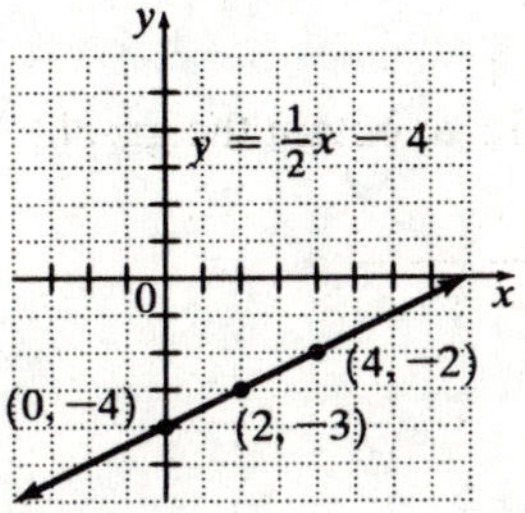

9. C

11. D

13. $x - y = 3$

Let $x = 0$ Let $y = 0$ Let $x = 2$

$0 - y = 3$ $x - 0 = 3$ $2 - y = 3$

$y = -3$ $x = 3$ $y = -1$

x	0	3	2
y	−3	0	−1

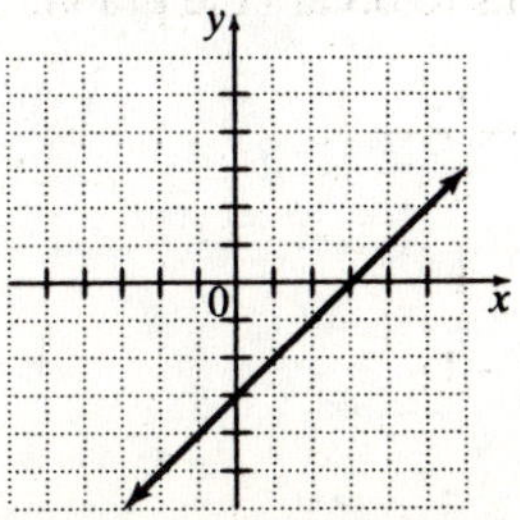

15. $x = 5y$

Let $x = 0$ Let $x = 5$ Let $x = -5$

$0 = 5y$ $5 = 5y$ $-5 = 5y$

$y = 0$ $y = 1$ $y = -1$

x	0	5	−5
y	0	1	−1

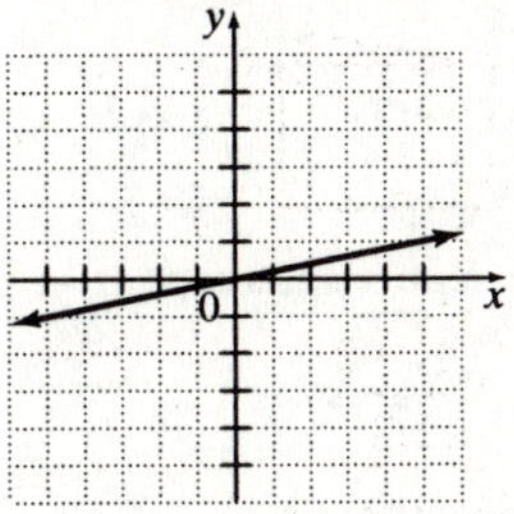

17. $-x + 2y = 6$

Let $x = 0$ Let $y = 0$

$-0 + 2y = 6$ $-x + 2(0) = 6$

$y = 3$ $x = -6$

Let $x = -4$

$-(-4) + 2y = 6$

$y = 1$

x	0	−6	−4
y	3	0	1

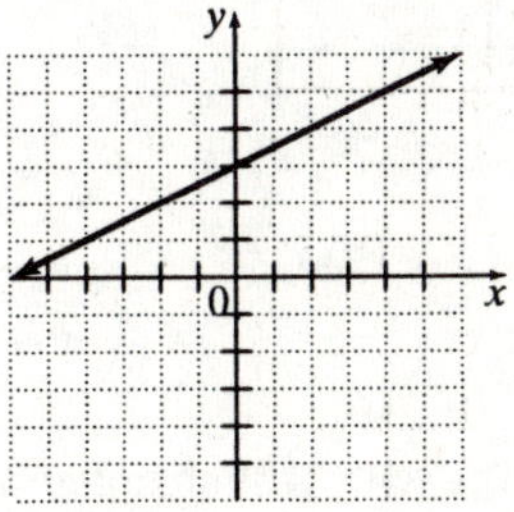

19. $2x - 4y = 8$

Let $x = 0$ Let $y = 0$

$2(0) - 4y = 8$ $2x - 4(0) = 8$

$y = -2$ $x = 4$

Let $x = 2$

$2(2) - 4y = 8$

$y = -1$

x	0	4	2
y	−2	0	−1

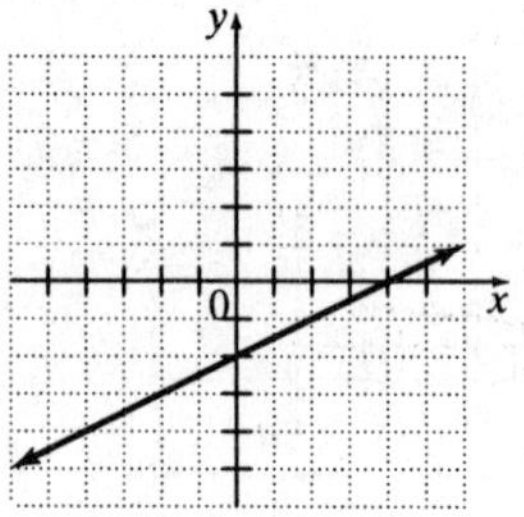

21. Answers may vary.

23. $x = -1$
Vertical line with x-intercept at -1

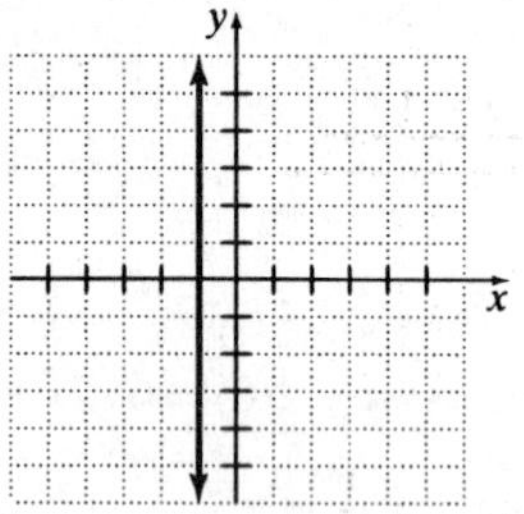

25. $y = 0$
Horizontal line with y-intercept at 0

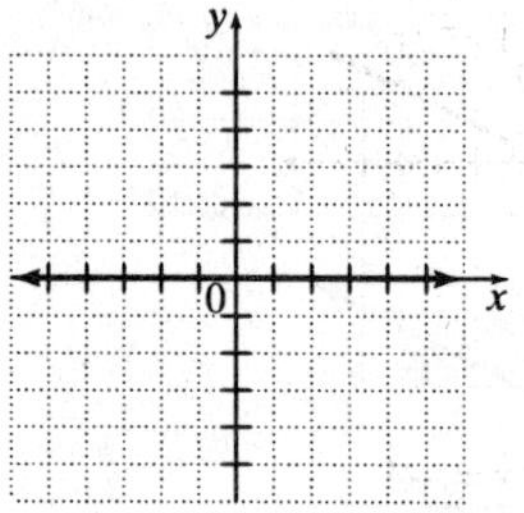

27. $y + 7 = 0$
$y = -7$
Horizontal line with y-intercept at -7

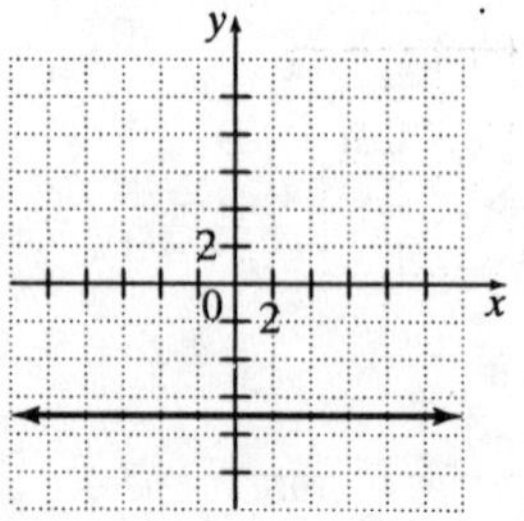

29. C

31. A

33. The vertical line $x = 0$ has y-intercepts.

For exercises 35–59, find ordered pair solutions, or find the x-and y-intercepts, or find the slope and y-intercept with the equation in the form $y = mx + b$.

35. $x + 2y = 8$

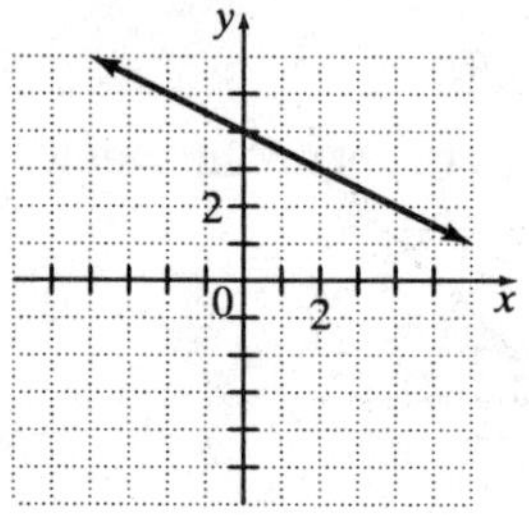

37. $f(x) = \frac{3}{4}x + 2$ or $y = \frac{3}{4}x + 2$

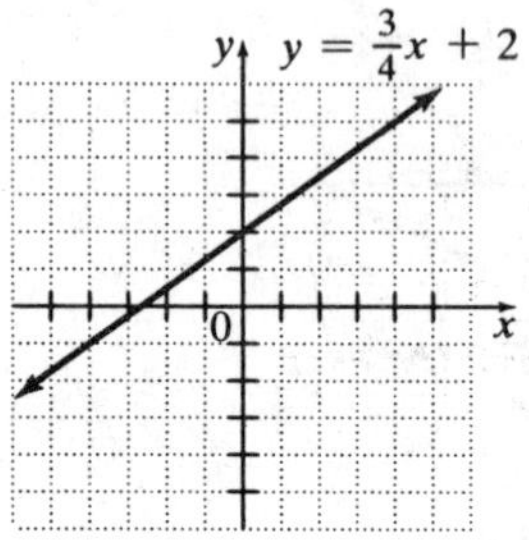

39. $x = -3$

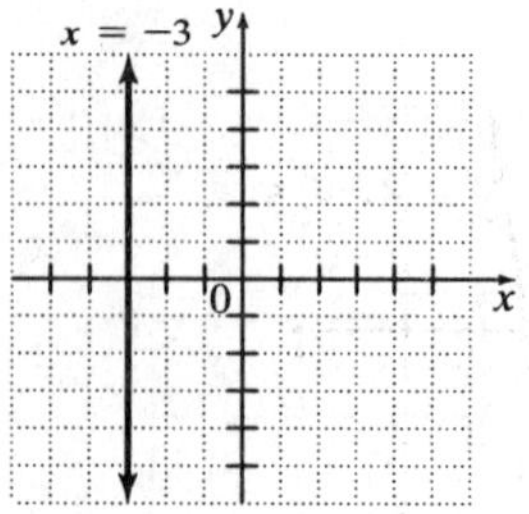

41. $3x + 5y = 7$

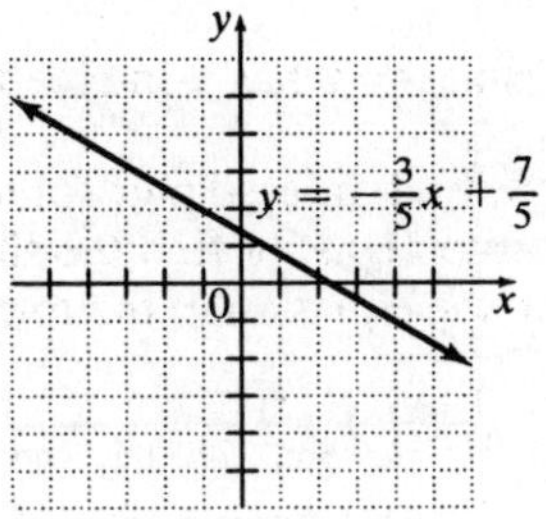

43. $f(x) = x$ or $y = x$

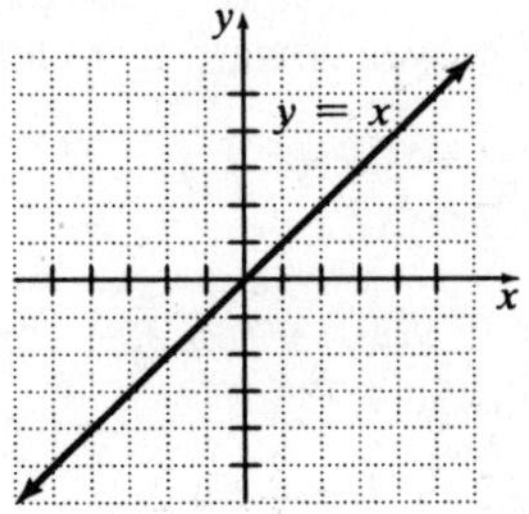

45. $x + 8y = 8$

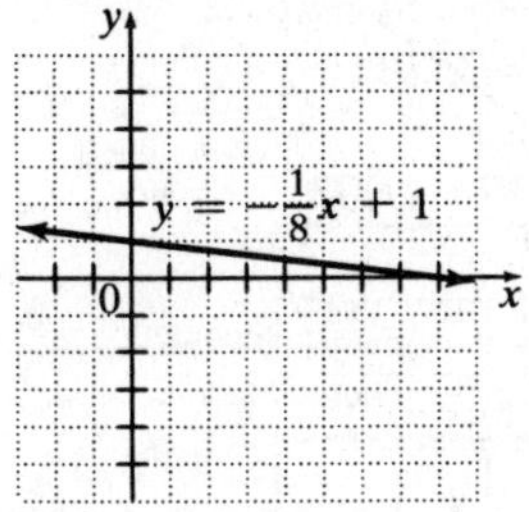

47. $5 = 6x - y$

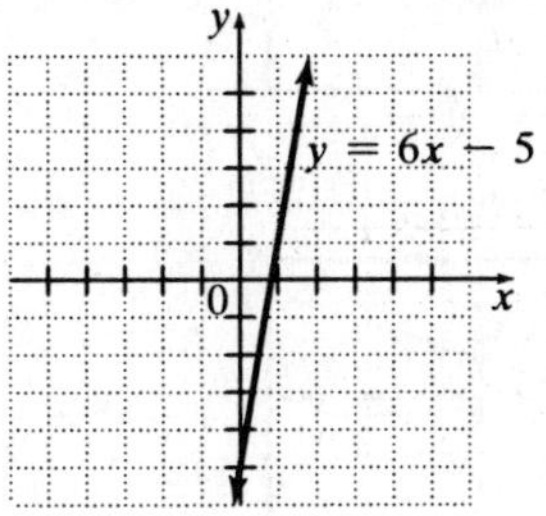

49. $-x + 10y = 11$

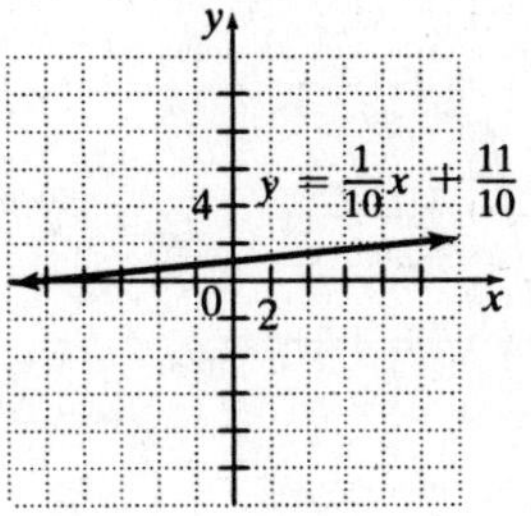

51. $y = 1$

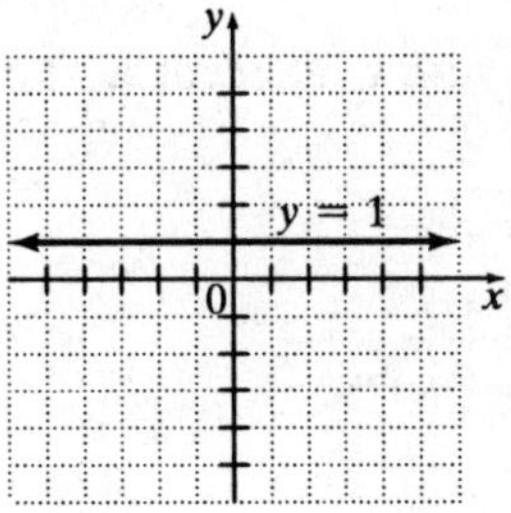

53. $f(x) = \frac{1}{2}x$ or $y = \frac{1}{2}x$

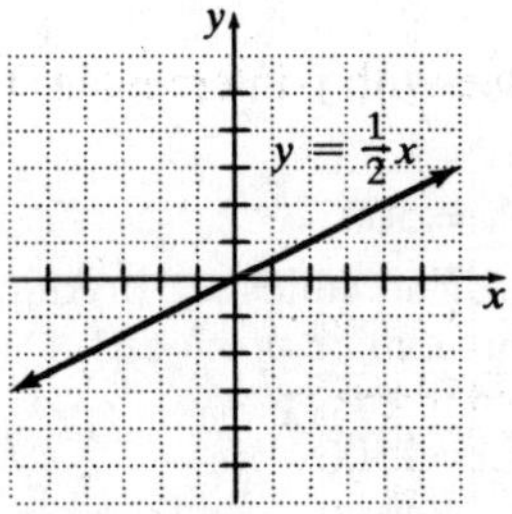

55. $x + 3 = 0$ or $x = -3$

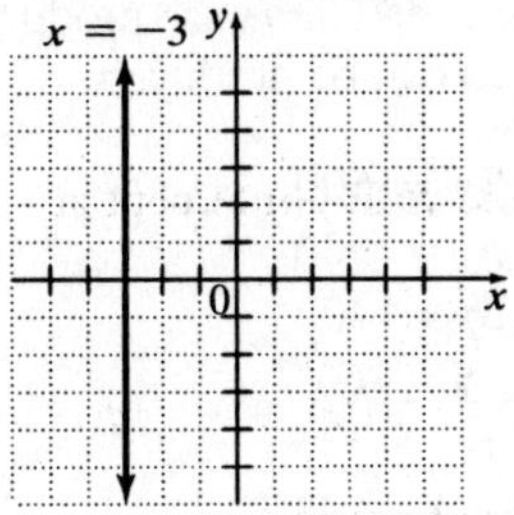

57. $f(x) = 4x - \frac{1}{3}$ or $y = 4x - \frac{1}{3}$

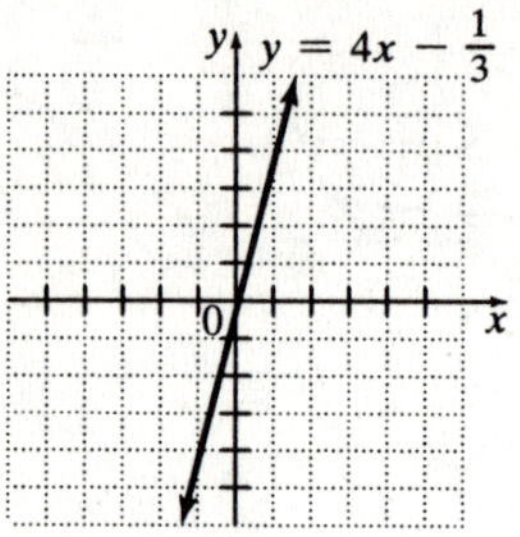

59. $2x + 3y = 6$

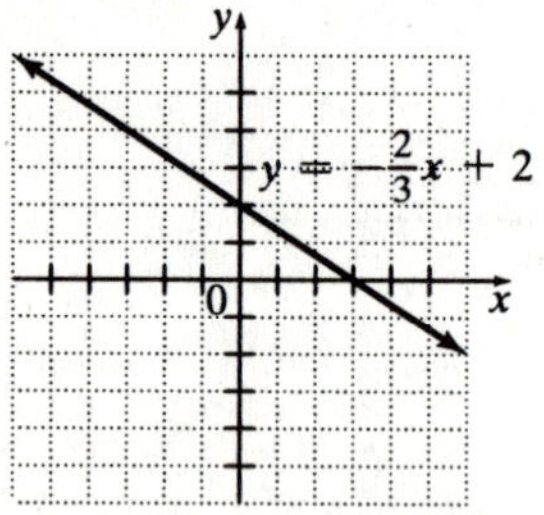

61. $2x + 3y = 1500$

a. $2(0) + 3y = 1500$
$3y = 1500$
$y = 500$
(0, 500); If no tables are produced, 500 chairs can be produced.

b. $2x + 3(0) = 1500$
$2x = 1500$
$x = 750$
(750, 0); If no chairs are produced, 750 tables can be produced.

c. $2(50) + 3y = 1500$
$100 + 3y = 1500$
$3y = 1400$
$y = 466.7$
466 chairs

63. $C(x) = 0.2x + 24$

a. $C(200) = 0.2(200) + 24$
$= 40 + 24$
$= 64$
\$64

b.

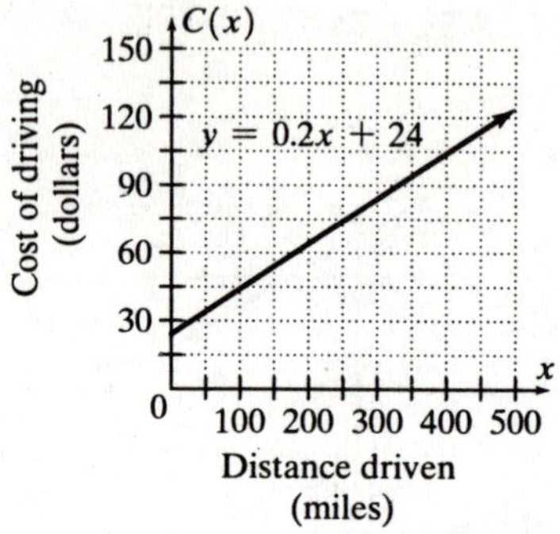

c. The line moves upward from left to right.

65. $f(t) = 72.9t + 785.2$

a. $f(20) = 72.9(20) + 785.2$
$= 1458 + 785.2$
$= 2243.2$
\$2243.20

b. $2000 = 72.9t + 785.2$
$1214.8 = 72.9t$
$16.66 = t$
1990 + 17 = 2007

c. Answers may vary.

67.

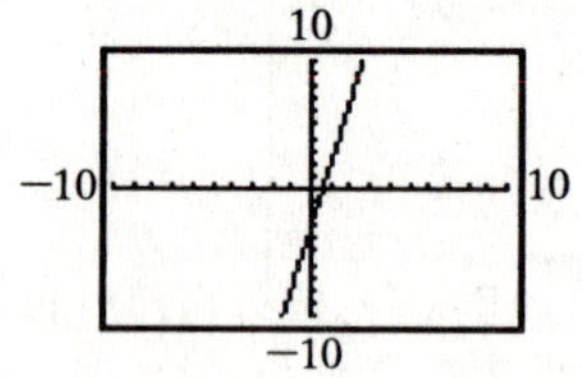

69.

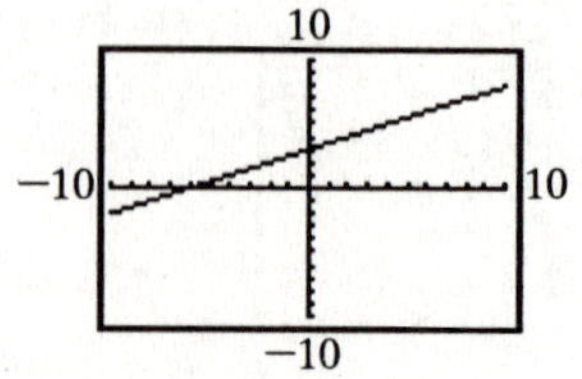

71. $|x - 3| = 6$
$x - 3 = 6$ or $x - 3 = -6$
$x = 9$ or $x = -3$
The solutions are –3 and 9.

73. $|2x+5|>3$
$2x+5<-3$ or $2x+5>3$
$2x<-8$ or $2x>-2$
$x<-4$ or $x>-1$
$(-\infty, -4)\cup(-1, \infty)$

75. $|3x-4|\le 2$
$-2\le 3x-4\le 2$
$2\le 3x\le 6$
$\frac{2}{3}\le x\le 2$
$\left[\frac{2}{3}, 2\right]$

77. $\frac{-6-3}{2-8}=\frac{-9}{-6}=\frac{3}{2}$

79. $\frac{-8-(-2)}{-3-(-2)}=\frac{-8+2}{-3+2}=\frac{-6}{-1}=6$

81. $\frac{0-6}{5-0}=-\frac{6}{5}$

Section 3.4

Graphing Calculator Explorations

1.

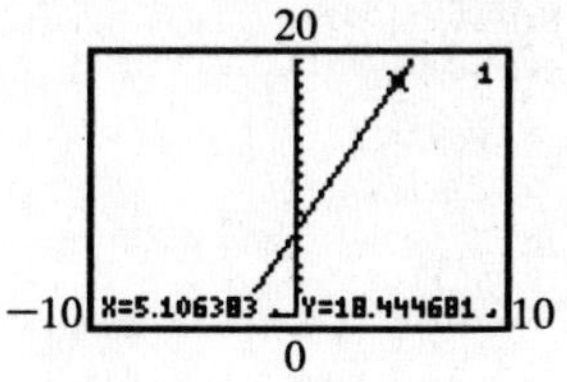

$y = 18.4$

3.

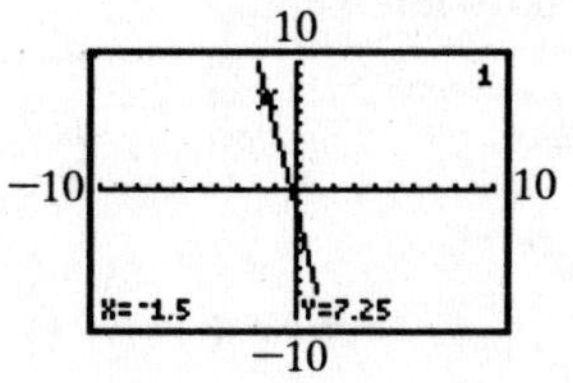

$x = -1.5$

5.

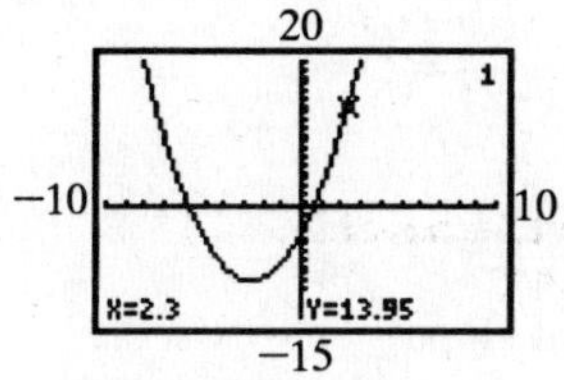

$y = 14.0$

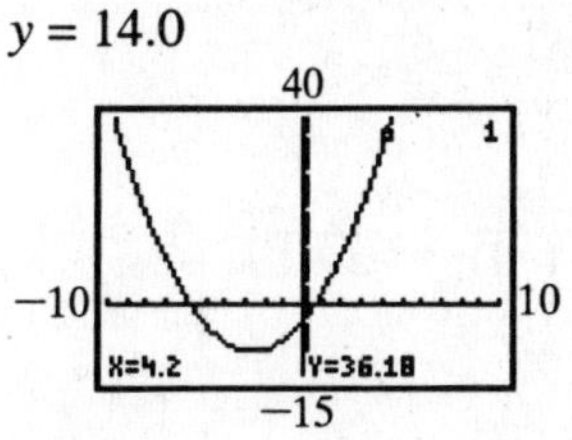

$x = 4.2$

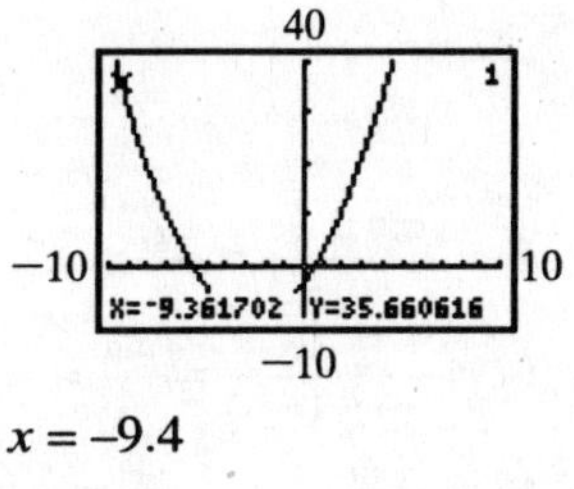

$x = -9.4$

Mental Math

1. $m=\frac{7}{6}$ slants upward.

2. $m = -3$ slants downward.

3. $m = 0$ slants horizontally.

4. m is undefined, slants vertically.

Exercise Set 3.4

1. $m=\frac{11-2}{8-3}=\frac{9}{5}$

3. $m=\frac{8-1}{1-3}=\frac{7}{-2}=-\frac{7}{2}$

5. $m=\frac{3-8}{4-(-2)}=-\frac{5}{6}$

7. $m=\frac{-4-(-6)}{4-(-2)}=\frac{2}{6}=\frac{1}{3}$

9. $m=\frac{11-(-1)}{-12-(-3)}=\frac{12}{-9}=-\frac{4}{3}$

11. $m = \frac{5-5}{3-(-2)} = \frac{0}{5} = 0$

13. $m = \frac{-5-1}{-1-(-1)} = -\frac{6}{0}$
undefined slope

15. $m = \frac{0-6}{-3-0} = \frac{-6}{-3} = 2$

17. $m = \frac{4-2}{-3-(-1)} = \frac{2}{-3+1} = \frac{2}{-2} = -1$

19. ℓ_2

21. ℓ_2

23. ℓ_2

25. a. $m \text{ for } \ell_1 = \frac{-2-4}{2-(-1)} = \frac{-6}{3} = -2$

$m \text{ for } \ell_2 = \frac{2-6}{-4-(-8)} = \frac{-4}{4} = -1$

$m \text{ for } \ell_3 = \frac{-4-0}{0-(-6)} = \frac{-4}{6} = -\frac{2}{3}$

b. Lesser

27. $f(x) = -2x + 6$
$m = -2, b = 6$

29. $-5x + y = 10$
$y = 5x + 10$
$m = 5, b = 10$

31. $-3x - 4y = 6$
$4y = -3x - 6$
$y = -\frac{3}{4}x - \frac{3}{2}$
$m = -\frac{3}{4},\ b = -\frac{3}{2}$

33. $f(x) = -\frac{1}{4}x$
$m = -\frac{1}{4},\ b = 0$

35. $f(x) = 2x - 3$
$m = 2, b = -3$
D

37. $f(x) = -2x - 3$
$m = -2, b = -3$
C

39. $y = -2$
$m = 0$

41. $x = 4$
m is undefined

43. $y - 7 = 0$
$y = 7$
$m = 0$

45. Answers may vary.

47. $f(x) = x + 2$
$m = 1, b = 2$

49. $4x - 7y = 28$
$7y = 4x - 28$
$y = \frac{4}{7}x - 4$
$m = \frac{4}{7},\ b = -4$

51. $2y - 7 = x$
$2y = x + 7$
$y = \frac{1}{2}x + \frac{7}{2}$
$m = \frac{1}{2},\ b = \frac{7}{2}$

53. $x = 7$
Slope is undefined.
There is no y-intercept.

55. $f(x) = \frac{1}{7}x$
$m = \frac{1}{7},\ b = 0$

57. $x - 7 = 0$
$x = 7$
Slope is undefined.
There is no y-intercept.

59. $2y+4=-7$

$2y=-11$

$y=-\frac{11}{2}$

$m=0,\ b=-\frac{11}{2}$

61. $f(x)=5x-6 \quad g(x)=5x+2$

$m=5 \quad m=5$

Parallel, since they have the same slope.

63. $2x-y=-10 \quad 2x+4y=2$

$y=2x+10 \quad y=-\frac{1}{2}x+\frac{1}{2}$

$m=2 \quad m=-\frac{1}{2}$

Perpendicular, since the product of their slopes is –1.

65. $x+4y=7 \quad 2x-5y=0$

$y=-\frac{1}{4}x+\frac{7}{4} \quad y=\frac{2}{5}x$

$m=-\frac{1}{4} \quad m=\frac{2}{5}$

Neither, since their slopes are not equal, nor does their product equal –1.

67. Answers may vary.

69. Two points on line: (1, 0), (0, 3)

$m=\frac{3-0}{0-1}=\frac{3}{-1}=-3$

71. Two points on line: (–3, –1), (2, 4)

$m=\frac{4-(-1)}{2-(-3)}=\frac{4+1}{2+3}=\frac{5}{5}=1$

73. $m=\frac{3}{25}$

75. $m=\frac{15}{100}=\frac{3}{20}$

77. $y=1054.7x+23{,}285.9$

a. $x=1996-1991=5$

$y=1054.7(5)+23{,}285.9$

$y=28{,}559.4$

The income is $28,559.40.

b. $m=1054.7$; the annual average income increases by $1054.70 for every one year.

c. $b=23{,}285.9$; at year $x=0$, or 1991, the annual average income was $23,285.90.

79. $-76x+10y=1130$

$10y=76x+1130$

$y=7.6x+113$

a. $m=7.6$; $b=113$

b. The number of people employed as paralegals increases by 7.6 thousand for every one year.

c. There were 113 thousand paralegals employed in 1996.

81. $f(x)=72.9x+785.2$

a. $m=72.9$; the yearly cost of tuition increases by $72.90 every one year.

b. $b=785.2$; the yearly cost of tuition in 1990 was $785.20.

83. $f(x)=x$

$m=1$

The slope of a parallel line is 1.

85. $f(x)=x$

$m=1$

The slope of a perpendicular line is –1.

87. $-3x+4y=0$

$y=\frac{3}{4}x$

$m=\frac{3}{4}$

The slope of a parallel line is $\frac{3}{4}$.

89. **a.** (6, 20)

b. (10, 13)

c. $m = \frac{13-20}{10-6} = -\frac{7}{4}$, or -1.75 yards per second

d. $m = \frac{8-2}{26-22} = \frac{6}{4} = \frac{3}{2}$, or 1.5 yards per second

91. $-4x + 2y = 5$
$2x - y = 7$

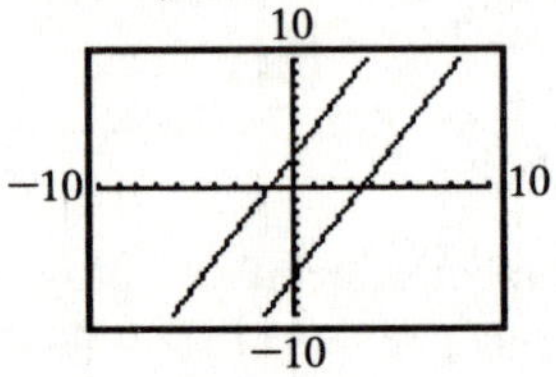

93. a. $y = \frac{1}{2}x + 1$
$y = x + 1$
$y = 2x + 1$

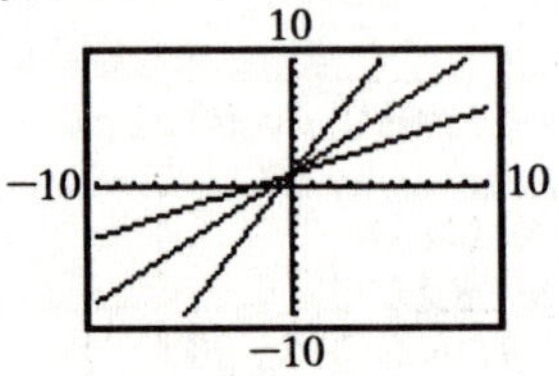

b. $y = -\frac{1}{2}x + 1$
$y = -x + 1$
$y = -2x + 1$

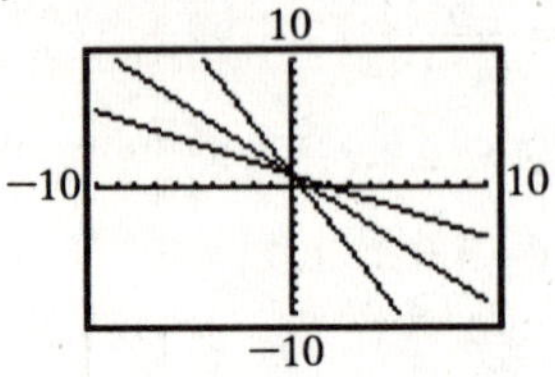

c. True

95. $P(B) = \frac{2}{11}$

97. $P(I \text{ or } T) = \frac{3}{11}$

99. $P(\text{vowel}) = \frac{4}{11}$

101. $y - 0 = -3[x - (-10)]$
$y = -3(x + 10)$
$y = -3x - 30$

103. $y - 9 = -8[x - (-4)]$
$y - 9 = -8(x + 4)$
$y - 9 = -8x - 32$
$y = -8x - 23$

Section 3.5

Mental Math

1. $m = -4, b = 12$

2. $m = \frac{2}{3}, b = -\frac{7}{2}$

3. $m = 5, b = 0$

4. $m = -1, b = 0$

5. $m = \frac{1}{2}, b = 6$

6. $m = -\frac{2}{3}, b = 5$

7. Parallel

8. Parallel

9. Neither

10. Neither

Exercise Set 3.5

1. $y = -x + 1$

3. $y = 2x + \frac{3}{4}$

5. $y = \frac{2}{7}x$

7. $y = 5x$

possible points: (0, 0), (1, 5)

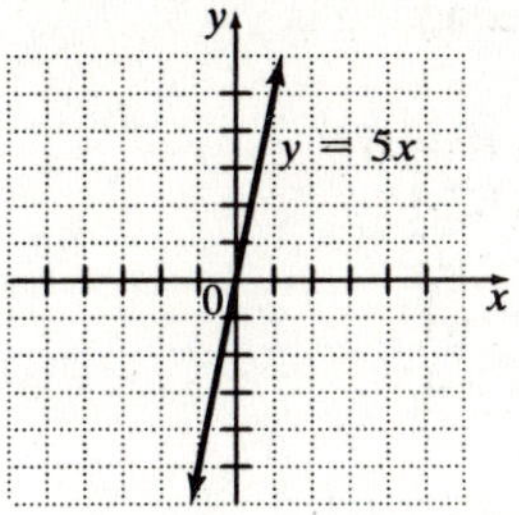

9. $x + y = 7$

$y = -x + 7$

possible points: (0, 7), (1, 6)

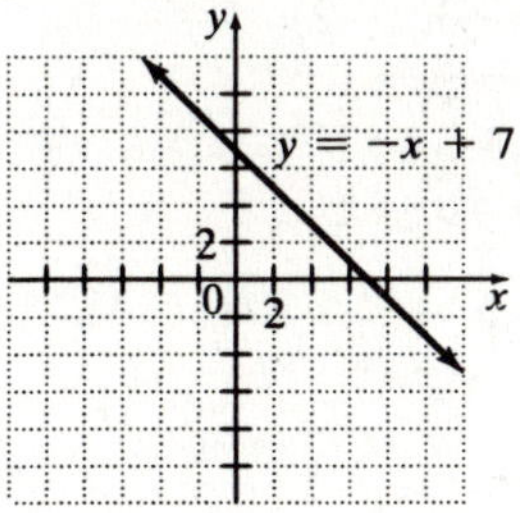

11. $-3x + 2y = 3$

$2y = 3x + 3$

$y = \frac{3}{2}x + \frac{3}{2}$

possible points: $\left(0, \frac{3}{2}\right), \left(2, \frac{9}{2}\right)$

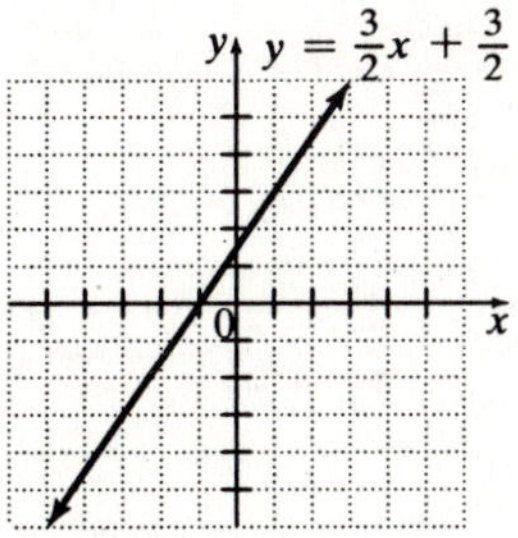

13. $y - y_1 = m(x - x_1)$

$y - 2 = 3(x - 1)$

$y - 2 = 3x - 3$

$y = 3x - 1$

15. $y - y_1 = m(x - x_1)$

$y - (-3) = -2(x - 1)$

$y + 3 = -2x + 2$

$y = -2x - 1$

17. $y - y_1 = m(x - x_1)$

$y - 2 = \frac{1}{2}[x - (-6)]$

$y - 2 = \frac{1}{2}(x + 6)$

$y - 2 = \frac{1}{2}x + 3$

$y = \frac{1}{2}x + 5$

19. $y - y_1 = m(x - x_1)$

$y - 0 = -\frac{9}{10}[x - (-3)]$

$y = -\frac{9}{10}(x + 3)$

$y = -\frac{9}{10}x - \frac{27}{10}$

21. (0, 3), (1, 1)

$m = \frac{1-3}{1-0} = -\frac{2}{1} = -2$

$b = 3$

$y = -2x + 3$

$2x + y = 3$

23. (−2, 1), (4, 5)

$m = \frac{5-1}{4-(-2)} = \frac{4}{6} = \frac{2}{3}$

$y - 1 = \frac{2}{3}(x + 2)$

$3y - 3 = 2x + 4$

$2x - 3y = -7$

25. $m = \frac{6-0}{4-2} = \frac{6}{2} = 3$

$y - 0 = 3(x - 2)$

$y = 3x - 6$

$f(x) = 3x - 6$

27. $m = \dfrac{13-5}{-6-(-2)} = \dfrac{8}{-4} = -2$
$y-5 = -2(x-(-2))$
$y-5 = -2(x+2)$
$y-5 = -2x-4$
$y = -2x+1$
$f(x) = -2x+1$

29. $m = \dfrac{-3-(-4)}{-4-(-2)} = \dfrac{1}{-2} = -\dfrac{1}{2}$
$y-(-4) = -\dfrac{1}{2}(x-(-2))$
$2(y+4) = -(x+2)$
$2y+8 = -x-2$
$y = -\dfrac{1}{2}x - \dfrac{10}{2}$
$f(x) = -\dfrac{1}{2}x-5$

31. $m = \dfrac{-9-(-8)}{-6-(-3)} = \dfrac{-1}{-3} = \dfrac{1}{3}$
$y-(-8) = \dfrac{1}{3}(x-(-3))$
$3(y+8) = x+3$
$3y+24 = x+3$
$3y = x-21$
$f(x) = \dfrac{1}{3}x-7$

33. Answers may vary.

35. $f(0) = -2$

37. $f(2) = 2$

39. $f(x) = -6$
$f(-2) = -6$
$x = -2$

41. $y = mx + b$
$-4 = 0(-2) + b$
$-4 = b$
$y = -4$

43. Every vertical line is in the form $x = c$. Since the line passes through the point (4, 7), its equation is $x = 4$.

45. Every horizontal line is in the form $y = c$. Since the line passes through the point (0, 5), its equation is $y = 5$.

47. $y = 4x - 2$ so $m = 4$
$y-8 = 4(x-3)$
$y-8 = 4x-12$
$y = 4x-4$
$f(x) = 4x-4$

49. $3y = x-6$ or $y = \dfrac{1}{3}x-2$

So $m_{\perp} = \dfrac{1}{3}$ and $m = -3$.

Now,
$y-(-5) = -3(x-2)$
$y+5 = -3x+6$
$y = -3x+1$
$f(x) = -3x+1$

51. $3x+2y = 5$
$2y = -3x+5$
$y = -\dfrac{3}{2}x+\dfrac{5}{2}$

so $m = \dfrac{-3}{2}$

Now,
$y-(-3) = -\dfrac{3}{2}(x-(-2))$
$2(y+3) = -3(x+2)$
$2y+6 = -3x-6$
$y = -\dfrac{3}{2}x-6$
$f(x) = -\dfrac{3}{2}x-6$

53. $y-3 = 2(x-(-2))$
$y-3 = 2(x+2)$
$y-3 = 2x+4$
$2x-y = -7$

55. $m = \dfrac{2-6}{5-1} = \dfrac{-4}{4} = -1$
$y-6 = -1(x-1)$
$y-6 = -x+1$
$y = -x+7$
$f(x) = -x+7$

57. $y = -\frac{1}{2}x + 11$
$2y = -x + 22$
$x + 2y = 22$

59. $m = \frac{-6-(-4)}{0-(-7)} = \frac{-2}{7}$
$y = -\frac{2}{7}x - 6$
$7y = -2x - 42$
$2x + 7y = -42$

61. $y - 0 = -\frac{4}{3}(x - (-5))$
$3y = -4(x+5)$
$3y = -4x - 20$
$4x + 3y = -20$

63. $x = -2$

65. $2x + 4y = 8$
$4y = -2x + 8$
$y = -\frac{1}{2}x + 2,$
so $m = -\frac{1}{2}$. Now,
$y - (-2) = -\frac{1}{2}(x - 6)$
$2(y+2) = -x + 6$
$2y + 4 = -x + 6$
$x + 2y = 2$

67. $y = 12$

69. $8x - y = 9$
$8x - 9 = y$ or $y = 8x - 9$
so $m = 8$
$y - 1 = 8(x - 6)$
$y - 1 = 8x - 48$
$47 = 8x - y$ or $8x - y = 47$

71. $x = 5$

73. $m = \frac{-5-(-8)}{-6-2} = \frac{3}{-8} = -\frac{3}{8}$
$y - (-8) = -\frac{3}{8}(x - 2)$
$8(y+8) = -3(x-2)$
$8y + 64 = -3x + 6$
$y = -\frac{3}{8}x - \frac{29}{4}$
$f(x) = -\frac{3}{8}x - \frac{29}{4}$

75. a. (1, 30,000), (4, 66,000)
$m = \frac{66,000 - 30,000}{4-1} = 12,000$
$y - 30,000 = 12,000(x-1)$
$y = 12,000x + 18,000$
$P(x) = 12,000x + 18,000$

b. $P(7) = 12,000(7) + 18,000$
$= \$102,000$

c. $126,000 = 12,000x + 18,000$
$x = \frac{126,000 - 18,000}{12,000}$
$x = 9$ years

77. a. (3, 10,000), (5, 8000)
$m = \frac{8000 - 10,000}{5-3} = -1000$
$y - 10,000 = -1000(x - 3)$
$y = -1000x + 13,000$

b. $y = -1000(3.5) + 13,000$
$y = 9500$
9500 Fun Noodles

79. a. (0, 109,900), (4, 128,400)
$m = \frac{128,400 - 109,900}{4-0}$
$= \frac{18,500}{4}$
$= 4625$
$y - 109,900 = 4625(x - 0)$
$y = 4625x + 109,900$

b. $x = 2008 - 1994 = 14$
$y = 4625(14) + 109,900$
$y = \$174,650$

81. a. (0, 225), (10, 391)

$$m = \frac{391-225}{10-0} = 16.6$$

$$y - 225 = 16.6(x - 0)$$

$$y = 16.6x + 225$$

b. $x = 2004 - 1996 = 8$

$y = 16.6(8) + 225$

$y = 357.8$ thousand people

83. $f(x) = -x + 7$

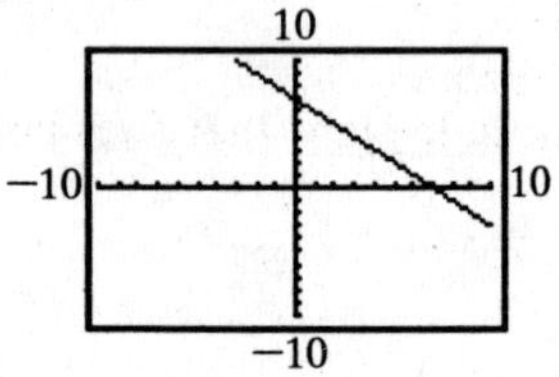

85. $4x + 3y = -20$

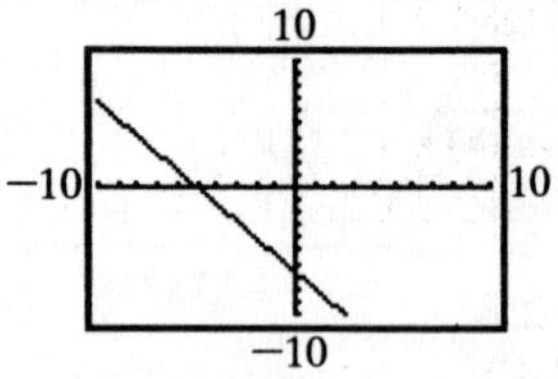

87. $2x - 7 \le 21$

$$2x \le 28$$

$$x \le 14$$

$(-\infty, 14]$

89. $5(x-2) \ge 3(x-1)$

$$5x - 10 \ge 3x - 3$$

$$2x \ge 7$$

$$x \ge \frac{7}{2}$$

$\left[\frac{7}{2}, \infty\right)$

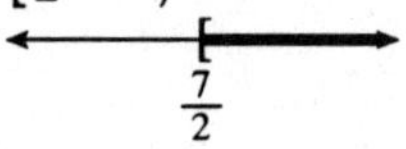

91. $\frac{x}{2} + \frac{1}{4} < \frac{1}{8}$

$$4x + 2 < 1$$

$$4x < -1$$

$$x < -\frac{1}{4}$$

$4x + 2 < 1$

$4x < -1$

$x < -\frac{1}{4}$

$\left(-\infty, -\frac{1}{4}\right)$

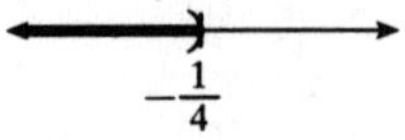

93. $m_\perp = \frac{1-(-1)}{-5-3} = \frac{2}{-8} = -\frac{1}{4}$, so $m = 4$

$$m((3, -1), (-5, 1)) = \left(\frac{3-5}{2}, \frac{-1+1}{2}\right) = (-1, 0)$$

so

$$y - 0 = 4(x - (-1))$$

$$y = 4(x + 1)$$

$$y = 4x + 4$$

$$-4x + y = 4$$

95. $m_\perp = \frac{-4-6}{-22-(-2)} = \frac{-10}{-20} = \frac{1}{2}$

so $m = -2$

$$m((-2, 6), (-22, -4)) = \left(\frac{-2-22}{2}, \frac{6-4}{2}\right) = (-12, 1)$$

Now, $y - 1 = -2(x - (-12)) = -2(x + 12)$

so $y - 1 = -2x - 24$ or $2x + y = -23$

97. $m_\perp = \frac{7-3}{-4-2} = \frac{4}{-6} = \frac{-2}{3}$, so $m = \frac{3}{2}$

$m((2, 3), (-4, 7)) = \left(\frac{2-4}{2}, \frac{3+7}{2}\right)$

$= (-1, 5)$

Now, $y - 5 = \frac{3}{2}(x - (-1))$

$2y - 10 = 3(x + 1) = 3x + 3$

or $3x - 2y = -13$

Exercise Set 3.6

1. $x < 2$

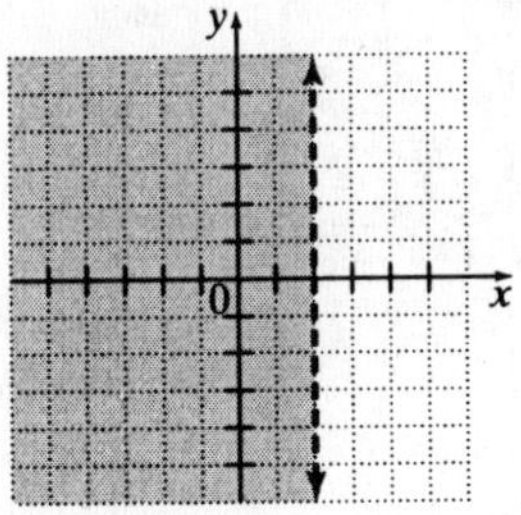

3. $x - y \geq 7$

Test: (0, 0)

$0 - 0 \geq 7$

$0 \geq 7$ False

Shade the half-plane that does not contain (0, 0).

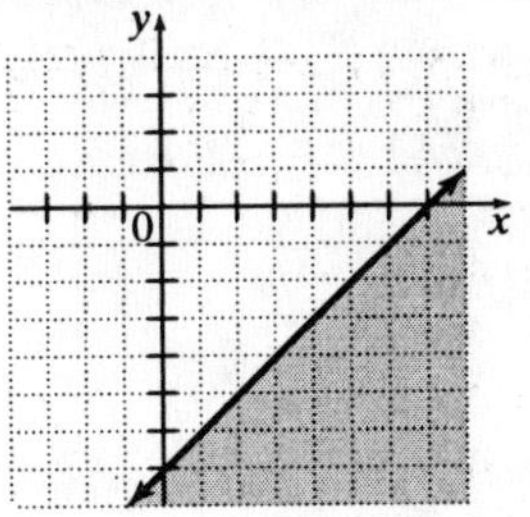

5. $3x + y > 6$

Test: (0, 0)

$3(0) + 0 > 6$

$0 > 6$ False

Shade the half-plane that does not contain (0, 0).

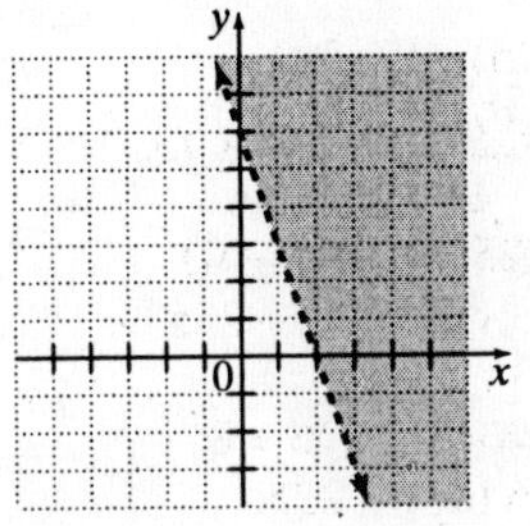

7. $y \leq -2x$

Test: (1, 1)

$1 \leq -2(1)$

$1 \leq -2$ False

Shade the half-plane that does not contain (1, 1).

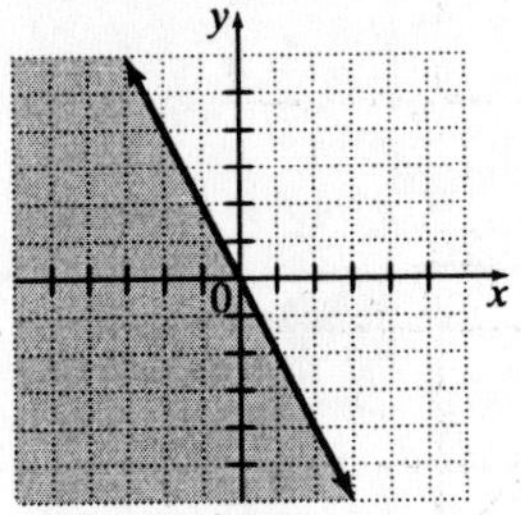

9. $2x + 4y \geq 8$

Test: (0, 0)

$2(0) + 4(0) \geq 8$

$0 \geq 8$ False

Shade the half-plane that does not contain (0, 0).

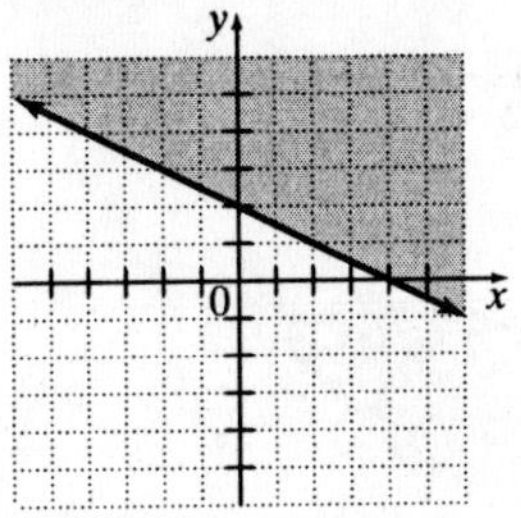

11. $5x + 3y > -15$

Test: (0, 0)

$5(0) + 3(0) > -15$

$0 > -15$ True

Shade the half-plane that contains (0, 0).

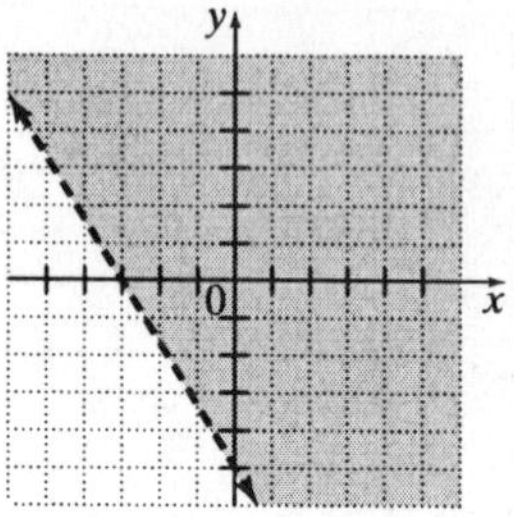

13. Answers may vary. A dashed boundary line should be used when the inequality contains a < or >.

15. $x \geq 3$ and $y \leq -2$

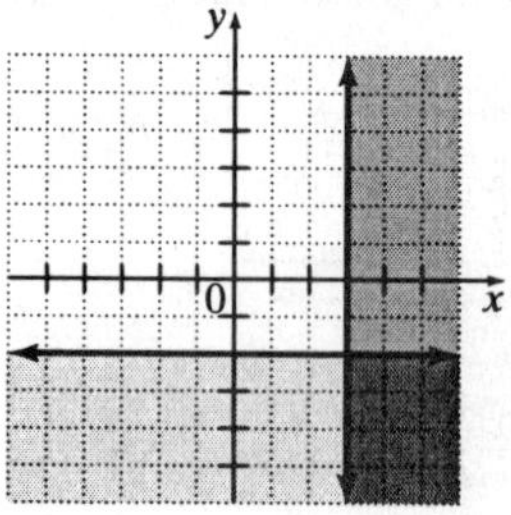

17. $x \leq -2$ or $y \geq 4$

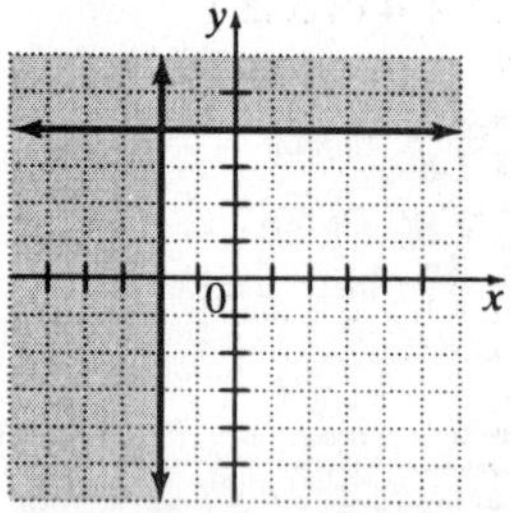

19. $x - y < 3$ and $x > 4$

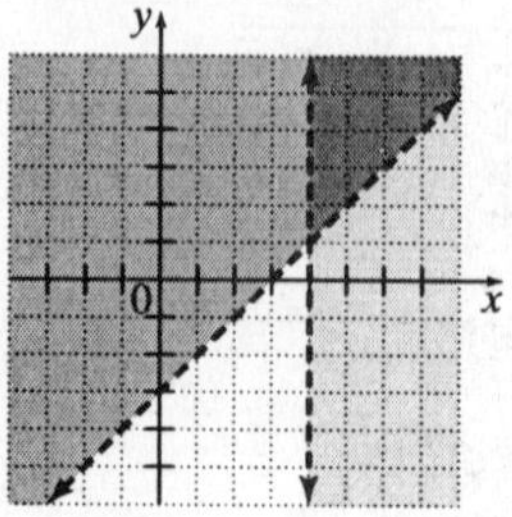

21. $x + y \leq 3$ or $x - y \geq 5$

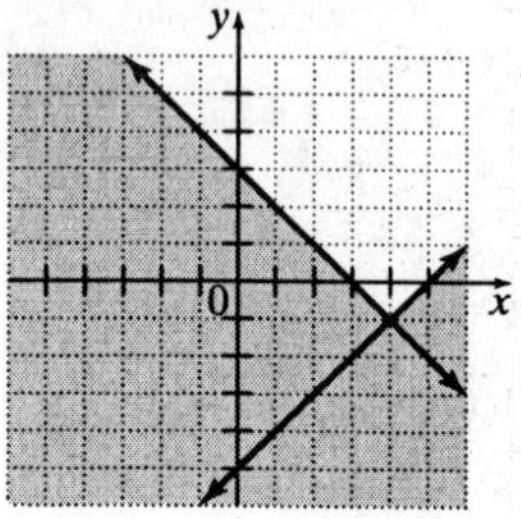

23. $y \geq -2$

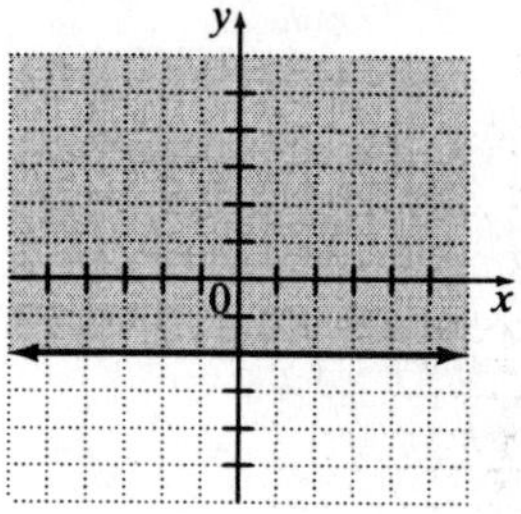

25. $x - 6y < 12$

$x - 12 < 6y$

$y > \frac{1}{6}x - 2$

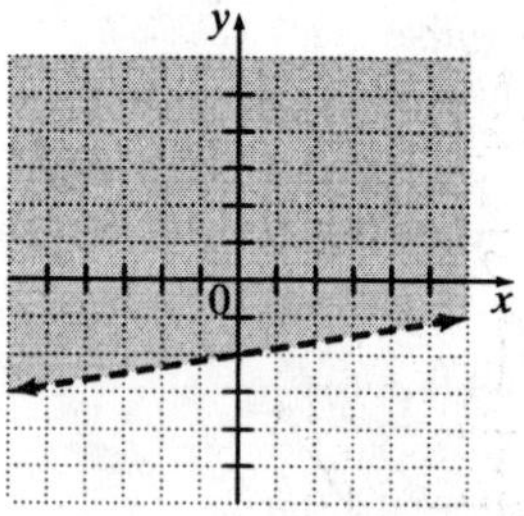

27. $x > 5$

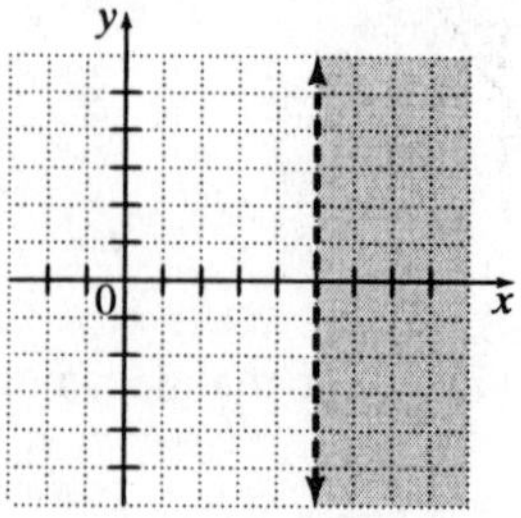

29. $-2x + y \leq 4$

$y \leq 2x + 4$

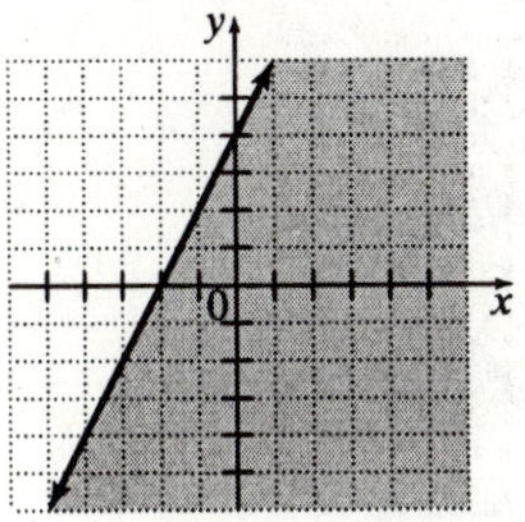

31. $x - 3y < 0$

$x < 3y$

$y > \dfrac{x}{3}$

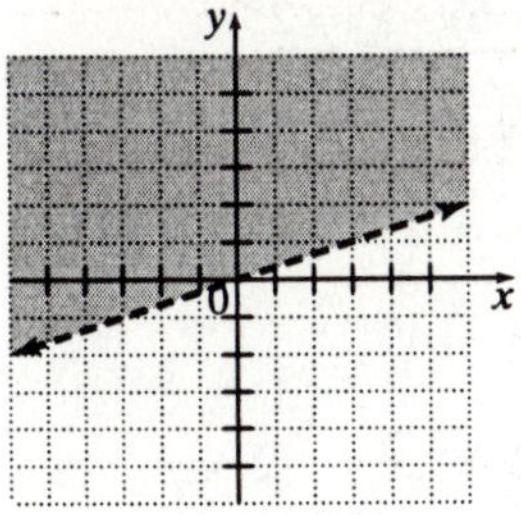

33. $3x - 2y \leq 12$

$3x - 12 \leq 2y$

$y \geq \dfrac{3}{2}x - 6$

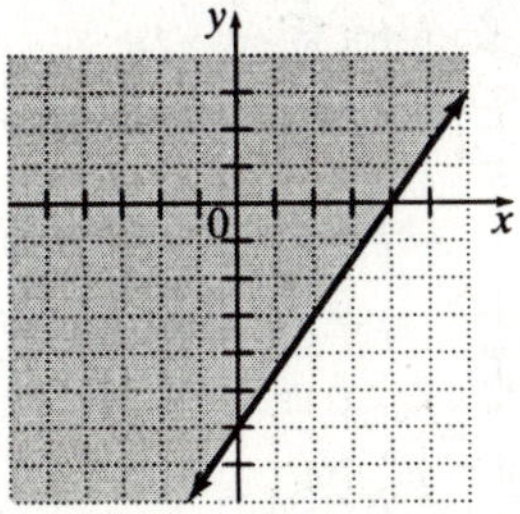

35. $x - y \geq 2 \quad \text{or } y < 5$

$y \leq x - 2 \text{ or } y < 5$

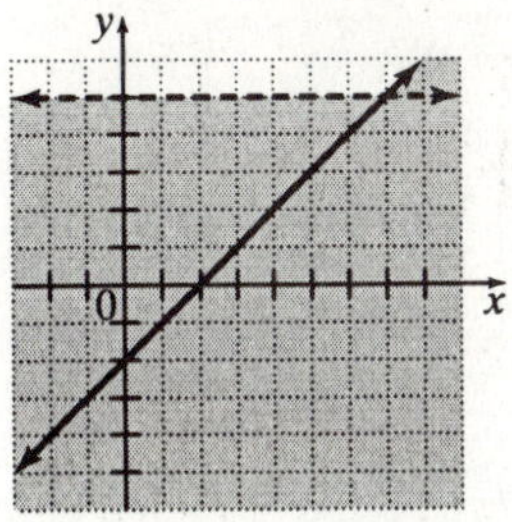

37. $x + y \leq 1 \quad \text{and } y \leq -1$

$y \leq -x + 1 \text{ and } y \leq -1$

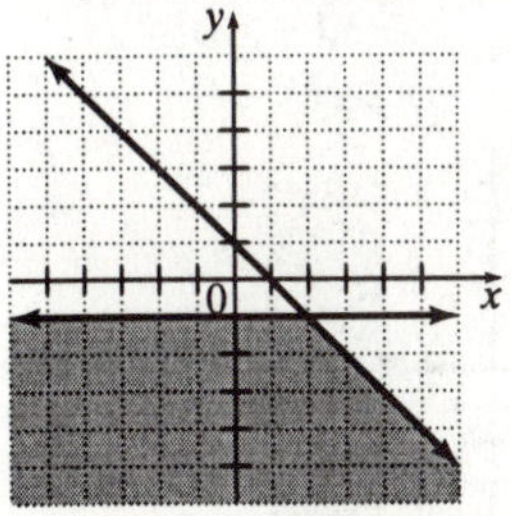

39. $2x + y > 4 \quad \text{or } x \geq 1$

$y > -2x + 4 \text{ or } x \geq 1$

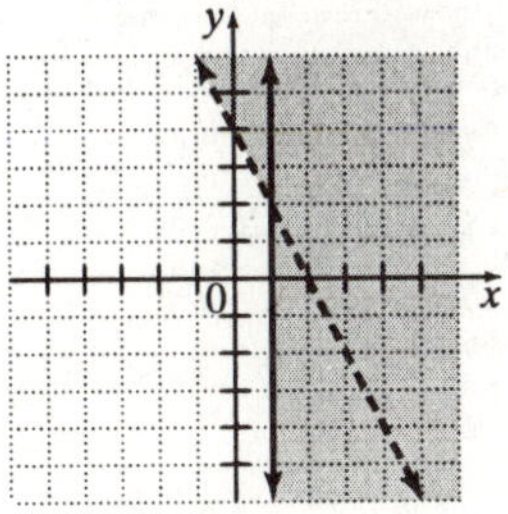

41. $-2 \leq x \leq 1$

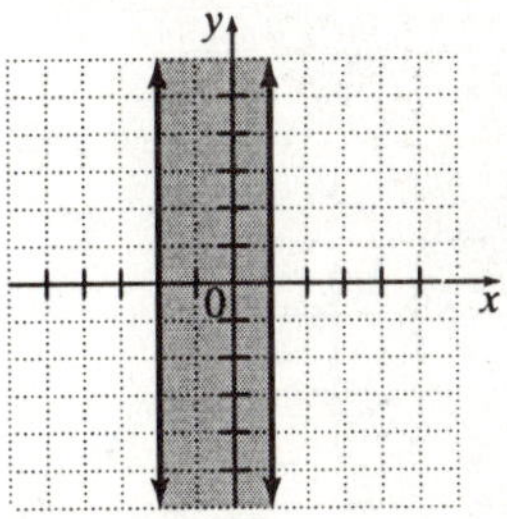

43. $x+y\le 0$ or $3x-6y\ge 12$
$y\le -x$ or $3x-12\ge 6y$
$y\le -x$ or $y\le \dfrac{1}{2}x-2$

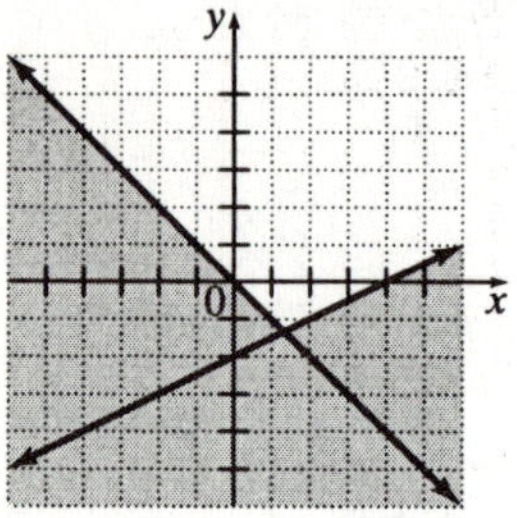

45. $2x-y>3$ and $x\ge 0$
$y<2x-3$ and $x\ge 0$

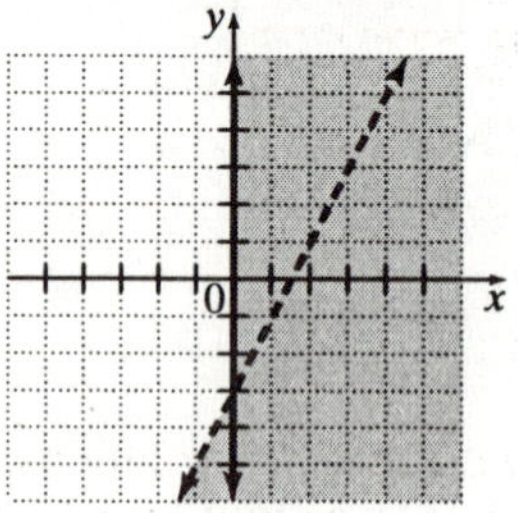

47. D

49. A

51. $x\ge 2$

53. $y\le -3$

55. $y>4$

57. $x<1$

59. $x\le 20$ and $y\ge 10$

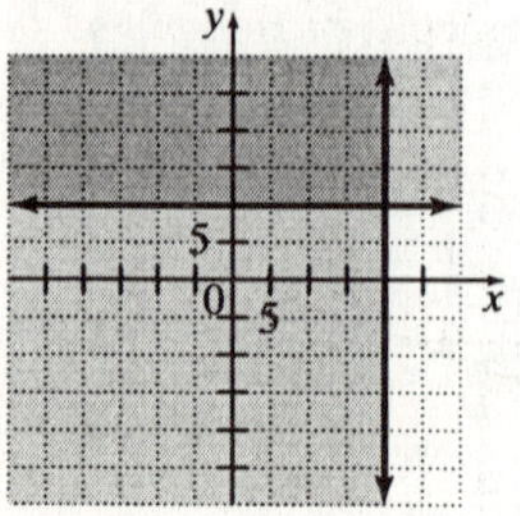

61. $\begin{cases} x\ge 0 \\ y\ge 0 \\ 2x+4y\le 40 \end{cases}$

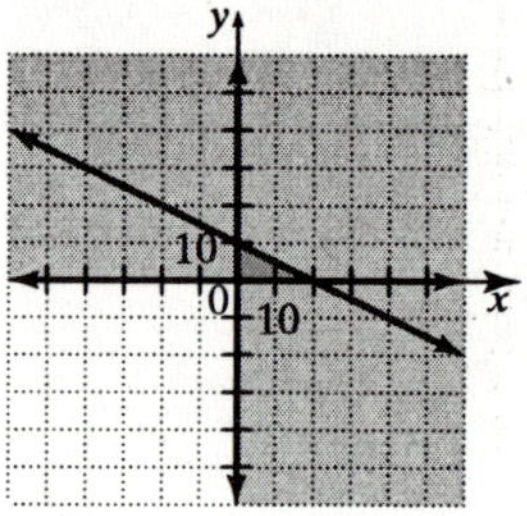

63. $3^2=3\cdot 3=9$

65. $(-5)^2=(-5)(-5)=25$

67. $-2^4=-(2)(2)(2)(2)=-16$

69. $\left(\frac{2}{7}\right)^2=\frac{2^2}{7^2}=\frac{4}{49}$

71. Domain: $(-\infty, 2]\cup[2, \infty)$
Range: $(-\infty, \infty)$
This relation is not a function.

Chapter 3 Review Exercises

1. $A(2, -1)$, quadrant IV
$B(-2, 1)$, quadrant II
$C(0, 3)$, y-axis
$D(-3, -5)$, quadrant III

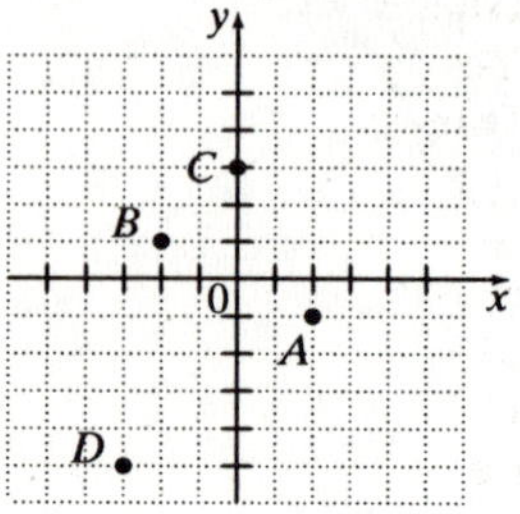

2. $A(-3, 4)$, quadrant II
$B(4, -3)$, quadrant IV
$C(-2, 0)$, x-axis
$D(-4, 1)$, quadrant II

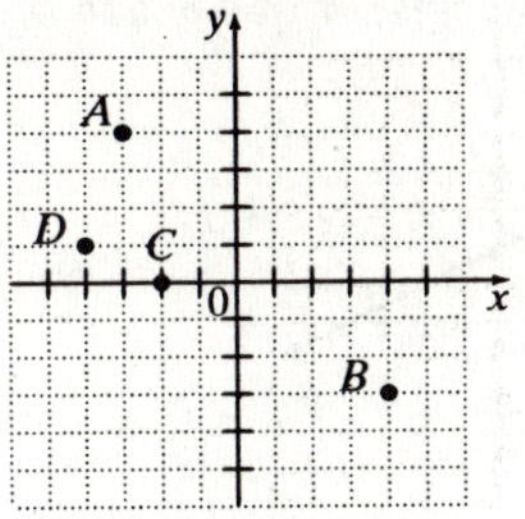

3. $7x - 8y = 56$
$(0, 56)$; No
$$7(0) - 8(56) = 56$$
$$-448 = 56, \text{ False}$$
$(8, 0)$; Yes
$$7(8) - 8(0) = 56$$
$$56 = 56, \text{ True}$$

4. $-2x + 5y = 10$
$(-5, 0)$; Yes
$$-2(-5) + 5(0) = 10$$
$$10 = 10, \text{ True}$$
$(1, 1)$; No
$$-2(1) + 5(1) = 10$$
$$-2 + 5 = 10$$
$$3 = 10, \text{ False}$$

5. $x = 13$
$(13, 5)$; Yes
$13 = 13$, True
$(13, 13)$; Yes
$13 = 13$, True

6. $y = 2$
$(7, 2)$; Yes
$2 = 2$, True
$(2, 7)$; No
$7 = 2$, False

7. Linear

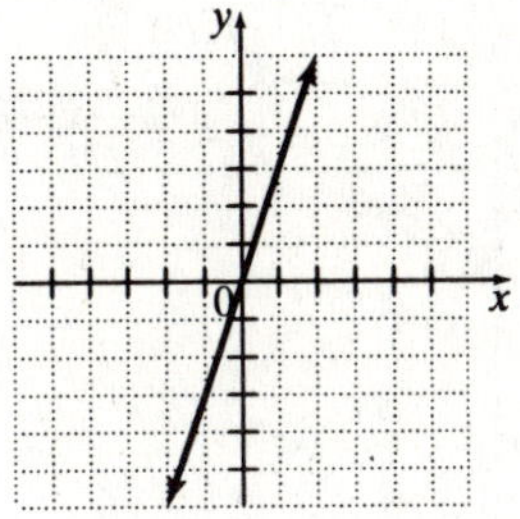

8. Linear

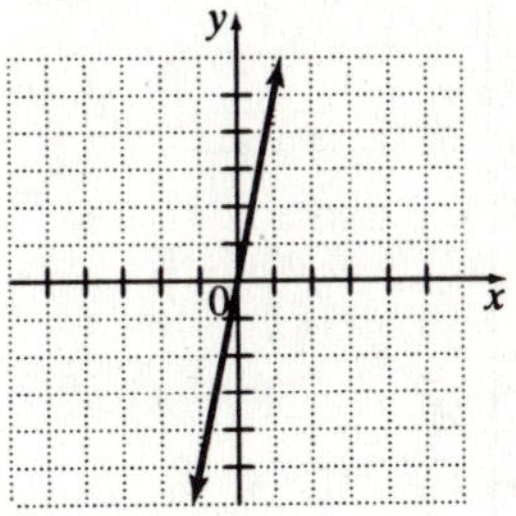

9. $3x - y = 4$
Linear
Find three ordered pair solutions, or find x- and y-intercepts, or find m and b.

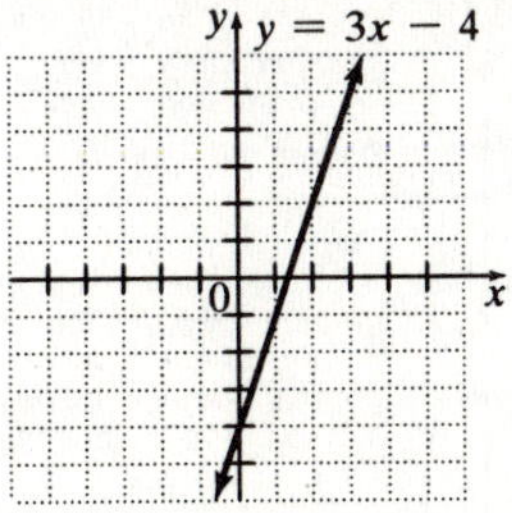

10. $x - 3y = 2$
Linear
Find three ordered pair solutions, or find x- and y-intercepts, or find m and b.

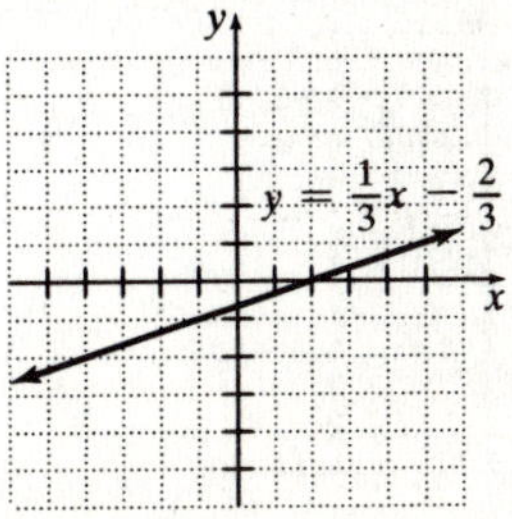

11. $y = |x| + 4$
Nonlinear

x	y
–3	7
–2	6
–1	5
0	4
1	5
2	6
3	7

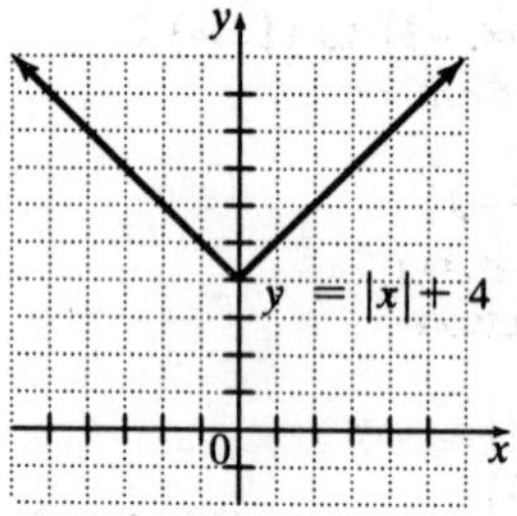

12. $y = x^2 + 4$
Nonlinear

x	y
–3	13
–2	8
–1	5
0	4
1	5
2	8
3	13

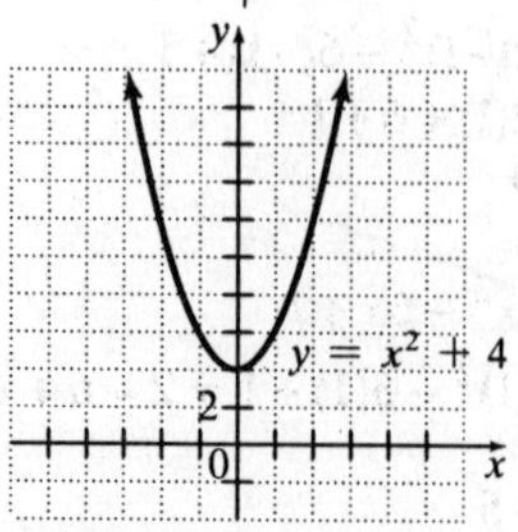

13. $y = -\frac{1}{2}x + 2$
Linear
Find three ordered pair solutions, or find x- and y-intercepts, or find m and b.

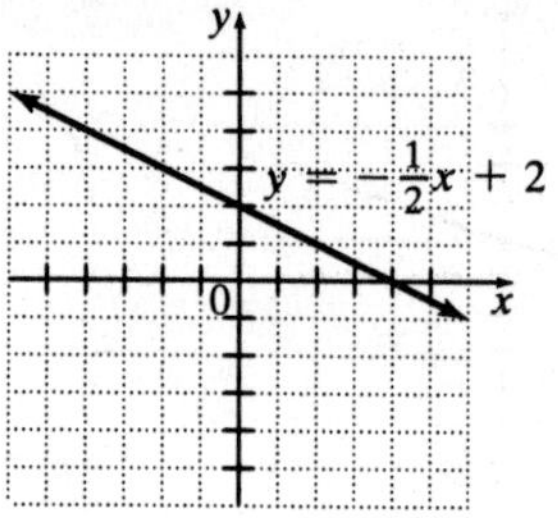

14. $y = -x + 5$
Linear
Find three ordered pair solutions, or find x- and y-intercepts, or find m and b.

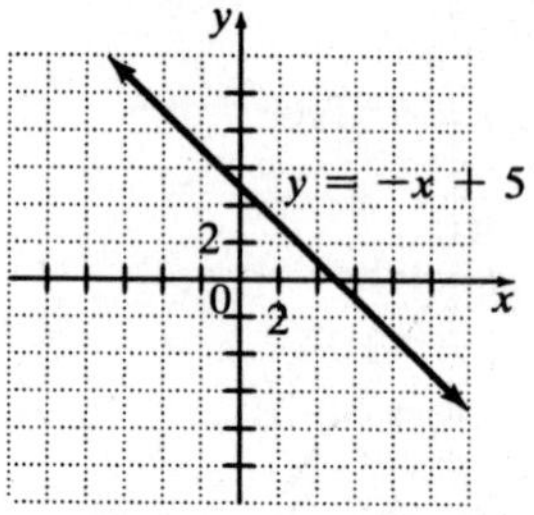

15. $y = 2x - 1$
Linear
Find three ordered pair solutions, or find x- and y-intercepts, or find m and b.

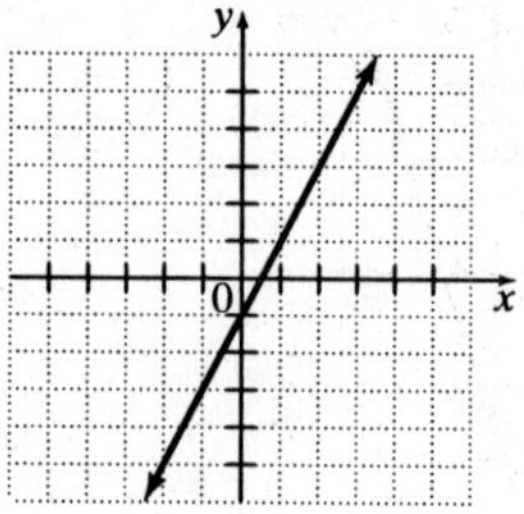

16. $y = \frac{1}{3}x + 1$
Linear
Find three ordered pair solutions, or find *x*- and *y*-intercepts, or find *m* and 6.

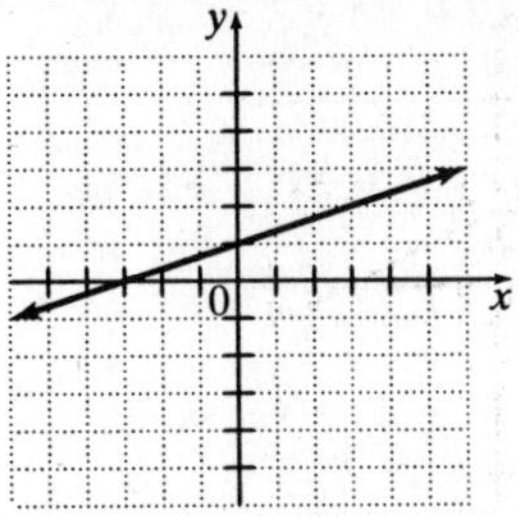

17. $y = -1.36x$
Linear
Find three ordered pair solutions, or find *x*- and *y*-intercepts, or find *m* and *b*.

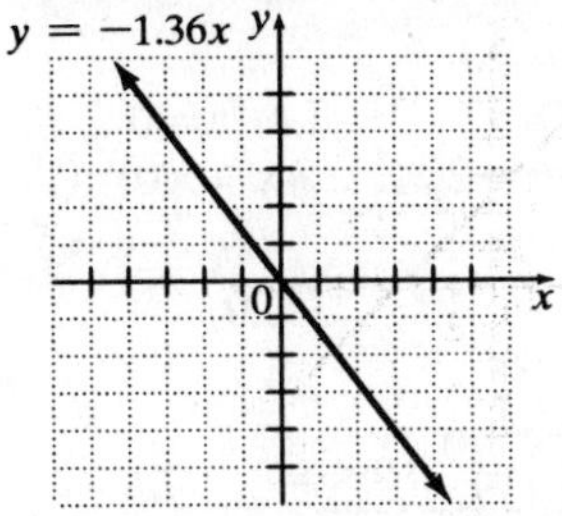

18. $y = 2.1x + 5.9$
Linear
Find three ordered pair solutions, or find *x*- and *y*-intercepts, or find *m* and *b*.

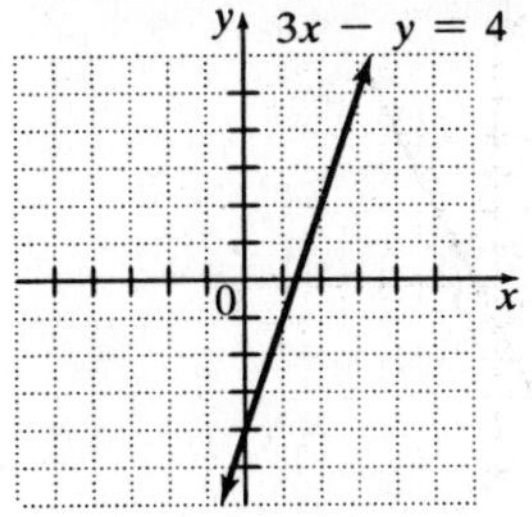

19. Domain: $\left\{-\frac{1}{2}, 6, 0, 25\right\}$
Range: $\left\{\frac{3}{4}, -12, 25\right\}$
Function

20. Domain: $\left\{\frac{3}{4}, -12, 25\right\}$
Range: $\left\{-\frac{1}{2}, 6, 0, 25\right\}$
Not a function

21. Domain: $\{2, 4, 6, 8\}$
Range: $\{2, 4, 5, 6\}$
Not a function

22. Domain: {Triangle, Square, Rectangle, Parallelogram}
Range: $\{3, 4\}$
Function

23. Domain: $(-\infty, \infty)$
Range: $(-\infty, -1] \cup [1, \infty)$
Not a function

24. Domain: $\{-3\}$
Range: $(-\infty, \infty)$
Not a function

25. Domain: $(-\infty, \infty)$
Range: $\{4\}$
Function

26. Domain: $[-1, 1]$
Range: $[-1, 1]$
Not a function

27. $f(x) = x - 5$
$f(2) = 2 - 5 = -3$

28. $g(x) = -3x$
$g(0) = -3(0) = 0$

29. $g(x) = -3x$
$g(-6) = -3(-6) = 18$

30. $h(x) = 2x^2 - 6x + 1$

$$\begin{aligned} h(-1) &= 2(-1)^2 - 6(-1) + 1 \\ &= 2(1) + 6 + 1 \\ &= 9 \end{aligned}$$

31. $h(x) = 2x^2 - 6x + 1$
$h(1) = 2(1)^2 - 6(1) + 1 = 2 - 6 + 1 = -3$

32. $f(x) = x - 5$
$f(5) = 5 - 5 = 0$

33. $J(x) = 2.54x$
$J(150) = 2.54(150) = 381$ pounds

34. $J(x) = 2.54x$
$J(2000) = 2.54(2000) = 5080$ pounds

35. 0

36. –2

37. $f(x) = 1$
$f(-2) = f(4) = 1$
–2, 4

38. $f(x) = -1$
$f(0) = f(2) = -1$
0, 2

39. $f(x) = x$ or $y = x$
$m = 1, b = 0$

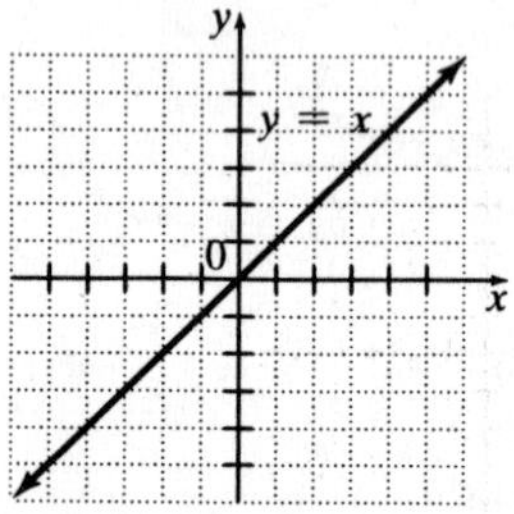

40. $f(x) = -\frac{1}{3}x$ or $y = -\frac{1}{3}x$
$m = -\frac{1}{3}, b = 0$

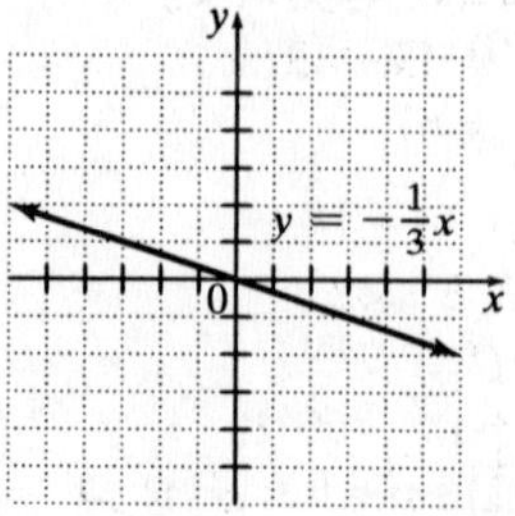

41. $g(x) = 4x - 1$ or $y = 4x - 1$
$m = 4, b = -1$

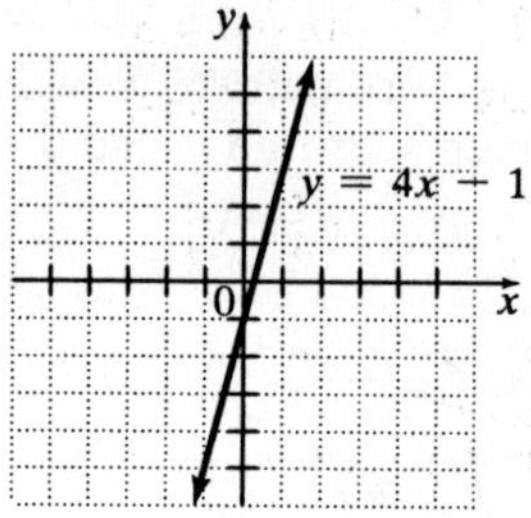

42. C

43. A

44. B

45. D

46. $4x + 5y = 20$

Let $x = 0$	Let $y = 0$
$4(0) + 5y = 20$	$4x + 5(0) = 20$
$y = 4$	$x = 5$
(0, 4)	(5, 0)

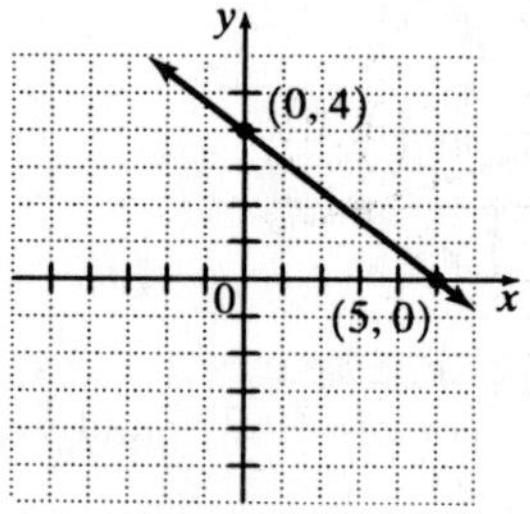

47. $3x - 2y = -9$

Let $x = 0$	Let $y = 0$
$3(0) - 2y = -9$	$3x - 2(0) = -9$
$y = \frac{9}{2}$	$x = -3$
$\left(0, \frac{9}{2}\right)$	(–3, 0)

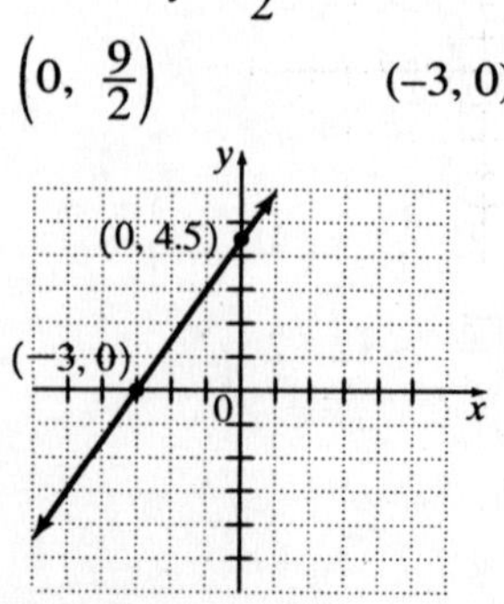

48. $4x - y = 3$

Let $x = 0$	Let $y = 0$
$4(0) - y = 3$	$4x - 0 = 3$
$y = -3$	$x = \frac{3}{4}$
$(0, -3)$	$\left(\frac{3}{4}, 0\right)$

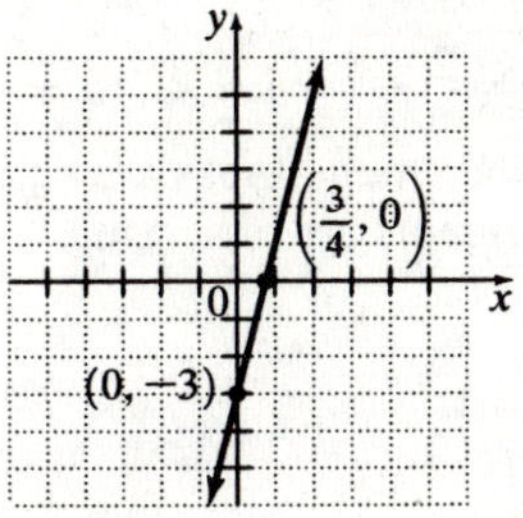

49. $2x + 6y = 9$

Let $x = 0$	Let $y = 0$
$2(0) + 6y = 9$	$2x + 6(0) = 9$
$y = \frac{3}{2}$	$x = \frac{9}{2}$
$\left(0, \frac{3}{2}\right)$	$\left(\frac{9}{2}, 0\right)$

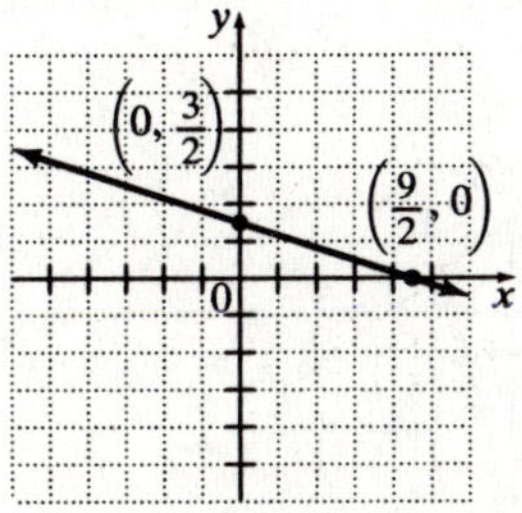

50. $y = 5$
Horizontal line with y-intercept 5.

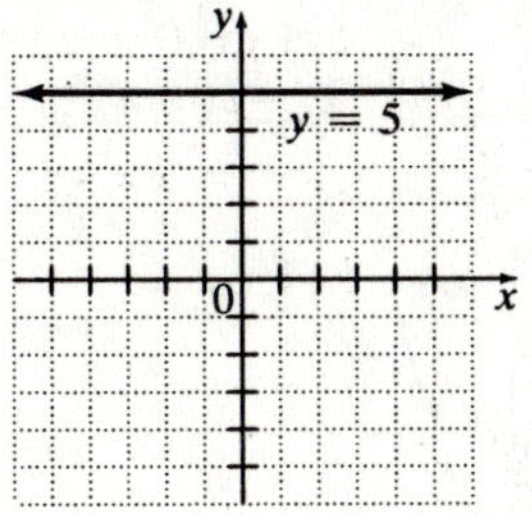

51. $x = -2$
Vertical line with x-intercept -2.

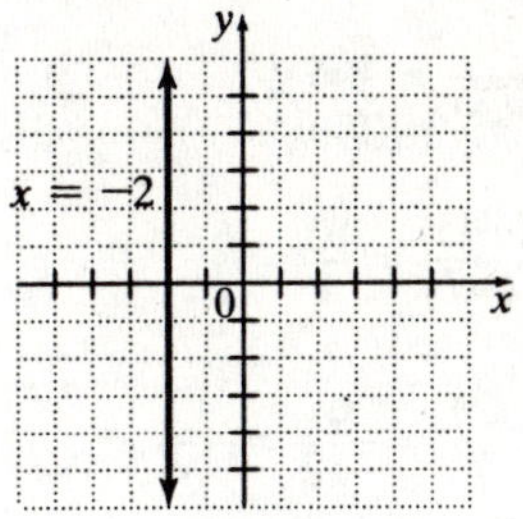

52.

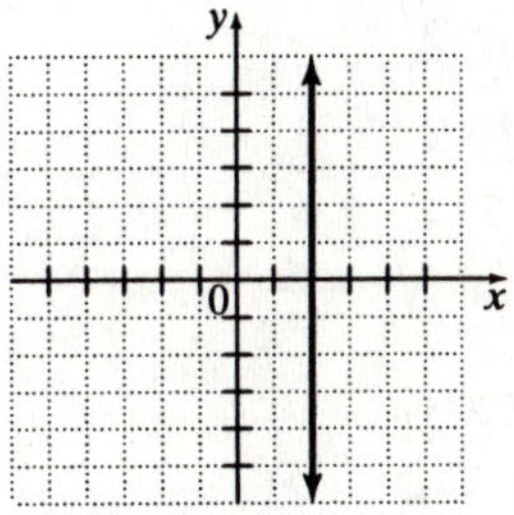

53.

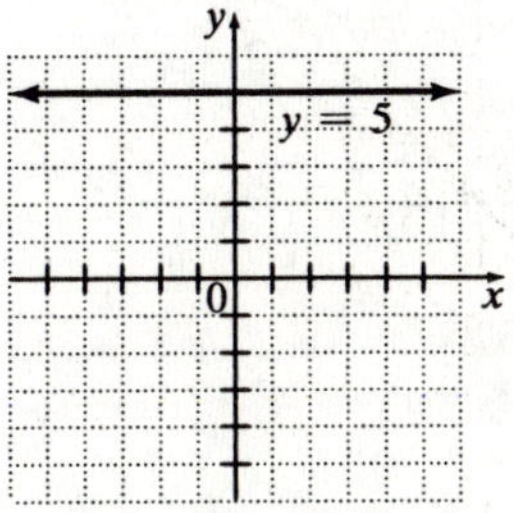

54. $C(x) = 0.3x + 42$

a. $C(150) = 0.3(150) + 42$
$= 45 + 42$
$= 87$

$87

b. $m = 0.3$, $b = 42$

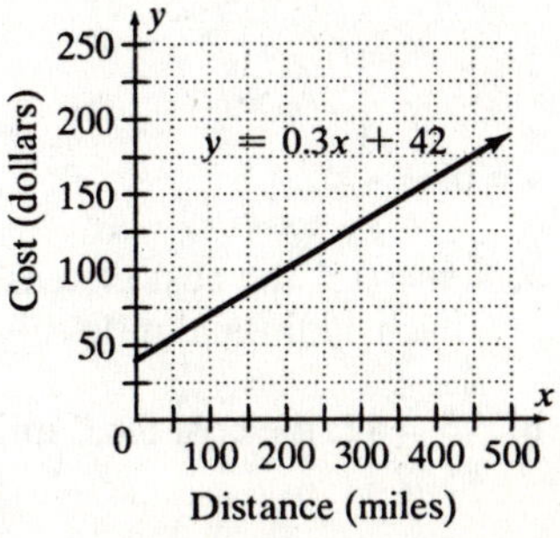

55. $m = \frac{-4-8}{6-2} = \frac{-12}{4} = -3$

56. $m = \frac{13-9}{5-(-3)} = \frac{4}{8} = \frac{1}{2}$

57. $m = \frac{6-(-4)}{-3-(-7)} = \frac{10}{4} = \frac{5}{2}$

58. $m = \frac{7-(-2)}{-5-7} = \frac{9}{-12} = -\frac{3}{4}$

59. $6x - 15y = 20$
$6x - 20 = 15y$
$y = \frac{2}{5}x - \frac{4}{3}$
$m = \frac{2}{5}$ and $b = -\frac{4}{3}$

60. $4x + 14y = 21$
$14y = -4x + 21$
$y = -\frac{2}{7}x + \frac{3}{2}$
$m = -\frac{2}{7},\ b = \frac{3}{2}$

61. $y - 3 = 0$
$y = 3$
Slope = 0

62. $x = -5$
Vertical line
Slope is undefined.

63. l_2

64. l_2

65. l_2

66. l_1

67. $y = 0.3x + 42$

a. $m = 0.3$; the cost increases by \$0.30 for each additional mile driven.

b. $b = 42$; the cost for 0 miles driven is \$42.

68. $f(x) = -2x + 6$ $g(x) = 2x - 1$
$m = -2$ $m = 2$
Neither; The slopes are not the same and their product is not -1.

69. $-x + 3y = 2$ $6x - 18y = 3$
$y = \frac{1}{3}x + \frac{2}{3}$ $y = \frac{1}{3}x - \frac{1}{6}$
$m = \frac{1}{3},\ b = \frac{2}{3}$ $m = \frac{1}{3},\ b = -\frac{1}{6}$
Parallel, since their slopes are equal and their y-intercepts are different.

70. $y = -x + 1$
$m = -1, b = 1$

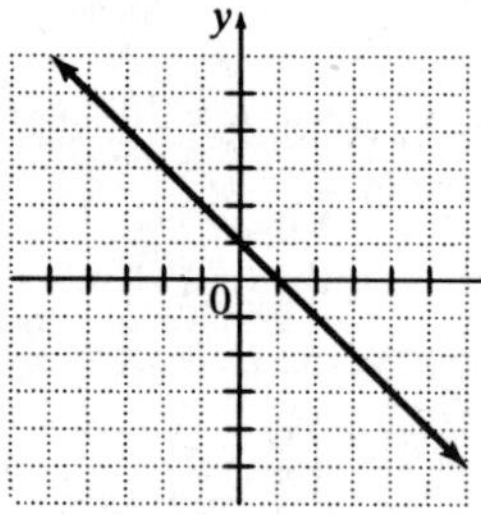

71. $y = 4x - 3$
$m = 4, b = -3$

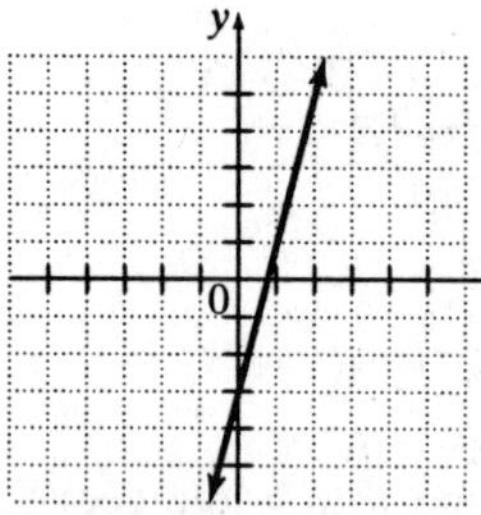

72. $3x - y = 6$
$y = 3x - 6$
$m = 3, b = -6$

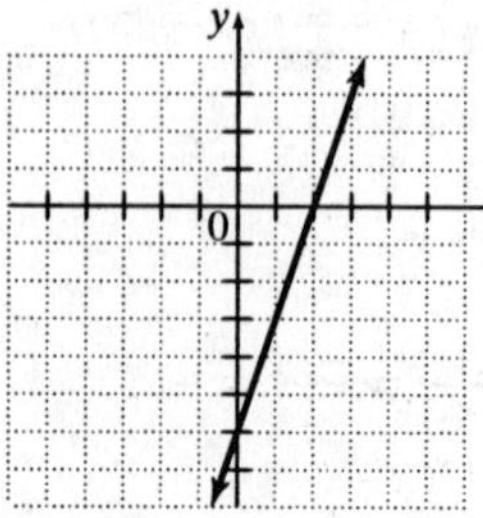

73. $y = -5x$
$m = -5, b = 0$

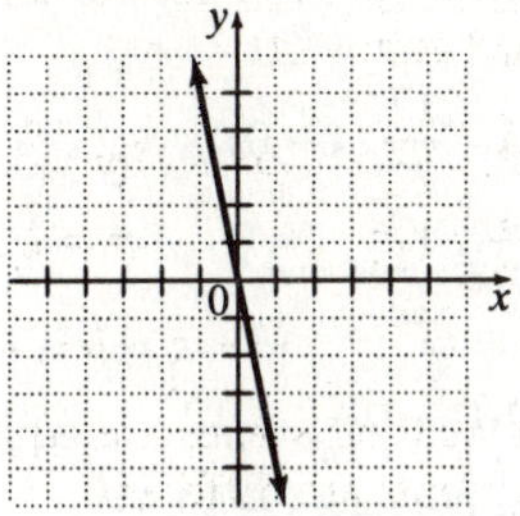

74. Horizontal lines have slope = 0.
The y-intercept is –1.
$y = -1$

75. Vertical lines have undefined slope.
The x-intercept is –2.
$x = -2$

76. The slope is undefined.
The x-intercept is –4.
$x = -4$

77. Horizontal line with y-intercept = 5
$y = 5$

78.
$$y - 5 = 3(x - (-3))$$
$$y - 5 = 3(x + 3)$$
$$y - 5 = 3x + 9$$
$$3x - y = -14$$

79.
$$y - (-2) = 2(x - 5)$$
$$y + 2 = 2x - 10$$
$$2x - y = 12$$

80. $m = \frac{-2 - (-1)}{-4 - (-6)} = -\frac{1}{2}$

Now, $y - (-1) = -\frac{1}{2}(x - (-6))$
$$2(y + 1) = -(x + 6)$$
$$2y + 2 = -x - 6$$
$$x + 2y = -8$$

81. $m = \frac{-8 - 3}{-4 - (-5)} = \frac{-11}{1} = -11$
$$y - 3 = -11(x - (-5))$$
$$y - 3 = -11x - 55$$
$$11x + y = -52$$

82. $x = 4$ has undefined slope
A line perpendicular to $x = 4$ has slope = 0.
y-intercept at 3
$y = 3$

83. Slope = 0, y-intercept = –5
$y = -5$

84.
$$y = -\frac{2}{3}x + 4$$
$$f(x) = -\frac{2}{3}x + 4$$

85.
$$y = -x - 2$$
$$f(x) = -x - 2$$

86.
$$6x + 3y = 5$$
$$3y = -6x + 5$$
$$y = -2x + \frac{5}{3},$$ so $m = -2$. Now,
$$y - (-6) = -2(x - 2)$$
$$y + 6 = -2x + 4$$
$$y = -2x - 2$$
$$f(x) = -2x - 2$$

87.
$$3x + 2y = 8$$
$$2y = -3x + 8$$
$$y = -\frac{3}{2}x + 4$$
$m = -\frac{3}{2}$ Now,
$$y - (-2) = -\frac{3}{2}(x - (-4))$$
$$y + 2 = -\frac{3}{2}x - 6$$
$$y = -\frac{3}{2}x - 8$$
$$f(x) = -\frac{3}{2}x - 8$$

88. $4x+3y=5$

$$3y=-4x+5$$
$$y=-\frac{4}{3}x+\frac{5}{3},$$

so $m_{\perp}=-\frac{4}{3}$ and $m=\frac{3}{4}$

Now,

$$y-(-1)=\frac{3}{4}(x-(-6))$$
$$4(y+1)=3(x+6)$$
$$4y+4=3x+18$$
$$y=\frac{3}{4}x+\frac{7}{2}$$
$$f(x)=\frac{3}{4}x+\frac{7}{2}$$

89. $2x-3y=6$

$$3y=2x-6$$
$$y=\frac{2}{3}x-2$$

$m=\frac{2}{3}$, so the slope of the perpendicular line is $-\frac{3}{2}$.

$$y-5=-\frac{3}{2}(x-(-4))$$
$$y-5=-\frac{3}{2}x-6$$
$$y=-\frac{3}{2}x-1$$
$$f(x)=-\frac{3}{2}x-1$$

90. a. Use ordered pairs (0, 42) and (3, 58).

$$m=\frac{58-42}{3-0}=\frac{16}{3}$$
$$y-42=\frac{16}{3}(x-0)$$
$$y=\frac{16}{3}x+42$$

b. $x=2007-1996=11$

$$y=\frac{16}{3}(11)+42=100.7$$

There will be about 101 million subscribers.

91. a. Use ordered pairs (0, 43) and (22, 60).

$$m=\frac{60-43}{22-0}=\frac{17}{22}$$
$$y-43=\frac{17}{22}(x-0)$$
$$y=\frac{17}{22}x+43$$

b. $x=2010-1998=12$

$$y=\frac{17}{22}(12)+43=52.3$$

There will be about 52 million people reporting arthritis.

92. $3x+y>4$

$$y>-3x+4$$

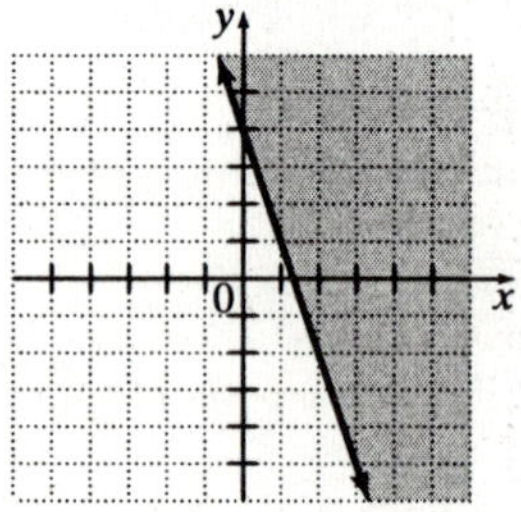

93. $\frac{1}{2}x-y<2$

$$y>\frac{1}{2}x-2$$

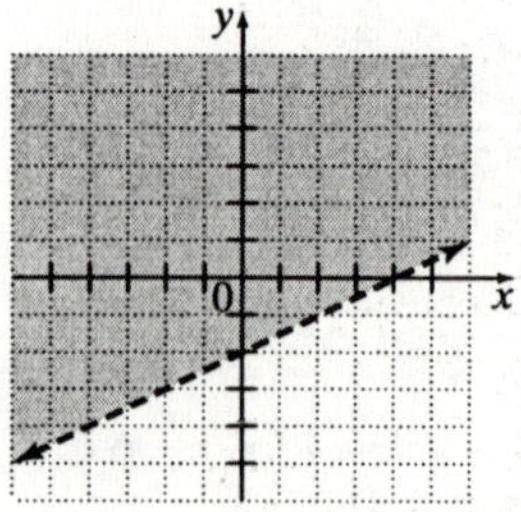

94. $5x - 2y \le 9$
$-2y \le -5x + 9$
$y \ge \frac{5}{2}x - \frac{9}{2}$

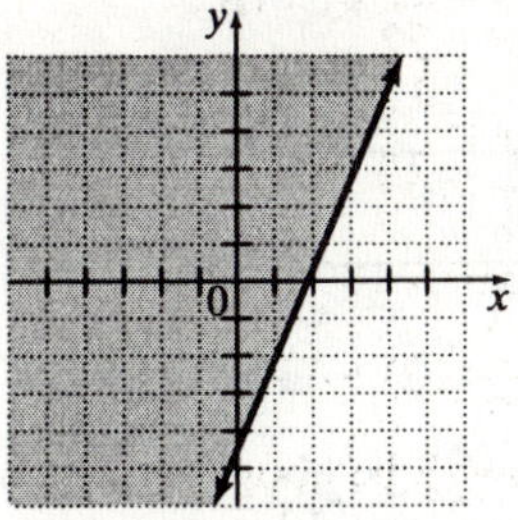

95. $3y \ge x$
$y \ge \frac{1}{3}x$

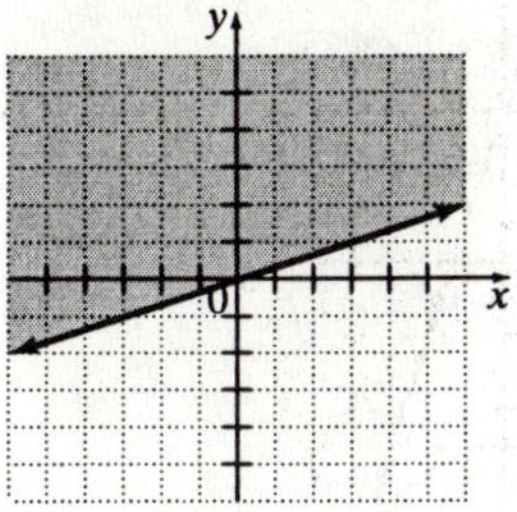

96. $y < 1$

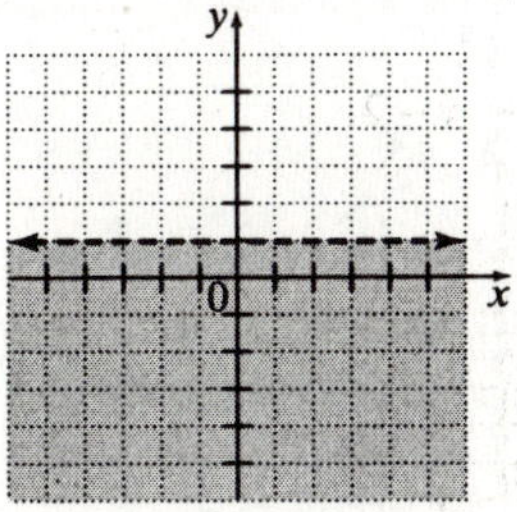

97. $x > -2$

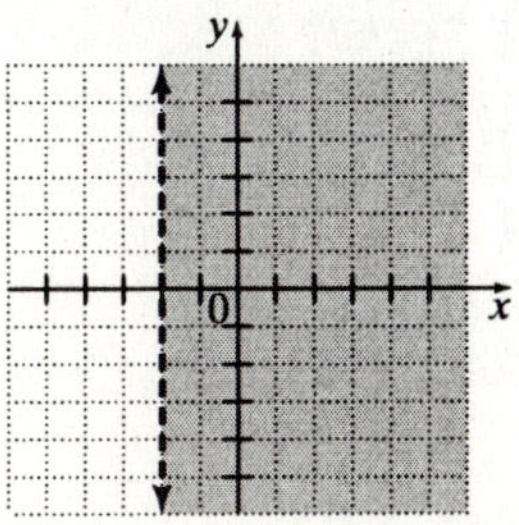

98. $y > 2x + 3$ or $x \le -3$

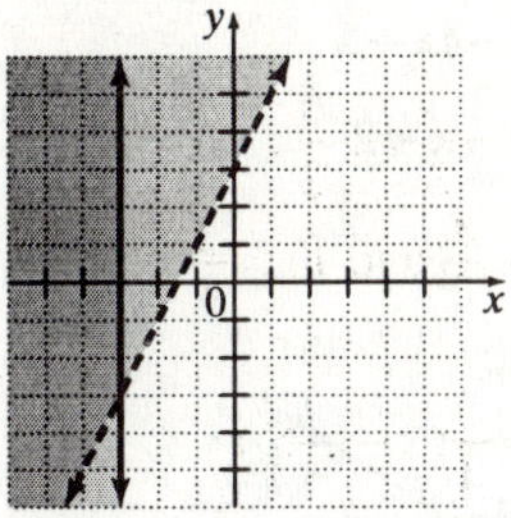

99. $2x < 3y + 8$ and $y \ge -2$

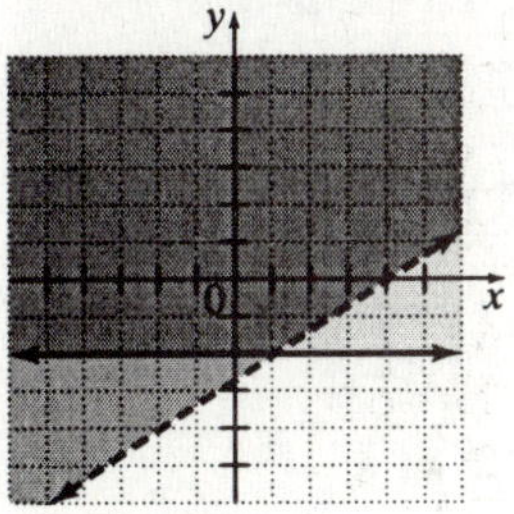

Chapter 3 Test

1.

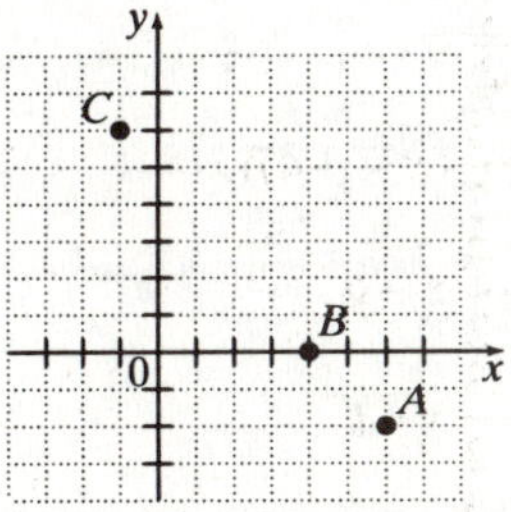

A is in quadrant IV.
B is on the *x*-axis, no quadrant.
C is in quadrant II.

2. $2y - 3x = 12$
$x = -6$:
$2y - 3(-6) = 12$
$2y + 18 = 12$
$2y = -6$
$y = -3$
$(-6, -3)$

3. $2x - 3y = -6$

$$-3y = -2x - 6$$

$$y = \frac{2}{3}x + 2$$

$m = \frac{2}{3}$, $b = 2$

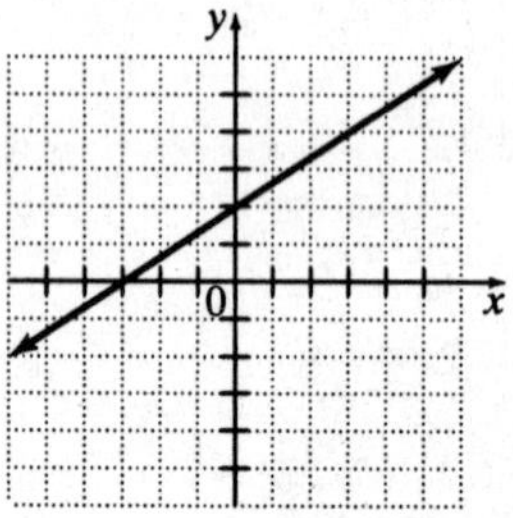

4. $4x + 6y = 7$

$$6y = -4x + 7$$

$$y = -\frac{2}{3}x + \frac{7}{6}$$

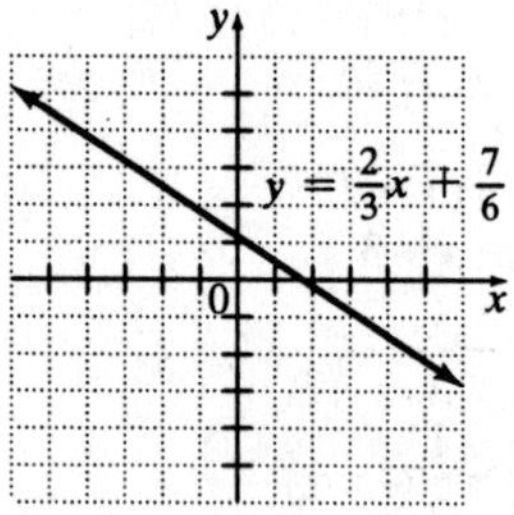

5. $y = \frac{2}{3}x$

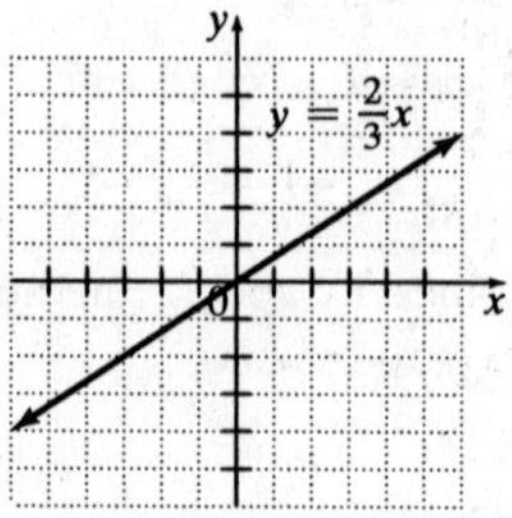

6. $y = -3$

Horizontal line with y-intercept at -3.

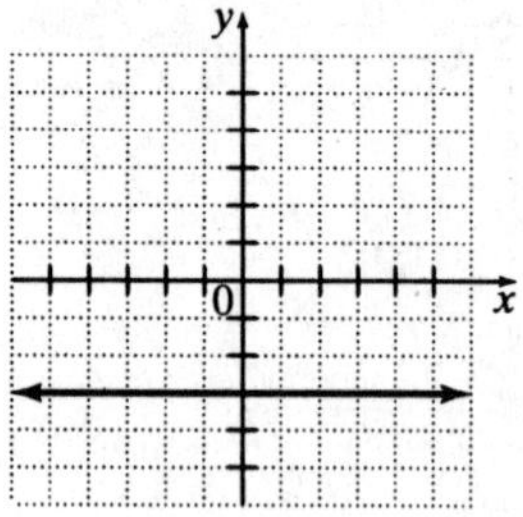

7. $m = \frac{10 - (-8)}{-7 - 5} = \frac{18}{-12} = -\frac{3}{2}$

8. $3x + 12y = 8$

$$12y = -3x + 8$$

$$y = -\frac{1}{4}x + \frac{2}{3}$$

$m = -\frac{1}{4}$ and $b = \frac{2}{3}$

9. $f(x) = (x - 1)^2$

x	$f(x)$
−2	9
−1	4
0	1
1	0
2	1
3	4
4	9

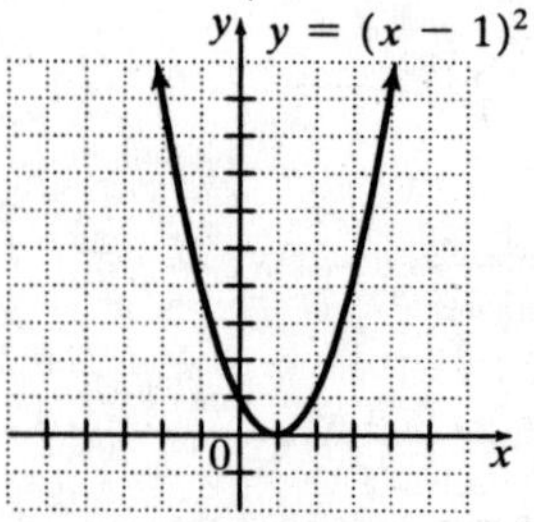

10. $g(x) = |x| + 2$

x	$g(x)$
−3	5
−2	4
−1	3
0	2
1	3
2	4
3	5

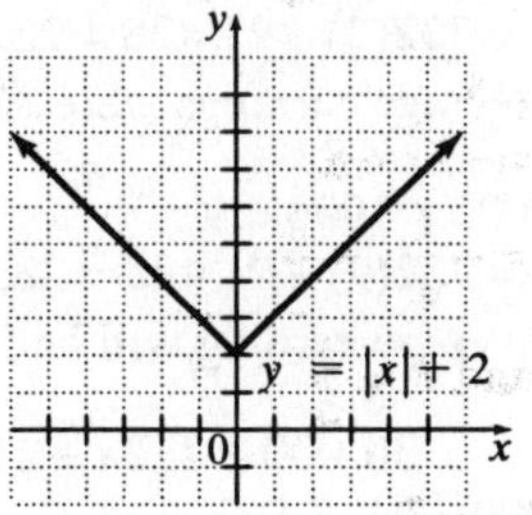

11. $y = -8$

12. $x = -4$

13. $y = -2$

14. $y - (-1) = -3(x - 4)$
$$y + 1 = -3x + 12$$
$$3x + y = 11$$

15. $y = 5x + (-2)$
$$5x - y = 2$$

16. $m = \frac{-3-(-2)}{6-4} = -\frac{1}{2}$

Now,
$$y - (-2) = -\frac{1}{2}(x - 4)$$
$$2(y + 2) = -(x - 4)$$
$$2y + 4 = -x + 4$$
$$y = -\frac{1}{2}x$$
$$f(x) = -\frac{1}{2}x$$

17. $3x - y = 4$
$$y = 3x - 4$$
so $m_{\perp} = 3$ and $m = -\frac{1}{3}$

Now,
$$y - 2 = -\frac{1}{3}(x - (-1))$$
$$3(y - 2) = -(x + 1)$$
$$3y - 6 = -x - 1$$
$$y = -\frac{1}{3}x + \frac{5}{3}$$
$$f(x) = -\frac{1}{3}x + \frac{5}{3}$$

18. $2y + x = 3$
$$2y = -x + 3$$
$$y = -\frac{1}{2}x + \frac{3}{2},$$
so $m = -\frac{1}{2}$

Now,
$$y - (-2) = -\frac{1}{2}(x - 3)$$
$$2(y + 2) = -(x - 3)$$
$$2y + 4 = -x + 3$$
$$y = -\frac{1}{2}x - \frac{1}{2}$$
$$f(x) = -\frac{1}{2}x - \frac{1}{2}$$

19. $2x - 5y = 8$
$$5y = 2x - 8$$
$$y = \frac{2}{5}x - \frac{8}{5},$$
so $m_1 = \frac{2}{5}$, $m_2 = \frac{-1-4}{-1-1} = \frac{-5}{-2} = \frac{5}{2}$

Therefore, lines L_1 and L_2 are neither parallel nor perpendicular.

20. $x \leq -4$

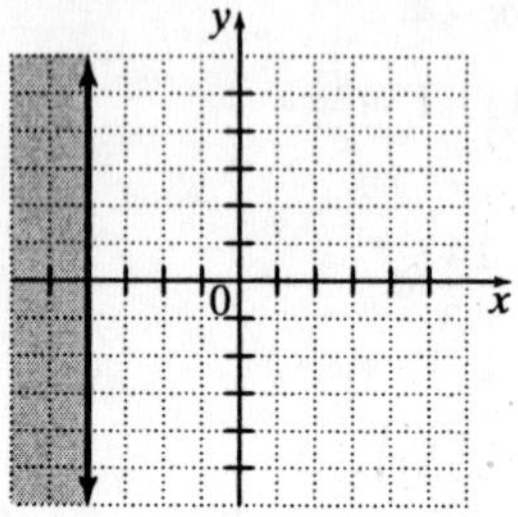

21. $y > -2$

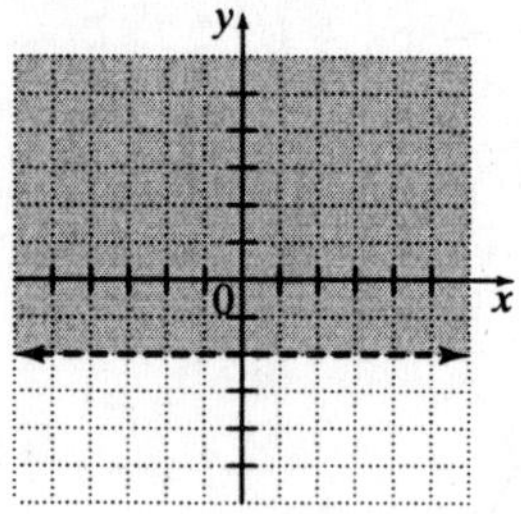

22. $2x - y > 5$

$y < 2x - 5$

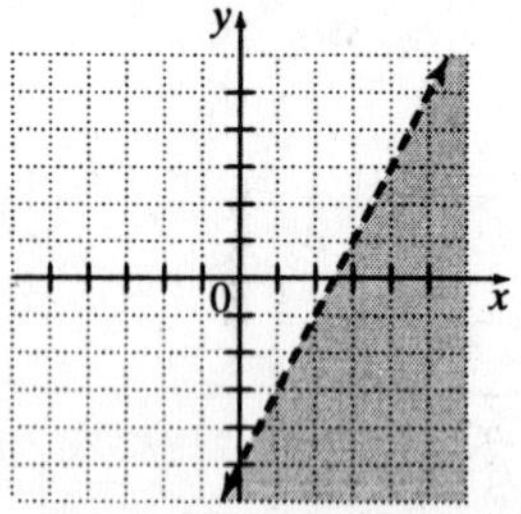

23. $2x + 4y < 6$ and $y \leq -4$

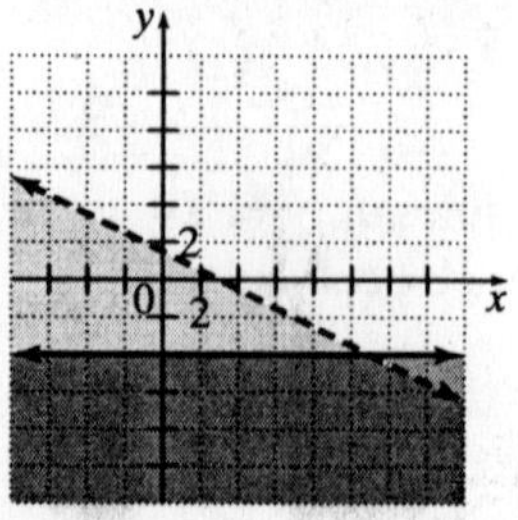

24. Domain: $(-\infty, \infty)$
Range: $\{5\}$
Function

25. Domain: $\{-2\}$
Range: $(-\infty, \infty)$
Not a function

26. Domain: $(-\infty, \infty)$
Range: $[0, \infty)$
Function

27. Domain: $(-\infty, \infty)$
Range: $(-\infty, \infty)$
Function

28. $f(x) = 732x + 21,428$

a. $f(2) = 732(2) + 21,428 = 22,892$
The average earnings were \$22,892.

b. $x = 2005 - 1996 = 9$
$f(9) = 732(9) + 21,428 = 28,016$
The average earning will be \$28,016.

c.
$$30,000 \geq 732x + 21,428$$
$$\frac{30,000 - 21,428}{732} \geq x$$
$$11.71 \geq x$$
It will be more than 11 years or in 2008.

d. $m = 732$; the yearly earnings for high school graduates increases by \$732 per year.

e. $b = 21,428$; the yearly earning for a high school graduate in 1996 was \$21,428.

Chapter 4

Section 4.1

Graphing Calculator Explorations

1.

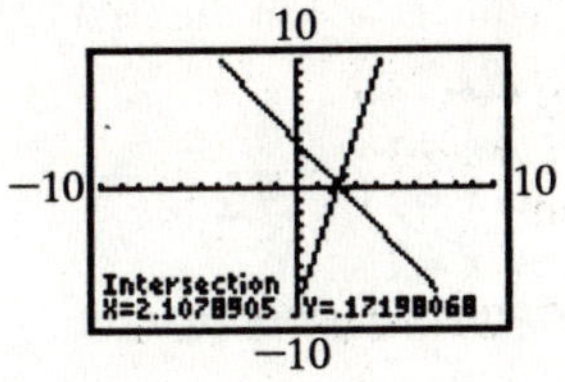

(2.11, 0.17)

3.

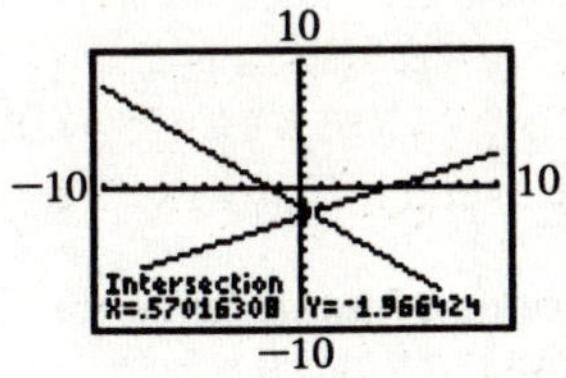

(0.57, −1.97)

Mental Math

1. B

2. C

3. A

4. D

Exercise Set 4.1

1. $\begin{cases} x - y = 3 \\ 2x - 4y = 8 \end{cases}$

$$\begin{aligned} x - y &= 3 \\ 2 - (-1) &= 3 \\ 2 + 1 &= 3 \\ 3 &= 3 \text{ True} \\ 2x - 4y &= 8 \\ 2(2) - 4(-1) &= 8 \\ 4 + 4 &= 8 \\ 8 &= 8 \text{ True} \end{aligned}$$

Yes, (2, −1) is a solution.

3. $\begin{cases} 2x - 3y = -9 \\ 4x + 2y = -2 \end{cases}$

$$\begin{aligned} 2x - 3y &= -9 \\ 2(3) - 3(5) &= -9 \\ 6 - 15 &= -9 \\ -9 &= -9 \text{ True} \\ 4x + 2y &= -2 \\ 4(3) + 2(5) &= -2 \\ 12 + 10 &= -2 \\ 22 &= -2 \text{ False} \end{aligned}$$

No, (3, 5) is not a solution.

5. $\begin{cases} y = -5x \\ x = -2 \end{cases}$

$$\begin{aligned} y &= -5x \\ 10 &= -5(-2) \\ 10 &= 10 \text{ True} \end{aligned} \qquad \begin{aligned} x &= -2 \\ -2 &= -2 \text{ True} \end{aligned}$$

Yes, (−2, 10) is a solution.

7. $\begin{cases} x + y = 1 \\ x - 2y = 4 \end{cases}$

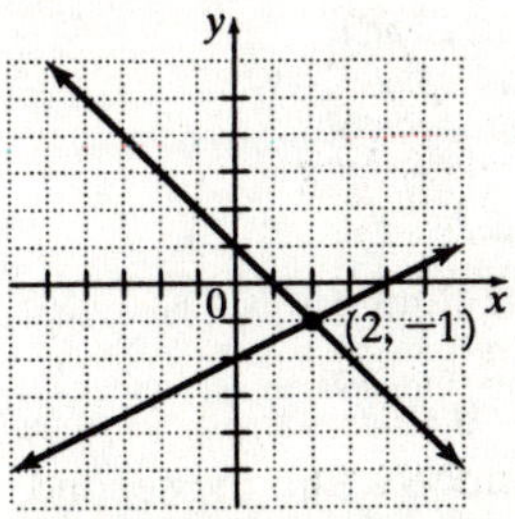

The solution is (2, −1).

9. $\begin{cases} 2y - 4 = 0 \\ x + 2y = 5 \end{cases}$

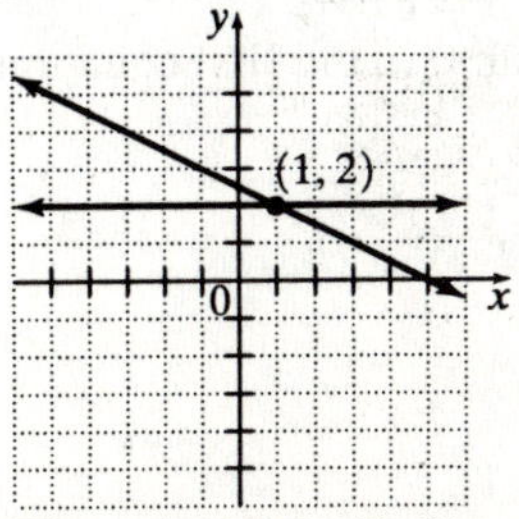

The solution is (1, 2).

11. $\begin{cases} 3x - y = 4 \\ 6x - 2y = 4 \end{cases}$

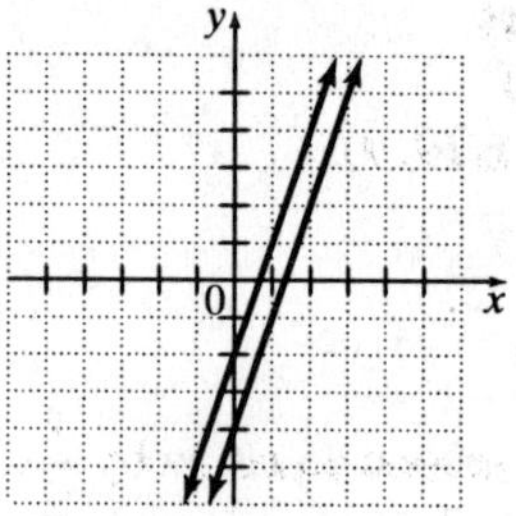

The solution is $\varnothing$.

13. No; answers may vary.

15. $\begin{cases} x + y = 10 \\ y = 4x \end{cases}$

Replace y with $4x$ in the first equation.
$x + 4x = 10$
$5x = 10$
$x = 2$
Replace x with 2 in the second equation.
$y = 4(2)$
$y = 8$
The solution is (2, 8).

17. $\begin{cases} 4x - y = 9 \\ 2x + 3y = -27 \end{cases}$

Solve the first equation for y.
$4x - y = 9$
$y = 4x - 9$
Replace y with $4x - 9$ in the second equation.
$2x + 3(4x - 9) = -27$
$2x + 12x - 27 = -27$
$14x = 0$
$x = 0$
Replace x with 0 in the first equation.
$4(0) - y = 9$
$y = -9$
The solution is (0, –9).

19. $\begin{cases} \frac{1}{2}x + \frac{3}{4}y = -\frac{1}{4} \\ \frac{3}{4}x - \frac{1}{4}y = 1 \end{cases}$

Clear fractions by multiplying each equation by 4.

$\begin{cases} 2x + 3y = -1 \\ 3x - y = 4 \end{cases}$

Now solve the second equation for y.
$3x - y = 4$
$y = 3x - 4$
Replace y with $3x - 4$ in the first equation.
$2x + 3(3x - 4) = -1$
$2x + 9x - 12 = -1$
$11x = 11$
$x = 1$
Replace x with 1 in the equation $y = 3x - 4$.
$y = 3(1) - 4$
$y = -1$
The solution is (1, –1).

21. $\begin{cases} \frac{x}{3} + y = \frac{4}{3} \\ -x + 2y = 11 \end{cases}$

Clear fractions by multiplying the first equation by 3.

$\begin{cases} x + 3y = 4 \\ -x + 2y = 11 \end{cases}$

Solve the second equation for x.
$2y - 11 = x$
$x = 2y - 11$
Replace x with $2y - 11$ in the first equation.
$2y - 11 + 3y = 4$
$5y = 15$
$y = 3$
Replace y with 3 in the equation $x = 2y - 11$.
$x = 2(3) - 11$
$x = -5$
The solution is (–5, 3).

23. $\begin{cases} 2x - 4y = 0 \\ x + 2y = 5 \end{cases}$

Multiply the first equation by $\frac{1}{2}$.

$\begin{cases} x - 2y = 0 \\ x + 2y = 5 \end{cases}$

Add the equations.

$x - 2y = 0$
$x + 2y = 5$
$2x \quad = 5$
$x = \frac{5}{2}$

Replace x with $\frac{5}{2}$ in the first equation.

$\frac{5}{2} - 2y = 0$
$\frac{5}{2} = 2y$
$y = \frac{5}{4}$

The solution is $\left(\frac{5}{2}, \frac{5}{4}\right)$.

25. $\begin{cases} 5x + 2y = 1 \\ x - 3y = 7 \end{cases}$

Multiply the second equation by –5.

$\begin{cases} 5x + 2y = 1 \\ -5x + 15y = -35 \end{cases}$

Add the equations.

$5x + 2y = 1$
$-5x + 15y = -35$
$17y = -34$
$y = -2$

Replace y with –2 in the second equation.

$x - 3(-2) = 7$
$x + 6 = 7$
$x = 1$

The solution is (1, –2).

27. $\begin{cases} 5x - 2y = 27 \\ -3x + 5y = 18 \end{cases}$

Multiply the first equation by 3 and the second equation by 5.

$\begin{cases} 15x - 6y = 81 \\ -15x + 25y = 90 \end{cases}$

Add the equations.

$15x - 6y = 81$
$-15x + 25y = 90$
$19y = 171$
$y = 9$

Replace y with 9 in the first equation.

$5x - 2(9) = 27$
$5x - 18 = 27$
$5x = 45$
$x = 9$

The solution is (9, 9).

29. $\begin{cases} 3x - 5y = 11 \\ 2x - 6y = 2 \end{cases}$

Multiply the second equation by $\frac{1}{2}$.

$\begin{cases} 3x - 5y = 11 \\ x - 3y = 1 \end{cases}$

Multiply the second equation by –3.

$\begin{cases} 3x - 5y = 11 \\ -3x + 9y = -3 \end{cases}$

Add the equations.

$3x - 5y = 11$
$-3x + 9y = -3$
$4y = 8$
$y = 2$

Replace y with 2 in the equation $x - 3y = 1$.

$x - 3(2) = 1$
$x - 6 = 1$
$x = 7$

The solution is (7, 2).

31. $\begin{cases} x - 2y = 4 \\ 2x - 4y = 4 \end{cases}$

Multiply the first equation by –2.

$\begin{cases} -2x + 4y = -8 \\ 2x - 4y = 4 \end{cases}$

Add the equations.

$-2x + 4y = -8$
$2x - 4y = 4$
$0 = -4$ False

Inconsistent system

The solution is $\varnothing$.

33. $\begin{cases} 3x + y = 1 \\ 2y = 2 - 6x \end{cases}$

$\begin{cases} 3x + y = 1 \\ 6x + 2y = 2 \end{cases}$

Multiply the first equation by –2.

$\begin{cases} -6x - 2y = -2 \\ 6x + 2y = 2 \end{cases}$

Add the equations.

$$\begin{array}{r} -6x - 2y = -2 \\ 6x + 2y = 2 \\ \hline 0 = 0 \end{array} \quad \text{True}$$

Dependent system

The solution is $\{(x, y) | 3x + y = 1\}$

35. Answers may vary.

37. $\begin{cases} 2x + 5y = 8 \\ 6x + y = 10 \end{cases}$

Multiply the first equation by –3.

$\begin{cases} -6x - 15y = -24 \\ 6x + y = 10 \end{cases}$

Add the equations.

$$\begin{array}{r} -6x - 15y = -24 \\ 6x + y = 10 \\ \hline -14y = -14 \\ y = 1 \end{array}$$

Replace y with 1 in the second equation.

$6x + 1 = 10$

$6x = 9$

$x = \frac{9}{6} = \frac{3}{2}$

The solution is $\left(\frac{3}{2}, 1\right)$.

39. $\begin{cases} x + y = 1 \\ x - 2y = 4 \end{cases}$

Multiply the second equation by –1.

$\begin{cases} x + y = 1 \\ -x + 2y = -4 \end{cases}$

Add the equations.

$$\begin{array}{r} x + y = 1 \\ -x + 2y = -4 \\ \hline 3y = -3 \\ y = -1 \end{array}$$

Replace y with –1 in the first equation.

$x + (-1) = 1$

$x - 1 = 1$

$x = 2$

The solution is (2, –1).

41. $\begin{cases} \frac{1}{3}x + y = \frac{4}{3} \\ -\frac{1}{4}x - \frac{1}{2}y = -\frac{1}{4} \end{cases}$

Clear fractions by multiplying the first equation by 3 and the second equation by 4.

$\begin{cases} x + 3y = 4 \\ -x - 2y = -1 \end{cases}$

Add the equations.

$$\begin{array}{r} x + 3y = 4 \\ -x - 2y = -1 \\ \hline y = 3 \end{array}$$

Replace y with 3 in the equation $x + 3y = 4$.

$x + 3(3) = 4$

$x + 9 = 4$

$x = -5$

The solution is (–5, 3).

43. $\begin{cases} 2x + 6y = 8 \\ 3x + 9y = 12 \end{cases}$

Multiply the first equation by –3 and the second equation by 2.

$\begin{cases} -6x - 18y = -24 \\ 6x + 18y = 24 \end{cases}$

Add the equations.

$$\begin{array}{r} -6x - 18y = -24 \\ 6x + 18y = 24 \\ \hline 0 = 0 \end{array} \quad \text{True}$$

Dependent system

The solution is $\{(x, y) | 3x + 9y = 12\}$.

45. $\begin{cases} 4x + 2y = 5 \\ 2x + y = -1 \end{cases}$

Multiply the second equation by –2.

$$\begin{cases} 4x+2y=5 \\ -4x-2y=2 \end{cases}$$

Add the equations.

$$\begin{array}{r} 4x+2y=5 \\ -4x-2y=2 \\ \hline 0=7 \end{array} \text{ True}$$

Inconsistent system

The solution is $\varnothing$.

47. $\begin{cases} 10y-2x=1 \\ 5y \quad =4-6x \end{cases}$

Multiply the second equation by 2.

$$\begin{cases} 10y= \quad 2x+1 \\ 10y=-12x+8 \end{cases}$$

Replace 10*y* with 2*x* + 1 in the second equation.

$2x+1=-12x+8$

$14x=7$

$x=\frac{7}{14}=\frac{1}{2}$

Replace *x* with $\frac{1}{2}$ in the first equation.

$10y-2\left(\frac{1}{2}\right)=1$

$10y-1=1$

$10y=2$

$y=\frac{2}{10}=\frac{1}{5}$

The solution is $\left(\frac{1}{2}, \frac{1}{5}\right)$.

49. $\begin{cases} \frac{3}{4}x+\frac{5}{2}y=11 \\ \frac{1}{16}x-\frac{3}{4}y=-1 \end{cases}$

Clear fractions by multiplying the first equation by 4 and the second equation by 16.

$$\begin{cases} 3x+10y=44 \\ x-12y=-16 \end{cases}$$

Multiply the second equation by –3.

$$\begin{cases} 3x+10y=44 \\ -3x+36y=48 \end{cases}$$

Add the equations.

$$\begin{array}{r} 3x+10y=44 \\ -3x+36y=48 \\ \hline 46y=92 \end{array}$$

$y=2$

$x-12(2)=-16$

$x-24=-16$

$x=8$

The solution is (8, 2).

51. $\begin{cases} x \quad =3y+2 \\ 5x-15y=10 \end{cases}$

Replace *x* with 3*y* + 2 in the second equation.

$5(3y+2)-15y=10$

$15y+10-15y=10$

$10=10$ True

The system is dependent.

The solution is $\{(x,\ y)\,|\,x=3y+2\}$

53. $\begin{cases} 2x-y=-1 \\ y=-2x \end{cases}$

Replace *y* with –2*x* in the first equation.

$2x-(-2x)=-1$

$4x=-1$

$x=-\frac{1}{4}$

Replace *x* with $-\frac{1}{4}$ in the second equation.

$y=-2\left(-\frac{1}{4}\right)=\frac{1}{2}$

The solution is $\left(-\frac{1}{4}, \frac{1}{2}\right)$.

55. $\begin{cases} 2x=6 \\ y=5-x \end{cases}$

The first equation yields $x=3$.

Replace *x* with 3 in the second equation.

$y=5-3=2$

The solution is (3, 2).

57. $\begin{cases} \frac{x+5}{2}=\frac{6-4y}{3} \\ \frac{3x}{5}=\frac{21-7y}{10} \end{cases}$

Multiply the first equation by 6 and the

second equation by 10.

$$\begin{cases} 3x+15 = 12-8y \\ 6x = 21-7y \end{cases}$$

$$\begin{cases} 3x+8y=-3 \\ 6x+7y=21 \end{cases}$$

Multiply the first equation by –2.

$$\begin{cases} -6x-16y=6 \\ 6x+7y=21 \end{cases}$$

Add the equations.

$$\begin{array}{r} -6x-16y=6 \\ 6x+7y=21 \\ \hline -9y=27 \\ y=-3 \end{array}$$

Replace y with –3 in the equation $3x+8y=-3$.

$$\begin{aligned} 3x+8(-3)&=-3 \\ 3x-24&=-3 \\ 3x&=21 \\ x&=7 \end{aligned}$$

The solution is (7, –3).

59. $\begin{cases} 4x-7y=7 \\ 12x-21y=24 \end{cases}$

Multiply the second equation by $\frac{1}{3}$.

$$\begin{cases} 4x-7y=7 \\ 4x-7y=8 \end{cases}$$

Inconsistent system

The solution is $\varnothing$.

61. $\begin{cases} \frac{2}{3}x-\frac{3}{4}y=-1 \\ -\frac{1}{6}x+\frac{3}{8}y=1 \end{cases}$

Multiply the first equation by 12 and the second equation by 24.

$$\begin{cases} 8x-9y=-12 \\ -4x+9y=24 \end{cases}$$

Add the equations.

$$\begin{aligned} 4x&=12 \\ x&=3 \end{aligned}$$

Replace x with 3 in the equation $-4x+9y=24$.

$$\begin{aligned} -4(3)+9y&=24 \\ -12+9y&=24 \\ 9y&=36 \\ y&=4 \end{aligned}$$

The solution is (3, 4).

63. $\begin{cases} 0.7x-0.2y=-1.6 \\ 0.2x-y=-1.4 \end{cases}$

Multiply the first equation by 10 and the second equation by 5.

$$\begin{cases} 7x-2y=-16 \\ x-5y=-7 \end{cases}$$

Multiply the second equation by –7.

$$\begin{cases} 7x-2y=-16 \\ -7x+35y=49 \end{cases}$$

Add the equations.

$$\begin{aligned} 33y&=33 \\ y&=1 \end{aligned}$$

Replace y with 1 in the equation $x-5y=-7$.

$$\begin{aligned} x-5(1)&=-7 \\ x-5&=-7 \\ x&=-2 \end{aligned}$$

The solution is (–2, 1).

65. $\begin{cases} 4x-1.5y=10.2 \\ 2x+7.8y=-25.68 \end{cases}$

Multiply the second equation by –2.

$$\begin{cases} 4x-1.5y=10.2 \\ -4x-15.6y=51.36 \end{cases}$$

Add the equations.

$$\begin{aligned} -17.1y&=61.56 \\ y&=-3.6 \end{aligned}$$

Replace y with –3.6 in the first equation.

$$\begin{aligned} 4x-1.5(-3.6)&=10.2 \\ 4x+5.4&=10.2 \\ 4x&=4.8 \\ x&=1.2 \end{aligned}$$

The solution is (1.2, –3.6).

67. Point of intersection (5, 21)
5000 ties; $21

69. Supply is greater than demand.

71. $\begin{cases} y = 2.5x \\ y = 0.9x + 3000 \end{cases}$

Substitute.

$2.5x = 0.9x + 3000$
$1.6x = 3000$
$x = 1875$
$y = 2.5(1875) = 4687.5$

The point of intersection is (1875, 4687.5).

73. Makes money because revenue is greater than cost at $x = 2000$.

75. For values of $x > 1875$ because the x-value at the intersection is 1875.

77. a. Consumption of red meat is decreasing while consumption of poultry is increasing.

b. $\begin{cases} y = -0.6x + 121.2 \\ y = 1.7x + 88 \end{cases}$

Replace y with $-0.6x + 121.2$ in the second equation.

$-0.6x + 121.2 = 1.7x + 88$
$33.2 = 2.3x$
$14 \approx x$

Replace x with 14 in the first equation.

$y = -0.6(14) + 121.2$
$y \approx 113$

The solution is (14, 113).

c. $x = 14$, $1995 + 14 = 2009$

In the year 2009, red meat and poultry consumption will each be about 113 pounds per person.

79. $\begin{cases} 107x + y = 11{,}096 \\ 399x + y = 15{,}149 \end{cases}$

Multiply the first equation by –1.

$\begin{cases} -107x - y = -11{,}096 \\ 399x + y = 15{,}149 \end{cases}$

Add the two new equations:

$292x = 4053$
$x \approx 13.88$

The year is 1993.

81.
$$\begin{aligned} x + 2y - z &= 7 \\ 2 + 2(-3) - 3 &= 7 \\ 2 - 6 - 3 &= 7 \\ -7 &= 7 \end{aligned}$$

False

83.
$$\begin{aligned} -4x + y - 8z &= 4 \\ -4(1) + 0 - 8(-1) &= 4 \\ -4 + 8 &= 4 \\ 4 &= 4 \end{aligned}$$

True

85.
$$\begin{array}{rl} & x + 4y - 5z = 20 \\ + & 2x - 4y - 2z = -17 \\ \hline & 3x \quad\;\; - 7z = 3 \end{array}$$

87.
$$\begin{array}{rl} & -9x - 8y - z = 31 \\ + & 9x + 4y - z = 12 \\ \hline & -4y - 2z = 43 \end{array}$$

89. Replacing $\frac{1}{y}$ with a, we have the following:

$\begin{cases} x + 2a = 7 \\ 3x + 3a = 6 \end{cases}$

Multiply the first equation by –3.

$\begin{cases} -3x - 6a = -21 \\ 3x + 3a = 6 \end{cases}$

Add the two new equations:

$-3a = -15$
$a = 5$
$x + 2a = 7$
$x + 2(5) = 7$
$x + 10 = 7$
$x = -3$

Since $a = 5$, $y = \frac{1}{5}$. The solution is $\left(-3, \frac{1}{5}\right)$.

91. Replacing $\frac{1}{x}$ with a, and $\frac{1}{y}$ with b, we have the following:

$\begin{cases} 2a + 3b = 5 \\ 5a - 3b = 2 \end{cases}$

Add the two new equations:

$$\begin{aligned}7a&=7\\a&=1\\2a+3b&=5\\2+3b&=5\\3b&=3\\b&=1\end{aligned}$$

Since $a = 1$, $x=\frac{1}{1}=1$. Similarly, $y = 1$.
The solution is (1, 1).

93. Replacing $\frac{1}{x}$ with a, and $\frac{1}{y}$ with b, we have the following:

$$\begin{cases}3a-2b=-18\\2a+3b=1\end{cases}$$

Multiplying the first equation by 3 and the second equation by 2 yields the following:

$$\begin{cases}9a-6b=-54\\4a+6b=2\end{cases}$$

Add the two new equations:

$$\begin{aligned}13a&=-52\\a&=-4\\3a-2b&=-18\\-12-2b&=-18\\-2b&=-6\\b&=3\end{aligned}$$

Since $a = -4$, $x=-\frac{1}{4}$. Similarly, $y=\frac{1}{3}$.

The solution is $\left(-\frac{1}{4},\frac{1}{3}\right)$.

95. Replacing $\frac{1}{x}$ with a, and $\frac{1}{y}$ with b, we have the following:

$$\begin{cases}5a+7b=1\\-10a-14b=0\end{cases}$$

Multiplying the first equation by 2 yields the following:

$$\begin{cases}10a+14b=2\\-10a-14b=0\end{cases}$$

Add the two new equations:
$0=2$
This is a false statement. The solution is $\varnothing$.

Section 4.2

Exercise Set 4.2

1. $$\begin{cases}x+y=3 & (1)\\2y=10 & (2)\\3x+2y-3z=1 & (3)\end{cases}$$

Solve the second equation for y.
$y=5$
Replace y with 5 in the first equation.

$$\begin{aligned}x+5&=3\\x&=-2\end{aligned}$$

Replace x with -2 and y with 5 in the third equation.

$$\begin{aligned}3(-2)+2(5)-3z&=1\\-6+10-3z&=1\\4-3z&=1\\-3z&=-3\\z&=1\end{aligned}$$

The solution is $(-2, 5, 1)$.

3. $$\begin{cases}2x+2y+z=1 & (1)\\-x+y+2z=3 & (2)\\x+2y+4z=0 & (3)\end{cases}$$

Add equations (2) and (3).
$3y + 6z = 3$ or $y + 2z = 1$
Add twice equation (2) to equation (1).

$$\begin{aligned}-2x+2y+4z&=6\\2x+2y+z&=1\\\hline 4y+5z&=7\end{aligned}$$

Solve the new system:

$$\begin{cases}y+2z=1\\4y+5z=7\end{cases}$$

Multiply the first equation by -4.

$$\begin{cases}-4y-8z=-4\\4y+5z=7\end{cases}$$

Add the equations.

$$\begin{aligned}-4y-8z&=-4\\4y+5z&=7\\\hline -3z&=3\end{aligned}$$

$z=-1$
Replace z with -1 in the equation $y + 2z = 1$.

$y+2(-1)=1$
$y-2=1$
$y=3$

Replace y with 3 and z with -1 in equation (3).

$x+2(3)+4(-1)=0$
$x+6-4=0$
$x+2=0$
$x=-2$

The solution is $(-2, 3, -1)$.

5. $\begin{cases} x-2y+z=-5 & (1) \\ -3x+6y-3z=15 & (2) \\ 2x-4y+2z=-10 & (3) \end{cases}$

Multiply equation (2) by $-\frac{1}{3}$ and equation (3) by $\frac{1}{2}$.

$\begin{cases} x-2y+z=-5 \\ x-2y+z=-5 \\ x-2y+z=-5 \end{cases}$

All three equations are identical. There are infinitely many solutions.
The solution is $\{(x, y, z) \mid x-2y+z=-5\}$.

7. $\begin{cases} 4x-y+2z=5 & (1) \\ 2y+z=4 & (2) \\ 4x+y+3z=10 & (3) \end{cases}$

Multiply equation (1) by -1 and add to equation (3).

$-4x+y-2z=-5$
$\underline{4x+y+3z=10}$
$2y+z=5 \quad (4)$

Multiply equation (4) by -1 and add to equation (2).

$-2y-z=-5$
$\underline{2y+z=4}$
$0=-1$ False

Inconsistent system
The solution is $\varnothing$.

9. Answers may vary.

11. $\begin{cases} x+5z=0 & (1) \\ 5x+y=0 & (2) \\ y-3z=0 & (3) \end{cases}$

Multiply equation (3) by -1 and add to equation (2).

$-y+3z=0$
$\underline{5x+y=0}$
$5x+3z=0 \quad (4)$

Multiply equation (1) by -5 and add to equation (4).

$-5x-25z=0$
$\underline{5x+3z=0}$
$-22z=0$
$z=0$

Replace z with 0 in equation (4).

$5x+3(0)=0$
$5x=0$
$x=0$

Replace x with 0 in equation (2).

$5(0)+y=0$
$y=0$

The solution is $(0, 0, 0)$.

13. $\begin{cases} 6x-5z=17 & (1) \\ 5x-y+3z=-1 & (2) \\ 2x+y=-41 & (3) \end{cases}$

Add equations (2) and (3).

$7x+3z=-42 \quad (4)$

Multiply equation (4) by 5, multiply equation (1) by 3, and add.

$35x+15z=-210$
$\underline{18x-15z=51}$
$53x=-159$
$x=-3$

Replace x with -3 in equation (1).

$6(-3)-5z=17$
$-18-5z=17$
$-5z=35$
$z=-7$

Replace x with -3 in equation (3).

$2(-3)+y=-41$
$-6+y=-41$
$y=-35$

The solution is $(-3, -35, -7)$.

15. $\begin{cases} x + y + z = 8 & (1) \\ 2x - y - z = 10 & (2) \\ x - 2y - 3z = 22 & (3) \end{cases}$

Add equations (1) and (2).
$3x = 18$ or $x = 6$
Add twice equation (1) to equation (3).

$$\begin{array}{r} 2x + 2y + 2z = 16 \\ x - 2y - 3z = 22 \\ \hline 3x - \quad z = 38 \end{array}$$

Replace x with 6 in this equation.

$$\begin{aligned} 3(6) - z &= 38 \\ 18 - z &= 38 \\ -z &= 20 \\ z &= -20 \end{aligned}$$

Replace x with 6 and z with -20 in equation (1).

$$\begin{aligned} 6 + y + (-20) &= 8 \\ y - 14 &= 8 \\ y &= 22 \end{aligned}$$

The solution is (6, 22, –20).

17. $\begin{cases} x + 2y - z = 5 & (1) \\ 6x + y + z = 7 & (2) \\ 2x + 4y - 2z = 5 & (3) \end{cases}$

Add equations (1) and (2).
$7x + 3y = 12$ (4)
Add twice equation (2) to equation (3).

$$\begin{array}{r} 12x + 2y + 2z = 14 \\ 2x + 4y - 2z = 5 \\ \hline 14x + 6y \quad = 19 \quad (5) \end{array}$$

Multiply equation (4) by –2 and add to equation (5).

$$\begin{array}{r} -14x - 6y = -24 \\ 14x + 6y = 19 \\ \hline 0 = -5 \quad \text{False} \end{array}$$

Inconsistent system
The solution is $\varnothing$.

19. $\begin{cases} 2x - 3y + z = 2 & (1) \\ x - 5y + 5z = 3 & (2) \\ 3x + y - 3z = 5 & (3) \end{cases}$

Add –2 times equation (2) to equation (1).

$$\begin{array}{r} 2x - 3y + z = 2 \\ -2x + 10y - 10z = -6 \\ \hline 7y - 9z = -4 \end{array}$$

Add –3 times equation (2) to equation (3).

$$\begin{array}{r} -3x + 15y - 15z = -9 \\ 3x + y - 3z = 5 \\ \hline 16y - 18z = -4 \end{array}$$

We now have the system:
$\begin{cases} 7y - 9z = -4 & (4) \\ 16y - 18z = -4 & (5) \end{cases}$

Multiply equation (4) by –2 and add to equation (5).

$$\begin{array}{r} -14y + 18z = 8 \\ 16y - 18z = -4 \\ \hline 2y \quad = 4 \end{array}$$

$y = 2$
Replace y with 2 in equation (4).

$$\begin{aligned} 7(2) - 9z &= -4 \\ -9z &= -18 \\ z &= 2 \end{aligned}$$

Replace y with 2 and z with 2 in equation (1).

$$\begin{aligned} 2x - 3(2) + 2 &= 2 \\ x &= 3 \end{aligned}$$

The solution is (3, 2, 2).

21. $\begin{cases} -2x - 4y + 6z = -8 & (1) \\ x + 2y - 3z = 4 & (2) \\ 4x + 8y - 12z = 16 & (3) \end{cases}$

Add 2 times equation (2) to equation (1).

$$\begin{array}{r} 2x + 4y - 6z = 8 \\ -2x - 4y + 6z = -8 \\ \hline 0 = 0 \end{array}$$

Add –4 times equation (2) to equation (3).

$$\begin{array}{r} -4x - 8y + 12z = -16 \\ 4x + 8y - 12z = 16 \\ \hline 0 = 0 \end{array}$$

The system is dependent.
The solution is $\{(x, y, z) \mid x + 2y - 3z = 4\}$.

23. $\begin{cases} 2x+2y-3z = 1 & (1) \\ y+2z=-14 & (2) \\ 3x-2y = -1 & (3) \end{cases}$

Add equations (1) and (3).
$5x-3z=0$ (4)
Add twice equation (2) to equation (3).

$$\begin{array}{r} 2y+4z=-28 \\ 3x-2y \quad = -1 \\ \hline 3x+ \quad 4z=-29 \end{array} \quad (5)$$

Multiply equation (4) by 4, multiply equation (5) by 3 and add.

$$\begin{array}{r} 20x-12z= \quad 0 \\ 9x+12z=-87 \\ \hline 29x \quad =-87 \end{array}$$

$x=-3$
Replace x with -3 in equation (4).
$5(-3)-3z=0$
$3z=-15$
$z=-5$
Replace z with -5 in equation (2).
$y+2(-5)=-14$
$y-10=-14$
$y=-4$
The solution is $(-3, -4, -5)$.

25. $\begin{cases} \frac{3}{4}x-\frac{1}{3}y+\frac{1}{2}z=9 & (1) \\ \frac{1}{6}x+\frac{1}{3}y-\frac{1}{2}z=2 & (2) \\ \frac{1}{2}x- \ y+\frac{1}{2}z=2 & (3) \end{cases}$

Multiply equation (1) by 12, multiply equation (2) by 6, and multiply equation (3) by 2.

$\begin{cases} 9x-4y+6z=108 & (4) \\ x+2y-3z= 12 & (5) \\ x-2y+ z= 4 & (6) \end{cases}$

Add twice equation (5) to equation (4).

$$\begin{array}{r} 2x+4y-6z= 24 \\ 9x-4y+6z=108 \\ \hline 11x \quad =132 \end{array}$$

$x=12$
Add equations (5) and (6).
$2x-2z=16$ or $x-z=8$
Replace x with 12 in this equation.
$12-z=8$
$z=4$
Replace x with 12 and z with 4 in equation (6).
$12-2y+4=4$
$12-2y=0$
$12=2y$
$y=6$
The solution is (12, 6, 4).

27. $\begin{cases} x+ y+ z= 1 & (1) \\ 2x- y+ z= 0 & (2) \\ -x+2y+2z=-1 & (3) \end{cases}$

Multiply equation (3) by 2 and add to equation (2).

$$\begin{array}{r} 2x- y + z= 0 \\ -2x+4y+4z=-2 \\ \hline 3y+5z=-2 \end{array} \quad (4)$$

Multiply equation (1) by -2 and add to equation (2).

$$\begin{array}{r} -2x-2y-2z=-2 \\ 2x- y+ z= 0 \\ \hline -3y- z=-2 \end{array} \quad (5)$$

Add equations (4) and (5).

$$\begin{array}{r} 3y+5z=-2 \\ -3y- z=-2 \\ \hline 4z=-4 \end{array}$$

$z=-1$
Replace z with -1 in equation (4).
$3y+5(-1)=-2$
$3y=3$
$y=1$
Replace y with 1 and z with -1 in equation (1).
$x+1+(-1)=1$
$x=1$
The solution is $(1, 1, -1)$.

$\frac{1}{24}=\frac{x}{8}+\frac{y}{4}+\frac{z}{3}$

$\frac{1}{24}=\frac{1}{8}+\frac{1}{4}-\frac{1}{3}$

$\frac{1}{24}=\frac{3}{24}+\frac{6}{24}-\frac{8}{24}$

$\frac{1}{24}=\frac{1}{24}$ True

29. Let x = first number.
Then $45 - x$ = second number.
$45 - x = 2x$
$45 = 3x$
$15 = x$
$45 - 15 = 30$
The numbers are 15 and 30.

31. $2(x-1) - 3x = x - 12$
$2x - 2 - 3x = x - 12$
$-2 - x = x - 12$
$10 = 2x$
$5 = x$
The solution is 5.

33. $-y - 5(y+5) = 3y - 10$
$-y - 5y - 25 = 3y - 10$
$-6y - 25 = 3y - 10$
$-15 = 9y$
$-\frac{15}{9} = y$
$y = -\frac{5}{3}$
The solution is $-\frac{5}{3}$.

35. $\begin{cases} x + y \qquad - w = 0 & (1) \\ y + 2z + w = 3 & (2) \\ x \qquad - z \qquad = 1 & (3) \\ 2x - y \qquad - w = -1 & (4) \end{cases}$
Add equations (2) and (4).
$2x + 2z = 2$ (5)
Add equation (5) to 2 times equation (3).
$2x - 2z = 2$
$\underline{2x + 2z = 2}$
$4x \qquad = 4$
$x = 1$
Replace x with 1 in equation (3).
$1 - z = 1$
$z = 0$
Replace x with 1 in equation (1).
$1 + y - w = 0$
$y - w = -1$ (6)
Replace x with 1 in equation (4).
$2 - y - w = -1$
$-y - w = -3$ (7)
Add equations (6) and (7).
$y - w = -1$
$\underline{-y - w = -3}$
$-2w = -4$
$w = 2$
Replace w with 2 in equation (6).
$y - 2 = -1$
$y = 1$
The solution is (1, 1, 0, 2).

37. $\begin{cases} x + y + z + w = 5 & (1) \\ 2x + y + z + w = 6 & (2) \\ x + y + z \qquad = 2 & (3) \\ x + y \qquad = 0 & (4) \end{cases}$
Multiply equation (1) by –1 and add to equation (2).
$-x - y - z - w = -5$
$\underline{2x + y + z + w = 6}$
$x \qquad = 1$
Replace x with 1 in equation (4).
$1 + y = 0$
$y = -1$
Replace x with 1 and y with –1 in equation (3).
$1 + (-1) + z = 2$
$z = 2$
Replace x with 1, y with –1 and z with 2 in equation (1).
$1 + (-1) + 2 + w = 5$
$2 + w = 5$
$w = 3$
The solution is (1, –1, 2, 3).

Section 4.3

Exercise Set 4.3

1. Let m = the first number
n = the second number
$\begin{cases} m = n + 2 \\ 2m = 3n - 4 \end{cases}$
Substitute $m = n + 2$ in the second equation.
$2(n+2) = 3n - 4$
$2n + 4 = 3n - 4$
$n = 8$

Replace n with 8 in the first equation.
$m = 8 + 2 = 10$
The numbers are 10 and 8.

3. Let p = the speed of the plane in still air
w = the speed of the wind
$$\begin{cases} p + w = 560 \\ p - w = 480 \end{cases}$$
Add the equations.
$2p = 1040$
$p = 520$
Replace p with 520 in the first equation.
$520 + w = 560$
$w = 40$
The speed of the plane in still air is 520 mph and the speed of the wind is 40 mph.

5. Let x = number of quarts of 4% butterfat milk
y = the number of quarts of 1% butterfat milk
$$\begin{cases} x + y = 60 \\ 0.04x + 0.01y = 0.02(60) \end{cases}$$
Multiply the second equation by –100 and add to the first equation.
$$\begin{array}{r} x + y = 60 \\ -4x - y = -120 \\ \hline -3x = -60 \end{array}$$
$x = 20$
Replace x with 20 in the first equation.
$20 + y = 60$
$y = 40$
20 quarts of 4% butterfat milk and 40 quarts of 1% butterfat milk should be used.

7. Let l = number of large frames
s = the number of small frames
$$\begin{cases} l + s = 22 \\ 15l + 8s = 239 \end{cases}$$
Multiply the first equation by –8 and add to the second equation.
$$\begin{array}{r} -8l - 8s = -176 \\ 15l + 8s = 239 \\ \hline 7l = 63 \end{array}$$
$l = 9$
Replace l with 9 in the first equation.
$9 + s = 22$
$s = 13$
She bought 9 large frames and 13 small frames.

9. Let m = the first number
n = the second number
$$\begin{cases} m = n - 2 \\ 2m = 3n + 4 \end{cases}$$
Substitute $m = n - 2$ in the second equation.
$2(n - 2) = 3n + 4$
$2n - 4 = 3n + 4$
$n = -8$
Replace n with –8 in the first equation.
$m = -8 - 2 = -10$
The numbers are –10 and –8.

11. Let x = the price of each tablet
y = the price of each pen
$$\begin{cases} 7x + 4y = 6.40 \\ 2x + 19y = 5.40 \end{cases}$$
Multiply the first equation by 2 and the second equation by –7 and add.
$$\begin{array}{r} 14x + 8y = 12.80 \\ -14x - 133y = -37.80 \\ \hline -125y = -25 \end{array}$$
$y = \frac{25}{125} = 0.20$
Replace y with 0.20 in the first equation.
$7x + 4(0.20) = 6.40$
$7x + 0.80 = 6.40$
$7x = 5.60$
$x = 0.80$
Tablets cost \$0.80 each and pens cost \$0.20 each.

13. Let p = the speed of the plane in still air
w = the speed of the wind
First note:
$\frac{2160 \text{ mi}}{3 \text{ hr}} = 720$ mph and
$\frac{2160 \text{ mi}}{4 \text{ hr}} = 540$ mph

Now,

$$\begin{cases} p+w=720 \\ p-w=540 \end{cases}$$

Add the equations.

$2p=1260$

$p=630$

Replace p with 630 in the first equation.

$630+w=720$

$w=90$

The speed of the plane in still air is 630 mph and the speed of the wind is 90 mph.

15. Let s = the length of the shortest side

l = the length of the longest side

m = the length of the other two sides

$$\begin{cases} s+2m+l=29 \\ l=2s \\ m=s+2 \end{cases}$$

Substitute $l = 2s$ and $m = s + 2$ in the first equation.

$s+2(s+2)+2s=29$

$s+2s+4+2s=29$

$5s+4=29$

$5s=25$

$s=5$

Replace s with 5 in the second and third equations.

$l = 2(5) = 10$

$m = 5 + 2 = 7$

The shortest side is 5 inches, the longest side is 10 inches, and the other two sides are 7 inches.

17. Let x = the first number

y = the second number

z = the third number

$$\begin{cases} x+y+z=40 \\ x=y+5 \\ x=2z \end{cases}$$

$$\begin{cases} x+y+z=40 \\ y=x-5 \\ z=\frac{1}{2}x \end{cases}$$

Substitute $y = x - 5$ and $z = \frac{1}{2}x$ in the first equation.

$x+x-5+\frac{1}{2}x=40$

$\frac{5}{2}x-5=40$

$\frac{5}{2}x=45$

$x=\frac{2}{5}(45)=18$

$y = 18 - 5 = 13$

$z=\frac{1}{2}(18)=9$

The three numbers are 18, 13, and 9.

19. Let x = total weekly sales

y = total weekly salary

$$\begin{cases} y=200+0.05x \\ y=0.15x \end{cases}$$

Substitute $y = 0.15x$ in the first equation.

$200+0.05x=0.15x$

$200=0.10x$

$x=\frac{200}{0.10}=2000$

Jack's salary would be the same, regardless of pay arrangement, at $2000 worth of sales.

21. Let x = price for template

y = price for pencil

z = price for paper

$$\begin{cases} 3x+y=6.45 \\ 2z+4y=7.50 \\ z=3y \end{cases}$$

Substitute $z = 3y$ in the second equation.

$2(3y)+4y=7.50$

$6y+4y=7.50$

$10y=7.50$

$y=0.75$

Replace y with 0.75 in the third equation.

$3(0.75) = z$

$2.25 = z$

Replace y with 0.75 in the first equation.

$3x + 0.75 = 6.45$
$3x = 5.70$
$x = 1.90$
\$1.90 for template; \$0.75 for pencil; \$2.25 for paper

23. $C(x) = 12x + 15{,}000$
$R(x) = 32x$
$12x + 15{,}000 = 32x$
$15{,}000 = 20x$
$750 = x$
750 units

25. $C(x) = 0.8x + 900$
$R(x) = 2x$
$0.8x + 900 = 2x$
$900 = 1.2x$
$750 = x$
750 units

27. $C(x) = 105x + 70{,}000$
$R(x) = 245x$
$105x + 70{,}000 = 245x$
$70{,}000 = 140x$
$500 = x$
500 units

29. a. $R(x) = 31x$

b. $C(x) = 15x + 500$

c. $R(x) = C(x)$
$31x = 15x + 500$
$16x = 500$
$x = 31.25$
32 baskets

31. $\begin{cases} x + 2y = 180 \\ 3x - 10 + y = 180 \end{cases}$

$\begin{cases} x = 180 - 2y \\ 3x + y = 190 \end{cases}$

Substitute $x = 180 - 2y$ in the second equation.
$3(180 - 2y) + y = 190$
$540 - 6y + y = 190$
$-5y = -350$
$y = 70$
Replace y with 70 in the equation
$x = 180 - 2y.$
$x = 180 - 2(70)$
$x = 40$
$x = 40$ and $y = 70$

33. Let x = ounces of 20% acid, and
y = ounces of 50% acid.
$\begin{cases} x + y = 60 & (1) \\ .2x + .5y = .3(60) = 18 & (2) \end{cases}$
Multiply the 2nd equation by 10:
$x + y = 60$
$2x + 5y = 180$
Multiply the 1st equation by –2 and add to the second:
$-2x - 2y = -120$
$2x + 5y = 180$
$3y = 60$
$y = 20$
Replace y with 20 in the first equation:
$x + 20 = 60$
$x = 40$
40 ounces of 20% acid solution
20 ounces of 50% acid solution

35.

Concentration	Amount	Solution
25%	$2x$	$0.25(2x) = 0.5x$
40%	x	$0.4x$
50%	y	$0.5y$
32%	200	$0.32(200) = 64$

$\begin{cases} 2x + x + y = 200 \\ 0.5x + 0.4x + 0.5y = 64 \end{cases}$

$\begin{cases} 3x + y = 200 \\ 0.9x + 0.5y = 64 \end{cases}$

Solve first equation for y and substitute into second equation.
$y = 200 - 3x$
$0.9x + 0.5(200 - 3x) = 64$
$0.9x + 100 - 1.5x = 64$
$-0.6x = -36$
$x = 60$
$2x = 120$

Replace x with 60 in the equation
$y = 200 - 3x$.
$y = 200 - 3(60) = 20$
120 liters of 25% solution
60 liters of 40% solution
20 liters of 50% solution

37. Let x = number of free throws
y = number of two-point field goals
z = number of three-point field goals

$$\begin{cases} x + 2y + 3z = 26 \\ x = z + 2 \\ y = x + 4 \end{cases}$$

$$\begin{cases} x + 2y + 3z = 26 & (1) \\ x - z = 2 & (2) \\ -x + y = 4 & (3) \end{cases}$$

Multiply equation (2) by −1 and add to equation (1).

$$\begin{array}{r} x + 2y + 3z = 26 \\ -x + z = -2 \\ \hline 2y + 4z = 24 \ (4) \end{array}$$

Add equation (3) to equation (1).

$$\begin{array}{r} x + 2y + 3z = 26 \\ -x + y \quad = 4 \\ \hline 3y + 3z = 30 \ (5) \end{array}$$

Multiply equation (5) by $-\frac{2}{3}$ and add to equation (4).

$$\begin{array}{r} 2y + 4z = 24 \\ -2y - 2z = -20 \\ \hline 2z = 4 \\ z = 2 \end{array}$$

Replace z with 2 in equation (2).
$x - (2) = 2$
$x = 4$
Replace x with 4 in equation (3).
$-(4) + y = 4$
$y = 8$
He scored 4 free throws, 8 two-point field goals and 2 three-point field goals.

39. No matter how much of each acid solution is used, the strength of the mixture should be somewhere between 25% and 60%. So you should suspect an error in the calculation if the result is 14%.

41. $y = ax^2 + bx + c$
$(1, 2): 2 = a + b + c \quad (1)$
$(2, 3): 3 = 4a + 2b + c \quad (2)$
$(-1, 6): 6 = a - b + c \quad (3)$
Add equations (1) and (3).

$$\begin{array}{r} 2 = a + b + c \\ 6 = a - b + c \\ \hline 8 = 2a + \quad 2c \ (4) \end{array}$$

Multiply equation (3) by 2 and add to equation (2).

$$\begin{array}{r} 3 = 4a + 2b + c \\ 12 = 2a - 2b + 2c \\ \hline 15 = 6a + \quad 3c \ (5) \end{array}$$

Multiply equation (4) by −3 and add to equation (5).

$$\begin{array}{r} -24 = -6a - 6c \\ 15 = 6a + 3c \\ \hline -9 = \quad -3c \end{array}$$

$c = 3$
Replace c with 3 in equation (4).
$8 = 2a + 2(3)$
$2 = 2a$
$a = 1$
Replace a with 1 and c with 3 in equation (1).
$2 = 1 + b + 3$
$b = -2$
$a = 1, \ b = -2, \$ and $c = 3$

43. $x = 180 - (z + 15)$
$x = 165 - z$

$y = 180 - (z - 13)$
$y = 193 - z$

$360 = x + y + z + 72$
$288 = x + y + z$

$$\begin{cases} z = 165 - x \\ z = 193 - y \\ x + y + z = 288 \end{cases}$$

Replace z with $165 - x$ in the third equation.
$x + y + 165 - x = 288$
$y = 123$

Replace z with $193 - y$ in the third equation.
$x + y + 193 - y = 288$
$x = 95$
Replace x with 95 in the first equation.
$z = 165 - 95 = 70$
$x = 95$, $y = 123$, and $z = 70$

45. $y = ax^2 + bx + c$
(4, 2.47): $2.47 = 16a + 4b + c$
(7, 0.6): $0.6 = 49a + 7b + c$
(8, 1.1): $1.1 = 64a + 8b + c$
Multiply the second equation by –1 and add to the first equation.

$$\begin{array}{r} 2.47 = 16a + 4b + c \\ \underline{-0.6 = -49a - 7b - c} \\ 1.87 = -33a - 3b \end{array}$$

Multiply the second equation by –1 and add to the third equation.

$$\begin{array}{r} -0.6 = -49a - 7b - c \\ \underline{1.1 = 64a + 8b + c} \\ 0.5 = 15a + b \end{array}$$

$b = 0.5 - 15a$
Replace b with $0.5 - 15a$.
$1.87 = -33a - 3(0.5 - 15a)$
$1.87 = -33a - 1.5 + 45a$
$3.37 = 12a$
$0.28 = a$
Replace a with 0.28.

$b = 0.5 - 15\ (0.28)$
$b = -3.71$

$c = 2.47 - 16a - 4b$
$c = 2.47 - 16(0.28) - 4(-3.71)$
$c = 12.83$
$a = 0.28$, $b = -3.71$, $c = 12.83$
September, $x = 9$
$y = 0.28(9)^2 - 3.71(9) + 12.83$
$= 2.12$ inches

47. $\begin{cases} 2x + y + 3z = 7 \\ -4x + y + 2z = 4 \end{cases}$

$$\begin{array}{r} 4x + 2y + 6z = 14 \\ \underline{-4x + y + 2z = 4} \\ 3y + 8z = 18 \end{array}$$

49. $\begin{cases} 2x - 3y + 2z = 5 \\ x - 9y + z = -1 \end{cases}$

$$\begin{array}{r} -6x + 9y - 6z = -15 \\ \underline{x - 9y + z = -1} \\ -5x \qquad - 5z = -16 \end{array}$$

51. $P(\text{green}) = \frac{3}{8}$

53. $P(\text{red or blue}) = \frac{5}{8}$

Section 4.4

Exercise Set 4.4

1. $\begin{cases} x + y = 1 \\ x - 2y = 4 \end{cases}$

$\left[\begin{array}{cc|c} 1 & 1 & 1 \\ 1 & -2 & 4 \end{array}\right]$

Multiply row 1 by –1 and add to row 2.

$\left[\begin{array}{cc|c} 1 & 1 & 1 \\ 0 & -3 & 3 \end{array}\right]$

Divide row 2 by –3.

$\left[\begin{array}{cc|c} 1 & 1 & 1 \\ 0 & 1 & -1 \end{array}\right]$

This corresponds to $\begin{cases} x + y = 1 \\ y = -1 \end{cases}$.

$x + (-1) = 1$
$x - 1 = 1$
$x = 2$
The solution is (2, –1).

3. $\begin{cases} x + 3y = 2 \\ x + 2y = 0 \end{cases}$

$\left[\begin{array}{cc|c} 1 & 3 & 2 \\ 1 & 2 & 0 \end{array}\right]$

Multiply row 1 by –1 and add to row 2.

$\left[\begin{array}{cc|c} 1 & 3 & 2 \\ 0 & -1 & -2 \end{array}\right]$

Multiply row 2 by –1.

$\left[\begin{array}{cc|c} 1 & 3 & 2 \\ 0 & 1 & 2 \end{array}\right]$

This corresponds to $\begin{cases} x+3y=2 \\ y=2 \end{cases}$

$x+3(2)=2$
$x+6=2$
$x=-4$

The solution is (–4, 2).

5. $\begin{cases} x-2y=4 \\ 2x-4y=4 \end{cases}$

$\left[\begin{array}{cc|c} 1 & -2 & 4 \\ 2 & -4 & 4 \end{array}\right]$

Multiply row 1 by –2 and add to row 2.

$\left[\begin{array}{cc|c} 1 & -2 & 4 \\ 0 & 0 & -4 \end{array}\right]$

This corresponds to $\begin{cases} x-2y=4 \\ 0=-4 \end{cases}$.

This is an inconsistent system.
The solution is $\varnothing$.

7. $\begin{cases} 3x-3y=9 \\ 2x-2y=6 \end{cases}$

$\left[\begin{array}{cc|c} 3 & -3 & 9 \\ 2 & -2 & 6 \end{array}\right]$

Divide row 1 by 3.

$\left[\begin{array}{cc|c} 1 & -1 & 3 \\ 2 & -2 & 6 \end{array}\right]$

Multiply row 1 by –2 and add to row 2.

$\left[\begin{array}{cc|c} 1 & -1 & 3 \\ 0 & 0 & 0 \end{array}\right]$

This corresponds to $\begin{cases} x-y=3 \\ 0=0 \end{cases}$.

This is a dependent system.
The solution is $\{(x, y) | x-y=3\}$.

9. $\begin{cases} x+y=3 \\ 2y=10 \\ 3x+2y-4z=12 \end{cases}$

$\left[\begin{array}{ccc|c} 1 & 1 & 0 & 3 \\ 0 & 2 & 0 & 10 \\ 3 & 2 & -4 & 12 \end{array}\right]$

Multiply row 1 by –3 and add to row 3.

$\left[\begin{array}{ccc|c} 1 & 1 & 0 & 3 \\ 0 & 2 & 0 & 10 \\ 0 & -1 & -4 & 3 \end{array}\right]$

Divide row 2 by 2.

$\left[\begin{array}{ccc|c} 1 & 1 & 0 & 3 \\ 0 & 1 & 0 & 5 \\ 0 & -1 & -4 & 3 \end{array}\right]$

Add row 2 to row 3.

$\left[\begin{array}{ccc|c} 1 & 1 & 0 & 3 \\ 0 & 1 & 0 & 5 \\ 0 & 0 & -4 & 8 \end{array}\right]$

Divide row 3 by –4.

$\left[\begin{array}{ccc|c} 1 & 1 & 0 & 3 \\ 0 & 1 & 0 & 5 \\ 0 & 0 & 1 & -2 \end{array}\right]$

This corresponds to

$\begin{cases} x+y=3 \\ y=5 \\ z=-2 \end{cases}$.

$x+5=3$
$x=-2$

The solution is (–2, 5, –2).

11. $\begin{cases} 2y-z=-7 \\ x+4y+z=-4 \\ 5x-y+2z=13 \end{cases}$

$\left[\begin{array}{ccc|c} 0 & 2 & -1 & -7 \\ 1 & 4 & 1 & -4 \\ 5 & -1 & 2 & 13 \end{array}\right]$

Interchange rows 1 and 2.

$\left[\begin{array}{ccc|c} 1 & 4 & 1 & -4 \\ 0 & 2 & -1 & -7 \\ 5 & -1 & 2 & 13 \end{array}\right]$

Multiply row 1 by –5 and add to row 3.

$\left[\begin{array}{ccc|c} 1 & 4 & 1 & -4 \\ 0 & 2 & -1 & -7 \\ 0 & -21 & -3 & 33 \end{array}\right]$

Divide row 2 by 2.

$$\left[\begin{array}{ccc|c} 1 & 4 & 1 & -4 \\ 0 & 1 & -\frac{1}{2} & -\frac{7}{2} \\ 0 & -21 & -3 & 33 \end{array}\right]$$

Multiply row 2 by 21 and add to row 3.

$$\left[\begin{array}{ccc|c} 1 & 4 & 1 & -4 \\ 0 & 1 & -\frac{1}{2} & -\frac{7}{2} \\ 0 & 0 & -\frac{27}{2} & -\frac{81}{2} \end{array}\right]$$

Multiply row 3 by $-\frac{2}{27}$.

$$\left[\begin{array}{ccc|c} 1 & 4 & 1 & -4 \\ 0 & 1 & -\frac{1}{2} & -\frac{7}{2} \\ 0 & 0 & 1 & 3 \end{array}\right]$$

This corresponds to

$$\begin{cases} x+4y+z=-4 \\ y-\frac{1}{2}z=-\frac{7}{2} \\ z=3 \end{cases}.$$

$$y-\frac{1}{2}(3)=-\frac{7}{2}$$
$$y-\frac{3}{2}=-\frac{7}{2}$$
$$y=-2$$
$$x+4(-2)+3=-4$$
$$x-8+3=-4$$
$$x=1$$

The solution is (1, –2, 3).

13. $\begin{cases} x-4=0 \\ x+y=1 \end{cases}$ or $\begin{cases} x=4 \\ x+y=1 \end{cases}$

$$\left[\begin{array}{cc|c} 1 & 0 & 4 \\ 1 & 1 & 1 \end{array}\right]$$

Multiply row 1 by –1 and add to row 2.

$$\left[\begin{array}{cc|c} 1 & 0 & 4 \\ 0 & 1 & -3 \end{array}\right]$$

This corresponds to $\begin{cases} x=4 \\ y=-3 \end{cases}$.

The solution is (4, –3).

15. $\begin{cases} x+y+z=2 \\ 2x-z=5 \\ 3y+z=2 \end{cases}$

$$\left[\begin{array}{ccc|c} 1 & 1 & 1 & 2 \\ 2 & 0 & -1 & 5 \\ 0 & 3 & 1 & 2 \end{array}\right]$$

Multiply row 1 by –2 and add to row 2.

$$\left[\begin{array}{ccc|c} 1 & 1 & 1 & 2 \\ 0 & -2 & -3 & 1 \\ 0 & 3 & 1 & 2 \end{array}\right]$$

Divide row 2 by –2.

$$\left[\begin{array}{ccc|c} 1 & 1 & 1 & 2 \\ 0 & 1 & \frac{3}{2} & -\frac{1}{2} \\ 0 & 3 & 1 & 2 \end{array}\right]$$

Multiply row 2 by –3 and add to row 3.

$$\left[\begin{array}{ccc|c} 1 & 1 & 1 & 2 \\ 0 & 1 & \frac{3}{2} & -\frac{1}{2} \\ 0 & 0 & -\frac{7}{2} & \frac{7}{2} \end{array}\right]$$

Multiply row 3 by $-\frac{2}{7}$.

$$\left[\begin{array}{ccc|c} 1 & 1 & 1 & 2 \\ 0 & 1 & \frac{3}{2} & -\frac{1}{2} \\ 0 & 0 & 1 & -1 \end{array}\right]$$

This corresponds to

$$\begin{cases} x+y+z=2 \\ y+\frac{3}{2}z=-\frac{1}{2} \\ z=-1 \end{cases}.$$

$$y+\frac{3}{2}(-1)=-\frac{1}{2}$$
$$y-\frac{3}{2}=-\frac{1}{2}$$
$$y=1$$

$$x+1+(-1)=2$$
$$x=2$$

The solution is (2, 1, –1).

17. $\begin{cases} 5x-2y=27 \\ -3x+5y=18 \end{cases}$

$\left[\begin{array}{cc|c} 5 & -2 & 27 \\ -3 & 5 & 18 \end{array}\right]$

Divide row 1 by 5.

$\left[\begin{array}{cc|c} 1 & -\frac{2}{5} & \frac{27}{5} \\ -3 & 5 & 18 \end{array}\right]$

Multiply row 1 by 3 and add to row 2.

$\left[\begin{array}{cc|c} 1 & -\frac{2}{5} & \frac{27}{5} \\ 0 & \frac{19}{5} & \frac{171}{5} \end{array}\right]$

Multiply row 2 by $\frac{5}{19}$.

$\left[\begin{array}{cc|c} 1 & -\frac{2}{5} & \frac{27}{5} \\ 0 & 1 & 9 \end{array}\right]$

Multiply row 2 by $\frac{2}{5}$ and add to row 1.

$\left[\begin{array}{cc|c} 1 & 0 & 9 \\ 0 & 1 & 9 \end{array}\right]$

This corresponds to

$\begin{cases} x=9 \\ y=9 \end{cases}.$

The solution is (9, 9).

19. $\begin{cases} 4x-7y=7 \\ 12x-21y=24 \end{cases}$

$\left[\begin{array}{cc|c} 4 & -7 & 7 \\ 12 & -21 & 24 \end{array}\right]$

Divide row 1 by 4.

$\left[\begin{array}{cc|c} 1 & -\frac{7}{4} & \frac{7}{4} \\ 12 & -21 & 24 \end{array}\right]$

Multiply row 1 by –12 and add to row 2.

$\left[\begin{array}{cc|c} 1 & -\frac{7}{4} & \frac{7}{4} \\ 0 & 0 & 3 \end{array}\right]$

This corresponds to

$\begin{cases} x-\frac{7}{4}y=\frac{7}{4}. \\ 0=3 \end{cases}$

This is an inconsistent system.
The solution is $\varnothing$.

21. $\begin{cases} 4x-y+2z=5 \\ 2y+z=4 \\ 4x+y+3z=10 \end{cases}$

$\left[\begin{array}{ccc|c} 4 & -1 & 2 & 5 \\ 0 & 2 & 1 & 4 \\ 4 & 1 & 3 & 10 \end{array}\right]$

Divide row 1 by 4.

$\left[\begin{array}{ccc|c} 1 & -\frac{1}{4} & \frac{1}{2} & \frac{5}{4} \\ 0 & 2 & 1 & 4 \\ 4 & 1 & 3 & 10 \end{array}\right]$

Multiply row 1 by –4 and add to row 3.

$\left[\begin{array}{ccc|c} 1 & -\frac{1}{4} & \frac{1}{2} & \frac{5}{4} \\ 0 & 2 & 1 & 4 \\ 0 & 2 & 1 & 5 \end{array}\right]$

Divide row 2 by 2.

$\left[\begin{array}{ccc|c} 1 & -\frac{1}{4} & \frac{1}{2} & \frac{5}{4} \\ 0 & 1 & \frac{1}{2} & 2 \\ 0 & 2 & 1 & 5 \end{array}\right]$

Multiply row 2 by –2 and add to row 3.

$\left[\begin{array}{ccc|c} 1 & -\frac{1}{4} & \frac{1}{2} & \frac{5}{4} \\ 0 & 1 & \frac{1}{2} & 2 \\ 0 & 0 & 0 & 1 \end{array}\right]$

This corresponds to

$\begin{cases} x-\frac{1}{4}y+\frac{1}{2}z=\frac{5}{4} \\ y+\frac{1}{2}z=2\,. \\ 0=1 \end{cases}$

This is an inconsistent system.
The solution is $\varnothing$.

23. $\begin{cases} 4x+y+z=3 \\ -x+y-2z=-11 \\ x+2y+2z=-1 \end{cases}$

$\left[\begin{array}{ccc|c} 4 & 1 & 1 & 3 \\ -1 & 1 & -2 & -11 \\ 1 & 2 & 2 & -1 \end{array}\right]$

Interchange rows 1 and 3.

$$\left[\begin{array}{ccc|c} 1 & 2 & 2 & -1 \\ -1 & 1 & -2 & -11 \\ 4 & 1 & 1 & 3 \end{array}\right]$$

Multiply row 1 by 1 and add to row 2.
Multiply row 1 by –4 and add to row 3.

$$\left[\begin{array}{ccc|c} 1 & 2 & 2 & -1 \\ 0 & 3 & 0 & -12 \\ 0 & -7 & -7 & 7 \end{array}\right]$$

Divide row 2 by 3.

$$\left[\begin{array}{ccc|c} 1 & 2 & 2 & -1 \\ 0 & 1 & 0 & -4 \\ 0 & -7 & -7 & 7 \end{array}\right]$$

Multiply row 2 by 7 and add to row 3.

$$\left[\begin{array}{ccc|c} 1 & 2 & 2 & -1 \\ 0 & 1 & 0 & -4 \\ 0 & 0 & -7 & -21 \end{array}\right]$$

Divide row 3 by –7.

$$\left[\begin{array}{ccc|c} 1 & 2 & 2 & -1 \\ 0 & 1 & 0 & -4 \\ 0 & 0 & 1 & 3 \end{array}\right]$$

This corresponds to

$$\begin{cases} x+2y+2z=-1 \\ \quad\; y \qquad\; =-4. \\ \qquad\quad\; z=3 \end{cases}$$

$$\begin{aligned} x+2(-4)+2(3) &= -1 \\ x-8+6 &= -1 \\ x &= 1 \end{aligned}$$

The solution is (1, –4, 3).

25. The matrix should be $\left[\begin{array}{cc|c} 2 & -3 & 8 \\ 1 & 5 & -3 \end{array}\right]$.

27. Function

29. Not a function

31. $(-1)(-5) - (6)(3) = 5 - 18 = -13$

33. $(4)(-10) - (2)(-2) = -40 + 4 = -36$

35. $(-3)(-3) - (-1)(-9) = 9 - 9 = 0$

Section 4.5

Exercise Set 4.5

1. $\begin{vmatrix} 3 & 5 \\ -1 & 7 \end{vmatrix} = 3(7) - 5(-1)$
$= 21 + 5$
$= 26$

3. $\begin{vmatrix} 9 & -2 \\ 4 & -3 \end{vmatrix} = 9(-3) - 4(-2)$
$= -27 + 8$
$= -19$

5. $\begin{vmatrix} -2 & 9 \\ 4 & -18 \end{vmatrix} = -2(-18) - 9(4)$
$= 36 - 36$
$= 0$

7. $\begin{cases} 2y - 4 = 0 \\ x + 2y = 5 \end{cases}$ or $\begin{cases} 2y = 4 \\ x + 2y = 5 \end{cases}$

$D = \begin{vmatrix} 0 & 2 \\ 1 & 2 \end{vmatrix} = 0(2) - 2(1) = 0 - 2 = -2$

$D_x = \begin{vmatrix} 4 & 2 \\ 5 & 2 \end{vmatrix} = 4(2) - 2(5) = 8 - 10 = -2$

$D_y = \begin{vmatrix} 0 & 4 \\ 1 & 5 \end{vmatrix} = 0(5) - 4(1) = 0 - 4 = -4$

$x = \frac{-2}{-2} = 1$ and $y = \frac{-4}{-2} = 2$

The solution is (1, 2).

9. $\begin{cases} 3x + y = 1 \\ 2y = 2 - 6x \end{cases}$ or $\begin{cases} 3x + y = 1 \\ 6x + 2y = 2 \end{cases}$

$D = \begin{vmatrix} 3 & 1 \\ 6 & 2 \end{vmatrix} = 3(2) - 1(6) = 6 - 6 = 0$

Thus, this system cannot be solved using Cramer's rule. Since equation 2 is 2 times equation 1, the system is dependent.
The solution is $\{(x, y) | 3x + y = 1\}$.

11. $\begin{cases} 5x-2y=27 \\ -3x+5y=18 \end{cases}$

$D=\begin{vmatrix} 5 & -2 \\ -3 & 5 \end{vmatrix}$
$=5(5)-(-2)(-3)$
$=25-6$
$=19$

$D_x=\begin{vmatrix} 27 & -2 \\ 18 & 5 \end{vmatrix}$
$=27(5)-(-2)18$
$=135+36$
$=171$

$D_y=\begin{vmatrix} 5 & 27 \\ -3 & 18 \end{vmatrix}$
$=5(18)-27(-3)$
$=90+81$
$=171$

$x=\frac{171}{19}=9$ and $y=\frac{171}{19}=9$

The solution is (9, 9).

13. $\begin{vmatrix} 2 & 1 & 0 \\ 0 & 5 & -3 \\ 4 & 0 & 2 \end{vmatrix}$

$=2\begin{vmatrix} 5 & -3 \\ 0 & 2 \end{vmatrix}-1\begin{vmatrix} 0 & -3 \\ 4 & 2 \end{vmatrix}+0\begin{vmatrix} 0 & 5 \\ 4 & 0 \end{vmatrix}$
$=2[5(2)-(-3)(0)]-[0(2)-4(-3)]+0$
$=2(10)-12$
$=8$

15. $\begin{vmatrix} 4 & -6 & 0 \\ -2 & 3 & 0 \\ 4 & -6 & 1 \end{vmatrix}$

$=0\begin{vmatrix} -2 & 3 \\ 4 & -6 \end{vmatrix}-0\begin{vmatrix} 4 & -6 \\ 4 & -6 \end{vmatrix}+1\begin{vmatrix} 4 & -6 \\ -2 & 3 \end{vmatrix}$
$=0-0+[4(3)-(-6)(-2)]$
$=0$

17. $\begin{vmatrix} 3 & 6 & -3 \\ -1 & -2 & 3 \\ 4 & -1 & 6 \end{vmatrix}$

$=3\begin{vmatrix} -2 & 3 \\ -1 & 6 \end{vmatrix}-6\begin{vmatrix} -1 & 3 \\ 4 & 6 \end{vmatrix}+(-3)\begin{vmatrix} -1 & -2 \\ 4 & -1 \end{vmatrix}$
$=3[-2(6)-3(-1)]-6[-1(6)-3(4)]$
$-3[(-1)(-1)-(-2)4]$
$=3(-9)-6(-18)-3(9)$
$=-27+108-27$
$=54$

19. $\begin{cases} 3x \quad +z=-1 \\ -x-3y+z=7 \\ 3y+z=5 \end{cases}$

$D=\begin{vmatrix} 3 & 0 & 1 \\ -1 & -3 & 1 \\ 0 & 3 & 1 \end{vmatrix}$

$=3\begin{vmatrix} -3 & 1 \\ 3 & 1 \end{vmatrix}-0\begin{vmatrix} -1 & 1 \\ 0 & 1 \end{vmatrix}+1\begin{vmatrix} -1 & -3 \\ 0 & 3 \end{vmatrix}$
$=3[(-3)(1)-1(3)]-0+[(-1)(3)-(-3)0]$
$=3(-6)-3$
$=-21$

$D_x=\begin{vmatrix} -1 & 0 & 1 \\ 7 & -3 & 1 \\ 5 & 3 & 1 \end{vmatrix}$

$=-1\begin{vmatrix} -3 & 1 \\ 3 & 1 \end{vmatrix}-0\begin{vmatrix} 7 & 1 \\ 5 & 1 \end{vmatrix}+1\begin{vmatrix} 7 & -3 \\ 5 & 3 \end{vmatrix}$
$=-[(-3)(1)-1(3)]-0+[7(3)-(-3)5]$
$=6+36$
$=42$

$D_y=\begin{vmatrix} 3 & -1 & 1 \\ -1 & 7 & 1 \\ 0 & 5 & 1 \end{vmatrix}$

$=3\begin{vmatrix} 7 & 1 \\ 5 & 1 \end{vmatrix}-(-1)\begin{vmatrix} -1 & 1 \\ 0 & 1 \end{vmatrix}+1\begin{vmatrix} -1 & 7 \\ 0 & 5 \end{vmatrix}$
$=3[7(1)-1(5)]+1[(-1)1-1(0)]$
$+[(-1)(5)-7(0)]$
$=3(2)+(-1)+(-5)$
$=0$

$$D_z = \begin{vmatrix} 3 & 0 & -1 \\ -1 & -3 & 7 \\ 0 & 3 & 5 \end{vmatrix}$$
$$= 3\begin{vmatrix} -3 & 7 \\ 3 & 5 \end{vmatrix} - 0\begin{vmatrix} -1 & 7 \\ 0 & 5 \end{vmatrix} + (-1)\begin{vmatrix} -1 & -3 \\ 0 & 3 \end{vmatrix}$$
$$= 3[(-3)(5) - 7(3)] - 0 - [(-1)(3) - (-3)0]$$
$$= 3(-36) - (-3)$$
$$= -105$$

$x = \frac{42}{-21} = -2,\ y = \frac{0}{-21} = 0,$

$z = \frac{-105}{-21} = 5$

The solution is (–2, 0, 5).

21. $\begin{cases} x + y + z = 8 \\ 2x - y - z = 10 \\ x - 2y + 3z = 22 \end{cases}$

$$D = \begin{vmatrix} 1 & 1 & 1 \\ 2 & -1 & -1 \\ 1 & -2 & 3 \end{vmatrix}$$
$$= 1\begin{vmatrix} -1 & -1 \\ -2 & 3 \end{vmatrix} - 1\begin{vmatrix} 2 & -1 \\ 1 & 3 \end{vmatrix} + 1\begin{vmatrix} 2 & -1 \\ 1 & -2 \end{vmatrix}$$
$$= (-3-2) - [6-(-1)] + [-4-(-1)]$$
$$= -5 - 7 - 3$$
$$= -15$$

$$D_x = \begin{vmatrix} 8 & 1 & 1 \\ 10 & -1 & -1 \\ 22 & -2 & 3 \end{vmatrix}$$
$$= 8\begin{vmatrix} -1 & -1 \\ -2 & 3 \end{vmatrix} - 1\begin{vmatrix} 10 & -1 \\ 22 & 3 \end{vmatrix} + 1\begin{vmatrix} 10 & -1 \\ 22 & -2 \end{vmatrix}$$
$$= 8(-3-2) - [30-(-22)] + [-20-(-22)]$$
$$= 8(-5) - 52 + 2$$
$$= -40 - 52 + 2$$
$$= -90$$

$$D_y = \begin{vmatrix} 1 & 8 & 1 \\ 2 & 10 & -1 \\ 1 & 22 & 3 \end{vmatrix}$$
$$= 1\begin{vmatrix} 10 & -1 \\ 22 & 3 \end{vmatrix} - 8\begin{vmatrix} 2 & -1 \\ 1 & 3 \end{vmatrix} + 1\begin{vmatrix} 2 & 10 \\ 1 & 22 \end{vmatrix}$$
$$= [30-(-22)] - 8[6-(-1)] + [44-10]$$
$$= 52 - 8(7) + 34$$
$$= 52 - 56 + 34$$
$$= 30$$

$$D_z = \begin{vmatrix} 1 & 1 & 8 \\ 2 & -1 & 10 \\ 1 & -2 & 22 \end{vmatrix}$$
$$= 1\begin{vmatrix} -1 & 10 \\ -2 & 22 \end{vmatrix} - 1\begin{vmatrix} 2 & 10 \\ 1 & 22 \end{vmatrix} + 8\begin{vmatrix} 2 & -1 \\ 1 & -2 \end{vmatrix}$$
$$= [-22-(-20)] - (44-10) + 8[-4-(-1)]$$
$$= -2 - 34 + 8(-3)$$
$$= -36 - 24$$
$$= -60$$

$x = \frac{-90}{-15} = 6,\ y = \frac{30}{-15} = -2,$

$z = \frac{-60}{-15} = 4$

The solution is (6, –2, 4).

23. $\begin{vmatrix} 10 & -1 \\ -4 & 2 \end{vmatrix} = 10(2) - (-1)(-4)$
$$= 20 - 4$$
$$= 16$$

25. $\begin{vmatrix} 1 & 0 & 4 \\ 1 & -1 & 2 \\ 3 & 2 & 1 \end{vmatrix} = 1\begin{vmatrix} -1 & 2 \\ 2 & 1 \end{vmatrix} - 0\begin{vmatrix} 1 & 2 \\ 3 & 1 \end{vmatrix} + 4\begin{vmatrix} 1 & -1 \\ 3 & 2 \end{vmatrix}$
$$= 1[-1-4] - 0 + 4[2-(-3)]$$
$$= -5 + 4(5)$$
$$= -5 + 20$$
$$= 15$$

27. $\begin{vmatrix} \frac{3}{4} & \frac{5}{2} \\ -\frac{1}{6} & \frac{7}{3} \end{vmatrix} = \frac{3}{4}\left(\frac{7}{3}\right) - \frac{5}{2}\left(-\frac{1}{6}\right)$

$= \frac{21}{12} + \frac{5}{12}$

$= \frac{26}{12}$

$= \frac{13}{6}$

29. $\begin{vmatrix} 4 & -2 & 2 \\ 6 & -1 & 3 \\ 2 & 1 & 1 \end{vmatrix}$

$= 4\begin{vmatrix} -1 & 3 \\ 1 & 1 \end{vmatrix} - (-2)\cdot\begin{vmatrix} 6 & 3 \\ 2 & 1 \end{vmatrix} + 2\begin{vmatrix} 6 & -1 \\ 2 & 1 \end{vmatrix}$

$= 4[-1 - 3] + 2[6 - 6] + 2[6 - (-2)]$

$= 4(-4) + 2 \cdot 0 + 2(8)$

$= -16 + 0 + 16$

$= 0$

31. $\begin{vmatrix} -2 & 5 & 4 \\ 5 & -1 & 3 \\ 4 & 1 & 2 \end{vmatrix}$

$= -2\begin{vmatrix} -1 & 3 \\ 1 & 2 \end{vmatrix} - 5\begin{vmatrix} 5 & 3 \\ 4 & 2 \end{vmatrix} + 4\begin{vmatrix} 5 & -1 \\ 4 & 1 \end{vmatrix}$

$= -2[-2 - 3] - 5[10 - 12] + 4[5 - (-4)]$

$= -2(-5) - 5(-2) + 4(9)$

$= 10 + 10 + 36$

$= 56$

33. If the elements of a single row (or column) of a determinant are all zero, the value of the determinant will be zero. To see this, consider expanding on that row (or column) containing all zeros.

35. $\begin{vmatrix} 1 & x \\ 2 & 7 \end{vmatrix} = -3$

$(1)(7) - 2x = -3$

$7 - 2x = -3$

$2x = 10$

$x = 5$

37. $\begin{cases} 2x - 5y = 4 \\ x + 2y = -7 \end{cases}$

$D = \begin{vmatrix} 2 & -5 \\ 1 & 2 \end{vmatrix} = 2(2) - (-5)(1) = 4 + 5 = 9$

$D_x = \begin{vmatrix} 4 & -5 \\ -7 & 2 \end{vmatrix}$

$= 4(2) - (-5)(-7)$

$= 8 - 35$

$= -27$

$D_y = \begin{vmatrix} 2 & 4 \\ 1 & -7 \end{vmatrix} = 2(-7) - 4(1) = -14 - 4 = -18$

$x = \frac{-27}{9} = -3$, and $y = \frac{-18}{9} = -2$

The solution is (–3, –2).

39. $\begin{cases} 4x + 2y = 5 \\ 2x + y = -1 \end{cases}$

$D = \begin{vmatrix} 4 & 2 \\ 2 & 1 \end{vmatrix} = 4(1) - 2(2) = 4 - 4 = 0$

Thus, Cramer's rule cannot be used to solve the system. Multiply equation 2 by 2 yielding the new system:

$\begin{cases} 4x + 2y = 5 \\ 4x + 2y = -2 \end{cases}$

Therefore, the system is inconsistent. The solution is $\varnothing$.

41. $\begin{cases} 2x + 2y + z = 1 \\ -x + y + 2z = 3 \\ x + 2y + 4z = 0 \end{cases}$

$D = \begin{vmatrix} 2 & 2 & 1 \\ -1 & 1 & 2 \\ 1 & 2 & 4 \end{vmatrix}$

$= 2\begin{vmatrix} 1 & 2 \\ 2 & 4 \end{vmatrix} - 2\begin{vmatrix} -1 & 2 \\ 1 & 4 \end{vmatrix} + 1\begin{vmatrix} -1 & 1 \\ 1 & 2 \end{vmatrix}$

$= 2(4 - 4) - 2(-4 - 2) + (-2 - 1)$

$= 2(0) - 2(-6) + (-3)$

$= 0 + 12 - 3$

$= 9$

$$D_x = \begin{vmatrix} 1 & 2 & 1 \\ 3 & 1 & 2 \\ 0 & 2 & 4 \end{vmatrix}$$
$$= 1\begin{vmatrix} 1 & 2 \\ 2 & 4 \end{vmatrix} - 3\begin{vmatrix} 2 & 1 \\ 2 & 4 \end{vmatrix} + 0\begin{vmatrix} 2 & 1 \\ 1 & 2 \end{vmatrix}$$
$$= (4-4) - 3(8-2) + 0$$
$$= 0 - 3(6)$$
$$= -18$$

$$D_y = \begin{vmatrix} 2 & 1 & 1 \\ -1 & 3 & 2 \\ 1 & 0 & 4 \end{vmatrix}$$
$$= 1\begin{vmatrix} 1 & 1 \\ 3 & 2 \end{vmatrix} - 0\begin{vmatrix} 2 & 1 \\ -1 & 2 \end{vmatrix} + 4\begin{vmatrix} 2 & 1 \\ -1 & 3 \end{vmatrix}$$
$$= (2-3) - 0 + 4[6-(-1)]$$
$$= -1 + 4(7)$$
$$= -1 + 28$$
$$= 27$$

$$D_z = \begin{vmatrix} 2 & 2 & 1 \\ -1 & 1 & 3 \\ 1 & 2 & 0 \end{vmatrix}$$
$$= 1\begin{vmatrix} 2 & 1 \\ 1 & 3 \end{vmatrix} - 2\begin{vmatrix} 2 & 1 \\ -1 & 3 \end{vmatrix} + 0\begin{vmatrix} 2 & 2 \\ -1 & 1 \end{vmatrix}$$
$$= (6-1) - 2[6-(-1)] + 0$$
$$= 5 - 2(7)$$
$$= 5 - 14$$
$$= -9$$

$x = -\frac{18}{9} = -2,\ y = \frac{27}{9} = 3,$

$z = \frac{-9}{9} = -1$

The solution is (–2, 3, –1).

43. $\begin{cases} \frac{2}{3}x - \frac{3}{4}y = -1 \\ -\frac{1}{6}x + \frac{3}{4}y = \frac{5}{2} \end{cases}$

$$D = \begin{vmatrix} \frac{2}{3} & -\frac{3}{4} \\ -\frac{1}{6} & \frac{3}{4} \end{vmatrix}$$
$$= \frac{2}{3} \cdot \frac{3}{4} - \left(-\frac{3}{4}\right)\left(-\frac{1}{6}\right)$$
$$= \frac{1}{2} - \frac{1}{8}$$
$$= \frac{3}{8}$$

$$D_x = \begin{vmatrix} -1 & -\frac{3}{4} \\ \frac{5}{2} & \frac{3}{4} \end{vmatrix}$$
$$= (-1)\frac{3}{4} - \left(-\frac{3}{4}\right)\frac{5}{2}$$
$$= -\frac{3}{4} + \frac{15}{8}$$
$$= \frac{9}{8}$$

$$D_y = \begin{vmatrix} \frac{2}{3} & -1 \\ -\frac{1}{6} & \frac{5}{2} \end{vmatrix}$$
$$= \frac{2}{3} \cdot \frac{5}{2} - (-1)\left(-\frac{1}{6}\right)$$
$$= \frac{5}{3} - \frac{1}{6}$$
$$= \frac{3}{2}$$

$x = \frac{\frac{9}{8}}{\frac{3}{8}} = 3$ and $y = \frac{\frac{3}{2}}{\frac{3}{8}} = 4$

The solution is (3, 4).

45. $\begin{cases} .7x - .2y = -1.6 \\ .2x - y = -1.4 \end{cases}$

$$D = \begin{vmatrix} .7 & -.2 \\ .2 & -1 \end{vmatrix}$$
$$= .7(-1) - (-.2)(.2)$$
$$= -.7 + .04$$
$$= -.66$$

$$D_x = \begin{vmatrix} -1.6 & -.2 \\ -1.4 & -1 \end{vmatrix}$$
$$= (-1.6)(-1) - (-.2)(-1.4)$$
$$= 1.6 - .28$$
$$= 1.32$$

$$D_y = \begin{vmatrix} .7 & -1.6 \\ .2 & -1.4 \end{vmatrix}$$
$$= (.7)(-1.4) - (-1.6)(.2)$$
$$= -.98 + .32$$
$$= -.66$$

$$x = \frac{1.32}{-.66} = -2 \text{ and } y = \frac{-.66}{-.66} = 1$$

The solution is (–2, 1).

47. $\begin{cases} -2x + 4y - 2z = 6 \\ x - 2y + z = -3 \\ 3x - 6y + 3z = -9 \end{cases}$

$$D = \begin{vmatrix} -2 & 4 & -2 \\ 1 & -2 & 1 \\ 3 & -6 & 3 \end{vmatrix}$$
$$= -2\begin{vmatrix} -2 & 1 \\ -6 & 3 \end{vmatrix} - 4\begin{vmatrix} 1 & 1 \\ 3 & 3 \end{vmatrix} + (-2)\begin{vmatrix} 1 & -2 \\ 3 & -6 \end{vmatrix}$$
$$= -2[-6 - (-6)] - 4[3 - 3] - 2[-6 - (-6)]$$
$$= -2(0) - 4(0) - 2(0)$$
$$= 0$$

Therefore, Cramer's rule will not provide the solution. Note that equation 1 is –2 times equation 2 and that equation 3 is 3 times equation 2. Thus, the system is dependent.

The solution is

$\{(x, y, z) | x - 2y + z = -3\}$.

49. $\begin{cases} x - 2y + z = -5 \\ 3y + 2z = 4 \\ 3x - y = -2 \end{cases}$

$$D = \begin{vmatrix} 1 & -2 & 1 \\ 0 & 3 & 2 \\ 3 & -1 & 0 \end{vmatrix}$$
$$= 1\begin{vmatrix} 3 & 2 \\ -1 & 0 \end{vmatrix} - 0\begin{vmatrix} -2 & 1 \\ -1 & 0 \end{vmatrix} + 3\begin{vmatrix} -2 & 1 \\ 3 & 2 \end{vmatrix}$$
$$= [0 - (-2)] - 0 + 3[-4 - 3]$$
$$= 2 + 3(-7)$$
$$= 2 - 21$$
$$= -19$$

$$D_x = \begin{vmatrix} -5 & -2 & 1 \\ 4 & 3 & 2 \\ -2 & -1 & 0 \end{vmatrix}$$
$$= 1\begin{vmatrix} 4 & 3 \\ -2 & -1 \end{vmatrix} - 2\begin{vmatrix} -5 & -2 \\ -2 & -1 \end{vmatrix} + 0\begin{vmatrix} 5 & -2 \\ 4 & 3 \end{vmatrix}$$
$$= [-4 - (-6)] - 2[5 - 4] + 0$$
$$= 2 - 2(1)$$
$$= 0$$

$$D_y = \begin{vmatrix} 1 & -5 & 1 \\ 0 & 4 & 2 \\ 3 & -2 & 0 \end{vmatrix}$$
$$= 1\begin{vmatrix} 4 & 2 \\ -2 & 0 \end{vmatrix} - 0\begin{vmatrix} -5 & 1 \\ -2 & 0 \end{vmatrix} + 3\begin{vmatrix} -5 & 1 \\ 4 & 2 \end{vmatrix}$$
$$= [0 - (-4)] - 0 + 3[-10 - 4]$$
$$= 4 + 3(-14)$$
$$= 4 - 42$$
$$= -38$$

$$D_z = \begin{vmatrix} 1 & -2 & -5 \\ 0 & 3 & 4 \\ 3 & -1 & -2 \end{vmatrix}$$
$$= 1\begin{vmatrix} 3 & 4 \\ -1 & -2 \end{vmatrix} - 0\begin{vmatrix} -2 & -5 \\ -1 & -2 \end{vmatrix} + 3\begin{vmatrix} -2 & -5 \\ 3 & 4 \end{vmatrix}$$
$$= [-6 - (-4)] - 0 + 3[-8 - (-15)]$$
$$= -2 + 3(7)$$
$$= -2 + 21$$
$$= 19$$

$$x = \frac{0}{-19} = 0, \quad y = \frac{-38}{-19} = 2,$$
$$z = \frac{19}{-19} = -1$$

The solution is (0, 2, –1).

51. The array of signs for use with a 4×4 matrix is

$$\begin{matrix} + & - & + & - \\ - & + & - & + \\ + & - & + & - \\ - & + & - & + \end{matrix}$$

53. $5x - 6 + x - 12 = 6x - 18$

55. $2(3x-6)+3(x-1) = 6x - 12 + 3x - 3$
$= 9x - 15$

57. $f(x) = 5x - 6$ or $y = 5x - 6$

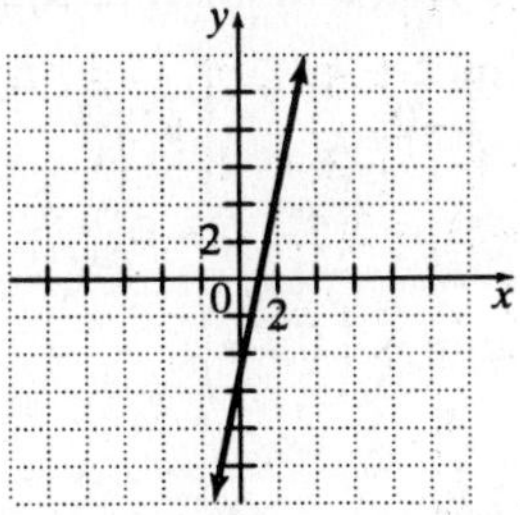

59. $h(x) = 3$ or $y = 3$

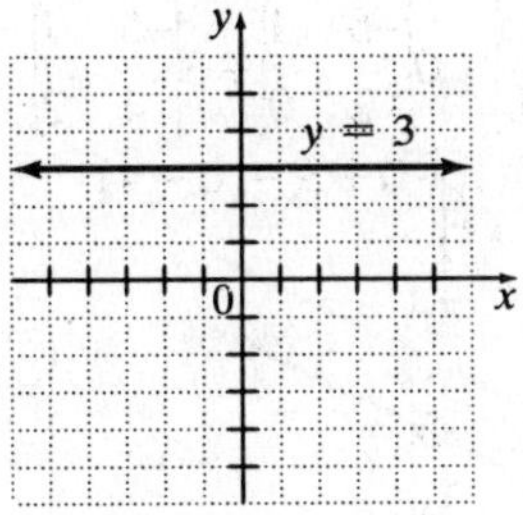

61. $$\begin{vmatrix} 5 & 0 & 0 & 0 \\ 0 & 4 & 2 & -1 \\ 1 & 3 & -2 & 0 \\ 0 & -3 & 1 & 2 \end{vmatrix} = 5\begin{vmatrix} 4 & 2 & -1 \\ 3 & -2 & 0 \\ -3 & 1 & 2 \end{vmatrix}$$
$$-0\begin{vmatrix} 0 & 2 & -1 \\ 1 & -2 & 0 \\ 0 & 1 & 2 \end{vmatrix} + 0\begin{vmatrix} 0 & 4 & -1 \\ 1 & 3 & 0 \\ 0 & -3 & 2 \end{vmatrix}$$
$$-0\begin{vmatrix} 0 & 4 & 2 \\ 1 & 3 & -2 \\ 0 & -3 & 1 \end{vmatrix}$$
$$= 5\left[(-1)\begin{vmatrix} 3 & -2 \\ -3 & 1 \end{vmatrix} - 0 \cdot \begin{vmatrix} 4 & 2 \\ -3 & 1 \end{vmatrix} + 2\begin{vmatrix} 4 & 2 \\ 3 & -2 \end{vmatrix}\right]$$
$= 5[-(3-6) - 0 + 2(-8-6)]$
$= 5[3 + 2(-14)]$
$= 5[3 - 28]$
$= 5[-25]$
$= -125$

63.

$$\begin{vmatrix} 4 & 0 & 2 & 5 \\ 0 & 3 & -1 & 1 \\ 0 & 0 & 2 & 0 \\ 0 & 0 & 0 & 1 \end{vmatrix} = 4\begin{vmatrix} 3 & -1 & 1 \\ 0 & 2 & 0 \\ 0 & 0 & 1 \end{vmatrix} - 0\begin{vmatrix} 0 & 2 & 5 \\ 0 & 2 & 0 \\ 0 & 0 & 1 \end{vmatrix}$$
$$+0\begin{vmatrix} 0 & 2 & 5 \\ 3 & -1 & 1 \\ 0 & 0 & 1 \end{vmatrix} - 0\begin{vmatrix} 0 & 2 & 5 \\ 3 & -1 & 1 \\ 0 & 2 & 0 \end{vmatrix}$$
$$= 4\left[3\begin{vmatrix} 2 & 0 \\ 0 & 1 \end{vmatrix} - 0\begin{vmatrix} -1 & 1 \\ 0 & 1 \end{vmatrix} + 0\begin{vmatrix} -1 & 1 \\ 2 & 0 \end{vmatrix}\right]$$
$= 4[3(2-0) - 0 + 0]$
$= 12(2)$
$= 24$

Chapter 4 Review Exercises

1. $\begin{cases} 3x + 10y = 1 \\ x + 2y = -1 \end{cases}$

(1)

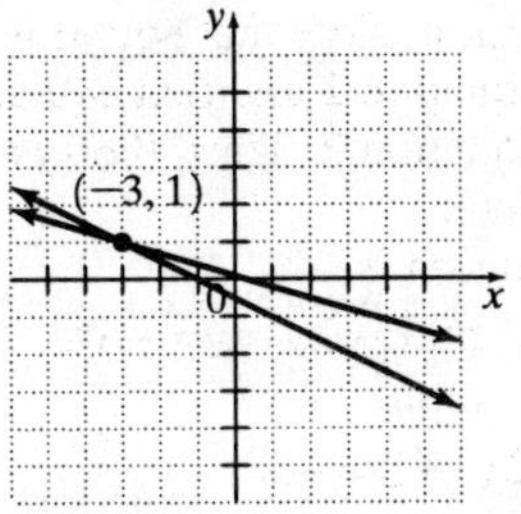

(2) $x = -2y - 1$
Replace x with $-2y - 1$ in the first equation.
$3(-2y-1)+10y = 1$
$-6y - 3 + 10y = 1$
$4y - 3 = 1$
$4y = 4$
$y = 1$
Replace y with 1 in the equation $x = -2y - 1$.

$x = -2(1) - 1 = -2 - 1 = -3$
The solution is (–3, 1).

(3) Multiply the second equation by –3.

$$\begin{cases} 3x + 10y = 1 \\ -3x - 6y = 3 \end{cases}$$

Add the equations.

$$\begin{array}{r} 3x + 10y = 1 \\ -3x - 6y = 3 \\ \hline 4y = 4 \\ y = 1 \end{array}$$

Replace y with 1 in the second equation.

$$x + 2(1) = -1$$
$$x + 2 = -1$$
$$x = -3$$

The solution is (–3, 1).

2. $$\begin{cases} y = \frac{1}{2}x + \frac{2}{3} \\ 4x + 6y = 4 \end{cases}$$

(1)

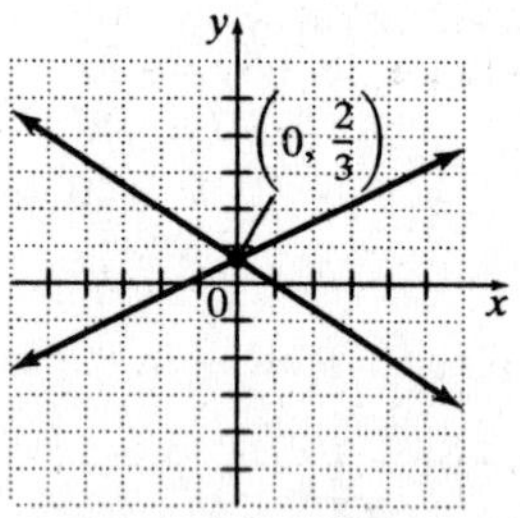

(2) Replace y with $\frac{1}{2}x + \frac{2}{3}$ in the second equation.

$$4x + 6\left(\frac{1}{2}x + \frac{2}{3}\right) = 4$$
$$4x + 3x + 4 = 4$$
$$7x = 0$$
$$x = 0$$

Replace x with 0 in the first equation.

$$y = \frac{1}{2}(0) + \frac{2}{3}$$
$$y = \frac{2}{3}$$

The solution is $\left(0, \frac{2}{3}\right)$.

(3) $$\begin{cases} -\frac{1}{2}x + y = \frac{2}{3} \\ 4x + 6y = 4 \end{cases}$$

Multiply the first equation by –6.

$$\begin{cases} 3x - 6y = -4 \\ 4x + 6y = 4 \end{cases}$$

Add the equations.

$$\begin{array}{l} 3x - 6y = -4 \\ 4x + 6y = 4 \\ \hline 7x \quad\; = 0 \\ x \quad\;\; = 0 \end{array}$$

Replace x with 0 in the first equation.

$$y = \frac{1}{2}(0) + \frac{2}{3}$$
$$y = \frac{2}{3}$$

The solution is $\left(0, \frac{2}{3}\right)$.

3. $$\begin{cases} 2x - 4y = 22 \\ 5x - 10y = 15 \end{cases}$$

(1)

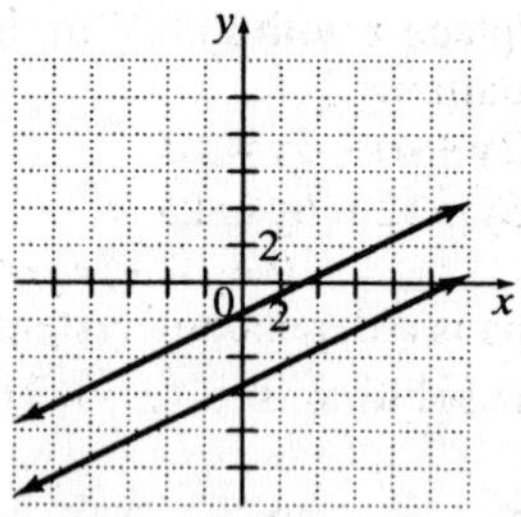

(2) Solve the first equation for x.

$$2x = 4y + 22$$
$$x = 2y + 11$$

Replace x with $2y + 11$ in the second equation.

$$5(2y + 1) - 10y = 15$$
$$10y + 55 - 10y = 15$$
$$55 = 15$$

This is an inconsistent system.
The solution is $\varnothing$.

(3) Divide the first equation by 2 and the second equation by –5.

$\begin{cases} x - 2y = 11 \\ -x + 2y = -3 \end{cases}$

Add the equations.

$$\begin{array}{r} x - 2y = 11 \\ -x + 2y = -3 \\ \hline 0 = 8 \end{array} \quad \text{False}$$

This is an inconsistent system.
The solution is $\varnothing$.

4. $\begin{cases} 3x - 6y = 12 \\ 2y = x - 4 \end{cases}$

(1)

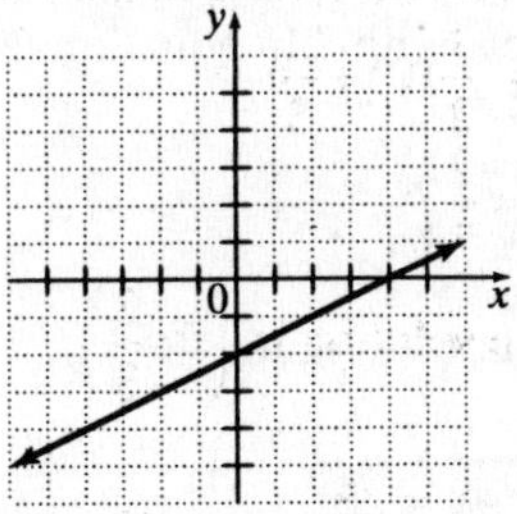

(2) Solve the second equation for x.
$x = 2y + 4$
Replace x with $2y + 4$ in the first equation.
$3(2y + 4) - 6y = 12$
$6y + 12 - 6y = 12$
$0 = 0$ True
This is a dependent system.
The solution is $\{(x, y) | 3x - 6y = 12\}$.

(3) $\begin{cases} 3x - 6y = 12 \\ -x + 2y = -4 \end{cases}$
Multiply the second equation by 3.
$\begin{cases} 3x - 6y = 12 \\ -3x + 6y = -12 \end{cases}$
Add the equations.

$$\begin{array}{r} 3x - 6y = 12 \\ -3x + 6y = -12 \\ \hline 0 = 0 \end{array} \quad \text{True}$$

This is a dependent system.
The solution is $\{(x, y) | 3x - 6y = 12\}$.

5. $\begin{cases} \frac{1}{2}x - \frac{3}{4}y = -\frac{1}{2} \\ \frac{1}{8}x + \frac{3}{4}y = \frac{19}{8} \end{cases}$

(1)

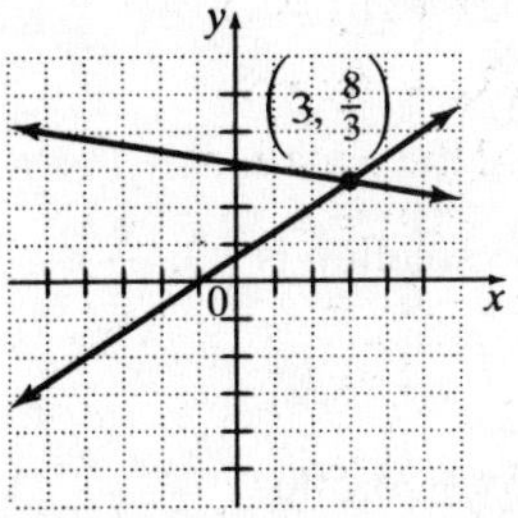

(2) Clear fractions by multiplying the first equation by 4 and the second equation by 8.
$\begin{cases} 2x - 3y = -2 \\ x + 6y = 19 \end{cases}$
Solve the new second equation for x.
$x = -6y + 19$
Replace x with $-6y + 19$ in the first equation.
$2(-6y + 19) - 3y = -2$
$-12y + 38 - 3y = -2$
$-15y = -40$
$y = \frac{-40}{-15} = \frac{8}{3}$

Replace y with $\frac{8}{3}$ in the equation
$x = -6y + 19$.
$x = -6\left(\frac{8}{3}\right) + 19$
$x = -16 + 19$
$x = 3$
The solution is $\left(3, \frac{8}{3}\right)$.

(3) Add the equations.

$$\begin{array}{r} \frac{1}{2}x - \frac{3}{4}y = -\frac{1}{2} \\ \frac{1}{8}x + \frac{3}{4}y = \frac{19}{8} \\ \hline \frac{5}{8}x \qquad = \frac{15}{8} \end{array}$$

$5x = 15$
$x = 3$
Replace x with 3 in the first equation.

$$\frac{1}{2}(3) - \frac{3}{4}y = -\frac{1}{2}$$
$$-\frac{3}{4}y = -2$$
$$-3y = -8$$
$$y = \frac{8}{3}$$

The solution is $\left(3, \frac{8}{3}\right)$.

6. $\begin{cases} y = 32x \\ y = 15x + 25{,}500 \end{cases}$

Multiply the first equation by –1 and add to the second equation.

$$\begin{array}{l} -y = -32x \\ \underline{\ \ y = 15x + 25{,}500} \\ \ \ 0 = -17x + 25{,}500 \end{array}$$
$17x = 25{,}500$
$x = 1500$

Replace x with 1500 in the first equation.
$y = 32(1500)$
$y = 48{,}000$

The number of backpacks that the company must sell is 1500.

7. $\begin{cases} x + \quad z = 4 & (1) \\ 2x - y \quad = 4 & (2) \\ x + y - z = 0 & (3) \end{cases}$

Adding (2) and (3) gives $3x - z = 4$ (4)
Adding (1) and (4) gives $4x = 8$ or $x = 2$
Replace x with 2 in equation (1).
$2 + z = 4$
$z = 2$
Replace x with 2 and z with 2 in equation (3).
$2 + y - 2 = 0$
$y = 0$
The solution is (2, 0, 2).

8. $\begin{cases} 2x + 5y \quad = 4 & (1) \\ x - 5y + z = -1 & (2) \\ 4x \quad - z = 11 & (3) \end{cases}$

Add equations 2 and 3.
$5x - 5y = 10$
We now have the system

$\begin{cases} 2x + 5y = 4 \\ 5x - 5y = 10 \end{cases}$

Add the equations.

$$\begin{array}{l} 2x + 5y = 4 \\ \underline{5x - 5y = 10} \\ 7x \qquad = 14 \end{array}$$
$x = 2$

Replace x with 2 in equation (1).
$2(2) + 5y = 4$
$4 + 5y = 4$
$5y = 0$
$y = 0$
Replace x with 2 in equation (3).
$4(2) - z = 11$
$8 - z = 11$
$-z = 3$
$z = -3$
The solution is (2, 0, –3).

9. $\begin{cases} \quad 4y + 2z = 5 & (1) \\ 2x + 8y \quad = 5 & (2) \\ 6x \quad + 4z = 1 & (3) \end{cases}$

Multiply equation (1) by –2 and add to equation (2).

$$\begin{array}{l} \quad -8y - 4z = -10 \\ \underline{2x + 8y \qquad = \ \ 5} \\ 2x \qquad - 4z = -5 \quad (4) \end{array}$$

Add equations (3) and (4).

$$\begin{array}{l} 6x + 4z = 1 \\ \underline{2x - 4z = -5} \\ 8x \qquad = -4 \end{array}$$
$x = -\frac{1}{2}$

Replace x with $-\frac{1}{2}$ in equation (2).

$2\left(-\frac{1}{2}\right) + 8y = 5$
$-1 + 8y = 5$
$8y = 6$
$y = \frac{3}{4}$

Replace x with $-\frac{1}{2}$ in equation (3).

$$6\left(-\frac{1}{2}\right)+4z=1$$
$$-3+4z=1$$
$$4z=4$$
$$z=1$$

The solution is $\left(-\frac{1}{2}, \frac{3}{4}, 1\right)$.

10. $\begin{cases} 5x+7y \quad = 9 & (1) \\ 14y-z=28 & (2) \\ 4x \quad +2z=-4 & (3) \end{cases}$

Solve equation (3) for z.

$$2z=-4x-4$$
$$z=-2x-2$$

Replace z with $-2x-2$ in equation (2).

$$14y-(-2x-2)=28$$
$$14y+2x+2=28$$
$$2x+14y=26$$
$$x+7y=13$$

Solve the new system.

$\begin{cases} 5x+7y=9 & (1) \\ x+7y=13 & (4) \end{cases}$

Multiply equation (4) by -1 and add to equation (1).

$$\begin{aligned} 5x+7y &= 9 \\ -x-7y &= -13 \\ \hline 4x &= -4 \\ y &= 1 \end{aligned}$$

Replace x with -1 in equation (4).

$$-1+7y=13$$
$$7y=14$$
$$y=2$$

Replace x with -1 in the equation $z=-2x-2$.

$$z=-2(-1)-2=2-2=0$$

The solution is $(-1, 2, 0)$.

11. $\begin{cases} 3x \; -2y+2z= 5 & (1) \\ -x \; +6y \; +z= 4 & (2) \\ 3x \; +14y+7z=20 & (3) \end{cases}$

Multiply equation (2) by 3 and add to equation (1).

$$\begin{aligned} 3x-2y+2z &= 5 \\ -3x+18y+3z &= 12 \\ \hline 16y+5z &= 17 \quad (4) \end{aligned}$$

Multiply equation (3) by -1 and add to equation (1).

$$\begin{aligned} 3x-\;2y+2z &= \;5 \\ -3x-14y-7z &= -20 \\ \hline -16y-5z &= -15 \quad (5) \end{aligned}$$

Add equations (4) and (5).

$$\begin{aligned} 16y+5z &= \;17 \\ -16y-5z &= -15 \\ \hline 0 &= \;2 \quad \text{False} \end{aligned}$$

The system is inconsistent.
The solution is $\varnothing$.

12. $\begin{cases} x+2y+3z=11 & (1) \\ y+2z= 3 & (2) \\ 2x \quad +2z=10 & (3) \end{cases}$

Multiply equation (2) by -2 and add to equation (1).

$$\begin{aligned} x+2y+3z &= 11 \\ -2y-4z &= -6 \\ \hline x- \quad z &= 5 \quad (4) \end{aligned}$$

Multiply equation (4) by 2 and add to equation (3).

$$\begin{aligned} 2x+2z &= 10 \\ 2x-2z &= 10 \\ \hline 4x \quad &= 20 \end{aligned}$$

$$x=5$$

Replace x with 5 in equation (3).

$$2(5)+2z=10$$
$$10+2z=10$$
$$2z=0$$
$$z=0$$

Replace z with 0 in equation (2).

$$y+2(0)=3$$
$$y+0=3$$
$$y=3$$

The solution is $(5, 3, 0)$.

13. $\begin{cases} 7x-3y+2z=0 & (1) \\ 4x-4y \; -z=2 & (2) \\ 5x+2y+3z=1 & (3) \end{cases}$

Multiply equation (2) by 2 and add to equation (1).

$$\begin{array}{r} 7x-\ \ 3y+2z=0 \\ 8x-\ \ 8y-2z=4 \\ \hline 15x-11y\qquad\ =4 \end{array}$$

Multiply equation (2) by 3 and add to equation (3).

$$\begin{array}{r} 12x-12y-3z=6 \\ 5x+\ \ 2y+3z=1 \\ \hline 17x-10y\qquad\ =7 \end{array}$$

Solve the new system.

$$\begin{cases} 15x-11y=4 & (4) \\ 17x-10y=7 & (5) \end{cases}$$

Multiply equation (4) by −10, multiply equation (5) by 11, and add.

$$\begin{array}{r} -150x+110y=-40 \\ 187x-110y=\ \ 77 \\ \hline 37x\qquad\quad\ =\ \ 37 \\ x\qquad\quad\ =\ \ \ 1 \end{array}$$

Replace x with 1 in equation (4).

$$\begin{aligned} 15(1)-11y&=4 \\ 15-11y&=4 \\ -11y&=-11 \\ y&=1 \end{aligned}$$

Replace x with 1 and y with 1 in equation (1).

$$\begin{aligned} 7(1)-3(1)+2z&=0 \\ 4+2z&=0 \\ 2z&=-4 \\ z&=-2 \end{aligned}$$

The solution is (1, 1, −2).

14. $$\begin{cases} x-3y-5z=-5 & (1) \\ 4x-2y+3z=13 & (2) \\ 5x+3y+4z=22 & (3) \end{cases}$$

Multiply equation (1) by −4 and add to equation (2).

$$\begin{array}{r} -4x+12y+20z=20 \\ 4x-\ \ 2y+\ \ 3z=13 \\ \hline 10y+23z=33 \end{array}$$

Multiply equation (1) by −5 and add to equation (3).

$$\begin{array}{r} -5x+15y+25z=25 \\ 5x+\ \ 3y+\ \ 4z=22 \\ \hline 18y+29z=47 \end{array}$$

Solve the new system.

$$\begin{cases} 10y+23z=33 & (4) \\ 18y+29z=47 & (5) \end{cases}$$

Multiply equation (4) by 9, multiply equation (5) by −5 and add.

$$\begin{array}{r} 90y+207z=\ \ 297 \\ -90y-145z=-235 \\ \hline 62z=\ \ \ 62 \\ z=1 \end{array}$$

Replace z with 1 in equation (4).

$$\begin{aligned} 10y+23(1)&=33 \\ 10y+23&=33 \\ 10y&=10 \\ y&=1 \end{aligned}$$

Replace y with 1 and z with 1 in equation (1).

$$\begin{aligned} x-3(1)-5(1)&=-5 \\ x-8&=-5 \\ x&=3 \end{aligned}$$

The solution is (3, 1, 1).

15. Let x = the first number
y = the second number
z = the third number

$$\begin{cases} x+y+z=98 & (1) \\ x+y=z+2 & (2) \\ y=4x & (3) \end{cases}$$

Replace y with $4x$ in equations (1) and (2).

$$\begin{aligned} x+4x+z&=98 \\ 5x+z&=98 \quad (4) \\ x+4x&=z+2 \\ 5x-z&=2 \quad (5) \end{aligned}$$

Add equations (4) and (5).

$$\begin{array}{r} 5x+z=98 \\ 5x-z=2 \\ \hline 10x=100 \\ x=10 \end{array}$$

Replace x with 10 in equation (3).

$y = 4(10) = 40$

Replace x with 10 and y with 40 in equation (2).

$$10+40=z+2$$
$$50=z+2$$
$$48=z$$

The numbers are 10, 40, and 48.

16. Let x = the first number
y = second number

$$\begin{cases} x=3y & (1) \\ 2(x+y)=168 & (2) \end{cases}$$

Replace x with $3y$ in equation (2).

$$2(3y+y)=168$$
$$8y=168$$
$$y=21$$

Replace y with 21 in equation (1).

$x=3(21)=63$

The numbers are 63 and 21.

17. Let x = speed of the first car
y = speed of the second car

$$\begin{cases} 4x+4y=492 & (1) \\ y=x+7 & (2) \end{cases}$$

Replace y with $x+7$ in equation (1).

$$4x+4(x+7)=492$$
$$8x+28=492$$
$$8x=464$$
$$x=58$$

Replace x with 58 in equation (2).

$y=58+7=65$

The cars are going 58 and 65 miles per hour.

18. Let w = the width of the foundation
l = the length of the foundation

$$\begin{cases} l=3w & (1) \\ 2w+2l=296 & (2) \end{cases}$$

Replace l with $3w$ in equation (2).

$$2w+2(3w)=296$$
$$2w+6w=296$$
$$8w=296$$
$$w=37$$

Replace w with 37 in equation (1).

$l=3(37)=111$

The foundation is 37 feet wide and 111 feet long.

19. Let x = liters of 10% solution
y = liters of 60% solution

$$\begin{cases} x+\quad y=50 & (1) \\ 0.10x+0.60y=0.40(50) & (2) \end{cases}$$

Solve equation (1) for y.

$y=50-x$

Replace y with $50-x$ in equation (2).

$$0.10x+0.60(50-x)=0.40(50)$$
$$10[0.10x+0.60(50-x)]=10[0.40(50)]$$
$$x+6(50-x)=4(50)$$
$$x+300-6x=200$$
$$-5x=-100$$
$$x=20$$

Replace x with 20 in the equation $y=50-x$.

$y=50-20=30$

He should use 20 liters of 10% solution and 30 liters of 60% solution.

20. Let c = pounds of chocolates used
n = pounds of nuts used
r = pounds of raisins used

$$\begin{cases} r=2n & (1) \\ c+n+r=45 & (2) \\ 3.00c+2.70n+2.25r=2.80(45) & (3) \end{cases}$$

Replace r with $2n$ in equation (2).

$$c+n+2n=45$$
$$c+3n=45$$
$$c=-3n+45$$

Replace r with $2n$ and c with $-3n+45$ in equation (3).

$$3.00(-3n+45)+2.70n+2.25(2n)=126$$
$$-9n+135+2.7n+4.5n=126$$
$$-1.8n+135=126$$
$$-1.8n=-9$$
$$n=5$$

Replace n with 5 in equation (1).

$r=2(5)=10$

Replace n with 5 and r with 10 in equation (2).

$$c+5+10=45$$
$$c+15=45$$
$$c=30$$

She should use 30 pounds of creme-filled chocolates, 5 pounds of chocolate-covered nuts, and 10 pounds of chocolate-covered raisins.

21. Let x = the number of pennies
y = the number of nickels
z = the number of dimes

$$\begin{cases} x+y+z=53 & (1)\\ 0.01x+0.05y+0.10z=2.77 & (2)\\ y=z+4 & (3)\end{cases}$$

Clear the decimals from equation (2) by multiplying by 100.
$x + 5y + 10z = 277$ (4)
Replace y with $z + 4$ in equation (1).

$$x+z+4+z=53$$
$$x+2z=49$$

Replace y with $z + 4$ in equation (4).

$$x+5(z+4)+10z=277$$
$$x+15z=257$$

Solve the new system.

$$\begin{cases} x+2z=49 & (5)\\ x+15z=257 & (6)\end{cases}$$

Multiply equation (5) by –1 and add to equation (6).

$$\begin{array}{r} -x-2z=-49\\ x+15z=257\\ \hline 13z=208 \end{array}$$
$$z=16$$

Replace z with 16 in equation (5).

$$x+2(16)=49$$
$$x+32=49$$
$$x=17$$

Replace z with 16 in equation (3).
$y = 16 + 4 = 20$
He has 17 pennies, 20 nickels, and 16 dimes in his jar.

22. Let l = the rate of interest on the larger investment and s = the rate of interest on the smaller investment, both expressed as decimals.

$$\begin{cases} 10{,}000l+4000s=1250 & (1)\\ l=s+0.02 & (2)\end{cases}$$

Replace l with $s + 0.02$ in equation (1).

$$10{,}000(s+0.02)+4000s=1250$$
$$10{,}000s+200+4000s=1250$$
$$14{,}000s+200=1250$$
$$14{,}000s=1050$$
$$s=\frac{1050}{14{,}000}=0.075$$

and $l = 0.075 + 0.02 = 0.095$.
The interest rate on the larger investment is 9.5% and the rate on the smaller investment is 7.5%.

23. Let x = length of the equal sides
y = length of the third side

$$\begin{cases} 2x+y=73 & (1)\\ y=x+7 & (2)\end{cases}$$

Replace y with $x + 7$ in equation (1).

$$2x+x+7=73$$
$$3x=66$$
$$x=22$$

Replace x with 22 in equation (2).
$y = 22 + 7 = 29$
Two sides of the triangle have length 22 cm, the third side has length 29 cm.

24. Let f = the first number
s = the second number
t = the third number

$$\begin{cases} f+s+t=295 & (1)\\ f=s+5 & (2)\\ f=2t & (3)\end{cases}$$

Solve equation (2) for s and equation (3) for t.

$$s=f-5$$
$$t=\frac{f}{2}$$

Replace s with $f – 5$ and t with $\frac{f}{2}$ in equation (1).

$$f+f-5+\frac{f}{2}=295$$
$$\frac{5}{2}f-5=295$$
$$\frac{5}{2}f=300$$
$$f=120$$

Replace f by 120 in the equation $s = f – 5$.
$s = 120 – 5 = 115$

Replace f by 120 in the equation $t = \frac{f}{2}$.

$t = \frac{120}{2} = 60$

The first number is 120, the second is 115, and the third is 60.

25. $\begin{cases} 3x + 10y = 1 \\ x + 2y = -1 \end{cases}$

$\left[\begin{array}{cc|c} 3 & 10 & 1 \\ 1 & 2 & -1 \end{array}\right]$

Interchange row 1 and row 2.

$\left[\begin{array}{cc|c} 1 & 2 & -1 \\ 3 & 10 & 1 \end{array}\right]$

Multiply row 1 by –3 and add to row 2.

$\left[\begin{array}{cc|c} 1 & 2 & -1 \\ 0 & 4 & 4 \end{array}\right]$

Divide row 2 by 4.

$\left[\begin{array}{cc|c} 1 & 2 & -1 \\ 0 & 1 & 1 \end{array}\right]$

This corresponds to

$\begin{cases} x + 2y = -1 \\ y = 1. \end{cases}$

$x + 2(1) = -1$

$x = -3$

The solution is (–3, 1).

26. $\begin{cases} 3x - 6y = 12 \\ 2y = x - 4 \end{cases}$

$\left[\begin{array}{cc|c} 3 & -6 & 12 \\ -1 & 2 & -4 \end{array}\right]$

Divide row 1 by 3.

$\left[\begin{array}{cc|c} 1 & -2 & 4 \\ -1 & 2 & -4 \end{array}\right]$

Add row 1 to row 2.

$\left[\begin{array}{cc|c} 1 & -2 & 4 \\ 0 & 0 & 0 \end{array}\right]$

This corresponds to

$\begin{cases} x - 2y = 4 \\ 0 = 0 \end{cases}$.

This is a dependent system.

The solution is $\{(x, y) \mid x - 2y = 4\}$.

27. $\begin{cases} 3x - 2y = -8 \\ 6x + 5y = 11 \end{cases}$

$\left[\begin{array}{cc|c} 3 & -2 & -8 \\ 6 & 5 & 11 \end{array}\right]$

Divide row 1 by 3.

$\left[\begin{array}{cc|c} 1 & -\frac{2}{3} & -\frac{8}{3} \\ 6 & 5 & 11 \end{array}\right]$

Multiply row 1 by –6 and add to row 2.

$\left[\begin{array}{cc|c} 1 & -\frac{2}{3} & -\frac{8}{3} \\ 0 & 9 & 27 \end{array}\right]$

Divide row 2 by 9.

$\left[\begin{array}{cc|c} 1 & -\frac{2}{3} & -\frac{8}{3} \\ 0 & 1 & 3 \end{array}\right]$

This corresponds to

$\begin{cases} x - \frac{2}{3}y = -\frac{8}{3} \\ y = 3 \end{cases}$.

$x - \frac{2}{3}(3) = -\frac{8}{3}$

$x - 2 = -\frac{8}{3}$

$x = -\frac{2}{3}$

The solution is $\left(-\frac{2}{3}, 3\right)$.

28. $\begin{cases} 6x - 6y = -5 \\ 10x - 2y = 1 \end{cases}$

$\left[\begin{array}{cc|c} 6 & -6 & -5 \\ 10 & -2 & 1 \end{array}\right]$

Divide row 1 by 6.

$\left[\begin{array}{cc|c} 1 & -1 & -\frac{5}{6} \\ 10 & -2 & 1 \end{array}\right]$

Multiply row 1 by –10 and add to row 2.

$\left[\begin{array}{cc|c} 1 & -1 & -\frac{5}{6} \\ 0 & 8 & \frac{28}{3} \end{array}\right]$

Divide row 2 by 8.

$\left[\begin{array}{cc|c} 1 & -1 & -\frac{5}{6} \\ 0 & 1 & \frac{7}{6} \end{array}\right]$

Add row 1 to row 2.

$$\left[\begin{array}{cc|c} 1 & 0 & \frac{1}{3} \\ 0 & 1 & \frac{7}{6} \end{array}\right]$$

This corresponds to

$$\begin{cases} x = \dfrac{1}{3} \\ y = \dfrac{7}{6} \end{cases}$$

The solution is $\left(\dfrac{1}{3}, \dfrac{7}{6}\right)$.

29. $\begin{cases} 3x - 6y = 0 \\ 2x + 4y = 5 \end{cases}$

$$\left[\begin{array}{cc|c} 3 & -6 & 0 \\ 2 & 4 & 5 \end{array}\right]$$

Divide row 1 by 3.

$$\left[\begin{array}{cc|c} 1 & -2 & 0 \\ 2 & 4 & 5 \end{array}\right]$$

Multiply row 1 by –2 and add to row 2.

$$\left[\begin{array}{cc|c} 1 & -2 & 0 \\ 0 & 8 & 5 \end{array}\right]$$

Divide row 2 by 8.

$$\left[\begin{array}{cc|c} 1 & -2 & 0 \\ 0 & 1 & \frac{5}{8} \end{array}\right]$$

This corresponds to

$$\begin{cases} x - 2y = 0 \\ y = \dfrac{5}{8}. \end{cases}$$

$$x - 2\left(\frac{5}{8}\right) = 0$$
$$x - \frac{5}{4} = 0$$
$$x = \frac{5}{4}$$

The solution is $\left(\dfrac{5}{4}, \dfrac{5}{8}\right)$.

30. $\begin{cases} 5x - 3y = 10 \\ -2x + y = -1 \end{cases}$

$$\left[\begin{array}{cc|c} 5 & -3 & 10 \\ -2 & 1 & -1 \end{array}\right]$$

Divide row 1 by 5.

$$\left[\begin{array}{cc|c} 1 & -\frac{3}{5} & 2 \\ -2 & 1 & -1 \end{array}\right]$$

Multiply row 1 by 2 and add to row 2.

$$\left[\begin{array}{cc|c} 1 & -\frac{3}{5} & 2 \\ 0 & -\frac{1}{5} & 3 \end{array}\right]$$

Multiply row 2 by –5.

$$\left[\begin{array}{cc|c} 1 & -\frac{3}{5} & 2 \\ 0 & 1 & -15 \end{array}\right]$$

This corresponds to

$$\begin{cases} x - \dfrac{3}{5}y = 2 \\ y = -15 \end{cases}.$$

$$x - \frac{3}{5}(-15) = 2$$
$$x + 9 = 2$$
$$x = -7$$

The solution is (–7, –15).

31. $\begin{cases} 0.2x - 0.3y = -0.7 \\ 0.5x + 0.3y = 1.4 \end{cases}$

$$\left[\begin{array}{cc|c} 0.2 & -0.3 & -0.7 \\ 0.5 & 0.3 & 1.4 \end{array}\right]$$

Multiply both rows by 10 to clear decimals.

$$\left[\begin{array}{cc|c} 2 & -3 & -7 \\ 5 & 3 & 14 \end{array}\right]$$

Divide row 1 by 2.

$$\left[\begin{array}{cc|c} 1 & -\frac{3}{2} & -\frac{7}{2} \\ 5 & 3 & 14 \end{array}\right]$$

Multiply row 1 by –5 and add to row 2.

$$\left[\begin{array}{cc|c} 1 & -\frac{3}{2} & -\frac{7}{2} \\ 0 & \frac{21}{2} & \frac{63}{2} \end{array}\right]$$

Multiply row 2 by $\dfrac{2}{21}$.

$$\left[\begin{array}{cc|c} 1 & -\frac{3}{2} & -\frac{7}{2} \\ 0 & 1 & 3 \end{array}\right]$$

This corresponds to

$$\begin{cases} x - \dfrac{3}{2}y = -\dfrac{7}{2} \\ y = 3 \end{cases}.$$

$$x-\frac{3}{2}(3)=-\frac{7}{2}$$
$$x-\frac{9}{2}=-\frac{7}{2}$$
$$x=1$$

The solution is (1, 3).

32. $\begin{cases}3x+2y=8\\3x-\ \ y=5\end{cases}$

$$\left[\begin{array}{cc|c}3 & 2 & 8\\3 & -1 & 5\end{array}\right]$$

Divide row 1 by 3.

$$\left[\begin{array}{cc|c}1 & \frac{2}{3} & \frac{8}{3}\\3 & -1 & 5\end{array}\right]$$

Multiply row 1 by –3 and add to row 2.

$$\left[\begin{array}{cc|c}1 & \frac{2}{3} & \frac{8}{3}\\0 & -3 & -3\end{array}\right]$$

Divide row 2 by –3.

$$\left[\begin{array}{cc|c}1 & \frac{2}{3} & \frac{8}{3}\\0 & 1 & 1\end{array}\right]$$

This corresponds to

$$\begin{cases}x+\frac{2}{3}y=\frac{8}{3}\\y=1\end{cases}.$$

$$x+\frac{2}{3}(1)=\frac{8}{3}$$
$$x=2$$

The solution is (2, 1).

33. $\begin{cases}x\quad\ +z=4\\2x-y\quad\ =0\\x+y-z=0\end{cases}$

$$\left[\begin{array}{ccc|c}1 & 0 & 1 & 4\\2 & -1 & 0 & 0\\1 & 1 & -1 & 0\end{array}\right]$$

Multiply row 1 by –2 and add to row 2.
Multiply row 1 by –1 and add to row 3.

$$\left[\begin{array}{ccc|c}1 & 0 & 1 & 4\\0 & -1 & -2 & -8\\0 & 1 & -2 & -4\end{array}\right]$$

Multiply row 2 by –1.

$$\left[\begin{array}{ccc|c}1 & 0 & 1 & 4\\0 & 1 & 2 & 8\\0 & 1 & -2 & -4\end{array}\right]$$

Multiply row 2 by –1 and add to row 3.

$$\left[\begin{array}{ccc|c}1 & 0 & 1 & 4\\0 & 1 & 2 & 8\\0 & 0 & -4 & -12\end{array}\right]$$

Divide row 3 by –4.

$$\left[\begin{array}{ccc|c}1 & 0 & 1 & 4\\0 & 1 & 2 & 8\\0 & 0 & 1 & 3\end{array}\right]$$

This corresponds to

$$\begin{cases}x+z=4\\y+2z=8\\z=3\end{cases}$$

$$y+2(3)=8$$
$$y+6=8$$
$$y=2$$
$$x+3=4$$
$$x=1$$

The solution is (1, 2, 3).

34. $\begin{cases}2x+5y\quad\ =4\\x-5y+z=-1\\4x\quad\ -z=11\end{cases}$

$$\left[\begin{array}{ccc|c}2 & 5 & 0 & 4\\1 & -5 & 1 & -1\\4 & 0 & -1 & 11\end{array}\right]$$

Interchange row 1 and row 2.

$$\left[\begin{array}{ccc|c}1 & -5 & 1 & -1\\2 & 5 & 0 & 4\\4 & 0 & -1 & 11\end{array}\right]$$

Multiply row 1 by –2 and add to row 2.
Multiply row 1 by –4 and add to row 3.

$$\left[\begin{array}{ccc|c}1 & -5 & 1 & -1\\0 & 15 & -2 & 6\\0 & 20 & -5 & 15\end{array}\right]$$

Divide row 2 by 15.

$$\left[\begin{array}{ccc|c} 1 & -5 & 1 & -1 \\ 0 & 1 & -\frac{2}{15} & \frac{2}{5} \\ 0 & 20 & -5 & 15 \end{array}\right]$$

Multiply row 2 by –20 and add to row 3.

$$\left[\begin{array}{ccc|c} 1 & -5 & 1 & -1 \\ 0 & 1 & -\frac{2}{15} & \frac{2}{5} \\ 0 & 0 & -\frac{7}{3} & 7 \end{array}\right]$$

Multiply row 3 by $-\dfrac{3}{7}$.

$$\left[\begin{array}{ccc|c} 1 & -5 & 1 & -1 \\ 0 & 1 & -\frac{2}{15} & \frac{2}{5} \\ 0 & 0 & 1 & -3 \end{array}\right]$$

This corresponds to

$$\begin{cases} x-5y \quad +z=-1 \\ \quad y-\dfrac{2}{15}z=\dfrac{2}{5} \\ \quad\quad z=-3 \end{cases}.$$

$$y-\frac{2}{15}(-3)=\frac{2}{5}$$
$$y+\frac{2}{5}=\frac{2}{5}$$
$$y=0$$
$$x-5(0)+(-3)=-1$$
$$x-3=-1$$
$$x=2$$

The solution is (2, 0, –3).

35. $\begin{cases} 3x-y \quad =11 \\ x+\ 2z=13 \\ \quad y-z=-7 \end{cases}$

$$\left[\begin{array}{ccc|c} 3 & -1 & 0 & 11 \\ 1 & 0 & 2 & 13 \\ 0 & 1 & -1 & -7 \end{array}\right]$$

Interchange row 1 and row 2.

$$\left[\begin{array}{ccc|c} 1 & 0 & 2 & 13 \\ 3 & -1 & 0 & 11 \\ 0 & 1 & -1 & -7 \end{array}\right]$$

Interchange row 2 and row 3.

$$\left[\begin{array}{ccc|c} 1 & 0 & 2 & 13 \\ 0 & 1 & -1 & -7 \\ 3 & -1 & 0 & 11 \end{array}\right]$$

Multiply row 1 by –3 and add to row 3.

$$\left[\begin{array}{ccc|c} 1 & 0 & 2 & 13 \\ 0 & 1 & -1 & -7 \\ 0 & -1 & -6 & -28 \end{array}\right]$$

Add row 2 to row 3.

$$\left[\begin{array}{ccc|c} 1 & 0 & 2 & 13 \\ 0 & 1 & -1 & -7 \\ 0 & 0 & -7 & -35 \end{array}\right]$$

Divide row 3 by –7.

$$\left[\begin{array}{ccc|c} 1 & 0 & 2 & 13 \\ 0 & 1 & -1 & -7 \\ 0 & 0 & 1 & 5 \end{array}\right]$$

This corresponds to

$$\begin{cases} x+2z=13 \\ y-z=-7 \\ z=5 \end{cases}$$

$$y-5=-7$$
$$y=-2$$
$$x+2(5)=13$$
$$x+10=13$$
$$x=3$$

The solution is (3, –2, 5).

36. $\begin{cases} 5x+7y+3z=\ 9 \\ \quad 14y-z=28 \\ 4x \quad\quad +2z=-4 \end{cases}$

$$\left[\begin{array}{ccc|c} 5 & 7 & 3 & 9 \\ 0 & 14 & -1 & 28 \\ 4 & 0 & 2 & -4 \end{array}\right]$$

Divide row 1 by 5.

$$\left[\begin{array}{ccc|c} 1 & \frac{7}{5} & \frac{3}{5} & \frac{9}{5} \\ 0 & 14 & -1 & 28 \\ 4 & 0 & 2 & -4 \end{array}\right]$$

Multiply row 1 by –4 and add to row 3.

$$\left[\begin{array}{ccc|c} 1 & \frac{7}{5} & \frac{3}{5} & \frac{9}{5} \\ 0 & 14 & -1 & 28 \\ 0 & -\frac{28}{5} & -\frac{2}{5} & -\frac{56}{5} \end{array}\right]$$

Divide row 2 by 14.

$$\left[\begin{array}{ccc|c} 1 & \frac{7}{5} & \frac{3}{5} & \frac{9}{5} \\ 0 & 1 & -\frac{1}{14} & 2 \\ 0 & -\frac{28}{5} & -\frac{2}{5} & -\frac{56}{5} \end{array}\right]$$

Multiply row 2 by $\frac{28}{5}$ and add to row 3.

$$\left[\begin{array}{ccc|c} 1 & \frac{7}{5} & \frac{3}{5} & \frac{9}{5} \\ 0 & 1 & -\frac{1}{14} & 2 \\ 0 & 0 & -\frac{4}{5} & 0 \end{array}\right]$$

Multiply row 3 by $-\frac{5}{4}$.

$$\left[\begin{array}{ccc|c} 1 & \frac{7}{5} & \frac{3}{5} & \frac{9}{5} \\ 0 & 1 & -\frac{1}{14} & 2 \\ 0 & 0 & 1 & 0 \end{array}\right]$$

This corresponds to

$$\begin{cases} x+\frac{7}{5}y+\frac{3}{5}z=\frac{9}{5} \\ y-\frac{1}{14}z=2 \\ z=0 \end{cases}.$$

$$y-\frac{1}{14}(0)=2$$
$$y=2$$
$$x+\frac{7}{5}(2)+\frac{3}{5}(0)=\frac{9}{5}$$
$$x+\frac{14}{5}=\frac{9}{5}$$
$$x=-1$$

The solution is (–1, 2, 0).

37. $\begin{cases} 7x-3y+2z=0 \\ 4x-4y-\ z=2 \\ 5x+2y+3z=1 \end{cases}$

$$\left[\begin{array}{ccc|c} 7 & -3 & 2 & 0 \\ 4 & -4 & -1 & 2 \\ 5 & 2 & 3 & 1 \end{array}\right]$$

Interchange row 1 and row 2.

$$\left[\begin{array}{ccc|c} 4 & -4 & -1 & 2 \\ 7 & -3 & 2 & 0 \\ 5 & 2 & 3 & 1 \end{array}\right]$$

Divide row 1 by 4.

$$\left[\begin{array}{ccc|c} 1 & -1 & -\frac{1}{4} & \frac{1}{2} \\ 7 & -3 & 2 & 0 \\ 5 & 2 & 3 & 1 \end{array}\right]$$

Multiply row 1 by –7 and add to row 2.
Multiply row 1 by –5 and add to row 3.

$$\left[\begin{array}{ccc|c} 1 & -1 & -\frac{1}{4} & \frac{1}{2} \\ 0 & 4 & \frac{15}{4} & -\frac{7}{2} \\ 0 & 7 & \frac{17}{4} & -\frac{3}{2} \end{array}\right]$$

Divide row 2 by 4.

$$\left[\begin{array}{ccc|c} 1 & -1 & -\frac{1}{4} & \frac{1}{2} \\ 0 & 1 & \frac{15}{16} & -\frac{7}{8} \\ 0 & 7 & \frac{17}{4} & -\frac{3}{2} \end{array}\right]$$

Multiply row 2 by –7 and add to row 3.

$$\left[\begin{array}{ccc|c} 1 & -1 & -\frac{1}{4} & \frac{1}{2} \\ 0 & 1 & \frac{15}{16} & -\frac{7}{8} \\ 0 & 0 & -\frac{37}{16} & \frac{37}{8} \end{array}\right]$$

Multiply row 3 by $-\frac{16}{37}$.

$$\left[\begin{array}{ccc|c} 1 & -1 & -\frac{1}{4} & \frac{1}{2} \\ 0 & 1 & \frac{15}{16} & -\frac{7}{8} \\ 0 & 0 & 1 & -2 \end{array}\right]$$

This corresponds to

$$\begin{cases} x-y-\frac{1}{4}z=\frac{1}{2} \\ y+\frac{15}{16}z=-\frac{7}{8} \\ z=-2 \end{cases}$$

$$y+\frac{15}{16}(-2)=-\frac{7}{8}$$
$$y-\frac{15}{8}=-\frac{7}{8}$$
$$y=1$$
$$x-1-\frac{1}{4}(-2)=\frac{1}{2}$$
$$x-1+\frac{1}{2}=\frac{1}{2}$$
$$x=1$$

The solution is (1, 1, –2).

38. $\begin{cases} x+2y+3z=14 \\ y+2z=3 \\ 2x \quad -2z=10 \end{cases}$

$\left[\begin{array}{ccc|c} 1 & 2 & 3 & 14 \\ 0 & 1 & 2 & 3 \\ 2 & 0 & -2 & 10 \end{array}\right]$

Multiply row 1 by –2 and add to row 3.

$\left[\begin{array}{ccc|c} 1 & 2 & 3 & 14 \\ 0 & 1 & 2 & 3 \\ 0 & -4 & -8 & -18 \end{array}\right]$

Multiply row 2 by 4 and add to row 3.

$\left[\begin{array}{ccc|c} 1 & 2 & 3 & 14 \\ 0 & 1 & 2 & 3 \\ 0 & 0 & 0 & -6 \end{array}\right]$

This corresponds to

$\begin{cases} x+2y+3z=14 \\ y+2z=3 \\ 0=-6 \end{cases}.$

The system is inconsistent.
The solution is $\varnothing$.

39. $\begin{vmatrix} -1 & 3 \\ 5 & 2 \end{vmatrix} = (-1)(2)-3(5) = -2-15 = -17$

40. $\begin{vmatrix} 3 & -1 \\ 2 & 5 \end{vmatrix} = 3(5)-(-1)2 = 15+2 = 17$

41. $\begin{vmatrix} 2 & -1 & -3 \\ 1 & 2 & 0 \\ 3 & -2 & 2 \end{vmatrix}$

$= 2\begin{vmatrix} 2 & 0 \\ -2 & 2 \end{vmatrix} - (-1)\begin{vmatrix} 1 & 0 \\ 3 & 2 \end{vmatrix} + (-3)\begin{vmatrix} 1 & 2 \\ 3 & -2 \end{vmatrix}$

$= 2(4-0)+(2-0)-3(-2-6)$

$= 2(4)+(2)-3(-8)$

$= 34$

42. $\begin{vmatrix} -2 & 3 & 1 \\ 4 & 4 & 0 \\ 1 & -2 & 3 \end{vmatrix}$

$= 1\begin{vmatrix} 4 & 4 \\ 1 & -2 \end{vmatrix} - 0\begin{vmatrix} -2 & 3 \\ 1 & -2 \end{vmatrix} + 3\begin{vmatrix} -2 & 3 \\ 4 & 4 \end{vmatrix}$

$= (-8-4)-0+3(-8-12)$

$= -12+3(-20)$

$= -12-60$

$= -72$

43. $\begin{cases} 3x-2y=-8 \\ 6x+5y=11 \end{cases}$

$D = \begin{vmatrix} 3 & -2 \\ 6 & 5 \end{vmatrix} = 15-(-12) = 27$

$D_x = \begin{vmatrix} -8 & -2 \\ 11 & 5 \end{vmatrix} = -40-(-22) = -18$

$D_y = \begin{vmatrix} 3 & -8 \\ 6 & 11 \end{vmatrix} = 33-(-48) = 81$

$x = \frac{D_x}{D} = \frac{-18}{27} = -\frac{2}{3}$

$y = \frac{D_y}{D} = \frac{81}{27} = 3$

The solution is $\left(-\frac{2}{3}, 3\right)$.

44. $\begin{cases} 6x-6y=-5 \\ 10x-2y=1 \end{cases}$

$D = \begin{vmatrix} 6 & -6 \\ 10 & -2 \end{vmatrix} = -12-(-60) = 48$

$D_x = \begin{vmatrix} -5 & -6 \\ 1 & -2 \end{vmatrix} = 10-(-6) = 16$

$D_y = \begin{vmatrix} 6 & -5 \\ 10 & 1 \end{vmatrix} = 6-(-50) = 56$

$x = \frac{D_x}{D} = \frac{16}{48} = \frac{1}{3}$ and $y = \frac{D_y}{D} = \frac{56}{48} = \frac{7}{6}$

The solution is $\left(\frac{1}{3}, \frac{7}{6}\right)$.

45. $\begin{cases} 3x+10y=1 \\ x+2y=-1 \end{cases}$

$D=\begin{vmatrix} 3 & 10 \\ 1 & 2 \end{vmatrix}=6-10=-4$

$D_x=\begin{vmatrix} 1 & 10 \\ -1 & 2 \end{vmatrix}=2-(-10)=12$

$D_y=\begin{vmatrix} 3 & 1 \\ 1 & -1 \end{vmatrix}=-3-1=-4$

$x=\frac{D_x}{D}=\frac{12}{-4}=-3$

$y=\frac{D_y}{D}=\frac{-4}{-4}=1$

The solution is (–3, 1).

46. $\begin{cases} y=\frac{1}{2}x+\frac{2}{3} \\ 4x+6y=4 \end{cases}$ or $\begin{cases} -\frac{1}{2}x+y=\frac{2}{3} \\ 4x+6y=4 \end{cases}$

$D=\begin{vmatrix} -\frac{1}{2} & 1 \\ 4 & 6 \end{vmatrix}=-3-4=-7$

$D_x=\begin{vmatrix} \frac{2}{3} & 1 \\ 4 & 6 \end{vmatrix}=4-4=0$

$D_y=\begin{vmatrix} -\frac{1}{2} & \frac{2}{3} \\ 4 & 4 \end{vmatrix}=-2-\frac{8}{3}=-\frac{14}{3}$

$x=\frac{D_x}{D}=\frac{0}{-7}=0$ and $y=\frac{D_y}{D}=\frac{-\frac{14}{3}}{-7}=\frac{2}{3}$

The solution is $\left(0, \frac{2}{3}\right)$.

47. $\begin{cases} 2x-4y=22 & (1) \\ 5x-10y=16 & (2) \end{cases}$

$D=\begin{vmatrix} 2 & -4 \\ 5 & -10 \end{vmatrix}=-20-(-20)=0$

This cannot be solved by Cramer's rule. Multiply equation (1) by –5, multiply equation (2) by 2, and add.

$$\begin{array}{r} -10x+20y=-110 \\ 10x-20y=\quad 32 \\ \hline 0=-78 \end{array}$$

The solution is $\varnothing$.

48. $\begin{cases} 3x-6y=12 \\ 2y=x-4 \end{cases}$ or $\begin{cases} 3x-6y=12 & (1) \\ -x+2y=-4 & (2) \end{cases}$

$D=\begin{vmatrix} 3 & -6 \\ -1 & 2 \end{vmatrix}=6-6=0$

Cramer's Rule cannot be used to solve this system. Since equation (1) is –3 times equation (2), the system is dependent. The solution is $\{(x, y) \mid x-2y=4\}$.

49. $\begin{cases} x+z=4 \\ 2x-y=0 \\ x+y-z=0 \end{cases}$

$D=\begin{vmatrix} 1 & 0 & 1 \\ 2 & -1 & 0 \\ 1 & 1 & -1 \end{vmatrix}$

$=1\begin{vmatrix} -1 & 0 \\ 1 & -1 \end{vmatrix}-0\begin{vmatrix} 2 & 0 \\ 1 & -1 \end{vmatrix}+1\begin{vmatrix} 2 & -1 \\ 1 & 1 \end{vmatrix}$

$=(1-0)-0+[2-(-1)]$

$=1+3$

$=4$

$D_x=\begin{vmatrix} 4 & 0 & 1 \\ 0 & -1 & 0 \\ 0 & 1 & -1 \end{vmatrix}$

$=4\begin{vmatrix} -1 & 0 \\ 1 & -1 \end{vmatrix}-0\begin{vmatrix} 0 & 1 \\ 1 & -1 \end{vmatrix}+0\begin{vmatrix} 0 & 1 \\ -1 & 0 \end{vmatrix}$

$=4(1-0)-0+0=4$

$D_y=\begin{vmatrix} 1 & 4 & 1 \\ 2 & 0 & 0 \\ 1 & 0 & -1 \end{vmatrix}$

$=-4\begin{vmatrix} 2 & 0 \\ 1 & -1 \end{vmatrix}+0\begin{vmatrix} 1 & 1 \\ 1 & -1 \end{vmatrix}-0\begin{vmatrix} 1 & 1 \\ 2 & 0 \end{vmatrix}$

$=(-4)(-2-0)$

$=8$

$D_z=\begin{vmatrix} 1 & 0 & 4 \\ 2 & -1 & 0 \\ 1 & 1 & 0 \end{vmatrix}$

$=4\begin{vmatrix} 2 & -1 \\ 1 & 1 \end{vmatrix}-0\begin{vmatrix} 1 & 0 \\ 1 & 1 \end{vmatrix}+0\begin{vmatrix} 1 & 0 \\ 2 & -1 \end{vmatrix}$

$=4[2-(-1)]$

$=12$

$$x = \frac{D_x}{D} = \frac{4}{4} = 1$$
$$y = \frac{D_y}{D} = \frac{8}{4} = 2$$
$$z = \frac{D_z}{D} = \frac{12}{4} = 3$$
The solution is (1, 2, 3).

50. $\begin{cases} 2x + 5y = 4 \\ x - 5y + z = -1 \\ 4x - z = 11 \end{cases}$

$$D = \begin{vmatrix} 2 & 5 & 0 \\ 1 & -5 & 1 \\ 4 & 0 & -1 \end{vmatrix}$$
$$= 0\begin{vmatrix} 1 & -5 \\ 4 & 0 \end{vmatrix} - 1\begin{vmatrix} 2 & 5 \\ 4 & 0 \end{vmatrix} + (-1)\begin{vmatrix} 2 & 5 \\ 1 & -5 \end{vmatrix}$$
$$= 0 - (0 - 20) - (-10 - 5)$$
$$= 20 + 15$$
$$= 35$$

$$D_x = \begin{vmatrix} 4 & 5 & 0 \\ -1 & -5 & 1 \\ 11 & 0 & -1 \end{vmatrix}$$
$$= 0\begin{vmatrix} -1 & -5 \\ 11 & 0 \end{vmatrix} - 1\begin{vmatrix} 4 & 5 \\ 11 & 0 \end{vmatrix} + (-1)\begin{vmatrix} 4 & 5 \\ -1 & -5 \end{vmatrix}$$
$$= 0 - (0 - 55) - [-20 - (-5)]$$
$$= 55 + 15$$
$$= 70$$

$$D_y = \begin{vmatrix} 2 & 4 & 0 \\ 1 & -1 & 1 \\ 4 & 11 & -1 \end{vmatrix}$$
$$= 0\begin{vmatrix} 1 & -1 \\ 4 & 11 \end{vmatrix} - 1\begin{vmatrix} 2 & 4 \\ 4 & 11 \end{vmatrix} + (-1)\begin{vmatrix} 2 & 4 \\ 1 & -1 \end{vmatrix}$$
$$= 0 - (22 - 16) - (-2 - 4)$$
$$= -6 + 6$$
$$= 0$$

$$D_z = \begin{vmatrix} 2 & 5 & 4 \\ 1 & -5 & -1 \\ 4 & 0 & 11 \end{vmatrix}$$
$$= 4\begin{vmatrix} 1 & -5 \\ 4 & 0 \end{vmatrix} - (-1)\begin{vmatrix} 2 & 5 \\ 4 & 0 \end{vmatrix} + 11\begin{vmatrix} 2 & 5 \\ 1 & -5 \end{vmatrix}$$
$$= 4[0 - (-20)] + 1(0 - 20) + 11(-10 - 5)$$
$$= 4(20) - 20 + 11(-15)$$
$$= 80 - 20 - 165$$
$$= -105$$

$$x = \frac{D_x}{D} = \frac{70}{35} = 2,\quad y = \frac{D_y}{D} = \frac{0}{35} = 0,$$
$$z = \frac{D_z}{D} = \frac{-105}{35} = -3$$
The solution is (2, 0, –3).

51. $\begin{cases} x + 3y - z = 5 \\ 2x - y - 2z = 3 \\ x + 2y + 3z = 4 \end{cases}$

$$D = \begin{vmatrix} 1 & 3 & -1 \\ 2 & -1 & -2 \\ 1 & 2 & 3 \end{vmatrix}$$
$$= 1\begin{vmatrix} -1 & -2 \\ 2 & 3 \end{vmatrix} - 3\begin{vmatrix} 2 & -2 \\ 1 & 3 \end{vmatrix} + (-1)\begin{vmatrix} 2 & -1 \\ 1 & 2 \end{vmatrix}$$
$$= [-3 - (-4)] - 3[6 - (-2)] - [4 - (-1)]$$
$$= 1 - 3(8) - 5$$
$$= 1 - 24 - 5$$
$$= -28$$

$$D_x = \begin{vmatrix} 5 & 3 & -1 \\ 3 & -1 & -2 \\ 4 & 2 & 3 \end{vmatrix}$$
$$= 5\begin{vmatrix} -1 & -2 \\ 2 & 3 \end{vmatrix} - 3\begin{vmatrix} 3 & -2 \\ 4 & 3 \end{vmatrix} + (-1)\begin{vmatrix} 3 & -1 \\ 4 & 2 \end{vmatrix}$$
$$= 5[-3 - (-4)] - 3[9 - (-8)] - [6 - (-4)]$$
$$= 5(1) - 3(17) - 10$$
$$= 5 - 51 - 10$$
$$= -56$$

$$D_y = \begin{vmatrix} 1 & 5 & -1 \\ 2 & 3 & -2 \\ 1 & 4 & 3 \end{vmatrix}$$
$$= 1\begin{vmatrix} 3 & -2 \\ 4 & 3 \end{vmatrix} - 5\begin{vmatrix} 2 & -2 \\ 1 & 3 \end{vmatrix} + (-1)\begin{vmatrix} 2 & 3 \\ 1 & 4 \end{vmatrix}$$
$$= [9-(-8)] - 5[6-(-2)] - (8-3)$$
$$= 17 - 5(8) - 5$$
$$= 17 - 40 - 5$$
$$= -28$$

$$D_z = \begin{vmatrix} 1 & 3 & 5 \\ 2 & -1 & 3 \\ 1 & 2 & 4 \end{vmatrix}$$
$$= 1\begin{vmatrix} -1 & 3 \\ 2 & 4 \end{vmatrix} - 3\begin{vmatrix} 2 & 3 \\ 1 & 4 \end{vmatrix} + 5\begin{vmatrix} 2 & -1 \\ 1 & 2 \end{vmatrix}$$
$$= (-4-6) - 3(8-3) + 5[4-(-1)]$$
$$= -10 - 3(5) + 5(5)$$
$$= -10 - 15 + 25$$
$$= 0$$

$$x = \frac{D_x}{D} = \frac{-56}{-28} = 2$$
$$y = \frac{D_y}{D} = \frac{-28}{-28} = 1$$
$$z = \frac{D_z}{D} = \frac{0}{-28} = 0$$

The solution is (2, 1, 0).

52. $\begin{cases} 2x \qquad\ - z = 1 \\ 3x - y + 2z = 3 \\ x + y + 3z = -2 \end{cases}$

$$D = \begin{vmatrix} 2 & 0 & -1 \\ 3 & -1 & 2 \\ 1 & 1 & 3 \end{vmatrix}$$
$$= 2\begin{vmatrix} -1 & 2 \\ 1 & 3 \end{vmatrix} - 0\begin{vmatrix} 3 & 2 \\ 1 & 3 \end{vmatrix} + (-1)\begin{vmatrix} 3 & -1 \\ 1 & 1 \end{vmatrix}$$
$$= 2(-3-2) - 0 - [3-(-1)]$$
$$= 2(-5) - 4$$
$$= -10 - 4$$
$$= -14$$

$$D_x = \begin{vmatrix} 1 & 0 & -1 \\ 3 & -1 & 2 \\ -2 & 1 & 3 \end{vmatrix}$$
$$= 1\begin{vmatrix} -1 & 2 \\ 1 & 3 \end{vmatrix} - 0\begin{vmatrix} 3 & 2 \\ -2 & 3 \end{vmatrix} + (-1)\begin{vmatrix} 3 & -1 \\ -2 & 1 \end{vmatrix}$$
$$= (-3-2) - 0 - (3-2)$$
$$= -5 - 1$$
$$= -6$$

$$D_y = \begin{vmatrix} 2 & 1 & -1 \\ 3 & 3 & 2 \\ 1 & -2 & 3 \end{vmatrix}$$
$$= 2\begin{vmatrix} 3 & 2 \\ -2 & 3 \end{vmatrix} - 1\begin{vmatrix} 3 & 2 \\ 1 & 3 \end{vmatrix} + (-1)\begin{vmatrix} 3 & 3 \\ 1 & -2 \end{vmatrix}$$
$$= 2[9-(-4)] - (9-2) - (-6-3)$$
$$= 2(13) - 7 + 9$$
$$= 26 + 2$$
$$= 28$$

$$D_z = \begin{vmatrix} 2 & 0 & 1 \\ 3 & -1 & 3 \\ 1 & 1 & -2 \end{vmatrix}$$
$$= 2\begin{vmatrix} -1 & 3 \\ 1 & -2 \end{vmatrix} - 0\begin{vmatrix} 3 & 3 \\ 1 & -2 \end{vmatrix} + 1\begin{vmatrix} 3 & -1 \\ 1 & 1 \end{vmatrix}$$
$$= 2(2-3) - 0 + [3-(-1)]$$
$$= 2(-1) + 4$$
$$= -2 + 4$$
$$= 2$$

$$x = \frac{D_x}{D} = \frac{-6}{-14} = \frac{3}{7},\quad y = \frac{D_y}{D} = \frac{28}{-14} = -2,$$
$$z = \frac{D_z}{D} = \frac{2}{-14} = -\frac{1}{7}$$

The solution is $\left(\frac{3}{7}, -2, -\frac{1}{7}\right)$.

53. $\begin{cases} x + 2y + 3z = 14 \quad (1) \\ \qquad y + 2z = 3 \quad (2) \\ 2x \qquad - 2z = 10 \quad (3) \end{cases}$

$$D = \begin{vmatrix} 1 & 2 & 3 \\ 0 & 1 & 2 \\ 2 & 0 & -2 \end{vmatrix}$$
$$= 1\begin{vmatrix} 1 & 2 \\ 0 & -2 \end{vmatrix} - 0\begin{vmatrix} 2 & 3 \\ 0 & -2 \end{vmatrix} + 2\begin{vmatrix} 2 & 3 \\ 1 & 2 \end{vmatrix}$$
$$= (-2-0) - 0 + 2(4-3)$$
$$= -2 + 2$$
$$= 0$$

This cannot be solved by Cramer's rule.
Solve equation (2) for y.
$y = -2z + 3$
Solve equation (3) for x.
$2x = 2z + 10$
$x = z + 5$
Replace x with $z + 5$ and y with $-2z + 3$ in equation (1).
$$z + 5 + 2(-2z + 3) + 3z = 14$$
$$z + 5 - 4z + 6 + 3z = 14$$
$$11 = 14 \quad \text{False}$$
The system is inconsistent.
The solution is $\varnothing$.

54. $\begin{cases} 5x + 7y = 9 \\ 14y - z = 28 \\ 4x + 2z = -4 \end{cases}$

$$D = \begin{vmatrix} 5 & 7 & 0 \\ 0 & 14 & -1 \\ 4 & 0 & 2 \end{vmatrix}$$
$$= 5\begin{vmatrix} 14 & -1 \\ 0 & 2 \end{vmatrix} - 7\begin{vmatrix} 0 & -1 \\ 4 & 2 \end{vmatrix} + 0\begin{vmatrix} 0 & 14 \\ 4 & 0 \end{vmatrix}$$
$$= 5(28-0) - 7[0-(-4)] + 0$$
$$= 140 - 28$$
$$= 112$$

$$D_x = \begin{vmatrix} 9 & 7 & 0 \\ 28 & 14 & -1 \\ -4 & 0 & 2 \end{vmatrix}$$
$$= 9\begin{vmatrix} 14 & -1 \\ 0 & 2 \end{vmatrix} - 7\begin{vmatrix} 28 & -1 \\ -4 & 2 \end{vmatrix} + 0\begin{vmatrix} 28 & 14 \\ -4 & 0 \end{vmatrix}$$
$$= 9(28-0) - 7(56-4) + 0$$
$$= 252 - 7(52)$$
$$= 252 - 364$$
$$= -112$$

$$D_y = \begin{vmatrix} 5 & 9 & 0 \\ 0 & 28 & -1 \\ 4 & -4 & 2 \end{vmatrix}$$
$$= 5\begin{vmatrix} 28 & -1 \\ -4 & 2 \end{vmatrix} - 9\begin{vmatrix} 0 & -1 \\ 4 & 2 \end{vmatrix} + 0\begin{vmatrix} 0 & 28 \\ 4 & -4 \end{vmatrix}$$
$$= 5(56-4) - 9[0-(-4)] + 0$$
$$= 5(52) - 9(4)$$
$$= 260 - 36$$
$$= 224$$

$$D_z = \begin{vmatrix} 5 & 7 & 9 \\ 0 & 14 & 28 \\ 4 & 0 & -4 \end{vmatrix}$$
$$= 5\begin{vmatrix} 14 & 28 \\ 0 & -4 \end{vmatrix} - 0\begin{vmatrix} 7 & 9 \\ 0 & -4 \end{vmatrix} + 4\begin{vmatrix} 7 & 9 \\ 14 & 28 \end{vmatrix}$$
$$= 5(-56-0) - 0 + 4(196-126)$$
$$= -280 + 4(70)$$
$$= -280 + 280$$
$$= 0$$

$$x = \frac{D_x}{D} = \frac{-112}{112} = -1,\quad y = \frac{D_y}{D} = \frac{224}{112} = 2,$$
$$z = \frac{D_z}{D} = \frac{0}{112} = 0$$
The solution is $(-1, 2, 0)$.

Chapter 4 Test

1. $\begin{vmatrix} 4 & -7 \\ 2 & 5 \end{vmatrix} = 4(5) - (-7)(2) = 20 + 14 = 34$

2. $\begin{vmatrix} 4 & 0 & 2 \\ 1 & -3 & 5 \\ 0 & -1 & 2 \end{vmatrix}$
$$= 4\begin{vmatrix} -3 & 5 \\ -1 & 2 \end{vmatrix} - 1\begin{vmatrix} 0 & 2 \\ -1 & 2 \end{vmatrix} + 0\begin{vmatrix} 0 & 2 \\ -3 & 5 \end{vmatrix}$$
$$= 4[-6-(-5)] - [0-(-2)] + 0$$
$$= 4(-1) - 2$$
$$= -4 - 2$$
$$= -6$$

3. $\begin{cases} 2x - y = -1 \\ 5x + 4y = 17 \end{cases}$

Graphically:

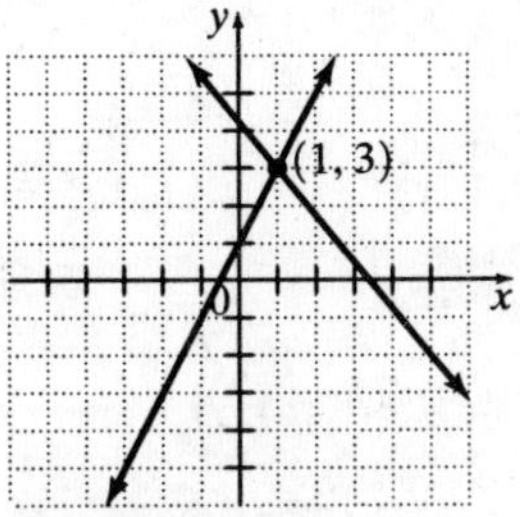

By elimination:
Multiply the first equation by 4 and add it to the second equation.

$$\begin{aligned} 8x - 4y &= -4 \\ 5x + 4y &= 17 \\ \hline 13x \quad &= 13 \\ x \quad &= 1 \end{aligned}$$

Replace x with 1 in the second equation.

$$\begin{aligned} 5(1) + 4y &= 17 \\ 4y &= 12 \\ y &= 3 \end{aligned}$$

The solution is (1, 3).

4. $\begin{cases} 7x - 14y = 5 \\ x = 2y \end{cases}$

Graphically:

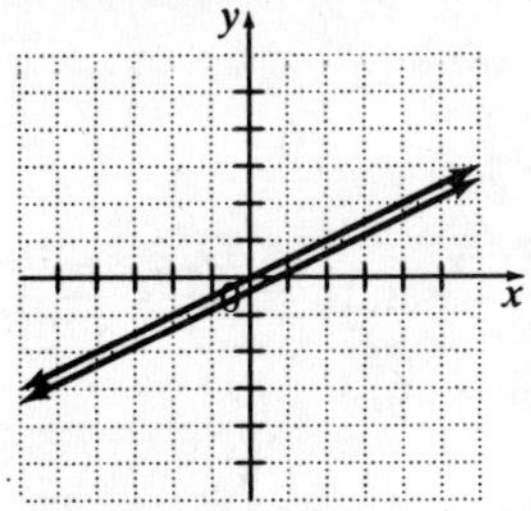

By substitution:
Replace x with $2y$ in the first equation.

$$\begin{aligned} 7(2y) - 14y &= 5 \\ 14y - 14y &= 5 \\ 0 &= 5 \quad \text{False} \end{aligned}$$

The system is inconsistent.
The solution is $\varnothing$.

5. $\begin{cases} 4x - 7y = 29 & (1) \\ 2x + 5y = -11 & (2) \end{cases}$

Multiply equation (2) by –2 and add to equation (1).

$$\begin{aligned} 4x - 7y &= 29 \\ -4x - 10y &= 22 \\ \hline -17y &= 51 \\ y &= -3 \end{aligned}$$

Replace y with –3 in equation (1).

$$\begin{aligned} 4x - 7(-3) &= 29 \\ 4x + 21 &= 29 \\ 4x &= 8 \\ x &= 2 \end{aligned}$$

The solution is (2, –3).

6. $\begin{cases} 15x + 6y = 15 \\ 10x + 4y = 10 \end{cases}$

Divide the first equation by 3 and the second equation by 2.

$\begin{cases} 5x + 2y = 5 \\ 5x + 2y = 5 \end{cases}$

The system is dependent.
The solution is $\{(x, y) | 10x + 4y = 10\}$.

7. $\begin{cases} 2x - 3y \quad = 4 & (1) \\ 3y + 2z = 2 & (2) \\ x \quad - z = -5 & (3) \end{cases}$

Add equations (1) and (2).
$2x + 2z = 6$
Solve the new system.

$\begin{cases} x - z = -5 & (3) \\ 2x + 2z = 6 & (4) \end{cases}$

Multiply equation (3) by 2 and add to equation (4).

$$\begin{aligned} 2x - 2z &= -10 \\ 2x + 2z &= 6 \\ \hline 4x \quad &= -4 \\ x \quad &= -1 \end{aligned}$$

Replace x with –1 in equation (3).

$$\begin{aligned} -1 - z &= -5 \\ -z &= -4 \\ z &= 4 \end{aligned}$$

Replace x with –1 in equation (1).

$$\begin{aligned} 2(-1) - 3y &= 4 \\ -2 - 3y &= 4 \\ -3y &= 6 \\ y &= -2 \end{aligned}$$

The solution is (–1, –2, 4).

8. $\begin{cases} 3x-2y-z=-1 & (1) \\ 2x-2y \quad = 4 & (2) \\ 2x \quad -2z=-12 & (3) \end{cases}$

Multiply equation (2) by –1 and add to equation (1).

$$\begin{array}{r} 3x-2y-z=-1 \\ -2x+2y \quad =-4 \\ \hline x \quad -z=-5 \quad (4) \end{array}$$

Multiply equation (4) by –2 and add to equation (3).

$$\begin{array}{r} 2x-2z=-12 \\ -2x+2z=10 \\ \hline 0=-2 \end{array} \quad \text{False}$$

The system is inconsistent.
The solution is $\varnothing$.

9. $\begin{cases} \frac{x}{2}+\frac{y}{4}=-\frac{3}{4} \\ x+\frac{3}{4}y=-4 \end{cases}$

Clear fractions by multiplying both equations by 4.

$\begin{cases} 2x+\ y=-3 & (1) \\ 4x+3y=-16 & (2) \end{cases}$

Multiply equation (1) by –2 and add to equation (2).

$$\begin{array}{r} -4x-2y=6 \\ 4x+3y=-16 \\ \hline y=-10 \end{array}$$

Replace y with –10 in equation (1).

$$2x+(-10)=-3$$
$$2x=7$$
$$x=\frac{7}{2}$$

The solution is $\left(\frac{7}{2}, -10\right)$.

10. $\begin{cases} 3x-y=7 \\ 2x+5y=-1 \end{cases}$

$$D=\begin{vmatrix} 3 & -1 \\ 2 & 5 \end{vmatrix}=3(5)-(-1)(2)$$
$$=15+2$$
$$=17$$

$$D_x=\begin{vmatrix} 7 & -1 \\ -1 & 5 \end{vmatrix}$$
$$=7(5)-(-1)(-1)$$
$$=35-1$$
$$=34$$

$$D_y=\begin{vmatrix} 3 & 7 \\ 2 & -1 \end{vmatrix}=3(-1)-7(2)$$
$$=-3-14=-17$$

$$x=\frac{D_x}{D}=\frac{34}{17}=2 \text{ and } y=\frac{D_y}{D}=\frac{-17}{17}=-1$$

The solution is (2, –1).

11. $\begin{cases} 4x-3y=-6 \\ -2x+y=0 \end{cases}$

$$D=\begin{vmatrix} 4 & -3 \\ -2 & 1 \end{vmatrix}=4(1)-(-3)(-2)$$
$$=4-6=-2$$

$$D_x=\begin{vmatrix} -6 & -3 \\ 0 & 1 \end{vmatrix}$$
$$=(-6)(1)-(-3)(0)$$
$$=-6-0$$
$$=-6$$

$$D_y=\begin{vmatrix} 4 & -6 \\ -2 & 0 \end{vmatrix}$$
$$=4(0)-(-6)(-2)$$
$$=0-12$$
$$=-12$$

$$x=\frac{D_x}{D}=\frac{-6}{-2}=3 \text{ and } y=\frac{D_y}{D}=\frac{-12}{-2}=6$$

The solution is (3, 6).

12. $\begin{cases} x+y+z=4 \\ 2x+5y \quad =1 \\ x-y-2z=0 \end{cases}$

$$D=\begin{vmatrix} 1 & 1 & 1 \\ 2 & 5 & 0 \\ 1 & -1 & -2 \end{vmatrix}$$
$$=1\begin{vmatrix} 2 & 5 \\ 1 & -1 \end{vmatrix}-0\begin{vmatrix} 1 & 1 \\ 1 & -1 \end{vmatrix}+(-2)\begin{vmatrix} 1 & 1 \\ 2 & 5 \end{vmatrix}$$
$$=(-2-5)-0-2(5-2)$$
$$=-7-2(3)$$
$$=-7-6$$
$$=-13$$

$$D_x = \begin{vmatrix} 4 & 1 & 1 \\ 1 & 5 & 0 \\ 0 & -1 & -2 \end{vmatrix}$$
$$= 1\begin{vmatrix} 1 & 5 \\ 0 & -1 \end{vmatrix} - 0\begin{vmatrix} 4 & 1 \\ 0 & -1 \end{vmatrix} + (-2)\begin{vmatrix} 4 & 1 \\ 1 & 5 \end{vmatrix}$$
$$= (-1-0) - 0 - 2(20-1)$$
$$= -1 - 2(19)$$
$$= -1 - 38$$
$$= -39$$
$$D_y = \begin{vmatrix} 1 & 4 & 1 \\ 2 & 1 & 0 \\ 1 & 0 & -2 \end{vmatrix}$$
$$= 1\begin{vmatrix} 2 & 1 \\ 1 & 0 \end{vmatrix} - 0\begin{vmatrix} 1 & 4 \\ 1 & 0 \end{vmatrix} + (-2)\begin{vmatrix} 1 & 4 \\ 2 & 1 \end{vmatrix}$$
$$= (0-1) - 0 - 2(1-8)$$
$$= -1 - 2(-7)$$
$$= -1 + 14$$
$$= 13$$
$$D_z = \begin{vmatrix} 1 & 1 & 4 \\ 2 & 5 & 1 \\ 1 & -1 & 0 \end{vmatrix}$$
$$= 1\begin{vmatrix} 1 & 4 \\ 5 & 1 \end{vmatrix} - (-1)\begin{vmatrix} 1 & 4 \\ 2 & 1 \end{vmatrix} + 0\begin{vmatrix} 1 & 1 \\ 2 & 5 \end{vmatrix}$$
$$= (1-20) + (1-8) + 0$$
$$= -19 - 7$$
$$= -26$$
$$x = \frac{D_x}{D} = \frac{-39}{-13} = 3,\ y = \frac{D_y}{D} = \frac{13}{-13} = -1,$$
$$z = \frac{D_z}{D} = \frac{-26}{-13} = 2$$
The solution is (3, –1, 2).

13. $\begin{cases} 3x + 2y + 3z = 3 \\ x \qquad\quad - z = 9 \\ \quad 4y + z = -4 \end{cases}$

$$D = \begin{vmatrix} 3 & 2 & 3 \\ 1 & 0 & -1 \\ 0 & 4 & 1 \end{vmatrix}$$
$$= -1\begin{vmatrix} 2 & 3 \\ 4 & 1 \end{vmatrix} + 0\begin{vmatrix} 3 & 3 \\ 0 & 1 \end{vmatrix} - (-1)\begin{vmatrix} 3 & 2 \\ 0 & 4 \end{vmatrix}$$
$$= -(2-12) + 0 + (12-0)$$
$$= -(-10) + 12$$
$$= 10 + 12$$
$$= 22$$
$$D_x = \begin{vmatrix} 3 & 2 & 3 \\ 9 & 0 & -1 \\ -4 & 4 & 1 \end{vmatrix}$$
$$= -9\begin{vmatrix} 2 & 3 \\ 4 & 1 \end{vmatrix} + 0\begin{vmatrix} 3 & 3 \\ -4 & 1 \end{vmatrix} - (-1)\begin{vmatrix} 3 & 2 \\ -4 & 4 \end{vmatrix}$$
$$= -9(2-12) + 0 + [12-(-8)]$$
$$= -9(-10) + 20$$
$$= 90 + 20$$
$$= 110$$
$$D_y = \begin{vmatrix} 3 & 3 & 3 \\ 1 & 9 & -1 \\ 0 & -4 & 1 \end{vmatrix}$$
$$= 3\begin{vmatrix} 9 & -1 \\ -4 & 1 \end{vmatrix} - 1\begin{vmatrix} 3 & 3 \\ -4 & 1 \end{vmatrix} + 0\begin{vmatrix} 3 & 3 \\ 9 & -1 \end{vmatrix}$$
$$= 3(9-4) - [3-(-12)] + 0$$
$$= 3(5) - (15)$$
$$= 15 - 15$$
$$= 0$$
$$D_z = \begin{vmatrix} 3 & 2 & 3 \\ 1 & 0 & 9 \\ 0 & 4 & -4 \end{vmatrix}$$
$$= 3\begin{vmatrix} 0 & 9 \\ 4 & -4 \end{vmatrix} - 1\begin{vmatrix} 2 & 3 \\ 4 & -4 \end{vmatrix} + 0\begin{vmatrix} 2 & 3 \\ 0 & 9 \end{vmatrix}$$
$$= 3(0-36) - (-8-12) + 0$$
$$= 3(-36) - (-20)$$
$$= -108 + 20$$
$$= -88$$
$$x = \frac{D_x}{D} = \frac{110}{22} = 5,\ y = \frac{D_y}{D} = \frac{0}{22} = 0,$$

$$z=\frac{D_z}{D}=\frac{-88}{22}=-4$$
The solution is (5, 0, –4).

14. $\begin{cases} x-\ y=-2 \\ 3x-3y=-6 \end{cases}$

$$\left[\begin{array}{cc|c} 1 & -1 & -2 \\ 3 & -3 & -6 \end{array}\right]$$
Multiply row 1 by –3 and add to row 2.
$$\left[\begin{array}{cc|c} 1 & -1 & -2 \\ 0 & 0 & 0 \end{array}\right]$$
This corresponds to
$$\begin{cases} x-y=-2 \\ 0=0 \end{cases}$$
This is a dependent system.
The solution is $\{(x,\ y)\,|\,x-y=-2\}$.

15. $\begin{cases} x+2y=-1 \\ 2x+5y=-5 \end{cases}$

$$\left[\begin{array}{cc|c} 1 & 2 & -1 \\ 2 & 5 & -5 \end{array}\right]$$
Multiply row 1 by –2 and add to row 2.
$$\left[\begin{array}{cc|c} 1 & 2 & -1 \\ 0 & 1 & -3 \end{array}\right]$$
Multiply row 2 by –2 and add to row 1.
$$\left[\begin{array}{cc|c} 1 & 0 & 5 \\ 0 & 1 & -3 \end{array}\right]$$
The solution is (5, –3).

16. $\begin{cases} x-\ y-z=\ 0 \\ 3x-y-5z=-2 \\ 2x+3y\qquad=-5 \end{cases}$

$$\left[\begin{array}{ccc|c} 1 & -1 & -1 & 0 \\ 3 & -1 & -5 & -2 \\ 2 & 3 & 0 & -5 \end{array}\right]$$
Multiply row 1 by –3 and add to row 2.
Multiply row 1 by –2 and add to row 3.
$$\left[\begin{array}{ccc|c} 1 & -1 & -1 & 0 \\ 0 & 2 & -2 & -2 \\ 0 & 5 & 2 & -5 \end{array}\right]$$
Divide row 2 by 2.
$$\left[\begin{array}{ccc|c} 1 & -1 & -1 & 0 \\ 0 & 1 & -1 & -1 \\ 0 & 5 & 2 & -5 \end{array}\right]$$
Multiply row 2 by –5 and add to row 3.
$$\left[\begin{array}{ccc|c} 1 & -1 & -1 & 0 \\ 0 & 1 & -1 & -1 \\ 0 & 0 & 7 & 0 \end{array}\right]$$
Divide row 3 by 7.
$$\left[\begin{array}{ccc|c} 1 & -1 & -1 & 0 \\ 0 & 1 & -1 & -1 \\ 0 & 0 & 1 & 0 \end{array}\right]$$
This corresponds to
$$\begin{cases} x-y-z=0 \\ y-z=-1. \\ z=0 \end{cases}$$
$$y-0=-1$$
$$y=-1$$
$$x-(-1)-0=0$$
$$x+1=0$$
$$x=-1$$
The solution is (–1, –1, 0).

17. $\begin{cases} 2x-\ y+3z=4 \\ 3x\qquad-3z=-2 \\ -5x+\ y\qquad=0 \end{cases}$

$$\left[\begin{array}{ccc|c} 2 & -1 & 3 & 4 \\ 3 & 0 & -3 & -2 \\ -5 & 1 & 0 & 0 \end{array}\right]$$
Divide row 1 by 2.
$$\left[\begin{array}{ccc|c} 1 & -\frac{1}{2} & \frac{3}{2} & 2 \\ 3 & 0 & -3 & -2 \\ -5 & 1 & 0 & 0 \end{array}\right]$$
Multiply row 1 by –3 and add to row 2.
Multiply row 1 by 5 and add to row 3.
$$\left[\begin{array}{ccc|c} 1 & -\frac{1}{2} & \frac{3}{2} & 2 \\ 0 & \frac{3}{2} & -\frac{15}{2} & -8 \\ 0 & -\frac{3}{2} & \frac{15}{2} & 10 \end{array}\right]$$
Multiply row 2 by $\frac{2}{3}$.

$$\left[\begin{array}{ccc|c} 1 & -\frac{1}{2} & \frac{3}{2} & 2 \\ 0 & 1 & -5 & -\frac{16}{3} \\ 0 & -\frac{3}{2} & \frac{15}{2} & 10 \end{array}\right]$$

Multiply row 2 by $\frac{3}{2}$ and add to row 3.

$$\left[\begin{array}{ccc|c} 1 & -\frac{1}{2} & \frac{3}{2} & 2 \\ 0 & 1 & -5 & -\frac{16}{3} \\ 0 & 0 & 0 & 2 \end{array}\right]$$

This corresponds to

$$\begin{cases} x - \frac{1}{2}y + \frac{3}{2}z = 2 \\ y - 5z = -\frac{16}{3} \\ 0 = 2 \end{cases}$$

This is an inconsistent system.
The solution is $\varnothing$.

18. $R(x) = 38x$
$C(x) = 18x + 5500$
$38x = 18x + 5500$
$20x = 5500$
$x = 275$
275 frames must be sold.

19. Let x = double occupancy rooms
y = single occupancy rooms

$$\begin{cases} x + y = 80 \\ 90x + 80y = 6930 \end{cases}$$

Multiply the first equation by –80 and add to the second equation.

$$\begin{array}{r} -80x - 80y = -6400 \\ 90x + 80y = 6930 \\ \hline 10x \quad\quad = 530 \\ x \quad\quad = 53 \end{array}$$

Replace x with 53 in the first equation.
$53 + y = 80$
$y = 27$
53 double-occupancy and
27 single-occupancy rooms are occupied.

20. Let x = number of gallons of 10% solution
y = number of gallons of 20% solution

$$\begin{cases} x + y = 20 \\ 0.10x + 0.20y = 0.175(20) \end{cases}$$

Multiply the first equation by –0.10 and add to the second equation.

$$\begin{array}{r} -0.10x - 0.10y = -2.0 \\ 0.10x + 0.20y = 3.5 \\ \hline 0.10y = 1.5 \\ y = 15 \end{array}$$

Replace y with 15 in the first equation.
$x + 15 = 20$
$x = 5$
They should use 5 gallons of 10% fructose solution and 15 gallons of the 20% solution.

21. $R(x) = 4x$ and $C(x) = 1.5x + 2000$
Break even occurs when $R(x) = C(x)$.
$4x = 1.5x + 2000$
$2.5x = 2000$
$x = 800$
The company must sell 800 packages to break even.

Chapter 5

Section 5.1

Calculator Explorations

1. $(3\times10^{11})(2\times10^{32})=6\times10^{43}$

3. $(5.2\times10^{23})(7.3\times10^{4})=3.796\times10^{28}$

Mental Math

1. $5x^{-1}y^{-2}=\frac{5}{xy^2}$

2. $7xy^{-4}=\frac{7x}{y^4}$

3. $a^2b^{-1}c^{-5}=\frac{a^2}{bc^5}$

4. $a^{-4}b^2c^{-6}=\frac{b^2}{a^4c^6}$

5. $\frac{y^{-2}}{x^{-4}}=\frac{x^4}{y^2}$

6. $\frac{x^{-7}}{z^{-3}}=\frac{z^3}{x^7}$

Exercise Set 5.1

1. $4^2\cdot4^3=4^{2+3}=4^5$

3. $x^5\cdot x^3=x^{5+3}=x^8$

5. $-7x^3\cdot20x^9=-7\cdot20x^{3+9}=-140x^{12}$

7. $(4xy)(-5x)=-20x^{1+1}y=-20x^2y$

9. $(-4x^3p^2)(4y^3x^3)=-16x^{3+3}y^3p^2$
$=-16x^6y^3p^2$

11. $-8^0=-(8^0)=-1$

13. $(4x+5)^0=1$

15. $(5x)^0+5x^0=1+5(1)=1+5=6$

17. Answers may vary.

19. $\frac{a^5}{a^2}=a^{5-2}=a^3$

21. $\frac{x^9y^6}{x^8y^6}=x^{9-8}y^{6-6}=x^1y^0=x$

23. $-\frac{26z^{11}}{2z^7}=-13z^{11-7}=-13z^4$

25. $\frac{-36a^5b^7c^{10}}{6ab^3c^4}=-6a^{5-1}b^{7-3}c^{10-4}$
$=-6a^4b^4c^6$

27. $4^{-2}=\frac{1}{4^2}=\frac{1}{16}$

29. $\frac{x^7}{x^{15}}=x^{7-15}=x^{-8}=\frac{1}{x^8}$

31. $5a^{-4}=5\frac{1}{a^4}=\frac{5}{a^4}$

33. $\frac{x^{-2}}{x^5}=x^{-2-5}=x^{-7}=\frac{1}{x^7}$

35. $\frac{8r^4}{2r^{-4}}=4r^{4-(-4)}=4r^8$

37. $\frac{x^{-9}x^4}{x^{-5}}=\frac{x^{-9+4}}{x^{-5}}$
$=\frac{x^{-5}}{x^{-5}}$
$=x^{-5-(-5)}$
$=x^0$
$=1$

39. $4^{-1}+3^{-2}=\frac{1}{4^1}+\frac{1}{3^2}$
$=\frac{1}{4}+\frac{1}{9}$
$=\frac{9}{36}+\frac{4}{36}$
$=\frac{13}{36}$

41. $4x^0+5=4(1)+5=4+5=9$

43. $x^7\cdot x^8=x^{7+8}=x^{15}$

45. $2x^3\cdot 5x^7=2\cdot 5x^{3+7}=10x^{10}$

47. $\frac{z^{12}}{z^{15}}=z^{12-15}=z^{-3}=\frac{1}{z^3}$

49. $\frac{y^{-3}}{y^{-7}}=y^{-3-(-7)}=y^4$

51. $3x^{-1}=3\cdot\frac{1}{x}=\frac{3}{x}$

53. $3^0-3t^0=1-3\cdot 1=1-3=-2$

55. $\frac{r^4}{r^{-4}}=r^{4-(-4)}=r^8$

57. $\frac{x^{-7}y^{-2}}{x^2y^2}=x^{-7-2}y^{-2-2}=x^{-9}y^{-4}=\frac{1}{x^9y^4}$

59. $\frac{2a^{-6}b^2}{18ab^{-5}}=\frac{a^{-6-1}b^{2-(-5)}}{9}=\frac{a^{-7}b^7}{9}=\frac{b^7}{9a^7}$

61. $\frac{(24x^8)x}{20x^{-7}}=\frac{6x^{8+1}}{5x^{-7}}=\frac{6x^{9-(-7)}}{5}=\frac{6x^{16}}{5}$

63. $31{,}250{,}000=3.125\times 10^7$

65. $0.016=1.6\times 10^{-2}$

67. $67{,}413=6.7413\times 10^4$

69. $0.0125=1.25\times 10^{-2}$

71. $0.000053=5.3\times 10^{-5}$

73. $3.6\times 10^{-9}=0.0000000036$

75. $9.3\times 10^7=93{,}000{,}000$

77. $1.278\times 10^6=1{,}278{,}000$

79. $7.35\times 10^{12}=7{,}350{,}000{,}000{,}000$

81. $4.03\times 10^{-7}=0.000000403$

83. Answers may vary.

85. *a*, *c*, and *d*

87. $778{,}300{,}000=7.783\times 10^8$

89. $43{,}141{,}000=4.3141\times 10^7$

91. $1{,}130{,}000{,}000=1.13\times 10^9$

93. $0.001=1\times 10^{-3}$

95. $x^5\cdot x^{7a}=x^{5+7a}$

97. $\frac{x^{3t-1}}{x^t}=x^{3t-1-t}=x^{2t-1}$

99. $x^{4a}\cdot x^7=x^{4a+7}$

101. $\frac{z^{6x}}{z^7}=z^{6x-7}$

103. $\frac{x^{3t}\cdot x^{4t-1}}{x^t}=\frac{x^{3t+4t-1}}{x^t}=x^{7t-1-t}=x^{6t-1}$

105. $x^{9+b}\cdot x^{3a-b}=x^{9+3a}$

107. 7^{13}

109. 7^{-11}

111. $(5\cdot 2)^2=(10)^2=100$

113. $\left(\frac{3}{4}\right)^3 = \frac{3^3}{4^3} = \frac{27}{64}$

115. $(2^3)^2 = 8^2 = 64$

117. $(2^{-1})^4 = \left(\frac{1}{2}\right)^4 = \frac{1^4}{2^4} = \frac{1}{16}$

Section 5.2

Mental Math

1. $(x^4)^5 = x^{4(5)} = x^{20}$

2. $(5^6)^2 = 5^{6(2)} = 5^{12}$

3. $x^4 \cdot x^5 = x^{4+5} = x^9$

4. $x^7 \cdot x^8 = x^{7+8} = x^{15}$

5. $(y^6)^7 = y^{6(7)} = y^{42}$

6. $(x^3)^4 = x^{3(4)} = x^{12}$

7. $(z^4)^5 = z^{4(5)} = z^{20}$

8. $(z^3)^7 = z^{3(7)} = z^{21}$

9. $(z^{-6})^{-3} = z^{-6(-3)} = z^{18}$

10. $(y^{-4})^{-2} = y^{-4(-2)} = y^8$

Exercise Set 5.2

1. $(3^{-1})^2 = 3^{-1(2)} = 3^{-2} = \frac{1}{3^2} = \frac{1}{9}$

3. $(x^4)^{-9} = x^{4(-9)} = x^{-36} = \frac{1}{x^{36}}$

5. $(y)^{-5} = \frac{1}{y^5}$

7. $$\begin{aligned}(3x^2y^3)^2 &= 3^2(x^2)^2(y^3)^2\\ &= 9x^{2(2)}y^{3(2)}\\ &= 9x^4y^6\end{aligned}$$

9. $$\begin{aligned}\left(\frac{2x^5}{y^{-3}}\right)^4 &= \frac{2^4(x^5)^4}{(y^{-3})^4}\\ &= \frac{16x^{5(4)}}{y^{-3(4)}}\\ &= \frac{16x^{20}}{y^{-12}}\\ &= 16x^{20}y^{12}\end{aligned}$$

11. $$\begin{aligned}(a^2bc^{-3})^{-6} &= (a^2)^{-6}b^{-6}(c^{-3})^{-6}\\ &= a^{2(-6)}b^{-6}c^{-3(-6)}\\ &= a^{-12}b^{-6}c^{18}\\ &= \frac{c^{18}}{a^{12}b^6}\end{aligned}$$

13. $$\begin{aligned}\left(\frac{x^7y^{-3}}{z^{-4}}\right)^{-5} &= \frac{(x^7)^{-5}(y^{-3})^{-5}}{(z^{-4})^{-5}}\\ &= \frac{x^{-35}y^{15}}{z^{20}}\\ &= \frac{y^{15}}{x^{35}z^{20}}\end{aligned}$$

15. $$\begin{aligned}\left(\frac{a^{-4}}{a^{-5}}\right)^{-2} &= \frac{(a^{-4})^{-2}}{(a^{-5})^{-2}}\\ &= \frac{a^8}{a^{10}}\\ &= a^{8-10}\\ &= a^{-2}\\ &= \frac{1}{a^2}\end{aligned}$$

17. $\left(\dfrac{2a^{-2}b^5}{4a^2b^7}\right)^{-2} = \left(\dfrac{1}{2a^4b^2}\right)^{-2}$
$= \dfrac{(1)^{-2}}{(2)^{-2}(a^4)^{-2}(b^2)^{-2}}$
$= \dfrac{2^2}{1^2a^{-8}b^{-4}}$
$= 4a^8b^4$

19. $\dfrac{4^{-1}x^2yz}{x^{-2}yz^3} = \dfrac{x^{2-(-2)}y^0}{4z^{3-1}} = \dfrac{x^4}{4z^2}$

21. Yes; $a = \pm 1$.

23. $(5^{-1})^3 = 5^{-3} = \dfrac{1}{5^3} = \dfrac{1}{125}$

25. $(x^7)^{-9} = x^{-63} = \dfrac{1}{x^{63}}$

27. $\left(\dfrac{7}{8}\right)^3 = \dfrac{7^3}{8^3} = \dfrac{343}{512}$

29. $(4x^2)^2 = 4^2(x^2)^2 = 16x^4$

31. $(-2^{-2}y)^3 = (-2^{-2})^3y^3$
$= -2^{-6}y^3$
$= -\dfrac{y^3}{2^6}$
$= -\dfrac{y^3}{64}$

33. $\left(\dfrac{4^{-4}}{y^3x}\right)^{-2} = \dfrac{(4^{-4})^{-2}}{(y^3)^{-2}x^{-2}}$
$= \dfrac{4^8}{y^{-6}x^{-2}}$
$= 4^8x^2y^6$

35. $\left(\dfrac{6p^6}{p^{12}}\right)^2 = \dfrac{6^2p^{12}}{p^{24}}$
$= 6^2p^{12-24}$
$= 36p^{-12}$
$= \dfrac{36}{p^{12}}$

37. $(-8y^3xa^{-2})^{-3} = (-8)^{-3}(y^3)^{-3}x^{-3}(a^{-2})^{-3}$
$= -\dfrac{y^{-9}a^6}{8^3x^3}$
$= -\dfrac{a^6}{512x^3y^9}$

39. $\left(\dfrac{x^{-2}y^{-2}}{a^{-3}}\right)^{-7} = \dfrac{(x^{-2})^{-7}(y^{-2})^{-7}}{(a^{-3})^{-7}}$
$= \dfrac{x^{14}y^{14}}{a^{21}}$

41. $\left(\dfrac{3x^5}{6x^4}\right)^4 = \left(\dfrac{x^{5-4}}{2}\right)^4 = \dfrac{(x^1)^4}{2^4} = \dfrac{x^4}{16}$

43. $\left(\dfrac{1}{4}\right)^{-3} = (4^{-1})^{-3} = 4^3 = 64$

45. $\dfrac{(y^3)^{-4}}{y^3} = y^{-12-3} = y^{-15} = \dfrac{1}{y^{15}}$

47. $\dfrac{8p^7}{4p^9} = 2p^{7-9} = 2p^{-2} = \dfrac{2}{p^2}$

49. $(4x^6y^5)^{-2}(6x^4y^3)$
$= 4^{-2}(x^6)^{-2}(y^5)^{-2}(6x^4y^3)$
$= \dfrac{1}{4^2}x^{-12}y^{-10}\cdot 6x^4y^3$
$= \dfrac{6}{16}x^{-12+4}y^{-10+3}$
$= \dfrac{3x^{-8}y^{-7}}{8}$
$= \dfrac{3}{8x^8y^7}$

51. $x^6(x^6bc)^{-6} = x^6(x^6)^{-6}b^{-6}c^{-6}$
$= \frac{x^6 \cdot x^{-36}}{b^6c^6}$
$= \frac{x^{-30}}{b^6c^6}$
$= \frac{1}{x^{30}b^6c^6}$

53. $\frac{2^{-3}x^2y^{-5}}{5^{-2}x^7y^{-1}} = \frac{5^2x^{2-7}y^{-5-(-1)}}{2^3}$
$= \frac{25x^{-5}y^{-4}}{8}$
$= \frac{25}{8x^5y^4}$

55. $\left(\frac{2x^2}{y^4}\right)^3 \cdot \left(\frac{2x^5}{y}\right)^{-2} = \frac{2^3(x^2)^3 2^{-2}(x^5)^{-2}}{(y^4)^3 y^{-2}}$
$= \frac{8x^6x^{-10}}{2^2y^{12}y^{-2}}$
$= \frac{8x^{6-10}}{4y^{12-2}}$
$= \frac{2x^{-4}}{y^{10}}$
$= \frac{2}{x^4y^{10}}$

57. $(5\times10^{11})(2.9\times10^{-3}) = 5\times2.9\times10^{11-3}$
$= 14.5\times10^8$
$= 1.45\times10^1\times10^8$
$= 1.45\times10^9$

59. $(2\times10^5)^3 = 2^3\times10^{5\cdot3} = 8\times10^{15}$

61. $\frac{3.6\times10^{-4}}{9\times10^2} = 0.4\times10^{-4-2}$
$= 0.4\times10^{-6}$
$= 4\times10^{-1}\times10^{-6}$
$= 4\times10^{-7}$

63. $\frac{0.0069}{0.023} = \frac{6.9\times10^{-3}}{2.3\times10^{-2}}$
$= 3\times10^{-3-(-2)}$
$= 3\times10^{-1}$

65. $\frac{18,200\times100}{91,000} = \frac{1.82\times10^4\times1\times10^2}{9.1\times10^4}$
$= \frac{1.82\times10^6}{9.1\times10^4}$
$= 0.2\times10^{6-4}$
$= 2\times10^{-1}\times10^2$
$= 2\times10^{2-1}$
$= 2\times10^1$

67. $\frac{6000\times0.006}{0.009\times400} = \frac{(6\times10^3)(6\times10^{-3})}{(9\times10^{-3})(4\times10^2)}$
$= \frac{36\times10^{3-3}}{36\times10^{-3+2}}$
$= \frac{10^0}{10^{-1}}$
$= 10^{0-(-1)}$
$= 10^1$
$= 1\times10^1$

69. $\frac{0.00064\times2000}{16,000} = \frac{(6.4\times10^{-4})(2\times10^3)}{1.6\times10^4}$
$= \frac{12.8\times10^{-4+3}}{1.6\times10^4}$
$= 8\times10^{-1-4}$
$= 8\times10^{-5}$

71. $\frac{66,000\times0.001}{0.002\times0.003} = \frac{(6.6\times10^4)(1\times10^{-3})}{(2\times10^{-3})(3\times10^{-3})}$
$= \frac{6.6\times10^{4-3}}{6\times10^{-3-3}}$
$= \frac{1.1\times10^1}{10^{-6}}$
$= 1.1\times10^{1-(-6)}$
$= 1.1\times10^7$

73. $\dfrac{8.25\times10^{15}}{(2.5\times10^{-2})(2.2\times10^{-5})}$
$=\dfrac{8.25\times10^{15}}{5.5\times10^{-7}}$
$=1.5\times10^{15-(-7)}$
$=1.5\times10^{22}$

75. $200{,}000\times10^{-8}=2\times10^{5}\times10^{-8}$
$=2\times10^{-3}$
$=0.002$
It would take 0.002 seconds.

77. $D=\dfrac{M}{V}$ or $M=DV$
$M=(3.12\times10^{-2})(4.269\times10^{14})$
$=13.31928\times10^{12}$
$=1.331928\times10^{13}$
The mass is 1.331928×10^{13} tons.

79. Recall $V=s^3$
$=\left(\dfrac{2x^{-2}}{y}\right)^3$
$=\dfrac{2^3(x^{-2})^3}{y^3}$
$=\dfrac{8x^{-6}}{y^3}$
$=\dfrac{8}{x^6y^3}$
The volume is $\dfrac{8}{x^6y^3}$ cubic meters.

81. $(4\times10^{-2})(6.452\times10^{-4})$
$=25.808\times10^{-6}$
$=2.5808\times10^{-5}$
The area is 2.5808×10^{-5} square meters.

83. Answers may vary.

85. $\dfrac{2.93\times10^{6}}{4.25\times10^{5}}\approx0.7\times10^{6-5}$
$=7.0\times10^{-1}\times10^{1}$
$=7\times10^{0}$
$=7$
China's fighting force numbers approximately 7 times Taiwan's.

87. $(x^{2b+7})^2=x^{(2b+7)2}=x^{4b+14}$

89. $\dfrac{x^{-5y+2}x^{2y}}{x}=x^{-5y+2+2y-1}=x^{-3y+1}$

91. $(c^{2a+3})^3=c^{(2a+3)3}=c^{6a+9}$

93. $\dfrac{(y^{4a})^7}{y^{2a-1}}=\dfrac{y^{28a}}{y^{2a-1}}$
$=y^{28a-(2a-1)}$
$=y^{28a-2a+1}$
$=y^{26a+1}$

95. $\left(\dfrac{3y^{5a}}{y^{-a+1}}\right)^2=(3y^{5a-(-a+1)})^2$
$=(3y^{6a-1})^2$
$=3^2y^{(6a-1)2}$
$=9y^{12a-2}$

97. $\dfrac{(y^{3-a})^b}{(y^{1-b})^a}=\dfrac{y^{3b-ab}}{y^{a-ab}}$
$=y^{3b-ab-(a-ab)}$
$=y^{3b-ab-a+ab}$
$=y^{3b-a}$

99. $\dfrac{x^{-5-3a}y^{-2a-b}}{x^{-5+3b}y^{-2b-a}}$
$= \left(\dfrac{x^{-5-3a}}{x^{-5+3b}}\right)\left(\dfrac{y^{-2a-b}}{y^{-2b-a}}\right)$
$= x^{-5-3a-(-5+3b)}y^{-2a-b-(-2b-a)}$
$= x^{-5-3a+5-3b}y^{-2a-b+2b+a}$
$= x^{-3a-3b}y^{b-a}$

101. $12m - 14 - 15m - 1 = -3m - 15$

103. $-9y - (5 - 6y) = -9y - 5 + 6y = -3y - 5$

105. $5(x-3) - 4(2x-5) = 5x - 15 - 8x + 20$
$= -3x + 5$

Section 5.3

Graphing Calculator Explorations

1. $(2x^2 + 7x + 6) + (x^3 - 6x^2 - 14)$
$= x^3 - 4x^2 + 7x - 8$

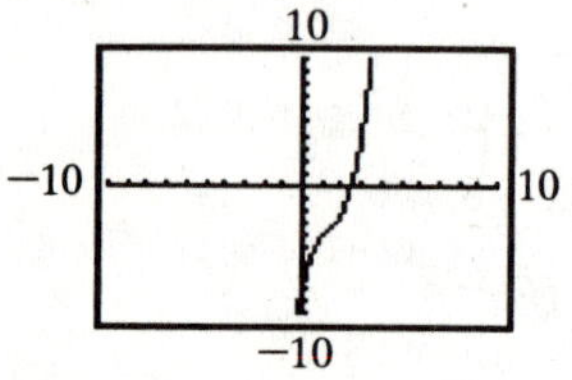

3. $(1.8x^2 - 6.8x - 1.7) - (3.9x^2 - 3.6x)$
$= -2.1x^2 - 3.2x - 1.7$

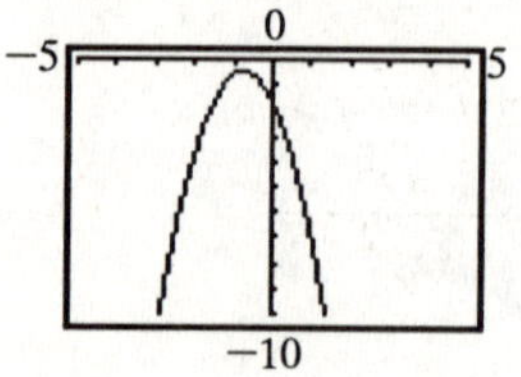

5. $(1.29x - 5.68)$
$+ (7.69x^2 - 2.55x + 10.98)$
$= 7.69x^2 - 1.26x + 5.3$

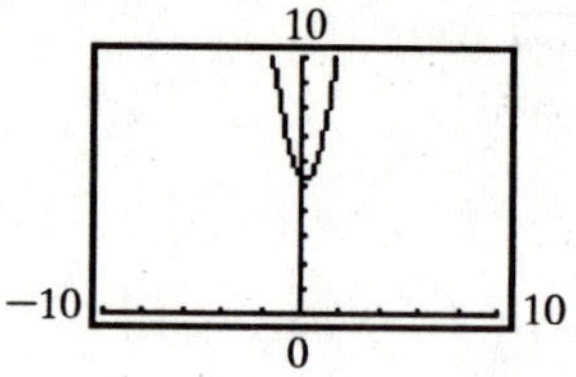

Exercise Set 5.3

1. 4 has degree 0.

3. $5x^2$ has degree 2.

5. $-3xy^2$ has degree $1 + 2 = 3$.

7. $6x + 3$ has degree 1 and is a binomial.

9. $3x^2 - 2x + 5$ has degree 2 and is a trinomial.

11. $-xyz$ has degree $1 + 1 + 1 = 3$ and is a monomial.

13. $x^2y - 4xy^2 + 5x + y$ has degree $2 + 1 = 3$ and is none of these.

15. Answers may vary.

17. $P(x) = x^2 + x + 1$
$P(7) = 7^2 + 7 + 1 = 49 + 7 + 1 = 57$

19. $Q(x) = 5x^2 - 1$
$Q(-10) = 5(-10)^2 - 1$
$= 5(100) - 1$
$= 500 - 1$
$= 499$

21. $P(x) = x^2 + x + 1$
$P(0) = 0^2 + 0 + 1 = 1$

23. $P(t) = -16t^2 + 1053$
$P(2) = -16(2)^2 + 1053$
$= -16(4) + 1053$
$= -64 + 1053$
$= 989$ ft

25. $P(t) = -16t^2 + 1053$
$P(6) = -16(6)^2 + 1053$
$= -16(36) + 1053$
$= -576 + 1053$
$= 477$ ft

27. $5y + y = 6y$

29. $4x + 7x - 3 = 11x - 3$

31. $4xy + 2x - 3xy - 1 = xy + 2x - 1$

33. $(9y^2 - 8) + (9y^2 - 9) = 18y^2 - 17$

35.
$$\begin{array}{r} x^2 + xy - y^2 \\ 2x^2 - 4xy + 7y^2 \\ \hline 3x^2 - 3xy + 6y^2 \end{array}$$

37.
$$\begin{array}{r} x^2 - 6x + 3 \\ + \quad (2x + 5) \\ \hline x^2 - 4x + 8 \end{array}$$

39.
$$\begin{array}{r} 9y^2 - 7y + 5 \\ - \; 8y^2 - 7y + 2 \\ \hline y^2 \qquad + 3 \end{array}$$

41.
$$\begin{array}{r} 4x^2 + 2x \\ - \; 6x^2 - 3x \\ \hline -2x^2 + 5x \end{array}$$

43.
$$\begin{array}{r} 3x^2 - 4x + 8 \\ -5x^2 \qquad - 7 \\ \hline -2x^2 - 4x + 15 \end{array}$$

45.
$$\begin{array}{r} 5x - 11 \\ + -x - 2 \\ \hline 4x - 13 \end{array}$$

47.
$$\begin{array}{r} 7x^2 + x + 1 \\ - \; 6x^2 + x - 1 \\ \hline x^2 \qquad + 2 \end{array}$$

49. $(7x^3 - 4x + 8) + (5x^3 + 4x + 8x)$
$= 7x^3 - 4x + 8 + 5x^3 + 12x$
$= 12x^3 + 8x + 8$

51.
$$\begin{array}{r} 9x^3 - 2x^2 + 4x - 7 \\ -2x^3 - 6x^2 - 4x + 3 \\ \hline 7x^3 + 4x^2 + 8x - 10 \end{array}$$

53.
$$\begin{array}{r} y^2 + 4yx + 7 \\ + -19y^2 + 7yx + 7 \\ \hline -18y^2 + 11yx + 14 \end{array}$$

55.
$$\begin{array}{r} 3x^3 - b + 2a - 6 \\ -4x^3 + b + 6a - 6 \\ \hline -x^3 \qquad + 8a - 12 \end{array}$$

57.
$$\begin{array}{r} 4x^2 - 6x + 2 \\ - -x^2 + 3x + 5 \\ \hline 5x^2 - 9x - 3 \end{array}$$

59.
$$\begin{array}{r} -3x + 8 \\ + -3x^2 + 3x - 5 \\ \hline -3x^2 \qquad + 3 \end{array}$$

61.
$$\begin{array}{r} 4x^2 + 7xy^2 - 3 \\ +2x^3 - x^2 + xy^2 \\ \hline 2x^3 + 3x^2 + 8xy^2 - 3 \end{array}$$

63.
$$\begin{array}{r} 6y^2 - 6y + 4 \\ - -y^2 - 6y + 7 \\ \hline 7y^2 \qquad - 3 \end{array}$$

65.
$$\begin{array}{r} 3x^2 + 15x + 8 \\ +(2x^2 + 7x + 8) \\ \hline 5x^2 + 22x + 16 \end{array}$$

67.
$$\begin{array}{r} 5q^4 - 2q^2 - 3q \\ + \underline{-6q^4 + 3q^2 \qquad +5} \\ -q^4 + q^2 - 3q + 5 \end{array}$$

69.
$$\begin{array}{r} 7x^2 + 4x + 9 \\ + \underline{8x^2 + 7x - 8} \\ 15x^2 + 11x + 1 \\ - \underline{\qquad 3x + 7} \\ 15x^2 + 8x - 6 \end{array}$$

71.
$$\begin{array}{r} 4x^4 - 7x^2 + 3 \\ + \underline{-3x^4 \qquad + 2} \\ x^4 - 7x^2 + 5 \end{array}$$

73.
$$\begin{array}{r} 8x^{2y} - 7x^y + 3 \\ + \underline{-4x^{2y} + 9x^y - 14} \\ 4x^{2y} + 2x^y - 11 \end{array}$$

75. $P(3) = -32(3) + 500$
$= -96 + 500$
$= 404$ ft per sec

77. $f(x) = 0.43x^2 + 164.6x + 949.3$

a. $f(5) = 0.43(5)^2 + 164.6(5) + 949.3$
$= \$1783.05$

b. $f(15) = 0.43(15)^2 + 164.6(15) + 949.3$
$= \$3515.05$

c. $f(30)$
$= 0.43(30)^2 + 164.6(30) + 949.3$
$= \$6274.30$

d. No, the amount of money spent is not rising at a steady rate. The first 15 years it increased \$3515.05 – \$949.3 = \$2565.75. The next 15 years it increased \$6274.30 – \$3515.05 = \$2759.25, which is more than the increase for the first 15 years.

79. $C(x) = 0.8x + 10{,}000$
$C(20{,}000) = 0.8(20{,}000) + 10{,}000$
$= \$26{,}000$

81. $P(t) = -16t^2 + 300t$

a. $P(1) = -16(1)^2 + 300(1) = 284$ feet

b. $P(2) = -16(2)^2 + 300(2) = 536$ feet

c. $P(3) = -16(3)^2 + 300(3) = 756$ feet

d. $P(4) = -16(4)^2 + 300(4) = 944$ feet

83.
$$\begin{aligned} 0 &= -16t^2 + 300t \\ 0 &= -4t(4t - 75) \\ 4t - 75 &= 0 \\ 4t &= 75 \\ t &= \frac{75}{4} = 18.75 \end{aligned}$$
19 sec

85. $f(x) = -0.85x^3 + 14.28x^2 - 49.38x + 574.16$

a. $f(5) = -0.85(5)^3 + 14.28(5)^2 - 49.38(5) + 574.16 \approx 578$

b. $f(8) = -0.85(8)^3 + 14.28(8)^2 - 49.38(8) + 574.16 \approx 658$

c. $f(13) = -0.85(13)^3 + 14.28(13)^2 - 49.38(13) + 574.16 \approx 478$

87. $P(x) + Q(x)$
$3x+3+4x^2-6x+3=4x^2-3x+6$

89. $Q(x) - R(x)$
$4x^2-6x+3-(5x^2-7)=-x^2-6x+10$

91. $2Q(x)-R(x)$
$=2(4x^2-6x+3)-(5x^2-7)$
$=8x^2-12x+6-5x^2+7$
$=3x^2-12x+13$

93. $3R(x)+4P(x)=3(5x^2-7)+4(3x+3)$
$=15x^2-21+12x+12$
$=15x^2+12x-9$

95. a. $P(x) = R(x) - C(x)$
$= 5.5x - (3x + 3000)$
$= 2.5x - 3000$

b. $P(2000) = 5000 - 3000 = \$2000$

97. $h(x)=5x^3-6x+2$; **B**

99. $g(x)=-2x^2-6x+2$; **C**

101. $(z+2)+2z^2+z+z^3-4z+1$
$=z^3+2z^2-2z+3$

103. $-7(2z-6y)=-14z+42y$

105. $-15y^2-10y+35$

107. $P(x) = 8x + 3$

a. $P(a) = 8a + 3$

b. $P(-x) = -8(-x) + 3 = -8x + 3$

c. $P(x + h) = 8(x + h) + 3 = 8x + 8h + 3$

109. $P(x) = -4x$

a. $P(a) = -4a$

b. $P(-x) = -4(-x) = 4x$

c. $P(x + h) = -4(x + h) = -4x - 4h$

111. $P(x) = 3x - 2$

a. $P(a) = 3a - 2$

b. $P(-x) = 3(-x) - 2 = -3x - 2$

c. $P(x + h) = 3(x + h) - 2 = 3x + 3h - 2$

Section 5.4

Graphing Calculator Explorations

1. $(x+4)(x-4)=x^2-16$

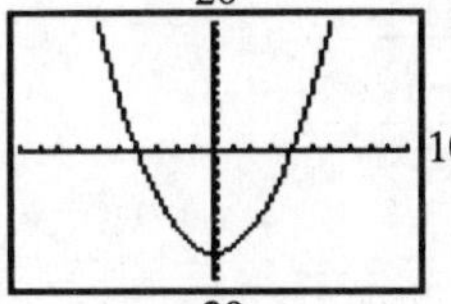

3. $(3x-7)^2=9x^2-42x+49$

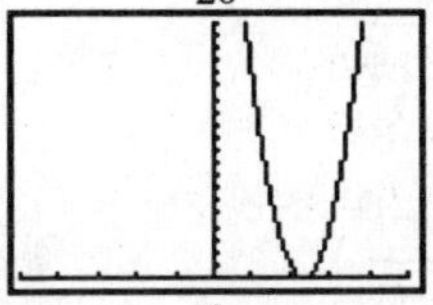

5. $(5x+1)(x^2-3x-2)$
$=5x^3-14x^2-13x-2$

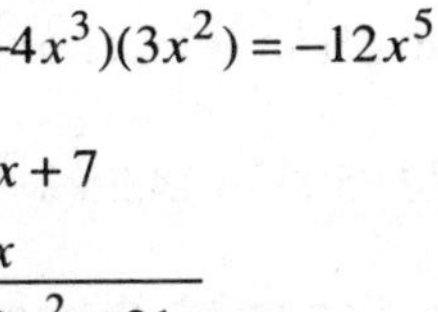

Exercise Set 5.4

1. $(-4x^3)(3x^2)=-12x^5$

3. $4x+7$
$\underline{3x}$
$12x^2+21x$

5. $-6xy(4x+y) = -6xy(4x) - 6xy(y)$
$= -24x^2y - 6xy^2$

7. $-4ab(xa^2 + ya^2 - 3)$
$= -4ab(xa^2) - 4ab(ya^2) - 4ab(-3)$
$= -4xa^3b - 4ya^3b + 12ab$

9.
$$\begin{array}{r} 2x+4 \\ \times \quad x-3 \\ \hline -6x-12 \\ 2x^2+4x \quad\;\; \\ \hline 2x^2-2x-12 \end{array}$$

11.
$$\begin{array}{r} x^3 \qquad\quad -x+2 \\ \times \qquad\qquad 2x+3 \\ \hline 3x^3 \qquad -3x+6 \\ 2x^4 \qquad -2x^2+4x \qquad\; \\ \hline 2x^4+3x^3-2x^2+x+6 \end{array}$$

13.
$$\begin{array}{r} 3x-2 \\ \times \quad 5x+1 \\ \hline 3x-2 \\ 15x^2-10x \quad\;\; \\ \hline 15x^2-7x-2 \end{array}$$

15.
$$\begin{array}{r} 3m^2+2m-1 \\ \times \qquad 5m+2 \\ \hline 6m^2+4m-2 \\ 15m^3+10m^2-5m \qquad \\ \hline 15m^3+16m^2-m-2 \end{array}$$

17. Answers may vary.

19.
$$\begin{array}{r} x-3 \\ \times \quad x+4 \\ \hline 4x-12 \\ x^2-3x \quad\;\; \\ \hline x^2+x-12 \end{array}$$

21.
$$\begin{array}{r} 5x+8y \\ \times \quad 2x-y \\ \hline -5xy-8y^2 \\ 10x^2+16xy \qquad\; \\ \hline 10x^2+11xy-8y^2 \end{array}$$

23.
$$\begin{array}{r} 3x-1 \\ \times \quad x+3 \\ \hline 9x-3 \\ 3x^2-x \quad\;\; \\ \hline 3x^2+8x-3 \end{array}$$

25.
$$\begin{array}{r} 3x+\frac{1}{2} \\ \times \quad 3x-\frac{1}{2} \\ \hline -\frac{3}{2}x-\frac{1}{4} \\ 9x^2+\frac{3}{2}x \qquad \\ \hline 9x^2 \qquad -\frac{1}{4} \end{array}$$

27. $(x+4)^2 = x^2 + 2(x)4 + 4^2 = x^2 + 8x + 16$

29. $(6y-1)(6y+1) = (6y)^2 - 1^2 = 36y^2 - 1$

31. $(3x-y)^2 = (3x)^2 - 2(3x)y + y^2$
$= 9x^2 - 6xy + y^2$

33. $(3b-6y)(3b+6y) = 9b^2 - 36y^2$

35. $[3+(4b+1)]^2$
$= 3^2 + 2(3)(4b+1) + (4b+1)^2$
$= 9 + 6(4b+1) + (4b)^2 + 2(4b)1 + 1^2$
$= 9 + 24b + 6 + 16b^2 + 8b + 1$
$= 16b^2 + 32b + 16$

37. $[(2s-3)-1][(2s-3)+1]$
$= (2s-3)^2 - 1^2$
$= (2s)^2 - 2(2s)3 + 3^2 - 1$
$= 4s^2 - 12s + 9 - 1$
$= 4s^2 - 12s + 8$

39. $[(xy+4)-6]^2$
$= (xy+4)^2 - 2(xy+4)6 + 6^2$
$= (xy)^2 + 2(xy)4 + 4^2 - 12(xy+4) + 36$
$= x^2y^2 + 8xy + 16 - 12xy - 48 + 36$
$= x^2y^2 - 4xy + 4$

41. Answers may vary.

43.
$$\begin{array}{r} 3x+1 \\ \times \quad 3x+5 \\ \hline 15x+5 \\ 9x^2+\ 3x \quad\; \\ \hline 9x^2+18x+5 \end{array}$$

45.
$$\begin{array}{r} 2x^3 \qquad\qquad +5 \\ \times \qquad 5x^2+\ 4x+1 \\ \hline 2x^3 \qquad\qquad +5 \\ 8x^4 \qquad\qquad +20x \quad \\ 10x^5 \qquad\qquad +25x^2 \qquad\qquad \\ \hline 10x^5+8x^4+2x^3+25x^2+20x+5 \end{array}$$

47. $(7x+3)(7x-3) = (7x)^2 - 3^2$
$= 49x^2 - 9$

49.
$$\begin{array}{r} 3x^2+\ 4x-\ 4 \\ \times \qquad 3x+\ 6 \\ \hline 18x^2+24x-24 \\ 9x^3+12x^2-12x \qquad \\ \hline 9x^3+30x^2+12x-24 \end{array}$$

51.
$$\begin{array}{r} 4x+\frac{1}{3} \\ \times \qquad 4x-\frac{1}{2} \\ \hline -2x-\frac{1}{6} \\ 16x^2+\frac{4}{3}x \qquad \\ \hline 16x^2-\frac{2}{3}x-\frac{1}{6} \end{array}$$

53. $(6x+1)^2 = (6x)^2 + 2(6x)1 + 1^2$
$= 36x^2 + 12x + 1$

55. $(x^2+2y)(x^2-2y) = (x^2)^2 - (2y)^2$
$= x^4 - 4y^2$

57.
$$\begin{array}{r} 5a^2b^2-6a-6b \\ \times \qquad -6a^2b^2 \\ \hline -30a^4b^4+36a^3b^2+36a^2b^3 \end{array}$$

59.
$$\begin{array}{r} a-4 \\ \times \qquad 2a-4 \\ \hline -4a+16 \\ 2a^2-\ 8a \qquad \\ \hline 2a^2-12a+16 \end{array}$$

61. $(7ab+3c)(7ab-3c) = (7ab)^2 - (3c)^2$
$= 49a^2b^2 - 9c^2$

63. $(m-4)^2 = m^2 - 2(m)4 + 4^2$
$= m^2 - 8m + 16$

65. $(3x+1)^2 = (3x)^2 + 2(3x)1 + 1^2$
$= 9x^2 + 6x + 1$

67.
$$\begin{array}{r} y-\ 4 \\ \times \qquad y-\ 3 \\ \hline -3y+12 \\ y^2-4y \qquad \\ \hline y^2-7y+12 \end{array}$$

69.
$$\begin{array}{r} x+y \\ \times \quad 2x-1 \\ \hline -x-y \\ 2x^2+2xy \quad\quad \\ \hline 2x^2+2xy-x-y \end{array}$$

Then,
$$\begin{array}{r} 2x^2+2xy-x-y \\ x+1 \\ \hline 2x^2+2xy-x-y \\ 2x^3+2x^2y+x^2-xy \quad\quad\quad \\ \hline 2x^3+2x^2y+3x^2+xy-x-y \end{array}$$

71.
$$\begin{array}{r} 3x^2+2x-1 \\ \times \quad 3x^2+2x-1 \\ \hline -3x^2-2x+1 \\ 6x^3+4x^2-2x \quad\quad \\ 9x^4+6x^3-3x^2 \quad\quad\quad\quad \\ \hline 9x^4+12x^3-2x^2-4x+1 \end{array}$$

73.
$$\begin{array}{r} 4x^2-2x+5 \\ \times \quad 3x+1 \\ \hline 4x^2-2x+5 \\ 12x^3-6x^2+15x \quad\quad \\ \hline 12x^3-2x^2+13x+5 \end{array}$$

75. $P(x) \cdot R(x) = (5x) \cdot (x+5)$
$$\begin{array}{r} x+5 \\ \times \quad 5x \\ \hline 5x^2+25x \end{array}$$

77.
$$\begin{aligned} [Q(x)]^2 &= (x^2-2)^2 \\ &= (x^2-2)(x^2-2) \\ &= (x^2)(x^2)-4(x^2)+(-2)(-2) \\ &= x^4-4x^2+4 \end{aligned}$$

79. $R(x) \cdot Q(x) = (x+5) \cdot (x^2-2)$
$$\begin{array}{r} x^2 \quad\quad -2 \\ \times \quad\quad x+5 \\ \hline 5x^2 \quad\quad -10 \\ x^3 \quad\quad -2x \quad\quad \\ \hline x^3+5x^2-2x-10 \end{array}$$

81. a.
$$\begin{aligned} (3x+5)+(3x+7) &= 3x+3x+5+7 \\ &= 6x+12 \end{aligned}$$

b.
$$\begin{array}{r} 3x+5 \\ \times \quad 3x+7 \\ \hline 21x+35 \\ 9x^2+15x \quad\quad \\ \hline 9x^2+36x+35 \end{array}$$

Part **a** involves adding polynomials while part **b** involves multiplying.

83.
$$\begin{aligned} A &= \pi(5x-2)^2 \\ &= \pi[(5x)^2-2\cdot 5x\cdot 2+2^2] \\ &= \pi(25x^2-20x+4) \text{ sq. km} \end{aligned}$$

85. $f(x) = x^2-3x$
$f(a) = a^2-3a$

87. $f(x) = x^2-3x$
$$\begin{aligned} f(a+h) &= (a+h)^2-3(a+h) \\ &= a^2+2ah+h^2-3a-3h \end{aligned}$$

89. $f(x) = x^2-3x$
$$\begin{aligned} f(b-2) &= (b-2)^2-3(b-2) \\ &= b^2-2\cdot b\cdot 2+2^2-3b+6 \\ &= b^2-4b+4-3b+6 \\ &= b^2-7b+10 \end{aligned}$$

91. a. $f(x) = x^2 + 3x + 2$
$F(a+h) = (a+h)^2 + 3(a+h) + 2$
$= a^2 + 2ah + h^2 + 3a + 3h + 2$

b. $F(a) = a^2 + 3a + 2$

c. $F(a+h) - F(a)$
$= a^2 + 2ah + h^2 + 3a + 3h + 2 - (a^2 + 3a + 2)$
$= 2ah + h^2 + 3h$

93. $5xy^n(6y^{n+1} - 2)$
$= 30x^2y^{2n+1} - 10x^2y^n$

95. $(x^a + 5)(x^{2a} - 3)$
$= x^a \cdot x^{2a} + x^a(-3) + (5)x^{2a} - 15$
$= x^{3a} - 3x^a + 5x^{2a} - 15$

97. $y = -2x + 7$
$m = -2$

99. $3x - 5y = 14$
$-5y = -3x + 14$
$y = \frac{3}{5}x - \frac{14}{5}$
$m = \frac{3}{5}$

101. Since any vertical line only crosses the graph once it is a function.

Section 5.5

Mental Math

1. $6 = 2 \cdot 3$
$12 = 2 \cdot 2 \cdot 3$
$\text{GCF} = 2 \cdot 3 = 6$

2. $9 = 3 \cdot 3$
$27 = 3 \cdot 3 \cdot 3$
$\text{GCF} = 3 \cdot 3 = 9$

3. $15x = 3 \cdot 5 \cdot x$
$10 = 2 \cdot 5$
$\text{GCF} = 5$

4. $9x = 3 \cdot 3 \cdot x$
$12 = 2 \cdot 2 \cdot 3$
$\text{GCF} = 3$

5. $13x = 13 \cdot x$
$2x = 2 \cdot x$
$\text{GCF} = x$

6. $4y = 2 \cdot 2 \cdot y$
$5y = 5 \cdot y$
$\text{GCF} = y$

7. $7x = 7 \cdot x$
$14x = 2 \cdot 7 \cdot x$
$\text{GCF} = 7x$

8. $8z = 2 \cdot 2 \cdot 2 \cdot z$
$4z = 2 \cdot 2 \cdot z$
$\text{GCF} = 2 \cdot 2 \cdot z = 4z$

Exercise Set 5.5

1. a^8, a^5, and a^3 have a GCF a^3.

3. $x^2y^3z^3$, y^2z^3, xy^2z^2 have GCF y^2z^2.

5. $6x^3y$, $9x^2y^2$, $12x^2y$ have GCF $3x^2y$.

7. $10x^3yz^3$, $20x^2z^5$, $45xz^3$ have GCF $5xz^3$.

9. $18x - 12 = 6(3x - 2)$

11. $4y^2 - 16xy^3 = 4y^2(1 - 4xy)$

13. $6x^5 - 8x^4 + 2x^3 = 2x^3(3x^2 - 4x + 1)$

15. $8a^3b^3 - 4a^2b^2 + 4ab + 16ab^2$
$= 4ab(2a^2b^2 - ab + 1 + 4b)$

17. $6(x + 3) + 5a(x + 3) = (6 + 5a)(x + 3)$

19. $2x(z + 7) + (z + 7) = (2x + 1)(z + 7)$

21. $3x(x^2 + 5) - 2(x^2 + 5) = (3x - 2)(x^2 + 5)$

23. Answers may vary.

25. $ab + 3a + 2b + 6 = a(b + 3) + 2(b + 3)$
$= (a + 2)(b + 3)$

27. $ac + 4a - 2c + 8 = a(c + 4) - 2(c + 4)$
$= (a - 2)(c + 4)$

29. $2xy - 3x - 4y + 6 = x(2y - 3) - 2(2y - 3)$
$= (x - 2)(2y - 3)$

31. $12xy - 8x - 3y + 2 = 4x(3y - 2) - (3y - 2)$
$= (4x - 1)(3y - 2)$

33. $6x^3 + 9 = 3(2x^3 + 3)$

35. $x^3 + 3x^2 = x^2(x + 3)$

37. $8a^3 - 4a = 4a(2a^2 - 1)$

39. $-20x^2y + 16x^3y = -4xy(5x - 4y^2)$

41. $10a^2b^3 + 5ab^2 - 15ab^3$
$= 5ab^2(2ab + 1 - 3b)$

43. $9abc^2 + 6a^2bc - 6ab + 3bc$
$= 3b(3ac^2 + 2a^2c - 2a + c)$

45. $4x(y - 2) - 3(y - 2) = (4x - 3)(y - 2)$

47. $6xy + 10x + 9y + 15$
$= 2x(3y + 5) + 3(3y + 5)$
$= (2x + 3)(3y + 5)$

49. $xy + 3y - 5x - 15 = y(x + 3) - 5(x + 3)$
$= (y - 5)(x + 3)$

51. $6ab - 2a - 9b + 3 = 2a(3b - 1) - 3(3b - 1)$
$= (2a - 3)(3b - 1)$

53. $12xy + 18x + 2y + 3$
$= 6x(2y + 3) + 1(2y + 3)$
$= (6x + 1)(2y + 3)$

55. $2m(n - 8) - (n - 8) = (2m - 1)(n - 8)$

57. $15x^3y^2 - 18x^2y^2 = 3x^2y^2(5x - 6)$

59. $2x^2 + 3xy + 4x + 6y$
$= x(2x + 3y) + 2(2x + 3y)$
$= (x + 2)(2x + 3y)$

61. $5x^2 + 5xy - 3x - 3y = 5x(x + y) - 3(x + y)$
$= (5x - 3)(x + y)$

63. $x^3 + 3x^2 + 4x + 12 = x^2(x + 3) + 4(x + 3)$
$= (x^2 + 4)(x + 3)$

65. $x^3 - x^2 - 2x + 2 = x^2(x - 1) - 2(x - 1)$
$= (x^2 - 2)(x - 1)$

67. $2\pi r^2 + 2\pi rh = 2\pi r(r + h)$

69. $A = P + PRT$
$A = P(1 + RT)$

71. $h(t) = -16t^2 + 64t$

a. $h(t) = -16t(t - 4)$

b. $h(1) = -16(1)^2 + 64(1) = 48$
$h(1) = -16(1)(1 - 4) = 48$
48 ft

c. Answers may vary.

73. None

a. $(2 - x)(3 - y) = 6 - 2y - 3x + xy$
$= xy - 3x - 2y + 6$

b. $(-2 + x)(-3 + y) = 6 - 2y - 3x + xy$
$= xy - 3x - 2y + 6$

c. $(x-2)(y-3) = xy - 3x - 2y + 6$

d. $(-x+2)(-y+3) = xy - 3x - 2y + 6$

75. **a** is correct.

77. $(7y)(-2y^3) = 7 \cdot -2 \cdot y \cdot y^3 = -14y^4$

79. $(-2y^3)^4 = (-2)^4(y^3)^4 = 16y^{3\cdot4} = 16y^{12}$

81. $(x-7)(x-1) = x^2 - x - 7x + 7$
$= x^2 - 8x + 7$

83. $(x-4)(x+2) = x^2 + 2x - 4x - 8$
$= x^2 - 2x - 8$

85. $(s+8)(s+10) = s^2 + 10s + 8s + 80$
$= s^2 + 18s + 80$

87. $3y^n + 3y^{2n} + 5y^{8n}$
$= 3y^n + 3y^n(y^n) + 5y^n(y^{7n})$
$= y^n(3 + 3y^n + 5y^{7n})$

89. $3x^{5a} - 6x^{3a} + 9x^{2a}$
$= 3x^{2a}(x^{3a}) - 3x^{2a}(2x^a) + 3x^{2a}(3)$
$= 3x^{2a}(x^{3a} - 2x^a + 3)$

Section 5.6

Mental Math

1. $10 = 2 \cdot 5$
$7 = 2 + 5$
2 and 5

2. $12 = 2 \cdot 2 \cdot 3 = 2 \cdot 6$
$8 = 2 + 6$
2 and 6

3. $24 = 2 \cdot 2 \cdot 2 \cdot 3 = 8 \cdot 3$
$11 = 8 + 3$
8 and 3

4. $30 = 2 \cdot 3 \cdot 5 = 10 \cdot 3$
$13 = 10 + 3$
10 and 3

Exercise Set 5.6

1. $x^2 + 9x + 18 = (x+6)(x+3)$

3. $x^2 - 12x + 32 = (x-4)(x-8)$

5. $x^2 + 10x - 24 = (x+12)(x-2)$

7. $x^2 - 2x - 24 = (x-6)(x+4)$

9. Note that the GCF is 3, so that
$3x^2 - 18x + 24 = 3(x^2 - 6x + 8)$
$= 3(x-2)(x-4).$

11. Note that the GCF is 4z, so that
$4x^2z + 28xz + 40z = 4z(x^2 + 7x + 10)$
$= 4z(x+2)(x+5).$

13. Note that the GCF is 2, so that
$2x^2 + 30x - 108 = 2(x^2 + 15x - 54)$
$= 2(x+18)(x-3).$

15. $x^2 + bx + 6$
$6 = 2 \cdot 3$ or $6 = (-2)(-3)$
$6 = 1 \cdot 6$ or $6 = (-1)(-6)$
$(x+2)(x+3) = x^2 + 5x + 6$
$(x-2)(x-3) = x^2 - 5x + 6$
$(x+1)(x+6) = x^2 + 7x + 6$
$(x-1)(x-6) = x^2 - 7x + 6$
$b = \pm 5$ and ± 7

17. $5x^2 + 16x + 3 = (5x+1)(x+3)$

19. $2x^2 - 11x + 12 = (2x-3)(x-4)$

21. $2x^2 + 25x - 20$ is prime.

23. $4x^2 - 12x + 9 = (2x-3)(2x-3)$
$= (2x-3)^2$

25. Note that the GCF is 2, so that
$12x^2 + 10x - 50 = 2(6x^2 + 5x - 25)$
$= 2(3x-5)(2x+5).$

27. Note that the GCF is y^2, so that
$3y^4 - y^3 - 10y^2 = y^2(3y^2 - y - 10)$
$= y^2(3y+5)(y-2).$

29. Note that the GCF is $2x$, so that
$6x^3 + 8x^2 + 24x = 2x(3x^2 + 4x + 12).$

31. $x^2 + 8xz + 7z^2 = (x+z)(x+7z)$

33. $2x^2 - 5xy - 3y^2 = (2x+y)(x-3y)$

35. $x^2 - x - 12$
$a = 1, b = -1, c = -12$
Now $ac = -12$. The two numbers are –4 and 3.
$x^2 - x - 12 = x^2 - 4x + 3x - 12$
$= x(x-4) + 3(x-4)$
$= (x+3)(x-4)$

37. Note that the GCF is 2, so that
$28y^2 + 22y + 4 = 2(14y^2 + 11y + 2).$
$a = 14$, $b = 11$, and $c = 2$
Now $ac = 28$. The two numbers are 4 and 7.
$14y^2 + 11y + 2 = 14y^2 + 7y + 4y + 2$
$= 7y(2y+1) + 2(2y+1)$
$= (7y+2)(2y+1)$
So, $28y^2 + 22y + 4 = 2(7y+2)(2y+1).$

39. $2x^2 + 15x - 27$
$a = 2, b = 15, c = -27$
Now $ac = -54$.
The two numbers are 18 and –3.
$2x^2 + 15x - 27 = 2x^2 + 18x - 3x - 27$
$= 2x(x+9) - 3(x+9)$
$= (2x-3)(x+9)$

41. $3x^2 + bx + 5$
$3 = 1 \cdot 3$ or $3 = (-1)(-3)$
$5 = 1 \cdot 5$ or $5 = (-1)(-5)$
$(3x+1)(x+5) = 3x^2 + 16x + 5$
$(3x-1)(x-5) = 3x^2 - 16x + 5$
$(-3x+1)(-x+5) = 3x^2 - 16x + 5$
$(-3x-1)(-x-5) = 3x^2 + 16x + 5$
$(3x+5)(x+1) = 3x^2 + 8x + 5$
$(3x-5)(x-1) = 3x^2 - 8x + 5$
$(-3x+5)(-x+1) = 3x^2 - 8x + 5$
$(-3x-5)(-x-1) = 3x^2 + 8x + 5$
$b = \pm 8$ and ± 16

43. First, let $y = x^2$. Then we have
$x^4 + x^2 - 6 = y^2 + y - 6 = (y+3)(y-2).$
This yields $(x^2+3)(x^2-2)$.

45. First, let $y = 5x + 1$. Then we have
$(5x+1)^2 + 8(5x+1) + 7 = y^2 + 8y + 7$
$= (y+1)(y+7).$
This yields
$[(5x+1)+1][(5x+1)+7] = (5x+2)(5x+8).$

47. First, let $y = x^3$. Then we have
$x^6 - 7x^3 + 12 = y^2 - 7y + 12 = (y-4)(y-3).$
This yields $(x^3-4)(x^3-3)$.

49. First, let $y = a + 5$. Then we have
$(a+5)^2 - 5(a+5) - 24 = y^2 - 5y - 24$
$= (y-8)(y+3).$
This yields
$[(a+5)-8][(a+5)+3] = (a-3)(a+8).$

51. Note that the GCF is x, so that
$v(x) = 3x^3 - 2x^2 - 8x = x(3x^2 - 2x - 8)$
$= x(3x+4)(x-2).$

53. $x^2 - 24x - 81 = (x-27)(x+3)$

55. $x^2 - 15x - 54 = (x-18)(x+3)$

57. $3x^2 - 6x + 3 = 3(x^2 - 2x + 1)$
$= 3(x-1)(x-1)$
$= 3(x-1)^2$

59. $3x^2 - 5x - 2 = (3x+1)(x-2)$

61. $8x^2 - 26x + 15 = (4x-3)(2x-5)$

63. $18x^4 + 21x^3 + 6x^2 = 3x^2(6x^2 + 7x + 2)$
$= 3x^2(3x+2)(2x+1)$

65. $3a^2 + 12ab + 12b^2 = 3(a^2 + 4ab + 4b^2)$
$= 3(a + 2b)(a + 2b)$
$= 3(a + 2b)^2$

67. $x^2 + 4x + 5$ is prime.

69. $2(x + 4)^2 + 3(x + 4) - 5$
$= [2(x + 4) + 5][(x + 4) - 1]$
$= [2x + 8 + 5][x + 3]$
$= (2x + 13)(x + 3)$

71. $6x^2 - 49x + 30 = (3x - 2)(2x - 15)$

73. $x^4 - 5x^2 - 6 = (x^2 - 6)(x^2 + 1)$

75. $6x^3 - x^2 - x = x(6x^2 - x - 1)$
$= x(3x + 1)(2x - 1)$

77. $12a^2 - 29ab + 15b^2 = (4a - 3b)(3a - 5b)$

79. $9x^2 + 30x + 25 = (3x + 5)(3x + 5)$
$= (3x + 5)^2$

81. $3x^2y - 11xy + 8y = y(3x^2 - 11x + 8)$
$= y(3x - 8)(x - 1)$

83. $2x^2 + 2x - 12 = 2(x^2 + x - 6)$
$= 2(x + 3)(x - 2)$

85. $(x - 4)^2 + 3(x - 4) - 18$
$= [(x - 4) + 6][(x - 4) - 3]$
$= (x + 2)(x - 7)$

87. $2x^6 + 3x^3 - 9 = (2x^3 - 3)(x^3 + 3)$

89. $72xy^4 - 24xy^2z + 2xz^2$
$= 2x(36y^4 - 12y^2z + z^2)$
$= 2x(6y^2 - z)(6y^2 - z)$
$= 2x(6y^2 - z)^2$

91. $x^4 + 6x^3 + 5x^2 = x^2(x^2 + 6x + 5)$
$= x^2(x + 5)(x + 1)$

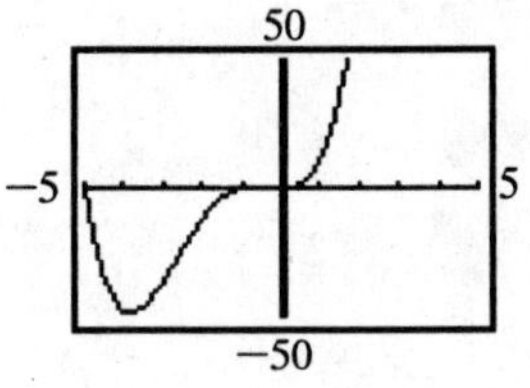

93. $30x^3 + 9x^2 - 3x = 3x(10x^2 + 3x - 1)$
$= 3x(5x - 1)(2x + 1)$

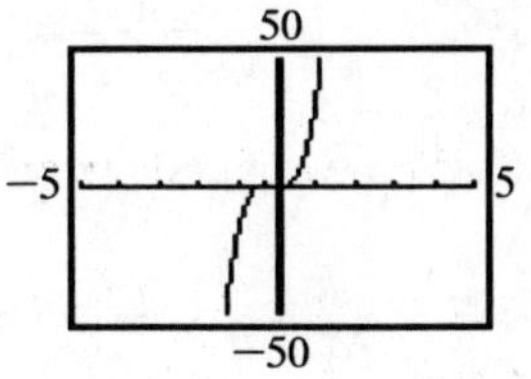

95.

$$\begin{array}{r} x^2 + 2x + 4 \\ x - 2 \\ \hline -2x^2 - 4x - 8 \\ x^3 + 2x^2 + 4x \phantom{{}-8} \\ \hline x^3 \phantom{{}+2x^2+4x} - 8 \end{array}$$

Answer is $x^3 - 8$.

97. $P(0) = 3(0)^2 + 2(0) - 9$
$= 0 + 0 - 9$
$= -9$

99. $P(-1) = 3(-1)^2 + 2(-1) - 9$
$= 3 - 2 - 9$
$= -8$

101. $x^{2n} + 10x^n + 16 = (x^n + 2)(x^n + 8)$

103. $x^{2n} - 3x^n - 18 = (x^n - 6)(x^n + 3)$

105. $2x^{2n} + 11x^n + 5 = (2x^n + 1)(x^n + 5)$

107. $4x^{2n} - 12x^n + 9 = (2x^n - 3)(2x^n - 3)$
$= (2x^n - 3)^2$

Exercise Set 5.7

1. $x^2+6x+9=(x+3)^2$

3. $4x^2-12x+9=(2x-3)^2$

5. $3x^2-24x+48=3(x^2-8x+16)$
$=3(x-4)^2$

7. $9y^2x^2+12yx^2+4x^2$
$=x^2(9y^2+12y+4)$
$=x^2(3y+2)^2$

9. $x^2-25=(x+5)(x-5)$

11. $9-4z^2=(3+2z)(3-2z)$

13. $(y+2)^2-49=[(y+2)-7][(y+2)+7]$
$=(y-5)(y+9)$

15. $64x^2-100=4(16x^2-25)$
$=4(4x+5)(4x-5)$

17. $x^3+27=x^3+3^3$
$=(x+3)(x^2-3x+9)$

19. $z^3-1=z^3-1^3=(z-1)(z^2+z+1)$

21. $m^3+n^3=(m+n)(m^2-mn+n^2)$

23. $x^3y^2-27y^2=(x^3-27)y^2$
$=(x^3-3^3)y^2$
$=(x-3)(x^2+3x+9)y^2$

25. $a^3b+8b^4=(a^3+8b^3)b$
$=(a^3+(2b)^3)b$
$=(a+2b)(a^2-2ab+4b^2)b$

27. $125y^3-8x^3$
$=(5y)^3-(2x)^3$
$=(5y-2x)(25y^2+10xy+4x^2)$

29. $(x^2+6x+9)-y^2=(x+3)^2-y^2$
$=(x+3+y)(x+3-y)$

31. $(x^2-10x+25)-y^2$
$=(x-5)^2-y^2$
$=(x-5-y)(x-5+y)$

33. $(4x^2+4x+1)-z^2$
$=(2x+1)^2-z^2$
$=(2x+1-z)(2x+1+z)$

35. $9x^2-49=(3x+7)(3x-7)$

37. $x^2-12x+36=(x-6)^2$

39. $x^4-81=(x^2+9)(x^2-9)$
$=(x^2+9)(x+3)(x-3)$

41. $(x^2+8x+16)-4y^2$
$=(x+4)^2-(2y)^2$
$=(x+4-2y)(x+4+2y)$

43. $(x+2y)^2-9=(x+2y-3)(x+2y+3)$

45. $x^3-216=x^3-6^3=(x-6)(x^2+6x+36)$

47. $x^3+125=x^3+5^3$
$=(x+5)(x^2-5x+25)$

49. $4x^2+25$ is prime.

51. $4a^2+12a+9=(2a+3)(2a+3)$
$=(2a+3)^2$

53. $18x^2y-2y=2y(9x^2-1)$
$=2y(3x-1)(3x+1)$

55. $8x^3+y^3=(2x)^3+y^3$
$=(2x+y)(4x^2-2xy+y^2)$

57. $x^6-y^3=(x^2)^3-y^3$
$=(x^2-y)(x^4+x^2y+y^2)$

59. $(x^2+16x+64)-x^4$
$=(x+8)^2-(x^2)^2$
$=(x+8-x^2)(x+8+x^2)$

61. $3x^6y^2+81y^2$
$=3y^2(x^6+27)$
$=3y^2[(x^2)^3+3^3]$
$=3y^2(x^2+3)(x^4-3x^2+9)$

63. $(x+y)^3+125$
$=(x+y)^3+5^3$
$=[(x+y)+5][(x+y)^2-5(x+y)+25]$
$=(x+y+5)(x^2+2xy+y^2-5x$
$-5y+25)$

65. $(2x+3)^3-64$
$=(2x+3)^3-4^3$
$=[(2x+3)-4][(2x+3)^2+4(2x+3)+16]$
$=(2x-1)(4x^2+12x+9+8x+12+16)$
$=(2x-1)(4x^2+20x+37)$

67. $A=\pi R^2-\pi r^2$
$=\pi(R^2-r^2)$
$=\pi(R-r)(R+r)$

69. The coating of the candy is the difference between the volume of the outer sphere $\frac{4}{3}\pi R^3$ and the inner sphere $\frac{4}{3}\pi(6)^3$.
Thus,
$V=\frac{4}{3}\pi R^3-\frac{4}{3}\pi(6)^3$
$V=\frac{4}{3}\pi(R^3-6^3)$
$V=\frac{4}{3}\pi(R-6)(R^2+6R+36)$

71. $1-y^3=1^3-y^3=(1-y)(1+y+y^2)$

73. $9x^2+6x+1=(3x+1)^2$

75. $x^2-8x+16-y^2=(x-4)^2-y^2$
$=(x-4-y)(x-4+y)$

77. $x^4-x=x(x^3-1)$
$=x(x-1)(x^2+x+1)$

79. $14x^2y-2xy=2xy(7x-1)$

81. $4x^2-16=4(x^2-4)=4(x+2)(x-2)$

83. $128a^3-2b^3$
$=2(64a^3-b^3)$
$=2([4a]^3-b^3)$
$=2(4a-b)(16a^2+4ab+b^2)$

85. $3x^2-8x-11=(3x-11)(x+1)$

87. $4x^2+8x-12=4(x^2+2x-3)$
$=4(x+3)(x-1)$

89. $4x^2+36x+81=(2x+9)^2$

91. $8x^3+27y^3$
$=(2x)^3+(3y)^3$
$=(2x+3y)(4x^2-6xy+9y^2)$

93. $64x^2y^3-8x^2$
$=8x^2(8y^3-1)$
$=8x^2[(2y)^3-(1)^3]$
$=8x^2(2y-1)(4y^2+2y+1)$

95. $(x+5)^3+y^3$
$=[(x+5)+y]\cdot[(x+5)^2-(x+5)y+y^2]$
$=(x+5+y)$
$\cdot(x^2+10x+25-xy-5y+y^2)$

97. $(5a-3)^2-6(5a-3)+9=[(5a-3)-3]^2$
$=(5a-6)^2$

99. $\left(\frac{6}{2}\right)^2=3^2=9$
$x^2+6x+9=(x+3)^2$

101. $\left(\frac{14}{2}\right)^2 = 7^2 = 49$

$m^2 - 14m + 49 = (m-7)^2$

103. $\pm 2\sqrt{16} = \pm 2 \cdot 4 = \pm 8$

$x^2 + 8x + 16 = (x+4)^2$

$x^2 - 8x + 16 = (x-4)^2$

105. a. $(x^3)^2 - 1^2$

$= (x^3+1)(x^3-1)$

$= (x+1)(x^2-x+1)(x-1)(x^2+x+1)$

b. $(x^2)^3 - 1^3 =$

$(x+1)(x-1)(x^4+x^2+1)$

c. Answers may vary.

107. $x + 7 = 0$

$x = -7$

109. $5x - 15 = 0$

$5x = 15$

$x = 3$

111. $3x = 0$

$x = 0$

113. $-4x - 16 = 0$

$-4x = 16$

$x = -4$

115. $x^{2n} - 36 = (x^n)^2 - 6^2 = (x^n - 6)(x^n + 6)$

117. $25x^{2n} - 81 = (5x^n)^2 - 9^2$

$= (5x^n - 9)(5x^n + 9)$

119. $x^{4n} - 625 = (x^{2n})^2 - (25)^2$

$= (x^{2n} - 25)(x^{2n} + 25)$

$= [(x^n)^2 - 5^2](x^{2n} + 25)$

$= (x^n - 5)(x^n + 5)(x^{2n} + 25)$

Section 5.8

Graphing Calculator Explorations

1.

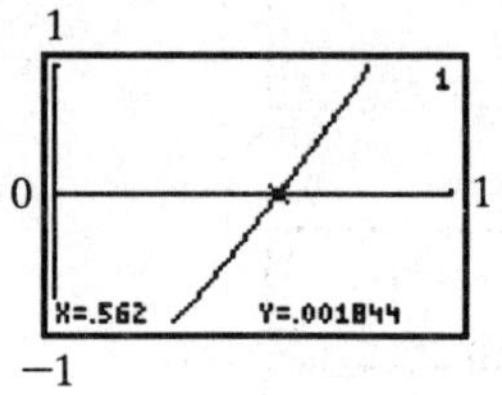

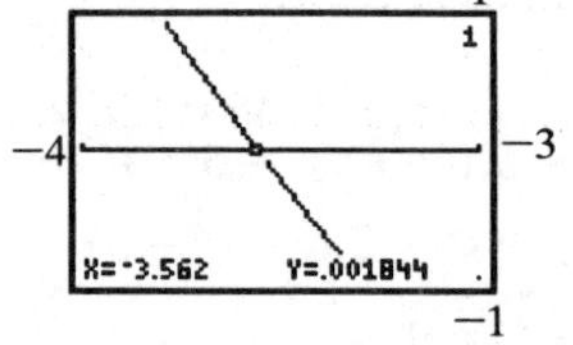

The intercepts are 0.562, −3.562.

3.

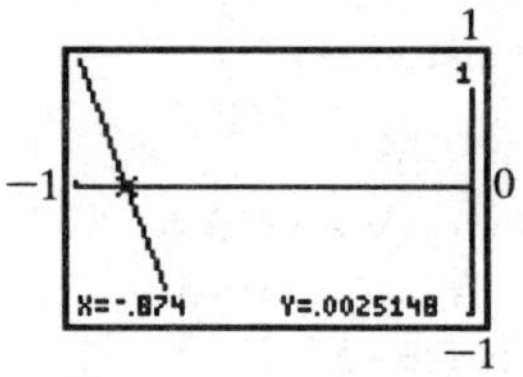

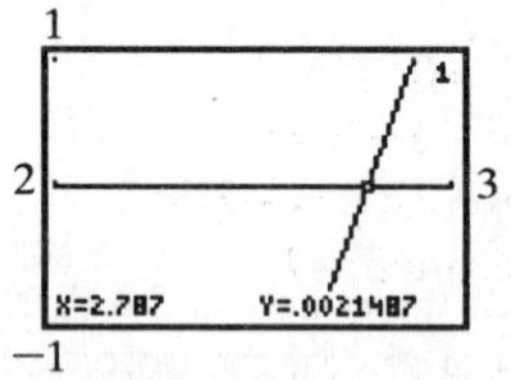

The intercepts are −0.874, 2.787.

5.

0.5

−1 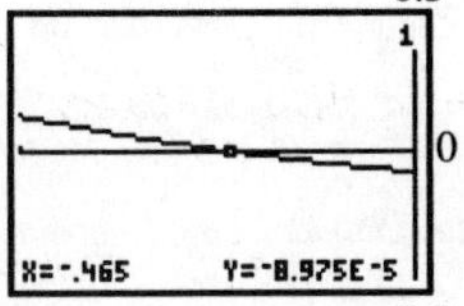 0

−0.5

0.5

1.5 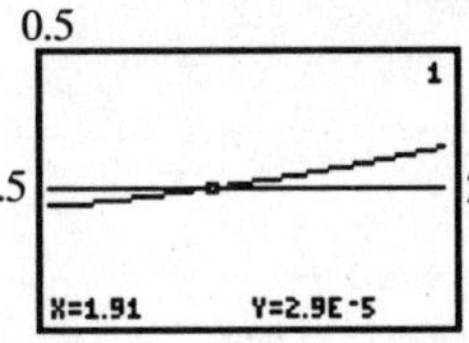 2.5

−0.5

The intercepts are –0.465, 1.910.

Mental Math

1. $(x-3)(x+5)=0$
$x-3=0$ or $x+5=0$
$x=3$ or $x=-5$
The solutions are 3, –5.

2. $(y+5)(y+3)=0$
$y+5=0$ or $y+3=0$
$y=-5$ or $y=-3$
The solutions are –5, –3.

3. $(z-3)(z+7)=0$
$z-3=0$ or $z+7=0$
$z=3$ or $z=-7$
The solutions are 3, –7.

4. $(c-2)(c-4)=0$
$c-2=0$ or $c-4=0$
$c=2$ or $c=4$
The solutions are 2, 4.

5. $x(x-9)=0$
$x=0$ or $x-9=0$
$x=9$
The solutions are 0, 9.

6. $w(w+7)=0$
$w=0$ or $w+7=0$
$w=-7$
The solutions are 0, –7.

Exercises 5.8

1. $(x+3)(3x-4)=0$
$x+3=0$ or $3x-4=0$
$x=-3$ or $3x=4$
$x=\frac{4}{3}$
The solutions are –3, $\frac{4}{3}$.

3. $3(2x-5)(4x+3)=0$
$2x-5=0$ or $4x+3=0$
$2x=5$ or $4x=-3$
$x=\frac{5}{2}$ or $x=-\frac{3}{4}$
The solutions are $\frac{5}{2}, -\frac{3}{4}$.

5. $x^2+11x+24=0$
$(x+3)(x+8)=0$
$x+3=0$ or $x+8=0$
$x=-3$ or $x=-8$
The solutions are –3, –8.

7. $12x^2+5x-2=0$
$(4x-1)(3x+2)=0$
$4x-1=0$ or $3x+2=0$
$4x=1$ or $3x=-2$
$x=\frac{1}{4}$ or $x=-\frac{2}{3}$
The solutions are $\frac{1}{4}, -\frac{2}{3}$.

9. $z^2+9=10z$
$z^2-10z+9=0$
$(z-1)(z-9)=0$
$z-1=0$ or $z-9=0$
$z=1$ or $z=9$
The solutions are 1, 9.

11. $x(5x+2)=3$
$5x^2+2x-3=0$
$(5x-3)(x+1)=0$
$5x-3=0$ or $x+1=0$
$5x=3$ or $x=-1$
$x=\frac{3}{5}$
The solutions are $\frac{3}{5}$, –1.

13. $x^2-6x=x(8+x)$
$x^2-6x=8x+x^2$
$0=14x$
$x=0$
The solution is 0.

15. $\frac{z^2}{6}-\frac{z}{2}-3=0$
$z^2-3z-18=0$
$(z-6)(z+3)=0$
$z-6=0$ or $z+3=0$
$z=6$ or $z=-3$
The solutions are 6, –3.

17. $\frac{x^2}{2}+\frac{x}{20}=\frac{1}{10}$
$10x^2+x=2$
$10x^2+x-2=0$
$(5x-2)(2x+1)=0$
$5x-2=0$ or $2x+1=0$
$5x=2$ or $2x=-1$
$x=\frac{2}{5}$ or $x=-\frac{1}{2}$
The solutions are $\frac{2}{5}$, $-\frac{1}{2}$.

19. $\frac{4t^2}{5}=\frac{t}{5}+\frac{3}{10}$
$8t^2=2t+3$
$8t^2-2t-3=0$
$(4t-3)(2t+1)=0$
$4t-3=0$ or $2t+1=0$
$4t=3$ or $2t=-1$
$t=\frac{3}{4}$ or $t=-\frac{1}{2}$
The solutions are $\frac{3}{4}$, $-\frac{1}{2}$.

21. $(x+2)(x-7)(3x-8)=0$
$x+2=0$ or $x-7=0$ or $3x-8=0$
$x=-2$ or $x=7$ or $3x=8$
$x=\frac{8}{3}$
The solutions are –2, 7, $\frac{8}{3}$.

23. $y^3=9y$
$y^3-9y=0$
$y(y^2-9)=0$
$y(y+3)(y-3)=0$
$y=0$ or $y+3=0$ or $y-3=0$
$y=-3$ or $y=3$
The solutions are 0, –3, 3.

25. $x^3-x=2x^2-2$
$x^3-2x^2-x+2=0$
$x^2(x-2)-(x-2)=0$
$(x^2-1)(x-2)=0$
$(x+1)(x-1)(x-2)=0$
$x+1=0$ or $x-1=0$ or $x-2=0$
$x=-1$ or $x=1$ or $x=2$
The solutions are –1, 1, 2.

27. Answers may vary.

29. $(2x+7)(x-10)=0$
$2x+7=0$ or $x-10=0$
$2x=-7$ or $x=10$
$x=-\frac{7}{2}$
The solutions are $-\frac{7}{2}$, 10.

31. $3x(x-5)=0$
$3x=0$ or $x-5=0$
$x=0$ or $x=5$
The solutions are 0, 5.

33. $x^2-2x-15=0$
$(x-5)(x+3)=0$
$x-5=0$ or $x+3=0$
$x=5$ or $x=-3$
The solutions are 5, –3.

35. $12x^2+2x-2=0$
$6x^2+x-1=0$
$(3x-1)(2x+1)=0$
$3x-1=0$ or $2x+1=0$
$3x=1$ or $2x=-1$
$x=\frac{1}{3}$ or $x=-\frac{1}{2}$
The solutions are $\frac{1}{3}, -\frac{1}{2}$.

37. $w^2-5w=36$
$w^2-5w-36=0$
$(w-9)(w+4)=0$
$w-9=0$ or $w+4=0$
$w=9$ or $w=-4$
The solutions are 9, –4.

39. $25x^2-40x+16=0$
$(5x-4)^2=0$
$5x-4=0$
$5x=4$
$x=\frac{4}{5}$
The solution is $\frac{4}{5}$.

41. $2r^3+6r^2=20r$
$r^3+3r^2=10r$
$r^3+3r^2-10r=0$
$r(r^2+3r-10)=0$
$r(r+5)(r-2)=0$
$r=0$ or $r+5=0$ or $r-2=0$
$r=-5$ or $r=2$
The solutions are 0, –5, 2.

43. $z(5z-4)(z+3)=0$
$z=0$ or $5z-4=0$ or $z+3=0$
$5z=4$ or $z=-3$
$z=\frac{4}{5}$
The solutions are 0, $\frac{4}{5}$, –3.

45. $2z(z+6)=2z^2+12z-8$
$2z^2+12z=2z^2+12z-8$
$0=-8$
No solution exists.

47. $(x-1)(x+4)=24$
$x^2+3x-4=24$
$x^2+3x-28=0$
$(x+7)(x-4)=0$
$x+7=0$ or $x-4=0$
$x=-7$ or $x=4$
The solutions are –7, 4.

49. $\frac{x^2}{4}-\frac{5}{2}x+6=0$
$x^2-10x+24=0$
$(x-4)(x-6)=0$
$x-4=0$ or $x-6=0$
$x=4$ or $x=6$
The solutions are 4, 6.

51. $y^2 + \frac{1}{4} = -y$

$4y^2 + 1 = -4y$

$4y^2 + 4y + 1 = 0$

$(2y + 1)^2 = 0$

$2y + 1 = 0$

$2y = -1$

$y = -\frac{1}{2}$

The solution is $-\frac{1}{2}$.

53. $y^3 + 4y^2 = 9y + 36$

$y^3 + 4y^2 - 9y - 36 = 0$

$y^2(y + 4) - 9(y + 4) = 0$

$(y^2 - 9)(y + 4) = 0$

$(y + 3)(y - 3)(y + 4) = 0$

$y + 3 = 0$ or $y - 3 = 0$ or $y + 4 = 0$

$y = -3$ or $y = 3$ or $y = -4$

The solutions are –3, 3, –4.

55. $2x^3 = 50x$

$x^3 = 25x$

$x^3 - 25x = 0$

$x(x^2 - 25) = 0$

$x(x + 5)(x - 5) = 0$

$x = 0$ or $x + 5 = 0$ or $x - 5 = 0$

$x = -5$ or $x = 5$

The solutions are 0, –5, 5.

57. $x^2 + (x + 1)^2 = 61$

$x^2 + x^2 + 2x + 1 = 61$

$2x^2 + 2x - 60 = 0$

$x^2 + x - 30 = 0$

$(x + 6)(x - 5) = 0$

$x + 6 = 0$ or $x - 5 = 0$

$x = -6$ or $x = 5$

The solutions are –6, 5.

59. $m^2(3m - 2) = m$

$3m^3 - 2m^2 = m$

$3m^3 - 2m^2 - m = 0$

$m(3m^2 - 2m - 1) = 0$

$m(3m + 1)(m - 1) = 0$

$m = 0$ or $3m + 1 = 0$ or $m - 1 = 0$

$3m = -1$ or $m = 1$

$m = -\frac{1}{3}$

The solutions are 0, $-\frac{1}{3}$, 1.

61. $3x^2 = -x$

$3x^2 + x = 0$

$x(3x + 1) = 0$

$x = 0$ or $3x + 1 = 0$

$3x = -1$

$x = -\frac{1}{3}$

The solutions are 0, $-\frac{1}{3}$.

63. $x(x - 3) = x^2 + 5x + 7$

$x^2 - 3x = x^2 + 5x + 7$

$-7 = 8x$

$x = -\frac{7}{8}$

The solution is $-\frac{7}{8}$.

65. $3(t - 8) + 2t = 7 + t$

$3t - 24 + 2t = 7 + t$

$5t - 24 = 7 + t$

$4t = 31$

$t = \frac{31}{4}$

The solution is $\frac{31}{4}$.

67. $-3(x - 4) + x = 5(3 - x)$

$-3x + 12 + x = 15 - 5x$

$-2x + 12 = 15 - 5x$

$3x = 3$

$x = 1$

The solution is 1.

69. **a** and **d** are incorrect because the right side of the equation is not zero.

71. Let n = the one number and $n + 5$ = the other number.
Then: $n(n + 5) = 66$

$$n^2 + 5n - 66 = 0$$
$$(n+11)(n-6) = 0$$
$$n+11=0 \quad \text{or} \quad n-6=0$$
$$n=-11 \quad \text{or} \quad n=6$$

There are two solutions: –11 and –6 and 6 and 11.

73. Let d = the amount of cable needed. Then from the Pythagorean theorem,
$d^2 = 45^2 + 60^2 = 5625.$
So, $d = \sqrt{5625} = 75$ ft.

75. $C(x) = x^2 - 15x + 50$

$$9500 = x^2 - 15x + 50$$
$$0 = x^2 - 15x - 9450$$
$$0 = (x-105)(x+90)$$
$$x-105=0 \quad \text{or} \quad x+90=0$$
$$x=105 \quad \text{or} \quad x=-90$$

Disregard the negative.
105 units

77. Let x = one leg of a right triangle and $x - 3$ = the other leg of the right triangle. Then from the Pythagorean theorem,

$$15^2 = x^2 + (x-3)$$
$$225 = x^2 + x^2 - 6x + 9$$
$$0 = 2x^2 - 6x - 216$$
$$0 = x^2 - 3x - 108$$
$$0 = (x-12)(x+9)$$
$$x-12=0 \quad \text{or} \quad x+9=0$$
$$x=12 \quad \text{or} \quad x=-9.$$

Discarding the extraneous solution of –9, we find that one leg of the right triangle is 12 cm and the other leg is 9 cm.

79. Note that the outer rectangle has lengths of $2x + 12$ and $2x + 16$. Thus, the area of the border is $(2x + 12)(2x + 16) - 12 \cdot 16$. We set this equal to 128 and solve for x.

$$(2x+12)(2x+16) - 12 \cdot 16 = 128$$
$$4x^2 + 56x = 128$$
$$4x^2 + 56x - 128 = 0$$
$$4(x^2 + 14x - 32) = 0$$
$$x^2 + 14x - 32 = 0$$
$$(x+16)(x-2) = 0$$
$$x+16=0 \quad \text{or} \quad x-2=0$$
$$x=-16 \quad \text{or} \quad x=2$$

Since x must be positive, we see that $x = 2$ inches.

81. The sunglasses will hit the ground when the height $h(t)$ equals 0.

$$-16t^2 + 1600 = 0$$
$$-16(t^2 - 100) = 0$$
$$-16(t-10)(t+10) = 0$$
$$t-10=0 \quad \text{or} \quad t+10=0$$
$$t=10 \quad \text{or} \quad t=-10$$

The sunglasses hit the ground 10 seconds after being dropped.

83. Let the width of the floor = w. Then the length is $w + 6$. So the area is

$$91 = w(w+6)$$
$$91 = w^2 + 6w$$
$$w^2 + 6w - 91 = 0$$
$$(w+13)(w-7) = 0$$
$$w+13=0 \quad \text{or} \quad w-7=0$$
$$w=-13 \quad \text{or} \quad w=7.$$

Since the width can't be negative, the width is 7ft and the length is 13 ft.

85.
$$0.5x^2 = 50$$
$$0.5x^2 - 50 = 0$$
$$5x^2 - 500 = 0$$
$$x^2 - 100 = 0$$
$$(x+10)(x-10) = 0$$
$$x+10=0 \quad \text{or} \quad x-10=0$$
$$x=-10 \quad \text{or} \quad x=10$$

Disregard the negative solution. A 10-inch square tier is needed, provided each person has one serving.

87. E

89. F

91. B

93. Answers may vary.
Ex.: $f(x) = (x-5)(x-3) = x^2 - 8x + 15$

95. Answers may vary.
Ex.: $f(x) = (x+1)(x-2) = x^2 - x - 2$

97. (–3, 0), (0, 2); function

99. (–4, 0), (4, 0), (0, –2), (0, 2); not a function

101. Answers may vary.

Section 5.9

Mental Math

1. $f(x) = 2x^2 + 7x + 10$ opens upward because 2 is positive.

2. $f(x) = -3x^2 - 5x$ opens downward because –3 is negative.

3. $f(x) = -x^2 + 5$ opens downward because –1 is negative.

4. $f(x) = x^2 + 3x + 7$ opens upward because 1 is positive.

Exercise Set 5.9

1. **a.** Domain, $(-\infty, \infty)$; Range, $(-\infty, 5]$

b. x-intercepts, (–2, 0), (6, 0); y-intercept, (0, 5)

c. (0, 5)

d. There is no such point.

e. –2, 6

f. $\{x \mid -2 < x < 6\}$

g. –2, 6

3. **a.** Domain, $(-\infty, \infty)$; Range, $[-4, \infty)$

b. x-intercepts, (–3, 0), (1, 0); y-intercept, (0, –3)

c. There is no such point.

d. (–1, –4)

e. –3, 1

f. $\{x \mid x < -3 \text{ or } x > 1\}$

g. –3, 1

5. **a.** Domain, $(-\infty, \infty)$; Range, $(-\infty, \infty)$

b. x-intercepts, (–2, 0), (0, 0), (2, 0); y-intercept, (0, 0)

c. There is no such point.

d. There is no such point.

e. –2, 0, 2

f. $\{x \mid -2 < x < 0 \text{ or } x > 2\}$

g. –2, 0, 2

7. $f(x) = 2x^2$

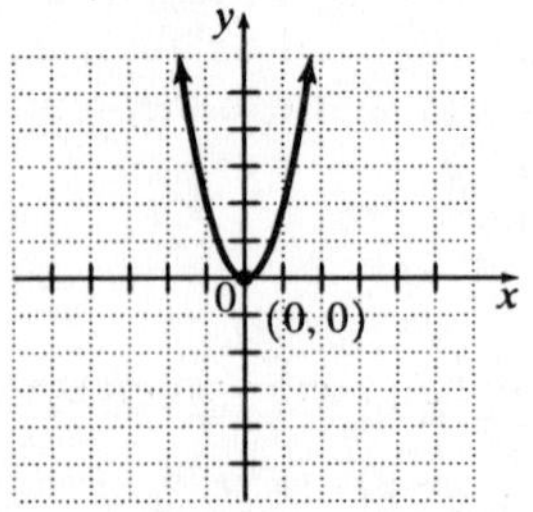

9. $f(x) = x^2 + 1$

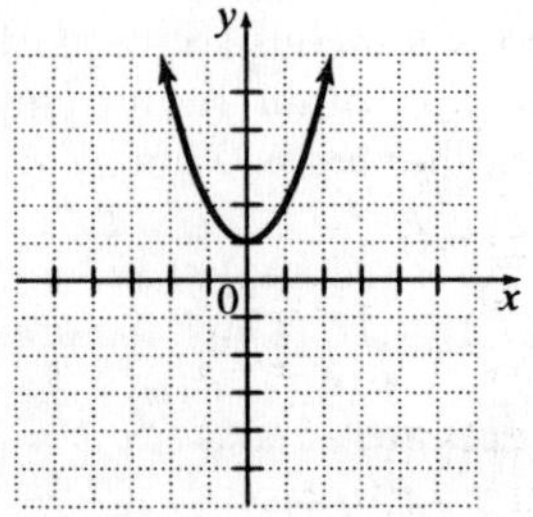

11. $f(x) = -x^2$

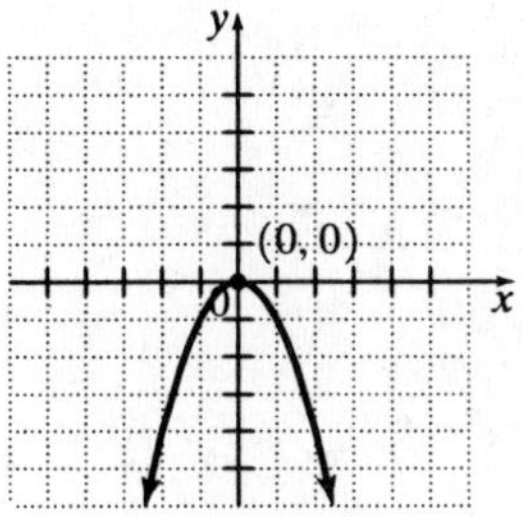

13. $f(x) = x^2 + 8x + 7$,
$a = 1$, and $b = 8$.
So, $h = \frac{-8}{2(1)} = -4$ and
$k = f(-4) = (-4)^2 + 8(-4) + 7 = -9$
thus, $V(-4, -9)$

15. $f(x) = 3x^2 + 6x + 4$, $a = 3$, and $b = 6$.
So, $h = -\frac{6}{2(3)} = -1$ and
$k = f(-1) = 3(-1)^2 + 6(-1) + 4 = 1$;
thus $V(-1, 1)$.

17. $f(x) = -x^2 + 10x + 5$, $a = -1$ and
$b = 10$. So, $h = -\frac{10}{2(-1)} = 5$ and
$k = f(5) = -5^2 + 10(5) + 5 = 30$;
thus $V(5, 30)$.

19. 2

21. 0

23. One x-intercept and one y-intercept

25. $f(x) = x^2 + 6x + 5$
$\frac{-b}{2a} = \frac{-6}{2(1)} = -3$
$f(-3) = 9 - 18 + 5 = -4$
The vertex is $(-3, -4)$.
If $x^2 + 6x + 5 = 0$
$(x + 1)(x + 5) = 0$
$x + 1 = 0$ or $x + 5 = 0$
$x = -1$ or $x = -5$
The x-intercepts are -1 and -5.
If $x = 0$, then $y = f(0) = 5$.
The y-intercept is 5.

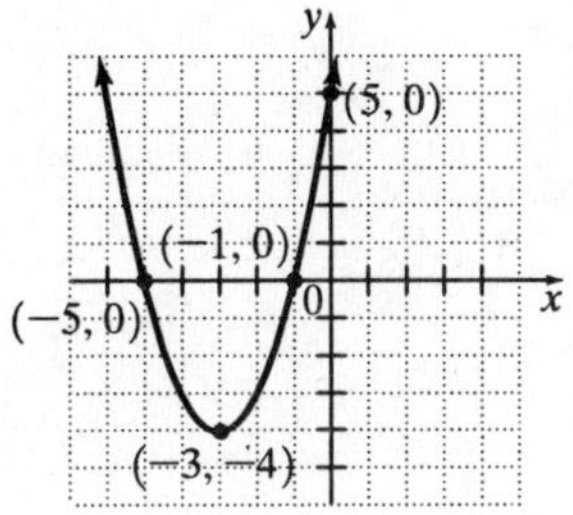

27. $f(x) = x^2 - 12x + 35$
$\frac{-b}{2a} = \frac{12}{2(1)} = 6$
$f(6) = 36 - 72 + 35 = -1$
The vertex is $(6, -1)$.
If $x^2 - 12x + 35 = 0$
$(x - 5)(x - 7) = 0$
$x - 5 = 0$ or $x - 7 = 0$
$x = 5$ or $x = 7$
The x-intercepts are 5 and 7. If $x = 0$, then $y = f(0) = 35$. The y-intercept is 35.

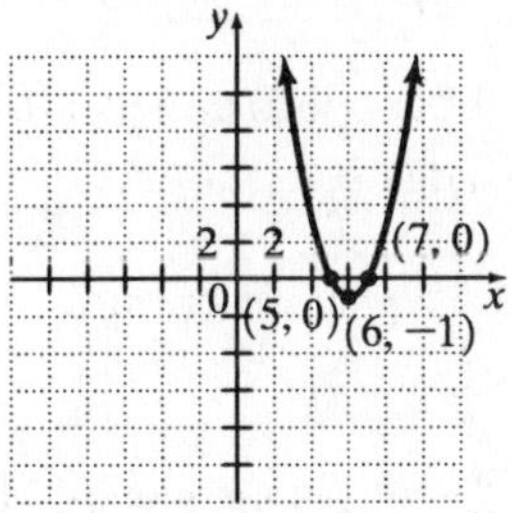

29. $f(x) = -3x^2 + 6x$
$\frac{-b}{2a} = \frac{-6}{2(-3)} = 1$
$f(1) = -3 + 6 = 3$
The vertex is $(1, 3)$.
If $-3x^2 + 6x = 0$
$3x(-x + 2) = 0$
$3x = 0$ or $-x + 2 = 0$
$x = 0$ or $x = 2$
The x-intercepts are 0 and 2. If $x = 0$, then

$y = f(0) = 0$. The y-intercept is 0.

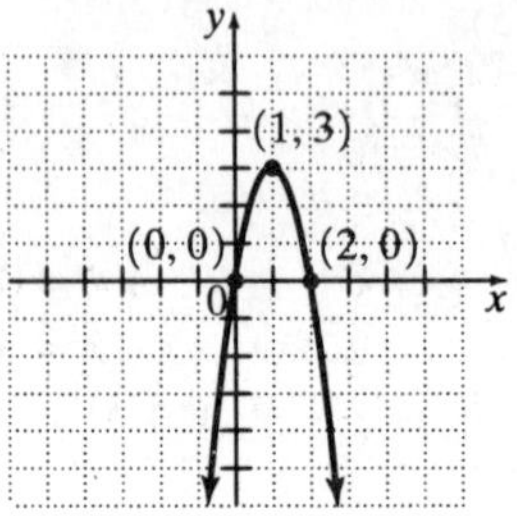

31. $f(x) = 2x^3 - 5x^2 - 3x$

If $2x^3 - 5x^2 - 3x = 0$

$x(2x^2 - 5x - 3) = 0$

$x(2x+1)(x-3) = 0$

$x = 0$ or $2x + 1 = 0$ or $x - 3 = 0$

$2x = -1$ or $x = 3$

$x = -\frac{1}{2}$

The x-intercepts are $-\frac{1}{2}$, 0, and 3.

The y-intercept is 0.

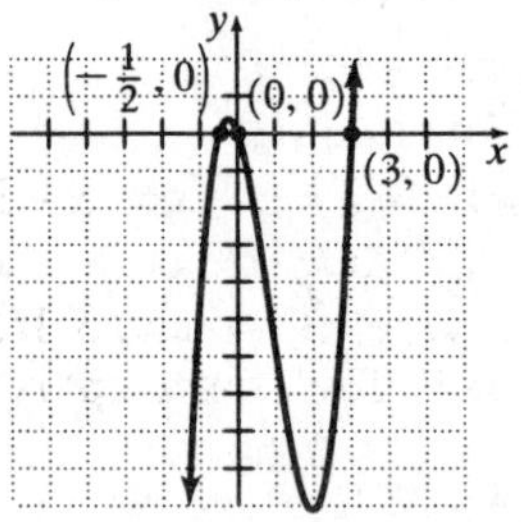

33. $f(x) = x^3 + x^2 - 4x - 4$

If $x^3 + x^2 - 4x - 4 = 0$

$x^2(x+1) - 4(x+1) = 0$

$(x+1)(x^2 - 4) = 0$

$(x+1)(x+2)(x-2) = 0$

$x + 1 = 0$ or $x + 2 = 0$ or $x - 2 = 0$

$x = -1$ or $x = -2$ or $x = 2$

The x-intercepts are –2, –1, and 2. If $x = 0$, then $y = f(0) = -4$. The y-intercept is –4.

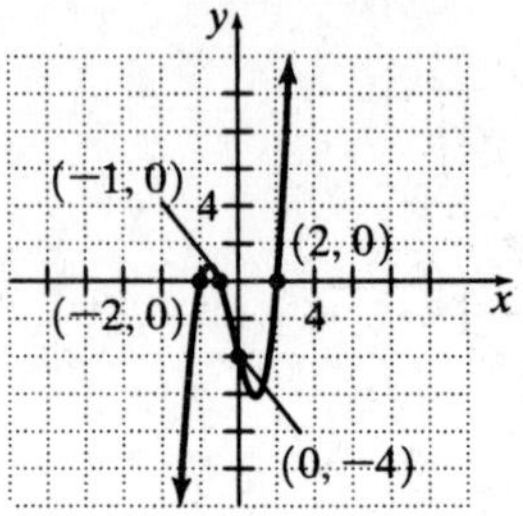

35. No

37. $f(x) = x^2 + 2x - 3$

$\frac{-b}{2a} = \frac{-2}{2(1)} = -1$

$f(-1) = 1 - 2 - 3 = -4$

The vertex is (–1, –4).

If $x^2 + 2x - 3 = 0$

$(x-1)(x+3) = 0$

$x - 1 = 0$ or $x + 3 = 0$

$x = 1$ or $x = -3$

The x-intercepts are –3 and 1. If $x = 0$, the $y = f(0) = -3$. The y-intercept is –3.

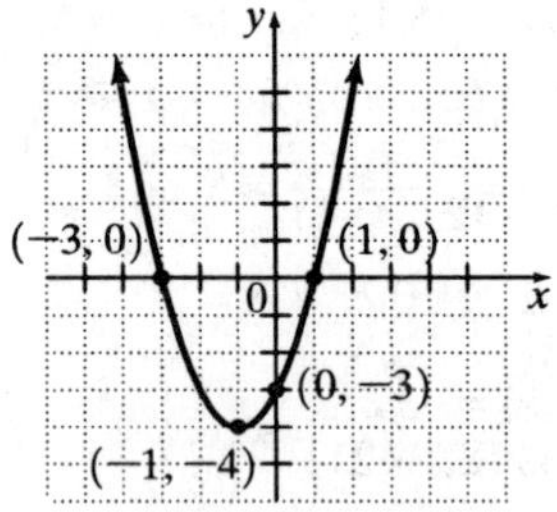

39. $f(x) = x^3 - 4x^2 + 3x$

If $x^3 - 4x^2 + 3x = 0$

$x(x^2 - 4x + 3) = 0$

$x(x-1)(x-3) = 0$

$x = 0$ or $x - 1 = 0$ or $x - 3 = 0$

$x = 1$ or $x = 3$

The x-intercepts are 0, 1, and 3.

The y-intercept is 0.

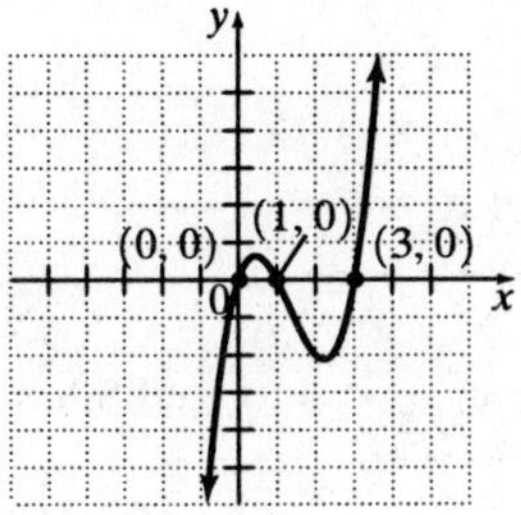

41. $f(x) = x^2 + 4$

$\frac{-b}{2a} = 0$

$f(0) = 4$

The vertex is (0, 4).

There are no x-intercepts.

The y-intercept is 4.

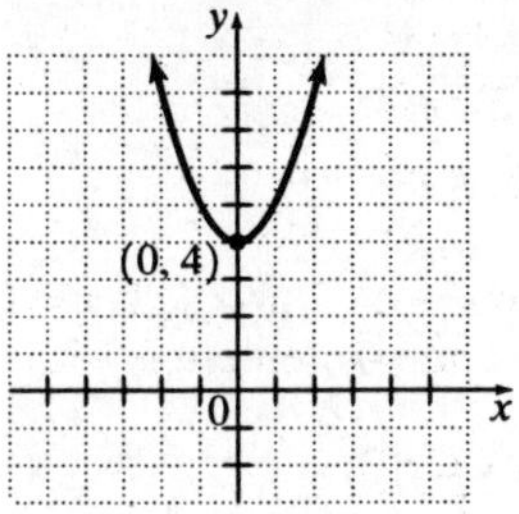

43. $f(x) = 3x^2 - 12x$

$\frac{-b}{2a} = \frac{12}{2(3)} = 2$

$f(2) = 12 - 24 = -12$

The vertex is (2, –12).

If $3x^2 - 12x = 0$

$3x(x - 4) = 0$

$3x = 0$ or $x - 4 = 0$

$x = 0$ or $x = 4$

The x-intercepts are 0 and 4.

The y-intercept is 0.

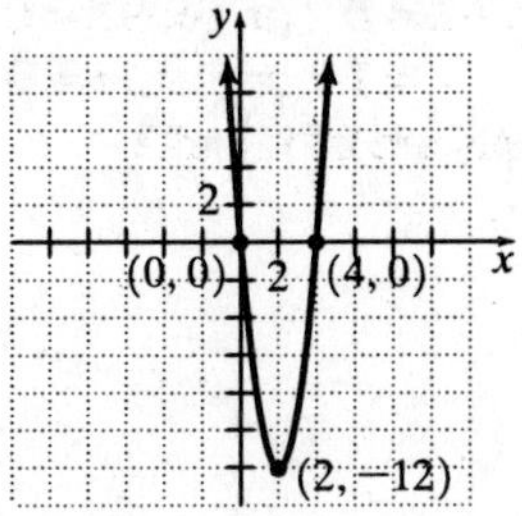

45. $f(x) = x^3 + x^2 - 12x$

If $x^3 + x^2 - 12x = 0$

$x(x^2 + x - 12) = 0$

$x(x - 3)(x + 4) = 0$

$x = 0$ or $x - 3 = 0$ or $x + 4 = 0$

$x = 3$ or $x = -4$

The x-intercepts are –4, 0, and 3.

The y-intercept is 0.

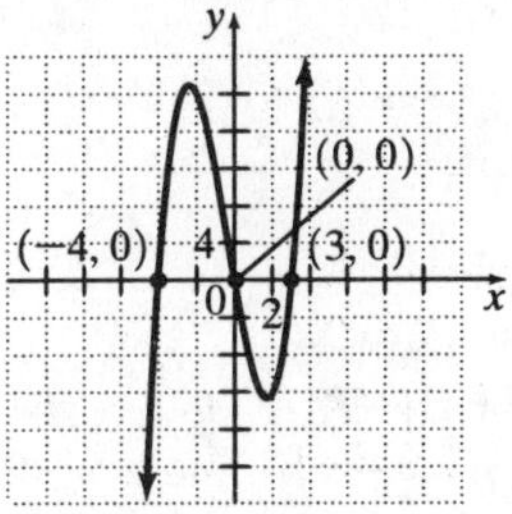

47. $f(x) = x^3 + x^2 - 9x - 9$

If $x^3 + x^2 - 9x - 9 = 0$

$x^2(x + 1) - 9(x + 1) = 0$

$(x + 1)(x^2 - 9) = 0$

$(x + 1)(x + 3)(x - 3) = 0$

$x + 1 = 0$ or $x + 3 = 0$ or $x - 3 = 0$

$x = -1$ or $x = -3$ or $x = 3$

The x-intercepts are –3, –1, and 3. If $x = 0$, then $y = f(0) = -9$. The y-intercept is –9.

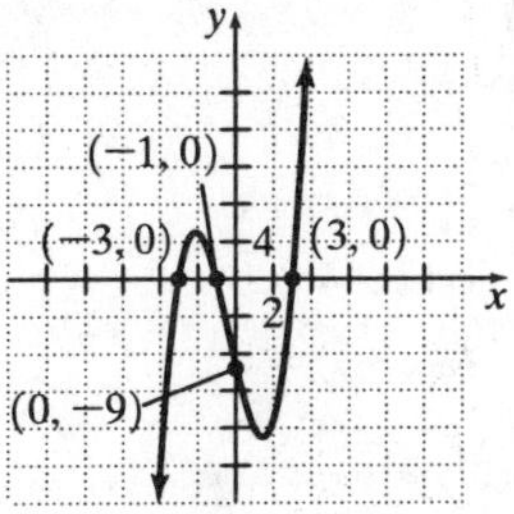

49. $f(x) = x^2 - 2x + 1$

$\frac{-b}{2a} = \frac{2}{2(1)} = 1$

$f(1) = 1 - 2 + 1 = 0$

The vertex is (1, 0).

If $x^2 - 2x + 1 = 0$

$(x - 1)(x - 1) = 0$

$x - 1 = 0$

$x = 1$

The x-intercept is 1. If $x = 0$, then $y = f(0) = 1$. The y-intercept is 1.

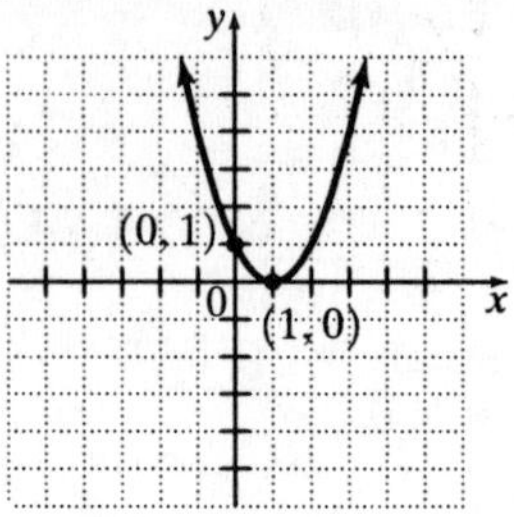

51. $f(x) = x^2 + 6x$

$\frac{-b}{2a} = \frac{-6}{2(1)} = -3$

$f(-3) = 9 - 18 = -9$

The vertex is $(-3, -9)$.

If $x^2 + 6x = 0$

$x(x+6) = 0$

$x = 0$ or $x + 6 = 0$

$x = -6$

The x-intercepts are -6 and 0.

The y-intercept is 0.

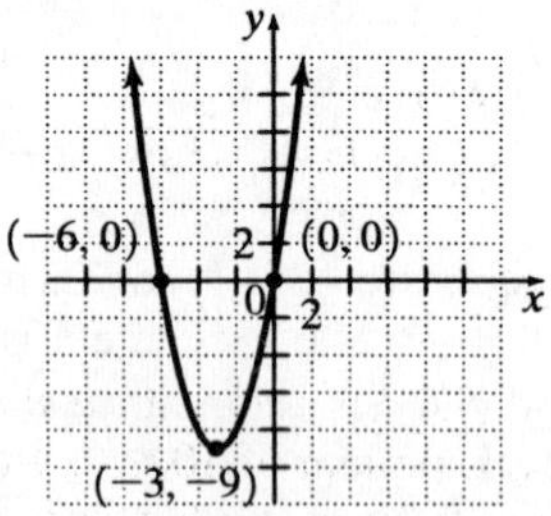

53. $f(x) = (x + 2)(x - 2)$

If $(x + 2)(x - 2) = 0$

$x + 2 = 0$ or $x - 2 = 0$

$x = -2$ or $x = 2$

The x-intercepts are -2 and 2.

If $x = 0$, then $y = f(0) = -4$.

The y-intercept is -4.

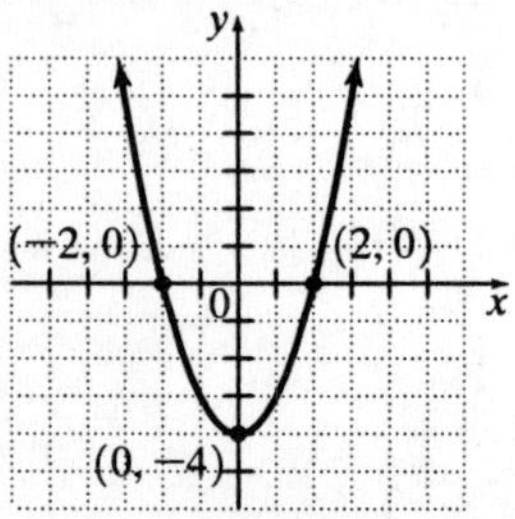

55. $f(x) = -x^3 + 25x$

If $-x^3 + 25x = 0$

$x(-x^2 + 25) = 0$

$x(5 - x)(5 + x) = 0$

$x = 0$ or $5 - x = 0$ or $5 + x = 0$

$x = 5$ or $x = -5$

The x-intercepts are -5, 0, and 5.

The y-intercept is 0.

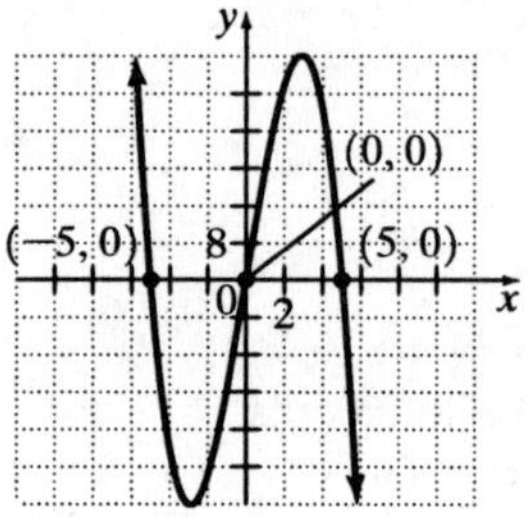

57. $f(x) = x^2 + 2x + 4$

$\frac{-b}{2a} = \frac{-2}{2(1)} = -1$

$f(-1) = 1 - 2 + 4 = 3$

The vertex is $(-1, 3)$.

There are no x-intercepts.

If $x = 0$, then $y = f(0) = 4$.

The y-intercept is 4.

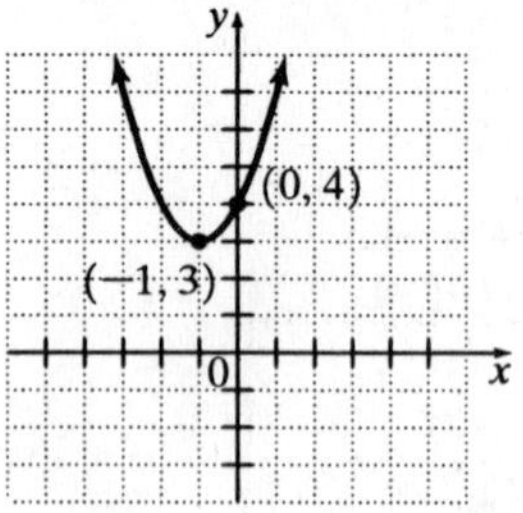

59. $f(x) = 3x(x - 3)(x + 5)$

If $3x(x - 3)(x + 5) = 0$

$3x = 0$ or $x - 3 = 0$ or $x + 5 = 0$

$x = 0$ or $x = 3$ or $x = -5$

The x-intercepts are -5, 0, and 3.

The y-intercept is 0.

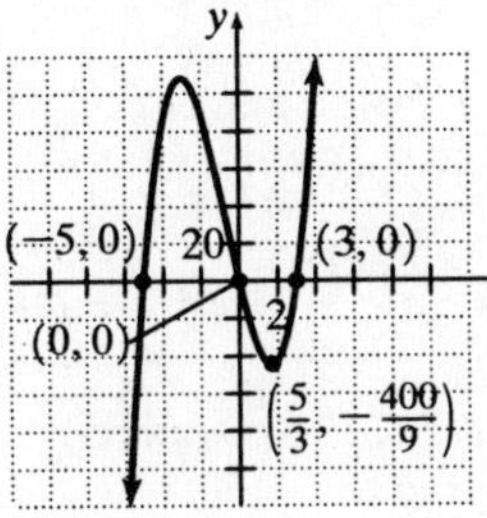

61. $h(x) = (x - 4)(x - 2)(2x + 1)(x + 3)$

$x - 4 = 0$ or $x - 2 = 0$
$x = 4$ or $x = 2$

or $2x + 1 = 0$ or $x + 3 = 0$
$x = -\frac{1}{2}$ or $x = -3$

The x-intercepts are 4, 2, $-\frac{1}{2}$, and -3.
If $x = 0$, then $y = h(0) = 24$.
The y-intercept is 24.

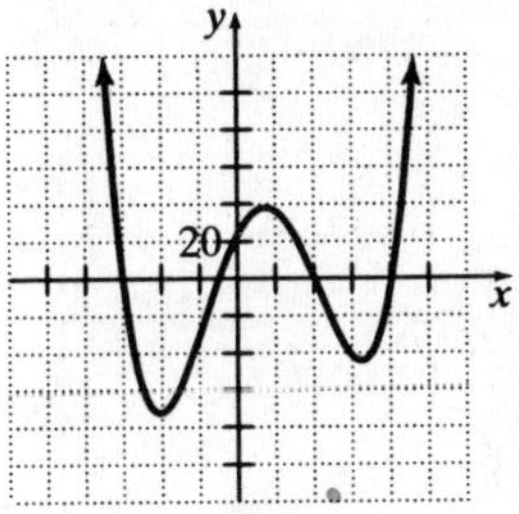

63. $f(x) = x^2 + 2x - 3$

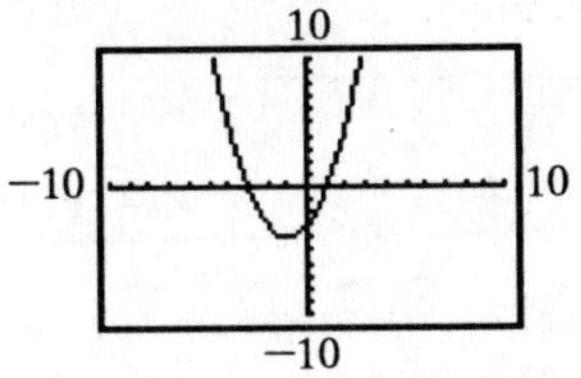

65. $f(x) = -x^3 + 25x$

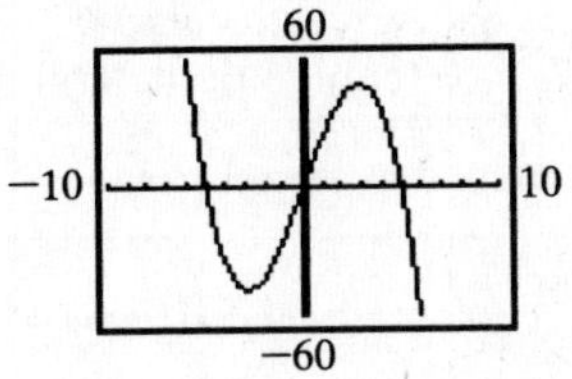

67. $G(x) = x^4 - 6.2x^2 - 6.2$
$x = -2.7, x = 2.7$

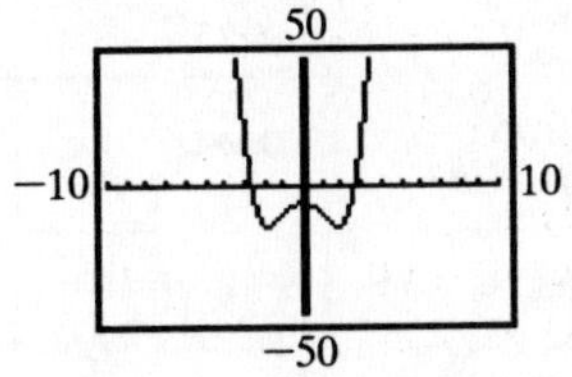

69. $-\frac{45}{100} = -\frac{9(5)}{20(5)} = -\frac{9}{20}$

71. $\frac{a^{14}b^2}{ab^4} = \frac{a^{14-1}}{b^{4-2}} = \frac{a^{13}}{b^2}$

73. $\frac{20x^{-3}y^5}{25y^{-2}x} = \frac{4(5)y^{5+2}}{5(5)x^{1+3}} = \frac{4y^7}{5x^4}$

Chapter 5 Project

1. Icecaps and glaciers account for more because the exponent of 10 is 7, while the exponent of 10 for groundwater is only 6.

2. The total volume of water that exists on planet Earth is 1.358204×10^9 cubic kilometers.

3.

	Water volume (cubic kilometers)	Percent
Atmosphere	1.3×10^4	0.00096%
Stream channels	1.0×10^3	0.00007%
Freshwater lakes	1.2×10^5	0.00884%
Groundwater	8.3×10^6	0.61110%
Icecaps and glaciers	2.9×10^7	2.13517%
Oceans	1.32×10^9	97.18717%
Saline lakes and inland seas	1.0×10^5	0.00736%
Water in soil above groundwater	6.7×10^5	0.04933%
TOTAL	1.358204×10^9	100%

4. $y = 0.0011(20)^2 + 0.1564(20) + 1.4545$
$= 5.0225$ cm

5. $y = 0.002x^2 + 0.2662x + 2.4621$

6. a. In 2025,
$y = 0.002(25)^2 + 0.2662(25) + 2.4621$
$= 10.3671$ cm

b. In 2050,
$y = 0.002(50)^2 + 0.2662(50) + 2.4621$
$= 20.7721$ cm

c. In 2075,
$y = 0.002(75)^2 + 0.2662(75) + 2.4621$
$= 33.6771$ cm

d. In 2100,
$y = 0.002(100)^2 + 0.2662(100) + 2.4621$
$= 49.0821$ cm

7. Answers may vary.

Chapter 5 Review

1. $(-2)^2 = (-2)(-2) = 4$

2. $(-3)^4 = (-3)\cdot(-3)\cdot(-3)\cdot(-3) = 81$

3. $-2^2 = -(2\cdot 2) = -4$

4. $-3^4 = -3\cdot 3\cdot 3\cdot 3 = -81$

5. $8^0 = 1$

6. $-9^0 = -1$

7. $-4^{-2} = -\frac{1}{4^2} = -\frac{1}{16}$

8. $(-4)^{-2} = \frac{1}{(-4)^2} = \frac{1}{16}$

9. $-xy^2y^3xy^2z = -x^{1+1}y^{2+3+2}z = -x^2y^7z$

10. $(-4xy)(-3xy^2b) = 12x^2y^3b$

11. $a^{-14}a^5 = a^{-14+5} = a^{-9} = \frac{1}{a^9}$

12. $\frac{a^{16}}{a^{17}} = a^{16-17} = a^{-1} = \frac{1}{a}$

13. $\frac{x^{-7}}{x^4} = \frac{1}{x^{4-(-7)}} = \frac{1}{x^{11}}$

14. $\frac{9a(a^{-3})}{18a^{15}} = \frac{a^{1-3-15}}{2} = \frac{a^{-17}}{2} = \frac{1}{2a^{17}}$

15.
$$\frac{y^{6p-3}}{y^{6p+2}} = \frac{1}{y^{(6p+2)-(6p-3)}}$$
$$= \frac{1}{y^{6p+2-6p+3}} = \frac{1}{y^5}$$

16. $36{,}890{,}000 = 3.689\times 10^7$

17. $-0.000362 = -3.62\times 10^{-4}$

18. $1.678\times 10^{-6} = 0.000001678$

19. $4.1\times 10^5 = 410{,}000$

20. $(8^5)^3 = 8^{5\cdot 3} = 8^{15}$

21. $\left(\frac{a}{4}\right)^2 = \frac{a^2}{4^2} = \frac{a^2}{16}$

22. $(3x^3) = 3^3x^3 = 27x^3$

23. $(-4x)^{-2} = \frac{1}{(-4x)^2} = \frac{1}{(-4)^2x^2} = \frac{1}{16x^2}$

24. $\left(\frac{6x}{5}\right)^2 = \frac{(6x)^2}{(5)^2} = \frac{36x^2}{25}$

25. $(8^6)^{-3} = 8^{-18} = \frac{1}{8^{18}}$

26. $\left(\frac{4}{3}\right)^{-2} = \frac{4^{-2}}{3^{-2}} = \frac{\frac{1}{4^2}}{\frac{1}{3^2}} = \frac{3^2}{4^2} = \frac{9}{16}$

27.
$$(-2x^3)^{-3} = \frac{1}{(-2x^3)^3}$$
$$= \frac{1}{(-2)^3(x^3)^3}$$
$$= \frac{1}{-8x^9}$$
$$= -\frac{1}{8x^9}$$

28. $\left(\frac{8p^6}{4p^4}\right)^{-2} = (2p^2)^{-2} = 2^{-2}p^{-4} = \frac{1}{4p^4}$

29.
$$(-3x^{-2}y^2)^3 = (-3)^3(x^{-2})^3(y^2)^3$$
$$= -27x^{-6}y^6$$
$$= \frac{-27y^6}{x^6}$$

30. $\left(\frac{x^{-5}y^{-3}}{z^3}\right)^{-5} = \frac{x^{25}y^{15}}{z^{-15}} = x^{25}y^{15}z^{15}$

31. $\frac{4^{-1}x^3yz}{x^{-2}yx^4} = \frac{x^{3-(-2)-4}z}{4} = \frac{x^{5-4}z}{4} = \frac{xz}{4}$

32. $(5xyz)^{-4}(x^{-2})^{-3} = \frac{1}{(5xyz)^4}x^6$
$= \frac{x^6}{625x^4y^4z^4}$
$= \frac{x^2}{625y^4z^4}$

33.
$\frac{2(3yz)^{-3}}{y^{-3}} = \frac{2(3)^{-3}y^{-3}z^{-3}}{y^{-3}}$
$= \frac{2}{3^3z^3} = \frac{2}{27z^3}$

34. $x^{4a}(3x^{5a})^3 = x^{4a}(3^3x^{15a})$
$= 27x^{4a+15a}$
$= 27x^{19a}$

35. $\frac{4y^{3x-3}}{2y^{2x+4}} = 2y^{(3x-3)-(2x+4)}$
$= 2y^{3x-3-2x-4}$
$= 2y^{x-7}$

36. $\frac{(0.00012)(144,000)}{0.0003}$
$= \frac{(1.2\times10^{-4})(1.44\times10^5)}{3\times10^{-4}}$
$= 0.576\times10^5$
$= 5.76\times10^4$

37. $\frac{(-0.00017)(0.00039)}{3,000}$
$= \frac{(-1.7\times10^{-4})(3.9\times10^{-4})}{3\times10^3}$
$= -2.21\times10^{-4-4-3}$
$= -2.21\times10^{-11}$

38. $\frac{27x^{-5}y^5}{18x^{-6}y^2}\cdot\frac{x^4y^{-2}}{x^{-2}y^3} = \frac{3x^{-5+4}y^{5-2}}{2x^{-6-2}y^{2+3}}$
$= \frac{3x^{-1}y^3}{2x^{-8}y^5}$
$= \frac{3}{2}x^{-1+8}y^{3-5}$
$= \frac{3x^7}{2y^2}$

39. $\frac{3x^5}{y^{-4}}\cdot\frac{(3xy^{-3})^{-2}}{(z^{-3})^{-4}} = \frac{3x^5\cdot3^{-2}x^{-2}(y^{-3})^{-2}}{y^{-4}z^{12}}$
$= \frac{3^{1-2}x^{5-2}y^{6-(-4)}}{z^{12}}$
$= \frac{3^{-1}x^3y^{10}}{z^{12}}$
$= \frac{x^3y^{10}}{3z^{12}}$

40. $\frac{(x^w)^2}{(x^{w-4})^{-2}} = x^{2w-(-2w+8)} = x^{4w-8}$

41. The degree of the polynomial $x^2y - 3xy^3z + 5x + 7y$ is the degree of the term $-3xy^3z$ which is 5.

42. $3x + 2$ has degree 1.

43. $4x + 8x - 6x^2 - 6x^2y$
$= (4+8)x - 6x^2 - 6x^2y$
$= 12x - 6x^2 - 6x^2y$

44. $-8xy^3 + 4xy^3 - 3x^3y$
$= (-8+4)xy^3 - 3x^3y$
$= -4xy^3 - 3x^3y$

45. $(3x+7y)+(4x^2-3x+7)+(y-1)$
$= 4x^2 + (3-3)x + (7+1)y + (7-1)$
$= 4x^2 + 8y + 6$

46.
$$\begin{array}{r} 4x^2-6xy+\ 9y^2 \\ -\ \ 8x^2-6xy-\ \ y^2 \\ \hline -4x^2 \qquad +10y^2 \end{array}$$

47.
$(3x^2-4b+28)+(9x^2-30)$
$-(4x^2-6b+20)$
$=3x^2-4b+28+9x^2-30$
$-4x^2+6b-20$
$=(3+9-4)x^2+(-4+6)b+(28-30-20)$
$=8x^2+2b-22$

48.
$$\begin{array}{r} 4x^2+\ 9xy \qquad +18 \\ +\ -4x^3 \qquad +\ 7xy-9x \qquad \\ \hline -4x^3+4x^2+16xy-9x+18 \end{array}$$

49. $(3x^2y-7xy-4)+(9x^2y+x)-(x-7)$
$=3x^2y-7xy-4+9x^2y+x-x+7$
$=(3+9)x^2y-7xy+(1-1)x+(-4+7)$
$=12x^2y-7xy+3$

50.
$$\begin{array}{r} x^2-5x+7 \\ -\qquad x+4 \\ \hline x^2-6x+3 \end{array}$$

51.
$$\begin{array}{r} x^3 \qquad +2xy^2-y \qquad \\ x-4xy^2 \qquad -7 \\ \hline x^3+x-2xy^2-y-7 \end{array}$$

52. $P(6)=9(6)^2-7(6)+8=290$

53. $P(-2)=9(-2)^2-7(-2)+8$
$=58$

54. $P(-3)=9(-3)^2-7(-3)+8=110$

55. $(2x-1)+(x^2+2x-5)=x^2+4x-6$

56. $2(2x-1)-(x^2+2x-5)$
$=4x-2-x^2-2x+5$
$=-x^2+2x+3$

57. $2(x^2y+5)+2(2x^2y-6x+1)$
$=2x^2y+10+4x^2y-12x+2$
$=(6x^2y-12x+12)$ cm

58.
$$\begin{array}{r} 4x^2-6x+1 \\ -6x \\ \hline -24x^3+36x^2-6x \end{array}$$

59. $-4ab^2(3ab^3+7ab+1)$
$=(-4ab^2)(3ab^3)-4ab^2(7ab)-4ab^2(1)$
$=-12a^2b^5-28a^2b^3-4ab^2$

60. $(x-4)(2x+9)=2x^2+9x-8x-36$
$=2x^2+x-36$

61. $(-3xa+4b)^2$
$=(-3xa)^2+2(-3xa)(4b)+(4b)^2$
$=9x^2a^2-24xab+16b^2$

62.
$$\begin{array}{r} 9x^2+4x+1 \\ 4x-3 \\ \hline -27x^2-12x-3 \\ 36x^3+16x^2+\ 4x \qquad \\ \hline 36x^3-11x^2-8x-3 \end{array}$$

63. $(5x-9y)(3x+9y)$
$=15x^2+45xy-27xy-81y^2$
$=15x^2+18xy-81y^2$

64. $\left(x-\frac{1}{3}\right)\left(x+\frac{2}{3}\right)$
$=(x)(x)+(x)\left(\frac{2}{3}\right)-\left(\frac{1}{3}\right)(x)-\left(\frac{1}{3}\right)\left(\frac{2}{3}\right)$
$=x^2+\frac{2}{3}x-\frac{1}{3}x-\frac{2}{9}$
$=x^2+\frac{1}{3}x-\frac{2}{9}$

65. $(x^2+9x+1)^2$
$=(x^2+9x+1)(x^2+9x+1)$
$=x^2(x^2)+x^2(9x)+x^2(1)$
$\quad +9x(x^2)+9x(9x)$
$\quad +9x(1)+1(x^2)+1(9x)+(1)(1)$
$=x^4+9x^3+x^2+9x^3+81x^2$
$\quad +9x+x^2+9x+1$
$=x^4+(9+9)x^3+(1+81+1)x^2$
$\quad +(9+9)x+1$
$=x^4+18x^3+83x^2+18x+1$

66. $(3x-y)^2=(3x)^2-2(3x)y+y^2$
$=9x^2-6xy+y^2$

67. $(4x+9)^2=(4x)^2+2(4x)(9)+(9)^2$
$=16x^2+72x+81$

68. $(x+3y)(x-3y)=x^2-(3y)^2=x^2-9y^2$

69. $[4+(3a-b)][4-(3a-b)]$
$=4^2-(3a-b)^2$
$=16-(9a^2-6ab+b^2)$
$=-9a^2+6ab-b^2+16$

70.

$$\begin{array}{r} x^2+\ 2x-5 \\ 2x-1 \\ \hline -x^2-\ 2x+5 \\ 2x^3+4x^2-10x \\ \hline 2x^3+3x^2-12x+5 \end{array}$$

71. $(3y-7z)(3y+7z)$
$=(3y)^2-(7z)^2$
$=(9y^2-49z^2)$ square units

72. $4a^b(3a^{b+2}-7)=4a^b(3a^{b+2})+4a^b(-7)$
$=12a^{b+b+2}-28a^b$
$=12a^{2b+2}-28a^b$

73. $(4xy^z-b)^2=(4xy^z)^2-2(4xy^z)b+b^2$
$=4^2x^2(y^z)^2-8xy^zb+b^2$
$=16x^2y^{2z}-8xy^zb+b^2$

74. $(3x^a-4)(3x^a+4)$
$=(3x^a)(3x^a)+(3x^a)(4)-4(3x^a)-4(4)$
$=9x^{2a}+12x^a-12x^a-16$
$=9x^{2a}-16$

75. $16x^3-24x^2=8x^2(2x-3)$

76. $36y-24y^2=12y(3-2y)$

77. $6ab^2+8ab-4a^2b^2=2ab(3b+4-2ab)$

78. $14a^2b^2-21ab^2+7ab$
$=7ab(2ab-3b+1)$

79. $6a(a+3b)-5(a+3b)=(6a-5)(a+3b)$

80. $4x(x-2y)-5(x-2y)=(x-2y)(4x-5)$

81. $xy-6y+3x-18=y(x-6)+3(x-6)$
$=(x-6)(y+3)$

82. $ab-8b+4a-32=b(a-b)+4(a-8)$
$=(a-8)(b+4)$

83. $pq-3p-5q+15=p(q-3)-5(q-3)$
$=(p-5)(q-3)$

84. $x^3-x^2-2x+2=x^2(x-1)-2(x-1)$
$=(x^2-2)(x-1)$

85. $2xy-x^2=x(2y-x)$

86. $x^2-14x-72=(x-18)(x+4)$

87. $x^2+16x-80=(x-4)(x+20)$

88. $2x^2-18x+28=2(x^2-9x+14)$
$=2(x-2)(x-7)$

89. $3x^2+33x+54=3(x^2+11x+18)$
$=3(x+2)(x+9)$

90. $2x^3-7x^2-9x=x(2x^2-7x-9)$
$=x(2x-9)(x+1)$

91. $3x^2+2x-16=(3x+8)(x-2)$

92. $6x^2+17x+10=(6x+5)(x+2)$

93. $15x^2-91x+6=(15x-1)(x-6)$

94. $4x^2+2x-12=2(2x^2+x-6)$
$=2(2x-3)(x+2)$

95. $9x^2-12x-12=3(3x^2-4x-4)$
$=3(x-2)(3x+2)$

96. $y^2(x+6)^2-2y(x+6)^2-3(x+6)^2$
$=(y^2-2y-3)(x+6)^2$
$=(y-3)(y+1)(x+6)^2$

97. Using the substitution $y=x+5$, we have
$(x+5)^2+6(x+5)+8$
$=y^2+6y+8=(y+2)(y+4)$
$=[(x+5)+2][(x+5)+4]$
$=(x+7)(x+9)$.

98. $x^4-6x^2-16=(x^2-8)(x^2+2)$

99. Using the substitution $y=x^2$, we have
$x^4+8x^2-20=y^2+8y-20$
$=(y-2)(y+10)$
$=(x^2-2)(x^2+10)$.

100. $x^2-100=(x+10)(x-10)$

101. $x^2-81=x^2-9^2=(x-9)(x+9)$

102. $2x^2-32=2(x^2-16)=2(x+4)(x-4)$

103. $6x^2-54=6(x^2-9)=6(x^2-3^2)$
$=6(x-3)(x+3)$

104. $81-x^4=(9+x^2)(9-x^2)$
$=(9+x^2)(3+x)(3-x)$

105. $16-y^4=4^2-(y^2)^2$
$=(4+y^2)(4-y^2)$
$=(4+y^2)(2^2-y^2)$
$=(4+y^2)(2-y)(2+y)$

106. $(y+2)^2-25=[(y+2)+5][(y+2)-5]$
$=(y+7)(y-3)$

107. $(x-3)^2-16=(x-3)^2-4^2$
$=[(x-3)-4][(x-3)+4]$
$=(x-7)(x+1)$

108. $x^3+216=x^3+6^3$
$=(x+6)(x^2-6\cdot x+36)$

109. $y^3+512=y^3+8^3$
$=(y+8)(y^2-8\cdot y+8^2)$
$=(y+8)(y^2-8y+64)$

110. $8-27y^3=2^3-(3y)^3$
$=(2-3y)(4+2\cdot 3y+9y^2)$
$=(2-3y)(4+6y+9y^2)$

111. $1-64y^3=1^3-(4y)^3$
$=(1-4y)(1+4y+(4y)^2)$
$=(1-4y)(1+4y+16y^2)$

112. $6x^4y+48xy=6xy(x^3+8)$
$=6xy(x^3+2^3)$
$=6xy(x+2)(x^2-2x+4)$

113. $2x^5+16x^2y^3$
$=2x^2(x^3+8y^3)$
$=2x^2(x^3+(2y)^3)$
$=2x^2(x+2y)(x^2-x\cdot 2y+(2y)^2)$
$=2x^2(x+2y)(x^2-2xy+4y^2)$

114. $(x^2-2x+1)-y^2=(x-1)^2-y^2$
$=(x-1-y)(x-1+y)$

115. $x^2-6x+9-4y^2$
$=(x-3)^2-4y^2$
$=(x-3)-(2y)^2$
$=(x-3-2y)(x-3+2y)$

116. $4x^2+12x+9=(2x+3)(2x+3)$
$=(2x+3)^2$

117. $16a^2-40ab+25b^2=(4a-5b)(4a-5b)$
$=(4a-5b)^2$

118. $\pi R^2h-\pi r^2h$
$=\pi h(R^2-r^2)$
$=\pi h(R-r)(R+r)$ cubic units

119. $(3x-1)(x+7)=0$
$3x-1=0$ or $x+7=0$
$3x=1$ or $x=-7$
$x=\frac{1}{3}$

The solutions are $\frac{1}{3}$, –7.

120. $3(x+5)(8x-3)=0$
$x+5=0$ or $8x-3=0$
$x=-5$ or $x=\frac{3}{8}$

The solutions are –5, $\frac{3}{8}$.

121. $5x(x-4)(2x-9)=0$
$5x=0$ or $x-4=0$ or $2x-9=0$
$x=0$ or $x=4$ or $2x=9$
$x=\frac{9}{2}$

The solutions are 0, 4, $\frac{9}{2}$.

122. $6(x+3)(x-4)(5x+1)=0$
$x+3=0$ or $x-4=0$ or $5x+1=0$
$x=-3$ or $x=4$ or $x=-\frac{1}{5}$

The solutions are –3, 4, $-\frac{1}{5}$.

123. $2x^2=12x$
$2x^2-12x=0$
$2x(x-6)=0$
$2x=0$ or $x-6=0$
$x=0$ or $x=6$
The solutions are 0, 6.

124. $4x^3-36x=0$
$4x(x^2-9)=0$
$4x(x-3)(x+3)=0$
$4x=0$ or $x-3=0$ or $x+3=0$
$x=0$ or $x=3$ or $x=-3$
The solutions are 0, 3, –3.

125. $(1-x)(3x+2)=-4x$
$3x+2-3x^2-2x=-4x$
$-3x^2+x+2=-4x$
$0=3x^2-5x-2$
$0=(3x+1)(x-2)$
$3x+1=0$ or $x-2=0$
$3x=-1$ or $x=2$
$x=-\frac{1}{3}$

The solutions are $-\frac{1}{3}$, 2.

126. $2x(x-12)=-40$
$2x^2-24x+40=0$
$2(x^2-12x+20)=0$
$2(x-2)(x-10)=0$
$x-2=0$ or $x-10=0$
$x=2$ or $x=10$
The solutions are 2, 10.

127.
$$3x^2 + 2x = 12 - 7x$$
$$3x^2 + 9x - 12 = 0$$
$$x^2 + 3x - 4 = 0$$
$$(x+4)(x-1) = 0$$
$$x+4=0 \quad \text{or} \quad x-1=0$$
$$x=-4 \quad \text{or} \quad x=1$$
The solutions are −4, 1.

128.
$$2x^2 + 3x = 35$$
$$2x^2 + 3x - 35 = 0$$
$$(2x-7)(x+5) = 0$$
$$2x-7=0 \quad \text{or} \quad x+5=0$$
$$x=\frac{7}{2} \quad \text{or} \quad x=-5$$
The solutions are $\frac{7}{2}$, −5.

129.
$$x^3 - 18x = 3x^2$$
$$x^3 - 3x^2 - 18x = 0$$
$$x(x^2 - 3x - 18) = 0$$
$$x(x-6)(x+3) = 0$$
$$x=0 \quad \text{or} \quad x-6=0 \quad \text{or} \quad x+3=0$$
$$x=6 \quad \text{or} \quad x=-3$$
The solutions are 0, 6, −3.

130.
$$19x^2 - 42x = -x^3$$
$$x^3 + 19x^2 - 42x = 0$$
$$x(x^2 + 19x - 42) = 0$$
$$x(x-2)(x+21) = 0$$
$$x=0 \quad \text{or} \quad x-2=0 \quad \text{or} \quad x+21=0$$
$$x=0 \quad \text{or} \quad x=2 \quad \text{or} \quad x=-21$$
The solutions are 0, 2, −21.

131.
$$12x = 6x^3 + 6x^2$$
$$6x^3 + 6x^2 - 12x = 0$$
$$6x(x^2 + x - 2) = 0$$
$$6x(x+2)(x-1) = 0$$
$$6x=0 \quad \text{or} \quad x+2=0 \quad \text{or} \quad x-1=0$$
$$x=0 \quad \text{or} \quad x=-2 \quad \text{or} \quad x=1$$
The solutions are 0, −2, 1.

132.
$$8x^3 + 10x^2 = 3x$$
$$8x^3 + 10x^2 - 3x = 0$$
$$x(8x^2 + 10x - 3) = 0$$
$$x(4x-1)(2x+3) = 0$$
$$x=0 \quad \text{or} \quad 4x-1=0 \quad \text{or} \quad 2x+3=0$$
$$x=\frac{1}{4} \quad \text{or} \quad x=-\frac{3}{2}$$
The solutions are 0, $\frac{1}{4}$, $-\frac{3}{2}$.

133. Let x be the number.
$$x + 2x^2 = 105$$
$$2x^2 + x - 105 = 0$$
$$(2x+15)(x-7) = 0$$
$$2x+15=0 \quad \text{or} \quad x-7=0$$
$$x=-\frac{15}{2} \quad \text{or} \quad x=7$$
The solutions are $-\frac{15}{2}$, 7.

134. Let x be the width. Then $5x - 2$ is the length.
$$x(5x-2) = 16$$
$$5x^2 - 2x - 16 = 0$$
$$(5x+8)(x-2) = 0$$
$$5x+8=0 \quad \text{or} \quad x-2=0$$
$$x=-\frac{8}{5} \quad \text{or} \quad x=2$$
Disregard the negative.
Width = 2 m
Length = 5(2) − 2 = 8 m

135.
$$h(t) = -16t^2 + 400$$
$$0 = -16t^2 + 400$$
$$0 = -16(t^2 - 25)$$
$$t^2 - 25 = 0$$
$$t^2 = 25$$
$$t = 5 \text{ seconds}$$

136. Domain: $(-\infty, \infty)$
Range: $(-\infty, 4]$

137. (−4, 0), (2, 0), (0, 3)

138. (−1, 4)

139. $\{x|-4 < x < 2\}$

140. $f(x) = x^2 + 6x + 9$

$f(x) = (x+3)^2$

Solve $(x+3)^2 = 0$

$x + 3 = 0$

$x = -3$

The x-intercept is -3. $h = \frac{-6}{2(1)} = -3$ and

$k = f(-3) = 0$ since -3 is an x-intercept.

Thus, $V(-3, 0)$.

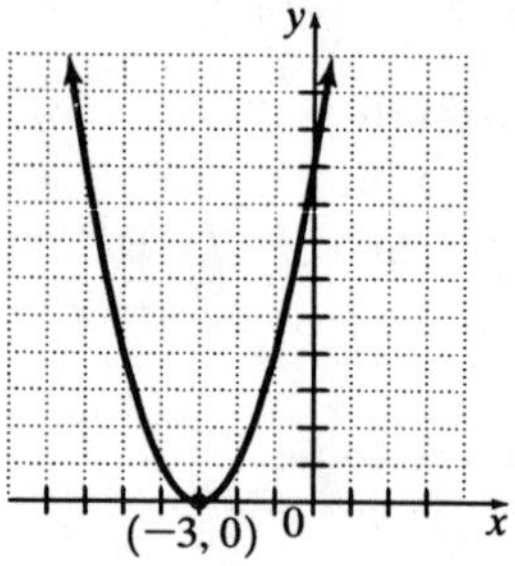

141. $f(x) = x^2 - 5x + 4$

$\frac{-b}{2a} = \frac{5}{2(1)} = \frac{5}{2}$

$f\left(\frac{5}{2}\right) = \left(\frac{5}{2}\right)^2 - 5\left(\frac{5}{2}\right) + 4 = -\frac{9}{4}$

The vertex is $\left(\frac{5}{2}, -\frac{9}{4}\right)$.

If $x^2 - 5x + 4 = 0$

$(x-1)(x-4) = 0$

$x - 1 = 0$ or $x - 4 = 0$

$x = 1$ or $x = 4$

The x-intercepts are 1 and 4. If $x = 0$, then $y = f(0) = 4$. The y-intercept is 4.

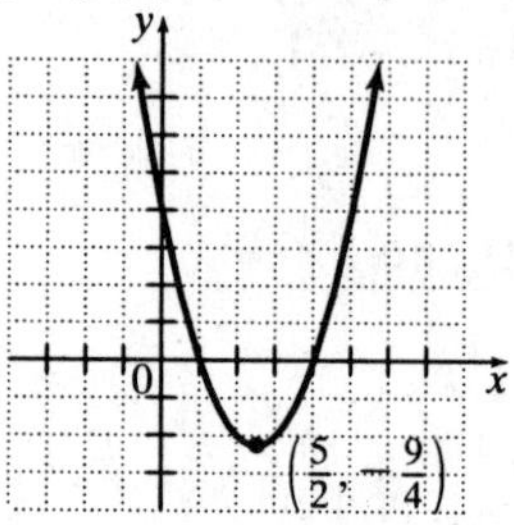

142. $f(x) = (x-1)(x^2 - 2x - 3)$

$= (x-1)(x-3)(x+1)$

Solve $(x-1)(x-3)(x+1) = 0$

$x - 1 = 0$ or $x - 3 = 0$ or $x + 1 = 0$

$x = 1$ or $x = 3$ or $x = -1$

If $x = 0$, then $y = f(0) = 3$. The y-intercept is 3. The x-intercepts are 1, 3, and -1.

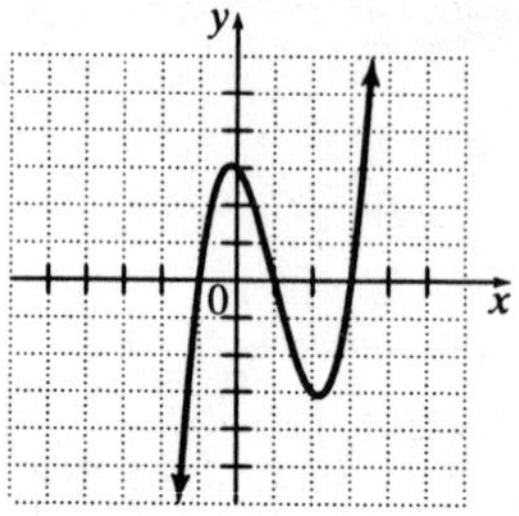

143. $f(x) = (x+3)(x^2 - 4x + 3)$

$f(x) = (x+3)(x-1)(x-3)$

If $(x+3)(x-1)(x-3) = 0$

$x + 3 = 0$ or $x - 1 = 0$ or $x - 3 = 0$

$x = -3$ or $x = 1$ or $x = 3$

The x-intercepts are -3, 1, and 3.

If $x = 0$, then $y = f(0) = (3)(3) = 9$.

The y-intercept is 9.

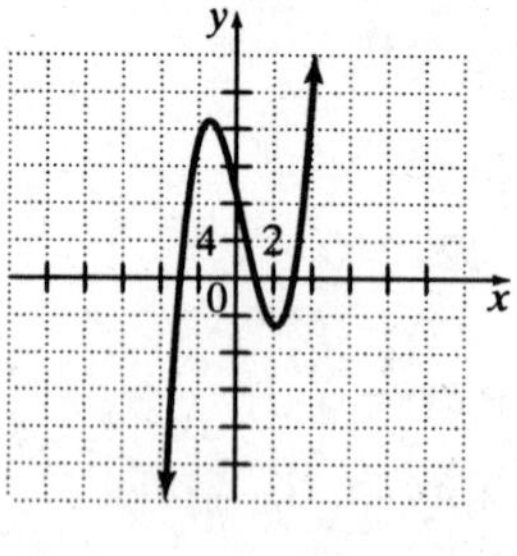

144. $f(x) = 2x^2 - 4x + 5$ does not factor, so the x-intercepts, if any, are not accessible.

$h = \frac{-(-4)}{2(2)} = 1$ and

$k = f(1) = 2(1)^2 - 4(1) + 5 = 3$, so $V(1, 3)$.

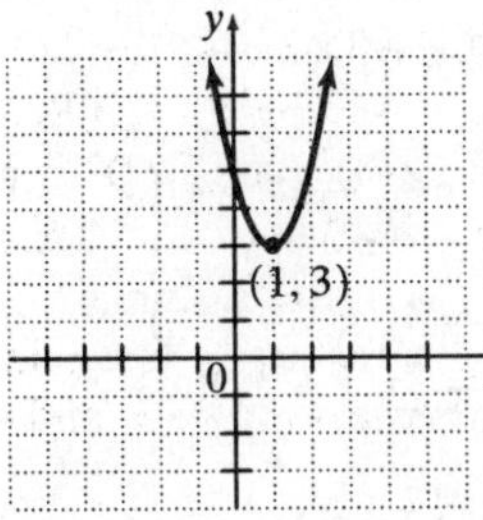

145. $f(x) = x^2 - 2x + 3$

$\frac{-b}{2a} = \frac{2}{2(1)} = 1$

$f(1) = 1 - 2 + 3 = 2$

The vertex is (1, 2). There are no x-intercepts. If $x = 0$, then $y = f(0) = 3$. The y-intercept is 3.

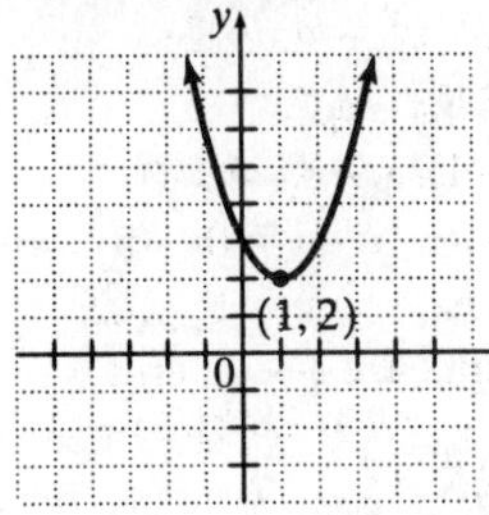

146. $f(x) = x^3 - 16x$

$= x(x^2 - 16)$

$= x(x + 4)(x - 4)$

Solve $x(x + 4)(x - 4) = 0$

$x = 0$ or $x + 4 = 0$ or $x - 4 = 0$

$x = -4$ or $x = 4$

The x-intercepts are 0, –4, and 4.
The y-intercept is 0.

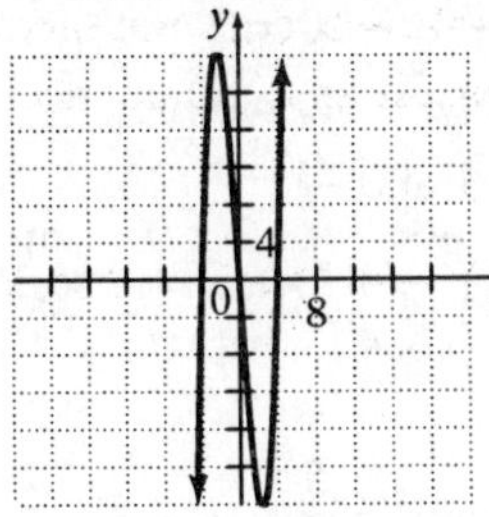

147. $f(x) = x^3 + 5x^2 + 6x$

$f(x) = x(x^2 + 5x + 6) = x(x + 3)(x + 2)$

If $x(x + 3)(x + 2) = 0$

$x = 0$ or $x + 3 = 0$ or $x + 2 = 0$

$x = 0$ or $x = -3$ or $x = -2$

The x-intercepts are –3, –2, and 0.
The y-intercept is 0.

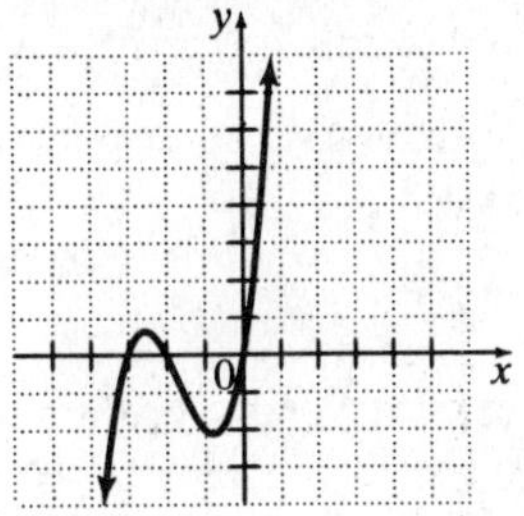

Chapter 5 Test

1. $(-9x)^{-2} = \frac{1}{(-9x)^2} = \frac{1}{81x^2}$

2. $-3xy^{-2}(4xy^2)z$

$= (-3)(4)x^{1+1}y^{-2+2}z$

$= -12x^2z$

3. $\frac{6^{-1}a^2b^{-3}}{3^{-2}a^{-5}b^2} = \frac{3^2a^{2+5}}{6^1b^{2+3}} = \frac{9a^7}{6b^5} = \frac{3a^7}{2b^5}$

4. $\left(\frac{-xy^{-5}z}{xy^3}\right)^{-5} = \frac{-x^{-5}y^{25}z^{-5}}{x^{-5}y^{-15}}$

$= \frac{-x^{-5+5}y^{25+15}}{z^5}$

$= -\frac{y^{40}}{z^5}$

5. $630{,}000{,}000 = 6.3 \times 10^8$

6. $0.01200 = 1.2 \times 10^{-2}$

7. $5.0 \times 10^{-6} = 0.000005$

8. $\dfrac{(2.4\times10^{-3})(1.2\times10^{-4})}{(3.2\times10^{-4})}$
$=\dfrac{(2.4)(1.2)}{(3.2)}\times10^{-3-4+4}$
$=0.9\times10^{-3}$
$=9\times10^{-4}$
$=0.0009$

9.
$$\begin{array}{r} 4x^3-3x-4 \\ -\ 9x^3+8x+5 \\ \hline -5x^3-11x-9 \end{array}$$

10. $-3xy(4x+y)=-12x^2y-3xy^2$

11.
$$\begin{array}{r} 3x+4 \\ \times\quad 4x-7 \\ \hline -21x-28 \\ 12x^2+16x\quad\ \ \\ \hline 12x^2-5x-28 \end{array}$$

12. $(5a-2b)(5a+2b)=(5a)^2-(2b)^2$
$=25a^2-4b^2$

13. $(6m+n)^2=(6m)^2+(6m)(n))2+(n)^2$
$=36m^2+12mn+n^2$

14.
$$\begin{array}{r} x^2-6x+4 \\ \times\quad 2x-1 \\ \hline -x^2+6x-4 \\ 2x^3-12x^2+8x\quad\ \ \\ \hline 2x^3-13x^2+14x-4 \end{array}$$

15. $16x^3y-12x^2y^4=4x^2y(4x-3y^3)$

16. $x^2-13x-30=(x-15)(x+2)$

17. $4y^2+20y+25=(2y+5)^2$

18. $6x^2-15x-9=3(2x^2-5x-3)$
$=3(2x+1)(x-3)$

19. $4x^2-25=(2x+5)(2x-5)$

20. $x^3+64=x^3+4^3=(x+4)(x^2-4x+16)$

21. $3x^2y-27y^3=3y(x^2-9y^2)$
$=3y(x+3y)(x-3y)$

22. $6x^2+24=6(x^2+4)$

23. $16y^3-2=2(8y^3-1)$
$=2[(2y)^3-(1)^3]$
$=2(2y-1)(4y^2+2y+1)$

24. $x^2y-9y-3x^2+27=(x^2-9)(y-3)$
$=(x+3)(x-3)(y-3)$

25. $3(n-4)(7n+8)=0$
$3(n-4)=0$ or $7n+8=0$
$n-4=0$ or $7n=-8$
$n=4$ or $n=-\dfrac{8}{7}$

The solutions are 4, $-\dfrac{8}{7}$.

26. $(x+2)(x-2)=5(x+4)$
$x^2-4=5x+20$
$x^2-5x-24=0$
$(x-8)(x+3)=0$
$x-8=0$ or $x+3=0$
$x=8$ or $x=-3$
The solutions are 8, –3.

27. $2x^3+5x^2-8x-20=0$
$(2x+5)(x^2-4)=0$
$(2x+5)(x+2)(x-2)=0$
$2x+5=0$ or $x+2=0$ or $x-2=0$
$2x=-5$ or $x=-2$ or $x=2$
$x=-\dfrac{5}{2}$

The solutions are $-\dfrac{5}{2}$, –2, 2.

28. $x^2-(2y)^2=(x-2y)(x+2y)$

29. $h(t) = -16t^2 + 96t + 880$

a. $-16(1)^2 + 96(1) + 880$
$= -16 + 96 + 880$
$= 960$
960 ft

b. $-16(5.1)^2 + 96(5.1) + 880$
$= -416.16 + 489.6 + 880$
$= 953.44$
953.44 ft

c. $0 = -16t^2 + 96t + 880$
$0 = (-16t - 80)(t - 11)$
$-16t - 80 = 0$ or $t - 11 = 0$
$t = -5$ or $t = 11$
Disregard the negative. 11 seconds

30. $f(x) = x^2 - 4x - 5$

$x = \frac{-b}{2a} = \frac{4}{2} = 2$

$f(2) = 2^2 - 4(2) - 5 = -9$

The vertex is (2, –9).

$f(x) = x^2 - 4x - 5 = (x+1)(x-5)$

The x-intercepts are (–1, 0), (5, 0).
The y-intercept is (0, –5).

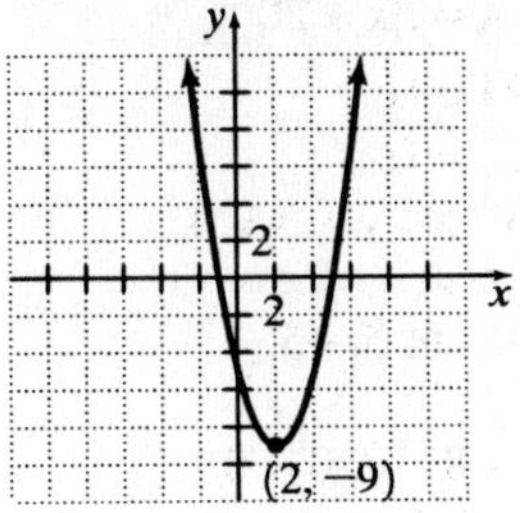

31. $f(x) = x^3 - 1$

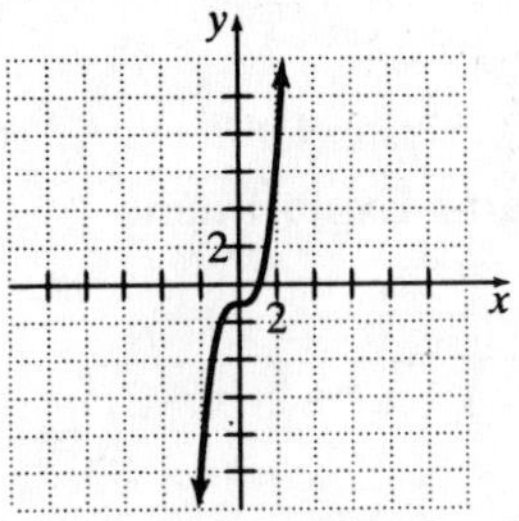

The y-intercept is (0, –1).

$f(x) = (x-1)(x^2 + x + 1)$

The x-intercept is (1, 0).

Chapter 6

Section 6.1

Graphing Calculator Explorations

1. $f(x) = \dfrac{x+1}{x^2-4} = \dfrac{x+1}{(x+2)(x-2)}$
Domain:
$\{x|x \text{ is a real number and } x \neq -2,\ x \neq 2\}$

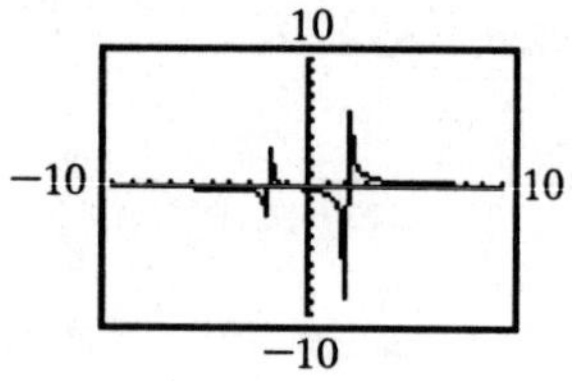

3. $h(x) = \dfrac{x^2}{2x^2+7x-4} = \dfrac{x^2}{(2x-1)(x+4)}$
Domain:
$\left\{x|x \text{ is a real number and } x \neq -4 \text{ and } x \neq \frac{1}{2}\right\}$

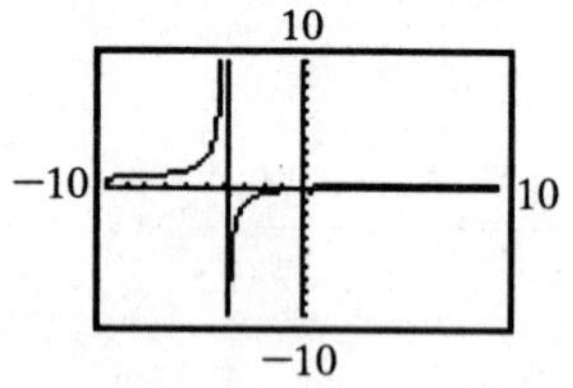

Exercise Set 6.1

1. $f(x) = \dfrac{x+8}{2x-1}$

$f(2) = \dfrac{2+8}{2(2)-1} = \dfrac{10}{4-1} = \dfrac{10}{3}$

$f(0) = \dfrac{0+8}{2(0)-1} = \dfrac{8}{-1} = -8$

$f(-1) = \dfrac{-1+8}{2(-1)-1} = \dfrac{7}{-3} = -\dfrac{7}{3}$

3. $g(x) = \dfrac{x^2+8}{x^3-25x}$

$g(3) = \dfrac{3^2+8}{3^3-25(3)} = \dfrac{9+8}{27-75} = \dfrac{17}{-48} = -\dfrac{17}{48}$

$g(-2) = \dfrac{(-2)^2+8}{(-2)^3-25(-2)} = \dfrac{4+8}{-8+50} = \dfrac{12}{42} = \dfrac{2}{7}$

$g(1) = \dfrac{1^2+8}{1^3-25(1)} = \dfrac{1+8}{1-25} = \dfrac{9}{-24} = -\dfrac{3}{8}$

5. $f(x) = \dfrac{5x-7}{4}$
Domain: $\{x|x \text{ is a real number}\}$

7. $s(t) = \dfrac{t^2+1}{2t}$
Undefined values when
$2t = 0$
$t = \dfrac{0}{2} = 0$
Domain: $\{t|t \text{ is a real number and } t \neq 0\}$

9. $f(x) = \dfrac{3x}{7-x}$
Undefined values when
$7 - x = 0$
$7 = x$
Domain: $\{x|x \text{ is a real number and } x \neq 7\}$

11. $R(x) = \frac{3+2x}{x^3+x^2-2x}$

Undefined values when

$x^3+x^2-2x=0$

$x(x^2+x-2)=0$

$x(x+2)(x-1)=0$

$x=0$ or $x+2=0$ or $x-1=0$

$x=-2$ or $x=1$

Domain: $\{x|x$ is a real number and $x \neq 0,\ x \neq -2,\ x \neq 1\}$

13. $C(x) = \frac{x+3}{x^2-4}$

Undefined values when

$x^2-4=0$

$(x+2)(x-2)=0$

$x+2=0$ or $x-2=0$

$x=-2$ or $x=2$

Domain: $\{x|x$ is a real number $x \neq 2,\ x \neq -2\}$

15. Answers may vary.

17. $\frac{4x-8}{3x-6} = \frac{4(x-2)}{3(x-2)} = \frac{4}{3}$

19. $\frac{2x-14}{7-x} = \frac{2(x-7)}{-(x-7)} = -2$

21. $\frac{x^2-2x-3}{x^2-6x+9} = \frac{(x-3)(x+1)}{(x-3)^2} = \frac{x+1}{x-3}$

23. $\frac{2x^2+12x+18}{x^2-9} = \frac{2(x^2+6x+9)}{x^2-3^2}$

$= \frac{2(x+3)^2}{(x+3)(x-3)}$

$= \frac{2(x+3)}{x-3}$

25. $\frac{3x+6}{x^2+2x} = \frac{3(x+2)}{x(x+2)} = \frac{3}{x}$

27. $\frac{2x^2-x-3}{2x^3-3x^2+2x-3} = \frac{(2x-3)(x+1)}{(2x-3)(x^2+1)}$

$= \frac{x+1}{x^2+1}$

29. $\frac{8q^2}{16q^3-16q^2} = \frac{8q^2}{16q^2(q-1)} = \frac{1}{2(q-1)}$

31. $\frac{x^2+6x-40}{10+x} = \frac{(x+10)(x-4)}{x+10} = x-4$

33. $\frac{x^3-125}{5-x} = \frac{x^3-5^3}{-(x-5)}$

$= \frac{(x-5)(x^2+5x+25)}{-(x-5)}$

$= -(x^2+5x+25)$

$= -x^2-5x-25$

35. $\frac{8x^3-27}{4x-6} = \frac{(2x)^3-3^3}{2(2x-3)}$

$= \frac{(2x-3)(4x^2+6x+9)}{2(2x-3)}$

$= \frac{4x^2+6x+9}{2}$

37. $\frac{3xy^3}{4x^3y^2} \cdot \frac{-8x^3y^4}{9x^4y^7} = \frac{-24x^4y^7}{36x^7y^9} = -\frac{2}{3x^3y^2}$

39. $\frac{8a}{3a^4b^2} \div \frac{4b^5}{6a^2b} = \frac{8a}{3a^4b^2} \cdot \frac{6a^2b}{4b^5}$

$= \frac{48a^3b}{12a^4b^7}$

$= \frac{4}{ab^6}$

41. $\frac{a^2b}{a^2-b^2} \cdot \frac{a+b}{4a^3b} = \frac{a^2b(a+b)}{4a^3b(a+b)(a-b)}$

$= \frac{1}{4a(a-b)}$

43. $\frac{x^2-9}{4} \div \frac{x^2-6x+9}{x^2-x-6}$

$= \frac{(x+3)(x-3)}{4} \cdot \frac{(x-3)(x+2)}{(x-3)^2}$

$= \frac{(x+3)(x+2)}{4}$

45. $\frac{9x+9}{4x+8} \cdot \frac{2x+4}{3x^2-3} = \frac{9(x+1) \cdot 2(x+2)}{4(x+2) \cdot 3(x^2-1)}$

$= \frac{18(x+1)}{12(x+1)(x-1)}$

$= \frac{3}{2(x-1)}$

47. $\frac{a+b}{ab} \div \frac{a^2-b^2}{4a^3b} = \frac{a+b}{ab} \cdot \frac{4a^3b}{(a+b)(a-b)}$

$= \frac{4a^2}{a-b}$

49. $\frac{2x^2-4x-30}{5x^2-40x-75} \div \frac{x^2-8x+15}{x^2-6x+9}$

$= \frac{2(x^2-2x-15)}{5(x^2-8x-15)} \cdot \frac{(x-3)^2}{(x-5)(x-3)}$

$= \frac{2(x-5)(x+3)(x-3)}{5(x^2-8x-15)(x-5)}$

$= \frac{2(x+3)(x-3)}{5(x^2-8x-15)}$

51. $\frac{2x^3-16}{6x^2+6x-36} \cdot \frac{9x+18}{3x^2+6x+12}$

$= \frac{2(x^3-8) \cdot 9(x+2)}{6(x^2+x-6) \cdot 3(x^2+2x+4)}$

$= \frac{18(x-2)(x^2+2x+4)(x+2)}{18(x+3)(x-2)(x^2+2x+4)}$

$= \frac{x+2}{x+3}$

53. $\frac{15b-3a}{b^2-a^2} \div \frac{a-5b}{ab+b^2}$

$= \frac{3(5b-a)}{(b+a)(b-a)} \cdot \frac{b(a+b)}{-(5b-a)}$

$= \frac{3}{b-a} \cdot \frac{b}{-1}$

$= \frac{3b}{a-b}$

55. $\frac{a^3+a^2b+a+b}{a^3+a} \cdot \frac{6a^2}{2a^2-2b^2}$

$= \frac{(a^2+1)(a+b)}{a(a^2+1)} \cdot \frac{6a^2}{2(a^2-b^2)}$

$= \frac{a+b}{1} \cdot \frac{3a}{(a+b)(a-b)}$

$= \frac{3a}{a-b}$

57. $\frac{5a}{12} \cdot \frac{2}{25a^2} \cdot \frac{15a}{2} = \frac{1}{6} \cdot \frac{1}{5a} \cdot \frac{15a}{2}$

$= \frac{1}{6} \cdot \frac{1}{1} \cdot \frac{3}{2}$

$= \frac{3}{12}$

$= \frac{1}{4}$

59. $\frac{3x-x^2}{x^3-27} \div \frac{x}{x^2+3x+9}$

$= \frac{x(3-x)}{(x-3)(x^2+3x+9)} \cdot \frac{x^2+3x+9}{x}$

$= \frac{3-x}{x-3}$

$= \frac{-(x-3)}{x-3}$

$= -1$

61. $\frac{4a}{7} \div \left(\frac{a^2}{14} \cdot \frac{3}{a}\right) = \frac{4a}{7} \div \frac{3a}{14}$

$= \frac{4a}{7} \cdot \frac{14}{3a}$

$= \frac{4}{1} \cdot \frac{2}{3}$

$= \frac{8}{3}$

63. $\frac{8b+24}{3a+6} \div \frac{ab-2b+3a-6}{a^2-4a+4} = \frac{8(b+3)}{3(a+2)} \cdot \frac{(a-2)^2}{(a-2)(b+3)} = \frac{8(a-2)}{3(a+2)}$

65. $\frac{4}{x} \div \frac{3xy}{x^2} \cdot \frac{6x^2}{x^4} = \frac{4}{x} \cdot \frac{x^2}{3xy} \cdot \frac{6}{x^2} = \frac{4}{3y} \cdot \frac{6}{x^2} = \frac{4}{y} \cdot \frac{2}{x^2} = \frac{8}{x^2 y}$

67. $\frac{3x^2-5x-2}{y^2+y-2} \cdot \frac{y^2+4y-5}{12x^2+7x+1} \div \frac{5x^2-9x-2}{8x^2-2x-1} = \frac{(3x+1)(x-2)(y+5)(y-1)(4x+1)(2x-1)}{(y+2)(y-1)(4x+1)(3x+1)(5x+1)(x-2)}$

$= \frac{(y+5)(2x-1)}{(y+2)(5x+1)}$

69. $\frac{5a^2-20}{3a^2-12a} \div \frac{a^3+2a^2}{2a^2-8a} \cdot \frac{9a^3+6a^2}{2a^2-4a} = \frac{5(a^2-4)}{3a(a-4)} \cdot \frac{2a(a-4)}{a^2(a+2)} \cdot \frac{3a^2(3a+2)}{2a(a-2)}$

$= \frac{5(a+2)(a-2)(3a+2)}{a(a+2)(a-2)} = \frac{5(3a+2)}{a}$

71. $\frac{5x^4+3x^2-2}{x-1} \cdot \frac{x+1}{x^4-1} = \frac{(5x^2-2)(x^2+1)(x+1)}{(x-1)(x^2+1)(x^2-1)} = \frac{(5x^2-2)(x+1)}{(x-1)(x+1)(x-1)} = \frac{5x^2-2}{(x-1)^2}$

73. Recall that $A = l \cdot w$. So,

$A = \frac{x+2}{x} \cdot \frac{5x}{x^2-4} = \frac{(x+2)5}{(x+2)(x-2)} = \frac{5}{x-2}$ sq. meter

75. Answers may vary.

77. Since $A = b \cdot h$, $b = \frac{A}{h}$. Now, $b = \frac{\frac{x^2+x-2}{x^3}}{\frac{x^2}{x-1}}$

$b = \frac{(x+2)(x-1)}{x^3} \cdot \frac{(x-1)}{x^2}$

$b = \frac{(x+2)(x-1)^2}{x^5}$ feet

79. $f(x)=\dfrac{20x}{100-x}$

x	0	10	30	50	70	90	95	99
y	0	$\frac{20}{9}$	$\frac{60}{7}$	20	$\frac{140}{3}$	180	380	1980

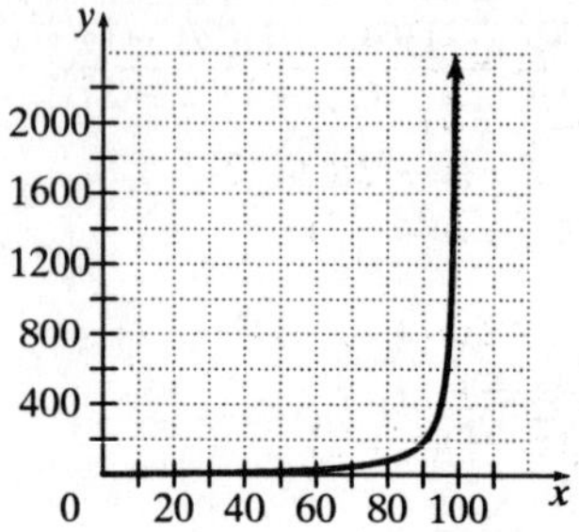

81. $f(x)=\dfrac{100,000x}{100-x}$

a. $\{x|0\le x<100\}$

b. $f(30)=\dfrac{100,000(30)}{100-30}=\$42,857.14$

c. $f(60)=\dfrac{100,000(60)}{100-60}=\$150,000$

$f(80)=\dfrac{100,000(80)}{100-80}=\$400,000$

d. $f(90)=\dfrac{100,000(90)}{100-90}=\$900,000$

$f(95)=\dfrac{100,000(95)}{100-95}=\$1,900,000$

$f(99)=\dfrac{100,000(99)}{100-99}=\$9,900,000$

83. $\frac{4}{5}+\frac{3}{5}=\frac{7}{5}$

85. $\frac{5}{28}-\frac{2}{21}$

The LCD is 84.

$$\frac{5}{28}\cdot\frac{3}{3}-\frac{2}{21}\cdot\frac{4}{4}=\frac{15}{84}-\frac{8}{84}$$
$$=\frac{7}{84}$$
$$=\frac{7\cdot 1}{7\cdot 12}$$
$$=\frac{1}{12}$$

87. $\frac{3}{8}+\frac{1}{2}-\frac{3}{16}$

The LCD is 16.

$$\frac{3}{8}\cdot\frac{2}{2}+\frac{1}{2}\cdot\frac{8}{8}-\frac{3}{16}=\frac{6}{16}+\frac{8}{16}-\frac{3}{16}$$
$$=\frac{14}{16}-\frac{3}{16}$$
$$=\frac{11}{16}$$

89. $\dfrac{x^{2n}-4}{7x}\cdot\dfrac{14x^3}{x^n-2}=\dfrac{(x^n+2)(x^n-2)2x^2}{x^n-2}$

$=2x^2(x^n+2)$

91. $\dfrac{y^{2n}+9}{10y}\cdot\dfrac{y^n-3}{y^{4n}-81}$

$$=\frac{(y^{2n}+9)(y^n-3)}{10y(y^{2n}+9)(y^{2n}-9)}$$
$$=\frac{y^n-3}{10y(y^n+3)(y^n-3)}$$
$$=\frac{1}{10y(y^n+3)}$$

93. $\dfrac{y^{2n}-y^n-2}{2y^n-4}\div\dfrac{y^{2n}-1}{1+y^n}$

$$=\frac{(y^n-2)(y^n+1)}{2(y^n-2)}\cdot\frac{y^n+1}{(y^n+1)(y^n-1)}$$
$$=\frac{y^n+1}{2(y^n-1)}$$

Section 6.2

Graphing Calculator Explorations

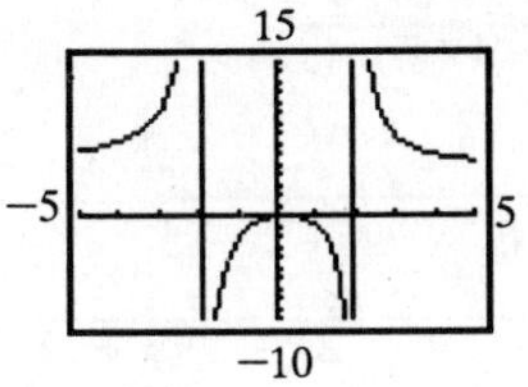

Exercise Set 6.2

1. $\frac{2}{x}-\frac{5}{x}=\frac{2-5}{x}=-\frac{3}{x}$

3. $\frac{2}{x-2}+\frac{x}{x-2}=\frac{2+x}{x-2}=\frac{x+2}{x-2}$

5. $\frac{x^2}{x+2}-\frac{4}{x+2}=\frac{x^2-4}{x+2}$
$=\frac{(x+2)(x-2)}{x+2}$
$=x-2$

7. $\frac{2x-6}{x^2+x-6}+\frac{3-3x}{x^2+x-6}=\frac{-x-3}{x^2+x-6}$
$=\frac{-(x+3)}{(x+3)(x-2)}$
$=\frac{-1}{x-2}$
$=\frac{1}{2-x}$

9. Recall that $P=4\cdot s$.
$P=4\left(\frac{x}{x+5}\right)=\frac{4x}{x+5}$ feet
Recall that $A=s^2$.
$A=\left(\frac{x}{x+5}\right)^2=\frac{x^2}{(x+5)^2}$
$=\frac{x^2}{x^2+10x+25}$ sq. feet

11. $7=7$
$5x=5\cdot x$
$\text{LCD}=7\cdot 5\cdot x=35x$

13. $x=x$
$x+1=x+1$
$\text{LCD}=x(x+1)$

15. $x+7=x+7$
$x-7=x-7$
$\text{LCD}=(x+7)(x-7)$

17. $3x+6=3(x+2)$
$2x-4=2(x-2)$
$\text{LCD}=2\cdot 3\cdot (x+2)(x-2)$
$=6(x+2)(x-2)$

19. $(3x-1)(x+2)=(3x-1)(x+2)$
$3x-1=3x-1$
$\text{LCD}=(3x-1)(x+2)$

21. $a^2-b^2=(a+b)(a-b)$
$a^2-2ab+b^2=(a-b)(a-b)$
$\text{LCD}=(a+b)(a-b)(a-b)$
$=(a+b)(a-b)^2$

23. $x^2-9=(x+3)(x-3)$
$x=x$
$12-4x=4(3-x)=-4(x-3)$
$\text{LCD}=-4x(x+3)(x-3)$

25. Answers may vary.

27. $\frac{4}{3x}+\frac{3}{2x}=\frac{(4)2}{(3x)2}+\frac{(3)3}{(2x)3}=\frac{8+9}{6x}=\frac{17}{6x}$

29. $\frac{5}{2y^2}-\frac{2}{7y}=\frac{(5)7}{(2y^2)7}-\frac{2(2y)}{(7y)2y}=\frac{35-4y}{14y^2}$

31. $\frac{x-3}{x+4}-\frac{x+2}{x-4}$
$=\frac{(x-3)(x-4)}{(x+4)(x-4)}-\frac{(x+2)(x+4)}{(x-4)(x+4)}$
$=\frac{(x^2-7x+12)-(x^2+6x+8)}{(x+4)(x-4)}$
$=\frac{-13x+4}{(x+4)(x-4)}$

33. $\frac{1}{x-5}+\frac{x}{x^2-x-20}$
$=\frac{1}{x-5}+\frac{x}{(x-5)(x+4)}$
$=\frac{1(x+4)}{(x-5)(x+4)}+\frac{x}{(x-5)(x+4)}$
$=\frac{x+4+x}{(x-5)(x+4)}$
$=\frac{2x+4}{(x-5)(x+4)}$

35. $\frac{1}{a-b}+\frac{1}{b-a}=\frac{1}{a-b}+\frac{1}{-(a-b)}$
$=\frac{1}{a-b}-\frac{1}{a-b}$
$=0$

37. $\frac{x+1}{1-x}+\frac{1}{x-1}=\frac{x+1}{-(x-1)}+\frac{1}{x-1}$
$=\frac{-(x+1)+1}{x-1}$
$=-\frac{x}{x-1}$

39. $\frac{5}{x-2}+\frac{x+4}{2-x}=\frac{5}{x-2}+\frac{x+4}{-(x-2)}$
$=\frac{5}{x-2}-\frac{x+4}{x-2}$
$=\frac{5-(x+4)}{x-2}$
$=\frac{5-x-4}{x-2}$
$=\frac{-x+1}{x-2}$

41. $\frac{y+1}{y^2-6y+8}-\frac{3}{y^2-16}$
$=\frac{y+1}{(y-2)(y-4)}-\frac{3}{(y+4)(y-4)}$
$=\frac{y^2+5y+4-(3y-6)}{(y-2)(y-4)(y+4)}$
$=\frac{y^2+5y+4-3y+6}{(y-2)(y-4)(y+4)}$
$=\frac{y^2+2y+10}{(y-2)(y-4)(y+4)}$

43. $\frac{x+4}{3x^2+11x+6}+\frac{x}{2x^2+x-15}$
$=\frac{x+4}{(3x+2)(x+3)}+\frac{x}{(2x-5)(x+3)}$
$+\frac{x(3x+2)}{(2x-5)(x+3)(3x+2)}$
$=\frac{2x^2+3x-20+3x^2+2x}{(3x+2)(x+3)(2x-5)}$
$=\frac{5x^2+5x-20}{(3x+2)(x+3)(2x-5)}$
$=\frac{5(x^2+x-4)}{(3x+2)(x+3)(2x-5)}$

45. $\frac{7}{x^2-x-2}+\frac{x}{x^2+4x+3}$
$=\frac{7}{(x-2)(x+1)}+\frac{x}{(x+1)(x+3)}$
$=\frac{7(x+3)}{(x-2)(x+1)(x+3)}$
$+\frac{x(x-2)}{(x+1)(x+3)(x-2)}$
$=\frac{7x+21+x^2-2x}{(x-2)(x+1)(x+3)}$
$=\frac{x^2+5x+21}{(x-2)(x+1)(x+3)}$

47. $\dfrac{2}{x+1}-\dfrac{3x}{3x+3}+\dfrac{1}{2x+2}$
$=\dfrac{2}{x+1}-\dfrac{3x}{3(x+1)}+\dfrac{1}{2(x+1)}$
$=\dfrac{2}{x+1}-\dfrac{x}{x+1}+\dfrac{1}{2(x+1)}$
$=\dfrac{2-x}{x+1}+\dfrac{1}{2(x+1)}$
$=\dfrac{2(2-x)}{2(x+1)}+\dfrac{1}{2(x+1)}$
$=\dfrac{4-2x+1}{2(x+1)}$
$=\dfrac{-2x+5}{2(x+1)}$

49. $\dfrac{3}{x+3}+\dfrac{5}{x^2+6x+9}-\dfrac{x}{x^2-9}$
$=\dfrac{3}{x+3}+\dfrac{5}{(x+3)^2}-\dfrac{x}{(x+3)(x-3)}$
$=\dfrac{3(x+3)(x-3)}{(x+3)^2(x-3)}+\dfrac{5(x-3)}{(x+3)^2(x-3)}$
$-\dfrac{x(x+3)}{(x+3)^2(x-3)}$
$=\dfrac{3x^2-27+5x-15-x^2-3x}{(x+3)^2(x-3)}$
$=\dfrac{2x^2+2x-42}{(x+3)^2(x-3)}$
$=\dfrac{2(x^2+x-21)}{(x+3)^2(x-3)}$

51. $\dfrac{4}{3x^2y^3}+\dfrac{5}{3x^2y^3}=\dfrac{4+5}{3x^2y^3}$
$=\dfrac{9}{3x^2y^3}$
$=\dfrac{3}{x^2y^3}$

53. $\dfrac{x-5}{2x}-\dfrac{x+5}{2x}=\dfrac{x-5-x-5}{2x}=\dfrac{-10}{2x}=-\dfrac{5}{x}$

55. $\dfrac{3}{2x+10}+\dfrac{8}{3x+15}=\dfrac{3}{2(x+5)}+\dfrac{8}{3(x+5)}$
$=\dfrac{9}{6(x+5)}+\dfrac{16}{6(x+5)}$
$=\dfrac{9+16}{6(x+5)}$
$=\dfrac{25}{6(x+5)}$

57. $\dfrac{-2}{x^2-3x}-\dfrac{1}{x^3-3x^2}$
$=\dfrac{-2}{x(x-3)}-\dfrac{1}{x^2(x-3)}$
$=\dfrac{-2x}{x^2(x-3)}-\dfrac{1}{x^2(x-3)}$
$=\dfrac{-2x-1}{x^2(x-3)}$

59. $\dfrac{ab}{a^2-b^2}+\dfrac{b}{a+b}$
$=\dfrac{ab}{(a+b)(a-b)}+\dfrac{b}{a+b}$
$=\dfrac{ab}{(a+b)(a-b)}+\dfrac{b(a-b)}{(a+b)(a-b)}$
$=\dfrac{ab+ab-b^2}{(a+b)(a-b)}$
$=\dfrac{2ab-b^2}{(a+b)(a-b)}$

61. $\dfrac{5}{x^2-4}-\dfrac{3}{x^2+4x+4}$
$=\dfrac{5}{(x+2)(x-2)}-\dfrac{3}{(x+2)^2}$
$=\dfrac{5(x+2)}{(x+2)^2(x-2)}-\dfrac{3(x-2)}{(x+2)^2(x-2)}$
$=\dfrac{5x+10-3x+6}{(x+2)^2(x-2)}$
$=\dfrac{2x+16}{(x+2)^2(x-2)}$

63. $\dfrac{2}{a^2+2a+1}+\dfrac{3}{a^2-1}$
$=\dfrac{2}{(a+1)^2}+\dfrac{3}{(a+1)(a-1)}$

$$= \frac{2(a-1)}{(a+1)^2(a-1)} + \frac{3(a+1)}{(a+1)^2(a-1)}$$
$$= \frac{2a-2+3a+3}{(a+1)^2(a-1)}$$
$$= \frac{5a+1}{(a+1)^2(a-1)}$$

65. Answers may vary.

67. Answers may vary.

69. $\left(\frac{2}{3} - \frac{1}{x}\right) \cdot \left(\frac{3}{x} + \frac{1}{2}\right)$
$$= \left(\frac{2x}{3x} - \frac{3}{3x}\right) \cdot \left(\frac{6}{2x} + \frac{x}{2x}\right)$$
$$= \frac{2x-3}{3x} \cdot \frac{x+6}{2x}$$
$$= \frac{2x^2+9x-18}{6x^2}$$

71. $\left(\frac{1}{x} + \frac{2}{3}\right) - \left(\frac{1}{x} - \frac{2}{3}\right)$
$$= \frac{1}{x} + \frac{2}{3} - \frac{1}{x} + \frac{2}{3}$$
$$= \frac{4}{3}$$

73. $\left(\frac{2a}{3}\right)^2 \div \left(\frac{a^2}{a+1} - \frac{1}{a+1}\right)$
$$= \frac{4a^2}{9} \div \frac{a^2-1}{a+1}$$
$$= \frac{4a^2}{9} \cdot \frac{a+1}{(a+1)(a-1)}$$
$$= \frac{4a^2}{9} \cdot \frac{1}{a-1}$$
$$= \frac{4a^2}{9(a-1)}$$

75. $\left(\frac{2x}{3}\right)^2 \div \left(\frac{x}{3}\right)^2 = \left(\frac{\frac{2x}{3}}{\frac{x}{3}}\right)^2 = 2^2 = 4$

77. $\frac{x}{x^2-9} + \frac{3}{x^2-6x+9} - \frac{1}{x+3}$
$$= \frac{x}{(x+3)(x-3)} + \frac{3}{(x-3)^2} - \frac{1}{(x+3)}$$
$$= \frac{x(x-3)+3(x+3)-(x-3)^2}{(x+3)(x-3)^2}$$
$$= \frac{x^2-3x+3x+9-x^2+6x-9}{(x-3)(x-3)^2}$$
$$= \frac{6x}{(x+3)(x-3)^2}$$

79. $\left(\frac{x}{x+1} - \frac{x}{x-1}\right) \div \frac{x}{2x+2}$
$$= \frac{x(x-1)-x(x+1)}{(x-1)(x+1)} \cdot \frac{2(x+1)}{x}$$
$$= \frac{x^2-x-x^2-x}{(x+1)(x-1)} \cdot \frac{2(x+1)}{x}$$
$$= \frac{-2x}{(x+1)(x-1)} \cdot \frac{2(x+1)}{x}$$
$$= \frac{-4}{x-1}$$

81. $\frac{4}{x} \cdot \left(\frac{2}{x+2} - \frac{2}{x-2}\right)$
$$= \frac{4 \cdot 2}{x}\left(\frac{1}{x+2} - \frac{1}{x-2}\right)$$
$$= \frac{8}{x}\left(\frac{x-2-(x+2)}{(x+2)(x-2)}\right)$$
$$= \frac{8}{x} \cdot \frac{x-2-x-2}{(x+2)(x-2)}$$
$$= \frac{8}{x} \cdot \frac{-4}{(x+2)(x-2)}$$
$$= -\frac{32}{x(x+2)(x-2)}$$

83.

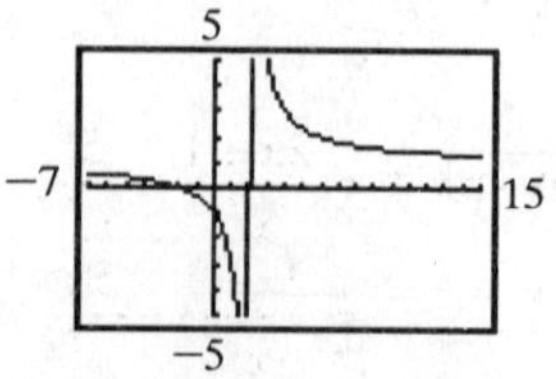

85.

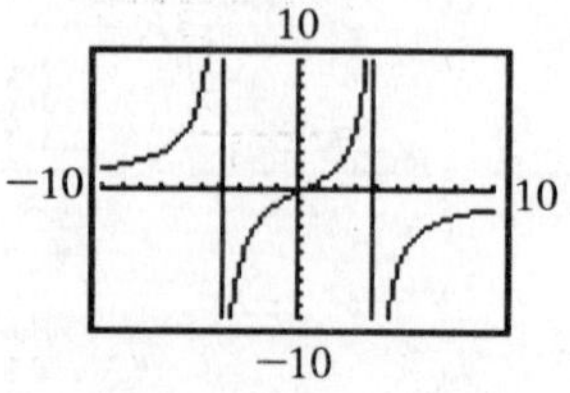

87. $12\left(\frac{2}{3}+\frac{1}{6}\right)=\frac{24}{3}+\frac{12}{6}=8+2=10$

89. $x^2\left(\frac{4}{x^2}+1\right)=x^2\cdot\frac{4}{x^2}+x^2\cdot 1=4+x^2$

91. $\sqrt{100}=10$

93. $\sqrt[3]{8}=2$

95. $\sqrt[4]{81}=3$

97. $a^2+b^2=c^2$
$3^2+4^2=c^2$
$9+16=c^2$
$25=c^2$
$c=5$ meters

99. $x^{-1}+(2x)^{-1}=\frac{1}{x}+\frac{1}{2x}=\frac{2}{2x}+\frac{1}{2x}=\frac{3}{2x}$

101. $4x^{-2}-3x^{-1}=\frac{4}{x^2}-\frac{3}{x}=\frac{4}{x^2}-\frac{3x}{x^2}=\frac{4-3x}{x^2}$

103. $x^{-3}(2x+1)-5x^{-2}=\frac{2x+1}{x^3}-\frac{5}{x^2}=\frac{2x+1}{x^3}-\frac{5x}{x^3}=\frac{2x+1-5x}{x^3}=\frac{1-3x}{x^3}$

Exercise Set 6.3

1. $\frac{\frac{1}{3}}{\frac{2}{5}}=\frac{1}{3}\cdot\frac{5}{2}=\frac{5}{6}$

3. $\frac{\frac{4}{x}}{\frac{5}{2x}}=\frac{4}{x}\cdot\frac{2x}{5}=\frac{4}{1}\cdot\frac{2}{5}=\frac{8}{5}$

5. $\frac{\frac{10}{3x}}{\frac{5}{6x}}=\frac{10}{3x}\cdot\frac{6x}{5}=\frac{2}{1}\cdot\frac{2}{1}=4$

7. $\frac{1+\frac{2}{5}}{2+\frac{3}{5}}=\frac{\left(1+\frac{2}{5}\right)5}{\left(2+\frac{3}{5}\right)5}=\frac{5+2}{10+3}=\frac{7}{13}$

9. $\frac{\frac{4}{x-1}}{\frac{x}{x-1}}=\frac{4}{x-1}\cdot\frac{x-1}{x}=\frac{4}{x}$

11. $\frac{1-\frac{2}{x}}{x-\frac{4}{9x}}=\frac{\left(1-\frac{2}{x}\right)9x}{\left(x-\frac{4}{9x}\right)9x}=\frac{9x-18}{9x^2-4}$

13. $\frac{\frac{1}{x+1}-1}{\frac{1}{x-1}+1}=\frac{\left(\frac{1}{x+1}-1\right)(x+1)(x-1)}{\left(\frac{1}{x-1}+1\right)(x-1)(x+1)}$
$=\frac{[1-(x+1)](x-1)}{[1+(x-1)](x+1)}$
$=\frac{(1-x-1)(x-1)}{(1+x-1)(x+1)}$
$=\frac{-x(x-1)}{x(x+1)}$
$=\frac{-(x-1)}{x+1}$
$=\frac{1-x}{x+1}$

15. $\dfrac{x^{-1}}{x^{-2}+y^{-2}}=\dfrac{\frac{1}{x}}{\frac{1}{x^2}+\frac{1}{y^2}}$

$=\dfrac{\frac{1}{x}(x^2y^2)}{\left(\frac{1}{x^2}+\frac{1}{y^2}\right)(x^2y^2)}$

$=\dfrac{xy^2}{y^2+x^2}$

17. $\dfrac{2a^{-1}+3b^{-2}}{a^{-1}-b^{-1}}=\dfrac{\frac{2}{a}+\frac{3}{b^2}}{\frac{1}{a}-\frac{1}{b}}$

$=\dfrac{\left(\frac{2}{a}+\frac{3}{b^2}\right)ab^2}{\left(\frac{1}{a}-\frac{1}{b}\right)ab^2}$

$=\dfrac{2b^2+3a}{b^2-ab}$

19. $\dfrac{1}{x-x^{-1}}=\dfrac{1}{x-\frac{1}{x}}$

$=\dfrac{1\cdot x}{\left(x-\frac{1}{x}\right)\cdot x}$

$=\dfrac{x}{x^2-1}$

21. $\dfrac{\frac{x+1}{7}}{\frac{x+2}{7}}=\dfrac{x+1}{7}\cdot\dfrac{7}{x+2}=\dfrac{x+1}{x+2}$

23. $\dfrac{\frac{1}{2}-\frac{1}{3}}{\frac{3}{4}+\frac{2}{5}}=\dfrac{\left(\frac{1}{2}-\frac{1}{3}\right)60}{\left(\frac{3}{4}+\frac{2}{5}\right)60}=\dfrac{30-20}{45+24}=\dfrac{10}{69}$

25. $\dfrac{\frac{x+1}{3}}{\frac{2x-1}{6}}=\dfrac{x+1}{3}\cdot\dfrac{6}{2x-1}$

$=\dfrac{x+1}{1}\cdot\dfrac{2}{2x-1}=\dfrac{2(x+1)}{2x-1}$

27. $\dfrac{\frac{x}{3}}{\frac{2}{x+1}}=\dfrac{x}{3}\cdot\dfrac{x+1}{2}=\dfrac{x(x+1)}{6}$

29. $\dfrac{\frac{2}{x}+3}{\frac{4}{x^2}-9}=\dfrac{\left(\frac{2}{x}+3\right)x^2}{\left(\frac{4}{x^2}-9\right)x^2}=\dfrac{2x+3x^2}{4-9x^2}$

$=\dfrac{x(2+3x)}{(2-3x)(2+3x)}=\dfrac{x}{2-3x}$

31. $\dfrac{1-\frac{x}{y}}{\frac{x^2}{y^2}-1}=\dfrac{\left(1-\frac{x}{y}\right)y^2}{\left(\frac{x^2}{y^2}-1\right)y^2}$

$=\dfrac{y^2-xy}{x^2-y^2}$

$=\dfrac{y(y-x)}{(x+y)(x-y)}$

$=\dfrac{y(-1)}{x+y}$

$=-\dfrac{y}{x+y}$

33. $\dfrac{\frac{-2x}{x-y}}{\frac{y}{x^2}}=\dfrac{-2x}{x-y}\cdot\dfrac{x^2}{y}=-\dfrac{2x^3}{y(x-y)}$

35. $\dfrac{\frac{2}{x}+\frac{1}{x^2}}{\frac{y}{x^2}}=\dfrac{\left(\frac{2}{x}+\frac{1}{x^2}\right)x^2}{\left(\frac{y}{x^2}\right)x^2}=\dfrac{2x+1}{y}$

37. $\dfrac{\frac{x}{9}-\frac{1}{x}}{1+\frac{3}{x}}=\dfrac{\left(\frac{x}{9}-\frac{1}{x}\right)9x}{\left(1+\frac{3}{x}\right)9x}=\dfrac{x^2-9}{9x+27}$

$=\dfrac{(x+3)(x-3)}{9(x+3)}=\dfrac{x-3}{9}$

39. $\dfrac{\frac{x-1}{x^2-4}}{1+\frac{1}{x-2}}=\dfrac{\frac{x-1}{(x+2)(x-2)}(x+2)(x-2)}{\left(1+\frac{1}{x-2}\right)(x-2)(x+2)}$

$=\dfrac{x-1}{(x-2+1)(x+2)}$

$=\dfrac{x-1}{(x-1)(x+2)}$

$=\dfrac{1}{x+2}$

41. $\dfrac{\frac{4}{5-x}+\frac{5}{x-5}}{\frac{2}{x}+\frac{3}{x-5}} = \dfrac{-\frac{4}{x-5}+\frac{5}{x-5}}{\frac{2(x-5)+3x}{x(x-5)}}$
$= \dfrac{\frac{1}{x-5}}{\frac{2x-10+3x}{x(x-5)}}$
$= \dfrac{1}{x-5}\cdot\dfrac{x(x-5)}{5x-10}$
$= \dfrac{x}{5x-10}$

43. $\dfrac{\frac{x+2}{x}-\frac{2}{x-1}}{\frac{x+1}{x}+\frac{x+1}{x-1}} = \dfrac{\frac{(x+2)(x-1)-2x}{x(x-1)}}{\frac{(x+1)(x-1)+(x+1)x}{x(x-1)}}$
$= \dfrac{x^2+x-2-2x}{x^2-1+x^2+x}$
$= \dfrac{x^2-x-2}{2x^2+x-1}$
$= \dfrac{(x-2)(x+1)}{(2x-1)(x+1)}$
$= \dfrac{x-2}{2x-1}$

45. $\dfrac{\frac{x-2}{x+2}+\frac{x+2}{x-2}}{\frac{x-2}{x+2}-\frac{x+2}{x-2}} = \dfrac{\frac{(x-2)(x-2)+(x+2)(x+2)}{(x+2)(x-2)}}{\frac{(x-2)(x-2)-(x+2)(x+2)}{(x+2)(x-2)}}$
$= \dfrac{x^2-4x+4+x^2+4x+4}{x^2-4x+4-x^2-4x-4}$
$= \dfrac{2x^2+8}{-8x}$
$= -\dfrac{x^2+4}{4x}$

47. $\dfrac{\frac{2}{y^2}-\frac{5}{xy}-\frac{3}{x^2}}{\frac{2}{y^2}+\frac{7}{xy}+\frac{3}{x^2}} = \dfrac{\frac{2x^2-5xy-3y^2}{x^2y^2}}{\frac{2x^2+7xy+3y^2}{x^2y^2}}$
$= \dfrac{(2x+y)(x-3y)}{(2x+y)(x+3y)}$
$= \dfrac{x-3y}{x+3y}$

49. $\dfrac{a^{-1}+1}{a^{-1}-1} = \dfrac{\left(\frac{1}{a}+1\right)a}{\left(\frac{1}{a}-1\right)a} = \dfrac{1+a}{1-a}$

51. $\dfrac{3x^{-1}+(2y)^{-1}}{x^{-2}} = \dfrac{\left(\frac{3}{x}+\frac{1}{2y}\right)\cdot 2x^2y}{\left(\frac{1}{x^2}\right)\cdot 2x^2y}$
$= \dfrac{6xy+x^2}{2y}$

53. $\dfrac{2a^{-1}+(2a)^{-1}}{a^{-1}+2a^{-2}} = \dfrac{\left(\frac{2}{a}+\frac{1}{2a}\right)\cdot 2a^2}{\left(\frac{1}{a}+\frac{2}{a^2}\right)\cdot 2a^2}$
$= \dfrac{4a+a}{2a+4}$
$= \dfrac{5a}{2a+4}$

55. $\dfrac{5x^{-1}+2y^{-1}}{x^{-2}y^{-2}} = \dfrac{\left(\frac{5}{x}+\frac{2}{y}\right)\cdot x^2y^2}{\left(\frac{1}{x^2}\cdot\frac{1}{y^2}\right)\cdot x^2y^2}$
$= \dfrac{5xy^2+2x^2y}{1}$
$= 5xy^2+2x^2y$

57. $\dfrac{5x^{-1}-2y^{-1}}{25x^{-2}-4y^{-2}}$
$= \dfrac{\left(\frac{5}{x}-\frac{2}{y}\right)\cdot x^2y^2}{\left(\frac{25}{x^2}-\frac{4}{y^2}\right)\cdot x^2y^2}$
$= \dfrac{5xy^2-2x^2y}{25y^2-4x^2}$
$= \dfrac{xy(5y-2x)}{(5y-2x)(5y+2x)}$
$= \dfrac{xy}{5y+2x}$

59. $(x^{-1}+y^{-1})^{-1} = \dfrac{1}{\frac{1}{x}+\frac{1}{y}}\cdot\dfrac{xy}{xy} = \dfrac{xy}{y+x}$

61. $\dfrac{x}{1-\frac{1}{1+\frac{1}{x}}}=\dfrac{x}{1-\frac{1}{x+1}}$
$=\dfrac{x}{\frac{x+1-x}{x+1}}$
$=x\cdot\dfrac{x+1}{1}$
$=x^2+x$

63. $\dfrac{a}{1-\frac{s}{770}}=\dfrac{a\cdot 770}{\left(1-\frac{s}{770}\right)\cdot 770}=\dfrac{770a}{770-s}$

65. $f(x)=\frac{1}{x}$

a. $f(a+h)=\dfrac{1}{a+h}$

b. $f(a)=\frac{1}{a}$

c. $\dfrac{f(a+h)-f(a)}{h}=\dfrac{\frac{1}{a+h}-\frac{1}{a}}{h}$

d. $\dfrac{f(a+h)-f(a)}{h}$
$=\dfrac{\frac{1}{a+h}\cdot\frac{a}{a}-\frac{1}{a}\cdot\frac{(a+h)}{(a+h)}}{h}$
$=\dfrac{\frac{a}{a(a+h)}-\frac{a+h}{a(a+h)}}{h}$
$=\dfrac{\frac{a-(a+h)}{a(a+h)}}{h}$
$=\dfrac{\frac{a-a-h}{a(a+h)}}{h}$
$=\dfrac{\frac{-h}{a(a+h)}}{h}$
$=\dfrac{-h}{a(a+h)}\cdot\dfrac{1}{h}$
$=\dfrac{-1}{a(a+h)}$

67. $f(x)=\dfrac{3}{x+1}$

a. $f(a+h)=\dfrac{3}{a+h+1}$

b. $f(a)=\dfrac{3}{a+1}$

c. $\dfrac{f(a+h)-f(a)}{h}=\dfrac{\frac{3}{a+h+1}-\frac{3}{a+1}}{h}$

d. $\dfrac{f(a+h)-f(a)}{h}$
$=\dfrac{h(a+h+1)(a+1)\left[\frac{3}{(a+h+1)}-\frac{3}{(a+1)}\right]}{h(a+h+1)(a+1)\cdot[h]}$
$=\dfrac{3h(a+1)-3h(a+h+1)}{h^2(a+h+1)(a+1)}$
$=\dfrac{3ah+3h-3ah-3h^2-3h}{h^2(a+h+1)(a+1)}$
$=\dfrac{-3h^2}{h^2(a+h+1)(a+1)}$
$=\dfrac{-3}{(a+h+1)(a+1)}$

69. $\dfrac{3x^3y^2}{12x}=\dfrac{x^{3-1}y^2}{4}=\dfrac{x^2y^2}{4}$

71. $\dfrac{144x^5y^5}{-16x^2y}=-9x^{5-2}y^{5-1}=-9x^3y^4$

73. $|x-5|=9$
$x-5=9$ or $x-5=-9$
$x=14$ or $x=-4$
$\{-4, 14\}$

75. $|x-5|<9$
$x-5<9$ and $x-5>-9$
$x<14$ and $x>-4$
$(-4, 14)$

77. $\dfrac{1}{1-(1-x)^{-1}} = \dfrac{1\cdot(1-x)}{\left(1-\frac{1}{1-x}\right)\cdot(1-x)}$
$= \dfrac{1-x}{1-x-1}$
$= \dfrac{1-x}{-x}$
$= \dfrac{x-1}{x}$

79. $\dfrac{(x+2)^{-1}+(x-2)^{-1}}{(x^2-4)^{-1}}$
$= \dfrac{\left(\frac{1}{x+2}+\frac{1}{x-2}\right)\cdot(x+2)(x-2)}{\left(\frac{1}{x^2-4}\right)\cdot(x+2)(x-2)}$
$= \dfrac{x-2+x+2}{1}$
$= 2x$

81. $\dfrac{3(a+1)^{-1}+4a^{-2}}{(a^3+a^2)^{-1}}$
$= \dfrac{\left(\frac{3}{a+1}+\frac{4}{a^2}\right)\cdot a^2(a+1)}{\left(\frac{1}{a^3+a^2}\right)\cdot a^2(a+1)}$
$= \dfrac{3a^2+4(a+1)}{1}$
$= 3a^2+4a+4$

Exercise Set 6.4

1. $\dfrac{4a^2+8a}{2a} = \dfrac{4a^2}{2a}+\dfrac{8a}{2a} = 2a+4$

3. $\dfrac{12a^5b^2+16a^4b}{4a^4b} = \dfrac{12a^5b^2}{4a^4b}+\dfrac{16a^4b}{4a^4b}$
$= 3ab+4$

5. $\dfrac{4x^2y^2+6xy^2-4y^2}{2x^2y}$
$= \dfrac{4x^2y^2}{2x^2y}+\dfrac{6xy^2}{2x^2y}-\dfrac{4y^2}{2x^2y}$
$= 2y+\dfrac{3y}{x}-\dfrac{2y}{x^2}$

7. $\dfrac{4x^2+8x+4}{4} = \dfrac{4x^2}{4}+\dfrac{8x}{4}+\dfrac{4}{4}$
$= x^2+2x+1$

9. $\dfrac{3x^4+6x-18}{3} = \dfrac{3x^4}{3}+\dfrac{6x}{3}-\dfrac{18}{3}$
$= (x^4+2x-6)$ meters

11.
$$\begin{array}{r|l} & x+1 \\ \hline x+2 & x^2+3x+2 \\ & \underline{x^2+2x} \\ & \quad x+2 \\ & \quad \underline{x+2} \\ & \quad\quad 0 \end{array}$$

Answer: $x + 1$

13.
$$\begin{array}{r|l} & 2x-8 \\ \hline x+1 & 2x^2-6x-8 \\ & \underline{2x^2+2x} \\ & \quad -8x-8 \\ & \quad \underline{-8x-8} \\ & \quad\quad 0 \end{array}$$

Answer: $2x - 8$

15.
$$\begin{array}{r|l} & x-\frac{1}{2} \\ \hline 2x+4 & 2x^2+3x-2 \\ & \underline{2x^2+4x} \\ & \quad -x-2 \\ & \quad \underline{-x-2} \\ & \quad\quad 0 \end{array}$$

Answer: $x-\frac{1}{2}$

17.
$$\begin{array}{r|l} & 2x^2 - \frac{1}{2}x + 5 \\ \hline 2x+4 & 4x^3 + 7x^2 + 8x + 20 \\ & \underline{4x^3 + 8x^2} \\ & -x^2 + 8x \\ & \underline{-x^2 - 2x} \\ & 10x + 20 \\ & \underline{10x + 20} \\ & 0 \end{array}$$

Answer: $2x^2 - \frac{1}{2}x + 5$

19. Recall that $A = \ell \cdot w$, so

$$w = \frac{A}{\ell} = \frac{15x^2 - 29x - 14}{5x + 2}.$$

$$\begin{array}{r|l} & 3x - 7 \\ \hline 5x+2 & 15x^2 - 29x - 14 \\ & \underline{15x^2 + 6x} \\ & -35x - 14 \\ & \underline{-35x - 14} \\ & 0 \end{array}$$

The width is $(3x - 7)$ in.

21. $\dfrac{25a^2b^{12}}{10a^5b^7} = \dfrac{5b^{12-7}}{2a^{5-2}} = \dfrac{5b^5}{2a^3}$

23. $\dfrac{x^6y^6 - x^3y^3}{x^3y^3} = \dfrac{x^6y^6}{x^3y^3} - \dfrac{x^3y^3}{x^3y^3} = x^3y^3 - 1$

25.
$$\begin{array}{r|l} & a + 3 \\ \hline a+1 & a^2 + 4a + 3 \\ & \underline{a^2 + a} \\ & 3a + 3 \\ & \underline{3a + 3} \\ & 0 \end{array}$$

Answer: $a + 3$

27.
$$\begin{array}{r|l} & 2x + 5 \\ \hline x-2 & 2x^2 + x - 10 \\ & \underline{2x^2 - 4x} \\ & 5x - 10 \\ & \underline{5x - 10} \\ & 0 \end{array}$$

Answer: $2x + 5$

29. $\dfrac{-16y^3 + 24y^4}{-4y^2} = \dfrac{-16y^3}{-4y^2} + \dfrac{24y^4}{-4y^2}$

$= 4y - 6y^2$

31.
$$\begin{array}{r|l} & 2x + 23 \\ \hline x-5 & 2x^2 + 13x + 15 \\ & \underline{2x^2 - 10x} \\ & 23x + 15 \\ & \underline{23x - 115} \\ & 130 \end{array}$$

Answer: $2x + 23 + \dfrac{130}{x - 5}$

33. $\dfrac{20x^2y^3 + 6xy^4 - 12x^3y^5}{2xy^3}$

$= \dfrac{20x^2y^3}{2xy^3} + \dfrac{6xy^4}{2xy^3} - \dfrac{12x^3y^5}{2xy^3}$

$= 10x + 3y - 6x^2y^2$

35.
$$\begin{array}{r|l} & 2x + 4 \\ \hline 3x+2 & 6x^2 + 16x + 8 \\ & \underline{6x^2 + 4x} \\ & 12x + 8 \\ & \underline{12x + 8} \\ & 0 \end{array}$$

Answer: $2x + 4$

37.
$$\begin{array}{r|l} & y + 5 \\ \hline 2y-3 & 2y^2 + 7y - 15 \\ & \underline{2y^2 - 3y} \\ & 10y - 15 \\ & \underline{10y - 15} \\ & 0 \end{array}$$

Answer: $y + 5$

39.
$$\begin{array}{r} 2x+3 \\ 2x-3\overline{)4x^2+0x-9} \\ \underline{4x^2-6x} \\ 6x-9 \\ \underline{6x-9} \\ 0 \end{array}$$

Answer: $2x+3$

41.
$$\begin{array}{r} 2x^2-8x+38 \\ x+4\overline{)2x^3+0x^2+6x-4} \\ \underline{2x^3+8x^2} \\ -8x^2+6x \\ \underline{-8x^2-32x} \\ 38x-4 \\ \underline{38x+152} \\ -156 \end{array}$$

Answer: $2x^2-8x+38-\dfrac{156}{x+4}$

43.
$$\begin{array}{r} 3x+3 \\ x-1\overline{)3x^2+0x-4} \\ \underline{3x^2-3x} \\ 3x-4 \\ \underline{3x-3} \\ -1 \end{array}$$

Answer: $3x+3-\dfrac{1}{x-1}$

45.
$$\begin{array}{r} -2x^3+3x^2-x+4 \\ -x+5\overline{)2x^4-13x^3+16x^2-9x+20} \\ \underline{2x^4-10x^3} \\ -3x^3+16x^2 \\ \underline{-3x^3+15x^2} \\ x^2-9x \\ \underline{x^2-5x} \\ -4x+20 \\ \underline{-4x+20} \\ 0 \end{array}$$

Answer: $-2x^3+3x^2-x+4$

47.
$$\begin{array}{r} 3x^3+5x+4 \\ x^2-2\overline{)3x^5-x^3+4x^2-12x-8} \\ \underline{3x^5-6x^3} \\ 5x^3+4x^2-12x \\ \underline{5x^3+0x^2-10x} \\ 4x^2-2x-8 \\ \underline{4x^2+0x-8} \\ -2x \end{array}$$

Answer: $3x^3+5x+4-\dfrac{2x}{x^2-2}$

49. $\dfrac{3x^3-5}{3x^2}=\dfrac{3x^3}{3x^2}-\dfrac{5}{3x^2}=x-\dfrac{5}{3x^2}$

51. $P(x)=3x^3+2x^2-4x+3$

$P(1)=3(1)^3+2(1)^2-4(1)+3$
$=3+2-4+3$
$=4$

$$\begin{array}{r} 3x^2+5x+1 \\ x-1\overline{)3x^3+2x^2-4x+3} \\ \underline{3x^3-3x^2} \\ 5x^2-4x \\ \underline{5x^2-5x} \\ x+3 \\ \underline{x-1} \\ 4 \end{array}$$

Remainder = 4

53. $P(x)=5x^4-2x^2+3x-6$

$P(-3)=5(-3)^4-2(-3)^2+3(-3)-6$
$=5(81)-2(9)-9-6$
$=405-18-9-6$
$=372$

$$\begin{array}{r}
5x^3 - 15x^2 + 43x - 126 \\
x+3\overline{\big)\,5x^4 + 0x^3 - 2x^2 + 3x - 6} \\
\underline{5x^4 + 15x^3} \phantom{{}- 2x^2 + 3x - 6} \\
-15x^3 - 2x^2 \phantom{{}+ 3x - 6} \\
\underline{-15x^3 - 45x^2} \phantom{{}+ 3x - 6} \\
43x^2 + 3x \phantom{{}- 6} \\
\underline{43x^2 + 129x} \phantom{{}- 6} \\
-126x - 6 \\
\underline{-126x - 378} \\
372
\end{array}$$

Remainder = 372

55. Answers may vary.

57.
$$\begin{array}{r}
3x^2 + 10x + 8 \phantom{{}+3x^3} \\
x-2\overline{\big)\,4x^2 - 12x - 12 + 3x^3} \\
\underline{-6x^2 \phantom{{}-12x-12} + 3x^3} \\
10x^2 - 12x \phantom{{}-12+3x^3} \\
\underline{10x^2 - 20x} \phantom{{}-12+3x^3} \\
8x - 12 \phantom{{}+3x^3} \\
\underline{8x - 16} \phantom{{}+3x^3} \\
4 \phantom{{}+3x^3}
\end{array}$$

$$\begin{array}{r}
3x^2 + 10x + 8 \\
x-2\overline{\big)\,3x^3 + 4x^2 - 12x - 12} \\
\underline{3x^3 - 6x^2} \phantom{{}-12x-12} \\
10x^2 - 12x \phantom{{}-12} \\
\underline{10x^2 - 20x} \phantom{{}-12} \\
8x - 12 \\
\underline{8x - 16} \\
4
\end{array}$$

Answer: $3x^2 + 10x + 8 + \dfrac{4}{x-2}$

59. $3^2 = (-3)^2$

61. $-2^3 = (-2)^3$

63. $|x+5| < 4$

$x+5 < 4$ or $x+5 > -4$

$x < -1$ or $x > -9$

$(-9, -1)$

65. $|2x+7| \ge 9$

$2x+7 \ge 9$ or $2x+7 \le -9$

$2x \ge 2$ or $2x \le -16$

$x \ge 1$ or $x \le -8$

$(-\infty, -8] \cup [1, \infty)$

67.
$$\begin{array}{r}
x^3 + \frac{5}{3}x^2 + \frac{5}{3}x + \frac{8}{3} \\
x-1\overline{\big)\,x^4 + \frac{2}{3}x^3 + 0x^2 + \phantom{\frac{5}{3}}x + 0} \\
\underline{x^4 - \phantom{\frac{2}{3}}x^3} \phantom{{}+0x^2+x+0} \\
\frac{5}{3}x^3 + 0x^2 \phantom{{}+x+0} \\
\underline{\frac{5}{3}x^3 - \frac{5}{3}x^2} \phantom{{}+x+0} \\
\frac{5}{3}x^2 + \phantom{\frac{5}{3}}x \phantom{{}+0} \\
\underline{\frac{5}{3}x^2 - \frac{5}{3}x} \phantom{{}+0} \\
\frac{8}{3}x + 0 \\
\underline{\frac{8}{3}x - \frac{8}{3}} \\
\frac{8}{3}
\end{array}$$

Answer: $x^3 + \dfrac{5}{3}x^2 + \dfrac{5}{3}x + \dfrac{8}{3} + \dfrac{8}{3(x-1)}$

69.
$$\begin{array}{r}
\frac{3}{2}x^3 + \frac{1}{4}x^2 + \frac{1}{8}x - \frac{7}{16} \\
2x-1\overline{\big)\,3x^4 - \phantom{\frac{3}{2}}x^3 + 0x^2 - \phantom{\frac{1}{8}}x + \frac{1}{2}} \\
\underline{3x^4 - \frac{3}{2}x^3} \phantom{{}+0x^2-x+\frac{1}{2}} \\
\frac{1}{2}x^3 + 0x^2 \phantom{{}-x+\frac{1}{2}} \\
\underline{\frac{1}{2}x^3 - \frac{1}{4}x^2} \phantom{{}-x+\frac{1}{2}} \\
\frac{1}{4}x^2 - \phantom{\frac{1}{8}}x \phantom{{}+\frac{1}{2}} \\
\underline{\frac{1}{4}x^2 - \frac{1}{8}x} \phantom{{}+\frac{1}{2}} \\
-\frac{7}{8}x + \frac{1}{2} \\
\underline{-\frac{7}{8}x + \frac{7}{16}} \\
\frac{1}{16}
\end{array}$$

Answer:
$\frac{3}{2}x^3+\frac{1}{4}x^2+\frac{1}{8}x-\frac{7}{16}+\frac{1}{16(2x-1)}$

71.
$$\begin{array}{r} x^3 \qquad\qquad -\tfrac{2}{5}x \\ 5x+10\overline{)5x^4+10x^3-2x^2-4x} \\ \underline{5x^4+10x^3} \qquad\qquad\quad \\ -2x^2-4x \\ \underline{-2x^2-4x} \\ 0 \end{array}$$

Answer: $x^3-\frac{2}{5}x$

Exercise Set 6.5

1.
$$\begin{array}{r|rrr} 5 & 1 & 3 & -40 \\ & & 5 & 40 \\ \hline & 1 & 8 & 0 \end{array}$$
$x+8$

3.
$$\begin{array}{r|rrr} -6 & 1 & 5 & -6 \\ & & -6 & 6 \\ \hline & 1 & -1 & 0 \end{array}$$
$x-1$

5.
$$\begin{array}{r|rrrr} 2 & 1 & -7 & -13 & 5 \\ & & 2 & -10 & -46 \\ \hline & 1 & -5 & -23 & -41 \end{array}$$
$x^2-5x-23-\frac{41}{x-2}$

7.
$$\begin{array}{r|rrr} 2 & 4 & 0 & -9 \\ & & 8 & 16 \\ \hline & 4 & 8 & 7 \end{array}$$
$4x+8+\frac{7}{x-2}$

9. a. $P(2)=3(2)^2-4(2)-1$
$=12-8-1=3$

b.
$$\begin{array}{r|rrr} 2 & 3 & -4 & -1 \\ & & 6 & 4 \\ \hline & 3 & 2 & 3 \end{array}$$
$P(2)=3$

11. a. $P(-2)=4(-2)^4+7(-2)^2+9(-2)-1$
$=64+28-18-1$
$=73$

b.
$$\begin{array}{r|rrrrr} -2 & 4 & 0 & 7 & 9 & -1 \\ & & -8 & 16 & -46 & 74 \\ \hline & 4 & -8 & 23 & -37 & 73 \end{array}$$
$P(-2)=73$

13. a. $P(-1)=(-1)^5+3(-1)^4+3(-1)-7$
$=-1+3-3-7$
$=-8$

b.
$$\begin{array}{r|rrrrrr} -1 & 1 & 3 & 0 & 0 & 3 & -7 \\ & & -1 & -2 & 2 & -2 & -1 \\ \hline & 1 & 2 & -2 & 2 & 1 & -8 \end{array}$$
$P(-1)=-8$

15.
$$\begin{array}{r|rrrr} 3 & 1 & -3 & 0 & 2 \\ & & 3 & 0 & 0 \\ \hline & 1 & 0 & 0 & 2 \end{array}$$
$x^2+\frac{2}{x-3}$

17.
$$\begin{array}{r|rrr} -1 & 6 & 13 & 8 \\ & & -6 & -7 \\ \hline & 6 & 7 & 1 \end{array}$$
$6x+7+\frac{1}{x+1}$

19.
$$\begin{array}{r|rrrrr} 5 & 2 & -13 & 16 & -9 & 20 \\ & & 10 & -15 & 5 & -20 \\ \hline & 2 & -3 & 1 & -4 & 0 \end{array}$$
$2x^3-3x^2+x-4$

21.
$$\begin{array}{r|rrr} -3 & 3 & 0 & -15 \\ & & -9 & 27 \\ \hline & 3 & -9 & 12 \end{array}$$
$3x-9+\frac{12}{x+3}$

23.
$$\begin{array}{r|rrrr} \frac{1}{2} & 3 & -6 & 4 & 5 \\ & & \frac{3}{2} & -\frac{9}{4} & \frac{7}{8} \\ \hline & 3 & -\frac{9}{2} & \frac{7}{4} & \frac{47}{8} \end{array}$$
$3x^2-\frac{9}{2}x+\frac{7}{4}+\frac{47}{8\left(x-\frac{1}{2}\right)}$

25.
$$\begin{array}{r|rrrr} \frac{1}{3} & 3 & 2 & -4 & 1 \\ & & 1 & 1 & -1 \\ \hline & 3 & 3 & -3 & 0 \end{array}$$
$3x^2+3x-3$

27.
$$\begin{array}{r|rrrr} -1 & 3 & 7 & -4 & 12 \\ & & -3 & -4 & 8 \\ \hline & 3 & 4 & -8 & 20 \end{array}$$
$3x^2+4x-8+\frac{20}{x+1}$

29.
$$\begin{array}{r|rrrr} 1 & 1 & 0 & 0 & -1 \\ & & 1 & 1 & 1 \\ \hline & 1 & 1 & 1 & 0 \end{array}$$
x^2+x+1

31.
$$\begin{array}{r|rrr} -6 & 1 & 0 & -36 \\ & & -6 & 36 \\ \hline & 1 & -6 & 0 \end{array}$$
$x-6$

33.
$$\begin{array}{r|rrrr} 1 & 1 & 3 & -7 & 4 \\ & & 1 & 4 & -3 \\ \hline & 1 & 4 & -3 & 1 \end{array}$$
Thus, $P(1)=1$.

35.
$$\begin{array}{r|rrrr} -3 & 3 & -7 & -2 & 5 \\ & & -9 & 48 & -138 \\ \hline & 3 & -16 & 46 & -133 \end{array}$$
Thus, $P(-3)=-133$.

37.
$$\begin{array}{r|rrrrr} -1 & 4 & 0 & 1 & 0 & -2 \\ & & -4 & 4 & -5 & 5 \\ \hline & 4 & -4 & 5 & -5 & 3 \end{array}$$
Thus, $P(-1)=3$.

39.
$$\begin{array}{r|rrrrr} \frac{1}{3} & 2 & 0 & -3 & 0 & -2 \\ & & \frac{2}{3} & \frac{2}{9} & -\frac{25}{27} & -\frac{25}{81} \\ \hline & 2 & \frac{2}{3} & -\frac{25}{9} & -\frac{25}{27} & -\frac{187}{81} \end{array}$$
Thus, $P\left(\frac{1}{3}\right)=-\frac{187}{81}$.

41.
$$\begin{array}{r|rrrrrr} \frac{1}{2} & 1 & 1 & -1 & 0 & 0 & 3 \\ & & \frac{1}{2} & \frac{3}{4} & -\frac{1}{8} & -\frac{1}{16} & -\frac{1}{32} \\ \hline & 1 & \frac{3}{2} & -\frac{1}{4} & -\frac{1}{8} & -\frac{1}{16} & \frac{95}{32} \end{array}$$
Thus, $P\left(\frac{1}{2}\right)=\frac{95}{32}$.

43. Answers may vary.

45.
$$\begin{array}{r|rrrr} -3 & 1 & 3 & 4 & 12 \\ & & -3 & 0 & -12 \\ \hline & 1 & 0 & 4 & 0 \end{array}$$
Remainder = 0

47. $P(x)$ is equal to the remainder when $P(x)$ is divided by $x - c$. Therefore, $P(c) = 0$.

49. Multiply $(x^2 - x + 10)$ by $(x + 3)$ and add the remainder, -2.
$$(x^2 - x + 10)(x + 3) - 2$$
$$= x^3 - x^2 + 10x + 3x^2 - 3x + 30 - 2$$
$$= x^3 + 2x^2 + 7x + 28$$

51. $V = lwh$
$$w = \frac{V}{lh}$$
$$= \frac{x^4 + 6x^3 - 7x^2}{x^2(x+7)}$$
$$= \frac{x^4 + 6x^3 - 7x^2}{x^3 + 7x^2}$$
$$\begin{array}{r} x - 1 \\ x^3 + 7x^2 \overline{\smash{)}\, x^4 + 6x^3 - 7x^2} \\ \underline{x^4 + 7x^3} \\ -1x^3 - 7x^2 \\ \underline{-1x^3 - 7x^2} \\ 0 \end{array}$$
The width is $(x - 1)$ meters.

53.
$$4 - 2x = 17 - 5x$$
$$5x - 2x = 17 - 4$$
$$3x = 13$$
$$x = \frac{13}{3}$$

55.
$$5x^2 + 10x = 15$$
$$5x^2 + 10x - 15 = 0$$
$$5(x^2 + 2x - 3) = 0$$
$$5(x+3)(x-1) = 0$$
$$x + 3 = 0 \quad \text{or} \quad x - 1 = 0$$
$$x = -3 \quad \text{or} \quad x = 1$$

57.
$$\frac{2x}{9} + 1 = \frac{7}{9}$$
$$2x + 9 = 7$$
$$2x = -2$$
$$x = -1$$

59. $8y^3 + 1 = (2y+1)(4y^2 - 2y + 1)$

61. $a^3 - 27 = (a-3)(a^2 + 3a + 9)$

63.
$$x^2 - x + xy - y = x(x-1) + y(x-1)$$
$$= (x+y)(x-1)$$

65.
$$2x^3 - 32x = 2x(x^2 - 16)$$
$$= 2x(x+4)(x-4)$$

Exercise Set 6.6

1.
$$\frac{x}{2} - \frac{x}{3} = 12$$
$$3x - 2x = 72$$
$$x = 72$$

3.
$$\frac{x}{3} = \frac{1}{6} + \frac{x}{4}$$
$$4x = 2 + 3x$$
$$x = 2$$

5.
$$\frac{2}{x} + \frac{1}{2} = \frac{5}{x}$$
$$\frac{1}{2} = \frac{3}{x}$$
$$x = 6$$

7.
$$\frac{x+3}{x} = \frac{5}{x}$$
$$x + 3 = 5$$
$$x = 2$$

9.
$$\frac{x+5}{x+3} = \frac{8}{x+3}$$
$$(x+3)\cdot\frac{x+5}{x+3} = (x+3)\cdot\frac{8}{x+3}$$
$$x + 5 = 8$$
$$x = 3$$

11. $\frac{1}{x-1}+\frac{1}{x+1}=\frac{2}{x^2-1}$

$\frac{x+1+x-1}{(x-1)(x+1)}=\frac{2}{(x+1)(x-1)}$

$2x=2$

$x=1$

which we discard as extraneous.
No solution.

13. $\frac{6}{x+3}=\frac{4}{x-3}$

$6(x-3)=4(x+3)$

$6x-18=4x+12$

$2x=30$

$x=15$

15. $\frac{3}{2x+3}-\frac{1}{2x-3}=\frac{4}{4x^2-9}$

$\frac{3(2x-3)-(2x+3)\cdot 1}{(2x+3)(2x-3)}=\frac{4}{(2x+3)(2x-3)}$

$6x-9-2x-3=4$

$4x-12=4$

$4x=16$

$x=4$

17. $\frac{2}{x^2-4}=\frac{1}{2x-4}$

$\frac{4}{2(x+2)(x-2)}=\frac{x+2}{2(x-2)(x+2)}$

$4=x+2$

$x=2$

which we discard as extraneous.
No solution.

19. $\frac{12}{3x^2+12x}=1-\frac{1}{x+4}$

$\frac{12}{3x(x+4)}=\frac{3x(x+4)-3x}{3x(x+4)}$

$12=3x^2+12x-3x$

$12=3x^2+9x$

$4=x^2+3x$

$x^2+3x-4=0$

$(x+4)(x-1)=0$

$x+4=0$ or $x-1=0$

$x=-4$ or $x=1$

We discard –4 as extraneous.
$x=1$

21. $\frac{2}{x}=\frac{10}{5}$

$10=10x$

$x=1$

23. $7+\frac{6}{a}=5$

$\frac{6}{a}=-2$

$6=-2a$

$a=-3$

25. $\frac{2}{x-5}+\frac{1}{2x}=\frac{5}{3x^2-15x}$

$\frac{2(2x)+x-5}{2x(x-5)}=\frac{5}{3x(x-5)}$

$\frac{3(4x+x-5)}{6x(x-5)}=\frac{2(5)}{6x(x-5)}$

$12x+3x-15=10$

$15x=25$

$x=\frac{25}{15}=\frac{5}{3}$

27. $\frac{x}{4}+\frac{5}{x}=3$

$\frac{x^2+20}{4x}=3$

$x^2+20=12x$

$x^2-12x+20=0$

$(x-10)(x-2)=0$

$x-10=0$ or $x-2=0$

$x=10$ or $x=2$

29. $1-\frac{5}{y+7}=\frac{4}{y+7}$

$1=\frac{9}{y+7}$

$y+7=9$

$y=2$

31. $\frac{6x+7}{2x+9} = \frac{5}{3}$
$3(6x+7) = 5(2x+9)$
$18x + 21 = 10x + 45$
$8x = 24$
$x = 3$

33. $\frac{2x+1}{4-x} = \frac{9}{4-x}$
$2x + 1 = 9$
$2x = 8$
$x = 4$
which we discard as extraneous.
No solution.

35. $\frac{12}{9-a^2} + \frac{3}{3+a} = \frac{2}{3-a}$
$\frac{12}{(3+a)(3-a)} + \frac{3(3-a)}{(3+a)(3-a)} = \frac{2(3+a)}{(3-a)(3+a)}$
$12 + 9 - 3a = 6 + 2a$
$21 - 3a = 6 + 2a$
$15 = 5a$
$3 = a$
which we discard as extraneous.
No solution.

37. $2 + \frac{3}{x} = \frac{2x}{x+3}$
$\frac{2x+3}{x} = \frac{2x}{x+3}$
$(2x+3)(x+3) = 2x \cdot x$
$2x^2 + 9x + 9 = 2x^2$
$9x + 9 = 0$
$9x = -9$
$x = -1$

39. $\frac{36}{x^2-9} + 1 = \frac{2x}{x+3}$
$\frac{36 + x^2 - 9}{(x+3)(x-3)} = \frac{2x(x-3)}{(x+3)(x-3)}$
$x^2 + 27 = 2x^2 - 6x$
$x^2 - 6x - 27 = 0$
$(x-9)(x+3) = 0$
$x - 9 = 0$ or $x + 3 = 0$
$x = 9$ or $x = -3$
We discard −3 as extraneous.
$x = 9$

41. $\dfrac{x^2-20}{x^2-7x+12}=\dfrac{3}{x-3}+\dfrac{5}{x-4}$

$\dfrac{x^2-20}{(x-3)(x-4)}=\dfrac{3(x-4)+5(x-3)}{(x-3)(x-4)}$

$x^2-20=3x-12+5x-15$

$x^2-20=8x-27$

$x^2-8x+7=0$

$(x-1)(x-7)=0$

$x-1=0$ or $x-7=0$

$x=1$ or $x=7$

43. $\dfrac{3}{2x-5}+\dfrac{2}{2x+3}=0$

$\dfrac{3(2x+3)+2(2x-5)}{(2x-5)(2x+3)}=0$

$6x+9+4x-10=0$

$10x-1=0$

$10x=1$

$x=\dfrac{1}{10}$

45. $f(x)=20+\dfrac{4000}{x}$

$25=20+\dfrac{4000}{x}$

$25-20=\dfrac{4000}{x}$

$5=\dfrac{4000}{x}$

$5x=4000$

$x=\dfrac{4000}{5}$

$x=800$

800 pencil sharpeners

47. $x^{-2}-5x^{-1}-36=0$

$\dfrac{1}{x^2}-\dfrac{5}{x}-36=0$

$1-5x-36x^2=0$

$36x^2+5x-1=0$

$(9x-1)(4x+1)=0$

$9x-1=0$ or $4x+1=0$

$9x=1$ or $4x=-1$

$x=\frac{1}{9}$ or $x=-\dfrac{1}{4}$

49. $6p^{-2}-5p^{-1}+1=0$

$\dfrac{6}{p^2}-\dfrac{5}{p}+1=0$

$6-5p+p^2=0$

$p^2-5p+6=0$

$(p-3)(p-2)=0$

$p-3=0$ or $p-2=0$

$p=3$ or $p=2$

51. $\dfrac{-8.5}{x+1.9}=\dfrac{5.7}{x-3.6}$

$-8.5(x-3.6)=5.7(x+1.9)$

$-8.5x+30.6=5.7x+10.83$

$30.6-10.83=5.7x+8.5x$

$19.77=14.2x$

$1.39\approx x$

53. $\dfrac{12.2}{x}+17.3=\dfrac{9.6}{x}-14.7$

$\dfrac{12.2}{x}-\dfrac{9.6}{x}=-14.7-17.3$

$\dfrac{12.2-9.6}{x}=-32$

$2.6=-32x$

$-0.08=x$

55. $Y_1=\dfrac{2}{x}$, $Y_2=2$

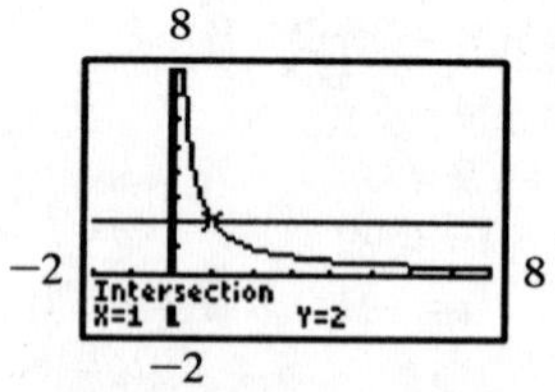

$x=1$

57. $Y_1=\dfrac{6x+7}{2x+9}$, $Y_2=\dfrac{5}{3}$

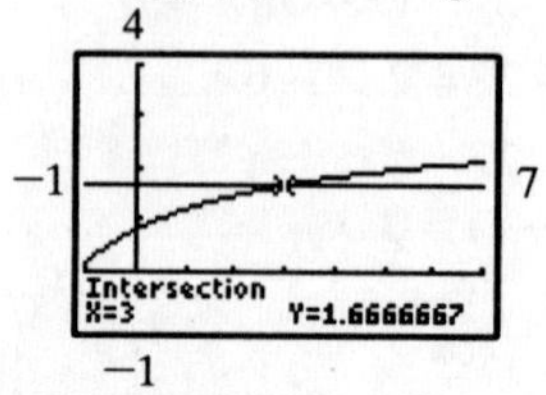

$x=3$

59. Let n = 1st integer.
$n + 1$ = 2nd integer.
$$n+(n+1)=147$$
$$2n+1=147$$
$$2n=146$$
$$n=73$$
The integers are 73 and 74.

61. Let x = the number.
$\frac{1}{x}$ = the reciprocal of the number.
$$x+\frac{1}{x}=\frac{5}{2}$$
$$\frac{x^2+1}{x}=\frac{5}{2}$$
$$2x^2+2=5x$$
$$2x^2-5x+2=0$$
$$(2x-1)(x-2)=0$$
$2x-1=0$ or $x-2=0$
$2x=1$ or $x=2$
$x=\frac{1}{2}$
The reciprocal pair is 2 and $\frac{1}{2}$.

63. 3% (reading from the graph)

65. Adding the percents for the three bars,
16% + 19% + 19% = 54%

67. $(x-1)^2+3(x-1)+2=0$
Let $u = x - 1$.
$$u^2+3u+2=0$$
$$(u+1)(u+2)=0$$
$u+1=0$ or $u+2=0$
$u=-1$ or $u=-2$
Substituting $x - 1$ for u,
$x-1=-1$ or $x-1=-2$
$x=0$ or $x=-1$

69. $\left(\frac{3}{x-1}\right)^2+2\left(\frac{3}{x-1}\right)+1=0$
Let $u=\frac{3}{x-1}$.
$$u^2+2u+1=0$$
$$(u+1)^2=0$$
$$u+1=0$$
$$u=-1$$
Substituting $\frac{3}{x-1}$ for u,
$$\frac{3}{x-1}=-1$$
$$(x-1)\left(\frac{3}{x-1}\right)=-1(x-1)$$
$$3=-x+1$$
$$x=-2$$

Supplementary Exercises

1. $\frac{x}{2}=\frac{1}{8}+\frac{x}{4}$
$$8\left(\frac{x}{2}\right)=8\left(\frac{1}{8}+\frac{x}{4}\right)$$
$$4x=1+2x$$
$$2x=1$$
$$x=\frac{1}{2}$$

3. $\frac{1}{8}+\frac{x}{4}=\frac{1}{8}+\frac{2\cdot x}{2\cdot 4}=\frac{1}{8}+\frac{2x}{8}=\frac{1+2x}{8}$

5. $\frac{4}{x+2}-\frac{2}{x-1}$
$$=\frac{4(x-1)}{(x+2)(x-1)}-\frac{2(x+2)}{(x-1)(x+2)}$$
$$=\frac{4(x-1)-2(x+2)}{(x+2)(x-1)}$$
$$=\frac{4x-4-2x-4}{(x+2)(x-1)}$$
$$=\frac{2x-8}{(x+2)(x-1)}$$
$$=\frac{2(x-4)}{(x+2)(x-1)}$$

7.

$$\frac{4}{x+2} = \frac{2}{x-1}$$

$$(x-1)(x+2)\left(\frac{4}{x+2}\right) = (x-1)(x+2)\left(\frac{2}{x-1}\right)$$

$$4(x-1) = 2(x+2)$$

$$4x-4 = 2x+4$$

$$2x = 8$$

$$x = 4$$

9.

$$\frac{2}{x^2-4} = \frac{1}{x+2} - \frac{3}{x-2}$$

$$\frac{2}{(x+2)(x-2)} = \frac{x-2-3(x+2)}{(x+2)(x-2)}$$

$$2 = (x-2)-3(x+2)$$

$$2 = x-2-3x-6$$

$$10 = -2x$$

$$-5 = x$$

11.

$$\frac{5}{x^2-3x} + \frac{4}{2x-6} = \frac{5}{x(x-3)} + \frac{4}{2(x-3)}$$

$$= \frac{5\cdot 2}{2x(x-3)} + \frac{4\cdot x}{2x(x-3)}$$

$$= \frac{10+4x}{2x(x-3)}$$

$$= \frac{2(5+2x)}{2x(x-3)}$$

$$= \frac{2x+5}{x(x-3)}$$

13.

$$\frac{x-1}{x+1} + \frac{x+7}{x-1} = \frac{4}{x^2-1}$$

$$(x+1)(x-1)\left(\frac{x-1}{x+1} + \frac{x+7}{x-1}\right) = (x+1)(x-1)\left[\frac{4}{(x+1)(x-1)}\right]$$

$$(x-1)^2 + (x+7)(x+1) = 4$$

$$x^2-2x+1+x^2+8x+7 = 4$$

$$2x^2+6x+4 = 0$$

$$2(x+2)(x+1) = 0$$

$x+2=0$ or $x+1=0$

$x=-2$ or $x=-1$

We discard –1 as extraneous.

15. $$\frac{a^2-9}{a-6}\cdot\frac{a^2-5a-6}{a^2-a-6}=\frac{(a+3)(a-3)}{a-6}\cdot\frac{(a-6)(a+1)}{(a-3)(a+2)}$$
$$=\frac{(a+3)(a+1)}{a+2}$$

17. $$\frac{2x+3}{3x-2}=\frac{4x+1}{6x+1}$$
$$(3x-2)(6x+1)\left(\frac{2x+3}{3x-2}\right)=(3x-2)(6x+1)\left(\frac{4x+1}{6x+1}\right)$$
$$(6x+1)(2x+3)=(3x-2)(4x+1)$$
$$12x^2+20x+3=12x^2-5x-2$$
$$25x=-5$$
$$x=-\frac{1}{5}$$

19. $$\frac{a}{9a^2-1}+\frac{2}{6a-2}$$
$$=\frac{a}{(3a-1)(3a+1)}+\frac{2}{2(3a-1)}$$
$$=\frac{2a+2(3a+1)}{2(3a-1)(3a+1)}$$
$$=\frac{2a+6a+2}{2(3a-1)(3a+1)}$$
$$=\frac{8a+2}{2(3a-1)(3a+1)}$$
$$=\frac{2(4a+1)}{2(3a-1)(3a+1)}$$
$$=\frac{4a+1}{(3a-1)(3a+1)}$$

21. $$-\frac{3}{x^2}-\frac{1}{x}+2=0$$
$$x^2\left(-\frac{3}{x^2}-\frac{1}{x}+2\right)=x^2\cdot 0$$
$$-3-x+2x^2=0$$
$$2x^2-x-3=0$$
$$(2x-3)(x+1)=0$$
$$2x-3=0 \quad\text{or}\quad x+1=0$$
$$x=\frac{3}{2} \quad\text{or}\quad x=-1$$

23. $$\frac{x-8}{x^2-x-2}+\frac{2}{x-2}$$
$$=\frac{x-8}{(x-2)(x+1)}+\frac{2}{x-2}$$
$$=\frac{x-8+2(x+1)}{(x-2)(x+1)}$$
$$=\frac{x-8+2x+2}{(x-2)(x+1)}$$
$$=\frac{3x-6}{(x-2)(x+1)}$$
$$=\frac{3(x-2)}{(x-2)(x+1)}$$
$$=\frac{3}{x+1}$$

25. $$\frac{3}{a}-5=\frac{7}{a}-1$$
$$a\left(\frac{3}{a}-5\right)=a\left(\frac{7}{a}-1\right)$$
$$3-5a=7-a$$
$$-4a=4$$
$$a=-1$$

Exercise Set 6.7

1. $$F=\frac{9}{5}C+32$$
$$F-32=\frac{9}{5}C$$
$$C=\frac{5}{9}(F-32)$$

3. $Q = \frac{A-I}{L}$

$QL = L\left(\frac{A-I}{L}\right)$

$QL = A - I$

$I = A - QL$

5. $\frac{1}{R} = \frac{1}{R_1} + \frac{1}{R_2}$

$\frac{1}{R} = \frac{R_2 + R_1}{R_1 R_2}$

$R = \frac{R_1 R_2}{R_1 + R_2}$

7. $S = \frac{n(a+L)}{2}$

$2S = n(a+L)$

$n = \frac{2S}{a+L}$

9. $A = \frac{h(a+b)}{2}$

$2A = ah + bh$

$2A - ah = bh$

$\frac{2A - ah}{h} = b$

11. $\frac{P_1 V_1}{T_1} = \frac{P_2 V_2}{T_2}$

$(T_1 T_2)\left(\frac{P_1 V_1}{T_1}\right) = \left(\frac{P_2 V_2}{T_2}\right)(T_1 T_2)$

$P_1 V_1 T_2 = P_2 V_2 T_1$

$T_2 = \frac{P_2 V_2 T_1}{P_1 V_1}$

13. $f = \frac{f_1 f_2}{f_1 + f_2}$

$(f_1 + f_2) f = \left(\frac{f_1 f_2}{f_1 + f_2}\right)(f_1 + f_2)$

$f_1 f + f_2 f = f_1 f_2$

$f_1 f = f_1 f_2 - f_2 f$

$f_1 f = f_2 (f_1 - f)$

$\frac{f_1 f}{f_1 - f} = f_2$

15. $\lambda = \frac{2L}{n}$

$n\lambda = \left(\frac{2L}{n}\right)(n)$

$n\lambda = 2L$

$\frac{n\lambda}{2} = L$

17. $\frac{\theta}{\omega} = \frac{2L}{c}$

$c\omega\left(\frac{\theta}{\omega}\right) = \frac{2L}{c}(c\omega)$

$c\theta = 2L\omega$

$c = \frac{2L\omega}{\theta}$

19. Let n = the number.

$\frac{1}{n}$ = the reciprocal of the number.

$n + 5\left(\frac{1}{n}\right) = 6$

$n + \frac{5}{n} = 6$

$\frac{n^2 + 5}{n} = 6$

$n^2 + 5 = 6n$

$n^2 - 6n + 5 = 0$

$(n-1)(n-5) = 0$

$n - 1 = 0$ or $n - 5 = 0$

$n = 1$ or $n = 5$

The number is either 1 or 5.

21. Let x = the number.

$$\frac{12+x}{41+2x}=\frac{1}{3}$$
$$3(12+x)=(41+2x)\cdot 1$$
$$36+3x=41+2x$$
$$x=5$$

The number is 5.

23. $R = 2$ ohms and $R_1 = 3$ ohms

$$\frac{1}{R}=\frac{1}{R_1}+\frac{1}{R_2}$$
$$\frac{1}{2}=\frac{1}{3}+\frac{1}{R_2}$$
$$\frac{1}{2}=\frac{R_2+3}{3R_2}$$
$$3R_2=2(R_2+3)$$
$$3R_2=2R_2+6$$
$$R_2=6$$

The other resistance is 6 ohms.

25. For three resistances R_1, R_2, and R_3 wired in a parallel circuit, the combined resistance R is given by

$$\frac{1}{R}=\frac{1}{R_1}+\frac{1}{R_2}+\frac{1}{R_3}$$
$$\frac{1}{R}=\frac{1}{5}+\frac{1}{6}+\frac{1}{2}$$
$$\frac{1}{R}=\frac{6+5+15}{30}$$
$$\frac{1}{R}=\frac{26}{30}=\frac{13}{15}$$
$$R=\frac{15}{13}$$

The combined resistance is $\frac{15}{13}$ ohms.

27. Let x = time for Steve to work alone.

$$\frac{1}{x}+\frac{1}{6}=\frac{1}{4}$$
$$12+2x=3x$$
$$12=x$$

Steve would take 12 hours to word process the paper alone.

29. Let x = time if both printers are operating.

$$\frac{1}{4}+\frac{2}{4}=\frac{1}{x}$$
$$x+2x=4$$
$$3x=4$$
$$x=\frac{4}{3}=1\frac{1}{3}$$

It will take $1\frac{1}{3}$ hours.

31. Let r = speed of truck, then $3r$ = speed of plane.

$d = rt$ or $t=\frac{d}{r}$

$$\frac{450}{r}-6=\frac{450}{3r}$$
$$3(450)-3r(6)=450$$
$$1350-18r=450$$
$$-18r=-900$$
$$r=50$$

The truck's speed is 50 mph.

33. Let r = speed of current, then $24 + r$ = speed of boat downstream, and $24 - r$ = speed of boat upstream.

$d = rt$ or $t=\frac{d}{r}$

$$\frac{54}{24-r}=\frac{90}{24+r}$$
$$54(24+r)=90(24-r)$$
$$1296+54r=2160-90r$$
$$144r=864$$
$$r=6$$

The speed of the current is 6 mph.

35. Let n = one integer, then $n + 2$ = other integer.

$$\frac{1}{n}+\frac{1}{n+2}=\frac{20}{99}$$
$$99(n+2)+99n=20n(n+2)$$
$$99n+198+99n=20n^2+40n$$
$$20n^2-158n-198=0$$
$$2(10n^2-79n-99)=0$$
$$2(10n+11)(n-9)=0$$
$$10n+11=0 \quad\text{or}\quad n-9=0$$
$$n=-\frac{11}{10} \quad\text{or}\quad n=9$$

Disregard the noninteger solution.
$n = 9$
$n + 2 = 11$
The integers are 9 and 11.

37. Let x = hours for Dick to do the job alone.

$$\frac{1}{5} + \frac{1}{x} = \frac{1}{2}$$
$$2x + 10 = 5x$$
$$10 = 3x$$
$$\frac{10}{3} = x$$
$$x = 3\frac{1}{3}$$

It takes Dick $3\frac{1}{3}$ hours to do the job alone.

39. Let r = speed of bicyclist, then $r - 10$ = speed of walker.

Recall $d = rt$ or $t = \frac{d}{r}$

$$\frac{26}{r} = \frac{6}{r-10}$$
$$26(r-10) = 6r$$
$$26r - 260 = 6r$$
$$20r = 260$$
$$r = 13$$

The bicyclist travels at 13 mph.

41. Let d = denominator, then $d - 4$ = numerator.

$$\frac{(d-4)+2}{d+2} = \frac{2}{3}$$
$$\frac{d-2}{d+2} = \frac{2}{3}$$
$$3(d-2) = 2(d+2)$$
$$3d - 6 = 2d + 4$$
$$d = 10$$
$$d - 4 = 6$$

The fraction is $\frac{6}{10}$.

43.

	distance	= rate	· time
First part	20	r	$\frac{20}{r}$
Cooldown	16	$r-2$	$\frac{16}{r-2}$

$$\frac{20}{r} = \frac{16}{r-2}$$
$$20(r-2) = 16r$$
$$20r - 40 = 16r$$
$$20r - 16r = 40$$
$$4r = 40$$
$$r = \frac{40}{4} = 10$$

The first part is at 10 mph, and the cool-down at 8 mph.

45. Let t = time to empty tank.

$$-\frac{1}{1.5} + \frac{1}{1} = \frac{1}{t}$$
$$\frac{1}{3} = \frac{1}{t}$$
$$t = 3$$

It will take 3 hours to empty the tank.

47. Let d = distance at which both rockets are an equal distance from earth.

Recall $d = rt$ or $t = \frac{d}{r}$

$$\frac{d}{9000} - \frac{1}{4} = \frac{d}{10{,}000}$$
$$90{,}000\left(\frac{d}{9000} - \frac{1}{4}\right) = 90{,}000\left(\frac{d}{10{,}000}\right)$$
$$10d - 22{,}500 = 9d$$
$$d = 22{,}500$$

The distance is 22,500 miles.

49.

	Time	In one hour
Experienced	4	$\frac{1}{4}$
Apprentice	5	$\frac{1}{5}$
Together	x	$\frac{1}{x}$

$$\frac{1}{4}+\frac{1}{5}=\frac{1}{x}$$
$$20x\left(\frac{1}{4}+\frac{1}{5}\right)=20x\left(\frac{1}{x}\right)$$
$$20x\left(\frac{1}{4}\right)+20x\left(\frac{1}{5}\right)=20$$
$$5x+4x=20$$
$$9x=20$$
$$x=\frac{20}{9}=2\frac{2}{9}\text{ hours}$$

51. Let x = time to complete job together.
$$\frac{1}{3}+\frac{1}{6}=\frac{1}{x}$$
$$2x+x=6$$
$$3x=6$$
$$x=2$$
It will take 2 hours.

53. Let t = time to complete job together.
$$\frac{1}{6}+\frac{1}{4}=\frac{1}{t}$$
$$2t+3t=12$$
$$5t=12$$
$$t=\frac{12}{5}$$
$$\$45\cdot\frac{12}{5}=\$108$$
The labor estimate should be $108.

55. Let t = time to complete job together.
$$\frac{1}{4}+\frac{1}{5}=\frac{1}{t}$$
$$5t+4t=20$$
$$9t=20$$
$$t=\frac{20}{9}=2\frac{2}{9}$$
It will take $2\frac{2}{9}$ days.

57. Let r = rowing rate, then
$r+6$ = rate of boat downstream, and
$r-6$ = rate of boat upstream.

Recall $d=rt$ or $t=\frac{d}{r}$.
$$\frac{9}{r+6}=\frac{3}{r-6}$$
$$9(r-6)=3(r+6)$$
$$9r-54=3r+18$$
$$6r=72$$
$$r=12$$
$$t=\frac{\text{distance downstream}}{\text{rate downstream}}+\frac{\text{distance upstream}}{\text{rate upstream}}$$
$$t=\frac{9}{12+6}+\frac{3}{12-6}$$
$$t=\frac{9}{18}+\frac{3}{6}$$
$$t=1$$
It will take 1 hour.

59.
$$B=\frac{705w}{h^2}$$
$$47=\frac{705(240)}{h^2}$$
$$h^2=\frac{705(240)}{47}$$
$$h^2=3600$$
$h=60$ or -60 (disregard)
The patient is 5 feet tall.

61.
$$\frac{x}{5}=\frac{x+2}{3}$$
$$3x=5(x+2)$$
$$3x=5x+10$$
$$-10=2x$$
$$x=-5$$

63.
$$\frac{x-3}{2}=\frac{x-5}{6}$$
$$6(x-3)=2(x-5)$$
$$6x-18=2x-10$$
$$4x=8$$
$$x=2$$

Exercise Set 6.8

1. $y = kx$
$4 = k(20)$
$k = \frac{1}{5}$

$y = \frac{1}{5}x$

3. $y = kx$
$6 = k(4)$
$k = \frac{3}{2}$

$y = \frac{3}{2}x$

5. $y = kx$
$7 = k\left(\frac{1}{2}\right)$
$k = 14$

$y = 14x$

7. $y = kx$
$0.2 = k(0.8)$
$k = 0.25$

$y = 0.25x$

9. $W = kr^3$
$1.2 = k \cdot 2^3$
$k = \frac{1.2}{8} = 0.15$

$W = 0.15r^3 = 0.15(3)^3$
$W = 0.15(27) = 4.05$ lb

11. $P = kN$
$260,000 = k(450,000)$
$k = \frac{26}{45}$

$P = \frac{26}{45}N = \frac{26}{45}(980,000)$
$P \approx 566,222$ tons

13. $y = \frac{k}{x}$
$6 = \frac{k}{5}$
$k = 30$

$y = \frac{30}{x}$

15. $y = \frac{k}{x}$
$100 = \frac{k}{7}$
$k = 700$

$y = \frac{700}{x}$

17. $y = \frac{k}{x}$
$\frac{1}{8} = \frac{k}{16}$
$k = 2$

$y = \frac{2}{x}$

19. $y = \frac{k}{x}$
$0.2 = \frac{k}{0.7}$
$k = 0.14$

$y = \frac{0.14}{x}$

21. $R = \frac{k}{T}$
$45 = \frac{k}{6}$
$k = 270$

$R = \frac{270}{T} = \frac{270}{5}$
$R = 54$ mph

23. $I = \dfrac{k}{R}$

$40 = \dfrac{k}{270}$

$k = 10{,}800$

$I = \dfrac{10{,}800}{R}$

$= \dfrac{10{,}800}{150}$

$= 72$ amps

25. $I_1 = \dfrac{k}{d^2}$

Replace d by $2d$.

$I_2 = \dfrac{k}{(2d)^2} = \dfrac{k}{4d^2} = \dfrac{1}{4}I_1$

Thus, the intensity is divided by 4.

27. $x = kyz$

29. $r = kst^3$

31. $W = \dfrac{kwh^2}{l}$

$12 = \dfrac{k\left(\frac{1}{2}\right)\left(\frac{1}{3}\right)^2}{10}$

$k = 2160$

$W = \dfrac{2160wh^2}{l}$

$= \dfrac{2160\left(\frac{2}{3}\right)\left(\frac{1}{2}\right)^2}{16}$

$= \dfrac{45}{2}$ tons or 22.5 tons

33. $V = kr^2h$

$32\pi = k(4)^2(6)$

$k = \dfrac{\pi}{3}$

$V = \dfrac{\pi}{3}r^2h$

$= \dfrac{\pi}{3}(3)^2(5)$

$= 15\pi$ cu. in.

35. $H = ksd^3$

$40 = k(120)(2)^3$

$k = \dfrac{1}{24}$

$H = \dfrac{1}{24}sd^3$

$= \dfrac{1}{24}(80)(3)^3$

$= 90$ hp

37. $y = \dfrac{k}{x}$

$400 = \dfrac{k}{8}$

$k = 3200$

$y = \dfrac{3200}{x} = \dfrac{3200}{4} = 800$ millibars

39. $V_1 = khr^2$

$V_2 = k\left(\dfrac{1}{2}h\right)(2r)^2$

$V_2 = 2khr^2 = 2V_1$

It is multiplied by 2.

41. $y_1 = kx^2$

$y_2 = k(2x)^2$

$y_2 = 4kx^2 = 4y_1$

It is multiplied by 4.

43.

x	$\frac{1}{4}$	$\frac{1}{2}$	1	2	4
$y=\frac{3}{x}$	12	6	3	$\frac{3}{2}$	$\frac{3}{4}$

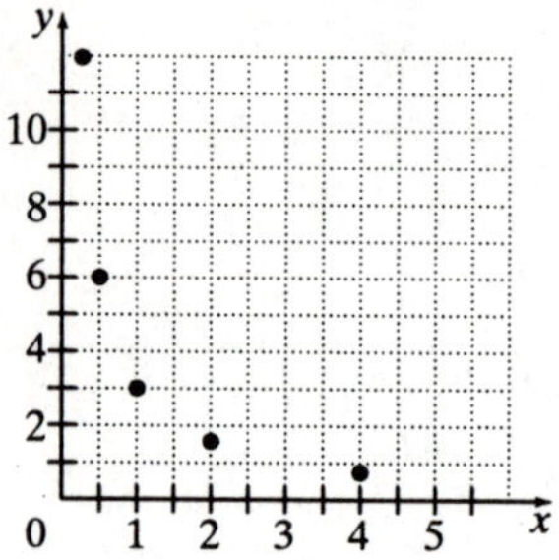

45.

x	$\frac{1}{4}$	$\frac{1}{2}$	1	2	4
$y=\frac{1}{2x}$	2	1	$\frac{1}{2}$	$\frac{1}{4}$	$\frac{1}{8}$

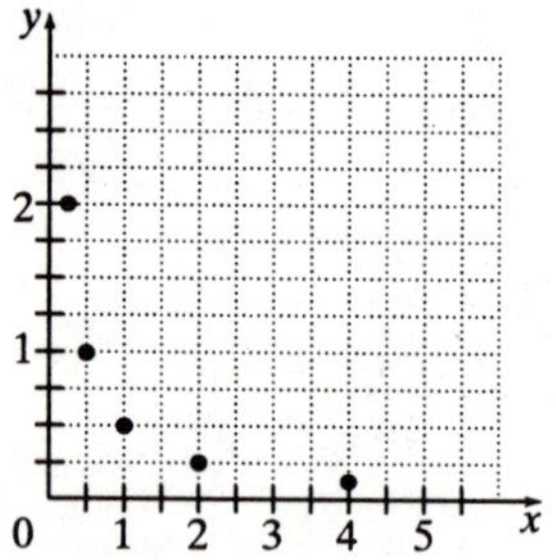

47. $r = 6$ cm
$C = 2\pi r = 2\pi(6) = 12\pi$ cm
$A = \pi r^2 = \pi(6^2) = 36\pi$ sq. cm

49. $r = 7$ m
$C = 2\pi r = 2\pi(7) = 14\pi$ m
$A = \pi r^2 = \pi(7^2) = 49\pi$ sq. m

51. $\sqrt{36} = 6$

53. $\sqrt{4} = 2$

55. $\sqrt{\frac{1}{25}} = \frac{1}{5}$

57. $\sqrt{\frac{25}{121}} = \frac{5}{11}$

Chapter 6 Review

1. $f(x) = \frac{3-5x}{7}$
Domain $\{x \mid x \text{ is a real number}\}$

2. $g(x) = \frac{2x+4}{11}$
Domain $\{x \mid x \text{ is a real number}\}$

3. $F(x) = \frac{-3x^2}{x-5}$
Undefined values when
$x - 5 = 0$
$x = 5$
Domain $\{x \mid x \text{ is a real number and } x \neq 5\}$

4. $h(x) = \frac{4x}{3x-12}$
Undefined values when
$3x - 12 = 0$
$3x = 12$
$x = 4$
Domain $\{x \mid x \text{ is a real number and } x \neq 4\}$

5. $f(x) = \frac{x^3+2}{x^2+8x}$
Undefined values when
$x^2 + 8x = 0$
$x(x+8) = 0$
$x = 0$ or $x + 8 = 0$
$x = -8$
Domain
$\{x \mid x \text{ is a real number and } x \neq 0, x \neq -8\}$

6. $G(x) = \frac{20}{3x^2-48}$
Undefined values when
$3x^2 - 48 = 0$
$3(x^2 - 16) = 0$
$3(x+4)(x-4) = 0$
$x + 4 = 0$ or $x - 4 = 0$
$x = -4$ or $x = 4$
Domain $\{x \mid x \text{ is a real number and } x \neq -4,$
$x \neq 4\}$

7. $\frac{15x^4}{45x^2} = \frac{15x^{4-2}}{15 \cdot 3} = \frac{x^2}{3}$

8. $\frac{x+2}{2+x} = \frac{x+2}{x+2} = 1$

9. $\frac{18m^6p^2}{10m^4p} = \frac{2 \cdot 9m^{6-4}p^{2-1}}{2 \cdot 5} = \frac{9m^2p}{5}$

10. $\frac{x-12}{12-x} = \frac{-1(12-x)}{12-x} = -1$

11. $\frac{5x-15}{25x-75} = \frac{5(x-3)}{5 \cdot 5(x-3)} = \frac{1}{5}$

12. $\frac{22x+8}{11x+4} = \frac{2(11x+4)}{11x+4} = 2$

13. $\frac{2x}{2x^2-2x} = \frac{2x}{2x(x-1)} = \frac{1}{x-1}$

14. $\frac{x+7}{x^2-49} = \frac{x+7}{(x+7)(x-7)} = \frac{1}{x-7}$

15. $\frac{2x^2+4x-30}{x^2+x-20} = \frac{2(x+5)(x-3)}{(x+5)(x-4)}$
$= \frac{2(x-3)}{x-4}$

16. $\frac{xy-3x+2y-6}{x^2+4x+4} = \frac{x(y-3)+2(y-3)}{(x+2)^2}$
$= \frac{(x+2)(y-3)}{(x+2)(x+2)}$
$= \frac{y-3}{x+2}$

17. $C(x) = \frac{35x+4200}{x}$

(a) $C(50) = \frac{35(50)+4200}{50} = \$119$

(b) $C(100) = \frac{35(100)+4200}{100} = \$77$

(c) Decrease

18. $\frac{5}{x^3} \cdot \frac{x^2}{15} = \frac{1}{3x}$

19. $\frac{3x^4yz^3}{15x^2y^2} \cdot \frac{10xy}{z^6} = \frac{30x^{4+1}y^{1+1}z^3}{15x^2y^2z^6}$
$= \frac{2x^{5-2}}{z^{6-3}} = \frac{2x^3}{z^3}$

20. $\frac{4-x}{5} \cdot \frac{15}{2x-8} = \frac{-1(x-4)}{5} \cdot \frac{15}{2(x-4)} = -\frac{3}{2}$

21. $\frac{x^2-6x+9}{2x^2-18} \cdot \frac{4x+12}{5x-15}$
$= \frac{(x-3)^2}{2(x^2-9)} \cdot \frac{4(x+3)}{5(x-3)}$
$= \frac{x-3}{(x+3)(x-3)} \cdot \frac{2(x+3)}{5}$
$= \frac{1}{1} \cdot \frac{2}{5}$
$= \frac{2}{5}$

22. $\frac{a-4b}{a^2+ab} \cdot \frac{b^2-a^2}{8b-2a}$
$= \frac{a-4b}{a(a+b)} \cdot \frac{(b-a)(b+a)}{2(4b-a)}$
$= \frac{-(b-a)}{2a}$
$= \frac{a-b}{2a}$

23. $\frac{x^2-x-12}{2x^2-32} \cdot \frac{x^2+8x+16}{3x^2+21x+36}$
$= \frac{(x-4)(x+3)}{2(x^2-16)} \cdot \frac{(x+4)^2}{3(x^2+7x+12)}$
$= \frac{(x-4)(x+3)(x+4)^2}{6(x+4)(x-4)(x+3)(x+4)}$
$= \frac{1}{6}$

24. $\dfrac{2x^3+54}{5x^2+5x-30}\cdot\dfrac{6x+12}{3x^2-9x+27}$
$=\dfrac{2(x^3+27)}{5(x^2+x-6)}\cdot\dfrac{6(x+2)}{3(x^2-3x+9)}$
$=\dfrac{2(x+3)(x^2-3x+9)}{5(x-2)(x+3)}\cdot\dfrac{6(x+2)}{3(x^2-3x+9)}$
$=\dfrac{4(x+2)}{5(x-2)}$

25. $\dfrac{3}{4x}\div\dfrac{8}{2x^2}=\dfrac{3}{4x}\cdot\dfrac{2x^2}{8}=\dfrac{6x^2}{32x}=\dfrac{3x}{16}$

26. $\dfrac{4x+8y}{3}\div\dfrac{5x+10y}{9}$
$=\dfrac{4(x+2y)}{3}\cdot\dfrac{9}{5(x+2y)}$
$=\dfrac{12}{5}$

27. $\dfrac{5ab}{14c^3}\div\dfrac{10a^4b^2}{6ac^5}=\dfrac{5ab}{14c^3}\cdot\dfrac{6ac^5}{10a^4b^2}$
$=\dfrac{30a^2bc^5}{140a^4b^2c^3}$
$=\dfrac{3c^{5-3}}{14a^{4-2}b^{2-1}}$
$=\dfrac{3c^2}{14a^2b}$

28. $\dfrac{2}{5x}\div\dfrac{4-18x}{6-27x}=\dfrac{2}{5x}\cdot\dfrac{3(2-9x)}{2(2-9x)}=\dfrac{3}{5x}$

29. $\dfrac{x^2-25}{3}\div\dfrac{x^2-10x+25}{x^2-x-20}$
$=\dfrac{(x+5)(x-5)}{3}\cdot\dfrac{(x-5)(x+4)}{(x-5)^2}$
$=\dfrac{(x+5)(x+4)}{3}$

30. $\dfrac{a-4b}{a^2+ab}\div\dfrac{20b-5a}{b^2-a^2}$
$=\dfrac{a-4b}{a(a+b)}\cdot\dfrac{(b-a)(b+a)}{5(4b-a)}$
$=\dfrac{-(b-a)}{5a}$
$=\dfrac{a-b}{5a}$

31. $\dfrac{7x+28}{2x+4}\div\dfrac{x^2+2x-8}{x^2-2x-8}$
$=\dfrac{7(x+4)}{2(x+2)}\cdot\dfrac{(x-4)(x+2)}{(x+4)(x-2)}$
$=\dfrac{7(x-4)}{2(x-2)}$

32. $\dfrac{3x+3}{x-1}\div\dfrac{x^2-6x-7}{x^2-1}$
$=\dfrac{3(x+1)}{x-1}\cdot\dfrac{(x-1)(x+1)}{(x+1)(x-7)}$
$=\dfrac{3(x+1)}{x-7}$

33. $\dfrac{2x-x^2}{x^3-8}\div\dfrac{x^2}{x^2+2x+4}$
$=\dfrac{x(2-x)}{(x-2)(x^2+2x+4)}\cdot\dfrac{x^2+2x+4}{x^2}$
$=\dfrac{-(x-2)}{x-2}\cdot\dfrac{1}{x}$
$=-1\cdot\dfrac{1}{x}$
$=-\dfrac{1}{x}$

34. $\dfrac{5a^2-20}{a^3+2a^2+a+2}\div\dfrac{7a}{a^3+a}$
$=\dfrac{5(a+2)(a-2)}{(a^2+1)(a+2)}\cdot\dfrac{a(a^2+1)}{7a}$
$=\dfrac{5(a-2)}{7}$

35. $\frac{2a}{21} \div \frac{3a^2}{7} \cdot \frac{4}{a} = \frac{2a}{21} \cdot \frac{7}{3a^2} \cdot \frac{4}{a}$
$= \frac{56a}{63a^3} = \frac{8}{9a^2}$

36. $\frac{5x-15}{3-x} \cdot \frac{x+2}{10x+20} \cdot \frac{x^2-9}{x^2-x-6}$
$= \frac{5(x-3)(x+2)(x-3)(x+3)}{(3-x)10(x+2)(x+2)(x-3)}$
$= -\frac{(x+3)}{2(x+2)}$

37. $\frac{4a+8}{5a^2-20} \cdot \frac{3a^2-6a}{a+3} \div \frac{2a^2}{5a+15}$
$= \frac{4(a+2)}{5(a^2-4)} \cdot \frac{3a(a-2)}{a+3} \cdot \frac{5(a+3)}{2a^2}$
$= \frac{60a(a+2)(a-2)(a+3)}{10a^2(a+2)(a-2)(a+3)}$
$= \frac{6}{a}$

38. The LCD is (9)(2) = 18.

39. $\frac{5}{4x^2y^5}, \frac{3}{10x^2y^4}, \frac{x}{6y^4}$
The LCD is $60x^2y^5$.

40. The LCD is $2x(x-2)$.

41. $\frac{3}{5x}, \frac{2}{x-5}$
The LCD is $5x(x-5)$.

42. The first denominator is $5x^3$.
The second denominator is $(x-4)(x+7)$.
The third denominator is $10x(x-3)$.
The LCD is $10x^3(x-4)(x+7)(x-3)$.

43. $\frac{2}{15} + \frac{4}{15} = \frac{2+4}{15} = \frac{6}{15} = \frac{2}{5}$

44. $\frac{4}{x-4} + \frac{x}{x-4} = \frac{4+x}{x-4}$

45. $\frac{4}{3x^2} + \frac{2}{3x^2} = \frac{4+2}{3x^2} = \frac{6}{3x^2} = \frac{2}{x^2}$

46. $\frac{1}{x-2} - \frac{1}{2(2-x)} = \frac{1(2)}{(x-2)(2)} + \frac{1}{2(x-2)}$
$= \frac{2+1}{2(x-2)}$
$= \frac{3}{2(x-2)}$

47. $\frac{2x+1}{x^2+x-6} + \frac{2-x}{x^2+x-6} = \frac{2x+1+2-x}{x^2+x-6}$
$= \frac{x+3}{(x+3)(x-2)}$
$= \frac{1}{x-2}$

48. $\frac{7}{2x} + \frac{5}{6x} = \frac{(7)\cdot(3)}{(2x)\cdot(3)} + \frac{5}{6x}$
$= \frac{21}{6x} + \frac{5}{6x}$
$= \frac{26}{6x}$
$= \frac{13}{3x}$

49. $\frac{1}{3x^2y^3} - \frac{1}{5x^4y} = \frac{5x^2}{15x^4y^3} - \frac{3y^2}{15x^4y^3}$
$= \frac{5x^2-3y^2}{15x^4y^3}$

50. $\frac{1}{10-x} + \frac{x-1}{x-10}$
$= \frac{-1}{x-10} + \frac{x-1}{x-10} = \frac{x-2}{x-10}$

51. $\frac{x-2}{x+1} - \frac{x-3}{x-1}$
$= \frac{(x-2)(x-1)-(x-3)(x+1)}{(x+1)(x-1)}$
$= \frac{x^2-3x+2-(x^2-2x-3)}{(x+1)(x-1)}$
$= \frac{x^2-3x+2-x^2+2x+3}{(x+1)(x-1)}$
$= \frac{-x+5}{x^2-1}$

52. $\frac{x}{9-x^2}-\frac{2}{5x-15}$
$=\frac{x}{(3-x)(3+x)}-\frac{2}{5(x-3)}$
$=\frac{-5x}{5(x-3)(x+3)}-\frac{2(x+3)}{5(x-3)(x+3)}$
$=\frac{-5x-2x-6}{5(x-3)(x+3)}$
$=\frac{-7x-6}{5(x-3)(x+3)}$

53. $2x+1-\frac{1}{x-3}=\frac{(2x+1)(x-3)-1}{x-3}$
$=\frac{2x^2-5x-3-1}{x-3}$
$=\frac{2x^2-5x-4}{x-3}$

54. $\frac{2}{a^2-2a+1}+\frac{3}{a^2-1}$
$=\frac{2}{(a-1)^2}+\frac{3}{(a+1)(a-1)}$
$=\frac{2(a+1)}{(a-1)^2(a+1)}+\frac{3(a-1)}{(a-1)^2(a+1)}$
$=\frac{2a+2+3a-3}{(a-1)^2(a+1)}$
$=\frac{5a-1}{(a-1)^2(a+1)}$

55. $\frac{x}{9x^2+12x+16}-\frac{3x+4}{27x^3-64}$
$=\frac{x}{9x^2+12x+16}$
$-\frac{3x+4}{(3x-4)(9x^2+12x+16)}$
$=\frac{(3x-4)x-(3x+4)}{(3x-4)(9x^2+12x+16)}$
$=\frac{3x^2-4x-3x-4}{(3x-4)(9x^2+12x+16)}$
$=\frac{3x^2-7x-4}{27x^3-64}$

56. $\frac{2}{x-1}-\frac{3x}{3x-3}+\frac{1}{2x-2}$
$=\frac{2}{x-1}-\frac{3x}{3(x-1)}+\frac{1}{2(x-1)}$
$=\frac{4}{2(x-1)}-\frac{2x}{2(x-1)}+\frac{1}{2(x-1)}$
$=\frac{5-2x}{2(x-1)}$

57. $\frac{3}{2x}\left(\frac{2}{x+1}-\frac{2}{x-3}\right)=\frac{3}{x}\left(\frac{1}{x+1}-\frac{1}{x-3}\right)$
$=\frac{3}{x}\left(\frac{x-3-(x+1)}{(x+1)(x-3)}\right)$
$=\frac{3}{x}\cdot\frac{x-3-x-1}{(x+1)(x-3)}$
$=\frac{3(-4)}{x(x+1)(x-3)}$
$=-\frac{12}{x(x+1)(x-3)}$

58. $\left(\frac{2}{x}-\frac{1}{5}\right)\cdot\left(\frac{2}{x}+\frac{1}{3}\right)=\left(\frac{10-x}{5x}\right)\cdot\left(\frac{6+x}{3x}\right)$
$=\frac{(10-x)(6+x)}{15x^2}$

59. $\frac{2}{x^2-16}-\frac{3x}{x^2+8x+16}+\frac{3}{x+4}$
$=\frac{2}{(x+4)(x-4)}-\frac{3x}{(x+4)^2}+\frac{3}{x+4}$
$=\frac{2(x+4)}{(x+4)^2(x-4)}-\frac{3x(x-4)}{(x+4)^2(x-4)}$
$+\frac{3(x^2-16)}{(x+4)^2(x-4)}$
$=\frac{2x+8-3x^2+12x+3x^2-48}{(x+4)^2(x-4)}$

60. $P = \frac{1}{x}+\frac{1}{x}+\frac{1}{x}+\frac{2}{x}+\frac{5}{2x}+\frac{2}{x}+\frac{3}{2x}$
$= \frac{1+1+1+2+2}{x}+\frac{5+3}{2x}$
$= \frac{7}{x}+\frac{8}{2x}$
$= \frac{14}{2x}+\frac{8}{2x}$
$= \frac{22}{2x}$
$= \frac{11}{x}$

61. $\frac{\frac{2}{5}}{\frac{3}{5}} = \frac{2}{5}\cdot\frac{5}{3} = \frac{2}{1}\cdot\frac{1}{3} = \frac{2}{3}$

62. $\frac{\left(1-\frac{3}{4}\right)(4)}{\left(2+\frac{1}{4}\right)(4)} = \frac{4-3}{8+1} = \frac{1}{9}$

63. $\frac{\frac{1}{x}-\frac{2}{3x}}{\frac{5}{2x}-\frac{1}{3}} = \frac{\frac{3x-2x}{3x^2}}{\frac{15-2x}{6x}}$
$= \frac{x}{3x^2}\cdot\frac{6x}{15-2x}$
$= \frac{6x^2}{3x^2(15-2x)}$
$= \frac{2}{15-2x}$

64. $\frac{\frac{x^2}{15}}{\frac{x+1}{5x}} = \frac{x^2}{15} \div \frac{x+1}{5x} = \frac{x^2}{15}\cdot\frac{5x}{x+1} = \frac{x^3}{3(x+1)}$

65. $\frac{\frac{3}{y^2}}{\frac{6}{y^3}} = \frac{3}{y^2}\cdot\frac{y^3}{6} = \frac{y}{2}$

66. $\frac{\frac{x+2}{3}}{\frac{5}{x-2}} = \frac{x+2}{3} \div \frac{5}{x-2}$
$= \frac{x+2}{3}\cdot\frac{x-2}{5}$
$= \frac{(x+2)(x-2)}{15}$

67. $\frac{2-\frac{3}{2x}}{x-\frac{2}{5x}} = \frac{\frac{4x-3}{2x}}{\frac{5x^2-2}{5x}}$
$= \frac{4x-3}{2x}\cdot\frac{5x}{5x^2-2}$
$= \frac{5(4x-3)}{2(5x^2-2)}$
$= \frac{20x-15}{10x^2-4}$

68. $\frac{\left(1+\frac{x}{y}\right)\cdot y^2}{\left(\frac{x^2}{y^2}-1\right)\cdot y^2} = \frac{y^2+xy}{x^2-y^2}$
$= \frac{y(y+x)}{(x+y)(x-y)}$
$= \frac{y}{x-y}$

69.
$\frac{\frac{5}{x}+\frac{1}{xy}}{\frac{3}{x^2}} = \frac{5y+1}{xy}\cdot\frac{x^2}{3} = \frac{x(5y+1)}{3y}$
$= \frac{5xy+x}{3y}$

70. $\frac{\left(\frac{x}{3}-\frac{3}{x}\right)\cdot(3x)}{\left(1+\frac{3}{x}\right)\cdot(3x)} = \frac{x^2-9}{3x+9}$
$= \frac{(x+3)(x-3)}{3(x+3)} = \frac{x-3}{3}$

71. $\frac{\frac{1}{x-1}+1}{\frac{1}{x+1}-1} = \frac{\frac{1+x-1}{x-1}}{\frac{1-(x+1)}{x+1}}$
$= \frac{x}{x-1}\cdot\frac{x+1}{1-x-1}$
$= \frac{x(x+1)}{-x(x-1)}$
$= \frac{x+1}{-(x-1)}$
$= \frac{1+x}{1-x}$

72. $\frac{(2)x}{\left(1-\frac{2}{x}\right)x} = \frac{2x}{x-2}$

73. $\dfrac{1}{1+\frac{2}{1-\frac{1}{x}}}=\dfrac{1}{1+\frac{2x}{x-1}}=\dfrac{1}{\frac{x-1+2x}{x-1}}=\dfrac{x-1}{3x-1}$

74. $\dfrac{\frac{x^2+5x-6}{4x+3}}{\frac{(x+6)^2}{8x+6}}=\dfrac{x^2+5x-6}{4x+3}\div\dfrac{(x+6)^2}{8x+6}$

$=\dfrac{(x-1)(x+6)}{4x+3}\cdot\dfrac{2(4x+3)}{(x+6)^2}$

$=\dfrac{2(x-1)}{x+6}$

75. $\dfrac{\frac{x-3}{x+3}+\frac{x+3}{x-3}}{\frac{x-3}{x+3}-\frac{x+3}{x-3}}=\dfrac{\frac{(x-3)^2+(x+3)^2}{(x+3)(x-3)}}{\frac{(x-3)^2-(x+3)^2}{(x+3)(x-3)}}$

$=\dfrac{x^2-6x+9+x^2+6x+9}{x^2-6x+9-(x^2+6x+9)}$

$=\dfrac{2x^2+18}{x^2-6x+9-x^2-6x-9}$

$=\dfrac{2x^2+18}{-12x}$

$=-\dfrac{x^2+9}{6x}$

76. $\dfrac{\frac{3}{x-1}-\frac{2}{1-x}}{\frac{2}{x-1}-\frac{2}{x}}=\dfrac{\frac{3}{x-1}+\frac{2}{x-1}}{\frac{2x}{x(x-1)}-\frac{2(x-1)}{x(x-1)}}$

$=\dfrac{\frac{5}{x-1}}{\frac{2}{x(x-1)}}$

$=\dfrac{5}{x-1}\div\dfrac{2}{x(x-1)}$

$=\dfrac{5}{x-1}\cdot\dfrac{x(x-1)}{2}$

$=\dfrac{5x}{2}$

77. $f(x)=\frac{3}{x}$

a. $f(a+h)=\dfrac{3}{a+h}$

b. $f(a)=\frac{3}{a}$

c. $\dfrac{f(a+h)-f(a)}{h}=\dfrac{\frac{3}{a+h}-\frac{3}{a}}{h}$

d. $\dfrac{\frac{3}{a+h}-\frac{3}{a}}{h}=\dfrac{\frac{3a}{a(a+h)}-\frac{3(a+h)}{a(a+h)}}{h}$

$=\dfrac{3a-3(a+h)}{a(a+h)}\cdot\dfrac{1}{h}$

$=\dfrac{3[a-(a+h)]}{ah(a+h)}$

$=\dfrac{3(-h)}{ah(a+h)}$

$=\dfrac{-3}{a(a+h)}$

78. $\dfrac{3x^5yb^9}{9xy^7}=\dfrac{x^{5-1}b^9}{3y^{7-1}}=\dfrac{x^4b^9}{3y^6}$

79. $\dfrac{-9xb^4z^3}{-4axb^2}=\dfrac{(-1)(9)x^{1-1}b^{4-2}z^3}{(-1)(4)a}=\dfrac{9b^2z^3}{4a}$

80. $\dfrac{4xy+2x^2-9}{4xy}=\dfrac{4xy}{4xy}+\dfrac{2x^2}{4xy}-\dfrac{9}{4xy}$

$=1+\dfrac{x^{2-1}}{2y}-\dfrac{9}{4xy}$

$=1+\dfrac{x}{2y}-\dfrac{9}{4xy}$

81. $\dfrac{12xb^2+16xb^4}{4xb^3}$

$=\dfrac{12xb^2}{4xb^3}+\dfrac{16xb^4}{4xb^3}$

$=\dfrac{3\cdot 4x^{1-1}}{4b^{3-2}}+\dfrac{4\cdot 4x^{1-1}b^{4-3}}{4}$

$=\dfrac{3}{b}+4b$

82.
$$\begin{array}{r}
3x^3+9x^2+2x+6 \\
x-3\overline{)3x^4+0x^3-25x^2+0x-20} \\
\underline{3x^4-9x^3} \qquad\qquad\qquad\quad \\
9x^3+25x^2 \qquad\qquad\quad \\
\underline{9x^3-27x^2} \qquad\qquad\quad \\
2x^2+0x \qquad\quad \\
\underline{2x^2-6x} \qquad\quad \\
6x-20 \\
\underline{6x-18} \\
-2
\end{array}$$

Answer: $3x^3+9x^2+2x+6-\frac{2}{x-3}$

83.
$$\begin{array}{r}
2x^3-4x^2+7x-9 \\
x+2\overline{)2x^4+0x^3-x^2+5x-12} \\
\underline{2x^4+4x^3} \qquad\qquad\qquad \\
-4x^3-x^2 \qquad\qquad \\
\underline{-4x^3-8x^2} \qquad\qquad \\
7x^2+5x \qquad \\
\underline{7x^2+14x} \qquad \\
-9x-12 \\
\underline{-9x-18} \\
6
\end{array}$$

Answer: $2x^3-4x^2+7x-9+\frac{6}{x+2}$

84.
$$\begin{array}{r}
2x^3 \qquad\quad +2x-2 \\
x-\tfrac{1}{2}\overline{)2x^4-x^3+2x^2-3x+1} \\
\underline{2x^4-x^3} \qquad\qquad\qquad \\
2x^2-3x \qquad \\
\underline{2x^2-x} \qquad \\
-2x+1 \\
\underline{-2x+1} \\
0
\end{array}$$

Answer: $2x^3+2x-2$

85.
$$\begin{array}{r}
2x^2 \qquad -2 \\
x+\tfrac{3}{2}\overline{)2x^3+3x^2-2x+2} \\
\underline{2x^3+3x^2} \qquad\qquad \\
-2x+2 \\
\underline{-2x-3} \\
5
\end{array}$$

Answer: $2x^2-2+\frac{5}{x+\frac{3}{2}}$

86.
$$\begin{array}{r}
3x^2+2x-1 \\
x^2+x+2\overline{)3x^4+5x^3+7x^2+3x-2} \\
\underline{3x^4+3x^3+6x^2} \qquad\qquad \\
2x^3+x^2+3x \qquad \\
\underline{2x^3+2x^2+4x} \qquad \\
-x^2-x-2 \\
\underline{-x^2-x-2} \\
0
\end{array}$$

Answer: $3x^2+2x-1$

87.
$$\begin{array}{r}
3x^2 \qquad\qquad +6 \\
3x^2-2x-5\overline{)9x^4-6x^3+3x^2-12x-30} \\
\underline{9x^4-6x^3-15x^2} \qquad\qquad\quad \\
18x^2-12x-30 \\
\underline{18x^2-12x-30} \\
0
\end{array}$$

Answer: $3x^2+6$

88.

2	3	0	12	–4
		6	12	48
	3	6	24	44

Answer: $3x^2+6x+24+\frac{44}{x-2}$

89.
$$\begin{array}{r|rrrr} -\frac{3}{2} & 3 & 2 & -4 & -1 \\ & & -\frac{9}{2} & \frac{15}{4} & \frac{3}{8} \\ \hline & 3 & -\frac{5}{2} & -\frac{1}{4} & -\frac{5}{8} \end{array}$$

Answer: $3x^2 - \frac{5}{2}x - \frac{1}{4} - \frac{5}{8\left(x+\frac{3}{2}\right)}$

90.
$$\begin{array}{r|rrrrrr} -1 & 1 & 0 & 0 & 0 & 0 & -1 \\ & & -1 & 1 & -1 & 1 & -1 \\ \hline & 1 & -1 & 1 & -1 & 1 & -2 \end{array}$$

Answer: $x^4 - x^3 + x^2 - x + 1 - \frac{2}{x+1}$

91.
$$\begin{array}{r|rrrr} 3 & 1 & 0 & 0 & -81 \\ & & 3 & 9 & 27 \\ \hline & 1 & 3 & 9 & -54 \end{array}$$

Answer: $x^2 + 3x + 9 - \frac{54}{x-3}$

92.
$$\begin{array}{r|rrrrr} 4 & 3 & 1 & -1 & 0 & -2 \\ & & 12 & 52 & 204 & 816 \\ \hline & 3 & 13 & 51 & 204 & 814 \end{array}$$

Answer: $3x^3 + 13x^2 + 51x + 204 + \frac{814}{x-4}$

93.
$$\begin{array}{r|rrrrr} -2 & 3 & 0 & -2 & 0 & 10 \\ & & -6 & 12 & -20 & 40 \\ \hline & 3 & -6 & 10 & -20 & 50 \end{array}$$

Answer: $3x^3 - 6x^2 + 10x - 20 + \frac{50}{x+2}$

94.
$$\begin{array}{r|rrrrrr} 4 & 3 & 0 & 0 & 0 & -9 & 7 \\ & & 12 & 48 & 192 & 768 & 3036 \\ \hline & 3 & 12 & 48 & 192 & 759 & 3043 \end{array}$$

Thus, $P(4) = 3043$

95.
$$\begin{array}{r|rrrrrr} -5 & 3 & 0 & 0 & 0 & -9 & 7 \\ & & -15 & 75 & -375 & 1875 & -9330 \\ \hline & 3 & -15 & 75 & -375 & 1866 & -9323 \end{array}$$

$P(-5) = -9323$

96.
$$\begin{array}{r|rrrrrr} \frac{2}{3} & 3 & 0 & 0 & 0 & -9 & 7 \\ & & 2 & \frac{4}{3} & \frac{8}{9} & \frac{16}{27} & -\frac{454}{81} \\ \hline & 3 & 2 & \frac{4}{3} & \frac{8}{9} & -\frac{227}{27} & \frac{113}{81} \end{array}$$

$P\left(\frac{2}{3}\right) = \frac{113}{81}$

97.
$$\begin{array}{r|rrrrrr} -\frac{1}{2} & 3 & 0 & 0 & 0 & -9 & 7 \\ & & -\frac{3}{2} & \frac{3}{4} & -\frac{3}{8} & \frac{3}{16} & \frac{141}{32} \\ \hline & 3 & -\frac{3}{2} & \frac{3}{4} & -\frac{3}{8} & -\frac{141}{16} & \frac{365}{32} \end{array}$$

$P\left(-\frac{1}{2}\right) = \frac{365}{32}$

98.
$$\begin{array}{r|rrrrr} 3 & 1 & -1 & -6 & -6 & 18 \\ & & 3 & 6 & 0 & -18 \\ \hline & 1 & 2 & 0 & -6 & 0 \end{array}$$

$(x^3 + 2x^2 - 6)$ miles

99.
$$\frac{2}{5} = \frac{x}{15}$$
$$30 = 5x$$
$$x = 6$$

100.
$$\frac{3}{x} + \frac{1}{3} = \frac{5}{x}$$
$$3x\left(\frac{3}{x} + \frac{1}{3}\right) = \left(\frac{5}{x}\right)(3x)$$
$$9 + x = 15$$
$$x = 6$$

101. $4+\frac{8}{x}=8$

$\frac{8}{x}=4$

$\frac{8}{4}=x$

$x=2$

102. $\frac{2x+3}{5x-9}=\frac{3}{2}$

$2(5x-9)\left(\frac{2x+3}{5x-9}\right)=\left(\frac{3}{2}\right)(2)(5x-9)$

$4x+6=15x-27$

$33=11x$

$3=x$

103. $\frac{1}{x-2}-\frac{3x}{x^2-4}=\frac{2}{x+2}$

$\frac{x+2-3x}{(x-2)(x+2)}=\frac{2(x-2)}{(x+2)(x-2)}$

$-2x+2=2x-4$

$6=4x$

$x=\frac{6}{4}=\frac{3}{2}$

104. $\frac{7}{x}-\frac{x}{7}=0$

$7x\left(\frac{7}{x}-\frac{x}{7}\right)=0(7x)$

$49-x^2=0$

$(7-x)(7+x)=0$

$7-x=0$ or $7+x=0$

$7=x$ or $x=-7$

105. $\frac{x-2}{x^2-7x+10}=\frac{1}{5x-10}-\frac{1}{x-5}$

$\frac{x-2}{(x-2)(x-5)}=\frac{1}{5(x-2)}-\frac{1}{x-5}$

$5(x-2)=x-5-5(x-2)$

$5x-10=x-5-5x+10$

$5x-10=-4x+5$

$9x=15$

$x=\frac{15}{9}=\frac{5}{3}$

106. $\frac{5}{x^2-7x}+\frac{4}{2x-14}=\frac{5}{x(x-7)}+\frac{4}{2(x-7)}$

$=\frac{10}{2x(x-7)}+\frac{4x}{2x(x-7)}$

$=\frac{4x+10}{2x(x-7)}$

$=\frac{2(2x+5)}{2x(x-7)}$

$=\frac{2x+5}{x(x-7)}$

107. $3-\frac{5}{x}-\frac{2}{x^2}=0$

$3x^2-5x-2=0$

$(3x+1)(x-2)=0$

$3x+1=0$ or $x-2=0$

$3x=-1$ or $x=2$

$x=-\frac{1}{3}$

108. $\frac{4}{3-x}-\frac{7}{2x-6}+\frac{5}{x}$

$=\frac{-4}{x-3}-\frac{7}{2(x-3)}+\frac{5}{x}$

$=\frac{-8x}{2x(x-3)}-\frac{7x}{2x(x-3)}+\frac{10(x-3)}{2x(x-3)}$

$=\frac{-5x-30}{2x(x-3)}$

109. $A=\frac{h(a+b)}{2}$

$\frac{2A}{h}=a+b$

$a=\frac{2A}{h}-b$

$=\frac{2A-hb}{h}$

110.
$$\frac{1}{R}=\frac{1}{R_1}+\frac{1}{R_2}$$
$$(RR_1R_2)\frac{1}{R}=\left(\frac{1}{R_1}+\frac{1}{R_2}\right)(RR_1R_2)$$
$$R_1R_2=RR_2+RR_1$$
$$R_1R_2-RR_2=RR_1$$
$$R_2=\frac{RR_1}{R_1-R}$$

111.
$$I=\frac{E}{R+r}$$
$$R+r=\frac{E}{I}$$
$$R=\frac{E}{I}-r$$
$$=\frac{E-Ir}{I}$$

112.
$$A=P+Prt$$
$$A-P=Prt$$
$$\frac{A-P}{Pt}=r$$

113.
$$H=\frac{kA(T_1-T_2)}{L}$$
$$HL=kA(T_1-T_2)$$
$$A=\frac{HL}{k(T_1-T_2)}$$

114. Let x = the number.
$$x+2\left(\frac{1}{x}\right)=3$$
$$x\left(x+\frac{2}{x}\right)=3(x)$$
$$x^2+2=3x$$
$$x^2-3x+2=0$$
$$(x-1)(x-2)=0$$
$$x-1=0 \quad \text{or} \quad x-2=0$$
$$x=1 \quad \text{or} \quad x=2$$

115. Let x = the number added to the numerator, so $2x$ = the number added to the denominator. Then
$$\frac{3+x}{7+2x}=\frac{10}{21}$$
$$21(3+x)=10(7+2x)$$
$$63+21x=70+20x$$
$$x=7$$
7 is the number to be added to the numerator.

116. Let x = the numerator of the fraction, so $x + 2$ = the denominator of the fraction. Then
$$\frac{x-3}{(x+2)+5}=\frac{2}{3}$$
$$\frac{x-3}{x+7}=\frac{2}{3}$$
$$3(x-3)=2(x+7)$$
$$3x-9=2x+14$$
$$x=23$$
$$x+2=25$$
The original fraction was $\frac{23}{25}$.

117. Let n and $n + 2$ represent two consecutive even integers. Then
$$\frac{1}{n}+\frac{1}{n+2}=-\frac{9}{40}$$
$$\frac{n+2+n}{n(n+2)}=-\frac{9}{40}$$
$$\frac{2n+2}{n^2+2n}=-\frac{9}{40}$$
$$40(2n+2)=-9(n^2+2n)$$
$$80n+80=-9n^2-18n$$
$$9n^2+98n+80=0$$
$$(9n+8)(n+10)=0$$
$$9n+8=0 \quad \text{or} \quad n+10=0$$
$$9n=-8 \quad \text{or} \quad n=-10$$
$$n=-\frac{8}{9}$$
Discarding the non-integer solution $-\frac{8}{9}$, we find the two consecutive even integers to be -10 and -8.

118. $\frac{1}{4}+\frac{1}{5}+\frac{1}{6}=\frac{1}{t}$

$120t\left[\frac{1}{4}+\frac{1}{5}+\frac{1}{6}\right]=\left(\frac{1}{t}\right)(120t)$

$30t+24t+20t=120$

$74t=120$

$t=\frac{120}{74}=1\frac{23}{37}$ hr

119. Let n = the number of hours required for Tom to type the mailing labels. Convert times to rates:

$\frac{1 \text{ task}}{6 \text{ hrs}}=\frac{1}{6}\frac{\text{task}}{\text{hr}}$

$\frac{1 \text{ task}}{n \text{ hrs}}=\frac{1}{n}\frac{\text{task}}{\text{hr}}$

$\frac{1 \text{ task}}{4 \text{ hrs}}=\frac{1}{4}\frac{\text{task}}{\text{hr}}$

Expressing their combined rate in two ways, we have

$\frac{1}{6}+\frac{1}{n}=\frac{1}{4}$

$\frac{1}{n}=\frac{1}{4}-\frac{1}{6}=\frac{6-4}{24}$

$\frac{1}{n}=\frac{2}{24}=\frac{1}{12}$

$n=12$

Therefore, Tom can type the mailing labels in 12 hours.

120. $\frac{1}{2}-\frac{1}{2.5}=\frac{1}{t}$

$10t\left[\frac{1}{2}-\frac{1}{2.5}\right]=\frac{1}{t}(10t)$

$5t-4t=10$

$t=10$ hours

121. Let r = the speed of the car, so $r + 430$ = the speed of the jet.

Recall that $d = rt$ or $t=\frac{d}{r}$. Using this relationship, we get:

$\frac{210}{r}=\frac{1715}{r+430}$

$1715r=210(r+430)$

$1715r=210r+90,300$

$1505r=90,300$

$r=60$

so $r+430=490$

Thus, the speed of the jet is 490 mph.

122. $\frac{1}{R}=\frac{1}{R_1}+\frac{1}{R_2}$

$\frac{1}{\frac{30}{11}}=\frac{1}{5}+\frac{1}{R_2}$

$\frac{11}{30}=\frac{1}{5}+\frac{1}{R_2}$

$30R_2\left[\frac{11}{30}\right]=\left(\frac{1}{5}+\frac{1}{R_2}\right)(30R_2)$

$11R_2=6R_2+30$

$5R_2=30$

$R_2=6$ ohms

123. Let r = the speed of the current. Recall $d = rt$ or $t=\frac{d}{r}$. Using this relationship, we get:

$\frac{72}{32-r}=\frac{120}{32+r}$

$72(32+r)=120(32-r)$

$2304+72r=3840-120r$

$192r=1536$

$r=8$

Therefore, the speed of the current is 8 mph.

124.

	distance = rate · time		
with wind	445	$400+x$	$\frac{445}{400+x}$
against wind	355	$400-x$	$\frac{355}{400-x}$

$$\frac{445}{400+x} = \frac{355}{400-x}$$
$$445(400-x) = 355(400+x)$$
$$178{,}000 - 445x = 142{,}000 + 355x$$
$$36{,}000 = 800x$$
$$45 = x$$

The speed of the wind is 45 mph.

125. Let r = the speed of the walker, so $r + 3$ = the speed of the jogger.

Recall that $d = rt$ or $t = \frac{d}{r}$. Using this relationship, we get:

$$\frac{14}{r+3} = \frac{8}{r}$$
$$14r = 8(r+3)$$
$$14r = 8r + 24$$
$$6r = 24$$
$$r = 4$$

Thus, the speed of the walker is 4 mph.

126.

	distance	= rate	· time
faster train	378	r	$\frac{378}{r}$
slower train	270	$r-18$	$\frac{270}{r-18}$

$$\frac{378}{r} = \frac{270}{r-18}$$
$$378(r-18) = 270r$$
$$378r - 6804 = 270r$$
$$108r = 6804$$
$$r = 63$$
$$r - 18 = 63 - 18$$
$$= 45$$

The speeds of the trains are 63 mph and 45 mph.

127. $A = kB$

$$6 = k(14)$$
$$k = \frac{6}{14} = \frac{3}{7}$$
$$A = \frac{3}{7}B = \frac{3}{7}(21) = 3(3) = 9$$

128. $C = \frac{k}{D}$

$$12 = \frac{k}{8}$$
$$96 = k$$
$$C = \frac{96}{D}$$
$$C = \frac{96}{24} = 4$$

129. $P = \frac{k}{V}$

$$1250 = \frac{k}{2}$$
$$k = 2500$$
$$P = \frac{2500}{V}$$
$$800 = \frac{2500}{V}$$
$$V = \frac{2500}{800} = 3.125 \text{ cu. ft}$$

130. $A = kr^2$

$$36\pi = k(3)^2$$
$$4\pi = k$$
$$A = 4\pi r^2$$
$$A = 4\pi(4)^2 = 64\pi \text{ sq. in.}$$

Chapter 6 Test

1. $f(x) = \frac{5x^2}{1-x}$

Undefined value when

$$1 - x = 0$$
$$1 = x$$

Domain $\{x \mid x \text{ is a real number and } x \neq 1\}$

2. $g(x) = \frac{9x^2 - 9}{x^2 + 4x + 3}$

Undefined values when

$$x^2 + 4x + 3 = 0$$
$$(x+3)(x+1) = 0$$
$$x + 3 = 0 \quad \text{or} \quad x + 1 = 0$$
$$x = -3 \quad \text{or} \quad x = -1$$

Domain

$\{x \mid x \text{ is a real number and } x \neq -3, x \neq -1\}$

3. $\frac{5x^7}{3x^4} = \frac{5x^{7-4}}{3} = \frac{5x^3}{3}$

4. $\frac{7x-21}{24-8x} = \frac{7(x-3)}{8(3-x)} = \frac{7(-1)}{8} = \frac{-7}{8} = -\frac{7}{8}$

5. $\frac{x^2-4x}{x^2+5x-36} = \frac{x(x-4)}{(x+9)(x-4)} = \frac{x}{x+9}$

6. $\frac{x}{x-2} \cdot \frac{x^2-4}{5x} = \frac{1}{x-2} \cdot \frac{(x-2)(x+2)}{5}$
$= \frac{1}{1} \cdot \frac{x+2}{5}$
$= \frac{x+2}{5}$

7. $\frac{2x^3+16}{6x^2+12x} \cdot \frac{5}{x^2-2x+4}$
$= \frac{2(x^3+8)\cdot 5}{6x(x+2)(x^2-2x+4)}$
$= \frac{5(x+2)(x^2-2x+4)}{3x(x+2)(x^2-2x+4)}$
$= \frac{5}{3x}$

8. $\frac{26ab}{7c} \div \frac{13a^2c^5}{14a^4b^3} = \frac{26ab}{7c} \cdot \frac{14a^4b^3}{13a^2c^5}$
$= \frac{2ab}{c} \cdot \frac{2a^4b^3}{a^2c^5}$
$= \frac{4a^5b^4}{a^2c^6}$
$= \frac{4a^3b^4}{c^6}$

9. $\frac{3x^2-12}{x^2+2x-8} \div \frac{6x+18}{x+4}$
$= \frac{3(x^2-4)}{(x+4)(x-2)} \cdot \frac{x+4}{6(x+3)}$
$= \frac{(x+2)(x-2)}{x-2} \cdot \frac{1}{2(x+3)}$
$= \frac{x+2}{2(x+3)}$

10. $\frac{4x-12}{2x-9} \div \frac{3-x}{4x^2-81} \cdot \frac{x+3}{5x+15}$
$= \frac{4(x-3)}{2x-9} \cdot \frac{(2x+9)(2x-9)}{-(x-3)} \cdot \frac{x+3}{5(x+3)}$
$= \frac{4}{1} \cdot \frac{2x+9}{-1} \cdot \frac{1}{5}$
$= \frac{-4(2x+9)}{5}$

11. $\frac{5}{4x^3} + \frac{7}{4x^3} = \frac{5+7}{4x^3} = \frac{12}{4x^3} = \frac{3}{x^3}$

12. $\frac{3+2x}{10-x} + \frac{13+x}{x-10} = \frac{3+2x-(13+x)}{10-x}$
$= \frac{3+2x-13-x}{10-x}$
$= \frac{-10+x}{-(-10+x)}$
$= \frac{1}{-1}$
$= -1$

13. $\frac{3}{x^2-x-6} + \frac{2}{x^2-5x+6}$
$= \frac{3}{(x-3)(x+2)} + \frac{2}{(x-3)(x-2)}$
$= \frac{3(x-2)+2(x+2)}{(x-3)(x+2)(x-2)}$
$= \frac{3x-6+2x+4}{(x-3)(x+2)(x-2)}$
$= \frac{5x-2}{(x-3)(x+2)(x-2)}$

14. $\frac{5}{x-7} - \frac{2x}{3x-21} + \frac{x}{2x-14}$
$= \frac{5}{x-7} - \frac{2x}{3(x-7)} + \frac{x}{2(x-7)}$
$= \frac{6\cdot 5 - 2\cdot 2x + 3\cdot x}{6(x-7)}$
$= \frac{30-4x+3x}{6(x-7)}$
$= \frac{30-x}{6(x-7)}$

15. $\frac{3x}{5}\cdot\left(\frac{5}{x}-\frac{5}{2x}\right)=3x\left(\frac{1}{x}-\frac{1}{2x}\right)$
$=3x\cdot\frac{2-1}{2x}$
$=3x\left(\frac{1}{2x}\right)$
$=\frac{3}{2}$

16. $\frac{\frac{4x}{13}}{\frac{20x}{13}}=\frac{4x}{20x}=\frac{1}{5}$

17. $\frac{\frac{5}{x}-\frac{7}{3x}}{\frac{9}{8x}-\frac{1}{x}}=\frac{\frac{15-7}{3x}}{\frac{9-8}{8x}}=\frac{8}{3x}\cdot\frac{8x}{1}=\frac{64}{3}$

18. $\frac{\frac{x^2-5x+6}{x+3}}{\frac{x^2-4x+4}{x^2-9}}=\frac{(x-3)(x-2)}{x+3}\cdot\frac{(x-3)(x+3)}{(x-2)(x-2)}$
$=\frac{(x-3)^2}{x-2}$

19. $\frac{4x^2y+9x+z}{3xz}=\frac{4x^2y}{3xz}+\frac{9x}{3xz}+\frac{z}{3xz}$
$=\frac{4xy}{3z}+\frac{3}{z}+\frac{1}{3x}$

20.

$$\begin{array}{r}
x^5+5x^4+8x^3+16x^2+33x+63 \\
x-2\,\overline{)\,x^6+3x^5-2x^4+0x^3+x^2-3x+2} \\
\underline{x^6-2x^5} \\
5x^5-2x^4 \\
\underline{5x^5-10x^4} \\
8x^4+0x^3 \\
\underline{8x^4-16x^3} \\
16x^3+x^2 \\
\underline{16x^3-32x^2} \\
33x^2-3x \\
\underline{33x^2-66x} \\
63x+2 \\
\underline{63x-126} \\
128
\end{array}$$

Answer:
$x^5+5x^4+8x^3+16x^2+33x+63+\frac{128}{x-2}$

21.

$$\begin{array}{r|rrrrr} -3 & 4 & -3 & 2 & -1 & -1 \\ & & -12 & 45 & -141 & 426 \\ \hline & 4 & -15 & 47 & -142 & 425 \end{array}$$

Answer: $4x^3 - 15x^2 + 47x - 142 + \dfrac{425}{x+3}$

22.

$$\begin{array}{r|rrrrr} -2 & 4 & 0 & 7 & -2 & -5 \\ & & -8 & 16 & -46 & 96 \\ \hline & 4 & -8 & 23 & -48 & 91 \end{array}$$

$P(-2) = 91$

23.

$$\frac{5x+3}{3x-7} = \frac{19}{7}$$
$$19(3x-7) = 7(5x+3)$$
$$57x - 133 = 35x + 21$$
$$22x = 154$$
$$x = 7$$

24.

$$\frac{5}{x-5} + \frac{x}{x+5} = -\frac{29}{21}$$
$$\frac{5(x+5) + x(x-5)}{(x-5)(x+5)} = -\frac{29}{21}$$
$$\frac{5x + 25 + x^2 - 5x}{x^2 - 25} = -\frac{29}{21}$$
$$\frac{x^2 + 25}{x^2 - 25} = -\frac{29}{21}$$
$$21x^2 + 525 = -29x^2 + 725$$
$$50x^2 = 200$$
$$x^2 = 4$$
$$x = \pm 2$$

25.

$$\frac{x}{x-4} = 3 - \frac{4}{x-4}$$
$$\frac{x}{x-4} + \frac{4}{x-4} = 3$$
$$\frac{x+4}{x-4} = 3$$
$$x + 4 = 3(x-4)$$
$$x + 4 = 3x - 12$$
$$16 = 2x$$
$$x = 8$$

26.

$$\frac{x+b}{a} = \frac{4x-7a}{b}$$
$$b(x+b) = a(4x-7a)$$
$$bx + b^2 = 4ax - 7a^2$$
$$7a^2 + b^2 = 4ax - bx$$
$$7a^2 + b^2 = (4a-b)x$$
$$x = \frac{7a^2 + b^2}{4a-b}$$

27. Let x = the number. Then

$$(x+1)\frac{2}{x} = \frac{12}{5}$$
$$\frac{2(x+1)}{x} = \frac{12}{5}$$
$$12x = 10(x+1)$$
$$12x = 10x + 10$$
$$2x = 10$$
$$x = 5$$

The number is 5.

28. Let n = the number of hours it takes Jan and her husband to weed the garden. Note that 1 hour and 30 minutes = $\frac{3}{2}$ hours.

Convert time to rates:

$$\frac{1 \text{ task}}{2 \text{ hrs}} = \frac{1}{2}\,\frac{\text{task}}{\text{hr}}$$
$$\frac{1 \text{ task}}{\frac{3}{2} \text{ hrs}} = \frac{2}{3}\,\frac{\text{task}}{\text{hr}}$$
$$\frac{1 \text{ task}}{n \text{ hrs}} = \frac{1}{n}\,\frac{\text{task}}{\text{hr}}$$

Expressing their combined rate in two ways, we get

$$\frac{1}{2}+\frac{2}{3}=\frac{1}{n}$$
$$\frac{7}{6}=\frac{1}{n}$$
$$n=\frac{6}{7}\text{ hr}$$

It takes the two of them $\frac{6}{7}$ hour to weed the garden.

29. $W=\frac{k}{V}$

$$20=\frac{k}{12}$$
$$k=240$$

so $W=\frac{240}{V}=\frac{240}{15}=16$

30. $Q=kRS^2$

$$24=k(3)4^2$$
$$24=48k$$
$$k=\frac{1}{2}$$

so $Q=\frac{1}{2}RS^2=\frac{1}{2}(2)3^2=9$

31. $S=k\sqrt{d}$

$$160=k\sqrt{400}$$
$$k=\frac{160}{20}=8$$

so

$$S=8\sqrt{d}$$
$$128=8\sqrt{d}$$
$$\sqrt{d}=16$$
$$d=256$$

The height of the cliff is 256 feet.

Chapter 7

Exercise Set 7.1

1. $\sqrt{100} = 10$ because $10^2 = 100$.

3. $\sqrt{\frac{1}{4}} = \frac{1}{2}$ because $\left(\frac{1}{2}\right)^2 = \frac{1}{4}$.

5. $\sqrt{0.0001} = 0.01$ because $(0.01)^2 = 0.0001$.

7. $-\sqrt{36} = -6$ because $(6)^2 = 36$.

9. $\sqrt{x^{10}} = x^5$ because $(x^5)^2 = x^{10}$.

11. $\sqrt{16y^6} = \sqrt{16}\sqrt{y^6} = 4y^3$ because $(4y^3)^2 = 16y^6$

13. $\sqrt{7} \approx 2.646$
Since $4 < 7 < 9$, then
$\sqrt{4} < \sqrt{7} < \sqrt{9}$, or $2 < \sqrt{7} < 3$. The approximation is between 2 and 3 and thus is reasonable.

15. $\sqrt{38} \approx 6.164$
Since $36 < 38 < 49$, then
$\sqrt{36} < \sqrt{38} < \sqrt{49}$, or $6 < \sqrt{38} < 7$. The approximation is between 6 and 7 and thus is reasonable.

17. $\sqrt{200} \approx 14.142$
Since $196 < 200 < 225$, then
$\sqrt{196} < \sqrt{200} < \sqrt{225}$, or
$14 < \sqrt{200} < 15$. The approximation is between 14 and 15 and thus is reasonable.

19. $\sqrt[3]{64} = 4$ because $(4)^3 = 64$.

21. $\sqrt[3]{\frac{1}{8}} = \frac{1}{2}$ because $\left(\frac{1}{2}\right)^3 = \frac{1}{8}$.

23. $\sqrt[3]{-1} = -1$ because $(-1)^3 = -1$.

25. $\sqrt[3]{x^{12}} = x^4$ because $(x^4)^3 = x^{12}$.

27. $\sqrt[3]{-27x^9} = -3x^3$ because $(-3x^3)^3 = -27x^9$.

29. $-\sqrt[4]{16} = -2$ because $(2)^4 = 16$.

31. $\sqrt[4]{-16}$ is not a real number. There is no real number that, when raised to the fourth power, is – 16.

33. $\sqrt[5]{-32} = -2$ because $(-2)^5 = -32$.

35. $\sqrt[5]{x^{20}} = x^4$ because $(x^4)^5 = x^{20}$.

37. $\sqrt[6]{64x^{12}} = 2x^2$ because $(2x^2)^6 = 64x^{12}$.

39. $\sqrt{81x^4} = 9x^2$ because $(9x^2)^2 = 81x^4$.

41. $\sqrt[4]{256x^8} = 4x^2$ because $(4x^2)^4 = 256x^8$.

43. $\sqrt{(-8)^2} = |-8| = 8$

45. $\sqrt[3]{(-8)^3} = -8$

47. $\sqrt{4x^2} = 2|x|$

49. $\sqrt[3]{x^3} = x$

51. $\sqrt[4]{(x-2)^4} = |x-2|$

53. $\sqrt{x^2+4x+4} = \sqrt{(x+2)^2} = |x+2|$

55. $-\sqrt{121} = -11$

57. $\sqrt[3]{8x^3} = 2x$

59. $\sqrt{y^{12}} = y^6$

61. $\sqrt{25a^2b^{20}} = 5ab^{10}$

63. $\sqrt[3]{-27x^9} = -3x^3$

65. $\sqrt[4]{a^{16}b^4} = a^4b$

67. $\sqrt[5]{-32x^{10}y^5} = \sqrt[5]{-32}\sqrt[5]{x^{10}}\sqrt[5]{y^5} = -2x^2y$

69. $\sqrt{\frac{25}{49}} = \frac{5}{7}$

71. $\sqrt{\frac{x^2}{4y^2}} = \frac{x}{2y}$

73. $-\sqrt[3]{\frac{z^{21}}{27x^3}} = -\frac{z^7}{3x}$

75. $\sqrt[4]{\frac{x^4}{16}} = \frac{x}{2}$

77. $f(x) = \sqrt{2x+3}$
$f(0) = \sqrt{2 \cdot 0 + 3} = \sqrt{3}$

79. $g(x) = \sqrt[3]{x-8}$
$g(7) = \sqrt[3]{7-8} = \sqrt[3]{-1} = -1$

81. $g(x) = \sqrt[3]{x-8}$
$g(-19) = \sqrt[3]{-19-8} = \sqrt[3]{-27} = -3$

83. $f(x) = \sqrt{2x+3}$
$f(2) = \sqrt{2 \cdot 2 + 3} = \sqrt{7}$

85. $f(x) = \sqrt{x} + 2$
Domain: $[0, \infty)$

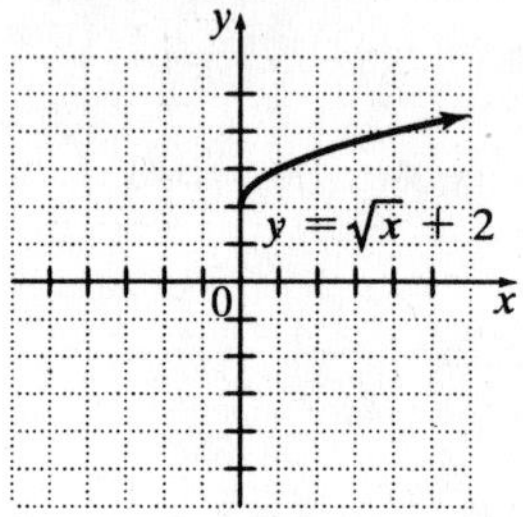

87. $f(x) = \sqrt{x-3}$
Domain: $[3, \infty)$

x	$f(x) = \sqrt{x-3}$
3	$\sqrt{3-3} = \sqrt{0} = 0$
4	$\sqrt{4-3} = \sqrt{1} = 1$
7	$\sqrt{7-3} = \sqrt{4} = 2$
12	$\sqrt{12-3} = \sqrt{9} = 3$

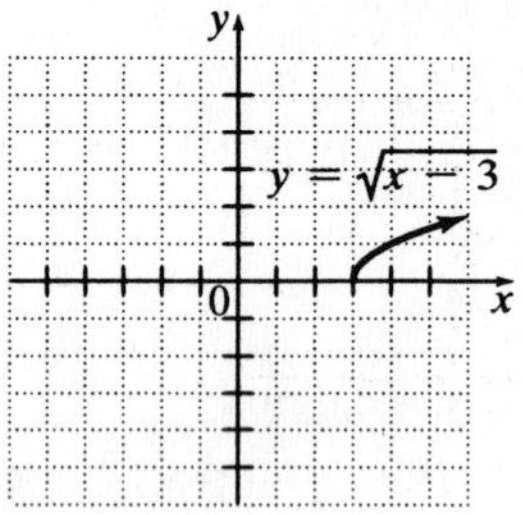

89. $f(x) = \sqrt[3]{x} + 1$
Domain: $(-\infty, \infty)$

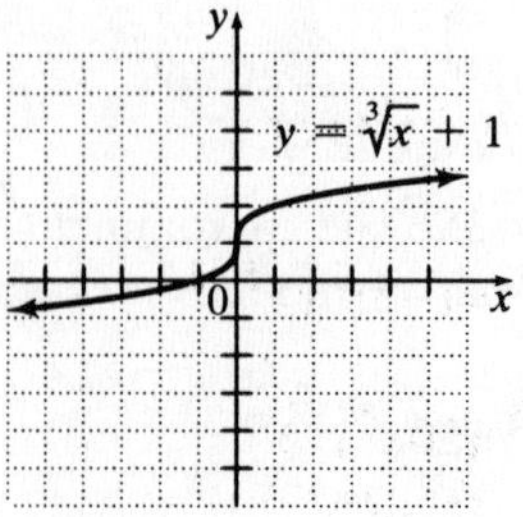

91. $g(x) = \sqrt[3]{x-1}$

Domain: $(-\infty, \infty)$

x	$g(x) = \sqrt[3]{x-1}$
1	$\sqrt[3]{1-1} = \sqrt[3]{0} = 0$
2	$\sqrt[3]{2-1} = \sqrt[3]{1} = 1$
0	$\sqrt[3]{0-1} = \sqrt[3]{-1} = -1$
9	$\sqrt[3]{9-1} = \sqrt[3]{8} = 2$
−7	$\sqrt[3]{-7-1} = \sqrt[3]{-8} = -2$

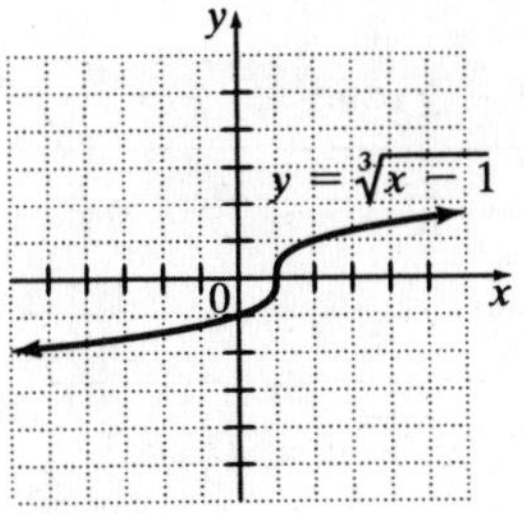

93. Answers may vary.

95.

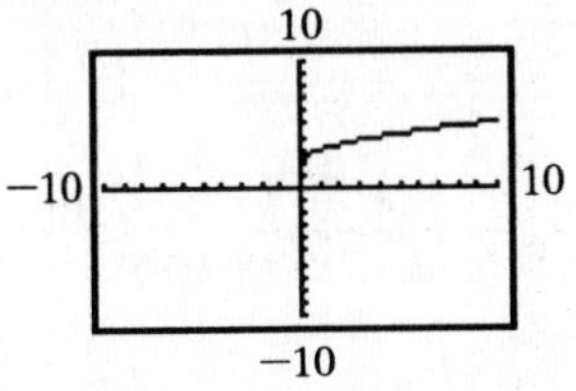

97.

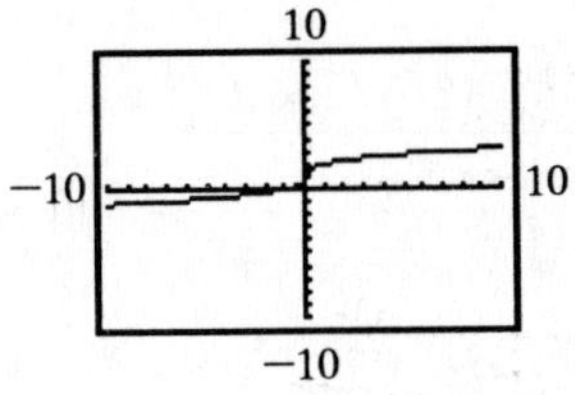

99. $(-2x^3y^2)^5 = (-2)^5 x^{3 \cdot 5} y^{2 \cdot 5} = -32x^{15}y^{10}$

101. $(-3x^2y^3z^5)(20x^5y^7)$
$= (-3)(20)x^{2+5}y^{3+7}z^5$
$= -60x^7y^{10}z^5$

103. $\dfrac{7x^{-1}y}{14(x^5y^2)^{-2}} = \dfrac{7x^{-1}y}{14x^{-10}y^{-4}} = \dfrac{x^9y^5}{2}$

Exercise Set 7.2

1. $49^{1/2} = \sqrt{49} = 7$

3. $27^{1/3} = \sqrt[3]{27} = 3$

5. $\left(\frac{1}{16}\right)^{1/4} = \sqrt[4]{\frac{1}{16}} = \frac{1}{2}$

7. $169^{1/2} = \sqrt{169} = 13$

9. $2m^{1/3} = 2\sqrt[3]{m}$

11. $(9x^4)^{1/2} = \sqrt{9x^4} = 3x^2$

13. $(-27)^{1/3} = \sqrt[3]{-27} = -3$

15. $-16^{1/4} = -\sqrt[4]{16} = -2$

17. $16^{3/4} = (\sqrt[4]{16})^3 = 2^3 = 8$

19. $(-64)^{2/3} = \left(\sqrt[3]{-64}\right)^2 = (-4)^2 = 16$

21. $(-16)^{3/4} = \left(\sqrt[4]{-16}\right)^3$ is not a real number.

23. $(2x)^{3/5} = \sqrt[5]{(2x)^3}$ or $\left(\sqrt[5]{2x}\right)^3$

25. $(7x+2)^{2/3} = \sqrt[3]{(7x+2)^2}$ or $\left(\sqrt[3]{7x+2}\right)^2$

27. $\left(\frac{16}{9}\right)^{3/2} = \left(\sqrt{\frac{16}{9}}\right)^3 = \left(\frac{4}{3}\right)^3 = \frac{64}{27}$

29. $8^{-4/3} = \dfrac{1}{8^{4/3}} = \dfrac{1}{(8^{1/3})^4} = \dfrac{1}{2^4} = \dfrac{1}{16}$

31. $(-64)^{-2/3} = \frac{1}{(-64)^{2/3}}$
$= \frac{1}{[(-64)^{1/3}]^2}$
$= \frac{1}{(-4)^2}$
$= \frac{1}{16}$

33. $(-4)^{-3/2} = \frac{1}{(-4)^{3/2}} = \frac{1}{\left[(-4)^{1/2}\right]^3}$ is not a real number.

35. $x^{-1/4} = \frac{1}{x^{1/4}}$

37. $\frac{1}{a^{-2/3}} = a^{2/3}$

39. $\frac{5}{7x^{-3/4}} = \frac{5x^{3/4}}{7}$

41. Answers may vary.

43. $a^{2/3}a^{5/3} = a^{2/3+5/3} = a^{7/3}$

45. $x^{-2/5} \cdot x^{7/5} = x^{-\frac{2}{5}+\frac{7}{5}} = x^{5/5} = x$

47. $3^{1/4} \cdot 3^{3/8} = 3^{\frac{1}{4}+\frac{3}{8}} = 3^{\frac{2}{8}+\frac{3}{8}} = 3^{5/8}$

49. $\frac{y^{1/3}}{y^{1/6}} = y^{\frac{1}{3}-\frac{1}{6}} = y^{\frac{2}{6}-\frac{1}{6}} = y^{1/6}$

51. $(4u^2)^{3/2} = 4^{3/2} \cdot u^{2\cdot 3/2}$
$= \left(\sqrt{4}\right)^3 \cdot u^3$
$= 2^3u^3$
$= 8u^3$

53. $\frac{b^{1/2}b^{3/4}}{-b^{1/4}} = -b^{\frac{1}{2}+\frac{3}{4}-\frac{1}{4}}$
$= -b^{\frac{1}{2}+\frac{1}{2}}$
$= -b^1$
$= -b$

55. $\frac{(3x^{1/4})^3}{x^{1/12}} = \frac{3^3x^{3/4}}{x^{1/12}}$
$= 27x^{\frac{3}{4}-\frac{1}{12}}$
$= 27x^{\frac{9}{12}-\frac{1}{12}}$
$= 27x^{8/12}$
$= 27x^{2/3}$

57. $y^{1/2}(y^{1/2} - y^{2/3}) = y^{1/2}y^{1/2} - y^{1/2}y^{2/3}$
$= y^{1/2+1/2} - y^{1/2+2/3}$
$= y^1 - y^{7/6}$
$= y - y^{7/6}$

59. $x^{2/3}(2x-2) = 2x^1x^{2/3} - 2x^{2/3}$
$= 2x^{5/3} - 2x^{2/3}$

61. $(2x^{1/3}+3)(2x^{1/3}-3) = (2x^{1/3})^2 - 3^2$
$= 2^2(x^{1/3})^2 - 9$
$= 4x^{2/3} - 9$

63. $x^{8/3} + x^{10/3} = x^{8/3}(1 + x^{2/3})$

65. $x^{2/5} - 3x^{1/5} = x^{1/5}(x^{1/5} - 3)$

67. $5x^{-1/3} + x^{2/3} = x^{-1/3}(5 + x^1)$
$= x^{-1/3}(5 + x)$

69. $\sqrt[6]{x^3} = x^{3/6} = x^{1/2} = \sqrt{x}$

71. $\sqrt[6]{4} = 4^{1/6} = (2^2)^{1/6} = 2^{2/6} = 2^{1/3} = \sqrt[3]{2}$

73. $\sqrt[4]{16x^2} = 16^{1/4}x^{2/4} = 2x^{1/2} = 2\sqrt{x}$

75. $\sqrt[8]{x^4y^4} = x^{4/8}y^{4/8} = x^{1/2}y^{1/2} = \sqrt{xy}$

77. $\sqrt[3]{y} \cdot \sqrt[5]{y^2} = y^{1/3} \cdot y^{2/5}$
$= y^{5/15} \cdot y^{6/15}$
$= y^{11/15}$
$= \sqrt[15]{y^{11}}$

79. $\frac{\sqrt[3]{b^2}}{\sqrt[4]{b}} = \frac{b^{2/3}}{b^{1/4}}$
$= b^{\frac{2}{3}-\frac{1}{4}}$
$= b^{\frac{8}{12}-\frac{3}{12}}$
$= b^{5/12}$
$= \sqrt[12]{b^5}$

81. $\frac{\sqrt[3]{a^2}}{\sqrt[6]{a}} = \frac{a^{2/3}}{a^{1/6}}$
$= a^{\frac{2}{3}-\frac{1}{6}}$
$= a^{\frac{4}{6}-\frac{1}{6}}$
$= a^{3/6}$
$= a^{1/2}$
$= \sqrt{a}$

83. $\sqrt{3}\cdot\sqrt[3]{4} = 3^{1/2}\cdot 4^{1/3}$
$= 3^{3/6}\cdot 4^{2/6}$
$= (3^3\cdot 4^2)^{1/6}$
$= (432)^{1/6}$
$= \sqrt[6]{432}$

85. $\sqrt[5]{7}\cdot\sqrt[3]{y} = 7^{1/5}\cdot y^{1/3}$
$= 7^{3/15}\cdot y^{5/15}$
$= (7^3\cdot y^5)^{1/15}$
$= (343y^5)^{1/15}$
$= \sqrt[15]{343y^5}$

87. $\frac{t^{1/2}}{u^{1/2}}$

89. $\text{BMR} = 70(50)^{3/4} \approx 1316$ calories

91. **a.** $f(x) = 6550x^{43/50}$
$f(1999) = 6550(9^{43/50})$
$= \$43,340$ million

b. $f(2004) = 6550(14^{43/50})$
$= \$63,374$ million

93. $x^{3/8}$
$x^{3/8}\cdot x^{1/8} = x^{4/8}$, or $x^{1/2}$

95. $y^{1/4}$
$\frac{y^{1/4}}{y^{-3/4}} = y^{1/4+3/4} = y^{4/4} = y$

97. $20^{1/5} \approx 1.8206$

99. $76^{5/7} \approx 22.0515$

101. $20 = 4\cdot 5$

103. $45 = 9\cdot 5$

105. $56 = 8\cdot 7$

107. $80 = 8\cdot 10$

Exercise Set 7.3

1. $\sqrt{7}\cdot\sqrt{2} = \sqrt{7\cdot 2} = \sqrt{14}$

3. $\sqrt[4]{8}\cdot\sqrt[4]{2} = \sqrt[4]{8\cdot 2} = \sqrt[4]{16} = 2$

5. $\sqrt[3]{4}\cdot\sqrt[3]{9} = \sqrt[3]{4\cdot 9} = \sqrt[3]{36}$

7. $\sqrt{2}\cdot\sqrt{3x} = \sqrt{2\cdot 3x} = \sqrt{6x}$

9. $\sqrt{\frac{7}{x}}\cdot\sqrt{\frac{2}{y}} = \sqrt{\frac{7\cdot 2}{x\cdot y}} = \sqrt{\frac{14}{xy}}$

11. $\sqrt[4]{4x^3}\cdot\sqrt[4]{5} = \sqrt[4]{4\cdot 5x^3} = \sqrt[4]{20x^3}$

13. $\sqrt{\frac{6}{49}} = \frac{\sqrt{6}}{\sqrt{49}} = \frac{\sqrt{6}}{7}$

15. $\sqrt{\frac{2}{49}} = \frac{\sqrt{2}}{\sqrt{49}} = \frac{\sqrt{2}}{7}$

17. $\sqrt[4]{\frac{x^3}{16}} = \frac{\sqrt[4]{x^3}}{\sqrt[4]{16}} = \frac{\sqrt[4]{x^3}}{2}$

19. $\sqrt[3]{\frac{4}{27}} = \frac{\sqrt[3]{4}}{\sqrt[3]{27}} = \frac{\sqrt[3]{4}}{3}$

21. $\sqrt[4]{\dfrac{8}{x^8}} = \dfrac{\sqrt[4]{8}}{\sqrt[4]{x^8}} = \dfrac{\sqrt[4]{8}}{x^2}$

23. $\sqrt[3]{\dfrac{2x}{81y^{12}}} = \dfrac{\sqrt[3]{2x}}{\sqrt[3]{81y^{12}}}$
$= \dfrac{\sqrt[3]{2x}}{\sqrt[3]{27y^{12}} \cdot \sqrt[3]{3}}$
$= \dfrac{\sqrt[3]{2x}}{3y^4\sqrt[3]{3}}$

25. $\sqrt{\dfrac{x^2y}{100}} = \dfrac{\sqrt{x^2} \cdot \sqrt{y}}{\sqrt{100}} = \dfrac{x\sqrt{y}}{10}$

27. $\sqrt{\dfrac{5x^2}{4y^2}} = \dfrac{\sqrt{5x^2}}{\sqrt{4y^2}} = \dfrac{\sqrt{5}\sqrt{x^2}}{\sqrt{4}\sqrt{y^2}} = \dfrac{\sqrt{5}x}{2y}$

29. $-\sqrt[3]{\dfrac{z^7}{27x^3}} = \dfrac{-\sqrt[3]{z^7}}{\sqrt[3]{27x^3}}$
$= \dfrac{-\sqrt[3]{z^6z}}{\sqrt[3]{27} \cdot \sqrt[3]{x^3}}$
$= \dfrac{-\sqrt[3]{z^6} \cdot \sqrt[3]{z}}{3x}$
$= \dfrac{-z^2\sqrt[3]{z}}{3x}$

31. $\sqrt{32} = \sqrt{16(2)} = \sqrt{16} \cdot \sqrt{2} = 4\sqrt{2}$

33. $\sqrt[3]{192} = \sqrt[3]{64(3)} = \sqrt[3]{64} \cdot \sqrt[3]{3} = 4\sqrt[3]{3}$

35. $5\sqrt{75} = 5\sqrt{25(3)}$
$= 5\sqrt{25} \cdot \sqrt{3}$
$= 5(5)\sqrt{3}$
$= 25\sqrt{3}$

37. $\sqrt{24} = \sqrt{4 \cdot 6} = \sqrt{4} \cdot \sqrt{6} = 2\sqrt{6}$

39. $\sqrt{100x^5} = \sqrt{(100x^4)x}$
$= \sqrt{100} \cdot \sqrt{x^4} \cdot \sqrt{x}$
$= 10x^2\sqrt{x}$

41. $\sqrt[3]{16y^7} = \sqrt[3]{(8y^6)(2y)}$
$= \sqrt[3]{8} \cdot \sqrt[3]{y^6} \cdot \sqrt[3]{2y}$
$= 2y^2\sqrt[3]{2y}$

43. $\sqrt[4]{a^8b^7} = \sqrt[4]{a^8b^4b^3}$
$= \sqrt[4]{a^8} \cdot \sqrt[4]{b^4} \cdot \sqrt[4]{b^3}$
$= a^2b\sqrt[4]{b^3}$

45. $\sqrt{y^5} = \sqrt{y^4y} + \sqrt{y^4} \cdot \sqrt{y} = y^2\sqrt{y}$

47. $\sqrt{25a^2b^3} = \sqrt{25} \cdot \sqrt{a^2} \cdot \sqrt{b^3}$
$= 5a\sqrt{b^2b}$
$= 5a\sqrt{b^2} \cdot \sqrt{b}$
$= 5ab\sqrt{b}$

49. $\sqrt[5]{-32x^{10}y} = \sqrt[5]{-32} \cdot \sqrt[5]{x^{10}} \cdot \sqrt[5]{y} = -2x^2\sqrt[5]{y}$

51. $\sqrt[3]{50x^{14}} = \sqrt[3]{x^{12}(50x^2)}$
$= \sqrt[3]{x^{12}} \cdot \sqrt[3]{50x^2}$
$= x^4\sqrt[3]{50x^2}$

53. $-\sqrt{32a^8b^7} = -\sqrt{16a^8b^6(2b)}$
$= -\sqrt{16} \cdot \sqrt{a^8} \cdot \sqrt{b^6} \cdot \sqrt{2b}$
$= -4a^4b^3\sqrt{2b}$

55. $\sqrt{9x^7y^9} = \sqrt{9x^6y^8 \cdot xy}$
$= \sqrt{9x^6y^8} \cdot \sqrt{xy}$
$= 3x^3y^4\sqrt{xy}$

57. $\sqrt[3]{125r^9s^{12}} = 5r^3s^4$

59. $\frac{\sqrt{14}}{\sqrt{7}} = \sqrt{\frac{14}{7}} = \sqrt{2}$

61. $\frac{\sqrt[3]{24}}{\sqrt[3]{3}} = \sqrt[3]{\frac{24}{3}} = \sqrt[3]{8} = 2$

63. $\frac{5\sqrt[4]{48}}{\sqrt[4]{3}} = 5\sqrt[4]{\frac{48}{3}} = 5\sqrt[4]{16} = 5 \cdot 2 = 10$

65. $\frac{\sqrt{x^5y^3}}{\sqrt{xy}} = \sqrt{\frac{x^5y^3}{xy}}$
$= \sqrt{x^4y^2}$
$= x^2y$

67. $\frac{8\sqrt[3]{54m^7}}{\sqrt[3]{2m}} = 8\sqrt[3]{\frac{54m^7}{2m}}$
$= 8\sqrt[3]{27m^6}$
$= 8 \cdot 3m^2$
$= 24m^2$

69. $\frac{3\sqrt{100x^2}}{2\sqrt{2x^{-1}}} = \frac{3}{2}\sqrt{\frac{100x^2}{2x^{-1}}}$
$= \frac{3}{2}\sqrt{50x^3}$
$= \frac{3}{2}\sqrt{25x^2 \cdot 2x}$
$= \frac{3}{2} \cdot 5x\sqrt{2x}$
$= \frac{15x}{2}\sqrt{2x}$

71. $\frac{\sqrt[4]{96a^{10}b^3}}{\sqrt[4]{3a^2b^3}} = \sqrt[4]{\frac{96a^{10}b^3}{3a^2b^3}}$
$= \sqrt[4]{32a^8}$
$= \sqrt[4]{16a^8 \cdot 2}$
$= 2a^2\sqrt[4]{2}$

73. $A = \pi r\sqrt{r^2 + h^2}$

a. $A = \pi \cdot 4\sqrt{4^2 + 3^2}$
$= 4\pi\sqrt{16+9}$
$= 4\pi\sqrt{25}$
$= 4 \cdot 5\pi$
$= 20\pi$ sq. centimeters

b. $A = \pi(6.8)\sqrt{(6.8)^2 + (7.2)^2}$
$= 6.8\pi\sqrt{46.24 + 51.84}$
$= 6.8\pi\sqrt{98.08}$
≈ 211.57 sq. feet

75. $F(x) = 0.6\sqrt{49 - x^2}$

a. $F(3) = 0.6\sqrt{49 - 3^2}$
$= 0.6\sqrt{49 - 9}$
$= 0.6\sqrt{40}$
≈ 3.8

b. $F(5) = 0.6\sqrt{49 - 5^2}$
$= 0.6\sqrt{49 - 25}$
$= 0.6\sqrt{24}$
≈ 2.9

c. Answers may vary.

77. $(6x)(8x) = (6)(8)x \cdot x = 48x^2$

79. $(2x+3) + (x-5) = 2x + 3 + x - 5$
$= (2x + x) + (3 - 5)$
$= 3x - 2$

81. $(9y^2)(-8y^2) = 9(-8)y^{2+2} = -72y^4$

83. $-3 + x + 5 = x + (-3 + 5) = x + 2$

85. $(2x+1)^2 = (2x)^2 + 2(2x)(1) + 1^2$
$= 4x^2 + 4x + 1$

Section 7.4

Mental Math

1. $2\sqrt{3} + 4\sqrt{3} = 6\sqrt{3}$

2. $5\sqrt{7}+3\sqrt{7}=8\sqrt{7}$

3. $8\sqrt{x}-5\sqrt{x}=3\sqrt{x}$

4. $3\sqrt{y}+10\sqrt{y}=13\sqrt{y}$

5. $7\sqrt[3]{x}+5\sqrt[3]{x}=12\sqrt[3]{x}$

6. $8\sqrt[3]{z}-2\sqrt[3]{z}=6\sqrt[3]{z}$

Exercise Set 7.4

1. $$\begin{aligned}\sqrt{8}-\sqrt{32}&=\sqrt{4(2)}-\sqrt{16(2)}\\&=\sqrt{4}\sqrt{2}-\sqrt{16}\sqrt{2}\\&=2\sqrt{2}-4\sqrt{2}\\&=-2\sqrt{2}\end{aligned}$$

3. $$\begin{aligned}&2\sqrt{2x^3}+4x\sqrt{8x}\\&=2\sqrt{x^2(2x)}+4x\sqrt{4(2x)}\\&=2\sqrt{x^2}\sqrt{2x}+4x\sqrt{4}\sqrt{2x}\\&=2x\sqrt{2x}+4x(2)\sqrt{2x}\\&=2x\sqrt{2x}+8x\sqrt{2x}\\&=10x\sqrt{2x}\end{aligned}$$

5. $$\begin{aligned}&2\sqrt{50}-3\sqrt{125}+\sqrt{98}\\&=2\sqrt{25(2)}-3\sqrt{25(5)}+\sqrt{49(2)}\\&=2\sqrt{25}\sqrt{2}-3\sqrt{25}\sqrt{5}+\sqrt{49}\sqrt{2}\\&=2(5)\sqrt{2}-3(5)\sqrt{5}+7\sqrt{2}\\&=10\sqrt{2}-15\sqrt{5}+7\sqrt{2}\\&=17\sqrt{2}-15\sqrt{5}\end{aligned}$$

7. $$\begin{aligned}\sqrt[3]{16x}-\sqrt[3]{54x}&=\sqrt[3]{8(2x)}-\sqrt[3]{27(2x)}\\&=\sqrt[3]{8}\sqrt[3]{2x}-\sqrt[3]{27}\sqrt[3]{2x}\\&=2\sqrt[3]{2x}-3\sqrt[3]{2x}\\&=-\sqrt[3]{2x}\end{aligned}$$

9. $$\begin{aligned}&\sqrt{9b^3}-\sqrt{25b^3}+\sqrt{49b^3}\\&=\sqrt{9b^2(b)}-\sqrt{25b^2(b)}+\sqrt{49b^2(b)}\\&=\sqrt{9b^2}\sqrt{b}-\sqrt{25b^2}\sqrt{b}+\sqrt{49b^2}\sqrt{b}\\&=3b\sqrt{b}-5b\sqrt{b}+7b\sqrt{b}\\&=5b\sqrt{b}\end{aligned}$$

11. $$\begin{aligned}\frac{5\sqrt{2}}{3}+\frac{2\sqrt{2}}{5}&=\frac{5(5\sqrt{2})+3(2\sqrt{2})}{3(5)}\\&=\frac{25\sqrt{2}+6\sqrt{2}}{15}\\&=\frac{31\sqrt{2}}{15}\end{aligned}$$

13. $$\begin{aligned}\sqrt[3]{\frac{11}{8}}-\frac{\sqrt[3]{11}}{6}&=\frac{\sqrt[3]{11}}{\sqrt[3]{8}}-\frac{\sqrt[3]{11}}{6}\\&=\frac{\sqrt[3]{11}}{2}-\frac{\sqrt[3]{11}}{6}\\&=\frac{3\sqrt[3]{11}-\sqrt[3]{11}}{6}\\&=\frac{2\sqrt[3]{11}}{6}\\&=\frac{\sqrt[3]{11}}{3}\end{aligned}$$

15. $$\begin{aligned}\frac{\sqrt{20x}}{9}+\sqrt{\frac{5x}{9}}&=\frac{\sqrt{4(5x)}}{9}+\frac{\sqrt{5x}}{\sqrt{9}}\\&=\frac{\sqrt{4}\sqrt{5x}}{9}+\frac{\sqrt{5x}}{3}\\&=\frac{2\sqrt{5x}+3\sqrt{5x}}{9}\\&=\frac{5\sqrt{5x}}{9}\end{aligned}$$

17. $$\begin{aligned}7\sqrt{9}-7+\sqrt{3}&=7(3)-7+\sqrt{3}\\&=21-7+\sqrt{3}\\&=14+\sqrt{3}\end{aligned}$$

19. $$\begin{aligned}2+3\sqrt{y^2}-6\sqrt{y^2}+5&=7-3\sqrt{y^2}\\&=7-3y\end{aligned}$$

21. $3\sqrt{108} - 2\sqrt{18} - 3\sqrt{48}$
$= 3\sqrt{36}\sqrt{3} - 2\sqrt{9}\sqrt{2} - 3\sqrt{16}\sqrt{3}$
$= 3(6)\sqrt{3} - 2(3)\sqrt{2} - 3(4)\sqrt{3}$
$= 18\sqrt{3} - 6\sqrt{2} - 12\sqrt{3}$
$= 6\sqrt{3} - 6\sqrt{2}$

23. $-5\sqrt[3]{625} + \sqrt[3]{40} = -5\sqrt[3]{125}\sqrt[3]{5} + \sqrt[3]{8}\sqrt[3]{5}$
$= -5(5)\sqrt[3]{5} + 2\sqrt[3]{5}$
$= -25\sqrt[3]{5} + 2\sqrt[3]{5}$
$= -23\sqrt[3]{5}$

25. $\sqrt{9b^3} - \sqrt{25b^3} + \sqrt{16b^3}$
$= \sqrt{9b^2}\sqrt{b} - \sqrt{25b^2}\sqrt{b} + \sqrt{16b^2}\sqrt{b}$
$= 3b\sqrt{b} - 5b\sqrt{b} + 4b\sqrt{b}$
$= (3 - 5 + 4)b\sqrt{b}$
$= 2b\sqrt{b}$

27. $5y\sqrt{8y} + 2\sqrt{50y^3}$
$= 5y\sqrt{4}\sqrt{2y} + 2\sqrt{25y^2}\sqrt{2y}$
$= 5y(2)\sqrt{2y} + 2(5y)\sqrt{2y}$
$= 10y\sqrt{2y} + 10y\sqrt{2y}$
$= 20y\sqrt{2y}$

29. $\sqrt[3]{54xy^3} - 5\sqrt[3]{2xy^3} + y\sqrt[3]{128x}$
$= \sqrt[3]{27y^3}\sqrt[3]{2x} - 5\sqrt[3]{y^3}\sqrt[3]{2x} + y\sqrt[3]{64}\sqrt[3]{2x}$
$= 3y\sqrt[3]{2x} - 5y\sqrt[3]{2x} + y(4)\sqrt[3]{2x}$
$= -2y\sqrt[3]{2x} + 4y\sqrt[3]{2x}$
$= 2y\sqrt[3]{2x}$

31. $6\sqrt[3]{11} + 8\sqrt{11} - 12\sqrt{11} = 6\sqrt[3]{11} - 4\sqrt{11}$

33. $-2\sqrt[4]{x^7} + 3\sqrt[4]{16x^7}$
$= -2\sqrt[4]{x^4}\sqrt[4]{x^3} + 3\sqrt[4]{16x^4}\sqrt[4]{x^3}$
$= -2x\sqrt[4]{x^3} + 3(2x)\sqrt[4]{x^3}$
$= -2x\sqrt[4]{x^3} + 6x\sqrt[4]{x^3}$
$= 4x\sqrt[4]{x^3}$

35. $\frac{4\sqrt{3}}{3} - \frac{\sqrt{12}}{3} = \frac{4\sqrt{3}}{3} - \frac{\sqrt{4}\sqrt{3}}{3}$
$= \frac{4\sqrt{3} - 2\sqrt{3}}{3}$
$= \frac{2\sqrt{3}}{3}$

37. $\frac{\sqrt[3]{8x^4}}{7} + \frac{3x\sqrt[3]{x}}{7} = \frac{\sqrt[3]{8x^3}\sqrt[3]{x} + 3x\sqrt[3]{x}}{7}$
$= \frac{2x\sqrt[3]{x} + 3x\sqrt[3]{x}}{7}$
$= \frac{5x\sqrt[3]{x}}{7}$

39. $\sqrt{\frac{28}{x^2}} + \sqrt{\frac{7}{4x^2}} = \frac{\sqrt{28}}{\sqrt{x^2}} + \frac{\sqrt{7}}{\sqrt{4x^2}}$
$= \frac{\sqrt{4}\sqrt{7}}{x} + \frac{\sqrt{7}}{2x}$
$= \frac{2\sqrt{7}}{x} + \frac{\sqrt{7}}{2x}$
$= \frac{4\sqrt{7} + \sqrt{7}}{2x}$
$= \frac{5\sqrt{7}}{2x}$

41. $\sqrt[3]{\frac{16}{27}} - \frac{\sqrt[3]{54}}{6} = \frac{\sqrt[3]{16}}{\sqrt[3]{27}} - \frac{\sqrt[3]{27}\sqrt[3]{2}}{6}$
$= \frac{\sqrt[3]{8}\sqrt[3]{2}}{3} - \frac{3\sqrt[3]{2}}{6}$
$= \frac{2(2)\sqrt[3]{2}}{6} - \frac{3\sqrt[3]{2}}{6}$
$= \frac{4\sqrt[3]{2} - 3\sqrt[3]{2}}{6}$
$= \frac{\sqrt[3]{2}}{6}$

43. $-\frac{\sqrt[3]{2x^4}}{9} + \sqrt[3]{\frac{250x^4}{27}}$
$= \frac{-\sqrt[3]{x^3}\sqrt[3]{2x}}{9} + \frac{\sqrt[3]{250x^4}}{\sqrt[3]{27}}$
$= \frac{-x\sqrt[3]{2x}}{9} + \frac{\sqrt[3]{125x^3}\sqrt[3]{2x}}{3}$
$= \frac{-x\sqrt[3]{2x}}{9} + \frac{5x\sqrt[3]{2x}}{3}$
$= \frac{-x\sqrt[3]{2x} + 15x\sqrt[3]{2x}}{9}$
$= \frac{14x\sqrt[3]{2x}}{9}$

45. $P = 2\sqrt{12} + \sqrt{12} + 2\sqrt{27} + 3\sqrt{3}$
$= 2\sqrt{4}\sqrt{3} + \sqrt{4}\sqrt{3} + 2\sqrt{9}\sqrt{3} + 3\sqrt{3}$
$= 2 \cdot 2\sqrt{3} + 2\sqrt{3} + 2 \cdot 3\sqrt{3} + 3\sqrt{3}$
$= (4 + 2 + 6 + 3)\sqrt{3}$
$= 15\sqrt{3}$
The perimeter of the trapezoid is $15\sqrt{3}$ inches.

47. $\sqrt{7}(\sqrt{5} + \sqrt{3}) = \sqrt{7}\sqrt{5} + \sqrt{7}\sqrt{3}$
$= \sqrt{35} + \sqrt{21}$

49. $(\sqrt{5} - \sqrt{2})^2 = \sqrt{5}^2 - 2\sqrt{5}\sqrt{2} + \sqrt{2}^2$
$= 5 - 2\sqrt{10} + 2$
$= 7 - 2\sqrt{10}$

51. $\sqrt{3x}(\sqrt{3} - \sqrt{x}) = \sqrt{3x}\sqrt{3} - \sqrt{3x}\sqrt{x}$
$= \sqrt{3}\sqrt{x}\sqrt{3} - \sqrt{3}\sqrt{x}\sqrt{x}$
$= 3\sqrt{x} - x\sqrt{3}$

53. $(2\sqrt{x} - 5)(3\sqrt{x} + 1)$
$= (2\sqrt{x})(3\sqrt{x}) + (2\sqrt{x})1 - 5(3\sqrt{x}) - 5 \cdot 1$
$= 6x + 2\sqrt{x} - 15\sqrt{x} - 5$
$= 6x - 13\sqrt{x} - 5$

55. $(\sqrt[3]{a} - 4)(\sqrt[3]{a} + 5)$
$= (\sqrt[3]{a})^2 + 5\sqrt[3]{a} - 4\sqrt[3]{a} - 4 \cdot 5$
$= \sqrt[3]{a^2} + \sqrt[3]{a} - 20$

57. $6(\sqrt{2} - 2) = 6\sqrt{2} - 6 \cdot 2 = 6\sqrt{2} - 12$

59. $\sqrt{2}(\sqrt{2} + x\sqrt{6}) = \sqrt{2}^2 + \sqrt{2}(x\sqrt{6})$
$= 2 + \sqrt{2}(x\sqrt{2}\sqrt{3})$
$= 2 + 2x\sqrt{3}$

61. $(2\sqrt{7} + 3\sqrt{5})(\sqrt{7} - 2\sqrt{5}) = 2\sqrt{7}^2 - (2\sqrt{7})(2\sqrt{5}) + (3\sqrt{5})\sqrt{7} - 3 \cdot 2\sqrt{5}^2$
$= 2 \cdot 7 - 4\sqrt{35} + 3\sqrt{35} - 6 \cdot 5$
$= 14 - \sqrt{35} - 30$
$= -16 - \sqrt{35}$

63. $(\sqrt{x} - y)(\sqrt{x} + y) = \sqrt{x}^2 - y^2 = x - y^2$

65. $\left(\sqrt{3}+x\right)^2 = \sqrt{3}^2 + 2\sqrt{3}x + x^2$
$= 3 + 2\sqrt{3}x + x^2$

67. $\left(\sqrt{5x}-3\sqrt{2}\right)\left(\sqrt{5x}-3\sqrt{3}\right) = \left(\sqrt{5x}\right)^2 - 3\sqrt{3}\sqrt{5x} - 3\sqrt{2}\sqrt{5x} + \left(3\sqrt{2}\right)\left(3\sqrt{3}\right)$
$= 5x - 3\sqrt{15x} - 3\sqrt{10x} + 9\sqrt{6}$

69. $\left(\sqrt[3]{4}+2\right)\left(\sqrt[3]{2}-1\right) = \sqrt[3]{4}\sqrt[3]{2} - \sqrt[3]{4}\cdot 1 + 2\sqrt[3]{2} - 2\cdot 1$
$= \sqrt[3]{8} - \sqrt[3]{4} + 2\sqrt[3]{2} - 2$
$= 2 - \sqrt[3]{4} + 2\sqrt[3]{2} - 2$
$= -\sqrt[3]{4} + 2\sqrt[3]{2}$

71. $\left(\sqrt[3]{x}+1\right)\left(\sqrt[3]{x}-4\sqrt{x}+7\right) = \sqrt[3]{x}\left(\sqrt[3]{x}-4\sqrt{x}+7\right) + 1\left(\sqrt[3]{x}-4\sqrt{x}+7\right)$
$= \sqrt[3]{x}^2 - \sqrt[3]{x}(4\sqrt{x}) + \sqrt[3]{x}(7) + \sqrt[3]{x} - 4\sqrt{x} + 7$
$= \sqrt[3]{x}^2 - x^{1/3}(4x^{1/2}) + 8\sqrt[3]{x} - 4\sqrt{x} + 7$
$= \sqrt[3]{x^2} - 4x^{5/6} + 8\sqrt[3]{x} - 4\sqrt{x} + 7$
$= \sqrt[3]{x^2} - 4\sqrt[6]{x^5} + 8\sqrt[3]{x} - 4\sqrt{x} + 7$

73. a. $P = 2(3\sqrt{20}) + 2\sqrt{125}$
$= 6\sqrt{4}\sqrt{5} + 2\sqrt{25}\sqrt{5}$
$= 6(2)\sqrt{5} + 2(5)\sqrt{5}$
$= 12\sqrt{5} + 10\sqrt{5}$
$= 22\sqrt{5}$ feet

b. $A = 3\sqrt{20}\cdot\sqrt{125}$
$= 3\sqrt{4}\sqrt{5}\sqrt{5}\sqrt{25}$
$= 3\cdot 2\cdot 5\cdot 5$
$= 150$ sq. feet

75. Answers may vary.

77. $\frac{2x-14}{2} = \frac{2(x-7)}{2} = x-7$

79. $\frac{7x-7y}{x^2-y^2} = \frac{7(x-y)}{(x-y)(x+y)} = \frac{7}{x+y}$

81. $\frac{6a^2b-9ab}{3ab} = \frac{3ab(2a-3)}{3ab} = 2a-3$

83. $\frac{-4+2\sqrt{3}}{6} = \frac{2(-2+\sqrt{3})}{2\cdot 3} = \frac{-2+\sqrt{3}}{3}$

Section 7.5

Mental Math

1. The conjugate of $\sqrt{2}+x$ is $\sqrt{2}-x$.
2. The conjugate of $\sqrt{3}+y$ is $\sqrt{3}-y$.
3. The conjugate of $5-\sqrt{a}$ is $5+\sqrt{a}$.
4. The conjugate of $6-\sqrt{b}$ is $6+\sqrt{b}$.
5. The conjugate of $7\sqrt{5}+8\sqrt{x}$ is $7\sqrt{5}-8\sqrt{x}$.
6. The conjugate of $9\sqrt{2}-6\sqrt{y}$ is $9\sqrt{2}+6\sqrt{y}$.

Exercise Set 7.5

1. $\frac{\sqrt{2}}{\sqrt{7}}=\frac{\sqrt{2}\cdot\sqrt{7}}{\sqrt{7}\cdot\sqrt{7}}=\frac{\sqrt{14}}{7}$

3. $\sqrt{\frac{1}{5}}=\sqrt{\frac{1}{5}\cdot\frac{5}{5}}=\sqrt{\frac{5}{25}}=\frac{\sqrt{5}}{\sqrt{25}}=\frac{\sqrt{5}}{5}$

5. $\sqrt[3]{\frac{3}{4}}=\sqrt[3]{\frac{3}{4}\cdot\frac{16}{16}}=\sqrt[3]{\frac{48}{64}}=\frac{\sqrt[3]{6\cdot 8}}{4}=\frac{\sqrt[3]{6}}{2}$

7. $\frac{4}{\sqrt[3]{3}}=\frac{4}{\sqrt[3]{3}}\cdot\frac{\sqrt[3]{9}}{\sqrt[3]{9}}=\frac{4\sqrt[3]{9}}{\sqrt[3]{27}}=\frac{4\sqrt[3]{9}}{3}$

9. $\frac{3}{\sqrt{8x}}=\frac{3}{\sqrt{8x}}\cdot\frac{\sqrt{2x}}{\sqrt{2x}}=\frac{3\sqrt{2x}}{\sqrt{16x^2}}=\frac{3\sqrt{2x}}{4x}$

11. $\frac{3}{\sqrt[3]{4x^2}}=\frac{3}{\sqrt[3]{4x^2}}\cdot\frac{\sqrt[3]{2x}}{\sqrt[3]{2x}}=\frac{3\sqrt[3]{2x}}{\sqrt[3]{8x^3}}=\frac{3\sqrt[3]{2x}}{2x}$

13. $\sqrt{\frac{4}{x}}=\sqrt{\frac{4}{x}\cdot\frac{x}{x}}=\frac{\sqrt{4x}}{\sqrt{x^2}}=\frac{\sqrt{4x}}{x}$

15. $\frac{9}{\sqrt{3a}}=\frac{9}{\sqrt{3a}}\cdot\frac{\sqrt{3a}}{\sqrt{3a}}=\frac{9\sqrt{3a}}{3a}=\frac{3\sqrt{3a}}{a}$

17. $\frac{3}{\sqrt[3]{2}}=\frac{3}{\sqrt[3]{2}}\cdot\frac{\sqrt[3]{4}}{\sqrt[3]{4}}=\frac{3\sqrt[3]{4}}{\sqrt[3]{8}}=\frac{3\sqrt[3]{4}}{2}$

19. $\frac{2\sqrt{3}}{\sqrt{7}}=\frac{2\sqrt{3}}{\sqrt{7}}\cdot\frac{\sqrt{7}}{\sqrt{7}}=\frac{2\sqrt{21}}{7}$

21. $\sqrt{\frac{2x}{5y}}=\frac{\sqrt{2x}}{\sqrt{5y}}=\frac{\sqrt{2x}\cdot\sqrt{5y}}{\sqrt{5y}\cdot\sqrt{5y}}=\frac{\sqrt{10xy}}{5y}$

23. $\sqrt[4]{\frac{81}{8}}=\frac{3}{\sqrt[4]{8}}\cdot\frac{\sqrt[4]{512}}{\sqrt[4]{512}}=\frac{3\sqrt[4]{512}}{\sqrt[4]{4096}}=\frac{3\sqrt[4]{512}}{8}$

25. $\sqrt[4]{\frac{16}{9x^7}}=\frac{\sqrt[4]{16}}{\sqrt[4]{9x^7}}$
$=\frac{2}{x\sqrt[4]{9x^3}}$
$=\frac{2\cdot\sqrt[4]{9x}}{x\sqrt[4]{9x^3}\cdot\sqrt[4]{9x}}$
$=\frac{2\sqrt[4]{9x}}{3x^2}$

27. $\frac{5a}{\sqrt[5]{8a^9b^{11}}}=\frac{5a}{ab^2\sqrt[5]{8a^4b}}$
$=\frac{5a\sqrt[5]{4ab^4}}{ab^2\sqrt[5]{8a^4b}\cdot\sqrt[5]{4ab^4}}$
$=\frac{5a\sqrt[5]{4ab^4}}{2a^2b^3}$
$=\frac{5\sqrt[5]{4ab^4}}{2ab^3}$

29. $\sqrt{\frac{5}{3}}=\frac{\sqrt{5}}{\sqrt{3}}\cdot\frac{\sqrt{5}}{\sqrt{5}}=\frac{\sqrt{25}}{\sqrt{15}}=\frac{5}{\sqrt{15}}$

31. $\sqrt{\frac{18}{5}}=\frac{\sqrt{18}}{\sqrt{5}}$
$=\frac{\sqrt{9}\sqrt{2}}{\sqrt{5}}$
$=\frac{3\sqrt{2}}{\sqrt{5}}$
$=\frac{3\sqrt{2}}{\sqrt{5}}\cdot\frac{\sqrt{2}}{\sqrt{2}}$
$=\frac{3\sqrt{4}}{\sqrt{10}}$
$=\frac{3(2)}{\sqrt{10}}$
$=\frac{6}{\sqrt{10}}$

33. $\frac{\sqrt{4x}}{7}=\frac{\sqrt{4x}}{7}\cdot\frac{\sqrt{4x}}{\sqrt{4x}}=\frac{\sqrt{16x^2}}{7\sqrt{4x}}=\frac{4x}{7\sqrt{4x}}$

35. $\dfrac{\sqrt[3]{5y^2}}{\sqrt[3]{4x}}=\dfrac{\sqrt[3]{5y^2}}{\sqrt[3]{4x}}\cdot\dfrac{\sqrt[3]{5^2y}}{\sqrt[3]{5^2y}}$
$=\dfrac{\sqrt[3]{5^3y^3}}{\sqrt[3]{4(5^2)xy}}$
$=\dfrac{5y}{\sqrt[3]{100xy}}$

37. $\sqrt{\dfrac{2}{5}}=\dfrac{\sqrt{2}}{\sqrt{5}}\cdot\dfrac{\sqrt{2}}{\sqrt{2}}=\dfrac{\sqrt{4}}{\sqrt{10}}=\dfrac{2}{\sqrt{10}}$

39. $\dfrac{\sqrt{2x}}{11}=\dfrac{\sqrt{2x}}{11}\cdot\dfrac{\sqrt{2x}}{\sqrt{2x}}=\dfrac{\sqrt{4x^2}}{11\sqrt{2x}}=\dfrac{2x}{11\sqrt{2x}}$

41. $\sqrt[3]{\dfrac{7}{8}}=\dfrac{\sqrt[3]{7}}{\sqrt[3]{8}}=\dfrac{\sqrt[3]{7}}{2}\cdot\dfrac{\sqrt[3]{7^2}}{\sqrt[3]{7^2}}=\dfrac{\sqrt[3]{7^3}}{2\sqrt[3]{7^2}}=\dfrac{7}{2\sqrt[3]{49}}$

43. $\dfrac{\sqrt[3]{3x^5}}{10}=\dfrac{\sqrt[3]{3x^5}}{10}\cdot\dfrac{\sqrt[3]{3^2x}}{\sqrt[3]{3^2x}}$
$=\dfrac{\sqrt[3]{3^3x^6}}{10\sqrt[3]{3^2x}}$
$=\dfrac{3x^2}{10\sqrt[3]{9x}}$

45. $\sqrt{\dfrac{18x^4y^6}{3z}}=\sqrt{\dfrac{6x^4y^6}{z}}$
$=\dfrac{x^2y^3\sqrt{6}}{\sqrt{z}}$
$=\dfrac{x^2y^3\sqrt{6}}{\sqrt{z}}\cdot\dfrac{\sqrt{6}}{\sqrt{6}}$
$=\dfrac{x^2y^3\sqrt{36}}{\sqrt{6z}}$
$=\dfrac{6x^2y^3}{\sqrt{6z}}$

47. Answers may vary.

49. $\dfrac{6}{2-\sqrt{7}}=\dfrac{6}{2-\sqrt{7}}\cdot\dfrac{2+\sqrt{7}}{2+\sqrt{7}}$
$=\dfrac{6\left(2+\sqrt{7}\right)}{2^2-\sqrt{7}^2}$
$=\dfrac{6\left(2+\sqrt{7}\right)}{4-7}$
$=-\dfrac{6\left(2+\sqrt{7}\right)}{3}$
$=-2\left(2+\sqrt{7}\right)$

51. $\dfrac{-7}{\sqrt{x}-3}=\dfrac{-7}{\sqrt{x}-3}\cdot\dfrac{\sqrt{x}+3}{\sqrt{x}+3}$
$=\dfrac{-7\left(\sqrt{x}+3\right)}{\sqrt{x}^2-3^2}$
$=\dfrac{-7\left(\sqrt{x}+3\right)}{x-9}$

53. $\dfrac{\sqrt{2}-\sqrt{3}}{\sqrt{2}+\sqrt{3}}=\dfrac{\sqrt{2}-\sqrt{3}}{\sqrt{2}+\sqrt{3}}\cdot\dfrac{\sqrt{2}-\sqrt{3}}{\sqrt{2}-\sqrt{3}}$
$=\dfrac{\sqrt{2}^2-2\sqrt{2}\sqrt{3}+\sqrt{3}^2}{\sqrt{2}^2-\sqrt{3}^2}$
$=\dfrac{2-2\sqrt{6}+3}{2-3}$
$=\dfrac{5-2\sqrt{6}}{-1}$
$=-5+2\sqrt{6}$

55. $\dfrac{\sqrt{a}+1}{2\sqrt{a}-\sqrt{b}}$
$=\dfrac{\sqrt{a}+1}{2\sqrt{a}-\sqrt{b}}\cdot\dfrac{2\sqrt{a}+\sqrt{b}}{2\sqrt{a}+\sqrt{b}}$
$=\dfrac{\sqrt{a}(2\sqrt{a})+\sqrt{a}\sqrt{b}+1(2\sqrt{a})+1\sqrt{b}}{(2\sqrt{a})^2-\sqrt{b}^2}$
$=\dfrac{2a+\sqrt{ab}+2\sqrt{a}+\sqrt{b}}{4a-b}$

57. $\frac{8}{1+\sqrt{10}} = \frac{8}{1+\sqrt{10}} \cdot \frac{1-\sqrt{10}}{1-\sqrt{10}}$
$= \frac{8\left(1-\sqrt{10}\right)}{1^2 - \sqrt{10}^2}$
$= \frac{8\left(1-\sqrt{10}\right)}{1-10}$
$= -\frac{8\left(1-\sqrt{10}\right)}{9}$

59. $\frac{\sqrt{x}}{\sqrt{x}+\sqrt{y}} = \frac{\sqrt{x}}{\sqrt{x}+\sqrt{y}} \cdot \frac{\sqrt{x}-\sqrt{y}}{\sqrt{x}-\sqrt{y}}$
$= \frac{\sqrt{x}\sqrt{x} - \sqrt{x}\sqrt{y}}{\sqrt{x}^2 - \sqrt{y}^2}$
$= \frac{x - \sqrt{xy}}{x - y}$

61. $\frac{2\sqrt{3}+\sqrt{6}}{4\sqrt{3}-\sqrt{6}} = \frac{2\sqrt{3}+\sqrt{6}}{4\sqrt{3}-\sqrt{6}} \cdot \frac{4\sqrt{3}+\sqrt{6}}{4\sqrt{3}+\sqrt{6}}$
$= \frac{8 \cdot 3 + 2\sqrt{18} + 4\sqrt{18} + 6}{16 \cdot 3 - 6}$
$= \frac{30 + 6\sqrt{18}}{42}$
$= \frac{30 + 18\sqrt{2}}{42}$
$= \frac{6(5+3\sqrt{2})}{6 \cdot 7}$
$= \frac{5+3\sqrt{2}}{7}$

63. $\frac{2-\sqrt{11}}{6} \cdot \frac{2+\sqrt{11}}{2+\sqrt{11}} = \frac{(2-\sqrt{11})(2+\sqrt{11})}{6(2+\sqrt{11})}$
$= \frac{4-\sqrt{121}}{12+6\sqrt{11}}$
$= \frac{4-11}{12+6\sqrt{11}}$
$= \frac{-7}{12+6\sqrt{11}}$

65. $\frac{2-\sqrt{7}}{-5} \cdot \frac{2+\sqrt{7}}{2+\sqrt{7}} = \frac{(2-\sqrt{7})(2+\sqrt{7})}{-5(2+\sqrt{7})}$
$= \frac{4-\sqrt{49}}{-10-5\sqrt{7}}$
$= \frac{4-7}{-10-5\sqrt{7}}$
$= \frac{-3}{-10-5\sqrt{7}}$
$= \frac{-1(3)}{-1(10+5\sqrt{7})}$
$= \frac{3}{10+5\sqrt{7}}$

67. $\frac{\sqrt{x}+3}{\sqrt{x}} \cdot \frac{\sqrt{x}-3}{\sqrt{x}-3} = \frac{(\sqrt{x}+3)(\sqrt{x}-3)}{\sqrt{x}(\sqrt{x}-3)}$
$= \frac{\sqrt{x^2}-9}{\sqrt{x^2}-3\sqrt{x}}$
$= \frac{x-9}{x-3\sqrt{x}}$

69. $\frac{\sqrt{2}-1}{\sqrt{2}+1} \cdot \frac{\sqrt{2}+1}{\sqrt{2}+1} = \frac{(\sqrt{2}-1)(\sqrt{2}+1)}{(\sqrt{2}+1)(\sqrt{2}+1)}$
$= \frac{\sqrt{4}-1}{\sqrt{4}+2\sqrt{2}+1}$
$= \frac{2-1}{2+2\sqrt{2}+1}$
$= \frac{1}{3+2\sqrt{2}}$

71. $\frac{\sqrt{x}+1}{\sqrt{x}-1} \cdot \frac{\sqrt{x}-1}{\sqrt{x}-1} = \frac{(\sqrt{x}+1)(\sqrt{x}-1)}{(\sqrt{x}-1)(\sqrt{x}-1)}$
$= \frac{\sqrt{x^2}-1}{\sqrt{x^2}-2\sqrt{x}+1}$
$= \frac{x-1}{x-2\sqrt{x}+1}$

73. $r = \sqrt{\dfrac{A}{4\pi}}$
$= \dfrac{\sqrt{A}}{\sqrt{4\pi}}$
$= \dfrac{\sqrt{A}}{2\sqrt{\pi}}$
$= \dfrac{\sqrt{A}\sqrt{\pi}}{2\sqrt{\pi}\sqrt{\pi}}$
$= \dfrac{\sqrt{A\pi}}{2\pi}$

75. Answers may vary.

77. $2x - 7 = 3(x - 4)$
$2x - 7 = 3x - 12$
$12 - 7 = 3x - 2x$
$5 = x$
The solution is 5.

79. $(x - 6)(2x + 1) = 0$
$x - 6 = 0$ or $2x + 1 = 0$
$x = 6$ or $x = -\dfrac{1}{2}$
The solutions are $-\dfrac{1}{2}$ and 6.

81. $x^2 - 8x = -12$
$x^2 - 8x + 12 = 0$
$(x - 6)(x - 2) = 0$
$x - 6 = 0$ or $x - 2 = 0$
$x = 6$ or $x = 2$
The solutions are 2 and 6.

Section 7.6

Graphing Calculator Explorations

1.

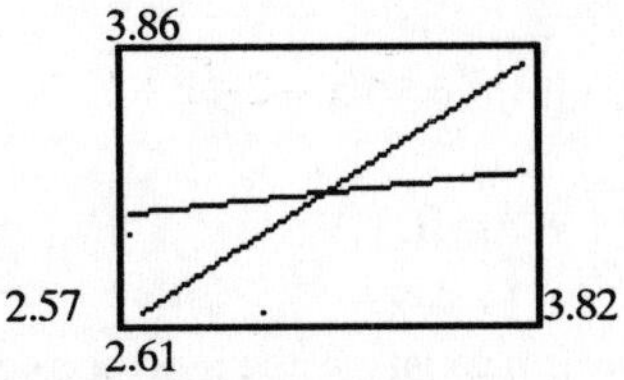

The solution is 3.19.

3.

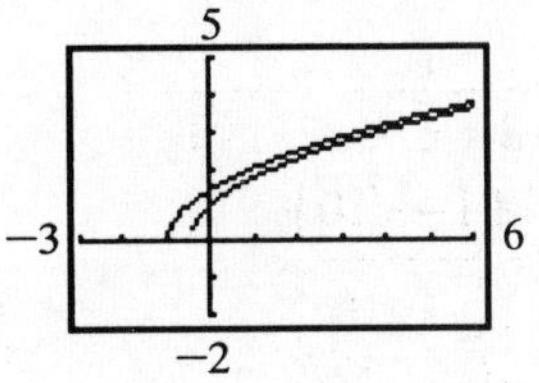

There is no solution.

5.

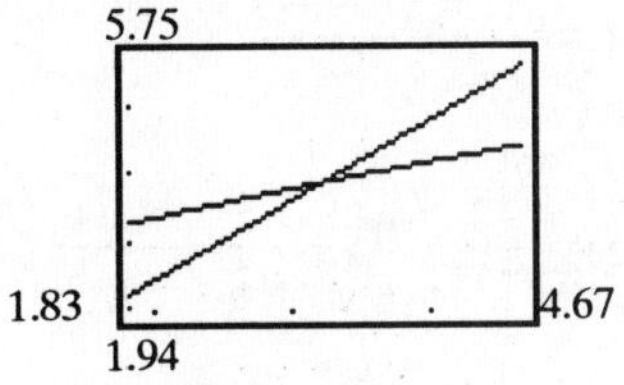

The solution is 3.23.

Exercise Set 7.6

1. $\sqrt{2x} = 4$
$2x = 4^2$
$2x = 16$
$x = 8$
The solution is 8.

3. $\sqrt{x - 3} = 2$
$x - 3 = 2^2$
$x - 3 = 4$
$x = 7$
The solution is 7.

5. $\sqrt{2x} = -4$
No solution since a principle square root does not yield a negative number.

7. $\sqrt{4x - 3} - 5 = 0$
$\sqrt{4x - 3} = 5$
$4x - 3 = 5^2$
$4x - 3 = 25$
$4x = 28$
$x = 7$
The solution is 7.

9. $\sqrt{2x-3}-2=1$
$\sqrt{2x-3}=3$
$2x-3=3^2$
$2x-3=9$
$2x=12$
$x=6$
The solution is 6.

11. $\sqrt[3]{6x}=-3$
$6x=(-3)^3$
$6x=-27$
$x=\frac{-27}{6}=-\frac{9}{2}$
The solution is $-\frac{9}{2}$.

13. $\sqrt[3]{x-2}-3=0$
$\sqrt[3]{x-2}=3$
$x-2=3^3$
$x-2=27$
$x=29$
The solution is 29.

15. $\sqrt{13-x}=x-1$
$13-x=(x-1)^2$
$13-x=x^2-2x+1$
$0=x^2-x-12$
$0=(x-4)(x+3)$
$x-4=0$ or $x+3=0$
$x=4$ or $x=-3$
We discard −3 as extraneous, leaving 4 as the only solution.

17. $x-\sqrt{4-3x}=-8$
$x+8=\sqrt{4-3x}$
$(x+8)^2=4-3x$
$x^2+16x+64=4-3x$
$x^2+19x+60=0$
$(x+4)(x+15)=0$
$x+4=0$ or $x+15=0$
$x=-4$ or $x=-15$
We discard −15 as extraneous, leaving −4 as the only solution.

19. $\sqrt{y+5}=2-\sqrt{y-4}$
$y+5=\left(2-\sqrt{y-4}\right)^2$
$y+5=4-4\sqrt{y-4}+y-4$
$y+5=y-4\sqrt{y-4}$
$5=-4\sqrt{y-4}$
$25=16(y-4)$
$\frac{25}{16}=y-4$
$y=\frac{89}{16}=5\frac{9}{16}$
which we discard as extraneous.
There is no solution.

21. $\sqrt{x-3}+\sqrt{x+2}=5$
$\sqrt{x-3}=5-\sqrt{x+2}$
$x-3=25-10\sqrt{x+2}+x+2$
$-3=27-10\sqrt{x+2}$
$-30=-10\sqrt{x+2}$
$3=\sqrt{x+2}$
$9=x+2$
$7=x$
The solution is 7.

23. $\sqrt{3x-2}=5$
$3x-2=5^2$
$3x-2=25$
$3x=27$
$x=9$
The solution is 9.

25. $-\sqrt{2x}+4=-6$
$10=\sqrt{2x}$
$10^2=2x$
$100=2x$
$x=50$
The solution is 50.

27. $\sqrt{3x+1}+2=0$
$\sqrt{3x+1}=-2$
No solution exists since a principle square root does not yield a negative number.

29. $\sqrt[4]{4x+1}-2=0$
$\sqrt[4]{4x+1}=2$
$4x+1=2^4$
$4x+1=16$
$4x=15$
$x=\frac{15}{4}$
The solution is $\frac{15}{4}$.

31. $\sqrt{4x-3}=7$
$4x-3=7^2$
$4x-3=49$
$4x=52$
$x=13$
The solution is 13.

33. $\sqrt[3]{6x-3}-3=0$
$\sqrt[3]{6x-3}=3$
$6x-3=3^3$
$6x-3=27$
$6x=30$
$x=5$
The solution is 5.

35. $\sqrt[3]{2x-3}-2=-5$
$\sqrt[3]{2x-3}=-3$
$2x-3=(-3)^3$
$2x-3=-27$
$2x=-24$
$x=-12$
The solution is –12.

37. $\sqrt{x+4}=\sqrt{2x-5}$
$x+4=2x-5$
$9=x$
$x=9$
The solution is 9.

39. $x-\sqrt{1-x}=-5$
$x+5=\sqrt{1-x}$
$(x+5)^2=1-x$
$x^2+10x+25=1-x$
$x^2+11x+24=0$
$(x+3)(x+8)=0$
$x+3=0$ or $x+8=0$
$x=-3$ or $x=-8$
We discard –8 as extraneous, leaving –3 as the only solution.

41. $\sqrt[3]{-6x-1}=\sqrt[3]{-2x-5}$
$-6x-1=-2x-5$
$4=4x$
$x=1$
The solution is 1.

43. $\sqrt{5x-1}-\sqrt{x}+2=3$
$\sqrt{5x-1}=\sqrt{x}+1$
$5x-1=x+2\sqrt{x}+1$
$4x-2=2\sqrt{x}$
$2x-1=\sqrt{x}$
$4x^2-4x+1=x$
$4x^2-5x+1=0$
$(4x-1)(x-1)=0$
$4x-1=0$ or $x-1=0$
$4x=1$ or $x=1$
$x=\frac{1}{4}$
We discard $\frac{1}{4}$ as extraneous, leaving 1 as the only solution.

45. $\sqrt{2x-1}=\sqrt{1-2x}$
$\sqrt{2x-1}=\sqrt{-(2x-1)}$
It follows that $2x-1=0$ (otherwise one of the radicands would be negative).
So $2x=1$
$x=\frac{1}{2}$
The solution is $\frac{1}{2}$.

47.
$$\sqrt{3x+4}-1=\sqrt{2x+1}$$
$$3x+4-2\sqrt{3x+4}+1=2x+1$$
$$x+4=2\sqrt{3x+4}$$
$$x^2+8x+16=4(3x+4)$$
$$x^2-4x=0$$
$$x(x-4)=0$$
$$x=0 \text{ or } x=4$$
The solutions are 0 and 4.

49. $\sqrt{y+3}-\sqrt{y-3}=1$
$$\sqrt{y+3}=1+\sqrt{y-3}$$
$$\left(\sqrt{y+3}\right)^2=\left(1+\sqrt{y-3}\right)^2$$
$$y+3=1+2\sqrt{y-3}+y-3$$
$$5=2\sqrt{y-3}$$
$$25=4(y-3)$$
$$\frac{25}{4}=y-3$$
$$\frac{25}{4}+\frac{12}{4}=y$$
$$\frac{37}{4}=y$$
The solution is $\frac{37}{4}$.

51. The left side of the equation on the third line should be
$2x+5+2\sqrt{(2x+5)(4-x)}+4-x.$

53. Let c = the length of the hypotenuse of the right triangle. By the Pythagorean theorem,
$c^2=6^2+3^2=36+9$
$c^2=45$
or $c=\sqrt{45}=\sqrt{9}\sqrt{5}=3\sqrt{5}$ feet

55. Let b = the length of the unknown leg of the right triangle. By the Pythagorean theorem,
$7^2=3^2+b^2$
$49=9+b^2$
$b^2=40$
$b=\sqrt{40}=\sqrt{4}\sqrt{10}=2\sqrt{10}$ meters

57. Let b = the length of the unknown leg of the right triangle. By the Pythagorean theorem,
$$\left(11\sqrt{5}\right)^2=9^2+b^2$$
$$605=81+b^2$$
$$b^2=524$$
$$b=\sqrt{524}$$
$$=\sqrt{4}\sqrt{131}$$
$$=2\sqrt{131}\approx 22.9 \text{ meters}$$

59. Let c = the length of the hypotenuse of the right triangle. By the Pythagorean theorem,
$c^2=7^2+(7.2)^2=100.84.$
So, $c=\sqrt{100.84}\approx 10.0$ millimeters.

61. Let c = the length of the hypotenuse of the right triangle as shown in the figure in the text. By the Pythagorean theorem,
$c^2=8^2+15^2=64+225=289.$
So, $c=\sqrt{289}=17$ feet.
Thus, 17 feet of cable are needed.

63. Let c = the length of the hypotenuse of the right triangle as shown in the figure in the text. By the Pythagorean theorem,
$c^2=12^2+5^2=144+25=169.$
So, $x=\sqrt{169}=13$ feet.
Thus, a 13 foot ladder is needed.

65.
$$r=\sqrt{\frac{A}{4\pi}}$$
$$1080=\sqrt{\frac{A}{4\pi}}$$
$$(1080)^2=\left(\sqrt{\frac{A}{4\pi}}\right)^2$$
$$1{,}166{,}400=\frac{A}{4\pi}$$
$$14{,}657{,}415\approx A$$
The surface area is 14,657,415 square miles.

67. $v = \sqrt{2gh}$
$80 = \sqrt{2(32)h}$
$(80)^2 = \left(\sqrt{2(32)h}\right)^2$
$6400 = 2(32) \cdot h$
$100 = h$
The object fell 100 feet.

69. $\sqrt{\sqrt{x+3} + \sqrt{x}} = \sqrt{3}$
$\left(\sqrt{\sqrt{x+3} + \sqrt{x}}\right)^2 = \left(\sqrt{3}\right)^2$
$\sqrt{x+3} + \sqrt{x} = 3$
$\sqrt{x+3} = 3 - \sqrt{x}$
$\left(\sqrt{x+3}\right)^2 = \left(3 - \sqrt{x}\right)^2$
$x + 3 = 9 - 6\sqrt{x} + x$
$\sqrt{x} = 1$
$x = 1$
The solution is 1.

71. We need to solve the equation:
$80\sqrt[3]{x} + 500 = 1620$
$80\sqrt[3]{x} = 1120$
$\sqrt[3]{x} = 14$
$x = 14^3$
$x = 2744$
Thus, the company needs to make fewer than 2744 deliveries per day, or 2743 deliveries.

73. Answers may vary.

75. It is not a function.

77. It is not a function.

79. It is not a function.

81. $\dfrac{\frac{1}{y} + \frac{4}{5}}{-\frac{3}{20}} = \dfrac{20y\left(\frac{1}{y} + \frac{4}{5}\right)}{20y\left(-\frac{3}{20}\right)}$
$= \dfrac{20y\left(\frac{1}{y}\right) + 20y\left(\frac{4}{5}\right)}{20y\left(-\frac{3}{20}\right)}$
$= \dfrac{20 + 16y}{-3y}$
$= -\dfrac{20 + 16y}{3y}$

83. $\dfrac{\frac{1}{y} + \frac{1}{x}}{\frac{1}{y} - \frac{1}{x}} = \dfrac{xy\left(\frac{1}{y} + \frac{1}{x}\right)}{xy\left(\frac{1}{y} - \frac{1}{x}\right)}$
$= \dfrac{xy\left(\frac{1}{y}\right) + xy\left(\frac{1}{x}\right)}{xy\left(\frac{1}{y}\right) - xy\left(\frac{1}{x}\right)}$
$= \dfrac{x + y}{x - y}$

85. $\sqrt{(x^2 - x) + 7} = 2(x^2 - x) - 1$
Let $u = x^2 - x$.
$\sqrt{u + 7} = 2u - 1$
$(\sqrt{u + 7})^2 = (2u - 1)^2$
$u + 7 = 4u^2 - 4u + 1$
$0 = 4u^2 - 5u - 6$
$0 = (4u + 3)(u - 2)$
$4u + 3 = 0$ or $u - 2 = 0$
$4u = -3$ or $u = 2$
$u = -\dfrac{3}{4}$ or $u = 2$
Substituting $x^2 - x = u$
$x^2 - x = -\dfrac{3}{4}$ or $x^2 - x = 2$
$4(x^2 - x) = 4\left(-\dfrac{3}{4}\right)$ or $x^2 - x - 2 = 0$
$4x^2 - 4x = -3$ or $(x - 2)(x + 1) = 0$

$4x^2 - 4x + 3 = 0$ or $x - 2 = 0$ or $x + 1 = 0$
Can't factor or $x = 2$ or $x = -1$
The solutions are –1 and 2.

87. $x^2+6x=4\sqrt{x^2+6x}$

Let $u=x^2+6x$.

$u=4\sqrt{u}$

$(u)^2=(4\sqrt{u})^2$

$u^2=16u$

$u^2-16u=0$

$u(u-16)=0$

$u=0$ or $u-16=0$

Substituting $x^2+6x=u$

$x^2+6x=0$ or $x^2+6x-16=0$

$x(x+6)=0$ or $(x+8)(x-2)=0$

$x=0$ or $x+6=0$ or $x+8=0$ or $x-2=0$

$x=0$ or $x=-6$ or $x=-8$ or $x=2$

All solutions check.

The solutions are –8, –6, 0, and 2.

Section 7.7

Mental Math

1. $\sqrt{-81}=9i$

2. $\sqrt{-49}=7i$

3. $\sqrt{-7}=i\sqrt{7}$

4. $\sqrt{-3}=i\sqrt{3}$

5. $-\sqrt{16}=-4$

6. $-\sqrt{4}=-2$

7. $\sqrt{-64}=8i$

8. $\sqrt{-100}=10i$

Exercise Set 7.7

1. $\sqrt{-24}=\sqrt{4}\sqrt{6}\sqrt{-1}=2i\sqrt{6}$

3. $-\sqrt{-36}=-\sqrt{36}\sqrt{-1}=-6i$

5. $8\sqrt{-63}=8\sqrt{9}\sqrt{7}\sqrt{-1}$
$=8\cdot 3i\sqrt{7}$
$=24i\sqrt{7}$

7. $-\sqrt{54}=-\sqrt{9}\sqrt{6}=-3\sqrt{6}$

9. $\sqrt{-2}\sqrt{-7}=(i\sqrt{2})(i\sqrt{7})$
$=i^2\sqrt{14}$
$=(-1)\sqrt{14}$
$=-\sqrt{14}$

11. $\sqrt{-5}\sqrt{-10}=(i\sqrt{5})(i\sqrt{10})$
$=i^2\sqrt{5}\sqrt{5}\sqrt{2}$
$=(-1)5\sqrt{2}$
$=-5\sqrt{2}$

13. $\sqrt{16}\sqrt{-1}=4i$

15. $\dfrac{\sqrt{-9}}{\sqrt{3}}=\dfrac{3i}{\sqrt{3}}=i\sqrt{3}$

17. $\dfrac{\sqrt{-80}}{\sqrt{-10}}=\dfrac{i\sqrt{80}}{i\sqrt{10}}$
$=\sqrt{\dfrac{80}{10}}$
$=\sqrt{8}$
$=\sqrt{4}\sqrt{2}$
$=2\sqrt{2}$

19. $(4-7i)+(2+3i)=(4+2)+(-7+3)i$
$=6-4i$

21. $(6+5i)-(8-i)=(6-8)+[5-(-1)]i$
$=-2+6i$

23. $6-(8+4i)=6-8-4i=-2-4i$

25. $6i(2-3i)=6i(2)-6i(3i)$
$=12i-18i^2$
$=12i-18(-1)$
$=18+12i$

27. $(\sqrt{3}+2i)(\sqrt{3}-2i) = (\sqrt{3})^2 - (2i)^2$
$= \sqrt{3^2} + 2^2$
$= 3+4$
$= 7$

29. $(4-2i)^2 = 16-16i+4i^2$
$= 16-16i+4(-1)$
$= 16-4-16i$
$= 12-16i$

31. $\frac{4}{i} = \frac{4}{i}\cdot\frac{-i}{-i} = \frac{-4i}{-i^2} = \frac{-4i}{-(-1)} = \frac{-4i}{1} = -4i$

33. $\frac{7}{4+3i} = \frac{7}{4+3i}\cdot\frac{4-3i}{4-3i}$
$= \frac{28-21i}{4^2+3^2}$
$= \frac{28-21i}{16+9}$
$= \frac{28-21i}{25}$
$= \frac{28}{25} - \frac{21}{25}i$

35. $\frac{3+5i}{1+i} = \frac{3+5i}{1+i}\cdot\frac{1-i}{1-i}$
$= \frac{3-3i+5i-5i^2}{1^2+1^2}$
$= \frac{3+2i-5(-1)}{1+1}$
$= \frac{3+5+2i}{2}$
$= \frac{8+2i}{2}$
$= 4+i$

37. $\frac{5-i}{3-2i} = \frac{5-i}{3-2i}\cdot\frac{3+2i}{3+2i}$
$= \frac{15+10i-3i-2i^2}{3^2+2^2}$
$= \frac{15+7i-2(-1)}{9+4}$
$= \frac{15+2+7i}{13}$
$= \frac{17+7i}{13}$
$= \frac{17}{13} + \frac{7}{13}i$

39. $(7i)(-9i) = -63i^2 = -63(-1) = 63$

41. $(6-3i)-(4-2i) = 6-3i-4+2i = 2-i$

43. $(6-2i)(3+i) = 2(3-i)(3+i)$
$= 2(3^2+1^2)$
$= 2(9+1)$
$= 2(10)$
$= 20$

45. $(8-3i)+(2+3i) = 8-3i+2+3i$
$= 10$

47. $(1-i)(1+i) = 1^2 - i^2 = 1^2 + 1^2 = 1+1 = 2$

49. $\frac{16+15i}{-3i} = \frac{16+5i}{-3i}\cdot\frac{i}{i}$
$= \frac{16i+15i^2}{-3i^2}$
$= \frac{16i+15(-1)}{-3(-1)}$
$= \frac{-15+16i}{3}$
$= \frac{-15}{3} + \frac{16}{3}i$
$= -5 + \frac{16}{3}i$

51. $(9+8i)^2 = 81+144i+64i^2$
$= 81+144i+64(-1)$
$= 81-64+144i$
$= 17+144i$

53. $\frac{2}{3+i} = \frac{2}{3+i} \cdot \frac{3-i}{3-i}$
$= \frac{2(3-i)}{3^2+1^2}$
$= \frac{6-2i}{9+1}$
$= \frac{6}{10} - \frac{2}{10}i$
$= \frac{3}{5} - \frac{1}{5}i$

55. $(5-6i) - 4i = 5 - 6i - 4i = 5 - 10i$

57. $\frac{2-3i}{2+i} = \frac{2-3i}{2+i} \cdot \frac{2-i}{2-i}$
$= \frac{4-2i-6i+3i^2}{2^2+1^2}$
$= \frac{4-8i+3(-1)}{4+1}$
$= \frac{4-3-8i}{5}$
$= \frac{1-8i}{5}$
$= \frac{1}{5} - \frac{8}{5}i$

59. $(2+4i)+(6-5i) = (2+6)+(4-5)i$
$= 8 - i$

61. $i^8 = (i^4)^2 = 1^2 = 1$

63. $i^{21} = i^{20}i = (i^4)^5 i = 1^5 i = 1i = i$

65. $i^{11} = i^{10}i = (i^2)^5 i = (-1)^5 i = (-1)i = -i$

67. $i^{-6} = (i^2)^{-3} = (-1)^{-3} = \frac{1}{(-1)^3} = \frac{1}{-1} = -1$

69. $i^3 + i^4 = -i + 1 = 1 - i$

71. $i^6 + i^8 = (i^4)(i^2) + (i^4)^2$
$= 1(-1) + 1^2$
$= -1 + 1$
$= 0$

73. $2 + \sqrt{-9} = 2 + 3i$

75. $\frac{6+\sqrt{-18}}{3} = \frac{6+3i\sqrt{2}}{3}$
$= \frac{6}{3} + \frac{3i\sqrt{2}}{3}$
$= 2 + i\sqrt{2}$

77. $\frac{5-\sqrt{-75}}{10} = \frac{5-5i\sqrt{3}}{10}$
$= \frac{5}{10} - \frac{5i\sqrt{3}}{10}$
$= \frac{1}{2} - \frac{\sqrt{3}}{2}i$

79. Answers may vary.

81. $\left(8-\sqrt{-4}\right) - \left(2+\sqrt{-16}\right)$
$= (8-2i) - (2+4i)$
$= (8-2) + (-2i-4i)$
$= 6 - 6i$

83.
$x^2 + 2x \stackrel{?}{=} -2$
$(-1+i)^2 + 2(-1+i) \stackrel{?}{=} -2$
$1 - 2i + i^2 - 2 + 2i \stackrel{?}{=} -2$
$1 - 2i - 1 - 2 + 2i \stackrel{?}{=} -2$
$-2 \stackrel{?}{=} -2$
Yes, –1 + i is a solution.

85. $x + 57° + 90° = 180°$
$x + 147° = 180°$
$x = 33°$

87.

-2	5	0	–3	0	2
		–10	20	–34	68
	5	–10	17	–34	70

$(5x^4 - 3x^2 + 2) \div (x+2)$
$= 5x^3 - 10x^2 + 17x - 34 + \frac{70}{x+2}$

89. 5 people reported an average checking balance of \$0 to \$100.

91. $6 + 2 + 3 = 11$
11 people reported an average checking balance of #301 or more.

93. $\frac{5}{30} = 0.166\overline{6}$
16.7% of the people reprted an average checking balance of \$0 to \$100.

Chapter 7 Review Exercises

1. $\sqrt{81} = 9$ because $9^2 = 81$.

2. $\sqrt[4]{81} = 3$ because $3^4 = 81$

3. $\sqrt[3]{-8} = -2$ because $(-2)^3 = -8$.

4. $\sqrt[4]{-16}$ is not a real number since there is no real number whose 4th power is negative.

5. $-\sqrt{\frac{1}{49}} = -\frac{1}{7}$ because $\left(\frac{1}{7}\right)^2 = \frac{1}{49}$

6. $\sqrt{x^{64}} = x^{32}$ because $(x^{32})^2 = x^{32 \cdot 2} = x^{64}$.

7. $-\sqrt{36} = -6$ because $(6)^2 = 36$

8. $\sqrt[3]{64} = 4$ because $4^3 = 64$.

9. $\sqrt[3]{-a^6b^9} = \sqrt[3]{-1}\sqrt[3]{a^6}\sqrt[3]{b^9}$
$= (-1)a^2b^3$
$= -a^2b^3$

10. $\sqrt{16a^4b^{12}} = \sqrt{16}\sqrt{a^4}\sqrt{b^{12}} = 4a^2b^6$

11. $\sqrt[5]{32a^5b^{10}} = \sqrt[5]{32}\sqrt[5]{a^5}\sqrt[5]{b^{10}} = 2ab^2$

12. $\sqrt[5]{-32x^{15}y^{20}} = \sqrt[5]{-32}\sqrt[5]{x^{15}}\sqrt[5]{y^{20}}$
$= -2x^3y^4$

13. $\sqrt{\frac{x^{12}}{36y^2}} = \frac{\sqrt{x^{12}}}{\sqrt{36y^2}} = \frac{x^6}{6y}$

14. $\sqrt[3]{\frac{27y^3}{z^{12}}} = \frac{\sqrt[3]{27y^3}}{\sqrt[3]{z^{12}}} = \frac{3y}{z^4}$

15. $\sqrt{(-x)^2} = |-x|$

16. $\sqrt[4]{(x^2-4)^4} = |x^2-4|$

17. $\sqrt[3]{(-27)^3} = -27$

18. $\sqrt[5]{(-5)^5} = -5$

19. $-\sqrt[5]{x^5} = -x$

20. $\sqrt[4]{16(2y+z)^{12}} = \sqrt[4]{16}\sqrt[4]{(2y+z)^{12}}$
$= 2|2y+z|^3$

21. $\sqrt{25(x-y)^{10}} = \sqrt{25}\sqrt{(x-y)^{10}}$
$= 5|(x-y)^5|$

22. $\sqrt[5]{-y^5} = \sqrt[5]{-1}\sqrt[5]{y^5} = (-1)y = -y$

23. $\sqrt[9]{-x^9} = \sqrt[9]{-1}\sqrt[9]{x^9} = -x$

24. $f(x) = \sqrt{x} + 3$
Domain: $[0, \infty)$

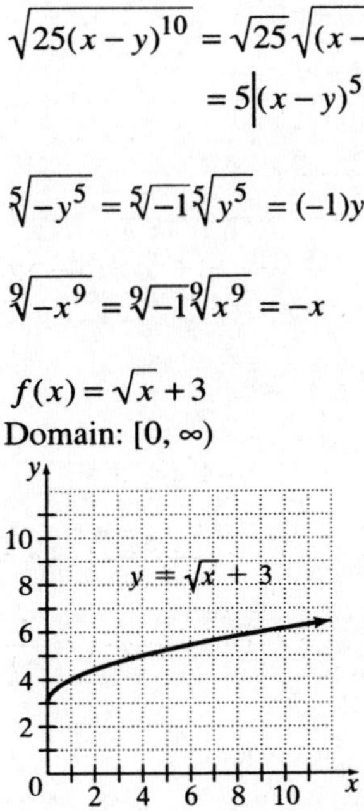

25. $g(x) = \sqrt[3]{x-3}$
Domain: $(-\infty, \infty)$

x	-5	2	3	4	11
$g(x)$	-2	-1	0	1	2

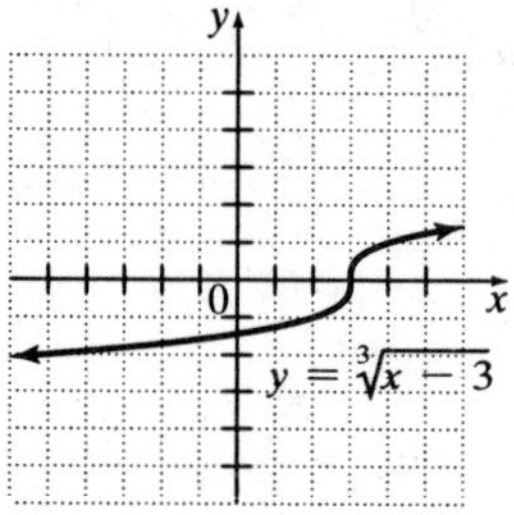

26. $\left(\frac{1}{81}\right)^{1/4} = \frac{1}{81^{1/4}} = \frac{1}{3}$

27. $\left(-\frac{1}{27}\right)^{1/3} = \frac{1}{(-27)^{1/3}} = \frac{1}{-3} = -\frac{1}{3}$

28. $(-27)^{-1/3} = \frac{1}{(-27)^{1/3}} = \frac{1}{-3} = -\frac{1}{3}$

29. $(-64)^{-1/3} = \frac{1}{(-64)^{1/3}} = \frac{1}{-4} = -\frac{1}{4}$

30. $-9^{3/2} = -(9^{1/2})^3 = -3^3 = -27$

31. $64^{-1/3} = \frac{1}{(64)^{1/3}} = \frac{1}{4}$

32. $(-25)^{5/2} = [(-25)^{1/2}]^5$ is not a real number, since there is no real number whose square is negative.

33.
$$\left(\frac{25}{49}\right)^{-3/2} = \frac{1}{\left(\frac{25}{49}\right)^{3/2}} = \frac{1}{\left(\left(\frac{25}{49}\right)^{1/2}\right)^3} = \frac{1}{\left(\frac{5}{7}\right)^3} = \frac{1}{\frac{125}{343}} = \frac{343}{125}$$

34.
$$\left(\frac{8}{27}\right)^{-2/3} = \frac{1}{\left(\frac{8}{27}\right)^{2/3}} = \frac{1}{\left[\left(\frac{8}{27}\right)^{1/3}\right]^2} = \frac{1}{\left(\frac{2}{3}\right)^2} = \frac{1}{\frac{4}{9}} = \frac{9}{4}$$

35. $\left(-\frac{1}{36}\right)^{-1/4} = \frac{1}{\left(-\frac{1}{36}\right)^{1/4}}$ is not a real number, since there is no real number whose 4th power is negative.

36. $\sqrt[3]{x^2} = x^{2/3}$

37.
$$\sqrt[5]{5x^2y^3} = (5x^2y^3)^{1/5} = 5^{1/5}(x^2)^{1/5}(y^3)^{1/5} = 5^{1/5}x^{2/5}y^{3/5}$$

38. $y^{4/5} = \sqrt[5]{y^4}$

39. $5(xy^2z^5)^{1/3} = 5\sqrt[3]{xy^2z^5}$

40. $(x+2y)^{-1/2} = \dfrac{1}{(x+2y)^{1/2}} = \dfrac{1}{\sqrt{x+2y}}$

41. $a^{1/3}a^{4/3}a^{1/2} = a^{(1/3+4/3+1/2)}$
$= a^{(5/3+1/2)}$
$= a^{(10/6+3/6)}$
$= a^{13/6}$

42. $\dfrac{b^{1/3}}{b^{4/3}} = b^{1/3-4/3} = b^{-1} = \dfrac{1}{b}$

43. $(a^{1/2}a^{-2})^3 = (a^{1/2-2})^3$
$= (a^{-3/2})^3$
$= a^{-9/2}$
$= \dfrac{1}{a^{9/2}}$

44.
$(x^{-3}y^6)^{1/3} = (x^{-3})^{1/3}(y^6)^{1/3}$
$= x^{-1}y^2 = \dfrac{y^2}{x}$

45. $\left(\dfrac{b^{3/4}}{a^{-1/2}}\right)^8 = \dfrac{(b^{3/4})^8}{(a^{-1/2})^8} = \dfrac{b^6}{a^{-4}} = a^4b^6$

46. $\dfrac{x^{1/4}x^{-1/2}}{x^{2/3}} = \dfrac{x^{1/4-1/2}}{x^{2/3}}$
$= \dfrac{x^{-1/4}}{x^{2/3}}$
$= \dfrac{1}{x^{2/3-(-1/4)}}$
$= \dfrac{1}{x^{8/12+3/12}}$
$= \dfrac{1}{x^{11/12}}$

47. $\left(\dfrac{49c^{5/3}}{a^{-1/4}b^{5/6}}\right)^{-1} = \dfrac{49^{-1}c^{-5/3}}{a^{1/4}b^{-5/6}}$
$= \dfrac{b^{5/6}}{49a^{1/4}c^{5/3}}$

48. $a^{-1/4}(a^{5/4} - a^{9/4})$
$= a^{-1/4+5/4} - a^{-1/4+9/4}$
$= a - a^2$

49. $\sqrt{20} \approx 4.472$

50. $\sqrt[3]{-39} \approx -3.391$

51. $\sqrt[4]{726} \approx 5.191$

52. $56^{1/3} \approx 3.826$

53. $-78^{3/4} \approx -26.246$

54. $105^{-2/3} \approx 0.045$

55. $\sqrt[3]{2} \cdot \sqrt{7} = 2^{1/3} \cdot 7^{1/2}$
$= 2^{2/6} \cdot 7^{3/6}$
$= (2^2)^{1/6} \cdot (7^3)^{1/6}$
$= 4^{1/6} \cdot 343^{1/6}$
$= (4 \cdot 343)^{1/6}$
$= \sqrt[6]{1372}$

56. $\sqrt[3]{3} \cdot \sqrt[4]{x} = 3^{1/3} \cdot x^{1/4}$
$= 3^{4/12} \cdot x^{3/12}$
$= (3^4)^{1/12}(x^3)^{1/12}$
$= (81)^{1/12}(x^3)^{1/12}$
$= (81x^3)^{1/12}$
$= \sqrt[12]{81x^3}$

57. $\sqrt{3} \cdot \sqrt{8} = \sqrt{24} = \sqrt{4 \cdot 6} = 2\sqrt{6}$

58. $\sqrt[3]{7y} \cdot \sqrt[3]{x^2z} = \sqrt[3]{7x^2yz}$

59. $\dfrac{\sqrt{44x^3}}{\sqrt{11x}} = \sqrt{\dfrac{44x^3}{11x}} = \sqrt{4x^2} = 2x$

60. $\dfrac{\sqrt[4]{a^6b^{13}}}{\sqrt[4]{a^2b}} = \sqrt[4]{\dfrac{a^6b^{13}}{a^2b}}$
$= \sqrt[4]{a^{6-2}b^{13-1}}$
$= \sqrt[4]{a^4b^{12}}$
$= ab^3$

61. $\sqrt{60} = \sqrt{4 \cdot 15} = 2\sqrt{15}$

62. $-\sqrt{75} = -\sqrt{25 \cdot 3} = -5\sqrt{3}$

63. $\sqrt[3]{162} = \sqrt[3]{27 \cdot 6} = 3\sqrt[3]{6}$

64. $\sqrt[3]{-32} = \sqrt[3]{(-8)(4)} = -2\sqrt[3]{4}$

65. $\sqrt{36x^7} = \sqrt{36x^6 \cdot x} = 6x^3\sqrt{x}$

66. $\sqrt[3]{24a^5b^7} = \sqrt[3]{8a^3b^6 \cdot 3a^2b} = 2ab^2\sqrt[3]{3a^2b}$

67. $\sqrt{\dfrac{p^{17}}{121}} = \dfrac{\sqrt{p^{16} \cdot p}}{\sqrt{121}} = \dfrac{p^8\sqrt{p}}{11}$

68. $\sqrt[3]{\dfrac{y^5}{27x^6}} = \dfrac{\sqrt[3]{y^3 \cdot y^2}}{\sqrt[3]{27x^6}} = \dfrac{y\sqrt[3]{y^2}}{3x^2}$

69. $\sqrt[4]{\dfrac{xy^6}{81}} = \dfrac{\sqrt[4]{y^4 \cdot xy^2}}{\sqrt[4]{81}} = \dfrac{y\sqrt[4]{xy^2}}{3}$

70. $\sqrt{\dfrac{2x^3}{49y^4}} = \dfrac{\sqrt{x^2 \cdot 2x}}{\sqrt{49y^4}} = \dfrac{x\sqrt{2x}}{7y^2}$

71. $r = \sqrt{\dfrac{A}{\pi}}$

a. $r = \sqrt{\dfrac{25}{\pi}} = \dfrac{\sqrt{25}}{\sqrt{\pi}} = \dfrac{5}{\sqrt{\pi}}$

$\dfrac{5}{\sqrt{\pi}}$ meters or $\dfrac{5\sqrt{\pi}}{\pi}$ meters

b. $r = \sqrt{\dfrac{104}{\pi}} \approx 5.75$ inches

72. $x\sqrt{75xy} - \sqrt{27x^3y}$
$= x\sqrt{25}\sqrt{3xy} - \sqrt{9x^2}\sqrt{3xy}$
$= 5x\sqrt{3xy} - 3x\sqrt{3xy}$
$= 2x\sqrt{3xy}$

73. $2\sqrt{32x^2y^3} - xy\sqrt{98y}$
$= 2\sqrt{16x^2y^2 \cdot 2y} - xy\sqrt{49 \cdot 2y}$
$= 2\sqrt{16x^2y^2}\sqrt{2y} - xy\sqrt{49}\sqrt{2y}$
$= 8xy\sqrt{2y} - 7xy\sqrt{2y}$
$= xy\sqrt{2y}$

74. $\sqrt[3]{128} + \sqrt[3]{250} = \sqrt[3]{64 \cdot 2} + \sqrt[3]{125 \cdot 2}$
$= \sqrt[3]{64}\sqrt[3]{2} + \sqrt[3]{125}\sqrt[3]{2}$
$= 4\sqrt[3]{2} + 5\sqrt[3]{2}$
$= 9\sqrt[3]{2}$

75. $3\sqrt[4]{32a^5} - a\sqrt[4]{162a}$
$= 3\sqrt[4]{16a^4 \cdot 2a} - a\sqrt[4]{81 \cdot 2a}$
$= 3\sqrt[4]{16a^4}\sqrt[4]{2a} - a\sqrt[4]{81}\sqrt[4]{2a}$
$= 6a\sqrt[4]{2a} - 3a\sqrt[4]{2a}$
$= 3a\sqrt[4]{2a}$

76. $\dfrac{5}{\sqrt{4}} + \dfrac{\sqrt{3}}{3} = \dfrac{5}{2} + \dfrac{\sqrt{3}}{3}$
$= \dfrac{5(3) + 2\sqrt{3}}{2 \cdot 3} = \dfrac{15 + 2\sqrt{3}}{6}$

77. $\sqrt{\dfrac{8}{x^2}} - \sqrt{\dfrac{50}{16x^2}} = \dfrac{2\sqrt{2}}{x} - \dfrac{5\sqrt{2}}{4x}$
$= \dfrac{8\sqrt{2} - 5\sqrt{2}}{4x}$
$= \dfrac{3\sqrt{2}}{4x}$

78. $2\sqrt{50}-3\sqrt{125}+\sqrt{98}$
$=2\sqrt{25}\sqrt{2}-3\sqrt{25}\sqrt{5}+\sqrt{49}\sqrt{2}$
$=2(5)\sqrt{2}-3(5)\sqrt{5}+7\sqrt{2}$
$=10\sqrt{2}-15\sqrt{5}+7\sqrt{2}$
$=17\sqrt{2}-15\sqrt{5}$

79. $2a\sqrt[4]{32b^5}-3b\sqrt[4]{162a^4b}+\sqrt[4]{2a^4b^5}$
$=2a\sqrt[4]{16b^4}\sqrt[4]{2b}-3b\sqrt[4]{81a^4}\sqrt[4]{2b}$
$+\sqrt[4]{a^4b^4}\sqrt[4]{2b}$
$=[(2a)(2b)-(3b)(3a)+ab]\sqrt[4]{2b}$
$=(4ab-9ab+ab)\sqrt[4]{2b}$
$=-4ab\sqrt[4]{2b}$

80. $\sqrt{3}(\sqrt{27}-\sqrt{3})=\sqrt{3}(\sqrt{9}\sqrt{3}-\sqrt{3})$
$=\sqrt{3}(3\sqrt{3}-\sqrt{3})$
$=\sqrt{3}\left(2\sqrt{3}\right)$
$=2(3)$
$=6$

81.
$(\sqrt{x}-3)^2=\left(\sqrt{x}\right)^2-2(3)\sqrt{x}+3^2$
$=x-6\sqrt{x}+9$

82. $(\sqrt{5}-5)(2\sqrt{5}+2)$
$=2\sqrt{25}+2\sqrt{5}-10\sqrt{5}-10$
$=10-8\sqrt{5}-10$
$=-8\sqrt{5}$

83. $(2\sqrt{x}-3\sqrt{y})(2\sqrt{x}+3\sqrt{y})$
$=(2\sqrt{x})^2-(3\sqrt{y})^2$
$=2^2\sqrt{x^2}-3^2\sqrt{y^2}$
$=4x-9y$

84. $(\sqrt{a}+3)(\sqrt{a}-3)=\left(\sqrt{a}\right)^2-3^2$
$=a-9$

85. $\left(\sqrt[3]{a}+2\right)^2=(\sqrt[3]{a})^2+2(2)\sqrt[3]{a}+2^2$
$=\sqrt[3]{a^2}+4\sqrt[3]{a}+4$

86. $(\sqrt[3]{5x}+9)(\sqrt[3]{5x}-9)=(\sqrt[3]{5x})^2-9^2$
$=\sqrt[3]{25x^2}-81$

87. $(\sqrt[3]{a}+4)(\sqrt[3]{a^2}-4\sqrt[3]{a}+16)=(\sqrt[3]{a})^3+4^3$
$=a+64$

88. $\frac{3}{\sqrt{7}}=\frac{3\sqrt{7}}{\sqrt{7}\cdot\sqrt{7}}=\frac{3\sqrt{7}}{7}$

89. $\sqrt{\frac{x}{12}}=\frac{\sqrt{x}}{\sqrt{12}}$
$=\frac{\sqrt{x}\cdot\sqrt{12}}{\sqrt{12}\cdot\sqrt{12}}$
$=\frac{\sqrt{12x}}{12}$
$=\frac{2\sqrt{3x}}{12}$
$=\frac{\sqrt{3x}}{6}$

90. $\frac{5}{\sqrt[3]{4}}=\frac{5\cdot\sqrt[3]{16}}{\sqrt[3]{4}\cdot\sqrt[3]{16}}$
$=\frac{5\sqrt[3]{8\cdot 2}}{\sqrt[3]{64}}$
$=\frac{10\sqrt[3]{2}}{4}$
$=\frac{5\sqrt[3]{2}}{2}$

91. $\sqrt{\frac{24x^5}{3y^2}}=\sqrt{\frac{8x^5}{3y^2}}$
$=\frac{\sqrt{4x^4\cdot 2x}}{\sqrt{y^2}}$
$=\frac{2x^2\sqrt{2x}}{y}$

92. $\sqrt[3]{\frac{15x^6y^7}{z^2}} = \frac{\sqrt[3]{15x^6y^7}}{\sqrt[3]{z^2}}$

$= \frac{\sqrt[3]{x^6y^6 \cdot 15y}\sqrt[3]{z}}{\sqrt[3]{z^2} \cdot \sqrt[3]{z}}$

$= \frac{x^2y^2\sqrt[3]{15yz}}{z}$

93. $\frac{5}{2-\sqrt{7}} = \frac{5}{2-\sqrt{7}} \cdot \frac{2+\sqrt{7}}{2+\sqrt{7}}$

$= \frac{5(2+\sqrt{7})}{2^2-(\sqrt{7})^2}$

$= \frac{10+5\sqrt{7}}{4-7}$

$= \frac{10+5\sqrt{7}}{-3}$

$= -\frac{10-5\sqrt{7}}{3}$

94. $\frac{3}{\sqrt{y}-2} \cdot \frac{\sqrt{y}+2}{\sqrt{y}+2} = \frac{3\sqrt{y}+6}{(\sqrt{y})^2-2^2}$

$= \frac{3\sqrt{y}+6}{y-4}$

95. $\frac{\sqrt{2}-\sqrt{3}}{\sqrt{2}+\sqrt{3}} = \frac{\sqrt{2}-\sqrt{3}}{\sqrt{2}+\sqrt{3}} \cdot \frac{\sqrt{2}-\sqrt{3}}{\sqrt{2}-\sqrt{3}}$

$= \frac{(\sqrt{2})^2-2\sqrt{2}\sqrt{3}+(\sqrt{3})^2}{(\sqrt{2})^2-(\sqrt{3})^2}$

$= \frac{2-2\sqrt{6}+3}{2-3}$

$= \frac{5-2\sqrt{6}}{-1}$

$= -5+2\sqrt{6}$

96. $\frac{\sqrt{11}}{3} = \frac{\sqrt{11} \cdot \sqrt{11}}{3 \cdot \sqrt{11}} = \frac{11}{3\sqrt{11}}$

97.

$\sqrt{\frac{18}{y}} = \frac{\sqrt{18}}{\sqrt{y}} = \frac{3\sqrt{2}}{\sqrt{y}}$

$= \frac{3\sqrt{2} \cdot \sqrt{2}}{\sqrt{y} \cdot \sqrt{2}} = \frac{3 \cdot 2}{\sqrt{2y}} = \frac{6}{\sqrt{2y}}$

98. $\frac{\sqrt[3]{9}}{7} = \frac{\sqrt[3]{9} \cdot \sqrt[3]{3}}{7 \cdot \sqrt[3]{3}} = \frac{3}{7\sqrt[3]{3}}$

99. $\sqrt{\frac{24x^5}{3y^2}} = \frac{\sqrt{4x^4 \cdot 6x}}{\sqrt{y^2 \cdot 3}}$

$= \frac{2x^2\sqrt{6x}}{y\sqrt{3}}$

$= \frac{2x^2\sqrt{6x} \cdot \sqrt{6x}}{y\sqrt{3} \cdot \sqrt{6x}}$

$= \frac{2x^2(6x)}{y\sqrt{18x}}$

$= \frac{12x^3}{3y\sqrt{2x}}$

$= \frac{4x^3}{y\sqrt{2x}}$

100. $\sqrt[3]{\frac{xy^2}{10z}} = \frac{\sqrt[3]{xy^2} \cdot \sqrt[3]{x^2y}}{\sqrt[3]{10z} \cdot \sqrt[3]{x^2y}} = \frac{xy}{\sqrt[3]{10x^2yz}}$

101. $\frac{\sqrt{x}+5}{-3} = \frac{(\sqrt{x}+5)(\sqrt{x}-5)}{-3(\sqrt{x}-5)} = \frac{x-25}{-3\sqrt{x}+15}$

102. $\sqrt{y-7} = 5$

$y-7 = 25$

$y = 32$

The solution is 32,

103. $\sqrt{2x}+1 = -4$

$\sqrt{2x} = -5$

No solution exists since the principle square root of a number is not negative.

104. $\sqrt[3]{2x-6}=4$
$(\sqrt[3]{2x-6})^3=(4)^3$
$2x-6=64$
$2x=70$
$x=35$
The solution is 35.

105. $\sqrt{x+6}=\sqrt{x+2}$
$x+6=x+2$
$6=2$
Since the last equation is never satisfied, there is no solution.

106. $2x-5\sqrt{x}=3$
$2x-3=5\sqrt{x}$
$(2x-3)^2=\left(5\sqrt{x}\right)^2$
$4x^2-12x+9=25x$
$4x^2-37x+9=0$
$(4x-1)(x-9)=0$
$4x-1=0$ or $x-9=0$
$4x=1$ or $x=9$
$x=\frac{1}{4}$

When $x=\frac{1}{4}$,
$2x-5\sqrt{x}=2\left(\frac{1}{4}\right)-5\sqrt{\frac{1}{4}}$
$=\frac{1}{2}-\frac{5}{2}$
$=-2\neq 3.$
When $x=9$,
$2x-5\sqrt{x}=2(9)-5\sqrt{9}$
$=18-5$
$=3.$
The solution is 3.

107. $\sqrt{x+9}=2+\sqrt{x-7}$
$x+9=4+4\sqrt{x-7}+x-7$
$12=4\sqrt{x-7}$
$3=\sqrt{x-7}$
$9=x-7$
$16=x$
The solution is 16.

108. Let c = the length of the hypotenuse of the right triangle. By the Pythagorean theorem,
$c^2=3^2+3^2=9+9=9(2)$
so $c=\sqrt{9(2)}=\sqrt{9}\sqrt{2}=3\sqrt{2}$

109. Let c = the length of the hypotenuse of the right triangle. By the Pythagorean theorem,
so $c^2=7^2+(8\sqrt{3})^2.$
$=49+192$
$=241$
$=\sqrt{241}$

110. $c^2=a^2+b^2$
$(65)^2=(40)^2+b^2$
$4225-1600=b^2$
$2625=b^2$
$\pm\sqrt{2625}=b$
$\pm 51.235=b$
51.2 feet

111. $c^2=3^2+3^2$
$c^2=9+9$
$c^2=18$
$c=\pm\sqrt{18}$
$c=\pm 4.243$
4.24 feet

112. $\sqrt{-8}=\sqrt{4}\sqrt{2}\sqrt{-1}=2i\sqrt{2}$

113. $-\sqrt{-6}=-\sqrt{6}\sqrt{-1}=-i\sqrt{6}$

114. $\sqrt{-4}+\sqrt{-16}=2i+4i=6i$

115. $\sqrt{-2}\sqrt{-5}=i\sqrt{2}\cdot i\sqrt{5}=i^2\sqrt{2\cdot 5}=-\sqrt{10}$

116. $(12-6i)+(3+2i)=(1+3)+(-6+2)i$
$=15-4i$

117. $(-8-7i)-(5-4i)=(-8-5)+(-7+4)i$
$=-13-3i$

118. $(\sqrt{3}+\sqrt{2})+(3\sqrt{2}-\sqrt{-8})$
$=\sqrt{3}+\sqrt{2}+3\sqrt{2}-\sqrt{4}\sqrt{2}\sqrt{-1}$
$=(\sqrt{3}+4\sqrt{2})-2i\sqrt{2}$

119. $2i(2-5i)=4i-10i^2$
$=4i-10(-1)$
$=10+4i$

120. $-3i(6-4i)=-18i+12i^2$
$=-18i+12(-1)$
$=-12-18i$

121. $(3+2i)(1+i)=3+2i+3i+2i^2$
$=3+5i+2(-1)$
$=(3-2)+5i$
$=1+5i$

122. $(2-3i)^2=(2-3i)(2-3i)$
$=(2)^2+2(2)(-3i)+(-3i)^2$
$=4-12i+9i^2$
$=4-12i-9$
$=-5-12i$

123. $(\sqrt{6}-9i)(\sqrt{6}+9i)=(\sqrt{6})^2-(9i)^2$
$=6-81i^2$
$=6+81$
$=87$

124. $\dfrac{2+3i}{2i}=\dfrac{(2+3i)i}{2i^2}$
$=\dfrac{2i+3i^2}{2(-1)}$
$=\dfrac{2i+3(-1)}{-2}$
$=\dfrac{-3+2i}{-2}$
$=\dfrac{3}{2}-i$

125. $\dfrac{1+i}{-3i}\cdot\dfrac{(3i)}{(3i)}=\dfrac{3i+3i^2}{-9i^2}$
$=\dfrac{3i+3(-1)}{-9(-1)}$
$=\dfrac{3(-1+i)}{9}$
$=\dfrac{-1+i}{3}$
$=-\dfrac{1}{3}+\dfrac{1}{3}i$

Chapter 7 Test

1. $\sqrt{216}=\sqrt{36\cdot 6}=6\sqrt{6}$

2. $-\sqrt[4]{x^{64}}=-(x^{64})^{1/4}=-x^{16}$

3. $\left(\dfrac{1}{125}\right)^{1/3}=\dfrac{1}{5}$

4. $\left[\dfrac{1}{125}\right]^{-1/3}=\left[\left(\dfrac{1}{125}\right)^{-1}\right]^{1/3}=125^{1/3}=5$

5. $\left[\dfrac{8x^3}{27}\right]^{2/3}=\dfrac{8^{2/3}(x^3)^{2/3}}{27^{2/3}}$
$=\dfrac{(8^{1/3})^2x^2}{(27^{1/3})^2}$
$=\dfrac{2^2x^2}{3^2}$
$=\dfrac{4x^2}{9}$

6. $\sqrt[3]{-a^{18}b^9}=\sqrt[3]{-1}\sqrt[3]{a^{18}}\sqrt[3]{b^9}$
$=(-1)a^6b^3$
$=-a^6b^3$

7. $\left[\dfrac{64c^{4/3}}{a^{-2/3}b^{5/6}}\right]^{1/2}$

$=\left[\dfrac{64a^{2/3}c^{4/3}}{b^{5/6}}\right]^{1/2}$

$=\dfrac{64^{1/2}\left(a^{2/3}\right)^{1/2}\left(c^{4/3}\right)^{1/2}}{\left(b^{5/6}\right)^{1/2}}$

$=\dfrac{8a^{1/3}c^{2/3}}{b^{5/12}}$

8. $a^{-2/3}(a^{5/4}-a^3)$

$=a^{-2/3+5/4}-a^{-2/3+3}$

$=a^{(-8+15/12)}-a^{(-2+9/3)}$

$=a^{7/12}-a^{7/3}$

9. $\sqrt[4]{(4xy)^4}=|4xy|$

10. $\sqrt[3]{(-27)^3}=-27$

11. $\sqrt{\dfrac{9}{y}}=\dfrac{\sqrt{9}}{\sqrt{y}}=\dfrac{3}{\sqrt{y}}\cdot\dfrac{\sqrt{y}}{\sqrt{y}}=\dfrac{3\sqrt{y}}{y}$

12. $\dfrac{4-\sqrt{x}}{4+2\sqrt{x}}=\dfrac{(4-\sqrt{x})(2-\sqrt{x})}{2(2+\sqrt{x})(2-\sqrt{x})}$

$=\dfrac{8-2\sqrt{x}-4\sqrt{x}+x}{2(2^2-\sqrt{x}^2)}$

$=\dfrac{8-6\sqrt{x}+x}{2(4-x)}$

13. $\dfrac{\sqrt[3]{ab}}{\sqrt[3]{ab^2}}=\sqrt[3]{\dfrac{ab}{ab^2}}$

$=\sqrt[3]{\dfrac{1}{b^1}}$

$=\dfrac{1}{\sqrt[3]{b}}$

$=\dfrac{\sqrt[3]{b^2}}{\sqrt[3]{b^3}}$

$=\dfrac{\sqrt[3]{b^2}}{b}$

14. $\dfrac{\sqrt{6}+x}{8}=\dfrac{(\sqrt{6}+x)(\sqrt{6}-x)}{8(\sqrt{6}-x)}$

$=\dfrac{(\sqrt{6})^2-x^2}{8(\sqrt{6}-x)}$

$=\dfrac{6-x^2}{8(\sqrt{6}-x)}$

15. $\sqrt{125x^3}-3\sqrt{20x^3}$

$=\sqrt{25x^2}\sqrt{5x}-3\sqrt{4x^2}\sqrt{5x}$

$=(5x-3(2x))\sqrt{5x}$

$=(5x-6x)\sqrt{5x}$

$=-x\sqrt{5x}$

16. $\sqrt{3}(\sqrt{16}-\sqrt{2})=\sqrt{3}(4-\sqrt{2})$

$=4\sqrt{3}-\sqrt{3}\sqrt{2}$

$=4\sqrt{3}-\sqrt{6}$

17. $(\sqrt{x}+1)^2=\sqrt{x}^2+2\sqrt{x}+1=x+2\sqrt{x}+1$

18. $(\sqrt{2}-4)(\sqrt{3}+1)=\sqrt{2}\sqrt{3}-4\sqrt{3}+\sqrt{2}-4$

$=\sqrt{6}-4\sqrt{3}+\sqrt{2}-4$

19. $(\sqrt{5}+5)(\sqrt{5}-5)=\sqrt{5}^2-5^2=5-25=$
-20

20. $\sqrt{561}=23.685$

21. $386^{-2/3}=0.019$

22.
$$\begin{aligned} x &= \sqrt{x-2}+2 \\ x-2 &= \sqrt{x-2} \\ (x-2)^2 &= x-2 \\ x^2-4x+4 &= x-2 \\ x^2-5x+6 &= 0 \\ (x-2)(x-3) &= 0 \end{aligned}$$
$x-2=0$ or $x-3=0$
$x=2$ or $x=3$
The solutions are 2 and 3.

23. $\sqrt{x^2-7}+3=0$
$\sqrt{x^2-7}=-3$
No solution exists since a principle square root is not negatively valued.

24. $\sqrt{x+5}=\sqrt{2x-1}$
$x+5=2x-1$
$6=x$
$x=6$
The solution is 6.

25. $\sqrt{-2}=i\sqrt{2}$

26. $-\sqrt{-8}=-\sqrt{4}\sqrt{2}\sqrt{-1}=-2i\sqrt{2}$

27. $(12-6i)-(12-3i)=12-6i-12+3i$
$=-3i$

28. $(6-2i)(6+2i)=6^2+2^2=36+4=40$

29.
$$\begin{aligned} (4+3i)^2 &= 16+24i+9i^2 \\ &= 16+24i+9(-1) \\ &= (16-9)+24i \\ &= 7+24i \end{aligned}$$

30.
$$\begin{aligned} \frac{1+4i}{1-i} &= \frac{(1+4i)(1+i)}{(1-i)(1+i)} \\ &= \frac{(1-4)+5i}{2} \\ &= -\frac{3}{2}+\frac{5}{2}i \end{aligned}$$

31. Let x = the length of a leg of the isosceles right triangle. By the Pythagorean theorem,
$$\begin{aligned} x^2+x^2 &= 5^2 \\ 2x^2 &= 25 \\ x^2 &= \frac{25}{2} \end{aligned}$$
$$x=\sqrt{\frac{25}{2}}=\frac{\sqrt{25}}{\sqrt{2}}=\frac{5}{\sqrt{2}}\cdot\frac{\sqrt{2}}{\sqrt{2}}=\frac{5\sqrt{2}}{2}$$

32. $g(x)=\sqrt{x+2}$
Domain: $[-2, \infty)$

x	-2	-1	2	7
$g(x)$	0	1	2	3

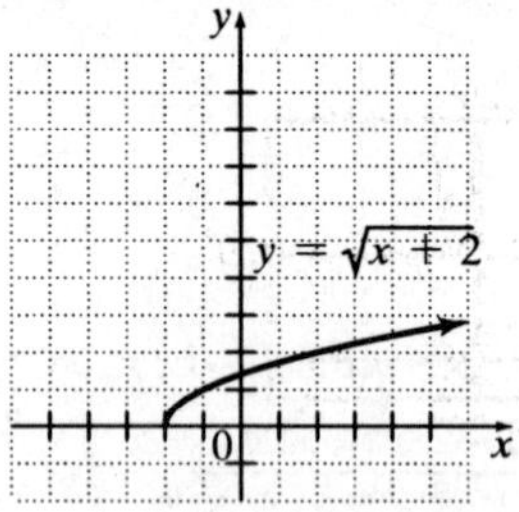

33. $V(300)=\sqrt{2.5(300)}\approx 27$ miles per hour

34. $V(r)=\sqrt{2.5r}$
$30=\sqrt{2.5r}$
$900=2.5r$
$360=r$
The radius of curvature is 360 feet.

Chapter 8

Section 8.1

Graphing Calculator Explorations

1. $-1.27, 6.27$

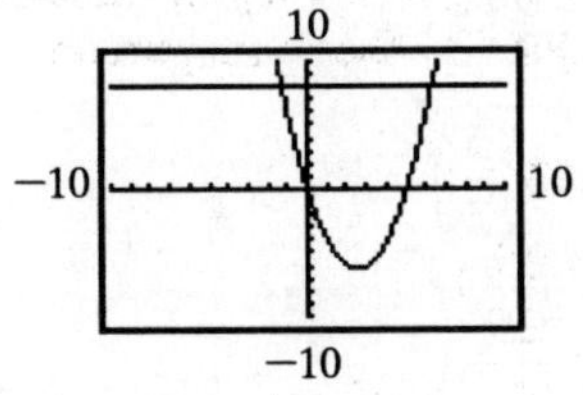

3. $-1.10, 0.90$

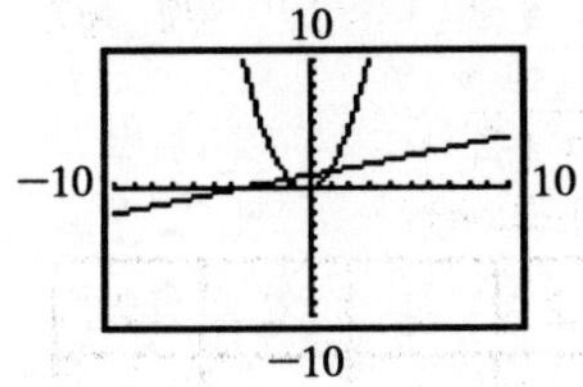

5. $\varnothing$

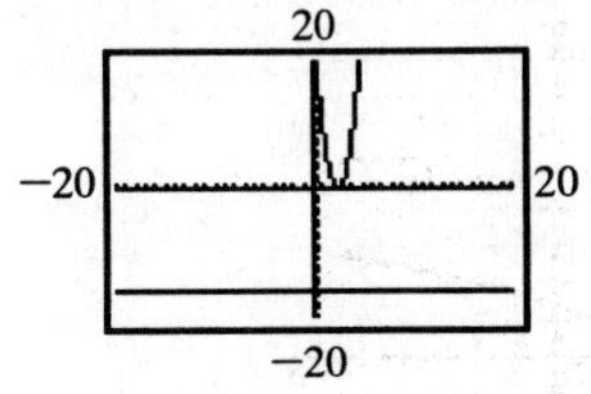

Exercise Set 8.1

1. $x^2 = 16$
   $x = \pm\sqrt{16}$
   $x = \pm 4$

3. $x^2 - 7 = 0$
   $x^2 = 7$
   $x = \pm\sqrt{7}$

5. $x^2 = 18$
   $x = \pm\sqrt{18}$
   $x = \pm\sqrt{9}\sqrt{2}$
   $x = \pm 3\sqrt{2}$

7. $3z^2 - 30 = 0$
   $3z^2 = 30$
   $z^2 = 10$
   $z = \pm\sqrt{10}$

9. $(x+5)^2 = 9$
   $x + 5 = \pm\sqrt{9}$
   $x + 5 = \pm 3$
   $x = -5 \pm 3$
   $x = -8$ or $x = -2$

11. $(z-6)^2 = 18$
    $z - 6 = \pm\sqrt{18}$
    $z - 6 = \pm\sqrt{9}\sqrt{2}$
    $z - 6 = \pm 3\sqrt{2}$
    $z = 6 \pm 3\sqrt{2}$

13. $(2x-3)^2 = 8$
    $2x - 3 = \pm\sqrt{8}$
    $2x - 3 = \pm\sqrt{4}\sqrt{2}$
    $2x - 3 = \pm 2\sqrt{2}$
    $2x = 3 \pm 2\sqrt{2}$
    $x = \dfrac{3 \pm 2\sqrt{2}}{2}$

15. $x^2 + 9 = 0$
    $x^2 = -9$
    $x = \pm\sqrt{-9}$
    $x = \pm 3i$

17. $x^2 - 6 = 0$
    $x^2 = 6$
    $x = \pm\sqrt{6}$

19. $2z^2+16=0$
$2z^2=-16$
$z^2=-8$
$z=\pm\sqrt{-8}$
$z=\pm\sqrt{4}\sqrt{2}\sqrt{-1}$
$z=\pm 2i\sqrt{2}$

21. $(x-1)^2=-16$
$x-1=\pm\sqrt{-16}$
$x-1=\pm 4i$
$x=1\pm 4i$

23. $(z+7)^2=5$
$z+7=\pm\sqrt{5}$
$z=-7\pm\sqrt{5}$

25. $(x+3)^2=-8$
$x+3=\pm\sqrt{-8}$
$x+3=\pm\sqrt{4}\sqrt{2}\sqrt{-1}$
$x+3=\pm 2i\sqrt{2}$
$x=-3\pm 2i\sqrt{2}$

27. $x^2+16x+\left(\frac{16}{2}\right)^2=x^2+16x+64$
$=(x+8)^2$

29. $z^2-12z+\left(\frac{12}{2}\right)^2=z^2-12z+36$
$=(z-6)^2$

31. $p^2+9p+\left(\frac{9}{2}\right)^2=p^2+9p+\frac{81}{4}$
$=\left(p+\frac{9}{2}\right)^2$

33. $x^2+x+\left(\frac{1}{2}\right)^2=x^2+x+\frac{1}{4}$
$=\left(x+\frac{1}{2}\right)^2$

35. $x^2+\_\_+16$
$\pm 2\sqrt{16}=\pm 2(4)=\pm 8$
$x^2-8x+16=(x-4)^2$
$x^2+8x+16=(x+4)^2$
Answer: $\pm 8x$

37. $z^2+\_\_+\frac{25}{4}$
$\pm\sqrt{\frac{25}{4}}=\pm 2\left(\frac{5}{2}\right)=\pm 5$
$z^2-5z+\frac{25}{4}=\left(z-\frac{5}{2}\right)^2$
$z^2+5z+\frac{25}{4}=\left(z+\frac{5}{2}\right)^2$
Answer: $\pm 5z$

39. $x^2+8x=-15$
$x^2+8x+\left(\frac{8}{2}\right)^2=-15+16$
$(x+4)^2=1$
$x+4=\pm\sqrt{1}$
$x=-4\pm 1$
$x=-5$ or $x=-3$

41. $x^2+6x+2=0$
$x^2+6x+\left(\frac{6}{2}\right)^2=-2+9$
$(x+3)^2=7$
$x+3=\pm\sqrt{7}$
$x=-3\pm\sqrt{7}$

43.
$$x^2+x-1=0$$
$$x^2+x+\left(\frac{1}{2}\right)^2=1+\left(\frac{1}{2}\right)^2$$
$$x^2+x+\frac{1}{4}=1+\frac{1}{4}$$
$$\left(x+\frac{1}{2}\right)^2=\frac{5}{4}$$
$$x+\frac{1}{2}=\pm\sqrt{\frac{5}{4}}$$
$$x=-\frac{1}{2}\pm\frac{\sqrt{5}}{2}$$
$$x=\frac{-1\pm\sqrt{5}}{2}$$

45.
$$x^2+2x-5=0$$
$$x^2+2x+\left(\frac{2}{2}\right)^2=5+1$$
$$(x+1)^2=6$$
$$x+1=\pm\sqrt{6}$$
$$x=-1\pm\sqrt{6}$$

47.
$$3p^2-12p+2=0$$
$$3p^2-12p=-2$$
$$p^2-4p=-\frac{2}{3}$$
$$p^2-4p+\left(\frac{4}{2}\right)^2=-\frac{2}{3}+4$$
$$(p-2)^2=\frac{10}{3}$$
$$p-2=\pm\sqrt{\frac{10}{3}}$$
$$p-2=\pm\sqrt{\frac{10\cdot 3}{3^2}}$$
$$p-2=\pm\frac{\sqrt{30}}{3}$$
$$p=2\pm\frac{\sqrt{30}}{3}$$
$$p=\frac{6\pm\sqrt{30}}{3}$$

49.
$$4y^2-12y-2=0$$
$$4y^2-12y=2$$
$$y^2-3y=\frac{1}{2}$$
$$y^2-3y+\left(\frac{3}{2}\right)^2=\frac{1}{2}+\frac{9}{4}$$
$$\left(y-\frac{3}{2}\right)^2=\frac{11}{4}$$
$$y-\frac{3}{2}=\pm\sqrt{\frac{11}{4}}$$
$$y-\frac{3}{2}=\pm\frac{\sqrt{11}}{2}$$
$$y=\frac{3}{2}\pm\frac{\sqrt{11}}{2}$$

51.
$$2x^2+7x=4$$
$$x^2+\frac{7}{2}x=2$$
$$x^2+\frac{7}{2}x+\left(\frac{\frac{7}{2}}{2}\right)^2=2+\frac{49}{16}$$
$$\left(x+\frac{7}{4}\right)^2=\frac{81}{16}$$
$$x+\frac{7}{4}=\pm\sqrt{\frac{81}{16}}$$
$$x=-\frac{7}{4}\pm\frac{9}{4}$$
$$x=-4,\ \frac{1}{2}$$

53.
$$x^2 - 4x - 5 = 0$$
$$x^2 - 4x = 5$$
$$x^2 - 4x + \left(\frac{4}{2}\right)^2 = 5 + 4$$
$$(x-2)^2 = 9$$
$$x - 2 = \pm\sqrt{9}$$
$$x - 2 = \pm 3$$
$$x = 2 \pm 3$$
$$x = -1,\ 5$$

55.
$$x^2 + 8x + 1 = 0$$
$$x^2 + 8x = -1$$
$$x^2 + 8x + \left(\frac{8}{2}\right)^2 = -1 + 16$$
$$(x+4)^2 = 15$$
$$x + 4 = \pm\sqrt{15}$$
$$x = -4 \pm \sqrt{15}$$

57.
$$3y^2 + 6y - 4 = 0$$
$$3y^2 + 6y = 4$$
$$y^2 + 2y = \frac{4}{3}$$
$$y^2 + 2y + \left(\frac{2}{2}\right)^2 = \frac{4}{3} + 1$$
$$(y+1)^2 = \frac{7}{3}$$
$$y + 1 = \pm\sqrt{\frac{7}{3}}$$
$$y + 1 = \pm\frac{\sqrt{7}}{\sqrt{3}} \cdot \frac{\sqrt{3}}{\sqrt{3}}$$
$$y + 1 = \pm\frac{\sqrt{21}}{3}$$
$$y = -1 \pm \frac{\sqrt{21}}{3}$$

59.
$$2x^2 - 3x - 5 = 0$$
$$2x^2 - 3x = 5$$
$$x^2 - \frac{3}{2}x = \frac{5}{2}$$
$$x^2 - \frac{3}{2}x + \left(\frac{\frac{3}{2}}{2}\right)^2 = \frac{5}{2} + \frac{9}{16}$$
$$\left(x - \frac{3}{4}\right)^2 = \frac{49}{16}$$
$$x - \frac{3}{4} = \pm\frac{7}{4}$$
$$x = \frac{3}{4} \pm \frac{7}{4}$$
$$x = -1,\ \frac{5}{2}$$

61.
$$y^2 + 2y + 2 = 0$$
$$y^2 + 2y + \left(\frac{2}{2}\right)^2 = -2 + 1$$
$$(y+1)^2 = -1$$
$$y + 1 = \pm\sqrt{-1}$$
$$y + 1 = \pm i$$
$$y = -1 \pm i$$

63.
$$x^2 - 6x + 3 = 0$$
$$x^2 - 6x + \left(\frac{6}{2}\right)^2 = -3 + 9$$
$$(x-3)^2 = 6$$
$$x - 3 = \pm\sqrt{6}$$
$$x = 3 \pm \sqrt{6}$$

65.

$$2a^2 + 8a = -12$$
$$a^2 + 4a = -6$$
$$a^2 + 4a = -6$$
$$a^2 + 4a + \left(\frac{4}{2}\right)^2 = -6 + 4$$
$$(a+2)^2 = -2$$
$$a + 2 = \pm\sqrt{-2}$$
$$a + 2 = \pm i\sqrt{2}$$
$$a = -2 \pm i\sqrt{2}$$

67.

$$5x^2 + 15x - 1 = 0$$
$$5x^2 + 15x = 1$$
$$x^2 + 3x = \frac{1}{5}$$
$$x^2 + 3x + \left(\frac{3}{2}\right)^2 = \frac{1}{5} + \frac{9}{4}$$
$$\left(x + \frac{3}{2}\right)^2 = \frac{49}{20}$$
$$x + \frac{3}{2} = \pm\sqrt{\frac{49}{20}}$$
$$= \frac{\pm\sqrt{49}}{\sqrt{4}\sqrt{5}}$$
$$= \pm\frac{7}{2\sqrt{5}}$$
$$= \pm\frac{7\sqrt{5}}{2\cdot\sqrt{5^2}}$$
$$= \pm\frac{7\sqrt{5}}{2\cdot 5}$$
$$x + \frac{3}{2} = \pm\frac{7\sqrt{5}}{10} \text{ so}$$
$$x = -\frac{3}{2} \pm \frac{7\sqrt{5}}{10}$$

69.

$$2x^2 - x + 6 = 0$$
$$2x^2 - x = -6$$
$$x^2 - \frac{1}{2}x = -3$$
$$x^2 - \frac{1}{2}x + \left(\frac{\frac{1}{2}}{2}\right)^2 = -3 + \frac{1}{16}$$
$$\left(x - \frac{1}{4}\right)^2 = -\frac{47}{16}$$
$$x - \frac{1}{4} = \pm\sqrt{-\frac{47}{16}}$$
$$= \pm\frac{\sqrt{47}\sqrt{-1}}{\sqrt{16}}$$
$$= \pm\frac{i\sqrt{47}}{4}$$
$$x = \frac{1 \pm i\sqrt{47}}{4}$$

71.

$$x^2 + 10x + 28 = 0$$
$$x^2 + 10x = -28$$
$$x^2 + 10x + \left(\frac{10}{2}\right)^2 = -28 + 25$$
$$(x+5)^2 = -3$$
$$x + 5 = \pm\sqrt{-3}$$
$$x + 5 = \pm i\sqrt{3}$$
$$x = -5 \pm i\sqrt{3}$$

73.

$$z^2 + 3z - 4 = 0$$
$$z^2 + 3z = 4$$
$$z^2 + 3z + \left(\frac{3}{2}\right)^2 = 4 + \frac{9}{4}$$
$$\left(z + \frac{3}{2}\right)^2 = \frac{25}{4}$$
$$z + \frac{3}{2} = \pm\sqrt{\frac{25}{4}}$$
$$z + \frac{3}{2} = \pm\frac{5}{2}$$
$$z = -\frac{3}{2} \pm \frac{5}{2}$$
$$z = -4 \text{ or } z = 1$$

75.
$$2x^2 - 4x + 3 = 0$$
$$2x^2 - 4x = -3$$
$$x^2 - 2x = -\frac{3}{2}$$
$$x^2 - 2x + \left(\frac{2}{2}\right)^2 = -\frac{3}{2} + 1$$
$$(x-1)^2 = -\frac{1}{2}$$
$$x - 1 = \pm\sqrt{-\frac{1}{2}}$$
$$x - 1 = \pm\frac{\sqrt{-1}}{\sqrt{2}}$$
$$x - 1 = \pm\frac{i}{\sqrt{2}}\cdot\frac{\sqrt{2}}{\sqrt{2}}$$
$$x - 1 = \pm\frac{\sqrt{2}}{2}i$$
$$x = 1 \pm \frac{\sqrt{2}}{2}i$$
$$x = \frac{2 \pm i\sqrt{2}}{2}$$

77.
$$3x^2 + 3x = 5$$
$$x^2 + x = \frac{5}{3}$$
$$x^2 + x + \left(\frac{1}{2}\right)^2 = \frac{5}{3} + \frac{1}{4}$$
$$\left(x + \frac{1}{2}\right)^2 = \frac{23}{12}$$
$$x + \frac{1}{2} = \pm\sqrt{\frac{23}{12}}$$
$$= \pm\frac{\sqrt{23}}{2\sqrt{3}}$$
$$= \pm\frac{\sqrt{23}\sqrt{3}}{2\sqrt{3^2}}$$
$$= \pm\frac{\sqrt{69}}{2\cdot 3}$$
$$x + \frac{1}{2} = \pm\frac{\sqrt{69}}{6}$$
$$x = -\frac{1}{2} \pm \frac{\sqrt{69}}{6}$$
$$x = \frac{-3 \pm \sqrt{69}}{6}$$

79.
$$A = P(1+r)^t$$
$$4320 = 3000(1+r)^2$$
$$\frac{4320}{3000} = (1+r)^2$$
$$1.44 = (1+r)^2$$
$$\pm\sqrt{1.44} = 1 + r$$
$$\pm 1.2 = 1 + r$$
$$-1 \pm 1.2 = r$$
$r = -1 + 1.2$ or $r = -1 - 1.2$
$r = 0.2$ or $r = -2.2$

Rate cannot be negative, so the rate is $r = 0.2 = 20\%$.

81. $A = P(1+r)^t$

$1000 = 810(1+r)^2$

$\frac{1000}{810} = (1+r)^2$

$\frac{100}{81} = (1+r)^2$

$\pm\sqrt{\frac{100}{81}} = 1+r$

$\pm\frac{10}{9} = 1+r$

$-1\pm\frac{10}{9} = r$

$r = -1+\frac{10}{9}$ or $r = -1-\frac{10}{9}$

$r = \frac{1}{9}$ or $r = -\frac{19}{9}$

Rate cannot be negative, so

$r = \frac{1}{9}$, or about 11%.

83. Answers may vary.

85. Simple

87. $s(t) = 16t^2$

$1053 = 16t^2$

$t = \pm\sqrt{\frac{1053}{16}}$

$t = 8.11$ or -8.11 (disregard)

It would take 8.11 seconds.

89. $s(t) = 16t^2$

$725 = 16t^2$

$t = \pm\sqrt{\frac{725}{16}}$

$t = 6.73$ or -6.73 (disregard)

It would take 6.73 seconds.

91. $A = \pi r^2$

$\frac{36\pi}{\pi} = \frac{\pi r^2}{\pi}$

$\sqrt{36} = \sqrt{r^2}$

$\pm 6 = r$

disregard –6

6 inches

93. $x^2 + x^2 = 27^2$

$\frac{2x^2}{2} = \frac{729}{2}$

$\sqrt{x^2} = \sqrt{\frac{729}{2}}$

$x = \pm\frac{27}{\sqrt{2}}\cdot\frac{\sqrt{2}}{\sqrt{2}}$

$x = \pm\frac{27\sqrt{2}}{\sqrt{4}}$

$x = \pm\frac{27\sqrt{2}}{2}$

Disregard the negative.

$\frac{27\sqrt{2}}{2}$ inches

95. $p = -x^2 + 15$

$7 = -x^2 + 15$

$x^2 = 8$

$x = \pm\sqrt{8}$

$x \approx 2.828$

Demand cannot be negative, therefore the demand is approximately 2.828 thousand units.

97. $\frac{3}{5} + \sqrt{\frac{16}{25}} = \frac{3}{5} + \frac{4}{5} = \frac{7}{5}$

99. $\frac{9}{10} - \sqrt{\frac{49}{100}} = \frac{9}{10} - \frac{7}{10} = \frac{2}{10} = \frac{1}{5}$

101. $\frac{10-20\sqrt{3}}{2} = \frac{10}{2} - \frac{20\sqrt{3}}{2} = 5-10\sqrt{3}$

103. $\dfrac{12-8\sqrt{7}}{16}=\dfrac{12}{16}-\dfrac{8\sqrt{7}}{16}$
$=\dfrac{3}{4}-\dfrac{\sqrt{7}}{2}$
$=\dfrac{3}{4}-\dfrac{2\sqrt{7}}{4}$
$=\dfrac{3-2\sqrt{7}}{4}$

105. $\sqrt{b^2-4ac}$
$a=1,\ b=6,\ c=2$
$\sqrt{(6)^2-4(1)(2)}=\sqrt{36-8}$
$=\sqrt{28}$
$=\sqrt{4}\sqrt{7}$
$=2\sqrt{7}$

107. $\sqrt{b^2-4ac}$
$a=1,\ b=-3,\ c=-1$
$\sqrt{(-3)^2-4(1)(-1)}=\sqrt{9+4}=\sqrt{13}$

Exercise Set 8.2

1. $m^2+5m-6=0$
$a=1,\ b=5,\ c=-6$
$m=\dfrac{-5\pm\sqrt{5^2-4(1)(-6)}}{2(1)}$
$m=\dfrac{-5\pm\sqrt{25+24}}{2}=\dfrac{-5\pm\sqrt{49}}{2}$
$m=\dfrac{-5\pm7}{2}$
$m=-6$ or $m=1$
The solutions are –6 and 1.

3. $2y=5y^2-3$
$5y^2-2y-3=0$
$a=5,\ b=-2,\ c=-3$
$y=\dfrac{2\pm\sqrt{(-2)^2-4(5)(-3)}}{2(5)}$
$y=\dfrac{2\pm\sqrt{4+60}}{10}=\dfrac{2\pm\sqrt{64}}{10}$
$y=\dfrac{2\pm8}{10}$
$y=-\dfrac{3}{5}$, or $y=1$
The solutions are $-\dfrac{3}{5}$ and 1.

5. $x^2-6x+9=0$
$a=1,\ b=-6,\ c=9$
$x=\dfrac{6\pm\sqrt{(-6)^2-4(1)(9)}}{2(1)}$
$x=\dfrac{6\pm\sqrt{36-36}}{2}=\dfrac{6\pm\sqrt{0}}{2}=\dfrac{6}{2}=3$
The solution is 3.

7. $x^2+7x+4=0$
$a=1,\ b=7,\ c=4$
$x=\dfrac{-7\pm\sqrt{7^2-4(1)(4)}}{2(1)}$
$x=\dfrac{-7\pm\sqrt{49-16}}{2}=\dfrac{-7\pm\sqrt{33}}{2}$
The solutions are $\dfrac{-7+\sqrt{33}}{2}$ and $\dfrac{-7-\sqrt{33}}{2}$.

9. $8m^2-2m=7$
$8m^2-2m-7=0$
$a=8,\ b=-2,\ c=-7$
$m=\dfrac{2\pm\sqrt{(-2)^2-4(8)(-7)}}{2(8)}$
$m=\dfrac{2\pm\sqrt{4+224}}{16}=\dfrac{2\pm\sqrt{228}}{16}$
$m=\dfrac{2\pm\sqrt{4}\sqrt{57}}{16}=\dfrac{2\pm2\sqrt{57}}{16}$
$m=\dfrac{1\pm\sqrt{57}}{8}$
The solutions are $\dfrac{1+\sqrt{57}}{8}$ and $\dfrac{1-\sqrt{57}}{8}$.

11. $3m^2 - 7m = 3$

$3m^2 - 7m - 3 = 0$

$a = 3, b = -7, c = -3$

$m = \dfrac{7 \pm \sqrt{(-7)^2 - 4(3)(-3)}}{2(3)}$

$m = \dfrac{7 \pm \sqrt{49+36}}{6} = \dfrac{7 \pm \sqrt{85}}{6}$

The solutions are $\dfrac{7+\sqrt{85}}{6}$ and $\dfrac{7-\sqrt{85}}{6}$.

13. $\dfrac{1}{2}x^2 - x - 1 = 0$

$x^2 - 2x - 2 = 0$

$a = 1, b = -2, c = -2$

$x = \dfrac{2 \pm \sqrt{(-2)^2 - 4(1)(-2)}}{2(1)}$

$x = \dfrac{2 \pm \sqrt{4+8}}{2} = \dfrac{2 \pm \sqrt{12}}{2}$

$x = \dfrac{2 \pm \sqrt{4}\sqrt{3}}{2} = \dfrac{2 \pm 2\sqrt{3}}{2}$

$x = 1 \pm \sqrt{3}$

The solutions are $1+\sqrt{3}$ and $1-\sqrt{3}$.

15. $\dfrac{2}{5}y^2 + \dfrac{1}{5}y = \dfrac{3}{5}$

$2y^2 + y = 3$

$2y^2 + y - 3 = 0$

$a = 2, b = 1, c = -3$

$y = \dfrac{-1 \pm \sqrt{1^2 - 4(2)(-3)}}{2(2)}$

$y = \dfrac{-1 \pm \sqrt{1+24}}{4} = \dfrac{-1 \pm \sqrt{25}}{4}$

$y = \dfrac{-1 \pm 5}{4}$

$y = -\dfrac{3}{2}$ or $y = 1$

The solutions are $-\dfrac{3}{2}$ and 1.

17. $\dfrac{1}{3}y^2 - y - \dfrac{1}{6} = 0$

$2y^2 - 6y - 1 = 0$

$a = 2, b = -6, c = -1$

$y = \dfrac{6 \pm \sqrt{(-6)^2 - 4(2)(-1)}}{2(2)}$

$y = \dfrac{6 \pm \sqrt{36+8}}{4} = \dfrac{6 \pm \sqrt{44}}{4}$

$y = \dfrac{6 \pm \sqrt{4}\sqrt{11}}{4} = \dfrac{6 \pm 2\sqrt{11}}{4}$

$y = \dfrac{3 \pm \sqrt{11}}{2}$

The solutions are $\dfrac{3+\sqrt{11}}{2}$ and $\dfrac{3-\sqrt{11}}{2}$.

19. $m^2 + 5m - 6 = 0$

$(m+6)(m-1) = 0$

$m + 6 = 0$ or $m - 1 = 0$

$m = -6$ or $m = 1$

The results are the same.

21. $6 = -4x^2 + 3x$

$4x^2 - 3x + 6 = 0$

$a = 4, b = -3, c = 6$

$x = \dfrac{3 \pm \sqrt{(-3)^2 - 4(4)(6)}}{2(4)}$

$x = \dfrac{3 \pm \sqrt{9-96}}{8}$

$x = \dfrac{3 \pm \sqrt{-87}}{8}$

$x = \dfrac{3 \pm \sqrt{87}i}{8}$

$x = \dfrac{3 \pm i\sqrt{87}}{8}$

The solutions are $\dfrac{3+i\sqrt{87}}{8}$ and $\dfrac{3-i\sqrt{87}}{8}$.

23. $(x+5)(x-1) = 2$

$x^2 + 4x - 5 = 2$

$x^2 + 4x - 7 = 0$

$a = 1, b = 4, c = -7$

$$x = \frac{-4 \pm \sqrt{4^2 - 4(1)(-7)}}{2(1)}$$
$$x = \frac{-4 \pm \sqrt{16+28}}{2}$$
$$x = \frac{-4 + \sqrt{44}}{2}$$
$$x = \frac{-4 \pm \sqrt{4}\sqrt{11}}{2}$$
$$x = \frac{-4 \pm 2\sqrt{11}}{2}$$
$$x = -2 \pm \sqrt{11}$$
The solutions are $-2+\sqrt{11}$ and $-2-\sqrt{11}$.

25. $10y^2 + 10y + 3 = 0$
$a = 10$, $b = 10$, $c = 3$
$$y = \frac{-10 \pm \sqrt{10^2 - 4(10)(3)}}{2(10)}$$
$$y = \frac{-10 \pm \sqrt{100-120}}{20}$$
$$y = \frac{-10 \pm \sqrt{-20}}{20}$$
$$y = \frac{-10 \pm \sqrt{4}\sqrt{5}\sqrt{-1}}{20}$$
$$y = \frac{-10 \pm 2i\sqrt{5}}{20}$$
$$y = \frac{-5 \pm i\sqrt{5}}{10}$$
The solutions are $\frac{-5+i\sqrt{5}}{10}$ and $\frac{-5-i\sqrt{5}}{10}$.

27. $$\frac{-b+\sqrt{b^2-4ac}}{2a} + \frac{-b-\sqrt{b^2-4ac}}{2a}$$
$$= \frac{-b+\sqrt{b^2-4ac}-b-\sqrt{b^2-4ac}}{2a}$$
$$= \frac{-2b}{2a}$$
$$= -\frac{b}{a}$$

29. $9x - 2x^2 + 5 = 0$
$-2x^2 + 9x + 5 = 0$
$a = -2$, $b = 9$, $c = 5$
$b^2 - 4ac = 9^2 - 4(-2)(5)$
$b^2 - 4ac = 81 + 40$
$b^2 - 4ac = 121 > 0$
Therefore, there are two real solutions.

31. $4x^2 + 12x = -9$
$4x^2 + 12x + 9 = 0$
$a = 4$, $b = 12$, $c = 9$
$b^2 - 4ac = 12^2 - 4(4)(9)$
$b^2 - 4ac = 144 - 144 = 0$
Therefore, there is 1 real solution.

33. $3x = -2x^2 + 7$
$2x^2 + 3x - 7 = 0$
$a = 2$, $b = 3$, $c = -7$
$b^2 - 4ac = 3^2 - 4(2)(-7)$
$b^2 - 4ac = 9 + 56$
$b^2 - 4ac = 65 > 0$
Therefore, there are two real solutions.

35. $6 = 4x - 5x^2$
$5x^2 - 4x + 6 = 0$
$a = 5$, $b = -4$, $c = 6$
$b^2 - 4ac = (-4)^2 - 4(5)(6)$
$b^2 - 4ac = 16 - 120$
$b^2 - 4ac = -104 < 0$
Therefore, there are two complex but not real solutions.

37. $x^2 + 5x = -2$
$x^2 + 5x + 2 = 0$
$a = 1$, $b = 5$, $c = 2$
$$x = \frac{-5 \pm \sqrt{5^2 - 4(1)(2)}}{2(1)}$$
$$x = \frac{-5 \pm \sqrt{25-8}}{2} = \frac{-5 \pm \sqrt{17}}{2}$$
The solutions are $\frac{-5+\sqrt{17}}{2}$ and $\frac{-5+\sqrt{17}}{2}$.

39. $(m+2)(2m-6)=5(m-1)-12$

$2m^2-2m-12=5m-5-12$

$2m^2-7m+5=0$

$a=2,\ b=-7,\ c=5$

$m=\frac{-(-7)\pm\sqrt{(-7)^2-4(2)(5)}}{2(2)}$

$m=\frac{7\pm\sqrt{49-40}}{4}$

$m=\frac{7\pm\sqrt{9}}{4}$

$m=\frac{7\pm3}{4}$

$m=\frac{10}{4}=\frac{5}{2}$, or $m=\frac{4}{4}=1$

The solutions are $\frac{5}{2}$ and 1.

41. $\frac{x^2}{3}-x=\frac{5}{3}$

$x^2-3x=5$

$x^2-3x-5=0$

$a=1,\ b=-3,\ c=-5$

$x=\frac{3\pm\sqrt{(-3)^2-4(1)(-5)}}{2(1)}$

$x=\frac{3\pm\sqrt{9+20}}{2}=\frac{3\pm\sqrt{29}}{2}$

The solutions are $\frac{3+\sqrt{29}}{2}$ and $\frac{3-\sqrt{29}}{2}$.

43. $x(6x+2)-3=0$

$6x^2+2x-3=0$

$a=6,\ b=2,\ c=-3$

$x=\frac{-2\pm\sqrt{2^2-4(6)(-3)}}{2(6)}$

$x=\frac{-2\pm\sqrt{4+72}}{12}=\frac{-2\pm\sqrt{76}}{12}$

$x=\frac{-2\pm\sqrt{4}\sqrt{19}}{12}=\frac{-2\pm2\sqrt{19}}{12}$

$x=\frac{-1\pm\sqrt{19}}{6}$

The solutions are $\frac{-1+\sqrt{19}}{6}$ and $\frac{-1-\sqrt{19}}{6}$.

45. $x^2+6x+13=0$

$a=1,\ b=6,\ c=13$

$x=\frac{-6\pm\sqrt{6^2-4(1)(13)}}{2(1)}$

$x=\frac{-6\pm\sqrt{36-52}}{2}=\frac{-6\pm\sqrt{-16}}{2}$

$x=\frac{-6\pm4i}{2}=-3\pm2i$

The solutions are $-3+2i$ and $-3-2i$.

47. $\frac{2}{5}y^2+\frac{1}{5}y+\frac{3}{5}=0$

$2y^2+y+3=0$

$a=2,\ b=1,\ c=3$

$y=\frac{-1\pm\sqrt{1^2-4(2)(3)}}{2(2)}$

$y=\frac{-1\pm\sqrt{1-24}}{4}=\frac{-1\pm\sqrt{-23}}{4}$

$y=\frac{-1\pm i\sqrt{23}}{4}$

The solutions are $\frac{-1+i\sqrt{23}}{4}$ and $\frac{-1-i\sqrt{23}}{4}$.

49. $\frac{1}{2}y^2=y-\frac{1}{2}$

$y^2=2y-1$

$y^2-2y+1=0$

$a=1,\ b=-2,\ c=1$

$y=\frac{2\pm\sqrt{(-2)^2-4(1)(1)}}{2(1)}$

$y=\frac{2\pm\sqrt{4-4}}{2}=\frac{2\pm\sqrt{0}}{2}=1$

The solution is 1.

51. $(n-2)^2=15n$

$n^2-4n+4=15n$

$n^2-19n+4=0$

$a=1,\ b=-19,\ c=4$

$n = \frac{19 \pm \sqrt{(-19)^2 - 4(1)(4)}}{2(1)}$

$n = \frac{19 \pm \sqrt{361 - 16}}{2}$

$n = \frac{19 \pm \sqrt{345}}{2}$

$n = \frac{19}{2} \pm \frac{\sqrt{345}}{2}$

The solutions are $\frac{19 + \sqrt{345}}{2}$ and $\frac{19 - \sqrt{345}}{2}$.

53. $(x+8)^2 + x^2 = 36^2$

$x^2 + 16x + 64 + x^2 = 1296$

$2x^2 + 16x - 1232 = 0$

$a = 2,\ b = 16,\ c = -1232$

$x = \frac{-16 \pm \sqrt{16^2 - 4(2)(-1232)}}{2(2)}$

$x = \frac{-16 \pm \sqrt{10,112}}{4}$

$x \approx 21$ or $x \approx -29$ (disregard)

$x + x + 8 = 2(21) + 8 = 50$

$50 - 36 = 14$

They save about 14 feet of walking distance.

55. Let x = length of leg

$x + 2$ = length of hypotenuse

$x^2 + x^2 = (x+2)^2$

$2x^2 = x^2 + 4x + 4$

$x^2 - 4x - 4 = 0$

$a = 1,\ b = -4,\ c = -4$

$x = \frac{4 \pm \sqrt{(-4)^2 - 4(1)(-4)}}{2(1)}$

$x = \frac{4 \pm \sqrt{32}}{2}$

$x = \frac{4 \pm 4\sqrt{2}}{2}$

$x = 2 \pm 2\sqrt{2}$

(disregard a negative value)

The sides measure $2 + 2\sqrt{2}$ cm, $2 + 2\sqrt{2}$ cm, and $4 + 2\sqrt{2}$ cm.

57. Let x = width, then $x + 10$ = length.

Area = length · width

$400 = (x+10)x$

$0 = x^2 + 10x - 400$

$a = 1,\ b = 10,\ c = -400$

$x = \frac{-10 \pm \sqrt{10^2 - 4(1)(-400)}}{2(1)}$

$x = \frac{-10 \pm \sqrt{1700}}{2}$

$x = \frac{-10 \pm 10\sqrt{17}}{2}$

$x = -5 \pm 5\sqrt{17}$

Disregard a negative length. The width is $-5 + 5\sqrt{17}$ ft and the length is $5 + 5\sqrt{17}$ ft.

59. a. Let x = length.

$x^2 + x^2 = 100^2$

$2x^2 - 10,000 = 0$

$a = 2,\ b = 0,\ c = -10,000$

$x = \frac{0 \pm \sqrt{0^2 - 4(2)(-10,000)}}{2(2)}$

$x = \frac{\pm\sqrt{80,000}}{4}$

$x = \frac{\pm 200\sqrt{2}}{4}$

$x = \pm 50\sqrt{2}$

Disregard a negative length. The side measures $50\sqrt{2}$ meters.

b. Area $= s^2$

$= \left(50\sqrt{2}\right)^2$

$= 50^2\left(\sqrt{2}\right)^2$

$= 2500 \cdot 2$

$= 5000$

The area is 5000 square meters.

61. $\frac{x-1}{1}=\frac{1}{x}$

$x(x-1)=1\cdot 1$

$x^2-x-1=0$

$a=1,\ b=-1,\ c=-1$

$x=\frac{1\pm\sqrt{(-1)^2-4(1)(-1)}}{2(1)}$

$x=\frac{1\pm\sqrt{5}}{2}$

(disregard a negative length)

The value is $\frac{1+\sqrt{5}}{2}$.

63. $h=-16t^2+20t+1100$

$0=-16t^2+20t+1100$

$a=-16,\ b=20,\ c=1100$

$t=\frac{-20\pm\sqrt{20^2-4(-16)(1100)}}{2(-16)}$

$t=\frac{-20\pm\sqrt{70,800}}{-32}$

$t\approx 8.9$ or $t\approx -7.7$ (disregard)

It will take about 8.9 seconds.

65. $h(t)=-16t^2-20t+180$

$0=-16t^2-20t+180$

$0=-4(4t^2+5t-45)$

$0=4t^2+5t-45$

$t=\frac{-5\pm\sqrt{5^2-4(4)(-45)}}{2(4)}$

$t=\frac{-5\pm\sqrt{745}}{8}$

$t\approx\frac{-5\pm 27.3}{8}$

$t\approx 2.8$ or $t\approx -4.0$

Time cannot be negative, therefore the time until it strikes the ground is 2.8 seconds.

67. From Sunday to Monday

69. Wednesday

71. $f(4)=3(4)^2-18(4)+56=32$

This answer appears to agree with the graph.

73. a. $f(x)=128.5x^2-69.5x+2681$

$x=1997-1995=2$

$f(2)=128.5(2)^2-69.5(2)+2681$

$f(2)=3056$

Their net income was \$3056 million.

b. $15,000=128.5x^2-69.5x+2681$

$0=128.5x^2-69.5x-12,319$

$a=128.5,\ b=-69.5,\ c=-12,319$

$x=\frac{69.5\pm\sqrt{(-69.5)^2-4(128.5)(-12,319)}}{2(128.5)}$

$x\approx 10$ or $x\approx -10$ (disregard)

$1995+10=2005$

10 years after 1995 or in 2005, the net income will be \$15,000 million.

75. Exercise 63:

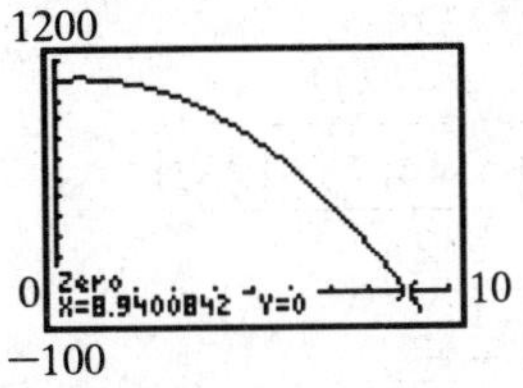

Exercise 65:

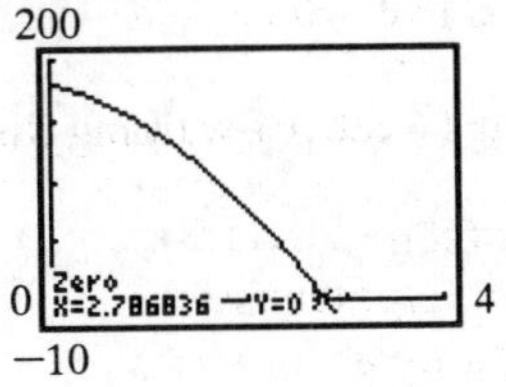

77.

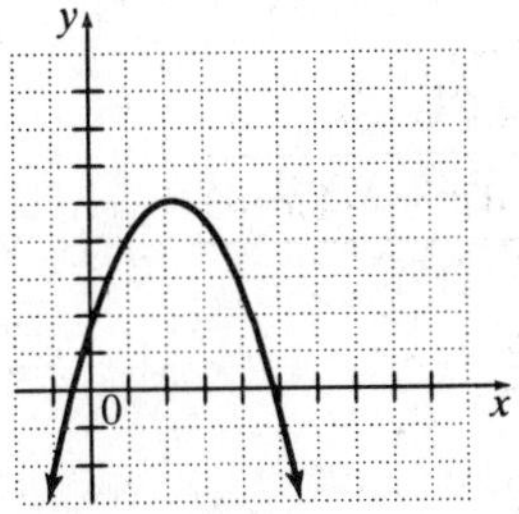

$y=9x-2x^2+5$

There are 2 intercepts.

79. $\sqrt{5x-2}=3$

$(\sqrt{5x-2})^2=3^2$

$5x-2=9$

$5x=11$

$x=\frac{11}{5}$

Checking:

$\sqrt{5\left(\frac{11}{5}\right)-2}\stackrel{?}{=}3$

$\sqrt{11-2}\stackrel{?}{=}3$

$\sqrt{9}\stackrel{?}{=}3$

$3=3$

The solution is $\frac{11}{5}$.

81. $\frac{1}{x}+\frac{2}{5}=\frac{7}{x}$

$5x\cdot\frac{1}{x}+5x\cdot\frac{2}{5}=5x\cdot\frac{7}{x}$

$5+2x=35$

$2x=30$

$x=15$

83. $x^4+x^2-20=(x^2+5)(x^2-4)$

$=(x^2+5)(x+2)(x-2)$

85. $z^4-13z^2+36=(z^2-9)(z^2-4)$

$=(z+3)(z-3)(z+2)(z-2)$

87. $3x^2-\sqrt{12x}+1=0$

$a=3,\ b=-\sqrt{12},\ c=1$

$x=\frac{\sqrt{12}\pm\sqrt{(-\sqrt{12})^2-4(3)(1)}}{2(3)}$

$x=\frac{\sqrt{12}\pm\sqrt{12-12}}{6}$

$x=\frac{\sqrt{4}\sqrt{3}\pm\sqrt{0}}{6}$

$x=\frac{2\sqrt{3}}{6}$

$x=\frac{\sqrt{3}}{3}$

The solution is $\frac{\sqrt{3}}{3}$.

89. $x^2+\sqrt{2}x+1=0$

$a=1,\ b=\sqrt{2},\ c=1$

$x=\frac{-\sqrt{2}\pm\sqrt{(\sqrt{2})^2-4(1)(1)}}{2(1)}$

$x=\frac{-\sqrt{2}\pm\sqrt{2-4}}{2}$

$x=\frac{-\sqrt{2}\pm\sqrt{-2}}{2}$

$x=\frac{-\sqrt{2}\pm i\sqrt{2}}{2}$

The solutions are $\frac{-\sqrt{2}+i\sqrt{2}}{2}$ and $\frac{-\sqrt{2}-i\sqrt{2}}{2}$.

91. $2x^2-\sqrt{3}x-1=0$

$a=2,\ b=-\sqrt{3},\ c=-1$

$x=\frac{\sqrt{3}\pm\sqrt{(-\sqrt{3})^2-4(2)(-1)}}{2(2)}$

$x=\frac{\sqrt{3}\pm\sqrt{3+8}}{4}$

$x=\frac{\sqrt{3}\pm\sqrt{11}}{4}$

The solutions are $\frac{\sqrt{3}+\sqrt{11}}{4}$ and $\frac{\sqrt{3}-\sqrt{11}}{4}$.

Exercise Set 8.3

1.
$$2x = \sqrt{10+3x}$$
$$4x^2 = 10+3x$$
$$4x^2 - 3x - 10 = 0$$
$$(4x+5)(x-2) = 0$$
$$4x+5=0 \quad \text{or} \quad x-2=0$$
$$x = -\frac{5}{4} \quad \text{or} \quad x = 2$$

Discard $x = -\frac{5}{4}$.

The solution is 2.

3. $x - 2\sqrt{x} = 8$
$$2\sqrt{x} = x-8$$
$$4x = x^2 - 16x + 64$$
$$0 = x^2 - 20x + 64$$
$$0 = (x-16)(x-4)$$
$$x-16=0 \quad \text{or} \quad x-4=0$$
$$x = 16 \quad \text{or} \quad x = 4 \text{ (discard)}$$

The solution is 16.

5. $\sqrt{9x} = x+2$
$$9x = x^2 + 4x + 4$$
$$0 = x^2 - 5x + 4$$
$$0 = (x-4)(x-1)$$
$$x-4=0 \quad \text{or} \quad x-1=0$$
$$x = 4 \quad \text{or} \quad x = 1$$

The solutions are 4 and 1.

7.
$$\frac{2}{x} + \frac{3}{x-1} = 1$$
$$\frac{2(x-1)+3x}{x(x-1)} = 1$$
$$\frac{2x-2+3x}{x^2-x} = 1$$
$$5x-2 = x^2 - x$$
$$x^2 - 6x + 2 = 0$$
$$x = \frac{6 \pm \sqrt{(-6)^2 - 4(1)(2)}}{2(1)}$$
$$= \frac{6 \pm \sqrt{28}}{2}$$
$$= \frac{6 \pm 2\sqrt{7}}{2}$$
$$= 3 \pm \sqrt{7}$$

The solutions are $3+\sqrt{7}$ and $3-\sqrt{7}$.

9.
$$\frac{3}{x} + \frac{4}{x+2} = 2$$
$$\frac{3(x+2)+4x}{x(x+2)} = 2$$
$$\frac{3x+6+4x}{x^2+2x} = 2$$
$$7x+6 = 2(x^2+2x)$$
$$7x+6 = 2x^2+4x$$
$$2x^2 - 3x - 6 = 0$$
$$x = \frac{3 \pm \sqrt{(-3)^2 - 4(2)(-6)}}{2(2)}$$
$$x = \frac{3 \pm \sqrt{57}}{4}$$

The solutions are $\frac{3+\sqrt{57}}{4}$ and $\frac{3-\sqrt{57}}{4}$.

11. $\dfrac{7}{x^2-5x+6}=\dfrac{2x}{x-3}-\dfrac{x}{x-2}$

$\dfrac{7}{(x-3)(x-2)}=\dfrac{2x(x-2)-x(x-3)}{(x-3)(x-2)}$

$7=2x^2-4x-x^2+3x$

$7=x^2-x$

$0=x^2-x-7$

$x=\dfrac{1\pm\sqrt{(-1)^2-4(1)(-7)}}{2(1)}$

$x=\dfrac{1\pm\sqrt{29}}{2}$

The solutions are $\dfrac{1+\sqrt{29}}{2}$ and $\dfrac{1-\sqrt{29}}{2}$.

13. $p^4-16=0$

$(p^2+4)(p^2-4)=0$

$(p+2i)(p-2i)(p+2)(p-2)=0$

$p+2i=0$ or $p-2i=0$ or $p+2=0$ or $p-2=0$

$p=-2i$ or $p=2i$ or $p=-2$ or $p=2$

The solutions are $-2i$, $2i$, -2, and 2.

15. $4x^4+11x^2=3$

$4x^4+11x^2-3=0$

$(4x^2-1)(x^2+3)=0$

$(2x+1)(2x-1)\left(x+i\sqrt{3}\right)\left(x-i\sqrt{3}\right)=0$

$2x+1=0$ or $2x-1=0$ or $x+i\sqrt{3}=0$ or $x-i\sqrt{3}=0$

$2x=-1$ or $2x=1$ or $x=-i\sqrt{3}$ or $x=i\sqrt{3}$

$x=-\dfrac{1}{2}$ or $x=\dfrac{1}{2}$

The solutions are $-\dfrac{1}{2}$, $\dfrac{1}{2}$, $-i\sqrt{3}$, and $i\sqrt{3}$.

17. $z^4-13z^2+36=0$

$(z^2-9)(z^2-4)=0$

$(z+3)(z-3)(z+2)(z-2)=0$

$z+3=0$ or $z-3=0$ or $z+2=0$ or $z-2=0$

$z=-3$ or $z=3$ or $z=-2$ or $z=2$

The solutions are -3, 3, -2, and 2.

19. $x^{2/3}-3x^{1/3}-10=0$

Let $y=x^{1/3}$.

$y^2-3y-10=0$

$(y-5)(y+2)=0$

$y-5=0$ or $y+2=0$
$y=5$ or $y=-2$
$x^{1/3}=5$ or $x^{1/3}=-2$
$x=125$ or $x=-8$
The solutions are 125 and −8.

21. $(5n+1)^2+2(5n+1)-3=0$
Let $y = 5n + 1$.
$y^2+2y-3=0$
$(y+3)(y-1)=0$
$y+3=0$ or $y-1=0$
$y=-3$ or $y=1$
$5n+1=-3$ or $5n+1=1$
$5n=-4$ or $5n=0$
$n=-\frac{4}{5}$ or $n=0$

The solutions are $-\frac{4}{5}$ and 0.

23. $2x^{2/3}-5x^{1/3}=3$
Let $y=x^{1/3}$.
$2y^2-5y=3$
$2y^2-5y-3=0$
$(2y+1)(y-3)=0$
$2y+1=0$ or $y-3=0$
$y=-\frac{1}{2}$ or $y=3$
$x^{1/3}=-\frac{1}{2}$ or $x^{1/3}=3$
$x=-\frac{1}{8}$ or $x=27$

The solutions are $-\frac{1}{8}$ and 27.

25. $1+\frac{2}{3t-2}=\frac{8}{(3t-2)^2}$

$(3t-2)^2+2(3t-2)-8=0$
Let $y = 3t - 2$.
$y^2+2y-8=0$
$(y+4)(y-2)=0$
$y+4=0$ or $y-2=0$
$y=-4$ or $y=2$
$3t-2=-4$ or $3t-2=2$
$3t=-2$ or $3t=4$
$t=-\frac{2}{3}$ or $t=\frac{4}{3}$

The solutions are $-\frac{2}{3}$ and $\frac{4}{3}$.

27. $20x^{2/3}-6x^{1/3}-2=0$
Let $y=x^{1/3}$.
$20y^2-6y-2=0$
$(5y+1)(4y-2)=0$
$5y=-1$ or $4y=2$
$y=-\frac{1}{5}$ or $y=\frac{1}{2}$
$x^{1/3}=-\frac{1}{5}$ or $x^{1/3}=\frac{1}{2}$
$x=-\frac{1}{125}$ or $x=\frac{1}{8}$

The solutions are $-\frac{1}{125}$ and $\frac{1}{8}$.

29. $a^4-5a^2+6=0$
$(a^2-3)(a^2-2)=0$
$a^2-3=0$ or $a^2-2=0$
$a^2=3$ or $a^2=2$
$a=\pm\sqrt{3}$ or $a=\pm\sqrt{2}$
The solutions are $-\sqrt{3}$, $\sqrt{3}$, $-\sqrt{2}$, and $\sqrt{2}$.

31. $\frac{2x}{x-2}+\frac{x}{x+3}=\frac{-5}{x+3}$
$\frac{2x}{x-2}=\frac{-x}{x+3}-\frac{5}{x+3}$
$\frac{2x}{x-2}=\frac{-x-5}{x+3}$
$2x(x+3)=(x-2)(-x-5)$
$2x^2+6x=-x^2+2x-5x+10$
$2x^2+6x=-x^2-3x+10$
$3x^2+9x-10=0$

$$x=\frac{-9\pm\sqrt{9^2-4(3)(-10)}}{2(3)}$$
$$x=\frac{-9\pm\sqrt{201}}{6}$$
The solutions are $\frac{-9+\sqrt{201}}{6}$ and $\frac{-9-\sqrt{201}}{6}$.

33. $(p+2)^2=9(p+2)-20$
$(p+2)^2-9(p+2)+20=0$
Let $x=p+2$.
$x^2-9x+20=0$
$(x-5)(x-4)=0$
$x=5$ or $x=4$
$p+2=5$ or $p+2=4$
$p=3$ or $p=2$
The solutions are 3 and 2.

35. $2x=\sqrt{11x+3}$
$4x^2=11x+3$
$4x^2-11x-3=0$
$(4x+1)(x-3)=0$
$x=-\frac{1}{4}$ (discard) or $x=3$
The solution is 3.

37. $x^{2/3}-8x^{1/3}+15=0$
Let $y=x^{1/3}$.
$y^2-8y+15=0$
$(y-5)(y-3)=0$
$y=5$ or $y=3$
$x^{1/3}=5$ or $x^{1/3}=3$
$x=5^3$ or $x=3^3$
$x=125$ or $x=27$
The solutions are 125 and 27.

39. $y^3+9y-y^2-9=0$
$y(y^2+9)-1(y^2+9)=0$
$(y^2+9)(y-1)=0$
$(y+3i)(y-3i)(y-1)=0$
$y+3i=0$ or $y-3i=0$ or $y-1=0$
$y=-3i$ or $y=3i$ or $y=1$
The solutions are $-3i$, $3i$, and 1.

41. $2x^{2/3}+3x^{1/3}-2=0$
Let $m=x^{1/3}$.
$2m^2+3m-2=0$
$(2m-1)(m+2)=0$
$m=\frac{1}{2}$ or $m=-2$
$x^{1/3}=\frac{1}{2}$ or $x^{1/3}=-2$
$x=\frac{1}{8}$ or $x=-8$
The solutions are $\frac{1}{8}$ and -8.

43. $x^{-2}-x^{-1}-6=0$
Let $y=x^{-1}$.
$y^2-y-6=0$
$(y-3)(y+2)=0$
$y=3$ or $y=-2$
$x^{-1}=3$ or $x^{-1}=-2$
$x=\frac{1}{3}$ or $x=-\frac{1}{2}$
The solutions are $\frac{1}{3}$ and $-\frac{1}{2}$.

45. $x-\sqrt{x}=2$
$x-2=\sqrt{x}$
$x^2-4x+4=x$
$x^2-5x+4=0$
$(x-4)(x-1)=0$
$x=4$ or $x=1$ (discard)
The solution is 4.

47. $\frac{x}{x-1}+\frac{1}{x+1}=\frac{2}{x^2-1}$

$\frac{x(x+1)+x-1}{(x-1)(x+1)}=\frac{2}{(x+1)(x-1)}$

$x^2+x+x-1=2$

$x^2+2x-3=0$

$(x+3)(x-1)=0$

$x+3=0$ or $x-1=0$

$x=-3$ or $x=1$ (discard)

The solution is –3.

49. $p^4-p^2-20=0$

$(p^2-5)(p^2+4)=0$

$p^2-5=0$ or $p^2+4=0$

$p^2=5$ or $(p+2i)(p-2i)=0$

$p=\pm\sqrt{5}$ or $p+2i=0$ or $p-2i=0$

$p=-2i$ or $p=2i$

The solutions are $-\sqrt{5}$, $\sqrt{5}$, $-2i$, and $2i$.

51. $2x^3=-54$

$x^3=-27$

$x^3+27=0$

$(x+3)(x^2-3x+9)=0$

$x+3=0$ or $x^2-3x+9=0$

$x=-3$ or

$x=\frac{3\pm\sqrt{(-3)^2-4(1)(9)}}{2(1)}$

$x=\frac{3\pm\sqrt{-27}}{2}$

$x=\frac{3\pm 3i\sqrt{3}}{2}$

The solutions are –3, $\frac{3+3i\sqrt{3}}{2}$, and $\frac{3-3i\sqrt{3}}{2}$.

53. $1=\frac{4}{x-7}+\frac{5}{(x-7)^2}$

$(x-7)^2=4(x-7)+5$

Let $y=x-7$.

$y^2-4y-5=0$

$(y-5)(y+1)=0$

$y=5$ or $y=-1$

$x-7=5$ or $x-7=-1$

$x=12$ or $x=6$

The solutions are 12 and 6.

55. $27y^4+15y^2=2$

$27y^4+15y^2-2=0$

$(9y^2-1)(3y^2+2)=0$

$(3y+1)(3y-1)(3y^2+2)=0$

$3y+1=0$ or $3y-1=0$ or $3y^2+2=0$

$3y=-1$ or $3y=1$ or $3y^2=-2$

$y=-\frac{1}{3}$ or $y=\frac{1}{3}$ or $y^2=-\frac{2}{3}$

$y=\pm\sqrt{-\frac{2}{3}}$

$y=\pm\frac{i\sqrt{6}}{3}$

The solutions are $-\frac{1}{3}, \frac{1}{3}, -\frac{i\sqrt{6}}{3}$, and $\frac{i\sqrt{6}}{3}$.

57. Let x = speed on first part.

$x-1$ = speed on second part

$D=r\cdot t$ or $t=\frac{D}{r}$, $1\frac{3}{5}=\frac{8}{5}$

$\frac{3}{x}+\frac{4}{x-1}=\frac{8}{5}$

$3\cdot 5(x-1)+4\cdot 5x=8\cdot x(x-1)$

$15x-15+20x=8x^2-8x$

$0=8x^2-43x+15$

$0=(8x-3)(x-5)$

$8x-3=0$ or $x-5=0$

$x=\frac{3}{8}$ or $x=5$

$x-1=4$

Her speeds were 5 mph and 4 mph.

59. Let x = time for hose alone, then

$x-1$ = time for inlet pipe alone.

$$\frac{1}{x}+\frac{1}{x-1}=\frac{1}{8}$$
$$8(x-1)+8x=x(x-1)$$
$$8x-8+8x=x^2-x$$
$$0=x^2-17x+8$$

$$x=\frac{17\pm\sqrt{(-17)^2-4(1)(8)}}{2(1)}$$
$$x=\frac{17\pm\sqrt{257}}{2}$$
$x \approx 0.5$ (discard) or $x \approx 16.5$

$x-1 \approx 15.5$

Inlet pipe: 15.5 hours
Hose: 16.5 hours

61. Let x = original speed.
$x + 11$ = return speed
$D = r \cdot t$ or $t = \frac{D}{r}$

$$\frac{330}{x}-\frac{330}{x+11}=1$$
$$330(x+11)-330x=x(x+11)$$
$$330x+3630-330x=x^2+11x$$
$$0=x^2+11x-3630$$
$$0=(x-55)(x+66)$$
$x=55$ or $x=-66$ (discard)

$x+11=66$

orginal speed: 55 mph
return speed: 66 mph

63.

	Hours	Part complete in one hour
Bill	$x-1$	$\frac{1}{x-1}$
Billy	x	$\frac{1}{x}$
Together	4	$\frac{1}{4}$

$$\frac{1}{x-1}+\frac{1}{x}=\frac{1}{4}$$
$$\frac{x+(x-1)}{x(x-1)}=\frac{1}{4}$$
$$4(2x-1)=x(x-1)$$
$$8x-4=x^2-x$$
$$0=x^2-9x+4$$
$$x=\frac{-(-9)\pm\sqrt{(-9)^2-4(1)(4)}}{2(1)}$$
$x=\frac{9-\sqrt{65}}{2}$ or $x=\frac{9+\sqrt{65}}{2}$

$x \approx 0.5$ (discard) or $x \approx 8.5$

It takes his son about 8.5 hours.

65. Let x = the number.
$$x(x-4)=96$$
$$x^2-4x=96$$
$$x^2-4x-96=0$$
$$(x-12)(x+8)=0$$
$x-12=0$ or $x+8=0$

$x=12$ or $x=-8$

The number is 12 or –8.

67. a. $x-3-3=x-6$

b. $V=l\cdot w\cdot h$

$300=(x-6)(x-6)\cdot 3$

c. $(x-6)(x-6)=100$

$x^2-12x+36=100$

$x^2-12x-64=0$

$(x-16)(x+4)=0$

$x = 16$ or $x = -4$ (discard)

The sheet is 16 cm by 16 cm.

69. a. Let x = Papis's fastest lap speed and $x + 3.8$ = Montoya's fastest lap speed.
Using time = distance ÷ speed,

$$\frac{7920}{x} = \frac{7920}{x+3.8} + 0.376$$

$$7920(x+3.8) = 7920x + 0.376x(x+3.8)$$

$$7920x + 30096 = 7920x + 0.376x^2 + 1.4288x$$

$$0 = 0.376x^2 + 1.4288x - 30096$$

$$x = \frac{-1.4288 \pm \sqrt{(1.4288)^2 - 4(0.376)(-30096)}}{2(0.376)}$$

Using the positive square root, $x \approx 281.0$ feet per second.

b. $x + 3.8 = 281.0 + 3.8 = 284.8$ feet per second.

c. There are 5,280 feet in one mile, and 3,600 seconds in one hour.

Papis: $\frac{281 \text{ ft}}{\text{sec}} \cdot \frac{3600 \text{ sec}}{1 \text{ hour}} \cdot \frac{1 \text{ mile}}{5280 \text{ ft}} \approx 191.6$ mph

Montoya: $\frac{284.8 \text{ ft}}{\text{sec}} \cdot \frac{3600 \text{ sec}}{1 \text{ hour}} \cdot \frac{1 \text{ mile}}{5280 \text{ ft}} \approx 194.2$ mph

71. Answers may vary.

73.

$$\frac{5x}{3} + 2 \le 7$$

$$\frac{5x}{3} \le 5$$

$$\frac{3}{5}\left(\frac{5x}{3}\right) \le \frac{3}{5}(5)$$

$$x \le 3$$

$\{x | x \le 3\}$

75.

$$\frac{y-1}{15} > -\frac{2}{5}$$

$$15\left(\frac{y-1}{15}\right) > 15\left(-\frac{2}{5}\right)$$

$$y - 1 > -6$$

$$y > -5$$

$\{y | y > -5\}$

77. Domain: $\{x | x \text{ is a real number}\}$
Range: $\{y | y \text{ is a real number}\}$
It is a function.

79. Domain: $\{x | x \text{ is a real number}\}$
Range: $\{y | y \ge -1\}$
It is a function.

Exercise Set 8.4

1. $(x + 1)(x + 5) > 0$

$x + 1 = 0$ or $x + 5 = 0$

$x = -1$ or $x = -5$

Region	Interval	Test Point
A	$(-\infty, -5)$	-6
B	$(-5, -1)$	-2
C	$(-1, \infty)$	0
$x = -6$	$(-6 + 1)(-6 + 5) > 0$	True
$x = -2$	$(-2 + 1)(-2 + 5) > 0$	False
$x = 0$	$(0 + 1)(0 + 5) > 0$	True

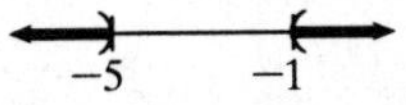

The solution is $(-\infty, -5) \cup (-1, \infty)$.

3. $(x-3)(x+4) \le 0$

$x-3=0$ or $x+4=0$

$x=3$ or $x=-4$

Region	Interval	Test Point
A	$(-\infty, -4)$	-5
B	$(-4, 3)$	0
C	$(3, \infty)$	4

$x=-5$	$(-5-3)(-5+4) \le 0$	False
$x=0$	$(0-3)(0+4) \le 0$	True
$x=4$	$(4-3)(4+4) \le 0$	False

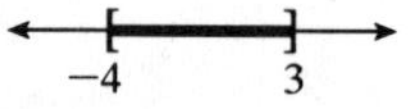

The solution is $[-4, 3]$.

5. $x^2-7x+10 \le 0$

$(x-5)(x-2) \le 0$

$(x-5)(x-2) = 0$

$x-5=0$ or $x-2=0$

$x=5$ or $x=2$

Region	Interval	Test Point
A	$(-\infty, 2)$	0
B	$(2, 5)$	3
C	$(5, \infty)$	6

$x=0$	$(0-5)(0-2) \le 0$	False
$x=3$	$(3-5)(3-2) \le 0$	True
$x=6$	$(6-5)(6-2) \le 0$	False

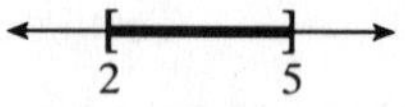

The solution is $[2, 5]$.

7. $3x^2+16x < -5$

$3x^2+16x+5 < 0$

$(3x+1)(x+5) < 0$

$(3x+1)(x+5) = 0$

$3x+1=0$ or $x+5=0$

$3x=-1$ or $x=-5$

$x=-\frac{1}{3}$

Region	Interval	Test Point
A	$(-\infty, -5)$	-6
B	$\left(-5, -\frac{1}{3}\right)$	-1
C	$\left(-\frac{1}{3}, \infty\right)$	0

$x=-6$	$[3(-6)+1](-6+5) < 0$	False
$x=-1$	$[3(-1)+1](-1+5) < 0$	True
$x=0$	$[3(0)+1](0+5) < 0$	False

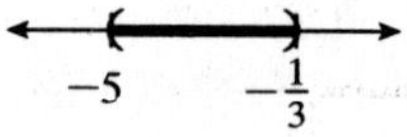

The solution is $\left(-5, -\frac{1}{3}\right)$.

9. $(x-6)(x-4)(x-2) > 0$

$(x-6)(x-4)(x-2) = 0$

$x-6=0$ or $x-4=0$ or $x-2=0$

$x=6$ or $x=4$ or $x=2$

Region	Interval	Test Point
A	$(-\infty, 2)$	1
B	$(2, 4)$	3
C	$(4, 6)$	5
D	$(6, \infty)$	7

$x=1$	$(1-6)(1-4)(1-2) > 0$	False
$x=3$	$(3-6)(3-4)(3-2) > 0$	True
$x=5$	$(5-6)(5-4)(5-2) > 0$	False
$x=7$	$(7-6)(7-4)(7-2) > 0$	True

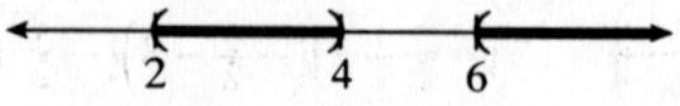

The solution is $(2, 4) \cup (6, \infty)$.

11. $x(x-1)(x+4) \le 0$

$x(x-1)(x+4) = 0$

$x=0$ or $x-1=0$ or $x+4=0$

$x=0$ or $x=1$ or $x=-4$

Region	Interval	Test Point
A	$(-\infty, -4)$	-5
B	$(-4, 0)$	-1
C	$(0, 1)$	$\frac{1}{2}$
D	$(1, \infty)$	2
$x=-5$	$-5(-5-1)(-5+4) \le 0$	True
$x=-1$	$-1(-1-1)(-1+4) \le 0$	False
$x=\frac{1}{2}$	$\frac{1}{2}\left(\frac{1}{2}-1\right)\left(\frac{1}{2}+4\right) \le 0$	True
$x=2$	$2(2-1)(2+4) \le 0$	False

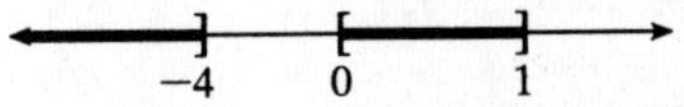

The solution is $(-\infty, -4] \cup [0, 1]$.

13.
$$(x^2-9)(x^2-4) > 0$$
$$(x+3)(x-3)(x+2)(x-2) > 0$$
$$(x+3)(x-3)(x+2)(x-2) = 0$$
$$x+3=0 \quad \text{or} \quad x-3=0 \quad \text{or} \quad x+2=0 \quad \text{or} \quad x-2=0$$
$$x=-3 \quad \text{or} \quad x=3 \quad \text{or} \quad x=-2 \quad \text{or} \quad x=2$$

Region	Interval	Test Point
A	$(-\infty, -3)$	-4
B	$(-3, -2)$	$-\frac{5}{2}$
C	$(-2, 2)$	0
D	$(2, 3)$	$\frac{5}{2}$
E	$(3, \infty)$	4
$x=-4$	$(-4+3)(-4-3)(-4+2)(-4-2) > 0$	True
$x=-\frac{5}{2}$	$\left(-\frac{5}{2}+3\right)\left(-\frac{5}{2}-3\right)\left(-\frac{5}{2}+2\right)\left(-\frac{5}{2}-2\right) > 0$	False
$x=0$	$(0+3)(0-3)(0+2)(0-2) > 0$	True
$x=\frac{5}{2}$	$\left(\frac{5}{2}+3\right)\left(\frac{5}{2}-3\right)\left(\frac{5}{2}+2\right)\left(\frac{5}{2}-2\right) > 0$	False
$x=4$	$(4+3)(4-3)(4+2)(4-2) > 0$	True

The solution is $(-\infty, -3) \cup (-2, 2) \cup (3, \infty)$.

15. $\frac{x+7}{x-2}<0$

$x+7=0$ or $x-2=0$

$x=-7$ or $x=2$

Region	Interval	Test Point
A	$(-\infty,-7)$	-8
B	$(-7,2)$	0
C	$(2,\infty)$	3
$x=-8$	$\frac{-8+7}{-8-2}<0$	False
$x=0$	$\frac{0+7}{0-2}<0$	True
$x=3$	$\frac{3+7}{3-2}<0$	False

The solution is $(-7, 2)$.

17. $\frac{5}{x+1}>0$

$x+1=0$

$x=-1$

Region	Interval	Test Point
A	$(-\infty,-1)$	-2
B	$(-1,\infty)$	0
$x=-2$	$\frac{5}{-2+1}>0$	False
$x=0$	$\frac{5}{0+1}>0$	True

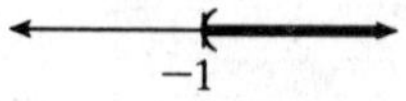

The solution is $(-1, \infty)$.

19. $\frac{x+1}{x-4}\ge 0$

$x+1=0$ or $x-4=0$

$x=-1$ or $x=4$

Region	Interval	Test Point
A	$(-\infty,-1)$	-2
B	$(-1,4)$	0
C	$(4,\infty)$	5
$x=-2$	$\frac{-2+1}{-2-4}\ge 0$	True
$x=0$	$\frac{0+1}{0-4}\ge 0$	False
$x=5$	$\frac{5+1}{5-4}\ge 0$	True

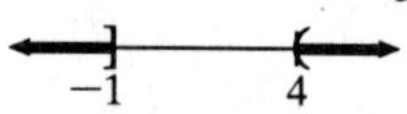

The solution is $(-\infty, -1] \cup (4, \infty)$.

21. Answers may vary.

23.
$$\frac{3}{x-2}<4$$
$$\frac{3}{x-2}-4<0$$
$$\frac{3-4(x-2)}{x-2}<0$$
$$\frac{3-4x+8}{x-2}<0$$
$$\frac{11-4x}{x-2}<0$$

$11-4x=0$ and $x-2=0$

$11=4x$ and $x=2$

$x=\frac{11}{4}$

Region	Interval	Test Point
A	$(-\infty, 2)$	1
B	$\left(2, \frac{11}{4}\right)$	$\frac{5}{2}$
C	$\left(\frac{11}{4}, \infty\right)$	3
$x = 1$	$\frac{11-4(1)}{1-2} < 0$	True
$x = \frac{5}{2}$	$\frac{11-4\left(\frac{5}{2}\right)}{\frac{5}{2}-2} < 0$	False
$x = 3$	$\frac{11-4(3)}{4-2} < 0$	True

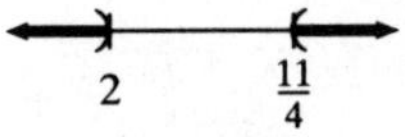

The solution is $(-\infty, 2) \cup \left(\frac{11}{4}, \infty\right)$.

25. $\frac{x^2+6}{5x} \geq 1$

$\frac{x^2+6}{5x} - 1 \geq 0$

$\frac{x^2+6-5x}{5x} \geq 0$

$\frac{(x-2)(x-3)}{5x} \geq 0$

$x - 2 = 0$ or $x - 3 = 0$ or $5x = 0$
$x = 2$ or $x = 3$ or $x = 0$

Region	Interval	Test Point
A	$(-\infty, 0)$	-1
B	$(0, 2)$	1
C	$(2, 3)$	$\frac{5}{2}$
D	$(3, \infty)$	4
$x = -1$	$\frac{(-1-2)(-1-3)}{5(-1)} \geq 0$	False
$x = 1$	$\frac{(1-2)(1-3)}{5(1)} \geq 0$	True
$x = \frac{5}{2}$	$\frac{\left(\frac{5}{2}-2\right)\left(\frac{5}{2}-3\right)}{5\left(\frac{5}{2}\right)} \geq 0$	False
$x = 4$	$\frac{(4-2)(4-3)}{5(4)} \geq 0$	True

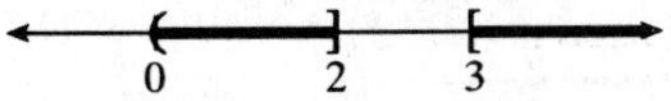

The solution is $(0, 2] \cup [3, \infty)$.

27. $(x-8)(x+7) > 0$
$(x-8)(x+7) = 0$
$x - 8 = 0$ or $x + 7 = 0$
$x = 8$ or $x = -7$

Region	Interval	Test Point
A	$(-\infty, -7)$	-8
B	$(-7, 8)$	0
C	$(8, \infty)$	9
$x = -8$	$(-8-8)(-8+7) > 0$	True
$x = 0$	$(0-8)(0+7) > 0$	False
$x = 9$	$(9-8)(9+7) > 0$	True

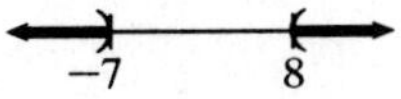

The solution is $(-\infty, -7) \cup (8, \infty)$.

29. $(2x-3)(4x+5) \le 0$
$(2x-3)(4x+5) = 0$
$2x-3=0$ or $4x+5=0$
$2x=3$ or $4x=-5$
$x=\frac{3}{2}$ or $x=-\frac{5}{4}$

Region	Interval	Test Point
A	$\left(-\infty, -\frac{5}{4}\right)$	–2
B	$\left(-\frac{5}{4}, \frac{3}{2}\right)$	0
C	$\left(\frac{3}{2}, \infty\right)$	2

$x=-2$	$[2(-2)-3][4(-2)+5] \le 0$	False
$x=0$	$[2(0)-3][4(0)+5] \le 0$	True
$x=2$	$[2(2)-3][4(2)+5] \le 0$	False

The solution is $\left[-\frac{5}{4}, \frac{3}{2}\right]$.

31. $x^2 > x$
$x^2 - x > 0$
$x(x-1) > 0$
$x(x-1) = 0$
$x=0$ or $x-1=0$
$x=1$

Region	Interval	Test Point
A	$(-\infty, 0)$	–1
B	$(0, 1)$	$\frac{1}{2}$
C	$(1, \infty)$	2

$x=-1$	$-1(-1-1) > 0$	True
$x=\frac{1}{2}$	$\frac{1}{2}\left(\frac{1}{2}-1\right) > 0$	False
$x=2$	$2(2-1) > 0$	True

The solution is $(-\infty, 0) \cup (1, \infty)$.

33. $(2x-8)(x+4)(x-6) \le 0$

$(2x-8)(x+4)(x-6) = 0$

$2x-8=0$ or $x+4=0$ or $x-6=0$

$2x=8$ or $x=-4$ or $x=6$

$x=4$

Region	Interval	Test Point
A	$(-\infty, -4)$	-5
B	$(-4, 4)$	0
C	$(4, 6)$	5
D	$(6, \infty)$	7

$x=-5$	$[2(-5)-8](-5+4)(-5-6) \le 0$	True
$x=0$	$[2(0)-8](0+4)(0-6) \le 0$	False
$x=5$	$[2(5)-8](5+4)(5-6) \le 0$	True
$x=7$	$[2(7)-8)](7+4)(7-6) \le 0$	False

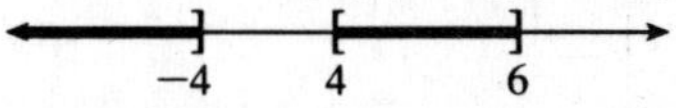

The solution is $(-\infty, -4] \cup [4, 6]$.

35. $6x^2 - 5x \ge 6$

$6x^2 - 5x - 6 \ge 0$

$(3x+2)(2x-3) \ge 0$

$(3x+2)(2x-3) = 0$

$3x+2=0$ or $2x-3=0$

$3x=-2$ or $2x=3$

$x=-\frac{2}{3}$ or $x=\frac{3}{2}$

Region	Interval	Test Point
A	$\left(-\infty, -\frac{2}{3}\right)$	-1
B	$\left(-\frac{2}{3}, \frac{3}{2}\right)$	0
C	$\left(\frac{3}{2}, \infty\right)$	2

$x=-1$	$[3(-1)+2][2(-1)-3] \ge 0$	True
$x=0$	$[3(0)+2][2(0)-3] \ge 0$	False
$x=2$	$[3(2)+2][2(2)-3] \ge 0$	True

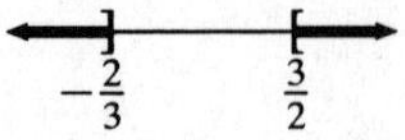

The solution is $\left(-\infty, -\frac{2}{3}\right] \cup \left[\frac{3}{2}, \infty\right)$.

37. $4x^3 + 16x^2 - 9x - 36 > 0$

$4x^2(x+4) - 9(x+4) > 0$

$(4x^2 - 9)(x+4) > 0$

$(2x-3)(2x+3)(x+4) > 0$

$2x - 3 = 0$ or $2x + 3 = 0$ or $x + 4 = 0$

$x = \frac{3}{2}$ or $x = -\frac{3}{2}$ or $x = -4$

Region	Interval	Test Point
A	$(-\infty, -4)$	-5
B	$\left(-4, -\frac{3}{2}\right)$	-3
C	$\left(-\frac{3}{2}, \frac{3}{2}\right)$	0
D	$\left(\frac{3}{2}, \infty\right)$	2
$x = -5$	$4(-5)^3 + 16(-5)^2 - 9(-5) - 36 = -91$ $-91 > 0$	False
$x = -3$	$4(-3)^3 + 16(-3)^2 - 9(-3) - 36 = 27$ $27 > 0$	True
$x = 0$	$4(0)^3 + 16(0)^2 - 9(0) - 36 = -36$ $-36 > 0$	False
$x = 2$	$4(2)^3 + 16(2)^2 - 9(2) - 36 = 42$ $42 > 0$	True

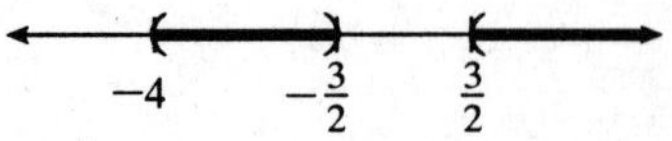

The solution is $\left(-4, -\frac{3}{2}\right) \cup \left(\frac{3}{2}, \infty\right)$.

39.
$$x^4 - 26x^2 + 25 \ge 0$$
$$(x^2 - 25)(x^2 - 1) \ge 0$$
$$(x-5)(x+5)(x-1)(x+1) \ge 0$$
$$(x-5)(x+5)(x-1)(x+1) = 0$$
$$x-5=0 \text{ or } x+5=0 \text{ or } x-1=0 \text{ or } x+1=0$$
$$x=5 \text{ or } x=-5 \text{ or } x=1 \text{ or } x=-1$$

Region	Interval	Test Point
A	$(-\infty, -5)$	-6
B	$(-5, -1)$	-3
C	$(-1, 1)$	0
D	$(1, 5)$	3
E	$(5, \infty)$	6

$x=-6$	$(-6-5)(-6+5)(-6-1)(-6+1) \ge 0$	True
$x=-3$	$(-3-5)(-3+5)(-3-1)(-3+1) \ge 0$	False
$x=0$	$(0-5)(0+5)(0-1)(0+1) \ge 0$	True
$x=3$	$(3-5)(3+5)(3-1)(3+1) \ge 0$	False
$x=6$	$(6-5)(6+5)(6-1)(6+1) \ge 0$	True

The solution is $(-\infty, -5] \cup [-1, 1] \cup [5, \infty)$.

41. $(2x-7)(3x+5) > 0$

$(2x-7)(3x+5) = 0$

$2x-7=0$ or $3x+5=0$

$2x=7$ or $3x=-5$

$x=\frac{7}{2}$ or $x=-\frac{5}{3}$

Region	Interval	Test Point
A	$\left(-\infty, -\frac{5}{3}\right)$	−2
B	$\left(-\frac{5}{3}, \frac{7}{2}\right)$	0
C	$\left(\frac{7}{2}, \infty\right)$	4

$x=-2$	$[2(-2)-7][3(-2)+5] > 0$	True
$x=0$	$[2(0)-7][3(0)+5] > 0$	False
$x=4$	$[2(4)-7][2(4)+5] > 0$	True

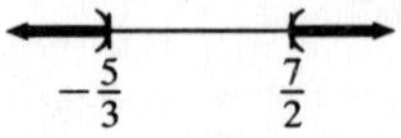

The solution is $\left(-\infty, -\frac{5}{3}\right) \cup \left(\frac{7}{2}, \infty\right)$.

43. $\frac{x}{x-10} < 0$

$x=0$ or $x-10=0$

$x=10$

Region	Interval	Test Point
A	$(-\infty, 0)$	−1
B	$(0, 10)$	1
C	$(10, \infty)$	11
$x=-1$	$\frac{-1}{-1-10} < 0$	False
$x=1$	$\frac{1}{1-10} < 0$	True
$x=11$	$\frac{11}{11-10} < 0$	False

The solution is $(0, 10)$.

45. $\frac{x-5}{x+4} \geq 0$

$x-5=0$ or $x+4=0$

$x=5$ or $x=-4$

Region	Interval	Test Point
A	$(-\infty, -4)$	−5
B	$(-4, 5)$	0
C	$(5, \infty)$	6
$x=-5$	$\frac{-5-5}{-5+4} \geq 0$	True
$x=0$	$\frac{0-5}{0+4} \geq 0$	False
$x=6$	$\frac{6-5}{6+4} \geq 0$	True

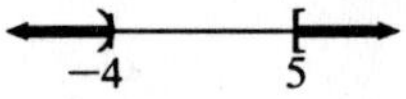

The solution is $(-\infty, -4) \cup [5, \infty)$.

47. $\dfrac{x(x+6)}{(x-7)(x+1)} \geq 0$

$x = 0$ or $x+6=0$ or $x-7=0$ or $x+1=0$

$x = 0$ or $x=-6$ or $x=7$ or $x=-1$

Region	Interval	Test Point
A	$(-\infty, -6)$	-7
B	$(-6, -1)$	-2
C	$(-1, 0)$	$-\frac{1}{2}$
D	$(0, 7)$	1
E	$(7, \infty)$	8
$x=-7$	$\frac{-7(-7+6)}{(-7-7)(-7+1)} \geq 0$	True
$x=-2$	$\frac{-2(-2+6)}{(-2-7)(-2+1)} \geq 0$	False
$x=-\frac{1}{2}$	$\frac{-\frac{1}{2}\left(-\frac{1}{2}+6\right)}{\left(-\frac{1}{2}-7\right)\left(-\frac{1}{2}+1\right)} \geq 0$	True
$x=1$	$\frac{1(1+6)}{(1-7)(1+1)} \geq 0$	False
$x=8$	$\frac{8(8+6)}{(8-7)(8+1)} \geq 0$	True

The solution is $(-\infty, -6] \cup (-1, 0] \cup (7, \infty)$.

49.
$$\frac{-1}{x-1} > -1$$
$$\frac{1}{x-1} < 1$$
$$\frac{1}{x-1} - 1 < 0$$
$$\frac{1-(x-1)}{x-1} < 0$$
$$\frac{1-x+1}{x-1} < 0$$
$$\frac{2-x}{x-1} < 0$$

$2-x=0$ or $x-1=0$

$2=x$ or $x=1$

Region	Interval	Test Point
A	$(-\infty, 1)$	0
B	$(1, 2)$	$\frac{3}{2}$
C	$(2, \infty)$	3
$x=0$	$\frac{2-0}{0-1} < 0$	True
$x=\frac{3}{2}$	$\frac{2-\frac{3}{2}}{\frac{3}{2}-1} < 0$	False
$x=3$	$\frac{2-3}{3-1} < 0$	True

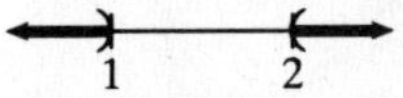

The solution is $(-\infty, 1) \cup (2, \infty)$.

51.
$$\frac{x}{x+4} \le 2$$
$$\frac{x}{x+4} - 2 \le 0$$
$$\frac{x-2(x+4)}{x+4} \le 0$$
$$\frac{-x-8}{x+4} \le 0$$
$$-x-8=0 \quad \text{or} \quad x+4=0$$
$$-x=8 \quad \text{or} \quad x=-4$$
$$x=-8$$

Region	Interval	Test Point
A	$(-\infty, -8)$	-9
B	$(-8, -4)$	-5
C	$(-4, \infty)$	0
$x=-9$	$\frac{-(-9)-8}{-9+4} \le 0$	True
$x=-5$	$\frac{-(-5)-8}{-5+4} \le 0$	False
$x=0$	$\frac{-0-8}{0+4} \le 0$	True

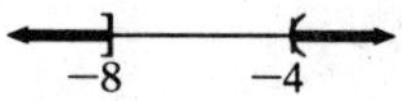

The solution is $(-\infty, -8] \cup (-4, \infty)$.

53.
$$\frac{z}{z-5} \ge 2z$$
$$\frac{z}{z-5} - 2z \ge 0$$
$$\frac{z-2z(z-5)}{z-5} \ge 0$$
$$\frac{z-2z^2+10z}{z-5} \ge 0$$
$$\frac{11z-2z^2}{z-5} \ge 0$$
$$\frac{z(11-2z)}{z-5} \ge 0$$

$$z=0 \quad \text{or} \quad 11-2z=0 \quad \text{or} \quad z-5=0$$
$$11=2z \quad \text{or} \quad z=5$$
$$\frac{11}{2}=z$$

Region	Interval	Test Point
A	$(-\infty, 0)$	-1
B	$(0, 5)$	1
C	$\left(5, \frac{11}{2}\right)$	$\frac{21}{4}$
D	$\left(\frac{11}{2}, \infty\right)$	6
$z=-1$	$\frac{-1[11-2(-1)]}{-1-5} \ge 0$	True
$z=1$	$\frac{1[11-2(1)]}{1-5} \ge 0$	False
$z=\frac{21}{4}$	$\frac{\frac{21}{4}\left[11-2\left(\frac{21}{4}\right)\right]}{\frac{21}{4}-5} \ge 0$	True
$z=6$	$\frac{6[11-2(6)]}{6-5} \ge 0$	False

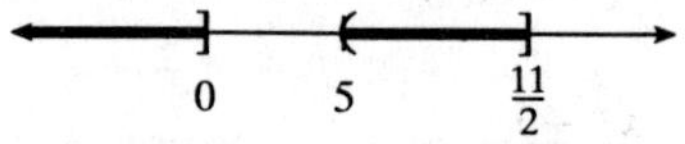

The solution is $(-\infty, 0] \cup \left(5, \frac{11}{2}\right]$.

55.
$$\frac{(x+1)^2}{5x} > 0$$
$$(x+1)^2 = 0 \quad \text{or} \quad 5x=0$$
$$x+1=0 \quad \text{or} \quad x=0$$
$$x=-1$$

Region	Interval	Test Point
A	$(-\infty, -1)$	-2
B	$(-1, 0)$	$-\frac{1}{2}$
C	$(-1, \infty)$	1
$x=-2$	$\frac{(-2+1)^2}{5(-2)} > 0$	False
$x=-\frac{1}{2}$	$\frac{\left(-\frac{1}{2}+1\right)^2}{5\left(-\frac{1}{2}\right)} > 0$	False
$x=1$	$\frac{(1+1)^2}{5(1)} > 0$	True

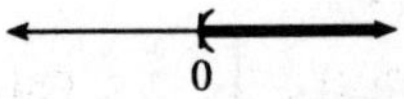

The solution is $(0, \infty)$.

57. Let x = the number.

$\frac{1}{x}$ = the reciprocal of the number

$$x - \frac{1}{x} < 0$$

$$\frac{x^2-1}{x} < 0$$

$$\frac{(x+1)(x-1)}{x} < 0$$

$$x+1=0 \quad \text{and} \quad x-1=0 \quad \text{and} \quad x=0$$

$$x=-1 \quad \text{and} \quad x=1 \quad \text{and} \quad x=0$$

Region	Interval	Test Point
A	$(-\infty, -1)$	-2
B	$(-1, 0)$	$-\frac{1}{2}$
C	$(0, 1)$	$\frac{1}{2}$
D	$(1, \infty)$	2
$x=-2$	$\frac{(-2+1)(-2-1)}{-2} < 0$	True
$x=-\frac{1}{2}$	$\frac{\left(-\frac{1}{2}+1\right)\left(-\frac{1}{2}-1\right)}{-\frac{1}{2}} < 0$	False
$x=\frac{1}{2}$	$\frac{\left(\frac{1}{2}+1\right)\left(\frac{1}{2}-1\right)}{\frac{1}{2}} < 0$	True
$x=2$	$\frac{(2+1)(2-1)}{2} < 0$	False

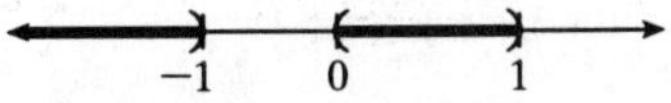

The solution is $(-\infty, -1) \cup (0, 1)$.

59. $P(x) = -2x^2 + 26x - 44$

$$-2x^2 + 26x - 44 > 0$$

$$-2(x^2 - 13x + 22) > 0$$

$$-2(x-11)(x-2) > 0$$

$$x-11=0 \quad \text{or} \quad x-2=0$$

$$x=11 \quad \text{or} \quad x=2$$

Region	Interval	Test Point
A	$(-\infty, 2)$	0
B	$(2, 11)$	3
C	$(11, \infty)$	12
$x=0$	$-2(0-11)(0-2) > 0$	False
$x=3$	$-2(3-11)(3-2) > 0$	True
$x=12$	$-2(12-11)(12-2) > 0$	False

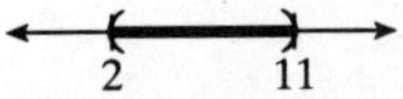

The solution is $(2, 11)$.

61.

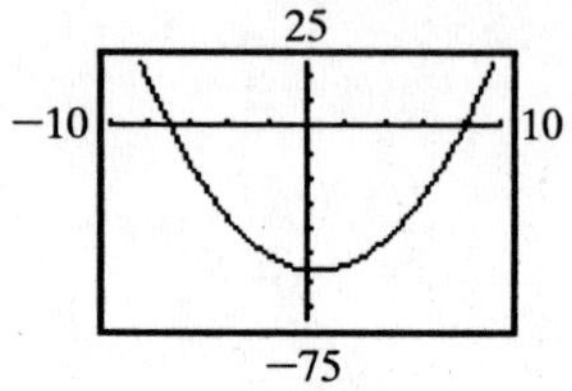

63.

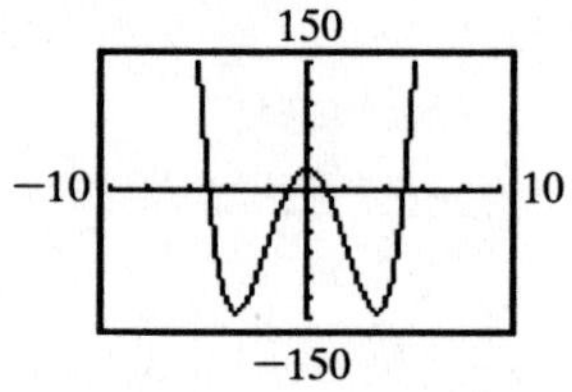

65. $g(x) = |x| + 2$

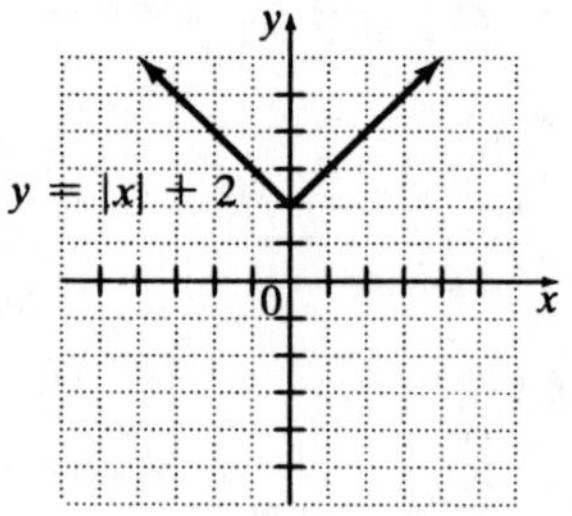

67. $F(x) = |x| - 1$

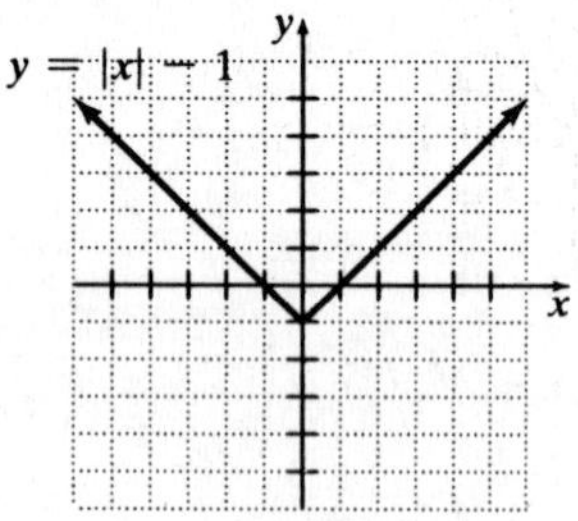

69. $F(x) = x^2 - 3$

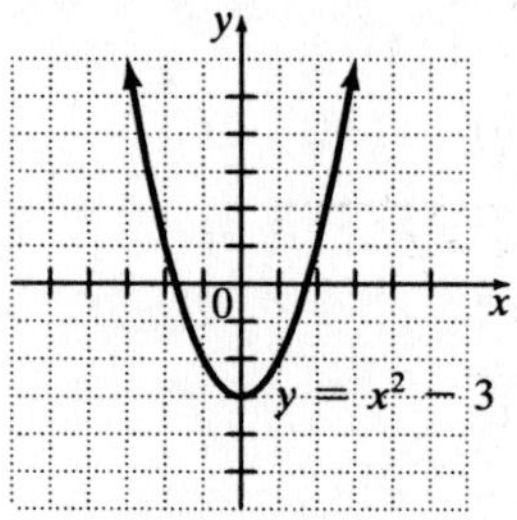

71. $H(x) = x^2 + 1$

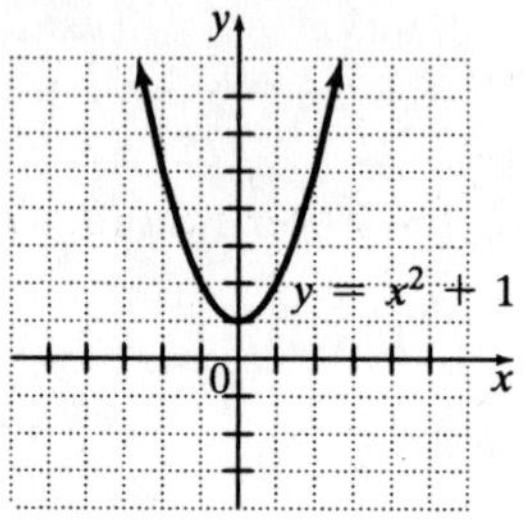

Section 8.5

Graphing Calculator Explorations

1.

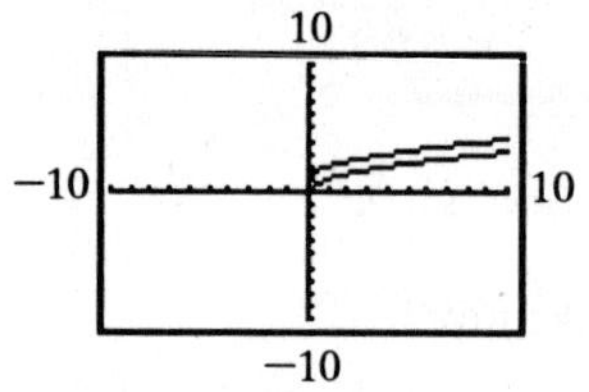

3.

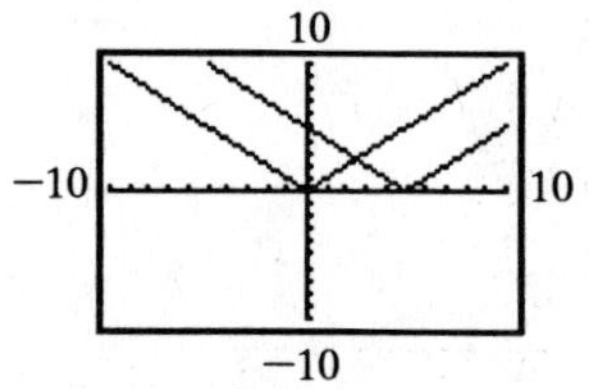

5.

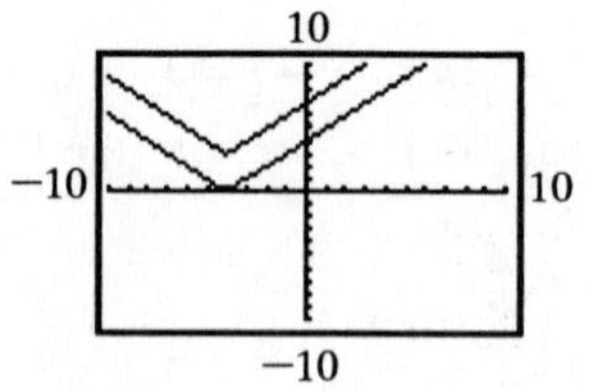

Mental Math

1. $f(x) = x^2$; vertex: (0, 0)

2. $f(x) = -5x^2$; vertex: (0, 0)

3. $g(x) = (x-2)^2$; vertex: (2, 0)

4. $g(x) = (x+5)^2$; vertex (−5, 0)

5. $f(x) = 2x^2 + 3$; vertex: (0, 3)

6. $h(x) = x^2 - 1$; vertex: $(0, -1)$

7. $g(x) = (x+1)^2 + 5$; vertex: $(-1, 5)$

8. $h(x) = (x-10)^2 - 7$; vertex: $(10, -7)$

Exercise Set 8.5

1. $f(x) = x^2 - 1$

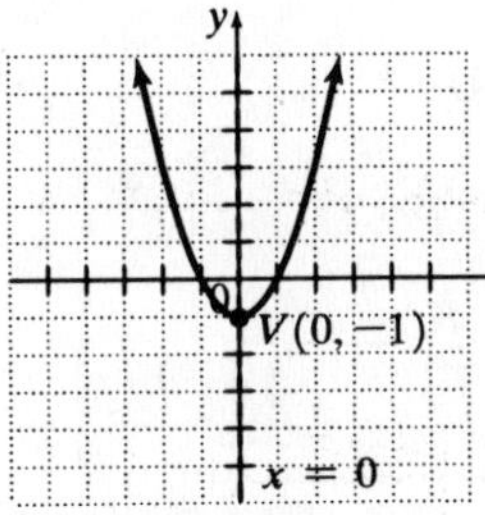

3. $h(x) = x^2 + 5$

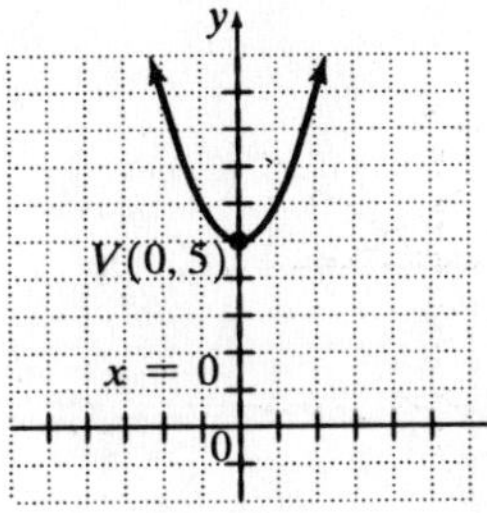

5. $g(x) = x^2 + 7$

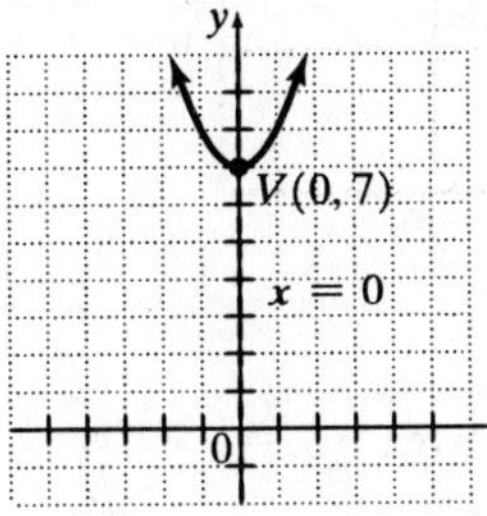

7. $f(x) = (x-5)^2$

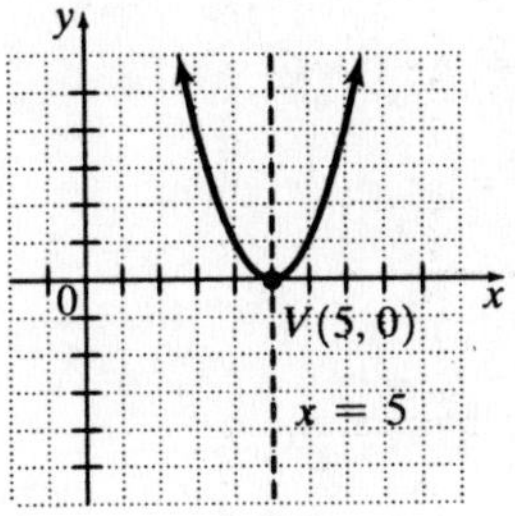

9. $h(x) = (x+2)^2$

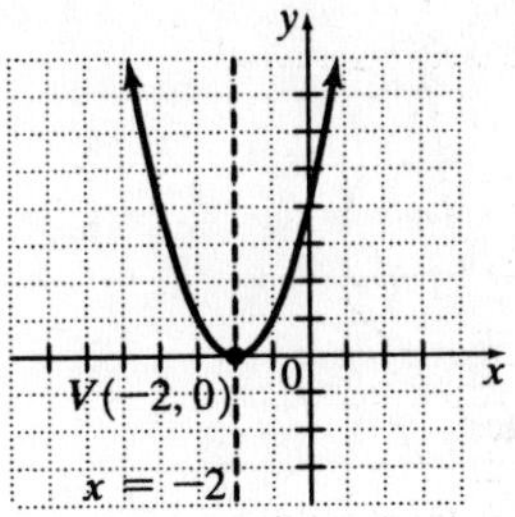

11. $G(x) = (x+3)^2$

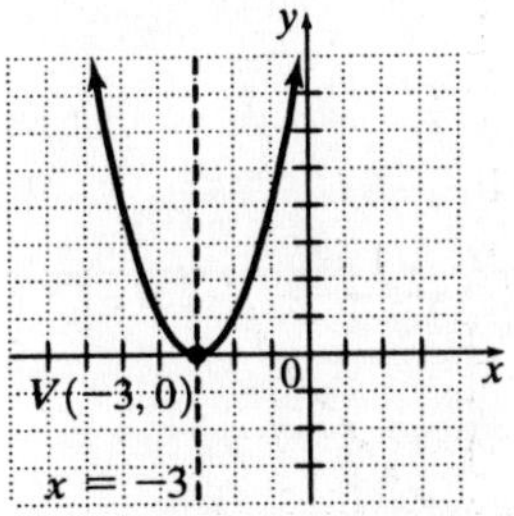

13. $f(x) = (x-2)^2 + 5$

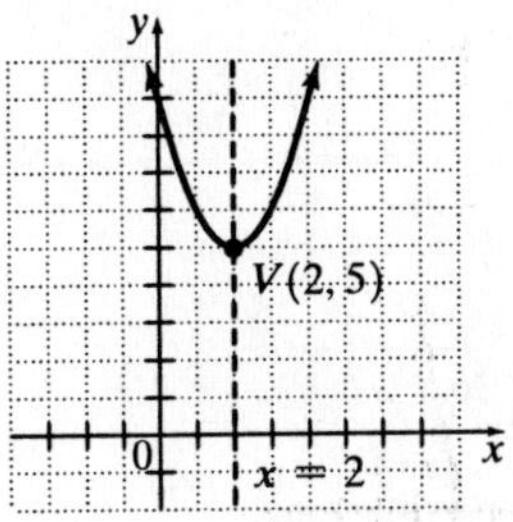

15. $h(x)=(x+1)^2+4$

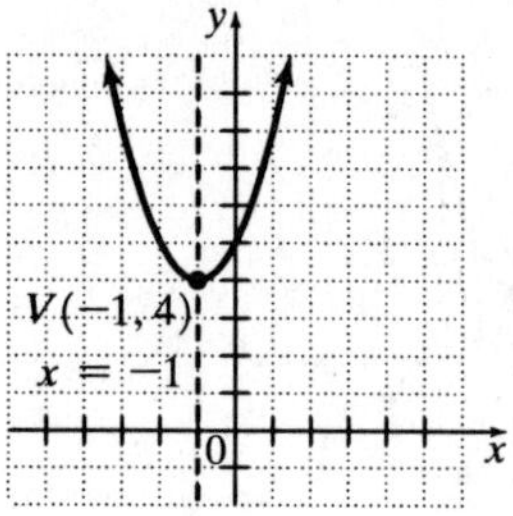

17. $g(x)=(x+2)^2-5$

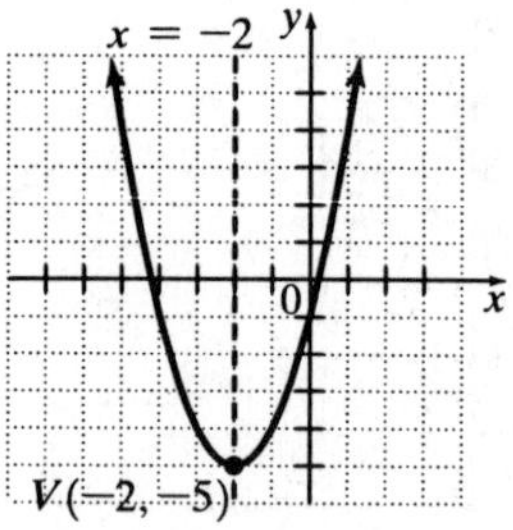

19. $g(x)=-x^2$

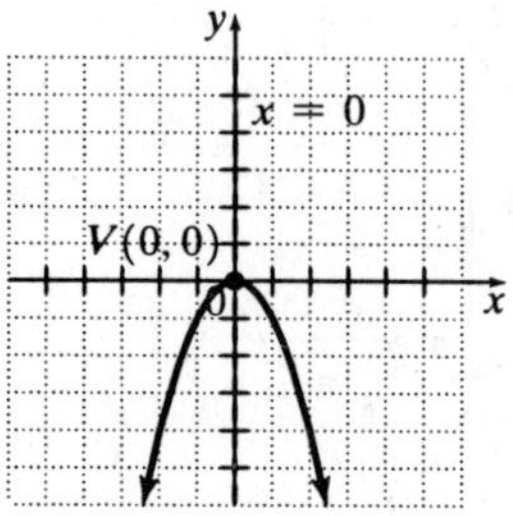

21. $h(x)=\frac{1}{3}x^2$

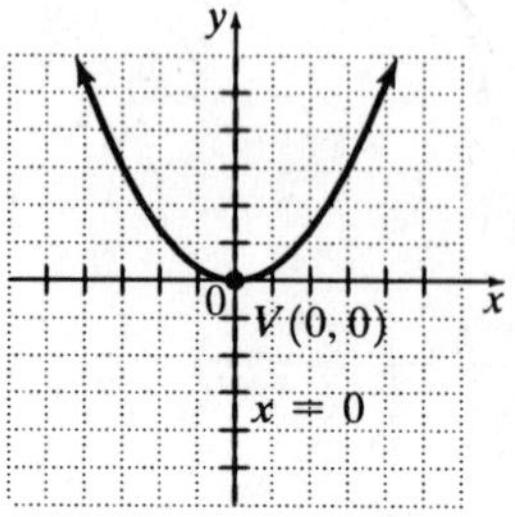

23. $H(x)=2x^2$

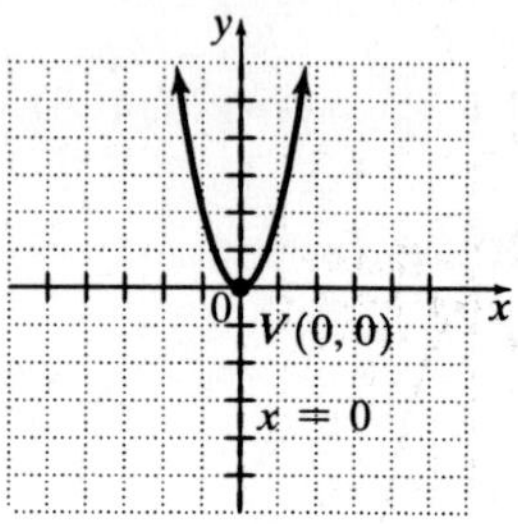

25. $f(x)=2(x-1)^2+3$

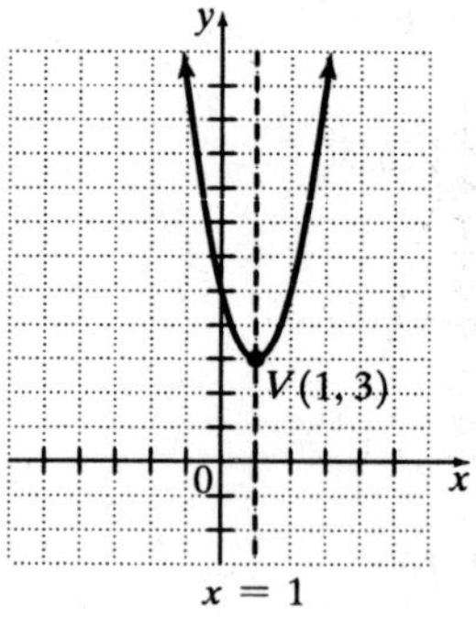

27. $h(x)=-3(x+3)^2+1$

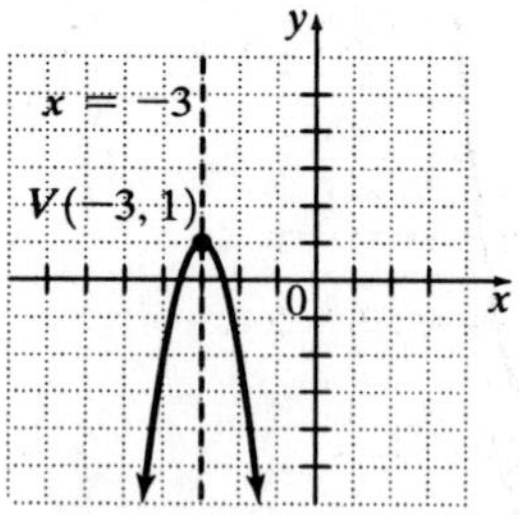

29. $H(x)=\frac{1}{2}(x-6)^2-3$

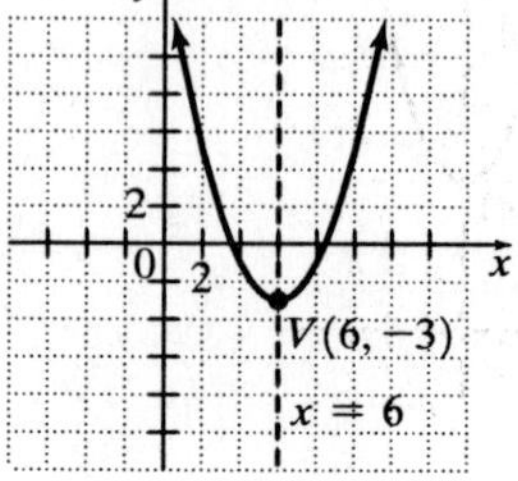

31. $f(x) = -(x-2)^2$

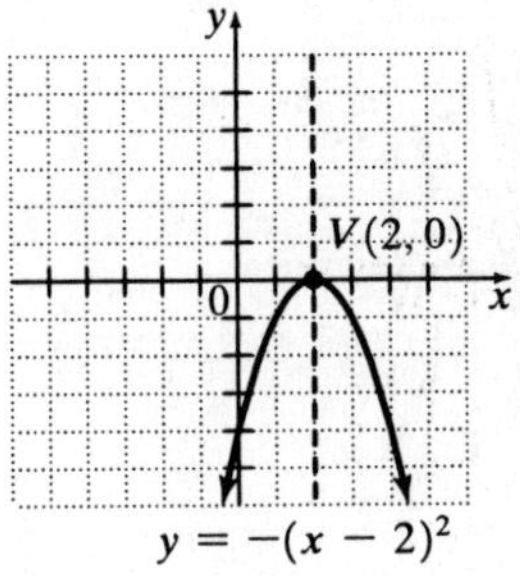

33. $F(x) = -x^2 + 4$

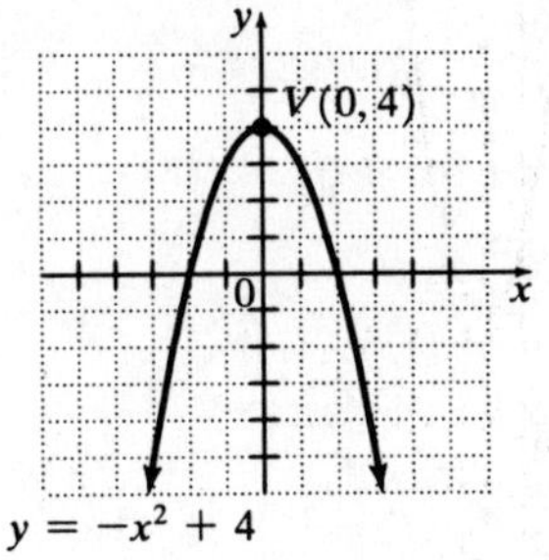

35. $F(x) = 2x^2 - 5$

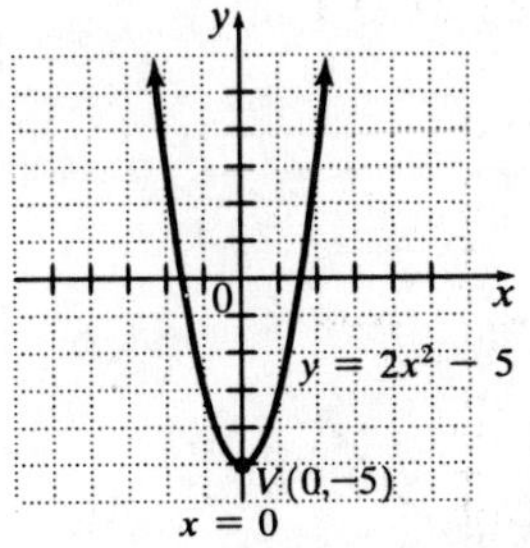

37. $h(x) = (x-6)^2 + 4$

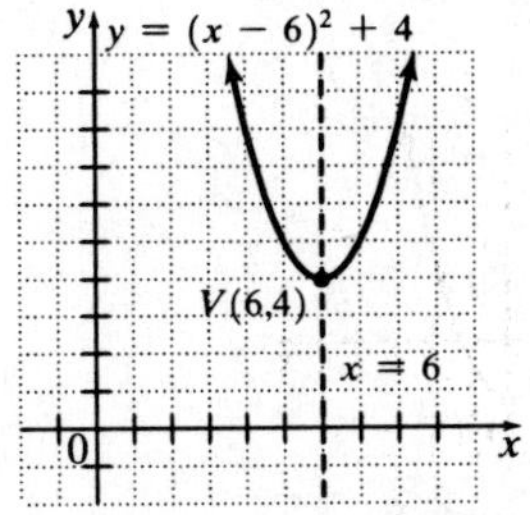

39. $F(x) = \left(x + \frac{1}{2}\right)^2 - 2$

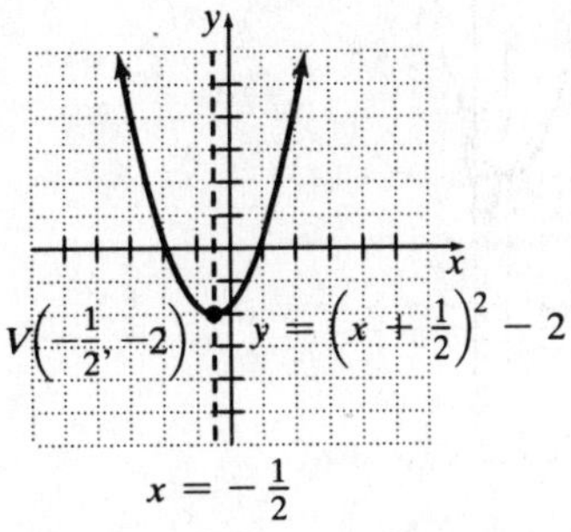

41. $F(x) = \frac{3}{2}(x+7)^2 + 1$

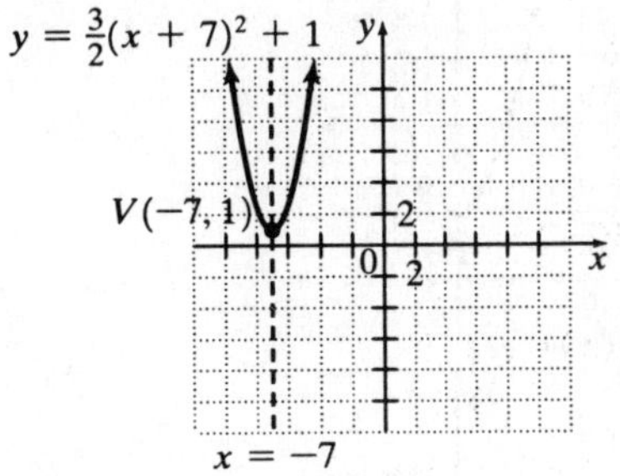

43. $f(x) = \frac{1}{4}x^2 - 9$

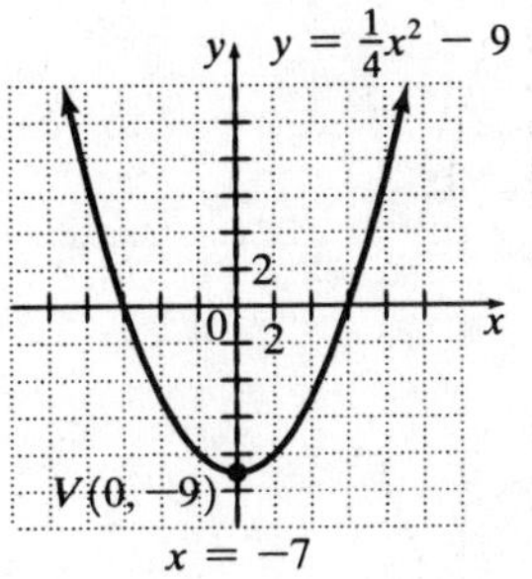

45. $G(x) = 5\left(x + \frac{1}{2}\right)^2$

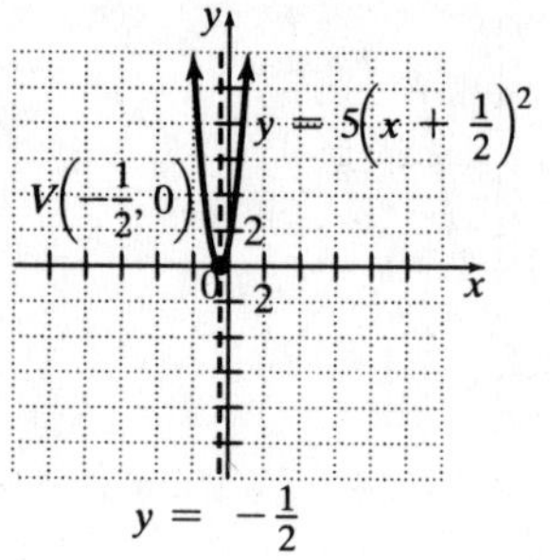

47. $f(x) = -(x-1)^2 - 1$

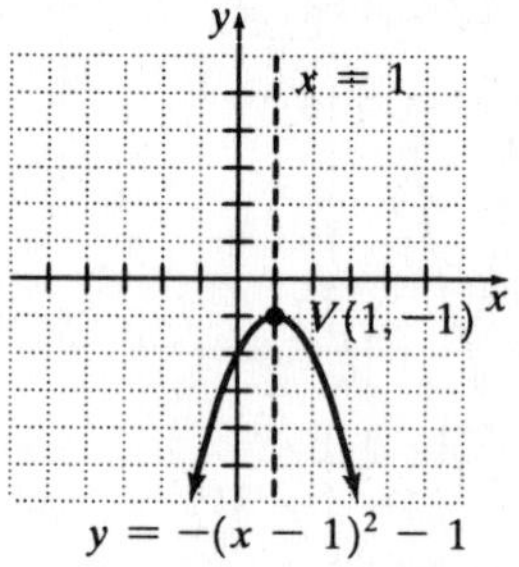

49. $g(x) = \sqrt{3}(x+5)^2 + \frac{3}{4}$

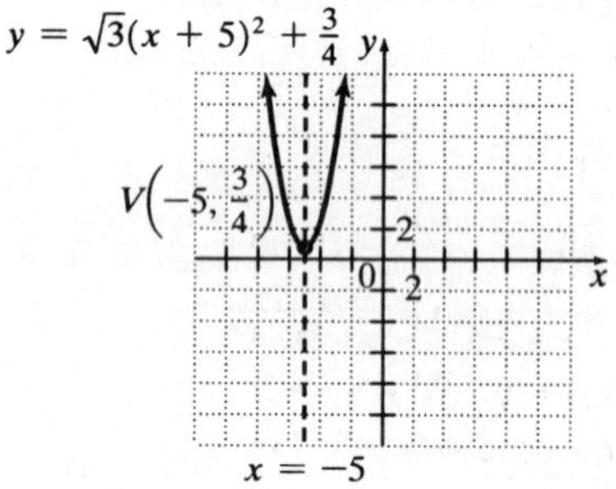

51. $h(x) = 10(x+4)^2 - 6$

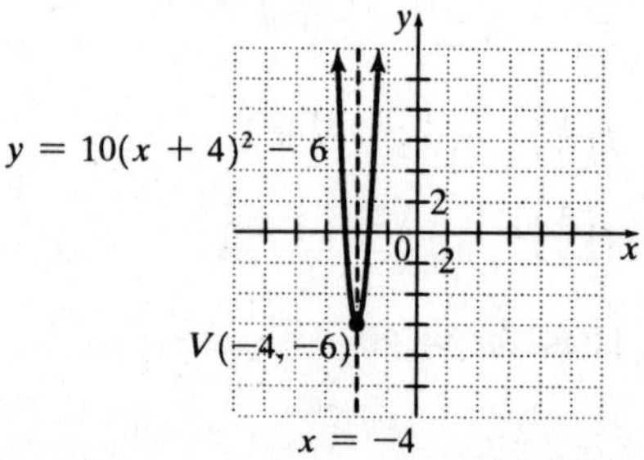

53. $f(x) = -2(x-4)^2 + 5$

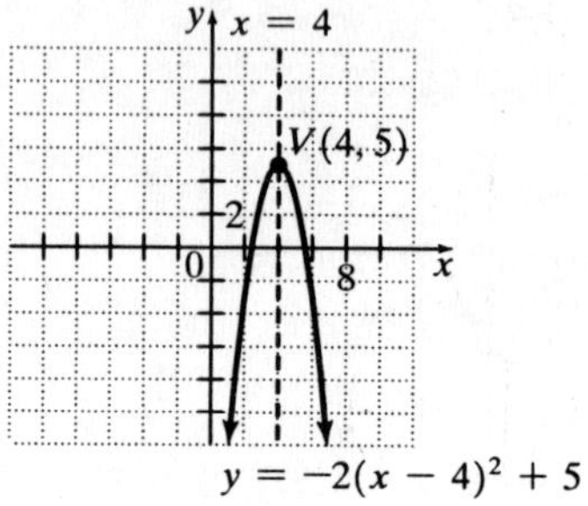

55. $f(x) = 5(x-2)^2 + 3$

57. $f(x) = 5[x-(-3)]^2 + 6$
$f(x) = 5(x+3)^2 + 6$

59. $y = f(x) + 1$

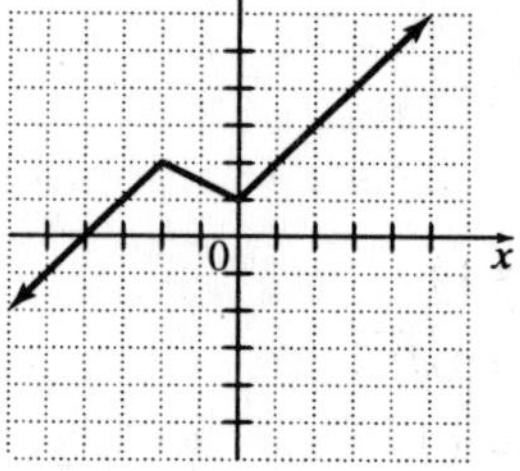

61. $y = f(x - 3)$

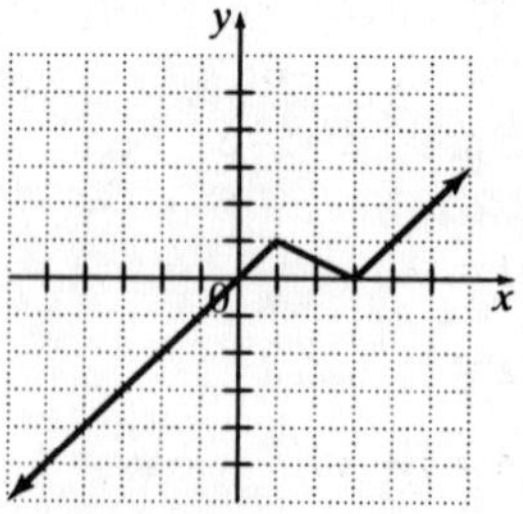

63. $y = f(x+2) + 2$

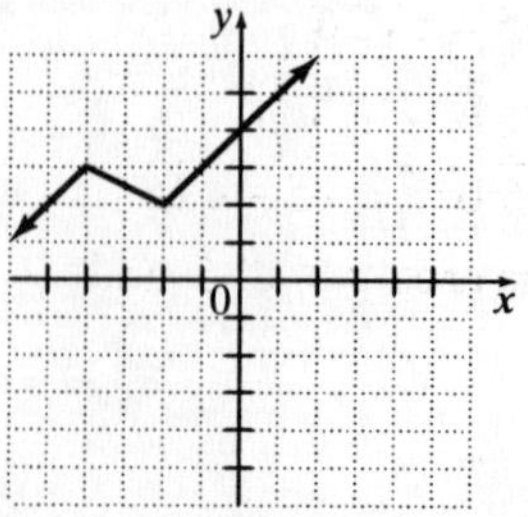

65. $x^2 + 8x$

$\left[\frac{1}{2}(8)\right]^2 = [4]^2 = 16$

$x^2 + 8x + 16$

67. $z^2 - 16z$

$\left[\frac{1}{2}(-16)\right]^2 = [-8]^2 = 64$

$z^2 - 16z + 64$

69. $y^2 + y$

$\left[\frac{1}{2}(1)\right]^2 = \left[\frac{1}{2}\right]^2 = \frac{1}{4}$

$y^2 + y + \frac{1}{4}$

71.
$$x^2 + 4x = 12$$
$$x^2 + 4x + 4 = 12 + 4$$
$$(x+2)^2 = 16$$
$$x + 2 = \pm\sqrt{16}$$
$$x + 2 = \pm 4$$
$$x = -2 \pm 4$$
$$x = -2 + 4 \quad \text{or} \quad x = -2 - 4$$
$$x = 2 \quad \text{or} \quad x = -6$$
The solutions are 2 and –6.

73.
$$z^2 + 10z - 1 = 0$$
$$z^2 + 10z = 1$$
$$z^2 + 10z + 25 = 1 + 25$$
$$(z+5)^2 = 26$$
$$z + 5 = \pm\sqrt{26}$$
$$z = -5 \pm \sqrt{26}$$
$$z = -5 + \sqrt{26} \text{ or } z = -5 - \sqrt{26}$$
The solutions are $-5 + \sqrt{26}$ and $-5 - \sqrt{26}$.

75.
$$z^2 - 8z = 2$$
$$z^2 - 8z + 16 = 2 + 16$$
$$(z-4)^2 = 18$$
$$z - 4 = \pm\sqrt{18}$$
$$z - 4 = \pm 3\sqrt{2}$$
$$z = 4 \pm 3\sqrt{2}$$
$$z = 4 + 3\sqrt{2} \text{ or } z = 4 - 3\sqrt{2}$$
The solutions are $4 + 3\sqrt{2}$ and $4 - 3\sqrt{2}$.

Exercise Set 8.6

1. $f(x) = x^2 + 8x + 7$

$-\frac{b}{2a} = \frac{-8}{2(1)} = -4$ and

$f(-4) = (-4)^2 + 8(-4) + 7$

$f(-4) = 16 - 32 + 7 = -9$

Thus, the vertex is (–4, –9).

3. $f(x) = -x^2 + 10x + 5$

$\frac{-b}{2a} = \frac{-10}{2(-1)} = 5$ and

$f(5) = -5^2 + 10(5) + 5$

$f(5) = -25 + 50 + 5 = 30$

Thus, the vertex is (5, 30).

5. $f(x) = 5x^2 - 10x + 3$

$\frac{-b}{2a} = \frac{-(-10)}{2(5)} = 1$ and

$f(1) = 5(1)^2 - 10(1) + 3$

$f(1) = 5 - 10 + 3 = -2$

Thus, the vertex is (1, –2).

7. $f(x) = -x^2 + x + 1$

$\frac{-b}{2a} = \frac{-1}{2(-1)} = \frac{1}{2}$ and

$f\left(\frac{1}{2}\right) = -\left(\frac{1}{2}\right)^2 + \frac{1}{2} + 1$

$f\left(\frac{1}{2}\right) = -\frac{1}{4} + \frac{1}{2} + 1 = \frac{5}{4}$

Thus, the vertex is $\left(\frac{1}{2}, \frac{5}{4}\right)$.

9. $f(x) = x^2 - 4x + 3$

$\frac{-b}{2a} = \frac{-(-4)}{2(1)} = 2$ and

$f(2) = 2^2 - 4(2) + 3 = -1$

The vertex is $(2, -1)$, so Graph D.

11. $f(x) = x^2 - 2x - 3$

$\frac{-b}{2a} = \frac{-(-2)}{2(1)} = 1$ and

$f(1) = 1^2 - 2(1) - 3 = -4$

The vertex is $(1, -4)$, so Graph B.

13. $f(x) = x^2 + 4x - 5$

$\frac{-b}{2a} = \frac{-4}{2(1)} = -2$ and

$f(-2) = (-2)^2 + 4(-2) - 5$

$f(-2) = 4 - 8 - 5 = -9$

Thus, the vertex is $(-2, -9)$.

The graph opens upward since $a = 1 > 0$.

$x^2 + 4x - 5 = 0$

$(x+5)(x-1) = 0$

$x + 5 = 0$ or $x - 1 = 0$

$x = -5$ or $x = 1$

The x-intercepts are $(-5, 0)$ and $(1, 0)$.

$f(0) = -5$, so the y-intercept is $(0, -5)$.

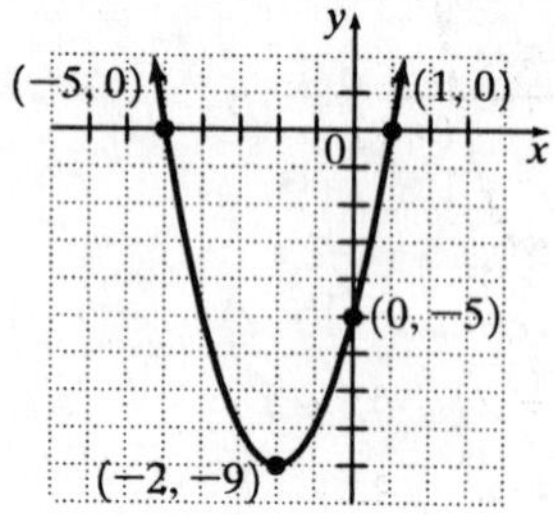

15. $f(x) = -x^2 + 2x - 1$

$\frac{-b}{2a} = \frac{-2}{2(-1)} = 1$ and

$f(1) = -1^2 + 2(1) - 1$

$f(1) = -1 + 2 - 1 = 0$

Thus, the vertex is $(1, 0)$.

The graph opens downward since $a = -1 < 0$.

$-x^2 + 2x - 1 = 0$

$-(x^2 - 2x + 1) = 0$

$-(x-1)^2 = 0$

$x - 1 = 0$

$x = 1$

The x-intercept is $(1, 0)$.

$f(0) = -1$, so the y-intercept is $(0, -1)$.

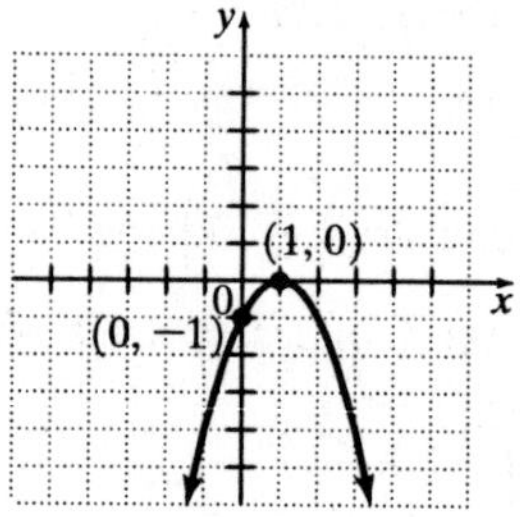

17. $f(x) = x^2 - 4$

$\frac{-b}{2a} = \frac{-0}{2(1)} = 0$ and

$f(0) = (0)^2 - 4$

$f(0) = 0 - 4 = -4$

Thus, the vertex is $(0, -4)$.

The graph opens upward since $a = 1 > 0$.

$x^2 - 4 = 0$

$(x+2)(x-2) = 0$

$x + 2 = 0$ or $x - 2 = 0$

$x = -2$ or $x = 2$

The x-intercepts are $(-2, 0)$ and $(2, 0)$.

$f(0) = -4$, so the y-intercept is $(0, -4)$.

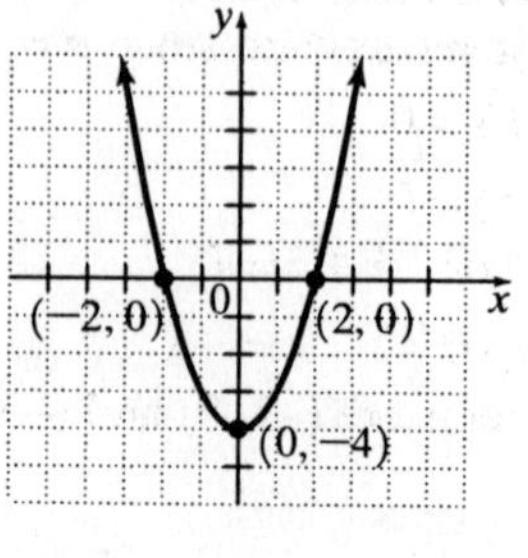

19. $f(x) = 4x^2 + 4x - 3$

$\frac{-b}{2a} = \frac{-4}{2(4)} = -\frac{1}{2}$ and

$f\left(-\frac{1}{2}\right) = 4\left(-\frac{1}{2}\right)^2 + 4\left(-\frac{1}{2}\right) - 3$

$f\left(-\frac{1}{2}\right) = 1 - 2 - 3 = -4$

Thus, the vertex is $\left(-\frac{1}{2}, -4\right)$.

The graph opens upward since $a = 4 > 0$.

$$4x^2 + 4x - 3 = 0$$
$$(2x - 1)(2x + 3) = 0$$
$$2x - 1 = 0 \quad \text{or} \quad 2x + 3 = 0$$
$$2x = 1 \quad \text{or} \quad 2x = -3$$
$$x = \frac{1}{2} \quad \text{or} \quad x = -\frac{3}{2}$$

The x-intercepts are $\left(\frac{1}{2}, 0\right)$ and $\left(-\frac{3}{2}, 0\right)$.

$f(0) = -3$, so the y-intercept is $(0, -3)$.

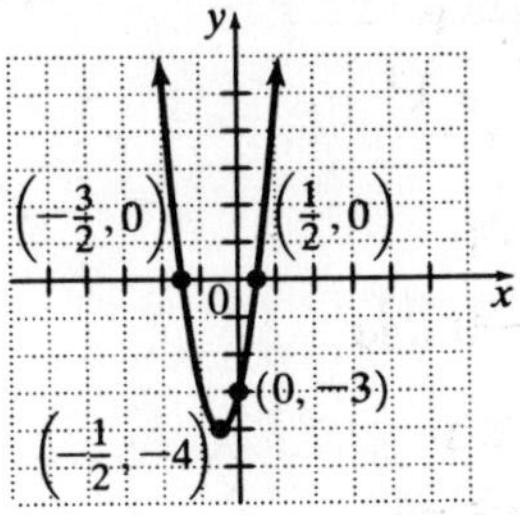

21.
$$y = x^2 + 8x + 15$$
$$y - 15 = x^2 + 8x$$
$$y - 15 + 16 = x^2 + 8x + 16$$
$$y + 1 = (x + 4)^2$$
$$f(x) = (x + 4)^2 - 1$$

Thus, the vertex is $(-4, -1)$.

The graph opens upward since $a = 1 > 0$.

$$x^2 + 8x + 15 = 0$$
$$(x + 3)(x + 5) = 0$$
$$x + 3 = 0 \quad \text{or} \quad x + 5 = 0$$
$$x = -3 \quad \text{or} \quad x = -5$$

The x-intercepts are $(-3, 0)$ and $(-5, 0)$.

$f(0) = 15$, so the y-intercept is $(0, 15)$.

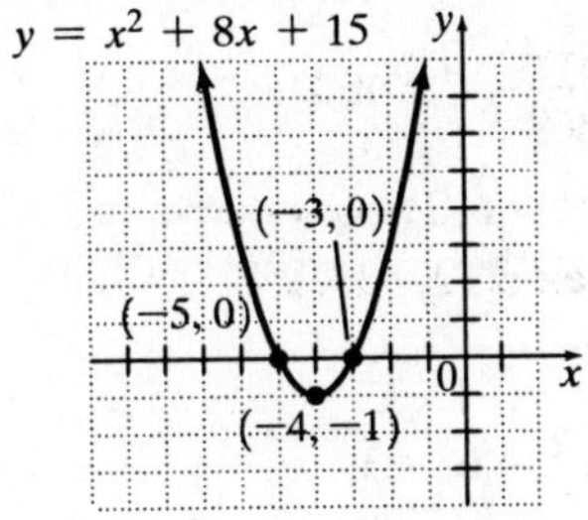

23.
$$y = x^2 - 6x + 5$$
$$y - 5 = x^2 - 6x$$
$$y - 5 + 9 = x^2 - 6x + 9$$
$$y + 4 = (x - 3)^2$$
$$f(x) = (x - 3)^2 - 4$$

Thus, the vertex is $(3, -4)$.

The graph opens upward since $a = 1 > 0$.

$$x^2 - 6x + 5 = 0$$
$$(x - 1)(x - 5) = 0$$
$$x - 1 = 0 \quad \text{or} \quad x - 5 = 0$$
$$x = 1 \quad \text{or} \quad x = 5$$

The x-intercepts are $(1, 0)$ and $(5, 0)$.

$f(0) = 5$, so the y-intercept is $(0, 5)$.

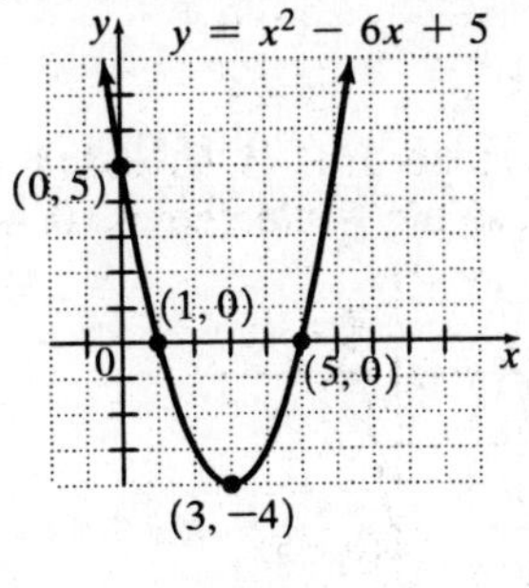

25.
$$y = x^2 - 4x + 5$$
$$y - 5 = x^2 - 4x$$
$$y - 5 + 4 = x^2 - 4x + 4$$
$$y - 1 = (x - 2)^2$$
$$f(x) = (x - 2)^2 + 1$$

Thus, the vertex is $(2, 1)$.

The graph opens upward since $a = 1 > 0$.

$x^2 - 4x + 5 = 0$

$(x-2)^2 + 1 = 0$

$(x-2)^2 = -1$

Hence, there are no x-intercepts.
$f(0) = 5$, so the y-intercept is (0, 5).

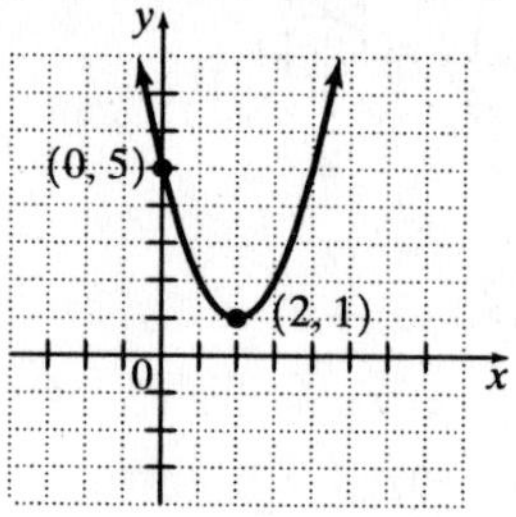

27. $y = 2x^2 + 4x + 5$

$y - 5 = 2(x^2 + 2x)$

$y - 5 + 2 = 2(x^2 + 2x + 1)$

$y - 3 = 2(x+1)^2$

$f(x) = 2(x+1)^2 + 3$

Thus, the vertex is (–1, 3).
The graph opens upward since $a = 2 > 0$.

$2x^2 + 4x + 5 = 0$

$2(x+1)^2 + 3 = 0$

$2(x+1)^2 = -3$

Hence, there are no x-intercepts.
$f(0) = 5$, so the y-intercept is (0, 5).

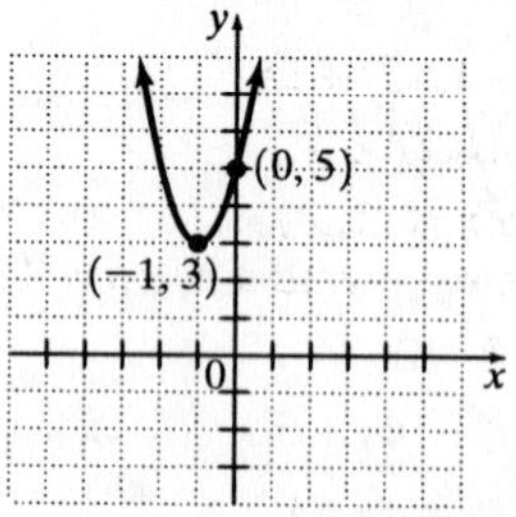

29. $y = -2x^2 + 12x$

$y = -2(x^2 + 6x)$

$y - 18 = -2(x^2 + 6x + 9)$

$y - 18 = -2(x+3)^2$

$f(x) = -2(x-3)^2 + 18$

Thus, the vertex is (3, 18).
The graph opens downward since $a = -2 < 0$.

$-2x^2 + 12x = 0$

$-2x(x-6) = 0$

$-2x = 0$ or $x - 6 = 0$

$x = 0$ or $x = 6$

The x-intercepts are (0, 0) and (6, 0).
$f(0) = 0$, so the y-intercept is (0, 0).

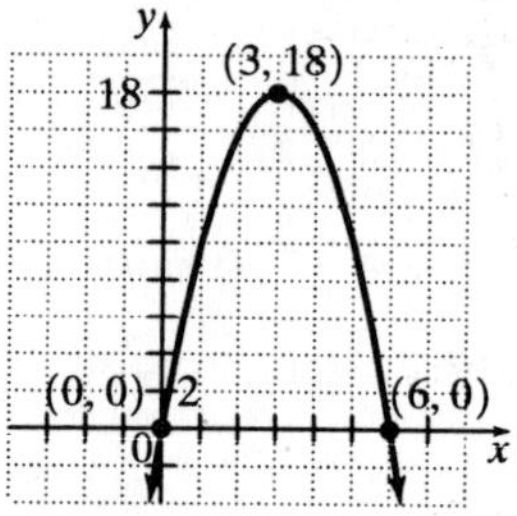

31. $f(x) = x^2 + 1$

$\frac{-b}{2a} = \frac{-0}{2(1)} = 0$ and

$f(0) = 0^2 + 1$

$f(0) = 0 + 1 = 1$

Thus, the vertex is (0, 1).
The graph opens upward since $a = 1 > 0$.

$x^2 + 1 = 0$

$x^2 = -1$

Hence, there are no x-intercepts.
$f(0) = 1$, so the y-intercept is (0, 1).

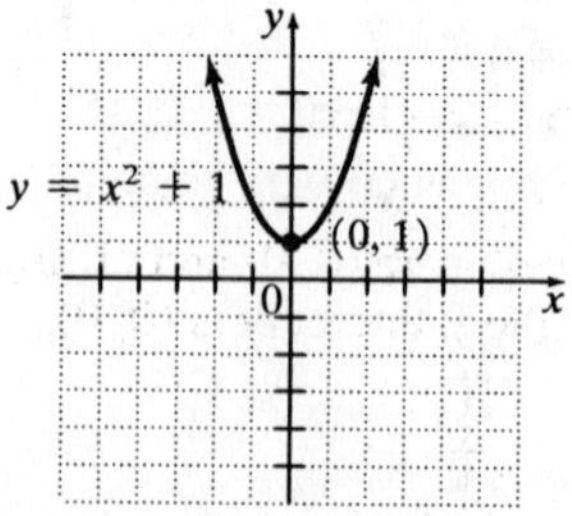

33. $f(x) = x^2 - 2x - 15$

$\frac{-b}{2a} = \frac{-(-2)}{2(1)} = 1$ and

$f(1) = 1^2 - 2(1) - 15$

$f(1) = 1 - 2 - 15 = -16$

Thus, the vertex is (1, –16).
The graph opens upward since $a = 1 > 0$.

$x^2 - 2x - 15 = 0$
$(x-5)(x+3) = 0$
$x - 5 = 0$ or $x + 3 = 0$
$x = 5$ or $x = -3$
The x-intercepts are (5, 0) and (–3, 0).
$f(0) = -15$, so the y-intercept is (0, –15).

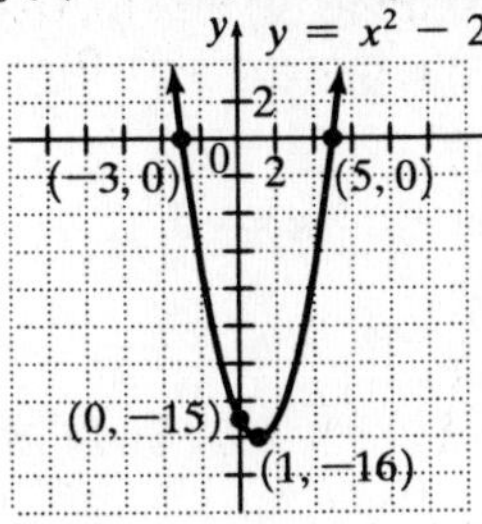

35. $f(x) = -5x^2 + 5x$
$\frac{-b}{2a} = \frac{-5}{2(-5)} = \frac{1}{2}$ and
$f\left(\frac{1}{2}\right) = -5\left(\frac{1}{2}\right)^2 + 5\left(\frac{1}{2}\right)$
$f\left(\frac{1}{2}\right) = -\frac{5}{4} + \frac{5}{2} = \frac{5}{4}$
Thus, the vertex is $\left(\frac{1}{2}, \frac{5}{4}\right)$.
The graph opens downward since $a = -5 < 0$.
$-5x^2 + 5x = 0$
$-5x(x-1) = 0$
$-5x = 0$ or $x - 1 = 0$
$x = 0$ or $x = 1$
The x-intercepts are (0, 0) and (1, 0).
$f(0) = 0$, so the y-intercept is (0, 0).

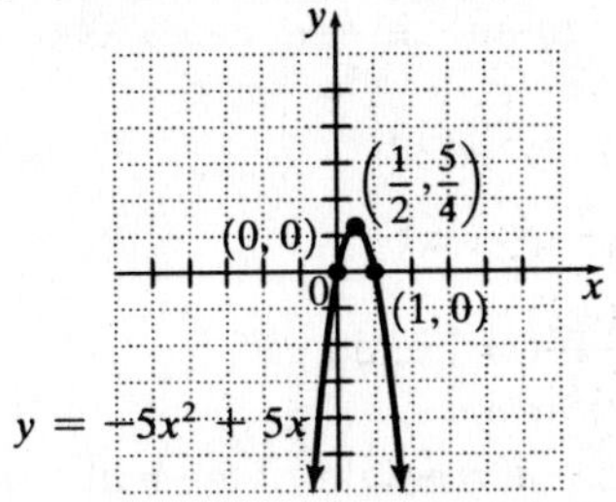

37. $f(x) = -x^2 + 2x - 12$
$\frac{-b}{2a} = \frac{-2}{2(-1)} = 1$ and
$f(1) = -1^2 + 2(1) - 12$
$f(1) = -1 + 2 - 12 = -11$
Thus, the vertex is (1, –11).
The graph opens downward since $a = -1 < 0$.
$-x^2 + 2x - 12 = 0$
$x^2 - 2x = -12$
$x^2 - 2x + 1 = -12 + 1$
$(x-1)^2 = -11$
Hence, there are no x-intercepts.
$f(0) = -12$, so the y-intercept is (0, –12).

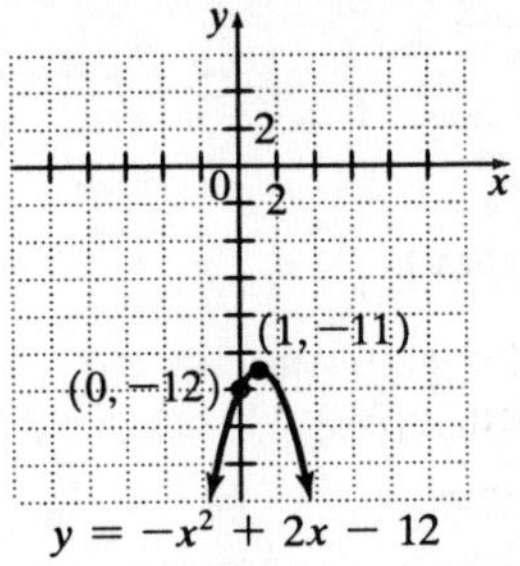

39. $f(x) = 3x^2 - 12x + 15$
$\frac{-b}{2a} = \frac{-(-12)}{2(3)} = 2$ and
$f(2) = 3(2)^2 - 12(2) + 15$
$f(2) = 12 - 24 + 15 = 3$
Thus, the vertex is (2, 3).
The graph opens upward since $a = 3 > 0$.
$3x^2 - 12x + 15 = 0$
$x^2 - 4x + 5 = 0$
$x^2 - 4x + 4 = -5 + 4$
$(x-2)^2 = -1$
Hence, there are no x-intercepts.

$f(0) = 15$, so the y-intercept is (0, 15).

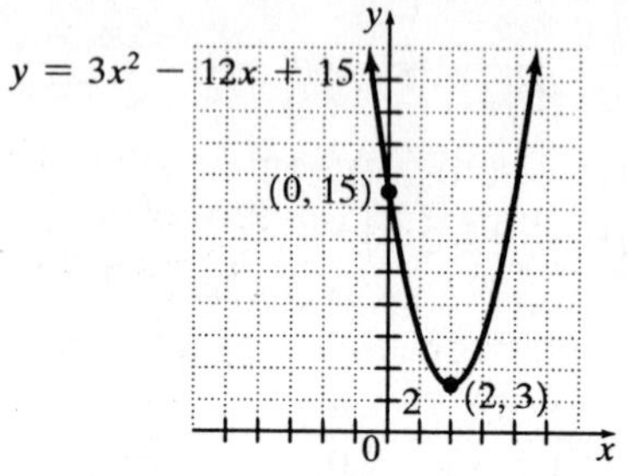

41. $f(x) = x^2 + x - 6$

$\frac{-b}{2a} = \frac{-1}{2(1)} = -\frac{1}{2}$ and

$f\left(-\frac{1}{2}\right) = \left(-\frac{1}{2}\right)^2 + \left(-\frac{1}{2}\right) - 6$

$f\left(-\frac{1}{2}\right) = \frac{1}{4} - \frac{1}{2} - 6 = -\frac{25}{4}$

Thus, the vertex is $\left(-\frac{1}{2}, -\frac{25}{4}\right)$.

The graph opens upward since $a = 1 > 0$.

$x^2 + x - 6 = 0$

$(x+3)(x-2) = 0$

$x + 3 = 0$ or $x - 2 = 0$

$x = -3$ or $x = 2$

The x-intercepts are (–3, 0) and (2, 0).

$f(0) = -6$, so the y-intercept is (0, –6).

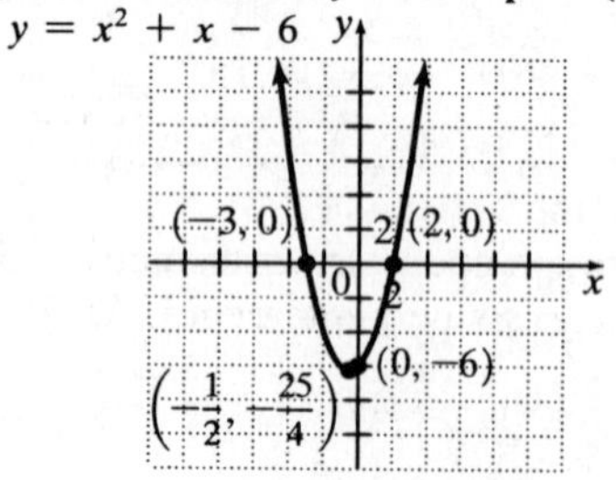

43. $f(x) = -2x^2 - 3x + 35$

$\frac{-b}{2a} = \frac{-(-3)}{2(-2)} = -\frac{3}{4}$ and

$f\left(-\frac{3}{4}\right) = -2\left(-\frac{3}{4}\right)^2 - 3\left(-\frac{3}{4}\right) + 35$

$f\left(-\frac{3}{4}\right) = -\frac{9}{8} + \frac{9}{4} + 35 = \frac{289}{8}$

Thus, the vertex is $\left(-\frac{3}{4}, \frac{289}{8}\right)$.

The graph opens downward since $a = -2 < 0$.

$-2x^2 - 3x + 35 = 0$

$2x^2 + 3x - 35 = 0$

$(2x-7)(x+5) = 0$

$2x - 7 = 0$ or $x + 5 = 0$

$2x = 7$ or $x = -5$

$x = \frac{7}{2}$

The x-intercepts are $\left(\frac{7}{2}, 0\right)$ and (–5, 0).

$f(0) = 35$, so the y-intercept is (0, 35).

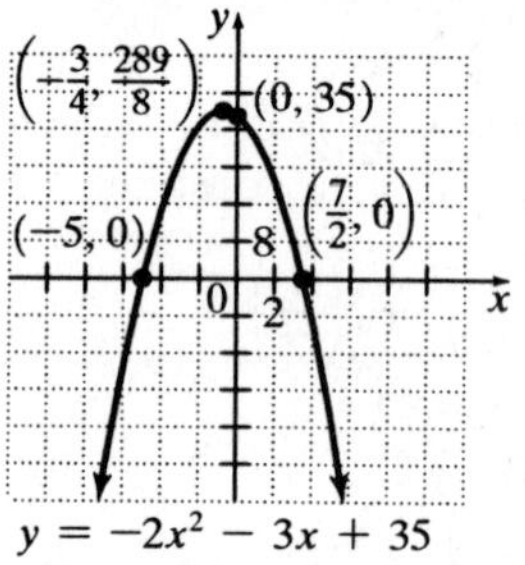

45. $C(x) = 2x^2 - 800x + 92{,}000$

a. $\frac{-b}{2a} = \frac{-(-800)}{2(2)} = 200$

200 bicycles

b. $C(200)$

$= 2(200)^2 - 800(200) + 92{,}000$

$= 12{,}000$

The minimum cost is $12,000.

47. $h(t) = -16t^2 + 32t$

$\frac{-b}{2a} = \frac{-32}{2(-16)} = 1$

$h(1) = -16(1)^2 + 32(1)$

$h(1) = 16$ feet

The maximum height of the ball is 16 feet.

49. Let x = one number.

$60 - x$ = other number

$f(x) = x(60 - x)$

$f(x) = 60x - x^2$
$f(x) = -x^2 + 60x$
$f(x) = -1(x^2 - 60x)$
$f(x) = -1(x^2 - 60x + 900) + 900$
$f(x) = -(x - 30)^2 + 900$
The maximum will occur at the vertex which is (30, 900). The numbers are 30 and 30.

51. Let x = one number
$10 + x$ = other number
$f(x) = x(10 + x)$
$f(x) = 10x + x^2$
$f(x) = (x^2 + 10x + 25) - 25$
$f(x) = (x + 5)^2 - 25$
The minimum will occur at the vertex which is (–5, –25). The numbers are –5 and 5.

53. Let x = width
$40 - x$ = the length
$f(x) = x(40 - x)$
$f(x) = 40x - x^2$
$f(x) = -x^2 + 40x$
$f(x) = -1(x^2 - 40x)$
$f(x) = -1(x^2 - 40x + 400) + 400$
$f(x) = -(x - 20)^2 + 400$
The maximum will occur at the vertex which is (20, 400). The width is 20 units and the length is 20 units.

55. a. $f(14) = -0.74(14)^2 + 8.66(14) + 159.07$
≈ 135.27 million metric tons

b. The maximum value occurs when
$x = \dfrac{-8.66}{2(-0.74)} \approx 6.$
Therefore emissions were maximized in 1996.

c. $f(6) = -0.74(6)^2 + 8.66(6) + 159.07$
≈ 184.39 million metric tons

57. $f(x) = x^2 + 10x + 15$
$\dfrac{-b}{2a} = \dfrac{-10}{2(1)} = -5$ and
$f(-5) = (-5)^2 + 10(-5) + 15 = -10$
Thus, the vertex is (–5, –10).
The graph opens upward since $a = 1 > 0$.
$x^2 + 10x + 15 = 0$
$$x = \frac{-10 \pm \sqrt{(10)^2 - 4(1)(15)}}{2(1)}$$
The x-intercepts are approximately (–8.2, 0) and (–1.8, 0).
$f(0) = 15$, so the y-intercept is (0, 15).

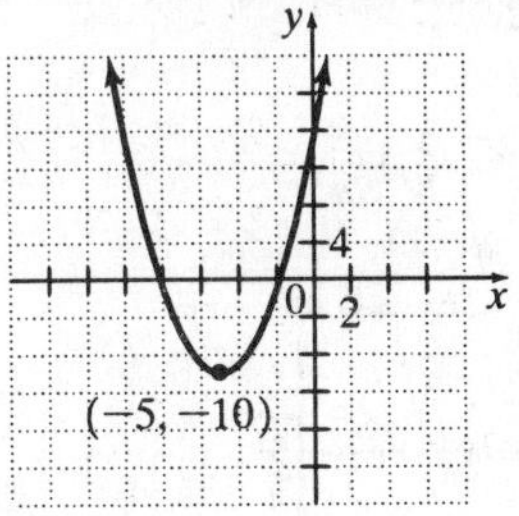

59. $f(x) = 3x^2 - 6x + 7$
$\dfrac{-b}{2a} = \dfrac{-(-6)}{2(3)} = 1$ and
$f(1) = 3(1)^2 - 6(1) + 7 = 4$
Thus the vertex is (1, 4).
The graph opens upward since $a = 3 > 0$.
$3x^2 - 6x + 7 = 0$
$$x = \frac{-(-6) \pm \sqrt{(-6)^2 - 4(3)(7)}}{2(3)}$$
There are no x-intercepts.
$f(0) = 7$, so the y-intercept is (0, 7).

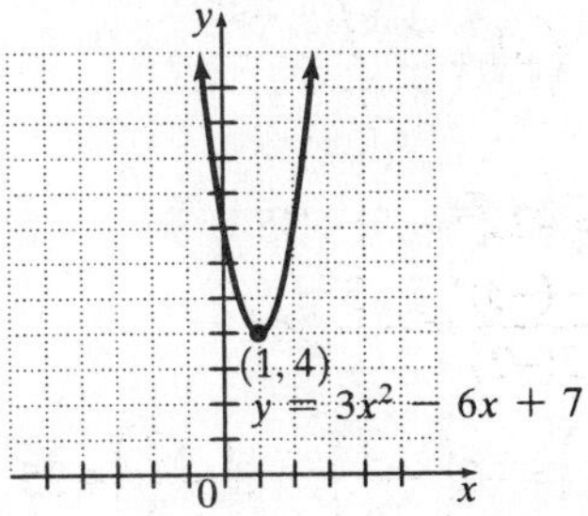

61.

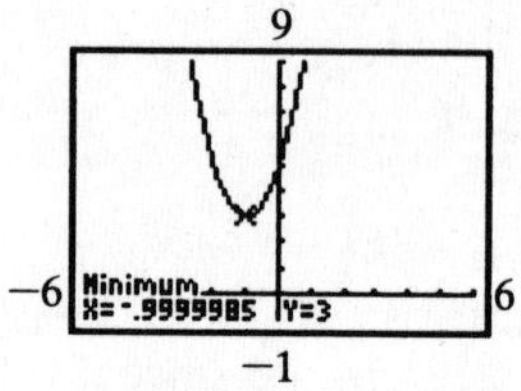

63.

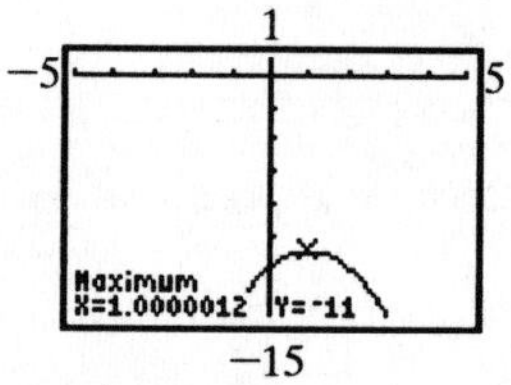

65. $f(x) = 2.3x^2 - 6.1x + 3.2$
minimum ≈ -0.84

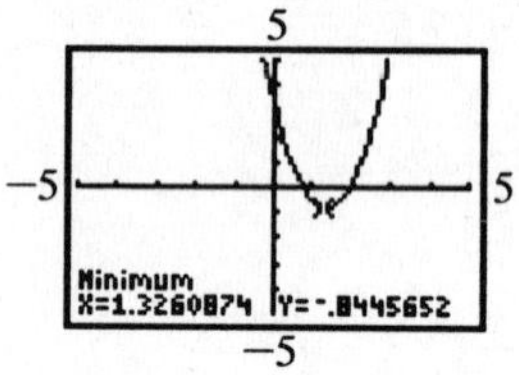

67. $f(x) = -1.9x^2 + 5.6x - 2.7$
maximum ≈ 1.43

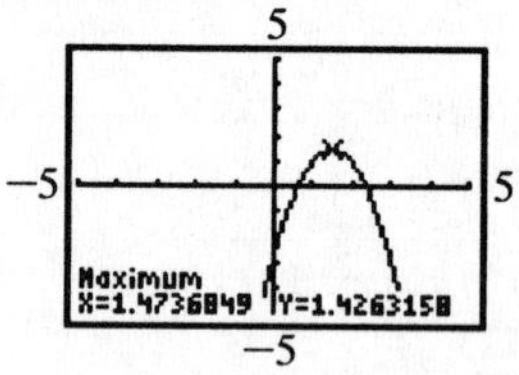

69.

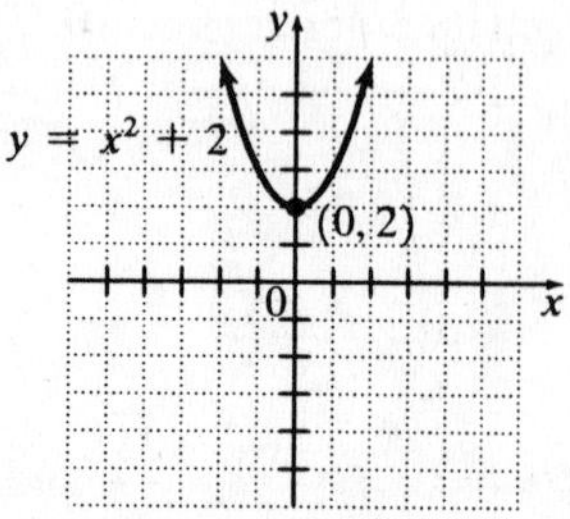

71.

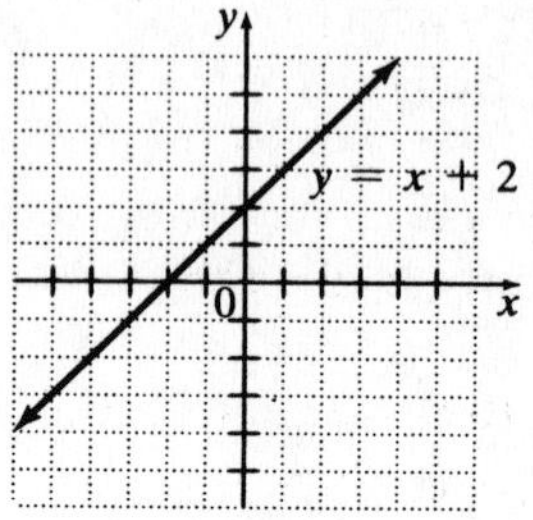

73.

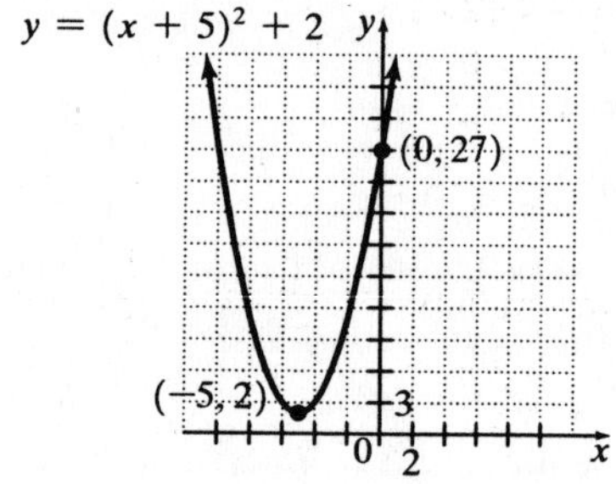

75.

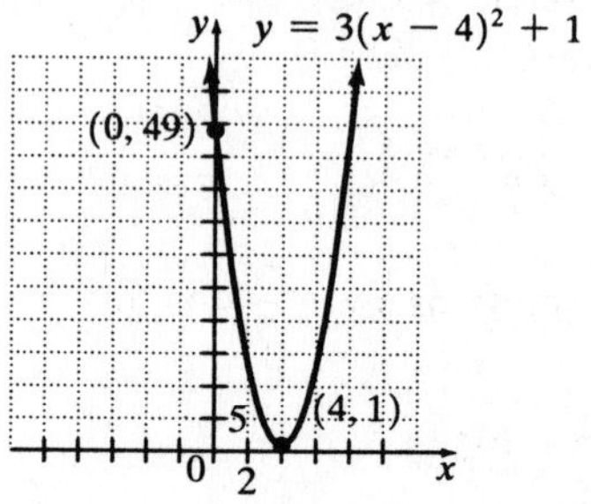

77.

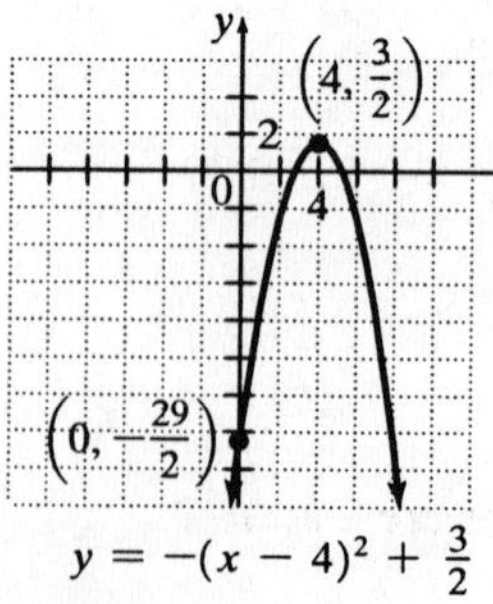

Chapter 8 Review Exercises

1. $x^2 - 15x + 14 = 0$
$(x-14)(x-1) = 0$
$x - 14 = 0$ or $x - 1 = 0$
$x = 14$ or $x = 1$
The solutions are 14 and 1.

2. $x^2 - x - 30 = 0$
$(x+5)(x-6) = 0$
$x+5=0$ or $x-6=0$
$x=-5$ or $x=6$
The solutions are –5 and 6.

3. $10x^2 = 3x + 4$
$10x^2 - 3x - 4 = 0$
$(5x-4)(2x+1) = 0$
$5x-4=0$ or $2x+1=0$
$5x=4$ or $2x=-1$
$x=\frac{4}{5}$ or $x=-\frac{1}{2}$
The solutions are $\frac{4}{5}$ and $-\frac{1}{2}$.

4. $7a^2 = 29a + 30$
$7a^2 - 29a - 30 = 0$
$(7a+6)(a-5) = 0$
$7a+6=0$ or $a-5=0$
$a=-\frac{6}{7}$ or $a=5$
The solutions are $-\frac{6}{7}$ and 5.

5. $4m^2 = 196$
$m^2 = 49$
$m = \pm\sqrt{49}$
$m = \pm 7$
The solutions are 7 and –7.

6. $9y^2 = 36$
$y^2 = 4$
$y = \pm 2$
The solutions are 2 and –2.

7. $(9n+1)^2 = 9$
$9n+1 = \pm\sqrt{9}$
$9n+1 = \pm 3$
$9n = -1 \pm 3$
$n = \frac{-1\pm 3}{9} = \frac{2}{9}, -\frac{4}{9}$
The solutions are $\frac{2}{9}$ and $-\frac{4}{9}$.

8. $(5x-2)^2 = 2$
$5x - 2 = \pm\sqrt{2}$
$5x = 2 \pm \sqrt{2}$
$x = \frac{2\pm\sqrt{2}}{5}$
The solutions are $\frac{2+\sqrt{2}}{5}$ and $\frac{2-\sqrt{2}}{5}$.

9. $z^2 + 3z + 1 = 0$
$z^2 + 3z = -1$
$z^2 + 3z + \left(\frac{3}{2}\right)^2 = -1 + \frac{9}{4}$
$\left(z + \frac{3}{2}\right)^2 = \frac{5}{4}$
$z + \frac{3}{2} = \pm\sqrt{\frac{5}{4}}$
$z + \frac{3}{2} = \frac{\pm\sqrt{5}}{\sqrt{4}}$
$z = -\frac{3}{2} \pm \frac{\sqrt{5}}{2}$
The solutions are $-\frac{3}{2} + \frac{\sqrt{5}}{2}$ and $-\frac{3}{2} - \frac{\sqrt{5}}{2}$.

10. $x^2 + x + 7 = 0$
$x^2 + x + \frac{1}{4} = -7 + \frac{1}{4}$
$\left(x + \frac{1}{2}\right)^2 = -\frac{27}{4}$
$x + \frac{1}{2} = \pm\sqrt{\frac{-27}{4}}$
$x = -\frac{1}{2} \pm \frac{3i\sqrt{3}}{2}$
$x = \frac{-1 \pm 3i\sqrt{3}}{2}$
The solutions are $\frac{-1+3i\sqrt{3}}{2}$ and $\frac{-1-3i\sqrt{3}}{2}$.

11.
$$(2x+1)^2 = x$$
$$4x^2+4x+1 = x$$
$$4x^2+3x = -1$$
$$x^2+\frac{3}{4}x = -\frac{1}{4}$$
$$x^2+\frac{3}{4}x+\left(\frac{\frac{3}{4}}{2}\right)^2 = -\frac{1}{4}+\frac{9}{64}$$
$$\left(x+\frac{3}{8}\right)^2 = -\frac{7}{64}$$
$$x+\frac{3}{8} = \pm\sqrt{\frac{-7}{64}}$$
$$x+\frac{3}{8} = \pm\frac{\sqrt{7}i}{8}$$
$$x = -\frac{3}{8}\pm\frac{\sqrt{7}}{8}i$$
The solutions are $-\frac{3}{8}+\frac{\sqrt{7}}{8}i$ and $-\frac{3}{8}-\frac{\sqrt{7}}{8}i$.

12.
$$(3x-4)^2 = 10x$$
$$9x^2-24x+16-10x = 0$$
$$9x^2-34x+16 = 0$$
$$x^2-\frac{34}{9}x+\frac{289}{81} = -\frac{16}{9}+\frac{289}{81}$$
$$\left(x-\frac{17}{9}\right)^2 = \frac{145}{81}$$
$$x-\frac{17}{9} = \pm\sqrt{\frac{145}{81}}$$
$$x = \frac{17\pm\sqrt{145}}{9}$$
The solutions are $\frac{17+\sqrt{145}}{9}$ and $\frac{17-\sqrt{145}}{9}$.

13. In this problem,
$P = 2500, A = 2717$
$A = P(1+r)^2$
$$2717 = 2500(1+r)^2$$
$$\frac{2717}{2500} = (1+r)^2$$
$$1.0868 = (1+r)^2$$
$$1.0425 = 1+r$$
$$0.0425 = r$$
The interest rate is 4.25%.

14. $c^2 = a^2+b^2$ where $a = b$
$$c^2 = a^2+a^2$$
$$(150)^2 = 2a^2$$
$$11,250 = a^2$$
$$\pm 75\sqrt{2} = a$$
Disregard the negative.
$75\sqrt{2}$ or 106.1 miles

15. Two complex but not real solutions exist.

16. Two real solutions exist.

17. Two distinct real solutions exist.

18. One real solution exists.

19. $x^2-16x+64 = 0$
$a = 1$, $b = -16$, and $c = 64$
$$x = \frac{16\pm\sqrt{(-16)^2-4(1)(64)}}{2(1)}$$
$$x = \frac{16\pm\sqrt{256-256}}{2} = \frac{16\pm\sqrt{0}}{2} = 8$$
The solution is 8.

20. $x^2+5x = 0$
$$x = \frac{-5\pm\sqrt{(5)^2-4(1)(0)}}{2(1)}$$
$$x = \frac{-5\pm\sqrt{25}}{2} = \frac{-5\pm 5}{2}$$
$x = 0$ or $x = -5$
The solutions are 0 and –5.

21. $x^2+11 = 0$
$a = 1$, $b = 0$, and $c = 11$

$$x = \frac{-0 \pm \sqrt{0^2 - 4(1)(11)}}{2(1)}$$
$$= \frac{\pm\sqrt{-44}}{2}$$
$$= \frac{\pm 2\sqrt{11}i}{2}$$
$$= \pm\sqrt{11}i$$

The solutions are $\sqrt{11}i$ and $-\sqrt{11}i$.

22. $2x^2 + 3x = 5$

$2x^2 + 3x - 5 = 0$

$$x = \frac{-3 \pm \sqrt{9 - 4(2)(-5)}}{2(2)}$$
$$x = \frac{-3 \pm \sqrt{49}}{4} = \frac{-3 \pm 7}{4}$$

$x = 1$ or $x = -\frac{5}{2}$

The solutions are 1 and $-\frac{5}{2}$.

23. $6x^2 + 7 = 5x$

$6x^2 - 5x + 7 = 0$

$a = 6$, $b = -5$, and $c = 7$

$$x = \frac{5 \pm \sqrt{(-5)^2 - 4(6)(7)}}{2(6)}$$
$$= \frac{5 \pm \sqrt{25 - 168}}{12}$$
$$= \frac{5 \pm \sqrt{-143}}{12}$$
$$= \frac{5 \pm \sqrt{143}i}{12}$$
$$= \frac{5}{12} \pm \frac{\sqrt{143}}{12}i$$

The solutions are $\frac{5}{12} + \frac{\sqrt{143}}{12}i$ and $\frac{5}{12} - \frac{\sqrt{143}}{12}i$.

24. $9a^2 + 4 = 2a$

$9a^2 - 2a + 4 = 0$

$$a = \frac{2 \pm \sqrt{4 - 4(9)(4)}}{2(9)}$$
$$= \frac{2 \pm \sqrt{-140}}{18}$$
$$= \frac{2 \pm 2i\sqrt{35}}{18}$$
$$= \frac{1 \pm i\sqrt{35}}{9}$$

The solutions are $\frac{1 + i\sqrt{35}}{9}$ and $\frac{1 - i\sqrt{35}}{9}$.

25. $(5a - 2)^2 - a = 0$

$25a^2 - 20a + 4 - a = 0$

$25a^2 - 21a + 4 = 0$

$a = 25$, $b = -21$ and $c = 4$

$$a = \frac{21 \pm \sqrt{(-21)^2 - 4(25)(4)}}{2(25)}$$
$$= \frac{21 \pm \sqrt{441 - 400}}{50}$$
$$= \frac{21 \pm \sqrt{41}}{50}$$

The solutions are $\frac{21 + \sqrt{41}}{50}$ and $\frac{21 - \sqrt{41}}{50}$.

26. $(2x - 3)^2 = x$

$4x^2 - 12x + 9 - x = 0$

$4x^2 - 13x + 9 = 0$

$$x = \frac{13 \pm \sqrt{169 - 4(4)(9)}}{2(4)}$$
$$x = \frac{13 \pm \sqrt{25}}{8} = \frac{13 \pm 5}{8}$$

$x = \frac{9}{4}$ or $x = 1$

The solutions are $\frac{9}{4}$ and 1.

27. $d(t) = -16t^2 + 30t + 6$

a. $d(1) = -16(1)^2 + 30(1) + 6$
$= -16 + 30 + 6$
$= 20$ feet

b. $-16t^2 + 30t + 6 = 0$
$8t^2 - 15t - 3 = 0$
$a = 8$, $b = -15$, and $c = -3$

$$x = \frac{15 \pm \sqrt{(-15)^2 - 4(8)(-3)}}{2(8)} = \frac{15 \pm \sqrt{225 + 96}}{16} = \frac{15 \pm \sqrt{321}}{16}$$

Discarding the negative value as extraneous, we find that $x = \frac{15 + \sqrt{321}}{16} \approx 2.1$ seconds.

28. Let x = length of leg.
$x + 6$ = length of hypotenuse

$$x^2 + x^2 = (x+6)^2$$
$$2x^2 = x^2 + 12x + 36$$
$$x^2 - 12x - 36 = 0$$
$$x = \frac{12 \pm \sqrt{144 - 4(1)(-36)}}{2}$$
$$x = \frac{12 \pm \sqrt{288}}{2}$$
$$x = \frac{12 \pm 12\sqrt{2}}{2} = 6 \pm 6\sqrt{2}$$

The length of each leg is $(6 + 6\sqrt{2})$ cm.

29.
$$x^3 = 27$$
$$x^3 - 27 = 0$$
$$(x-3)(x^2 + 3x + 9) = 0$$
$$x - 3 = 0 \quad \text{or} \quad x^2 + 3x + 9 = 0$$
$$x = 3$$

$a = 1$, $b = 3$, and $c = 9$

$$x = \frac{-3 \pm \sqrt{3^2 - 4(1)(9)}}{2(1)} = \frac{-3 \pm \sqrt{9 - 36}}{2} = \frac{-3 \pm \sqrt{-27}}{2} = \frac{-3 \pm 3\sqrt{3}i}{2} = -\frac{3}{2} \pm \frac{3\sqrt{3}i}{2}$$

The solutions are $3, -\frac{3}{2} + \frac{3\sqrt{3}i}{2}$ and $-\frac{3}{2} - \frac{3\sqrt{3}i}{2}$.

30.
$$y^3 = -64$$
$$y^3 + 64 = 0$$
$$(y+4)(y^2 - 4y + 16) = 0$$
$$y + 4 = 0 \quad \text{or} \quad y^2 - 4y + 16 = 0$$
$$y = -4 \quad \text{or} \quad y = \frac{4 \pm \sqrt{16 - 4(1)(16)}}{2}$$
$$y = \frac{4 \pm \sqrt{-48}}{2}$$
$$y = \frac{4 \pm 4i\sqrt{3}}{2}$$
$$y = 2 \pm 2i\sqrt{3}$$

The solutions are -4, $2 + 2i\sqrt{3}$, and $2 - 2i\sqrt{3}$.

31. $\frac{5}{x}+\frac{6}{x-2}=3$

$\frac{5(x-2)+6x}{x(x-2)}=3$

$5x-10+6x=3x(x-2)$

$11x-10=3x^2-6x$

$3x^2-17x+10=0$

$(3x-2)(x-5)=0$

$3x-2=0$ or $x-5=0$

$3x=2$ or $x=5$

$x=\frac{2}{3}$

The solutions are $\frac{2}{3}$ and 5.

32. $\frac{7}{8}=\frac{8}{x^2}$

$7x^2=64$

$x^2=\frac{64}{7}$

$x=\pm\sqrt{\frac{64}{7}}=\frac{\pm 8\sqrt{7}}{7}$

The solutions are $\frac{8\sqrt{7}}{7}$ and $-\frac{8\sqrt{7}}{7}$.

33. $x^4-21x^2-100=0$

$(x^2-25)(x^2+4)=0$

$x^2-25=0$ or $x^2+4=0$

$x^2=25$ or $x^2=-4$

$x=\pm\sqrt{25}$ or $x=\pm\sqrt{-4}$

$x=\pm 5$ or $x=\pm 2i$

The solutions are 5, -5, $2i$, and $-2i$.

34. $5(x+3)^2-19(x+3)=4$

Let $u=x+3$. Then

$5u^2-19u-4=0$

$(5u+1)(u-4)=0$

$5u+1=0$ or $u-4=0$

$u=-\frac{1}{5}$ or $u=4$

$x+3=-\frac{1}{5}$ or $x+3=4$

$x=-\frac{16}{5}$ or $x=1$

The solutions are $-\frac{16}{15}$ and 1.

35. $x^{2/3}-6x^{1/3}+5=0$

$(x^{1/3}-1)(x^{1/3}-5)=0$

$x^{1/3}-1=0$ or $x^{1/3}-5=0$

$x^{1/3}=1$ or $x^{1/3}=5$

$x=1^3=1$ or $x=5^3=125$

The solutions are 1 and 125.

36. $x^{2/3}-6x^{1/3}+8=0$

Let $m=x^{1/3}$. Then

$m^2-6m+8=0$

$(m-2)(m-4)=0$

$m-2=0$ or $m-4=0$

$m=2$ or $m=4$

$x^{1/3}=2$ or $x^{1/3}=4$

$x=2^3=8$ or $x=4^3=64$

The solutions are 8 and 64.

37.
$$a^6 - a^2 = a^4 - 1$$
$$a^2(a^4 - 1) - (a^4 - 1) = 0$$
$$(a^2 - 1)(a^4 - 1) = 0$$
$$(a^2 - 1)(a^2 - 1)(a^2 + 1) = 0$$
$$(a^2 - 1)^2(a^2 + 1) = 0$$
$$[(a+1)(a-1)]^2(a^2 + 1) = 0$$
$$(a+1)^2(a-1)^2(a^2 + 1) = 0$$
$$(a+1)^2 = 0 \quad \text{or} \quad (a-1)^2 = 0 \quad \text{or} \quad a^2 + 1 = 0$$
$$a+1 = 0 \quad \text{or} \quad a-1 = 0 \quad \text{or} \quad a^2 = -1$$
$$a = -1 \quad \text{or} \quad a = 1 \quad \text{or} \quad a = \pm\sqrt{-1} = \pm i$$
The solutions are 1, –1, i and $-i$.

38. $y^{-2} + y^{-1} = 20$
$$\frac{1}{y^2} + \frac{1}{y} = 20$$
$$1 + y = 20y^2$$
$$0 = 20y^2 - y - 1$$
$$0 = (5y+1)(4y-1)$$
$$5y + 1 = 0 \quad \text{or} \quad 4y - 1 = 0$$
$$y = -\frac{1}{5} \quad \text{or} \quad y = \frac{1}{4}$$
The solutions are $-\frac{1}{5}$ and $\frac{1}{4}$.

39.

	Hours	Job complete in one hour
Jerome	x	$\frac{1}{x}$
Tim	$x - 1$	$\frac{1}{x-1}$
Together	5	$\frac{1}{5}$

$$\frac{1}{x} + \frac{1}{x-1} = \frac{1}{5}$$
$$\frac{(x-1)+x}{x(x-1)} = \frac{1}{5}$$
$$\frac{2x-1}{x^2 - x} = \frac{1}{5}$$
$$10x - 5 = x^2 - x$$
$$0 = x^2 - 11x + 5$$
$a = 1$, $b = -11$, and $c = 5$
$$x = \frac{-(-11) \pm \sqrt{(-11)^2 - 4(1)(5)}}{2(1)}$$
$$x = \frac{11 \pm \sqrt{101}}{2}$$
$x \approx 0.475$ or 10.525
$x \approx 10.5,\ x - 1 \approx 9.5$
Jerome, 10.5 hours
Tim, 9.5 hours

40. Let x = the number so $\frac{1}{x}$ = the reciprocal of the number. Then
$$x - \frac{1}{x} = -\frac{24}{5}$$
$$\frac{x^2 - 1}{x} = -\frac{24}{5}$$
$$5(x^2 - 1) = -24x$$
$$5x^2 - 5 = -24x$$
$$5x^2 + 24x - 5 = 0$$
$$(5x - 1)(x + 5) = 0$$
$$5x - 1 = 0 \quad \text{or} \quad x + 5 = 0$$
$$5x = 1 \quad \text{or} \quad x = -5$$
$$x = \frac{1}{5}$$
Discarding the positive value as extraneous, we find the number to be –5.

41. $2x^2 - 50 \le 0$

$2x^2 \le 50$

$x^2 \le 25$

$-5 \le x \le 5$

$[-5, 5]$

42. $\frac{1}{4}x^2 < \frac{1}{16}$

$x^2 < \frac{1}{4}$

$x^2 - \frac{1}{4} < 0$

$\left(x + \frac{1}{2}\right)\left(x - \frac{1}{2}\right) < 0$

A	B	C

$-\frac{1}{2}$ $\quad \frac{1}{2}$

Region	Test Point	$\left(x - \frac{1}{2}\right)\left(x + \frac{1}{2}\right) < 0$
A	−1	$\left(-\frac{3}{2}\right)\left(-\frac{1}{2}\right) > 0$
B	0	$\left(-\frac{1}{2}\right)\left(\frac{1}{2}\right) < 0$
C	1	$\left(\frac{1}{2}\right)\left(\frac{3}{2}\right) > 0$

$\left(-\frac{1}{2}, \frac{1}{2}\right)$ satisfies the inequality.

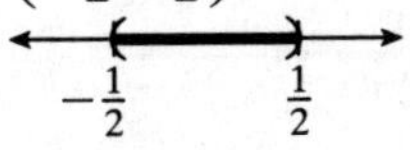

43. $(2x - 3)(4x + 5) \ge 0$

First solve

$(2x - 3)(4x + 5) = 0.$

$2x - 3 = 0$ or $4x + 5 = 0$

$2x = 3$ or $4x = -5$

$x = \frac{3}{2}$ or $x = -\frac{5}{4}$

Now select three appropriate test points.

$x = -2$: $[2(-2) - 3][4(-2) + 5] = (-)(-) = +$

$x = 2$: $[2 \cdot 2 - 3][4 \cdot 2 + 5] = (+)(+) = +$

$x = 0$: $[2(0) - 3][4(0) + 5] = (-)(+) = -$

Thus, the solutions are $x \ge \frac{3}{2}$ and $x \le -\frac{5}{4}$.

$\left(-\infty, -\frac{5}{4}\right] \cup \left[\frac{3}{2}, \infty\right)$

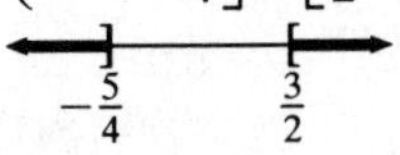

44. $(x^2 - 16)(x^2 - 1) > 0$

$(x + 4)(x - 4)(x + 1)(x - 1) > 0$

A	B	C	D	E

−4 $\quad$ −1 $\quad$ 1 $\quad$ 4

Region	Test Point	$(x^2 - 16)(x^2 - 1) > 0$
A	−5	$(9)(24) > 0$
B	−2	$(-12)(3) < 0$
C	0	$(-16)(-1) > 0$
D	2	$(-12)(3) < 0$
E	5	$(9)(24) > 0$

$(-\infty, -4) \cup (-1, 1) \cup (4, \infty)$ satisfies the inequality.

45. $\frac{x - 5}{x - 6} < 0$

First solve both

$x - 5 = 0$ and $x - 6 = 0$

$x = 5$ and $x = 6$.

Now select three appropriate test points.

$x = 4$: $\frac{4 - 5}{4 - 6} = \frac{-}{-} = +$

$x = \frac{11}{2}$: $\frac{\frac{11}{2} - 5}{\frac{11}{2} - 6} = \frac{+}{-} = -$

$x = 7$: $\frac{7 - 5}{7 - 6} = \frac{+}{+} = +$

Thus, the solution is $5 < x < 6$.

(5, 6)

46. $\dfrac{x(x+5)}{4x-3} \ge 0$

A	B	C	D

$-5 \quad 0 \quad \frac{3}{4}$

Region	Test Point	$\dfrac{x(x+5)}{4x-3} \ge 0$
A	-6	$\dfrac{(-6)(-1)}{-27} \le 0$
B	-1	$\dfrac{(-1)(4)}{-7} \ge 0$
C	$\frac{1}{2}$	$\dfrac{\frac{1}{2}\left(5\frac{1}{2}\right)}{(-1)} \le 0$
D	1	$\dfrac{1(6)}{1} \ge 0$

$[-5,\ 0] \cup \left(\dfrac{3}{4},\ \infty\right)$ satisfies the inequality.

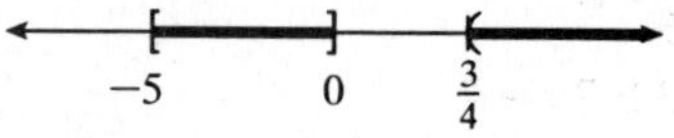

47. $\dfrac{(4x+3)(x-5)}{x(x+6)} > 0$

First solve both:

$(4x + 3)(x - 5) = 0$ and $x(x + 6) = 0$

$4x + 3 = 0$ or $x - 5 = 0$ and $x = 0$ or $x + 6 =$
0

$4x = -3$ or $x = 5$ and $x = 0$ or $x = -6$

$x = -\dfrac{3}{4}$

Now select five appropriate test points.

$x = -7$:

$\dfrac{[4(-7)+3][-7-5]}{-7(-7+6)} = \dfrac{(-)(-)}{(-)(-)} = \dfrac{+}{+} = +$

$x = -1$:

$\dfrac{[4(-1)+3][-1-5]}{-1(-1+6)} = \dfrac{(-)(-)}{(-)(+)} = \dfrac{+}{-} = -$

$x = -\dfrac{1}{2}$: $\dfrac{\left[4\left(-\frac{1}{2}\right)+3\right]\left[-\frac{1}{2}-5\right]}{-\frac{1}{2}\left(-\frac{1}{2}+6\right)} = \dfrac{(+)(-)}{(-)(+)}$

$= \dfrac{-}{-}$

$= +$

$x = 1$: $\dfrac{[4(1)+3][1-5]}{1(1+6)} = \dfrac{(+)(-)}{(+)(+)} = \dfrac{-}{+} = -$

$x = 6$: $\dfrac{[4(6)+3][6-5]}{6(6+6)} = \dfrac{(+)(+)}{(+)(+)} = \dfrac{+}{+} = +$

Thus, the solutions are $x < -6$,

$-\dfrac{3}{4} < x < 0$, and

$x > 5$.

$(-\infty,\ -6) \cup \left(-\dfrac{3}{4},\ 0\right) \cup (5,\ \infty)$

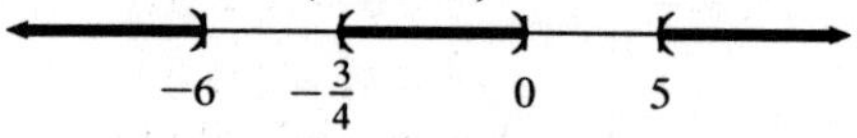

48. $(x + 5)(x - 6)(x + 2) \le 0$

A	B	C	D

$-5 \quad -2 \quad 6$

Region	Test Point	$(x + 5)(x - 6)(x + 2) \le 0$
A	-6	$(-1)(-12)(-4) \le 0$
B	-3	$(2)(-9)(-1) \ge 0$
C	0	$(5)(-6)(2) \le 0$
D	7	$(12)(1)(9) \ge 0$

$(-\infty,\ -5] \cup [-2,\ 6]$ satisfies the inequality.

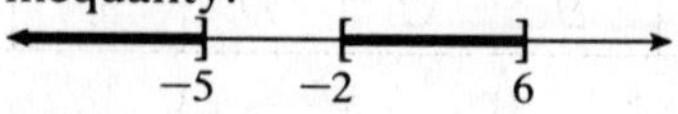

49. $x^3 + 3x^2 - 25x - 75 > 0$

$x^2(x+3) - 25(x+3) > 0$

$(x+3)(x^2 - 25) > 0$

$(x+3)(x+5)(x-5) > 0$

Region	A	B	C	D
x-values	$(-\infty,\ -5)$	$(-5, -3)$	$(-3, 5)$	$(5,\ \infty)$

Region	Test Point	$(x+3)(x+5)(x-5)>0$
A	–6	$(-3)(-1)(-11)<0$
B	–4	$(-1)(1)(-9)>0$
C	0	$(3)(5)(-5)<0$
D	6	$(9)(11)(1)>0$

$(-5,\ -3)\cup(5,\ \infty)$ satisfies the inequality.

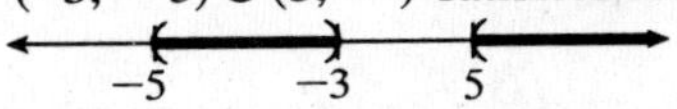

50. $\dfrac{x^2+4}{3x}\le 1$

First solve both

$3x=0$ and $\dfrac{x^2+4}{3x}=1.$

$x=0$

$x^2+4=3x$

$x^2-3x+4=0$

has no real solutions

Now select two appropriate test points.

$x=-5$: $\dfrac{(-5)^2+4}{3(-5)}=\dfrac{+}{-}=-$

$x=1$: $\dfrac{1^2+4}{3(1)}=\dfrac{+}{+}=+$

Thus the solutions are $x<0$.

$(-\infty,\ 0)$

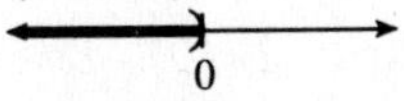

51. $\dfrac{(5x+6)(x-3)}{x(6x-5)}<0$

A	*B*	*C*	*D*	*E*

$-\frac{6}{5}$ 0 $\frac{5}{6}$ 3

Region	Test Point	$\dfrac{(5x+6)(x-3)}{x(6x-5)}<0$
A	–2	$\dfrac{(-1)(-5)}{(-2)(-17)}>0$
B	–1	$\dfrac{(1)(-4)}{(-1)(-11)}<0$
C	$\frac{1}{2}$	$\dfrac{\left(8\frac{1}{2}\right)\left(-2\frac{1}{2}\right)}{\frac{1}{2}(-2)}>0$
D	1	$\dfrac{(11)(-2)}{1(1)}<0$
E	4	$\dfrac{(26)(1)}{4(19)}>0$

$\left(-\dfrac{6}{5},\ 0\right)\cup\left(\dfrac{5}{6},\ 3\right)$ satisfies the inequality.

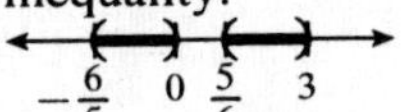

52. $\dfrac{3}{x-2}>2$

$3>2x-4$

$0>2x-7$

$x=2$ (undefined)

$x=\dfrac{7}{2}$

Region	A	B	C
x-values	$(-\infty,\ 2)$	$\left(2,\ \frac{7}{2}\right)$	$\left(\frac{7}{2},\ \infty\right)$

Region	Test Point	$\frac{3}{x-2} > 2$
A	0	$-\frac{3}{2} < 2$
B	3	$3 > 2$
C	4	$\frac{3}{2} < 2$

$\left(2, \frac{7}{2}\right)$ satisfies the inequality.

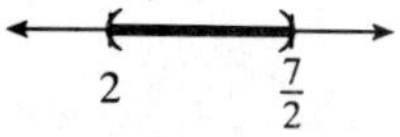

53. $f(x) = x^2 - 4$

$\frac{-b}{2a} = \frac{0}{2(1)} = 0$

$f(0) = 0^2 - 4 = -4$

Vertex: (0, –4)

Axis of symmetry: $x = 0$

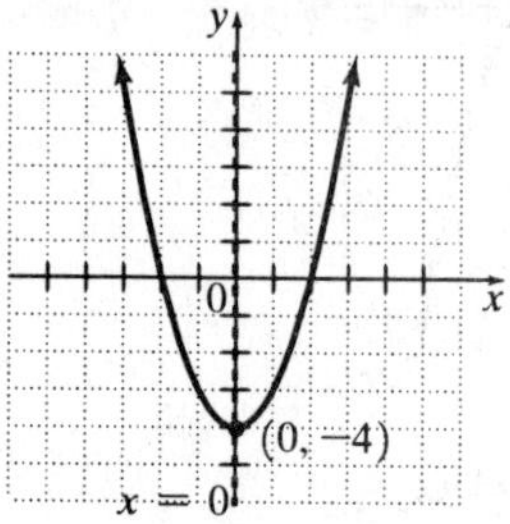

54. $g(x) = x^2 + 7$

$\frac{-b}{2a} = \frac{0}{2(1)} = 0$

$g(0) = 0^2 + 7$

Vertex: (0, 7)

Axis of symmetry: $x = 0$

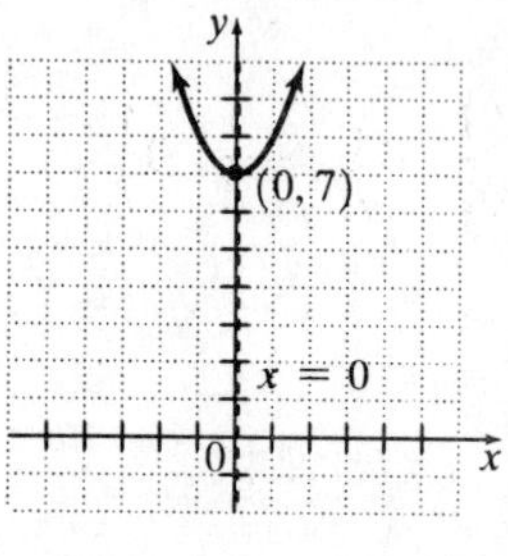

55. $H(x) = 2x^2$

Vertex: (0, 0)

Axis of symmetry: $x = 0$

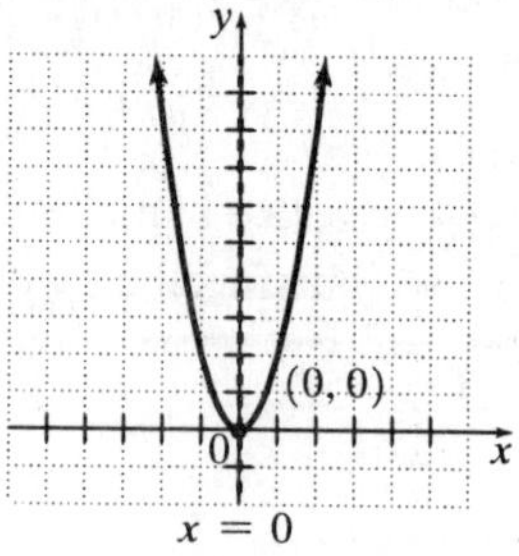

56. $h(x) = -\frac{1}{3}x^2$

Vertex: (0, 0)

Axis of symmetry: $x = 0$

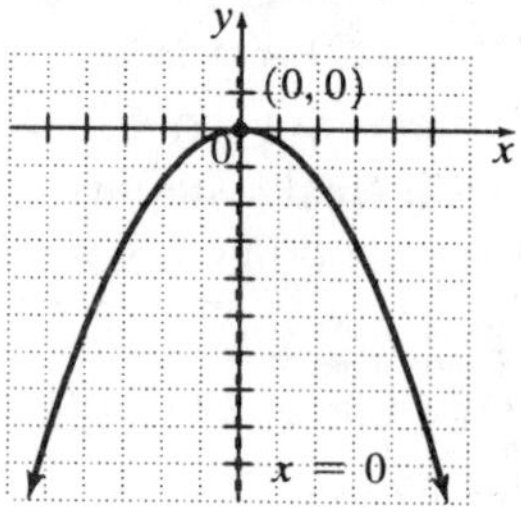

57. $F(x) = (x-1)^2$

$F(x) = x^2 - 2x + 1$

$\frac{-b}{2a} = \frac{-(-2)}{2(1)} = 1$

$F(1) = 0$

Vertex: (1, 0)

Axis of symmetry: $x = 1$

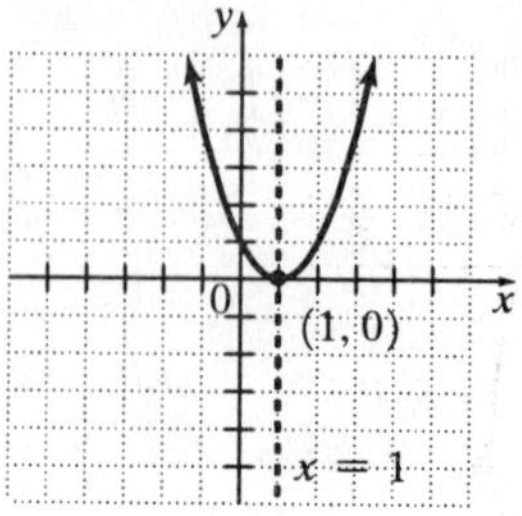

58. $G(x) = (x+5)^2$

$G(x) = x^2 + 10x + 25$

$\frac{-b}{2a} = \frac{-10}{2(1)} = -5$

$G(-5) = 0$

Vertex: (–5, 0)

Axis of symmetry: $x = -5$

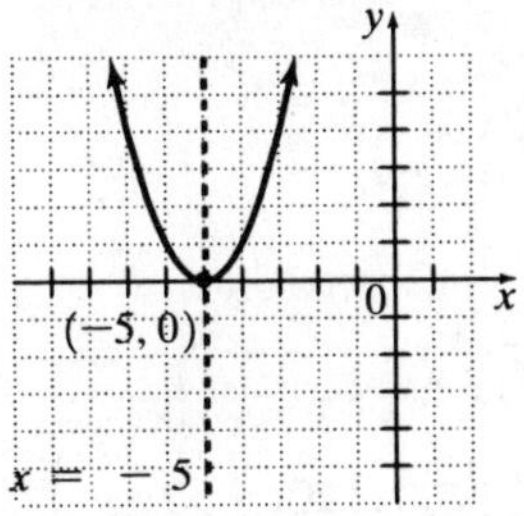

59. $f(x) = (x-4)^2 - 2$

Vertex: (4, –2)

Axis of symmetry: $x = 4$

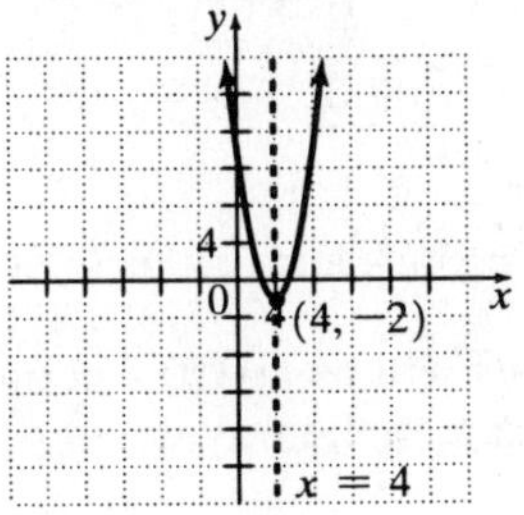

60. $y = -3(x-1)^2 + 1$

Vertex: (1, 1)

Axis of symmetry: $x = 1$

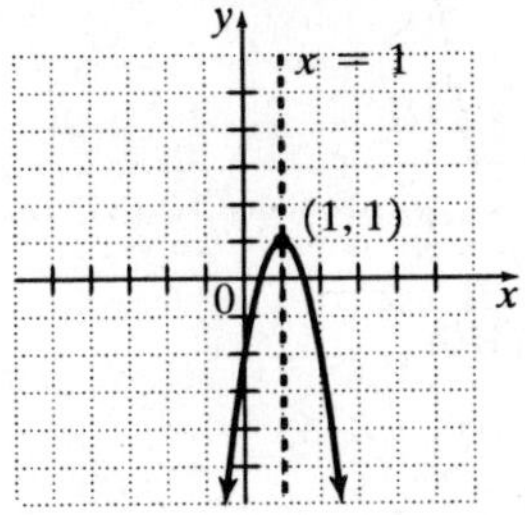

61. $f(x) = x^2 + 10x + 25$ or

$f(x) = (x+5)^2 + 0$

Vertex: (–5, 0).

$(x+5)^2 = 0 \Rightarrow x + 5 = 0$

x-intercept: (–5, 0).

y-intercept: (0, 25).

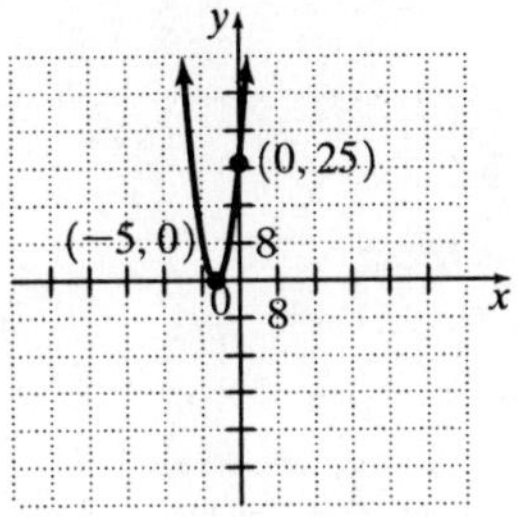

62. $y = -x^2 + 6x - 9$

$\frac{-b}{2a} = \frac{-6}{2(-1)} = 3$

$y = -(3)^2 + 6(3) - 9 = 0$

Vertex: (3, 0)

$y(0) = -9$

y-intercept: (0, –9)

x-intercept: (3, 0)

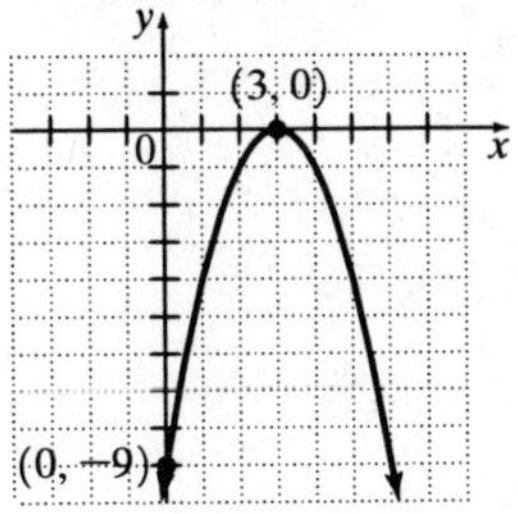

63. $f(x) = 4x^2 - 1$ or

$f(x) = 4(x-0)^2 - 1$

Vertex: (0, –1)

$4x^2 - 1 = 0$

$4x^2 = 1$

$x^2 = \frac{1}{4}$

$x = \pm\frac{1}{2}$

x-intercepts: $\left(\frac{1}{2}, 0\right), \left(-\frac{1}{2}, 0\right)$.

y-intercept: $(0, -1)$

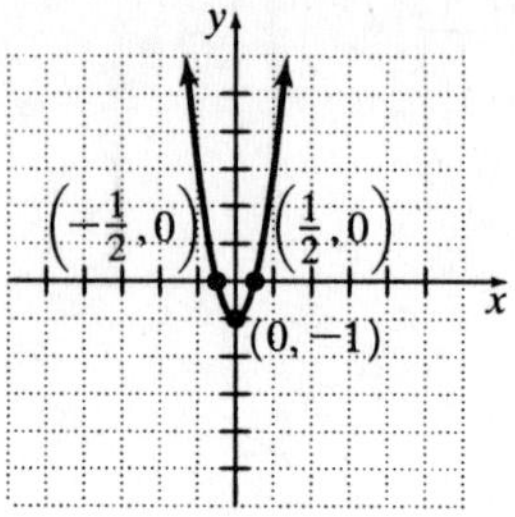

64. $y = -5x^2 + 5$

$\frac{-b}{2a} = 0$
$y = 0 + 5$
Vertex: $(0, 5)$
y-intercept: $(0, 5)$
x-intercepts: $(-1, 0), (1, 0)$

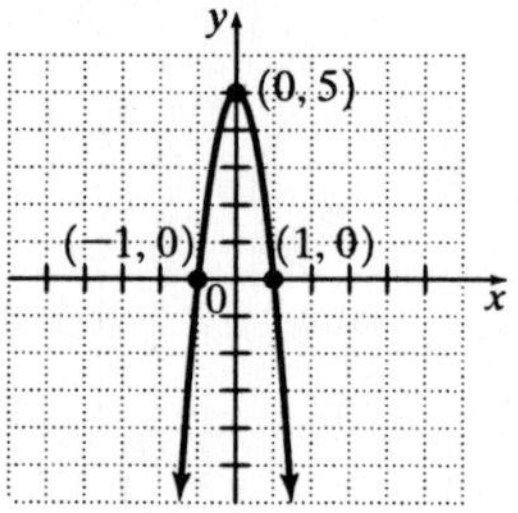

65. $f(x) = -3x^2 - 5x + 4$

$\frac{-b}{2a} = \frac{-(-5)}{2(-3)} = -\frac{5}{6}$

$f\left(\frac{5}{6}\right) = -3\left(-\frac{5}{6}\right)^2 - 5\left(-\frac{5}{6}\right) + 4 = \frac{73}{12}$

Vertex: $\left(-\frac{5}{6}, \frac{73}{12}\right)$

The graph opens down because $a = -3 < 0$.

y-intercept: $(0, 4)$.

$x = \frac{-(-5) \pm \sqrt{(-5)^2 - 4(-3)(4)}}{2(-3)}$

$x = \frac{5 \pm \sqrt{73}}{-6}$

The x-intercepts are approximately $(-2.26, 0)$ and $(0.59, 0)$.

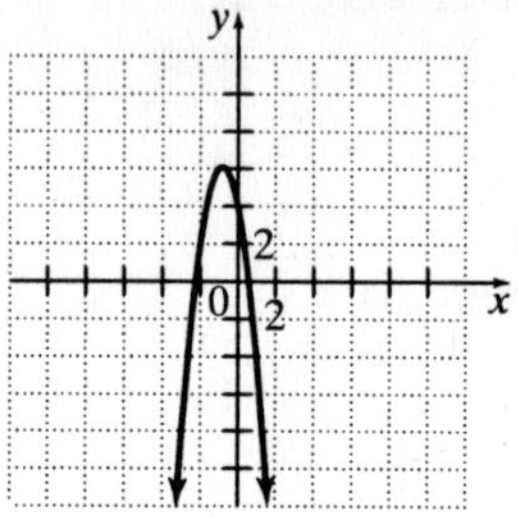

66. $h(t) = -16t^2 + 120t + 300$

a. $350 = -16t^2 + 120t + 300$

$0 = -16t^2 + 120t - 50$

$0 = -8t^2 + 60t - 25$

$a = -8, b = 60, c = -25$

$t = \frac{-60 \pm \sqrt{(60)^2 - 4(-8)(-25)}}{2(-8)}$

$t = \frac{-60 \pm \sqrt{2800}}{-16}$

$t \approx 0.4$ seconds and 7.1 seconds

b. The object will be at 350 feet on the way up and on the way down.

67. Let x = one number so $420 - x$ = the other number. Let $f(x)$ represent their product. Thus,

$f(x) = x(420 - x)$

$f(x) = -x^2 + 420x$

$\frac{-b}{2a} = \frac{-420}{2(-1)} = 210$

Therefore, the numbers are both 210.

68. $x = a(y - k)^2 + h$

$y = a(x - h)^2 + k$

$y = a(x + 3)^2 + 7$

$0 = a(0 + 3)^2 + 7$

$y - 7 = 9a$

$-\frac{7}{9} = a$

$y = -\frac{7}{9}(x + 3)^2 + 7$

Chapter 8 Test

1. $5x^2 - 2x = 7$

$5x^2 - 2x - 7 = 0$

$(5x - 7)(x + 1) = 0$

$5x - 7 = 0$ or $x + 1 = 0$

$5x = 7$ or $x = -1$

$x = \frac{7}{5}$

The solutions are $\frac{7}{5}$ and -1.

2. $(x + 1)^2 = 10$

$x + 1 = \pm\sqrt{10}$

$x = -1 \pm \sqrt{10}$

The solutions are $-1 + \sqrt{10}$ and $-1 - \sqrt{10}$.

3. $m^2 - m + 8 = 0$

$a = 1$, $b = -1$, and $c = 8$

$$m = \frac{1 \pm \sqrt{(-1)^2 - 4(1)(8)}}{2(1)}$$

$$= \frac{1 \pm \sqrt{1 - 32}}{2}$$

$$= \frac{1 \pm \sqrt{-31}}{2}$$

$$= \frac{1 \pm \sqrt{31}i}{2}$$

$$= \frac{1}{2} \pm \frac{\sqrt{31}}{2}i$$

The solutions are $\frac{1}{2} + \frac{\sqrt{31}}{2}i$ and $\frac{1}{2} - \frac{\sqrt{31}}{2}i$.

4. $u^2 - 6u + 2 = 0$

$a = 1$, $b = -6$, $c = 2$

$$u = \frac{6 \pm \sqrt{(-6)^2 - 4(1)(2)}}{2(1)}$$

$$= \frac{6 \pm \sqrt{36 - 8}}{2}$$

$$= \frac{6 \pm \sqrt{28}}{2}$$

$$= \frac{6 \pm 2\sqrt{7}}{2}$$

$$= 3 \pm \sqrt{7}$$

The solutions are $3 + \sqrt{7}$ and $3 - \sqrt{7}$.

5. $7x^2 + 8x + 1 = 0$

$(7x + 1)(x + 1) = 0$

$7x + 1 = 0$ or $x + 1 = 0$

$7x = -1$ or $x = -1$

$x = -\frac{1}{7}$

The solutions are $-\frac{1}{7}$ and -1.

6. $a^2 - 3a = 5$

$a^2 - 3a - 5 = 0$

$a = 1$, $b = -3$, and $c = -5$

$$a = \frac{3 \pm \sqrt{(-3)^2 - 4(1)(-5)}}{2(1)}$$

$$= \frac{3 \pm \sqrt{9 + 20}}{2}$$

$$= \frac{3 \pm \sqrt{29}}{2}$$

$$= \frac{3}{2} \pm \frac{\sqrt{29}}{2}$$

The solutions are $\frac{3}{2} + \frac{\sqrt{29}}{2}$ and $\frac{3}{2} - \frac{\sqrt{29}}{2}$.

7. $\frac{4}{x+2}+\frac{2x}{x-2}=\frac{6}{x^2-4}$

$\frac{4(x-2)+2x(x+2)}{(x+2)(x-2)}=\frac{6}{(x+2)(x-2)}$

$4x-8+2x^2+4x=6$

$2x^2+8x-8=6$

$2x^2+8x-14=0$

$x^2+4x-7=0$

$a=1$, $b=4$, and $c=-7$

$x=\frac{-4\pm\sqrt{4^2-4(1)(-7)}}{2(1)}$

$=\frac{-4\pm\sqrt{16+28}}{2}$

$=\frac{-4\pm\sqrt{44}}{2}$

$=\frac{-4\pm2\sqrt{11}}{2}$

$=-2\pm\sqrt{11}$

The solutions are $-2+\sqrt{11}$ and $-2-\sqrt{11}$.

8. $x^4-8x^2-9=0$

$(x^2-9)(x^2+1)=0$

$x^2-9=0$ or $x^2+1=0$

$x^2=9$ or $x^2=-1$

$x=\pm\sqrt{9}$ or $x=\pm\sqrt{-1}$

$x=\pm3$ or $x=\pm i$

The solutions are 3, –3, *i* and –*i*.

9. $x^6+1=x^4+x^2$

$x^6-x^4-x^2+1=0$

$x^4(x^2-1)-(x^2-1)=0$

$(x^4-1)(x^2-1)=0$

$(x^2+1)(x^2-1)(x^2-1)=0$

$(x^2+1)(x^2-1)^2=0$

$(x^2+1)[(x+1)(x-1)]^2=0$

$(x^2+1)(x+1)^2(x-1)^2=0$

$x^2+1=0$ or $(x+1)^2=0$ or $(x-1)^2=0$

$x^2=-1$ or $x+1=0$ or $x-1=0$

$x=\pm\sqrt{-1}$ or $x=-1$ or $x=1$

The solutions are 1, –1, *i*, and –*i*.

10. $(x+1)^2-15(x+1)+56=0$

Let $u=x+1$.

$u^2-15u+56=0$

$(u-7)(u-8)=0$

$u-7=0$ or $u-8=0$

$u=7$ or $u=8$

Substitute $x+1$ for u.

$x+1=7$ or $x+1=8$

$x=6$ or $x=7$

The solutions are 6 and 7.

11. $x^2-6x=-2$

$x^2-6x+\left(\frac{6}{2}\right)^2=-2+9$

$(x-3)^2=7$

$x-3=\pm\sqrt{7}$

$x=3\pm\sqrt{7}$

The solutions are $3+\sqrt{7}$ and $3-\sqrt{7}$.

12. $2a^2+5=4a$

$2a^2-4a=-5$

$a^2-2a=-\frac{5}{2}$

$a^2-2a+\left(\frac{2}{2}\right)^2=-\frac{5}{2}+1$

$(a-1)^2=-\frac{3}{2}$

$a-1=\pm\sqrt{\frac{-3}{2}}=\pm\frac{\sqrt{3}i}{\sqrt{2}}$

$a-1=\frac{\pm\sqrt{6}i}{2}$

$a=1\pm\frac{\sqrt{6}}{2}i$

The solutions are $1+\frac{\sqrt{6}}{2}i$ and $1-\frac{\sqrt{6}}{2}i$.

13. $2x^2 - 7x > 15$

$2x^2 - 7x - 15 > 0$

$(2x + 3)(x - 5) > 0$

$2x + 3 = 0 \quad \text{or } x - 5 = 0$

$x = -\frac{3}{2} \quad \text{or } x = 5$

Now select three appropriate test points.

$x = -2$: $2(-2)^2 - 7(-2) - 15 = 8 > 0$

$x = 0$: $2(0)^2 - 7(0) - 15 = -15 < 0$

$x = 6$: $2(6)^2 - 7(6) - 15 = 15 > 0$

Thus, the solutions are

$\left(-\infty, -\frac{3}{2}\right) \cup (5, \infty)$

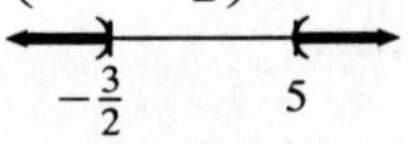

14. $(x^2 - 16)(x^2 - 25) > 0$

Let $u = x^2$.

$(u - 16)(u - 25) > 0$

First solve $(u - 16)(u - 25) = 0$

$u - 16 = 0 \quad \text{or } u - 25 = 0$

$u = 16 \quad \text{or } u = 25$

Now select three appropriate test points.

$u = 15$: $(15 - 16)(15 - 25) = (-)(-) = +$

$u = 20$: $(20 - 16)(20 - 25) = (+)(-) = -$

$u = 30$: $(30 - 16)(30 - 25) = (+)(+) = +$

Thus, the solutions are $u < 16$ or $u > 25$.

Now substitute back

$x^2 < 16$ or $x^2 > 25x$

$-4 < x < 4$ or $x < -5$ or $x > 5$

$(-\infty, -5) \cup (-4, 4) \cup (5, \infty)$

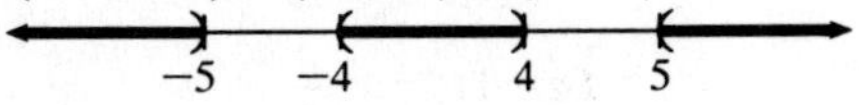

15. $\frac{5}{x+3} < 1$

$x = -3$ (undefined)

$5 = x + 3$

$x = 2$

Now select three appropriate test points.

$x = -4$: $\frac{5}{-4+3} < 1$

$x = 0$: $\frac{5}{0+3} > 1$

$x = 3$: $\frac{5}{3+3} < 1$

$(-\infty, -3) \cup (2, \infty)$

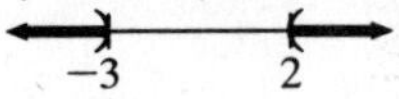

16. $\frac{7x - 14}{x^2 - 9} \le 0$

First solve both

$7x - 14 = 0$ and $x^2 - 9 = 0$.

$7x = 14$ and $x^2 = 9$

$x = 2$ and $x = \pm\sqrt{9} = \pm 3$

Now select four appropriate test points.

$x = -4$: $\frac{7(-4) - 14}{(-4)^2 - 9} = \frac{-}{+} = -$

$x = 0$: $\frac{7(0) - 14}{0^2 - 9} = \frac{-}{-} = +$

$x = \frac{5}{2}$: $\frac{7\left(\frac{5}{2}\right) - 14}{\left(\frac{5}{2}\right)^2 - 9} = \frac{+}{-} = -$

$x = 4$: $\frac{7(4) - 14}{4^2 - 9} = \frac{+}{+} = +$

Thus, the solutions are $x < -3$ and $2 \le x < 3$.

$(-\infty, -3) \cup [2, 3)$

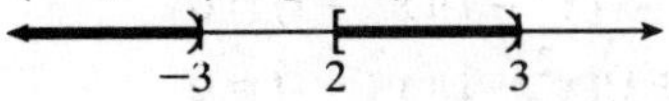

17. $f(x) = 3x^2$

vertex: (0, 0)

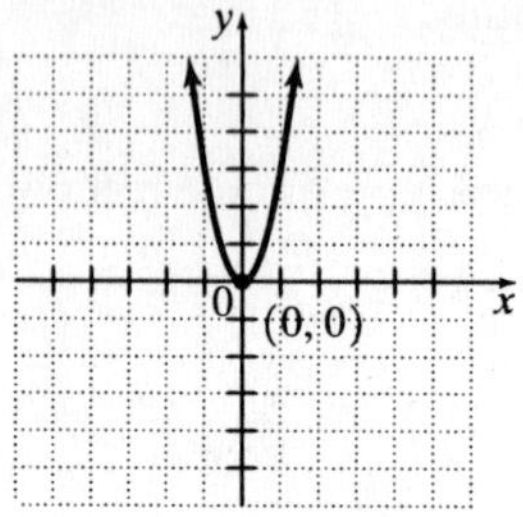

18. $G(x) = -2(x-1)^2 + 5$

vertex: (1, 5)

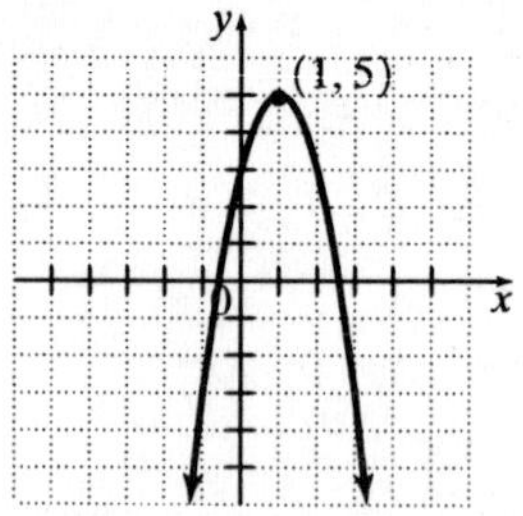

19. $h(x) = x^2 - 4x + 4$

$\frac{-b}{2a} = \frac{-(-4)}{2(1)} = 2$

$h(2) = 2^2 - 4(2) + 4 = 0$

vertex: (2, 0)

y-intercept: (0, 4).

$0 = x^2 - 4x + 4 = (x-2)^2$

x-intercept: (2, 0)

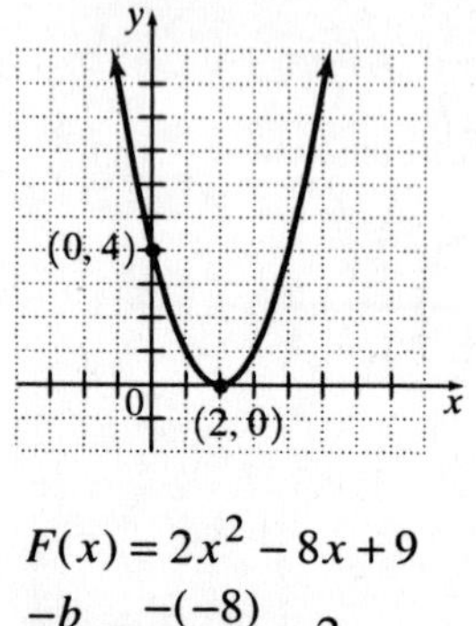

20. $F(x) = 2x^2 - 8x + 9$

$\frac{-b}{2a} = \frac{-(-8)}{2(2)} = 2$

$F(2) = 2(2)^2 - 8(2) + 9 = 1$

Vertex: (2, 1)

y-intercept: (0, 9)

There are no x-intercepts.

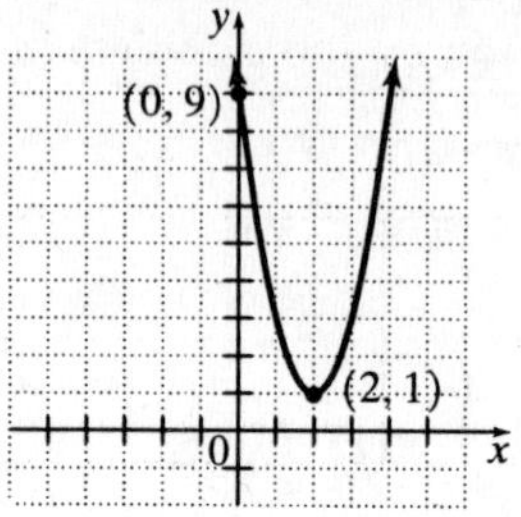

21.

$$c^2 = a^2 + b^2$$

$$(10)^2 = x^2 + (x-4)^2$$

$$100 = x^2 + x^2 - 8x + 16$$

$$0 = 2x^2 - 8x - 84$$

$$0 = x^2 - 4x - 42$$

$a = 1$, $b = -4$, $c = -42$

$$x = \frac{-(-4) \pm \sqrt{(-4)^2 - 4(1)(-42)}}{2(1)}$$

$$x = \frac{4 \pm \sqrt{16 + 168}}{2}$$

$$x = \frac{4 \pm \sqrt{184}}{2}$$

$$x = \frac{4 \pm 2\sqrt{46}}{2}$$

$$x = 2 \pm \sqrt{46}$$

Disregard the negative result.

$2 + \sqrt{46}$ or 8.8 ft

22.

	Hours	Job complete in 1 hr
Dave	$x-2$	$\frac{1}{x-2}$
Sandy	x	$\frac{1}{x}$
Together	4	$\frac{1}{4}$

$$\frac{1}{x-2}+\frac{1}{x}=\frac{1}{4}$$
$$\frac{x+(x-2)}{x(x-2)}=\frac{1}{4}$$
$$\frac{2x-2}{x^2-2x}=\frac{1}{4}$$
$$8x-8=x^2-2x$$
$$0=x^2-10x+8$$
$a=1, b=-10, c=8$
$$x=\frac{-(-10)\pm\sqrt{(-10)^2-4(1)(8)}}{2(1)}$$
$$x=\frac{10\pm\sqrt{68}}{2}$$
$$x=\frac{10\pm2\sqrt{17}}{2}$$
$$x=5\pm\sqrt{17}$$
Since $x-2$ must be positive,
$x=5+\sqrt{17}$.
$5+\sqrt{17}$ or 9.12 hours

23. $s(t)=-16t^2+32t+256$

a. $\frac{-b}{2a}=\frac{-32}{2(-16)}=1$

$s(1)=-16(1)^2+32(1)+256=272$
Vertex: (1, 272)
Maximum height = 272 feet

b. $0=-16t^2+32t+256$

$0=-t^2+2t+16$
$a=-1, b=2, c=16$
$$x=\frac{-2\pm\sqrt{2^2-4(-1)(16)}}{2(-1)}$$
$$x=\frac{-2\pm\sqrt{68}}{-2}$$
$$x=\frac{-2\pm2\sqrt{17}}{-2}$$
$$x=1\pm\sqrt{17}$$
$x\approx-3.12$ and 5.12
Disregard the negative.
5.12 seconds

24. $(x+8)^2+x^2=20^2$

$$x^2+16x+64+x^2=400$$
$$2x^2+16x-336=0$$
$$x^2+8x-168=0$$
$$x=\frac{-8\pm\sqrt{8^2-4(1)(-168)}}{2(1)}$$
$$x=\frac{-8\pm\sqrt{736}}{2}$$
$$x=-4\pm2\sqrt{46}$$
$x\approx10$
$(10+8)+10=28$
$28-20=8$
approximately 8 feet

Chapter 9

Section 9.1

Graphing Calculator Explorations

1. $x^2 + y^2 = 55$

$y^2 = 55 - x^2$

$y = \pm\sqrt{55 - x^2}$

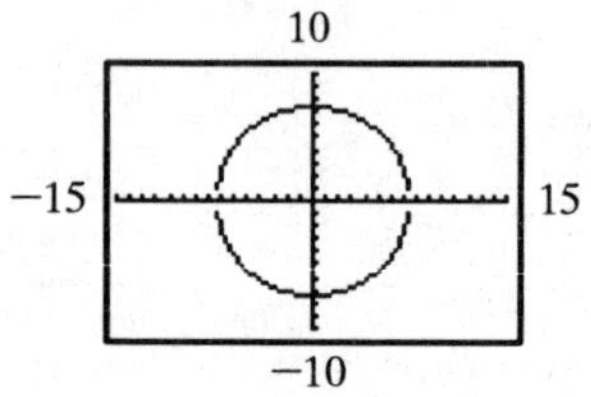

3. $5x^2 + 5y^2 = 50$

$5y^2 = 50 - 5x^2$

$y^2 = 10 - x^2$

$y = \pm\sqrt{10 - x^2}$

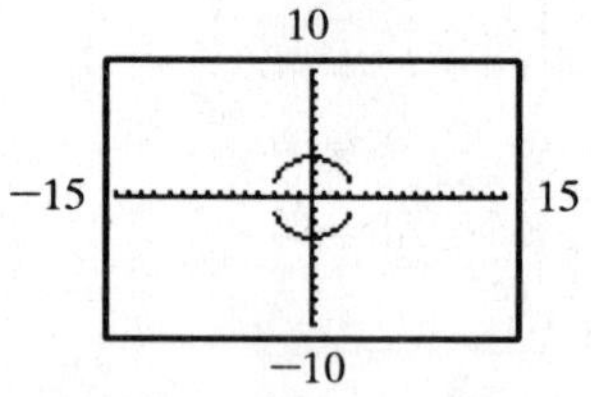

5. $2x^2 + 2y^2 - 34 = 0$

$2y^2 = 34 - 2x^2$

$y^2 = 17 - x^2$

$y = \pm\sqrt{17 - x^2}$

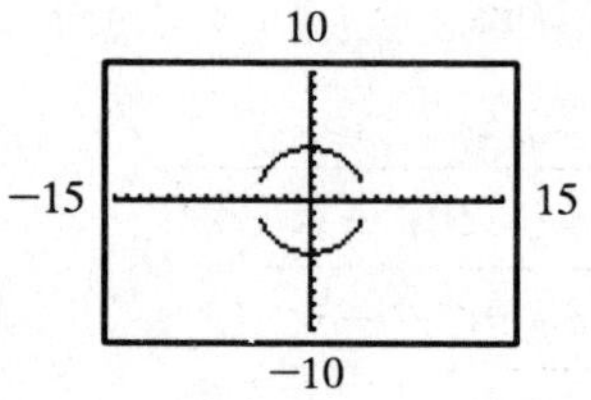

7. $7x^2 + 7y^2 - 89 = 0$

$7y^2 = 89 - 7x^2$

$y^2 = \dfrac{89 - 7x^2}{7}$

$y = \pm\sqrt{\dfrac{89 - 7x^2}{7}}$

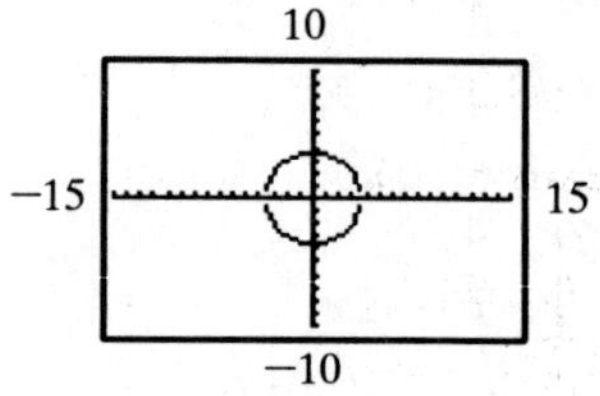

Mental Math

1. $y = x^2 - 7x + 5$; upward

2. $y = -x^2 + 16$; downward

3. $x = -y^2 - y + 2$; to the left

4. $x = 3y^2 + 2y - 5$; to the right

5. $y = -x^2 + 2x + 1$; downward

6. $x = -y^2 + 2y - 6$; to the left

Exercise Set 9.1

1. $x = 3y^2$

$x = 3(y - 0)^2$

The vertex is (0, 0).

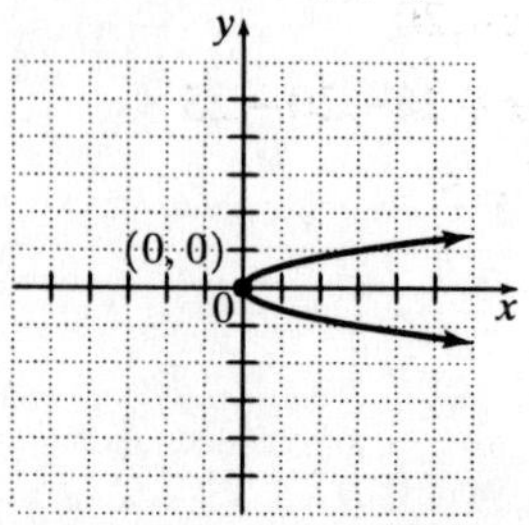

3. $x = (y-2)^2 + 3$

The vertex is (3, 2).

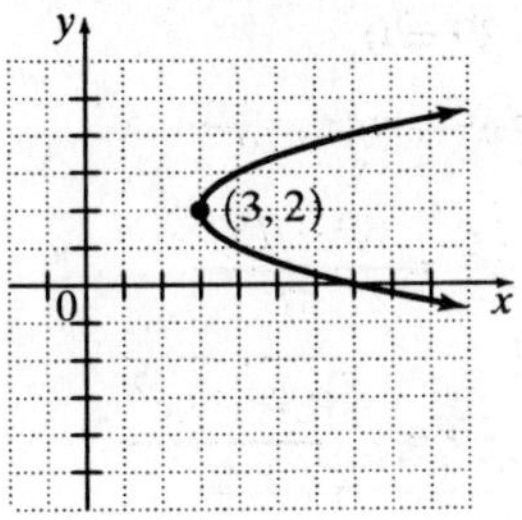

5. $y = 3(x-1)^2 + 5$

The vertex is (1, 5).

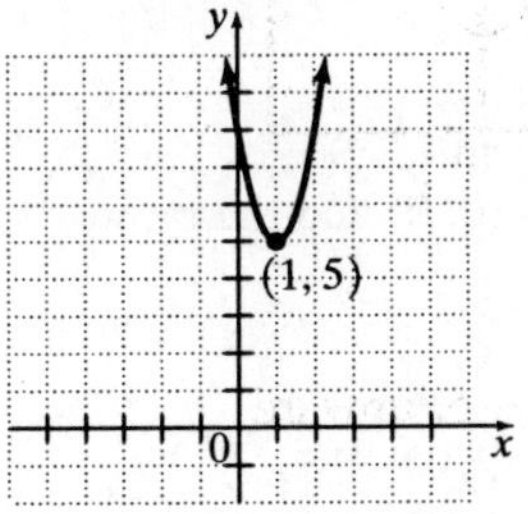

7. $x = y^2 + 6y + 8$

$x = y^2 + 6y + 9 + 8 - 9$

$x = (y+3)^2 - 1$

The vertex is (–1, –3).

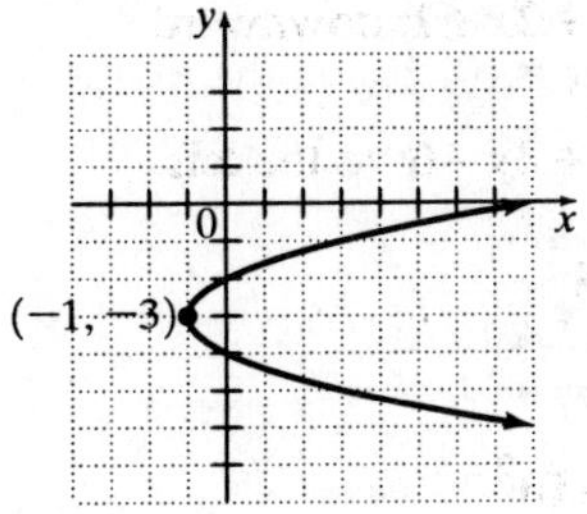

9. $y = x^2 + 10x + 20$

$y = x^2 + 10x + 25 + 20 - 25$

$y = (x+5)^2 - 5$

The vertex is (–5, –5).

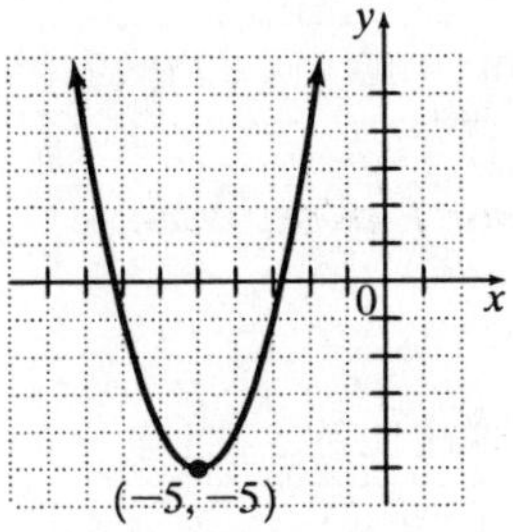

11. $x = -2y^2 + 4y + 6$

$x = -2(y^2 - 2y) + 6$

$x = -2(y^2 - 2y + 1) + 6 + 2$

$x = -2(y-1)^2 + 8$

Thus, the vertex is (8, 1).

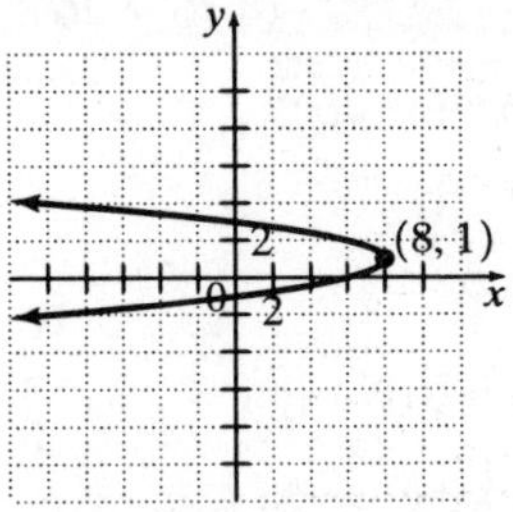

13. (5, 1), (8, 5)

$d = \sqrt{(8-5)^2 + (5-1)^2}$

$d = \sqrt{9+16}$

$d = \sqrt{25}$

$d = 5$ units

15. (–3, 2), (1, –3)

$d = \sqrt{[1-(-3)]^2 + (-3-2)^2}$

$d = \sqrt{16+25}$

$d = \sqrt{41}$ units

17. (–9, 4), (–8, 1)

$d = \sqrt{[-8-(-9)]^2 + (1-4)^2}$

$d = \sqrt{(-8+9)^2 + (-3)^2}$

$d = \sqrt{1^2 + 9}$

$d = \sqrt{10}$ units

19. $(0, -\sqrt{2}), (\sqrt{3}, 0)$

$d = \sqrt{(\sqrt{3}-0)^2 + [0-(-\sqrt{2})]^2}$

$d = \sqrt{(\sqrt{3})^2 + (\sqrt{2})^2}$

$d = \sqrt{3+2}$

$d = \sqrt{5}$ units

21. $(1.7, -3.6), (-8.6, 5.7)$

$d = \sqrt{(-8.6-1.7)^2 + [5.7-(-3.6)]^2}$

$d = \sqrt{(-10.3)^2 + (9.3)^2}$

$d = \sqrt{192.58}$

$d = 13.88$ units

23. $(2\sqrt{3}, \sqrt{6}), (-\sqrt{3}, 4\sqrt{6})$

$d = \sqrt{(-\sqrt{3}) - 2\sqrt{3})^2 + (4\sqrt{6} - \sqrt{6})^2}$

$d = \sqrt{(-3\sqrt{3})^2 + (3\sqrt{6})^2}$

$d = \sqrt{27+54}$

$d = \sqrt{81}$

$d = 9$ units

25. $(6, -8), (2, 4)$

$\left(\frac{6+2}{2}, \frac{-8+4}{2}\right)$

The midpoint of the segment is $(4, -2)$.

27. $(-2, -1), (-8, 6)$

$\left(\frac{-2+(-8)}{2}, \frac{-1+6}{2}\right)$

The midpoint of the segment is $\left(-5, \frac{5}{2}\right)$.

29. $(7, 3), (-1, -3)$

$\left(\frac{7+(-1)}{2}, \frac{3+(-3)}{2}\right)$

The midpoint of the segment is $(3, 0)$.

31. $\left(\frac{1}{2}, \frac{3}{8}\right), \left(-\frac{3}{2}, \frac{5}{8}\right)$

$\left(\frac{\frac{1}{2}+\left(-\frac{3}{2}\right)}{2}, \frac{\frac{3}{8}+\frac{5}{8}}{2}\right)$

The midpoint of the segment is $\left(-\frac{1}{2}, \frac{1}{2}\right)$.

33. $(\sqrt{2}, 3\sqrt{5}), (\sqrt{2}, -2\sqrt{5})$

$\left(\frac{\sqrt{2}+\sqrt{2}}{2}, \frac{3\sqrt{5}-2\sqrt{5}}{2}\right)$

The midpoint of the segment is $\left(\sqrt{2}, \frac{\sqrt{5}}{2}\right)$.

35. $(4.6, -3.5), (7.8, -9.8)$

$\left(\frac{4.6+7.8}{2}, \frac{-3.5+(-9.8)}{2}\right)$

The midpoint of the segment is (6.2, −6.65).

37. $x^2 + y^2 = 9$

$(x-0)^2 + (y-0)^2 = 3^2$

The center is (0, 0) and the radius is 3.

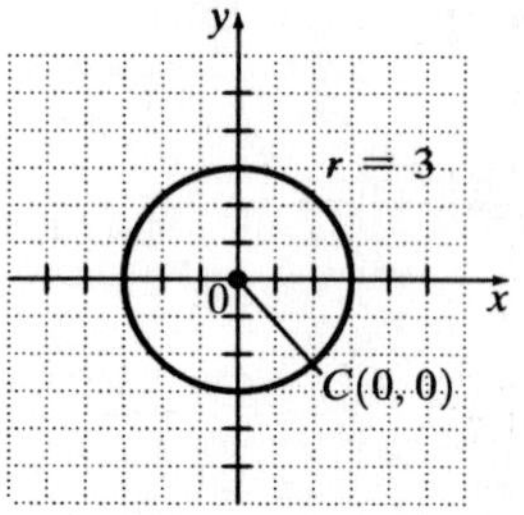

39. $x^2 + (y-2)^2 = 1$

$(x-0)^2 + (y-2)^2 = 1^2$

The center is (0, 2) and the radius is 1.

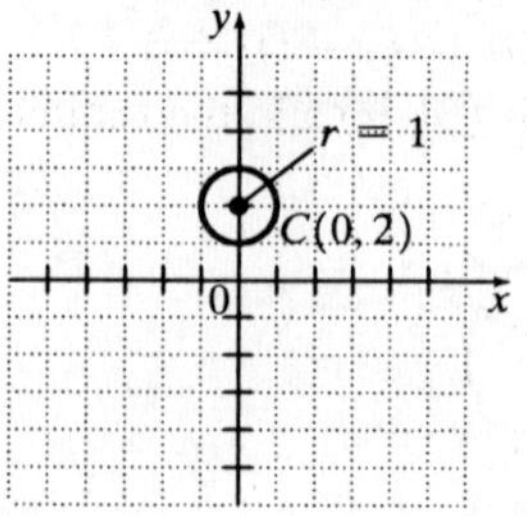

41. $(x-5)^2+(y+2)^2=1$

$(x-5)^2+(y+2)^2=1^2$

The center is (5, –2) and the radius is 1.

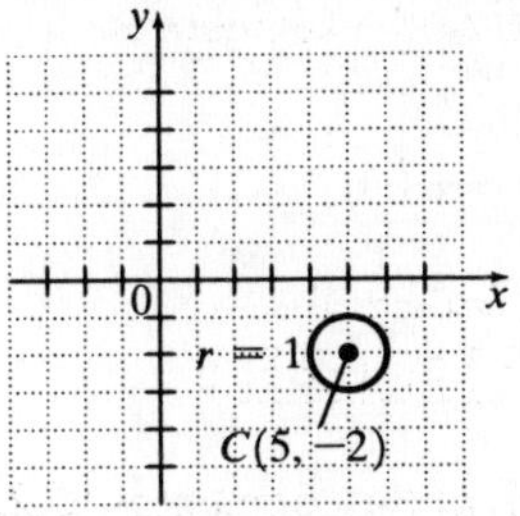

43. $x^2+y^2+6y=0$

$x^2+y^2+6y+9=9$

$(x-0)^2+(y+3)^2=3^2$

The center is (0, –3) and the radius is 3.

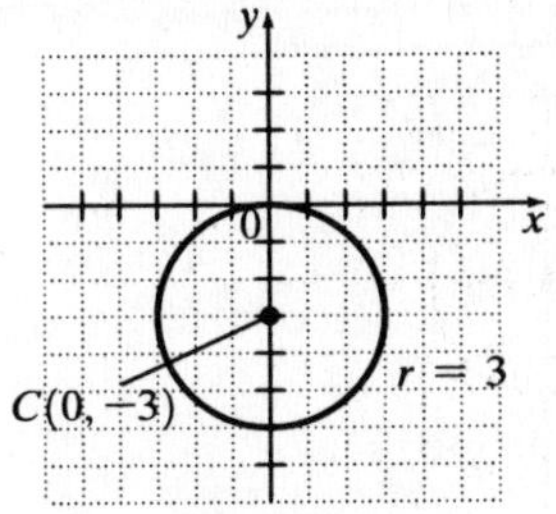

45. $x^2+y^2+2x-4y=4$

$x^2+2x+1+y^2-4y+4=4+1+4$

$(x+1)^2+(y-2)^2=9$

$(x+1)^2+(y-2)^2=3^2$

The center is (–1, 2) and the radius is 3.

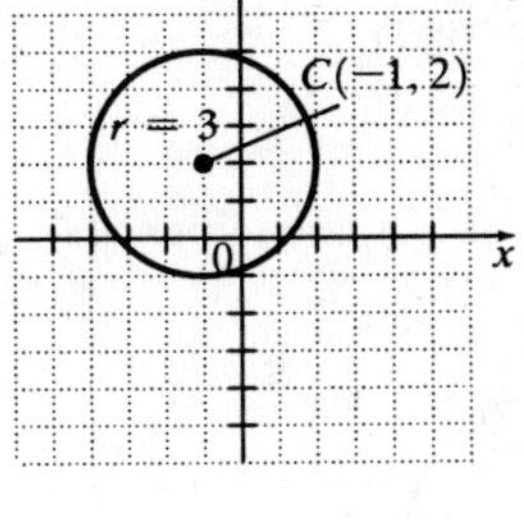

47. $x^2+y^2-4x-8y-2=0$

$(x^2-4x+4)+(y^2-8y+16)=2+4+16$

$(x-2)^2+(y-4)^2=22$

The center is (2, 4) and the radius is $\sqrt{22}$.

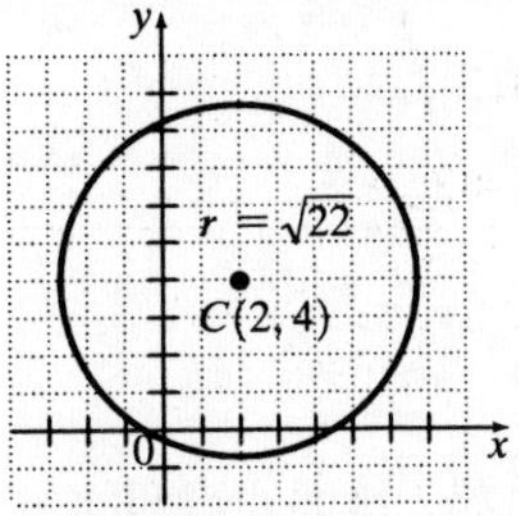

49. The center is (2, 3); the radius is 6.

$(x-2)^2+(y-3)^2=6^2$

$(x-2)^2+(y-3)^2=36$

51. The center is (0, 0); the radius is $\sqrt{3}$.

$(x-0)^2+(y-0)^2=(\sqrt{3})^2$

$x^2+y^2=3$

53. The center is (–5, 4); the radius is $3\sqrt{5}$.

$[x-(-5)]^2+(y-4)^2=(3\sqrt{5})^2$

$(x+5)^2+(y-4)^2=45$

55. Answers may vary.

57. $x=y^2-3$

The vertex is (–3, 0).

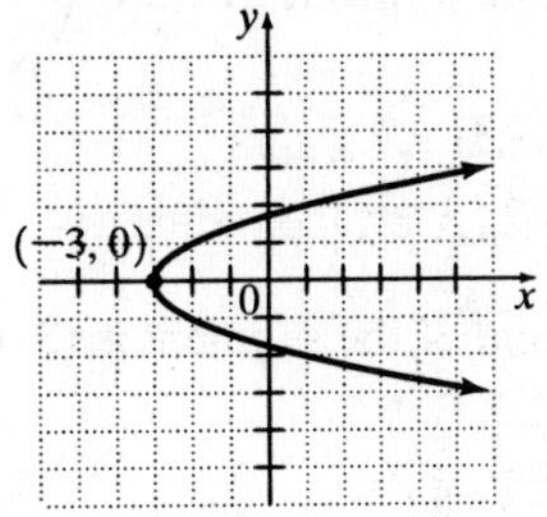

59. $y = (x-2)^2 - 2$

The vertex is (2, –2).

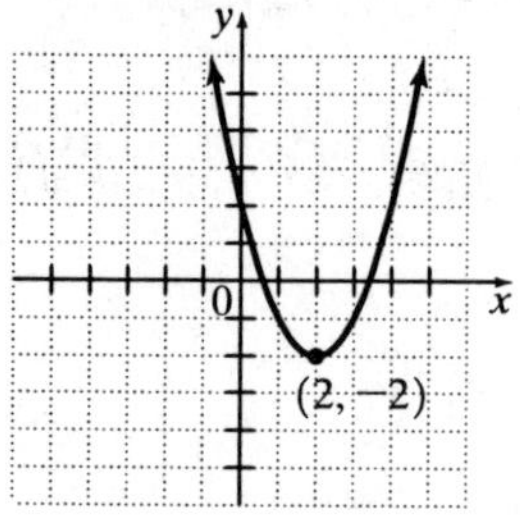

61. $x^2 + y^2 = 1$

The vertex is (0, 0) and the radius is 1.

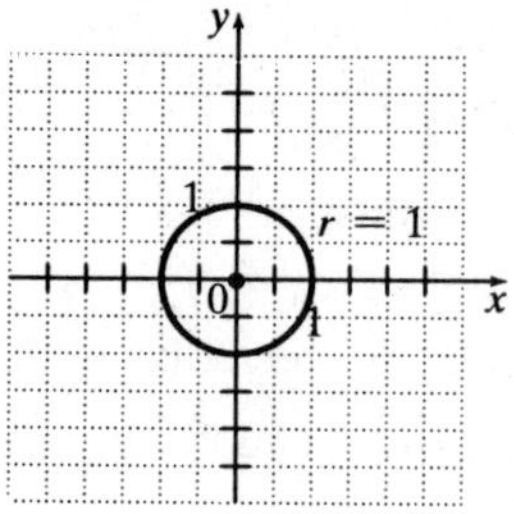

63. $x = (y+3)^2 - 1$

The vertex is (–1, –3).

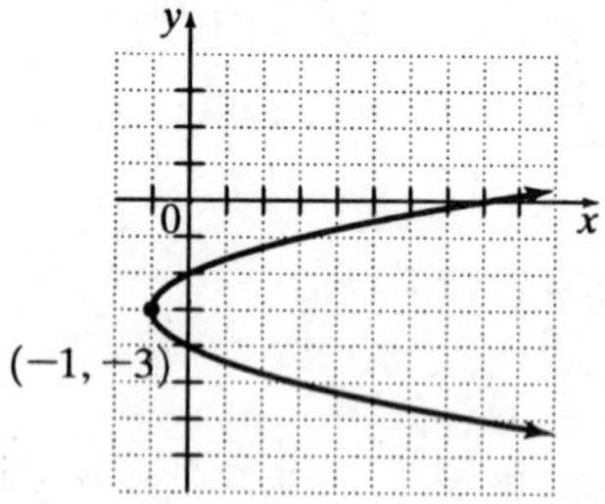

65. $(x-2)^2 + (y-2)^2 = 16$

The center is (2, 2); the radius is 4.

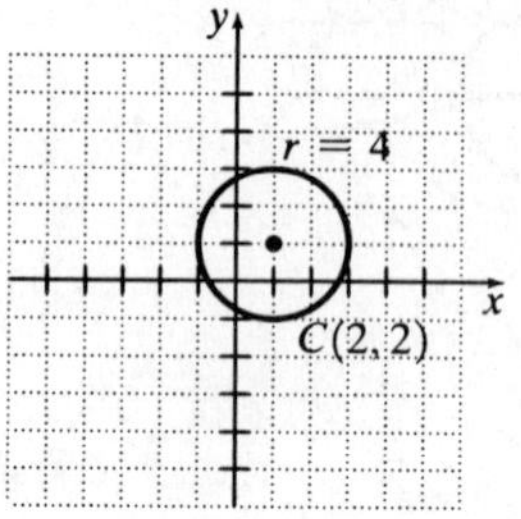

67. $x = -(y-1)^2$

The vertex is (0, 1).

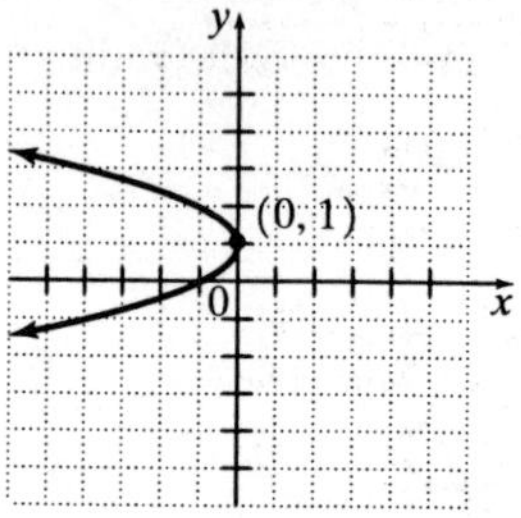

69. $(x-4)^2 + y^2 = 7$

The center is (4, 0); the radius is $\sqrt{7}$.

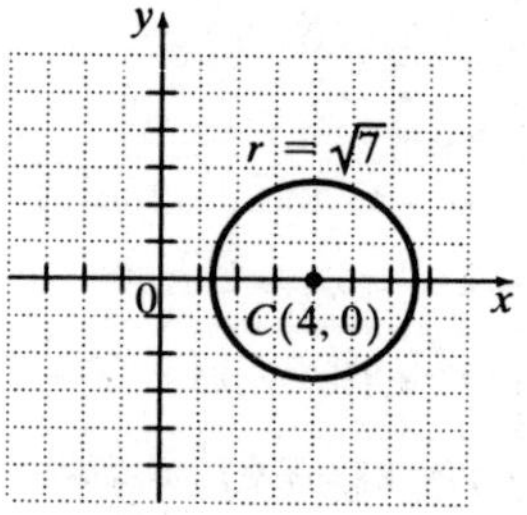

71. $y = 5(x+5)^2 + 3$

The vertex is (–5, 3).

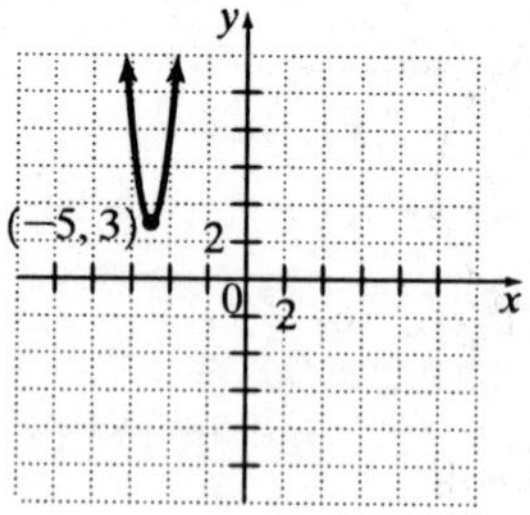

73. $\frac{x^2}{8} + \frac{y^2}{8} = 2$

The center is (0, 0); the radius is 4.

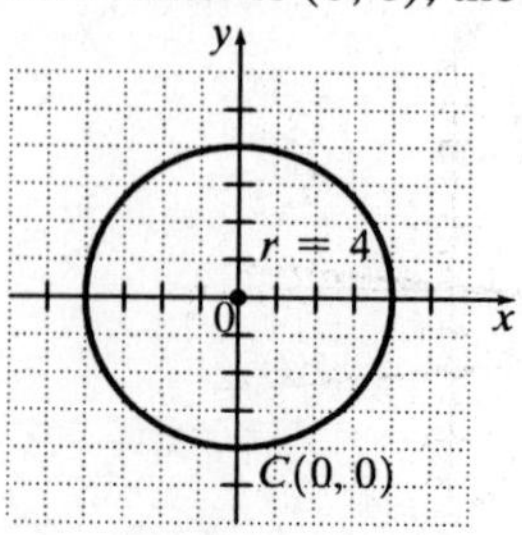

75. $y = x^2 + 7x + 6$

$y = x^2 + 7x + \frac{49}{4} + 6 - \frac{49}{4}$

$y = \left(x + \frac{7}{2}\right)^2 - \frac{25}{4}$

The vertex is $\left(-\frac{7}{2}, -\frac{25}{4}\right)$.

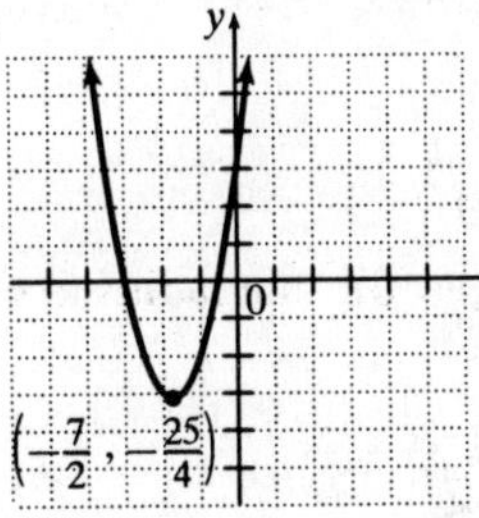

77. $x^2 + y^2 + 2x + 12y - 12 = 0$

$x^2 + 2x + 1 + y^2 + 12y + 36 = 12 + 1 + 36$

$(x+1)^2 + (y+6)^2 = 49$

The center is (–1, –6); the radius is 7.

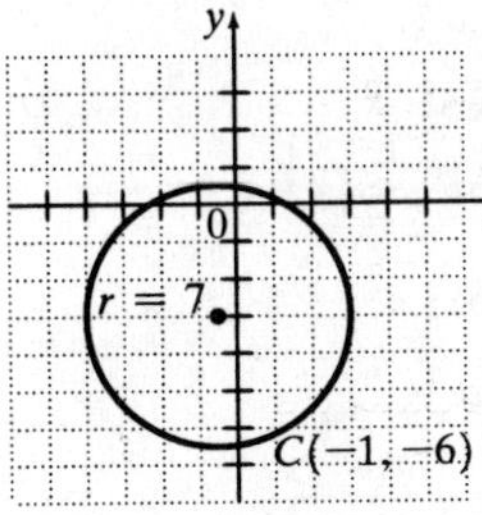

79. $x = y^2 + 8y - 4$

$x = y^2 + 8y + 16 - 4 - 16$

$x = (y+4)^2 - 20$

$V(-20, -4)$

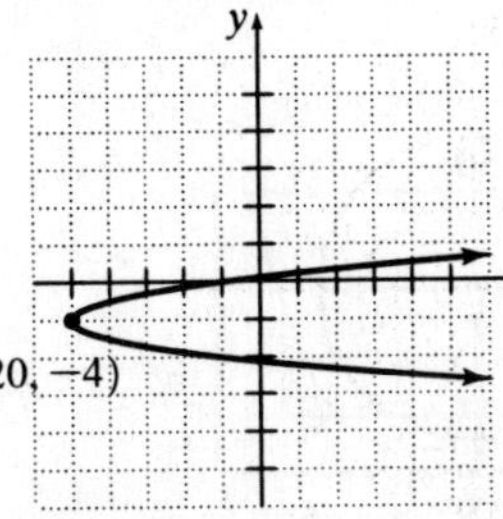

81. $x^2 - 10y + y^2 + 4 = 0$

$x^2 + y^2 - 10y + 25 = -4 + 25$

$x^2 + (y-5)^2 = 21$

The center is (0, 5); the radius is $\sqrt{21}$.

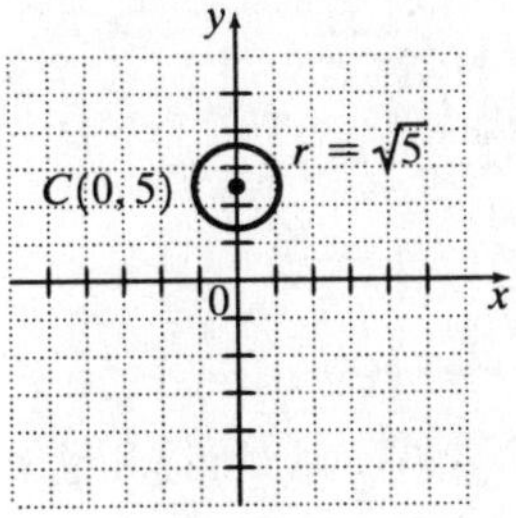

83. $x = -3y^2 + 30y$

$x = -3(y^2 - 10y + 25) + 75$

$x = -3(y-5)^2 + 75$

The vertex is (75, 5).

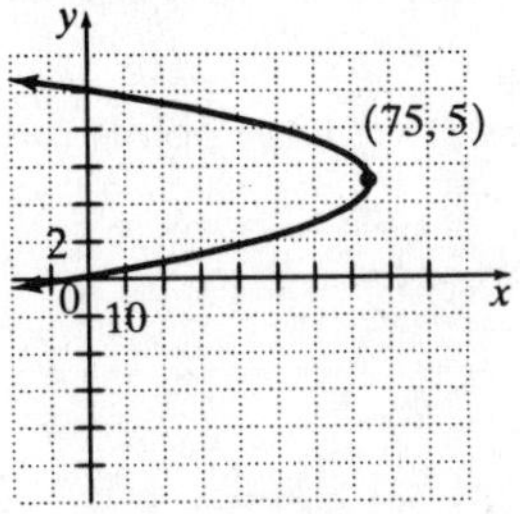

85. $5x^2 + 5y^2 = 25$

$x^2 + y^2 = 5$

The center is (0, 0); the radius is $\sqrt{5}$.

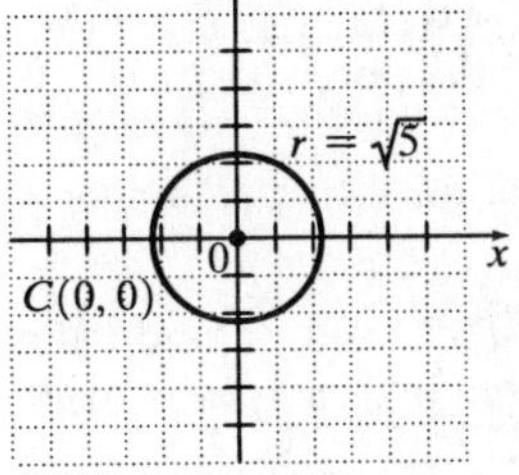

87. $y = 5x^2 - 20x + 16$

$y = 5(x^2 - 4x + 4) + 16 - 20$

$y = 5(x - 2)^2 - 4$

The vertex is (2, –4).

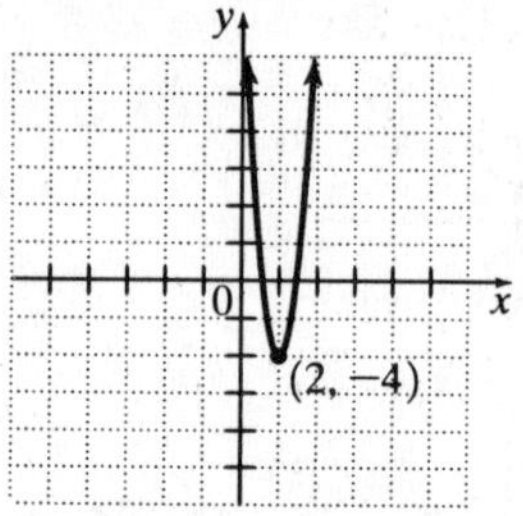

89. The distance between the points (5, 1) and (2, 6) is

$d = \sqrt{(5-2)^2 + (1-6)^2}$

$d = \sqrt{3^2 + (-5)^2}$

$d = \sqrt{9 + 25}$

$d = \sqrt{34}$ units.

The distance between the points (5, 1) and (0, –2) is

$d = \sqrt{(5-0)^2 + [1-(-2)]^2}$

$d = \sqrt{5^2 + 3^2}$

$d = \sqrt{25 + 9}$

$d = \sqrt{34}$ units.

Therefore, the triangle with vertices (2, 6), (0, –2) and (5, 1) is an isosceles triangle.

91. Setting up a coordinate system with the axis of symmetry as the *y*-axis and the base as the *x*-axis. The parabola would pass through the points (–50, 0), (0, 40), and (50, 0).

Use the equation for a parabola:

$y = ax^2 + bx + c$.

Substituting the *x* and *y* coordinates for the known points yields a system of equations.

$$\begin{cases} 0 = a(-50)^2 + b(-50) + c \\ 40 = a(0)^2 + b(0) + c \\ 0 = a(50)^2 + b(50) + c \end{cases}$$

$$\begin{cases} 0 = 2500a - 50b + c \\ 40 = c \\ 0 = 2500a + 50b + c \end{cases}$$

Solving $c = 40$:

$$\begin{cases} 0 = 2500a - 50b + 40 \\ 0 = 2500a + 50b + 40 \end{cases}$$

Adding the two equations:

$0 = 5000a + 80$

$-80 = 5000a$

$\frac{-80}{5000} = a$

$-\frac{2}{125} = a$

$0 = 2500\left(-\frac{2}{125}\right) - 50b + 40$

$0 = -40 - 50b + 40$

$0 = -50b$

$0 = b$

Thus, the equation of the parabola is

$y = -\frac{2}{125}x^2 + 40.$

93. $5x^2 + 5y^2 = 25$

$5y^2 = 25 - 5x^2$

$y^2 = 5 - x^2$

$y = \pm\sqrt{5 - x^2}$

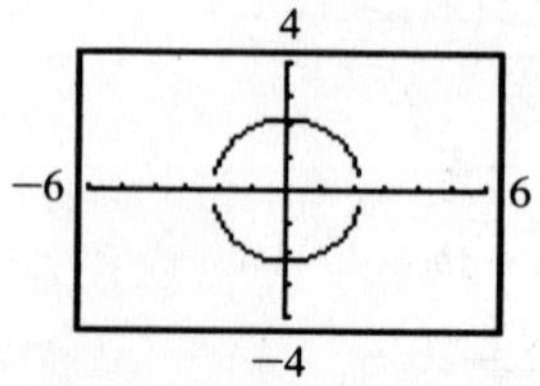

Answers are the same.

95. $y = 5x^2 - 20x + 16$

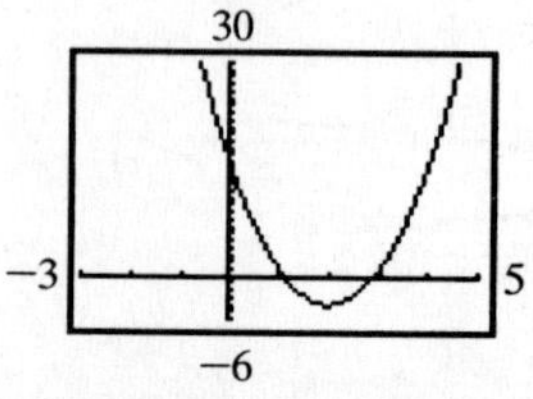

Answers are the same.

97. $y = -3x + 3$

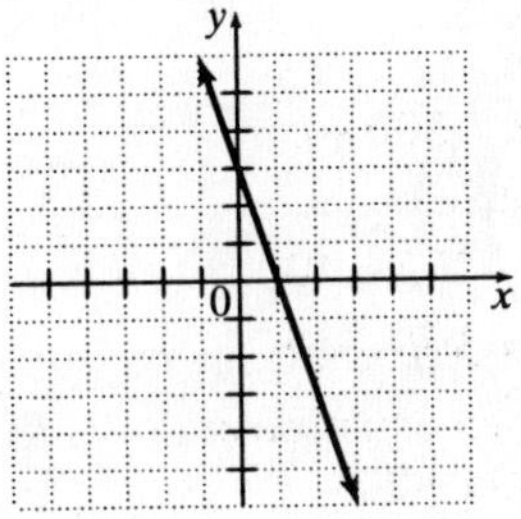

99. $x = -2$

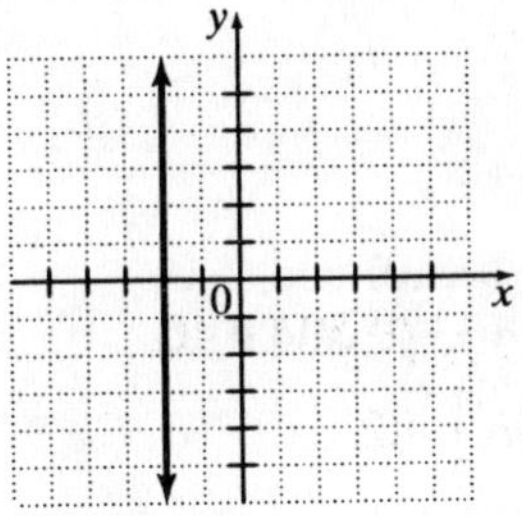

101. $\frac{\sqrt{5}}{\sqrt{8}} = \frac{\sqrt{5}}{2\sqrt{2}}$

$= \frac{\sqrt{5}}{2\sqrt{2}} \cdot \frac{\sqrt{2}}{\sqrt{2}}$

$= \frac{\sqrt{10}}{2\sqrt{4}}$

$= \frac{\sqrt{10}}{2(2)}$

$= \frac{\sqrt{10}}{4}$

103. $\frac{10}{\sqrt{5}} = \frac{10}{\sqrt{5}} \cdot \frac{\sqrt{5}}{\sqrt{5}}$

$= \frac{10\sqrt{5}}{\sqrt{25}}$

$= \frac{10\sqrt{5}}{5}$

$= 2\sqrt{5}$

Section 9.2

Graphing Calculator Explorations

1. $10x^2 + y^2 = 32$

$y^2 = 32 - 10x^2$

$y = \pm\sqrt{32 - 10x^2}$

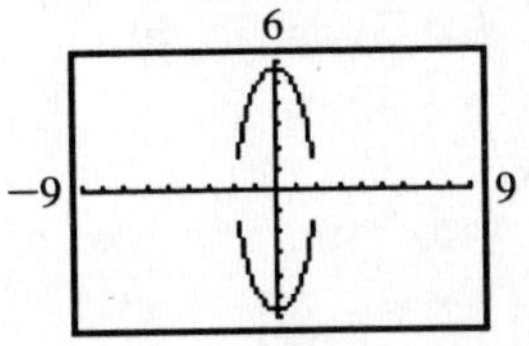

3. $20x^2 + 5y^2 = 100$

$5y^2 = 100 - 20x^2$

$y^2 = 20 - 4x^2$

$y = \pm\sqrt{20 - 4x^2}$

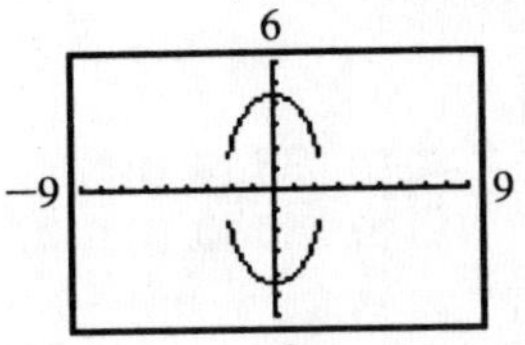

5. $7.3x^2 + 15.5y^2 = 95.2$

$15.5y^2 = 95.2 - 7.3x^2$

$y^2 = \frac{95.2 - 7.3x^2}{15.5}$

$y = \pm\sqrt{\frac{95.2 - 7.3x^2}{15.5}}$

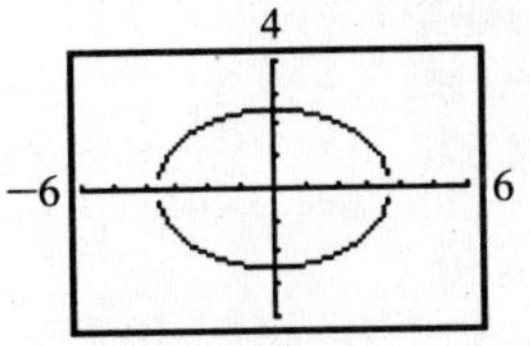

Exercise Set 9.2

1. $\frac{x^2}{4}+\frac{y^2}{25}=1$

$\frac{x^2}{2^2}+\frac{y^2}{5^2}=1$

The center is (0, 0).
x-intercepts: $(-2, 0)$ and $(2, 0)$
y-intercepts: $(0, -5)$ and $(5, 0)$

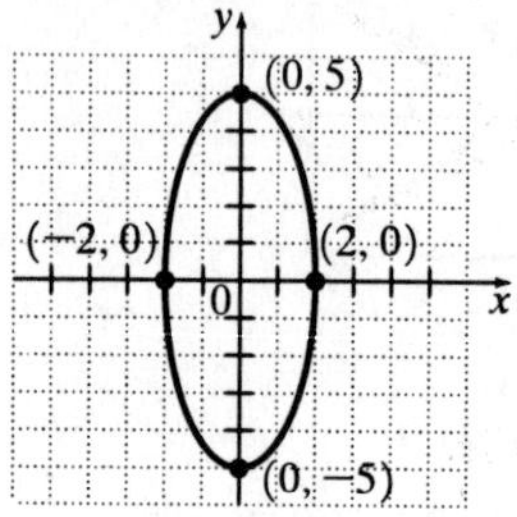

3. $\frac{x^2}{16}+\frac{y^2}{9}=1$

$\frac{x^2}{4^2}+\frac{y^2}{3^2}=1$

The center is (0, 0).
x-intercepts: $(-4, 0)$ and $(4, 0)$
y-intercepts: $(0, -3)$ and $(0, 3)$

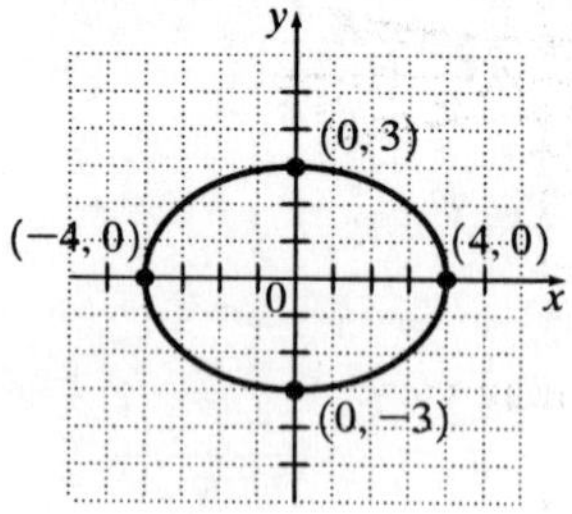

5. $9x^2+4y^2=36$

$\frac{x^2}{4}+\frac{y^2}{9}=1$

$\frac{x^2}{2^2}+\frac{y^2}{3^2}=1$

The center is (0, 0).
x-intercepts: $(-2, 0)$ and $(2, 0)$
y-intercepts: $(0, -3)$ and $(0, 3)$

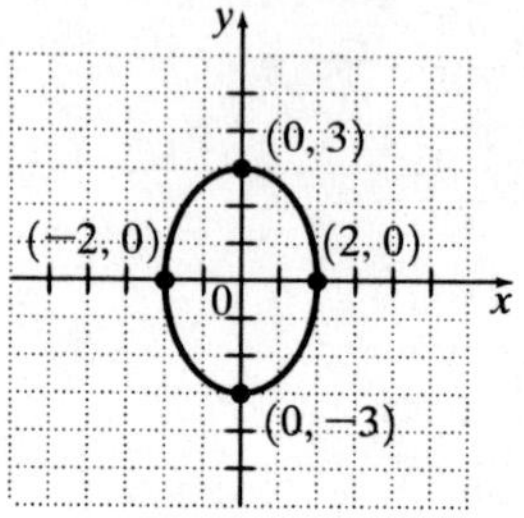

7. $4x^2+25y^2=100$

$\frac{x^2}{25}+\frac{y^2}{4}=1$

$\frac{x^2}{5^2}+\frac{y^2}{2^2}=1$

x-intercepts: $(-5, 0)$ and $(5, 0)$
y-intercepts: $(0, -2)$ and $(0, 2)$
The center is (0, 0).

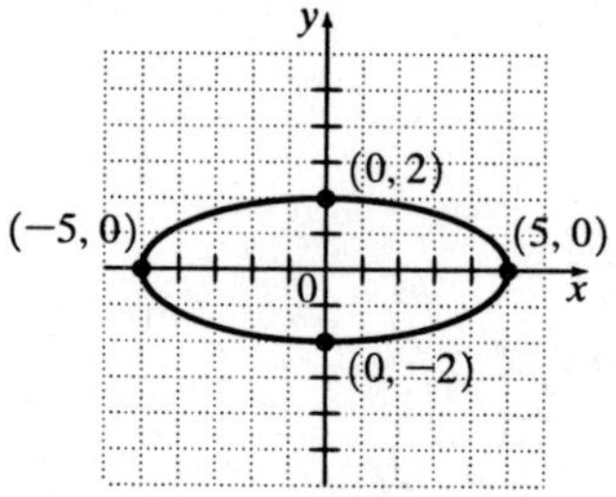

9. $\frac{(x+1)^2}{36}+\frac{(y-2)^2}{49}=1$

$\frac{(x+1)^2}{6^2}+\frac{(y-2)^2}{7^2}=1$

The center is $(-1, 2)$.
Other points:
$(-1-6, 2)$ or $(-7, 2)$
$(-1+6, 2)$ or $(5, 2)$
$(-1, 2-7)$ or $(-1, -5)$
$(-1, 2+7)$ or $(-1, 9)$

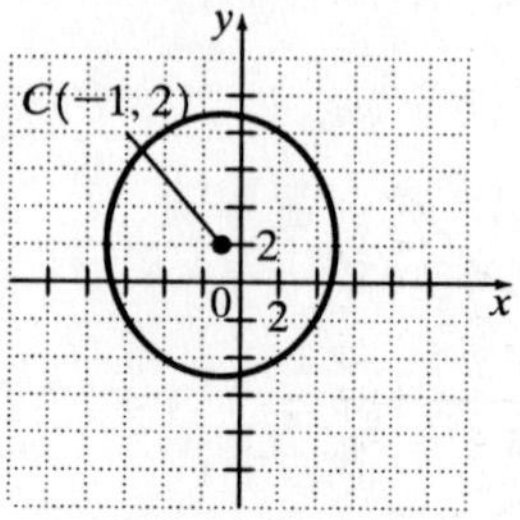

11. $\frac{(x-1)^2}{4}+\frac{(y-1)^2}{25}=1$

$\frac{(x-1)^2}{2^2}+\frac{(y-1)^2}{5^2}=1$

The center is (1, 1).
Other points:
(1 – 2, 1) or (–1, 1)
(1 + 2, 1) or (3, 1)
(1, 1 – 5) or (1, –4)
(1, 1 + 5) or (1, 6)

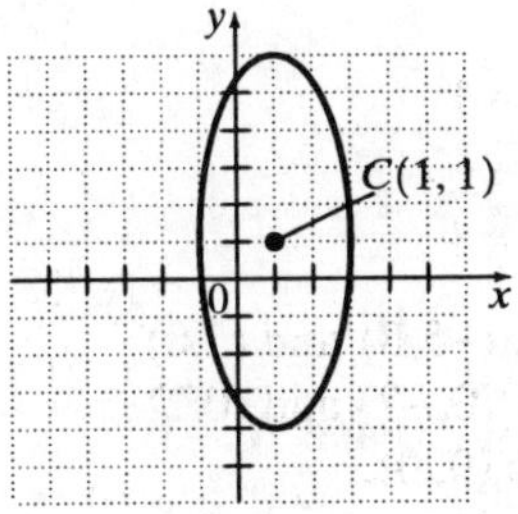

13. $\frac{x^2}{4}-\frac{y^2}{9}=1$

$\frac{x^2}{2^2}-\frac{y^2}{3^2}=1$

$a=2,\ b=3$

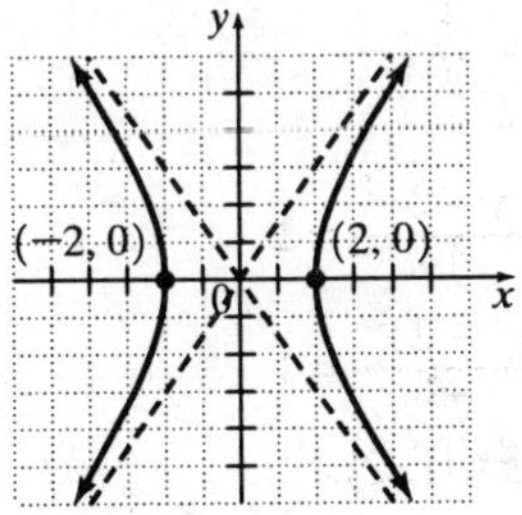

15. $\frac{y^2}{25}-\frac{x^2}{16}=1$

$\frac{y^2}{5^2}-\frac{x^2}{4^2}=1$

$a=4,\ b=5$

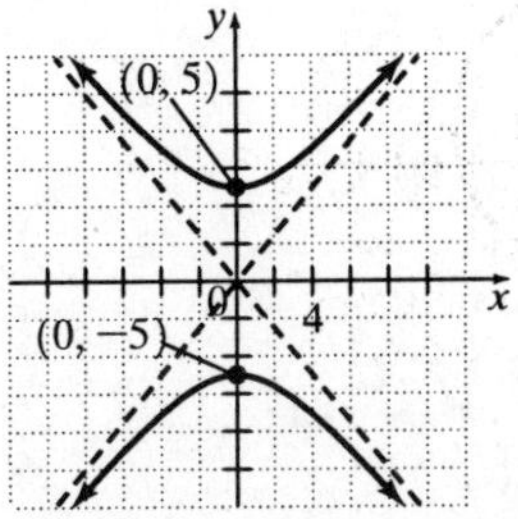

17. $x^2-4y^2=16$

$\frac{x^2}{16}-\frac{y^2}{4}=1$

$\frac{x^2}{4^2}-\frac{y^2}{2^2}=1$

$a=4,\ b=2$

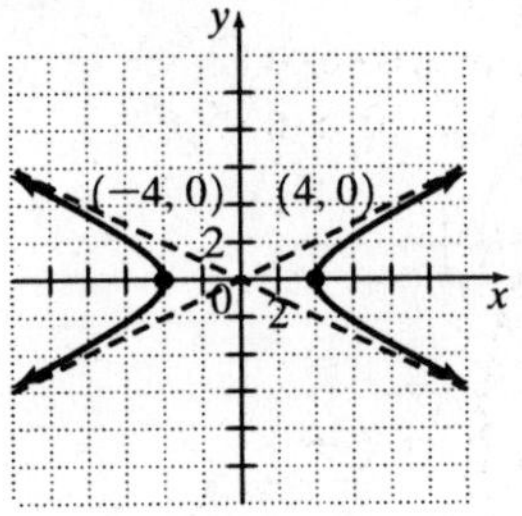

19. $16y^2-x^2=16$

$\frac{y^2}{1}-\frac{x^2}{16}=1$

$\frac{y^2}{1^2}-\frac{x^2}{4^2}=1$

$a=4,\ b=1$

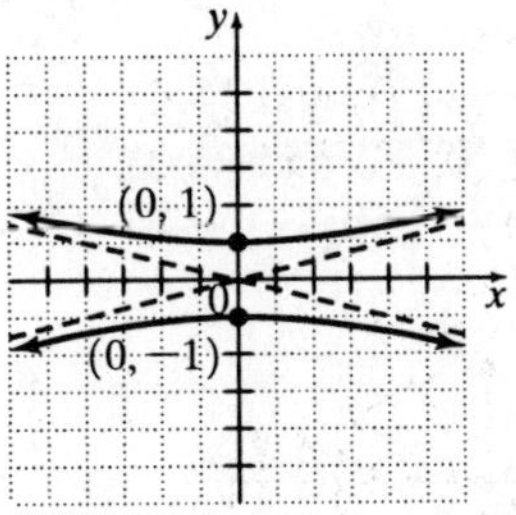

21. Answers may vary.

23. $y=x^2+4$

parabola: The vertex is (0, 4).
opens upward

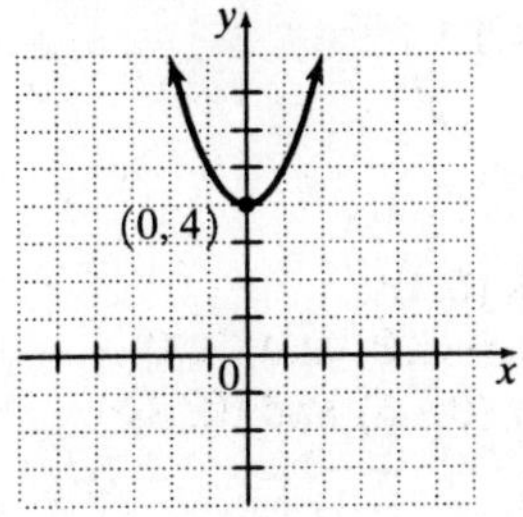

25. $\dfrac{x^2}{4}+\dfrac{y^2}{9}=1$

ellipse: center (0, 0)

$a = 3,\ b = 2$

y-intercepts:(0, 3) and (0, −3)

x-intercepts: (2, 0) and (−2, 0)

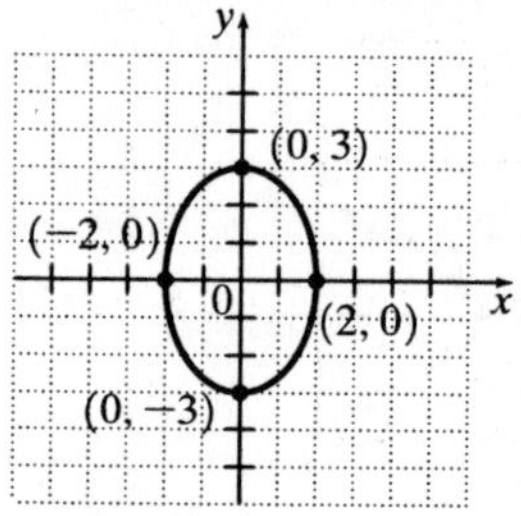

27. $\dfrac{x^2}{16}-\dfrac{y^2}{4}=1$

hyperbola: center (0, 0)

$a = 4,\ b = 2$

x-intercepts: (4, 0) and (−4, 0)

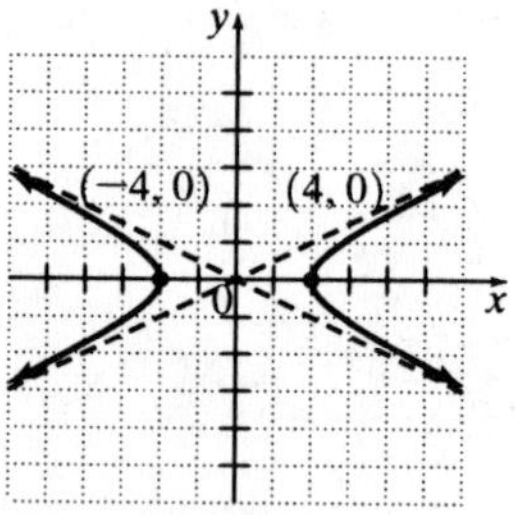

29. $x^2+y^2=16$

circle: center (0, 0)

radius = 4

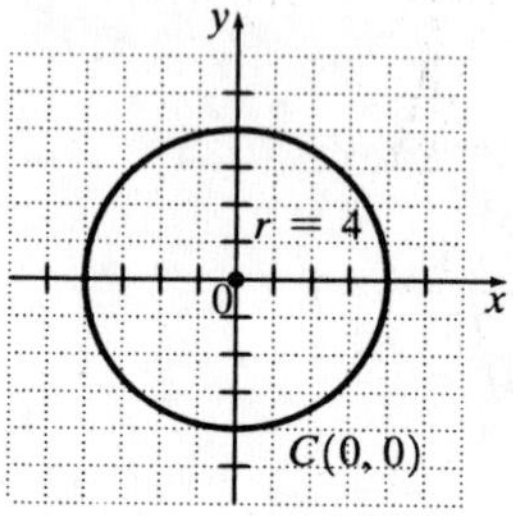

31. $x=-y^2+6y$

parabola: $\dfrac{-b}{2a}=\dfrac{-6}{2(-1)}=3$

$x = -9 + 18 = 9$

The vertex is (9, 3).

opens to the left

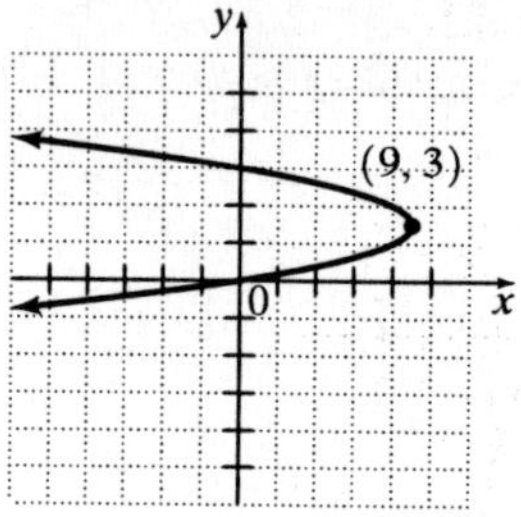

33. $9x^2+4y^2=36$

$\dfrac{x^2}{4}+\dfrac{y^2}{9}=1$

ellipse: center (0, 0)

$a = 3,\ b = 2$

y-intercepts: (0, 3) and (0, −3)

x-intercepts at: (2, 0) and (−2, 0)

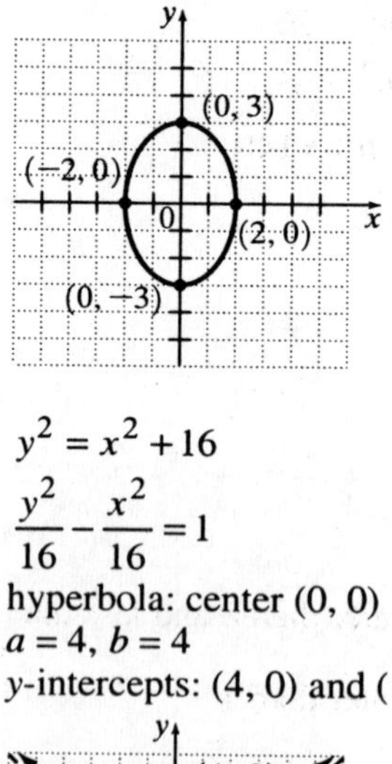

35. $y^2=x^2+16$

$\dfrac{y^2}{16}-\dfrac{x^2}{16}=1$

hyperbola: center (0, 0)

$a = 4,\ b = 4$

y-intercepts: (4, 0) and ((−4, 0)

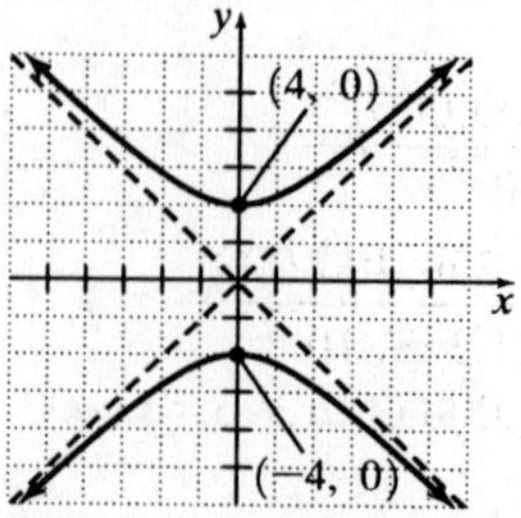

37. $y=-2x^2+4x-3$

parabola: $\dfrac{-b}{2a}=\dfrac{-4}{2a}=1$

$y = -2 + 4 - 3 = -1$
The vertex is $(1, -1)$.
opens downward

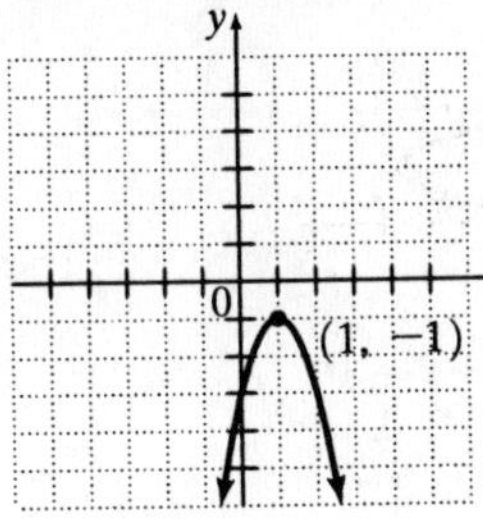

39. A: $a^2 = 36$, $b^2 = 13$
B: $a^2 = 4$, $b^2 = 4$
C: $a^2 = 25$, $b^2 = 16$
D: $a^2 = 39$, $b^2 = 25$
E: $a^2 = 17$, $b^2 = 81$
F: $a^2 = 36$, $b^2 = 36$
G: $a^2 = 16$, $b^2 = 65$
H: $a^2 = 144$, $b^2 = 140$

41. A: $d = 6$
B: $d = 2$
C: $d = 5$
D: $d = 5$
E: $d = 9$
F: $d = 6$
G: $d = 4$
H: $d = 12$

43. They are greater than 0 and less than 1.

45. They are greater than 1.

47. $$\frac{(x - 1{,}782{,}000{,}000)^2}{(3.42)(10^{23})} + \frac{(y - 356{,}400{,}000)^2}{(1.368)(10^{22})} = 1$$
center (1,782,000,000, 356,400,000)

49. $x^2 + 4y^2 = 16$
$$4y^2 = 16 - x^2$$
$$y^2 = \frac{16 - x^2}{4}$$
$$y = \pm\sqrt{\frac{16 - x^2}{4}}$$

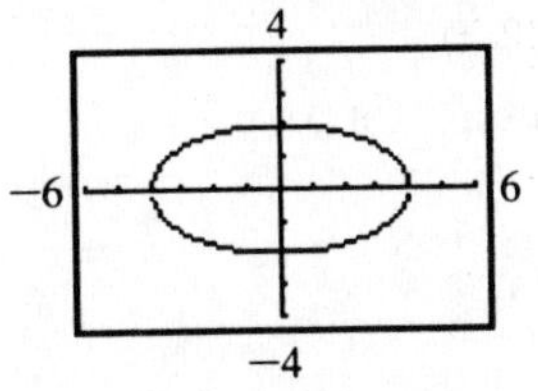

Answers are the same.

51. $x < 5$ or $x < 1$
$(-\infty, 5)$

53. $2x - 1 \ge 7$ and $-3x \le -6$
$x \ge 4$ and $x \ge 2$
$[4, \infty)$

55. $2x^3 - 4x^3 = -2x^3$

57. $(-5x^2)(x^2) = -5x^4$

59.

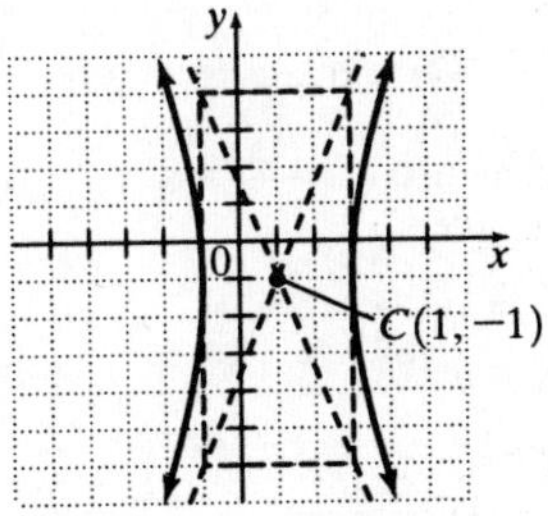

$$\frac{(x-1)^2}{4} - \frac{(y+1)^2}{25} = 1$$
center $(1, -1)$
$a = 2$, $b = 5$

61. $$\frac{y^2}{16} - \frac{(x+3)^2}{9} = 1$$
center $(-3, 0)$
$a = 4$, $b = 3$

63. $\dfrac{(x+5)^2}{16} - \dfrac{(y+2)^2}{25} = 1$

center $(-5, -2)$

$a = 4$, $b = 5$

Exercise Set 9.3

1. $\begin{cases} x^2 + y^2 = 25 \\ 4x + 3y = 0 \end{cases}$

Solve equation 2 for y.

$3y = -4x$

$y = \dfrac{-4x}{3}$

Substitute.

$$x^2 + \left(-\frac{4x}{3}\right)^2 = 25$$
$$x^2 + \frac{16x^2}{9} = 25$$
$$\frac{25}{9}x^2 = 25$$
$$\frac{x^2}{9} = 1$$
$$x^2 = 9$$
$$x = \pm\sqrt{9} = \pm 3$$

$x = 3$: $y = -\dfrac{4}{3}(3) = -4$

$x = -3$: $y = -\dfrac{4}{3}(-3) = 4$

The solutions are $(3, -4)$ and $(-3, 4)$.

3. $\begin{cases} x^2 + 4y^2 = 10 \\ y = x \end{cases}$

Substitute.

$$x^2 + 4x^2 = 10$$
$$5x^2 = 10$$
$$x^2 = 2$$
$$x = \pm\sqrt{2}$$

$x = \sqrt{2}$; $y = \sqrt{2}$

$x = -\sqrt{2}$; $y = -\sqrt{2}$

The solutions are

$\left(\sqrt{2}, \sqrt{2}\right)$ and $\left(-\sqrt{2}, -\sqrt{2}\right)$.

5. $\begin{cases} y^2 = 4 - x \\ x - 2y = 4 \end{cases}$

$-2y = 4 - x$

Substitute.

$$y^2 = -2y$$
$$y^2 + 2y = 0$$
$$y(y+2) = 0$$
$$y = 0 \quad \text{or} \quad y + 2 = 0$$
$$y = -2$$

$y = 0$: $x - 2(0) = 4$

$x = 4$

$y = -2$: $x - 2(-2) = 4$

$x + 4 = 4$

$x = 0$

The solutions are $(4, 0)$ and $(0, -2)$.

7. $\begin{cases} x^2 + y^2 = 9 \\ 16x^2 - 4y^2 = 64 \end{cases}$

Divide equation two by 4.

$\begin{cases} x^2 + y^2 = 9 \\ 4x^2 - y^2 = 16 \end{cases}$

Add.

$$5x^2 = 25$$
$$x^2 = 5$$
$$x = \pm\sqrt{5}$$

Substitute back.

$$5 + y^2 = 9$$
$$y^2 = 4$$
$$y = \pm 2$$

The solutions are

$\left(\sqrt{5}, 2\right)$, $\left(\sqrt{5}, -2\right)$, $\left(-\sqrt{5}, 2\right)$, and

$\left(-\sqrt{5}, -2\right)$.

9. $\begin{cases} x^2 + 2y^2 = 2 \\ x - y = 2 \end{cases}$

$x = y + 2$

Substitute.

$$(y+2)^2 + 2y^2 = 2$$
$$y^2 + 4y + 4 + 2y^2 = 2$$
$$3y^2 + 4y + 4 = 2$$
$$3y^2 + 4y + 2 = 0$$

$b^2 - 4ac = 4^2 - 4(3)(2) = 16 - 24 = -8 < 0$

Therefore, no real solutions exist.

The solution is $\varnothing$.

11. $\begin{cases} y = x^2 - 3 \\ 4x - y = 6 \end{cases}$

Substitute.

$$4x - (x^2 - 3) = 6$$
$$4x - x^2 + 3 = 6$$
$$0 = x^2 - 4x + 3$$
$$0 = (x-3)(x-1)$$

$x - 3 = 0$ or $x - 1 = 0$

$x = 3$ or $x = 1$

$x = 3$: $y = 3^2 - 3 = 9 - 3 = 6$

$x = 1$: $y = 1^2 - 3 = 1 - 3 = -2$

The solutions are (3, 6) and (1, –2).

13. $\begin{cases} y = x^2 \\ 3x + y = 10 \end{cases}$

Substitute.

$$3x + x^2 = 10$$
$$x^2 + 3x - 10 = 0$$
$$(x+5)(x-2) = 0$$

$x + 5 = 0$ or $x - 2 = 0$

$x = -5$ or $x = 2$

$x = -5$: $y = (-5)^2 = 25$

$x = 2$: $y = 2^2 = 4$

The solutions are (–5, 25) and (2, 4).

15. $\begin{cases} y = 2x^2 + 1 \\ x + y = -1 \end{cases}$

Substitute.

$$x + 2x^2 + 1 = -1$$
$$2x^2 + x + 1 = -1$$
$$2x^2 + x + 2 = 0$$

$b^2 - 4ac = 1^2 - 4(2)(2) = -15 < 0$

Therefore, no real solutions exist.

The solution is $\varnothing$.

17. $\begin{cases} y = x^2 - 4 \\ y = x^2 - 4x \end{cases}$

Substitute.

$$x^2 - 4 = x^2 - 4x$$
$$-4 = -4x$$
$$x = 1$$

$y = 1^2 - 4 = -3$

The solution is (1, –3).

19. $\begin{cases} 2x^2 + 3y^2 = 14 \\ -x^2 + y^2 = 3 \end{cases}$

$y^2 = x^2 + 3$

Substitute.

$$2x^2 + 3(x^2 + 3) = 14$$
$$2x^2 + 3x^2 + 9 = 14$$
$$5x^2 + 9 = 14$$
$$5x^2 = 5$$
$$x^2 = 1$$
$$x = \pm 1$$

Substitute back.

$$y^2 = 1 + 3$$
$$y^2 = 4$$
$$y = \pm 2$$

The solutions are (1, –2), (1, 2), (–1, –2), and (–1, 2).

21. $\begin{cases} x^2+y^2=1 \\ x^2+(y+3)^2=4 \end{cases}$

Subtract equation 1 from equation 2.

$$(y+3)^2-y^2=3$$
$$y^2+6y+9-y^2=3$$
$$6y+9=3$$
$$6y=-6$$
$$y=-1$$

Substitute back.

$$x^2+(-1)^2=1$$
$$x^2+1=1$$
$$x^2=0$$
$$x=0$$

The solution is (0, –1).

23. $\begin{cases} y=x^2+2 \\ y=-x^2+4 \end{cases}$

Substitute.

$$x^2+2=-x^2+4$$
$$2x^2=2$$
$$x^2=1$$
$$x=\pm 1$$

Substitute back.

$y=1+2=3$

The solutions are (1, 3) and (–1, 3).

25. $\begin{cases} 3x^2+y^2=9 \\ 3x^2-y^2=9 \end{cases}$

Subtract.

$$2y^2=0$$
$$y^2=0$$
$$y=0$$

Substitute back.

$$3x^2+0=9$$
$$3x^2=9$$
$$x^2=3$$
$$x=\pm\sqrt{3}$$

The solutions are $\left(\sqrt{3},\ 0\right)$ and $\left(-\sqrt{3},\ 0\right)$.

27. $\begin{cases} x^2+3y^2=6 \\ x^2-3y^2=10 \end{cases}$

Add.

$$2x^2=16$$
$$x^2=8$$
$$x=\pm\sqrt{8}$$
$$x=\pm 2\sqrt{2}$$

Substitute back.

$$8+3y^2=6$$
$$3y^2=-2$$
$$y^2=-\frac{2}{3}$$

Therefore, no real solutions exist.
The solution is $\varnothing$.

29. $\begin{cases} x^2+y^2=36 \\ y=\frac{1}{6}x^2-6 \end{cases}$

$$y+6=\frac{1}{6}x^2$$
$$x^2=6(y+6)$$

Substitute.

$$6(y+6)+y^2=36$$
$$6y+36+y^2=36$$
$$6y+y^2=0$$
$$y(6+y)=0$$
$$y=0 \quad \text{or} \quad 6+y=0$$
$$y=-6$$

$y=0$: $x^2+0^2=36$
$$x^2=36$$
$$x=\pm 6$$

$y=-6$: $x^2+(-6)^2=36$
$$x^2+36=36$$
$$x^2=0$$
$$x=0$$

The solutions are (6, 0), (–6, 0), and (0, –6).

31. There can be 0, 1, 2, 3, or 4 real solutions.
For the circle $x^2 + y^2 = 9$:

a. the parabola $y = -x^2 - 4$ does not intersect it;

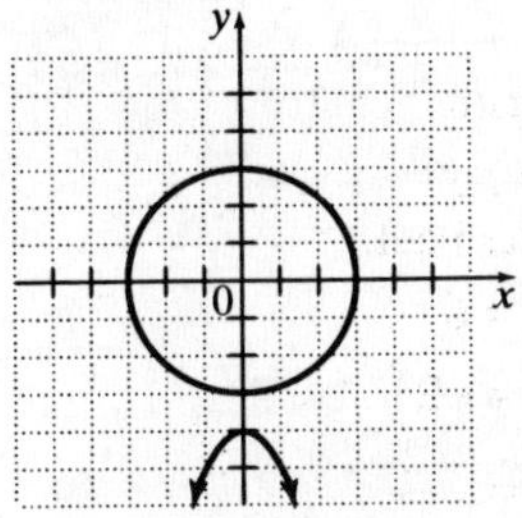

b. the parabola $y = x^2 + 3$ intersects it once;

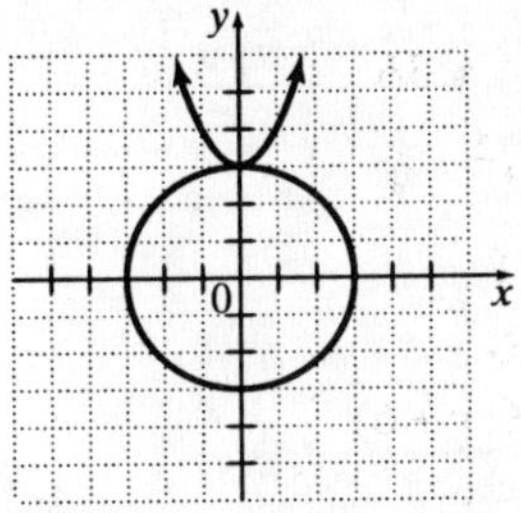

c. the parabola $y = x^2$ intersects it 2 times;

d. the parabola $y = x^2 - 3$ intersects it 3 times;

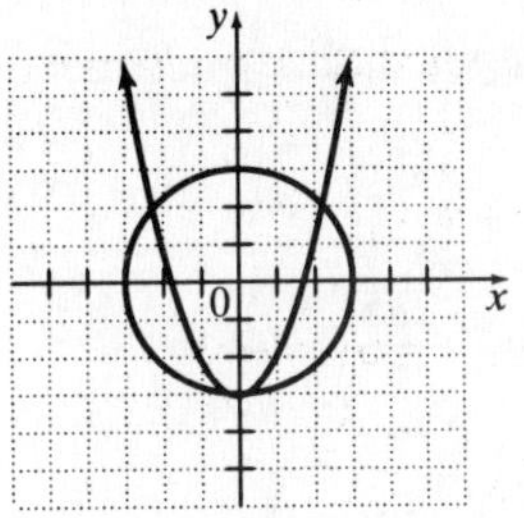

e. the parabola $y = x^2 - 4$ intersects it 4 times.

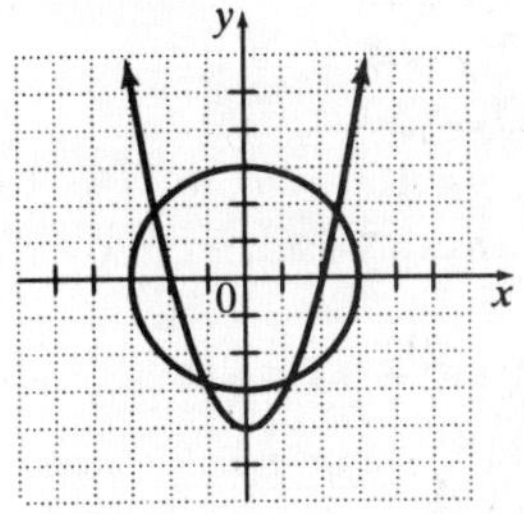

33. $\begin{cases} x^2 + y^2 = 130 \\ x^2 - y^2 = 32 \end{cases}$

Add.

$2x^2 = 162$

$x^2 = 81$

$x = \pm 9$

Substitute back.

$9^2 + y^2 = 130 \qquad (-9)^2 + y^2 = 130$

$y^2 = 49 \qquad y^2 = 49$

$y = \pm 7 \qquad y = \pm 7$

Answer:
9 and 7, 9 and –7, –9 and 7, –9 and –7

35. $\begin{cases} 2x + 2y = 68 \\ xy = 285 \end{cases}$

Substitute $y = \dfrac{285}{x}$.

$2x + 2\left(\dfrac{285}{x}\right) = 68$

$x + \dfrac{285}{x} = 34$

$x^2 - 34x + 285 = 0$

$(x - 19)(x - 15) = 0$

$x = 19$ or $x = 15$

Substitute back.

$2(19) + 2y = 68$

$38 + 2y = 68$

$2y = 30$

$y = 15$

The dimensions are 19 cm by 15 cm.

37. $p = -0.01x^2 - 0.2x + 9$

$p = 0.01x^2 - 0.1x + 3$

$-0.01x^2 - 0.2x + 9 = 0.01x^2 - 0.1x + 3$

$0 = 0.02x^2 + 0.1x - 6$

$0 = x^2 + 5x - 300$

$0 = (x + 20)(x - 15)$

$x + 20 = 0$ or $x - 15 = 0$

$x = -20$ or $x = 15$

Disregard the negative.

$p = -0.01(15)^2 - 0.2(15) + 9$

$p = 3.75$

15 thousand compact discs; price, $3.75

39. $\begin{cases} x^2 + 4y^2 = 10 \\ y = x \end{cases}$

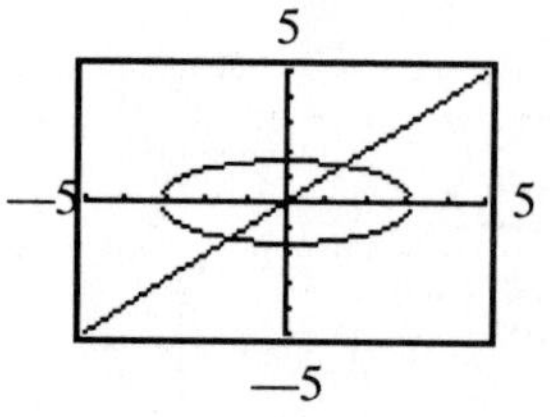

41. $\begin{cases} y = x^2 + 2 \\ y = -x^2 + 4 \end{cases}$

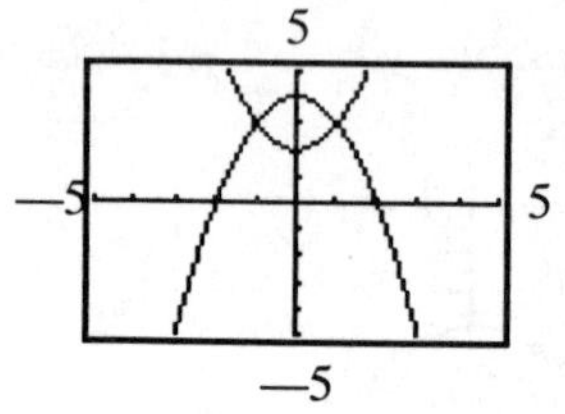

43. $x > -3$

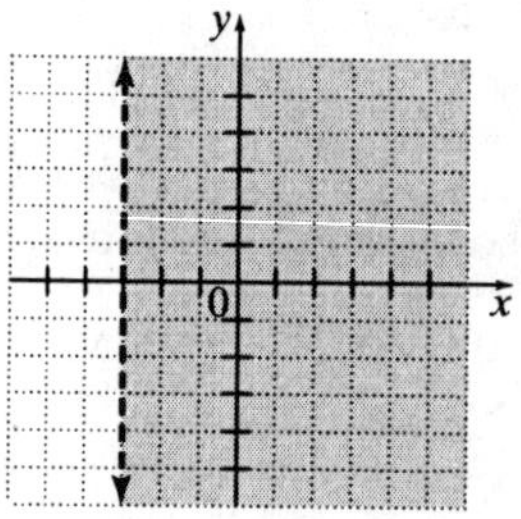

45. $y < 2x - 1$

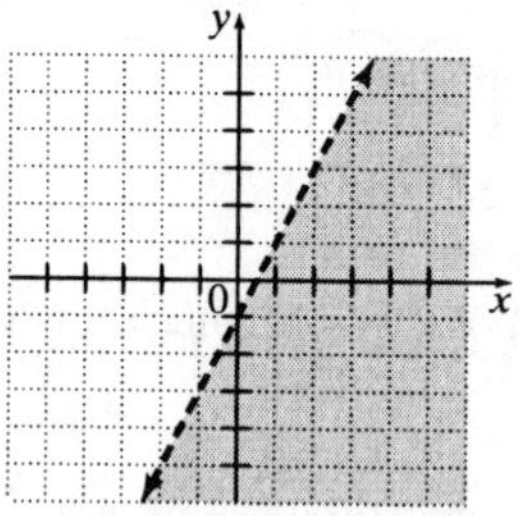

47. $P = x + (2x - 5) + (5x - 20)$

$P = (8x - 25)$ in.

49. $P = 2(x^2 + 3x + 1) + 2x^2$

$P = 2x^2 + 6x + 2 + 2x^2$

$P = (4x^2 + 6x + 2)$ m

Exercise Set 9.4

1. $y < x^2$

First, graph the parabola with a dashed line.

Does (0, 1) satisfy the inequality?

$1 < 0^2$

$1 < 0$ false

Shade portion of the graph which does not contain (0, 1).

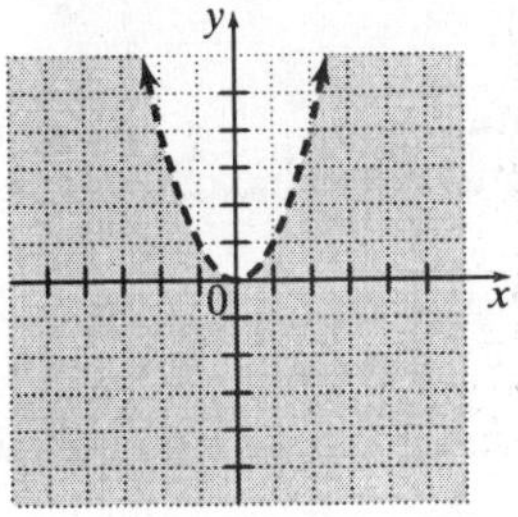

3. $x^2 + y^2 \geq 16$
First graph the circle with a solid line.
Does (0, 0) satisfy the inequality?
$0^2 + 0^2 \geq 16$
$0 \geq 16$ false
Shade the portion of the graph which does not contain (0, 0).

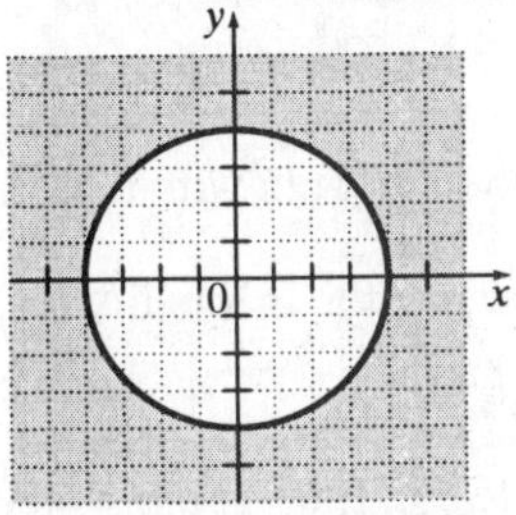

5. $\frac{x^2}{4} - y^2 < 1$
First graph the hyperbola with a dashed line.
Does (–4, 0) satisfy the inequality?
$\frac{(-4)^2}{4} - 0^2 < 1$
$4 < 1$ false
Does (0, 0) satisfy the inequality?
$\frac{0^2}{4} - 0^2 < 1$
$0 < 1$ true
Does (4, 0) satisfy the inequality?
$\frac{4^2}{4} - 0^2 < 1$
$4 < 1$ false

Shade the portion of the graph containing (0, 0).

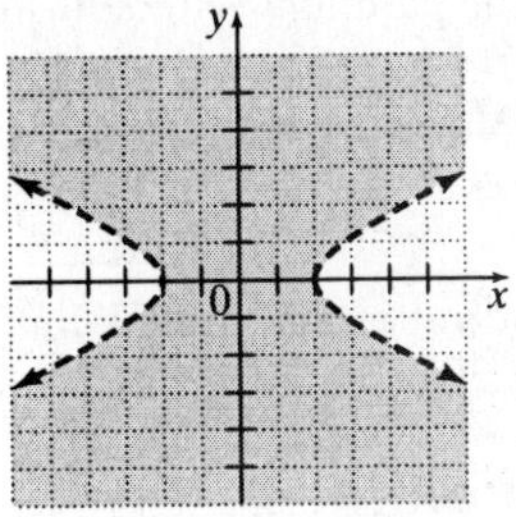

7. $y > (x-1)^2 - 3$
First graph the parabola with a dashed line.
Does (0, 0) satisfy the inequality?
$0 > (0-1)^2 - 3$
$0 > -2$ true
Shade the portion of the graph which does contain (0, 0).

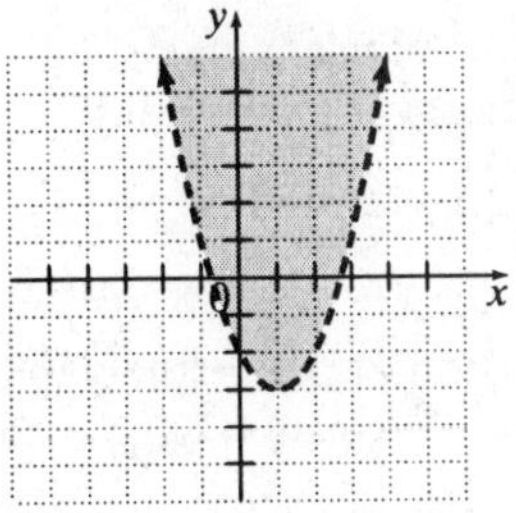

9. $x^2 + y^2 \leq 9$
First graph the circle with a solid line.
Does (0, 0) satisfy the inequality?
$0^2 + 0^2 \leq 9$
$0 \leq 9$ true
Shade the portion of the graph which does contain (0, 0).

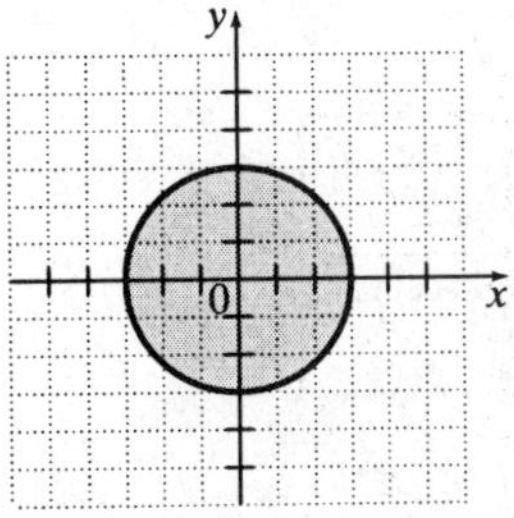

11. $y > -x^2 + 5$

First, graph the parabola with a dashed line.

Does (0, 0) satisfy the inequality?

$0 > -0^2 + 5$

$0 > 5$ false

Shade the portion of the graph which does not contain (0, 0).

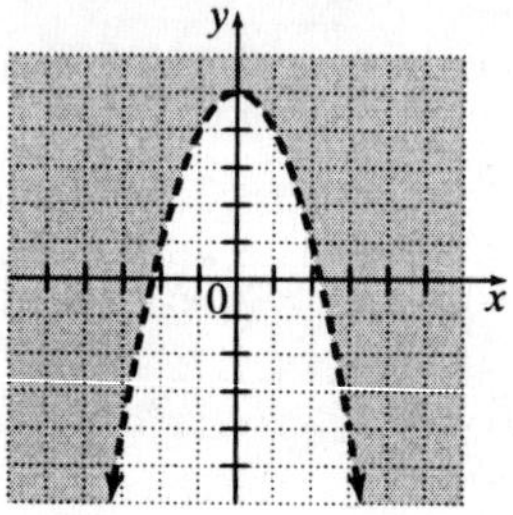

13. $\frac{x^2}{4} + \frac{y^2}{9} \le 1$

First, graph the ellipse with a solid line.

Does (0, 0) satisfy the inequality?

$\frac{0^2}{4} + \frac{0^2}{9} \le 1$

$0 \le 1$ true

Shade the portion of the graph which does contain (0, 0).

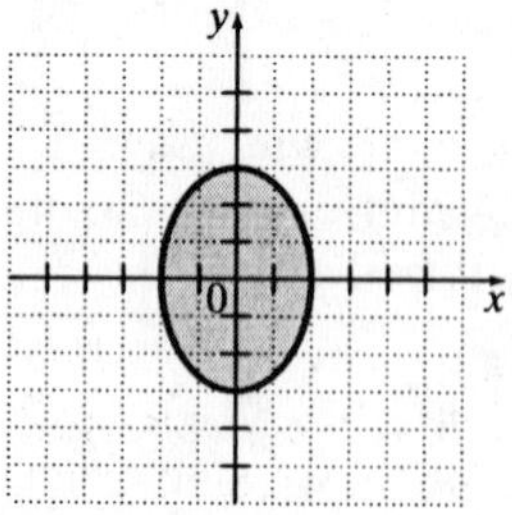

15. $\frac{y^2}{4} - x^2 \le 1$

First, graph the hyperbola with a solid line.

Does (0, –4) satisfy the inequality?

$\frac{(-4)^2}{4} - 0 \le 1$

$4 \le 1$ false

Does (0, 0) satisfy the inequality?

$\frac{0^2}{4} - 0 \le 1$

$0 \le 1$ true

Does (0, 4) satisfy the inequality?

$\frac{4^2}{4} - 0 \le 1$

$16 \le 1$ false

Shade the portion of the graph containing (0, 0).

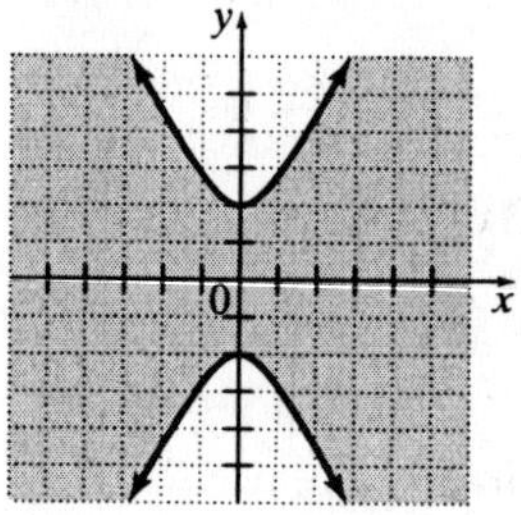

17. $y < (x-2)^2 + 1$

First, graph the parabola with a dashed line.

Does (2, 0) satisfy the inequality?

$0 < (2-2)^2 + 1$

$0 < 1$ true

Shade the portion of the graph which does contain (2, 0).

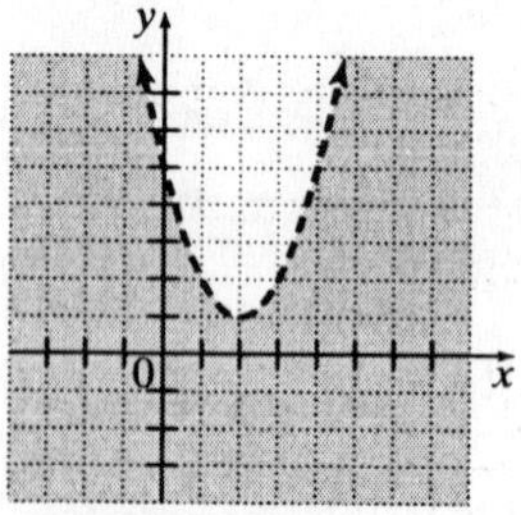

19. $y \le x^2 + x - 2$

First, graph the parabola with a solid line.

Does (0, 0) satisfy the inequality?

$0 \le 0^2 + 0 - 2$

$0 \le -2$ false

Shade the portion of the graph which does not contain (0, 0).

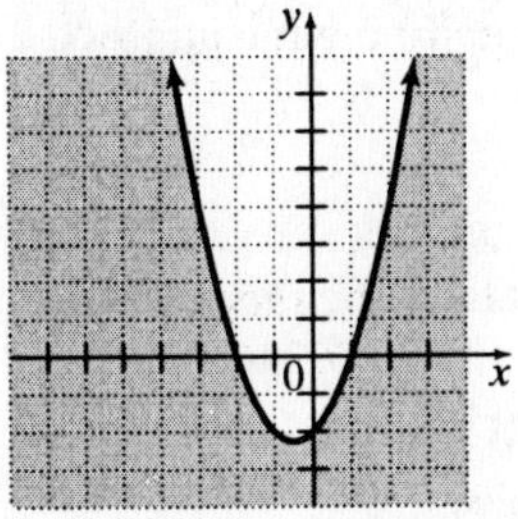

21. Answers may vary.

23. $\begin{cases} 2x - y < 2 \\ y \le -x^2 \end{cases}$

First, graph $2x - y < 2$ with a dotted line by solving for y.

$-y < -2x + 2$

$y > 2x - 2$

Does (0, 0) satisfy $2x - y < 2$?

$2(0) - 0 < 2$

$0 < 2$ true

Shade the portion of the graph which contains (0, 0).

Next, graph the parabola $y \le -x^2$ with a solid line.

Does (0, 1) satisfy $y \le -x^2$?

$1 \le -0^2$

$1 \le 0$ false

Shade the portion of the graph which does not contain (0, 1).

The solution to the system is the overlapping region.

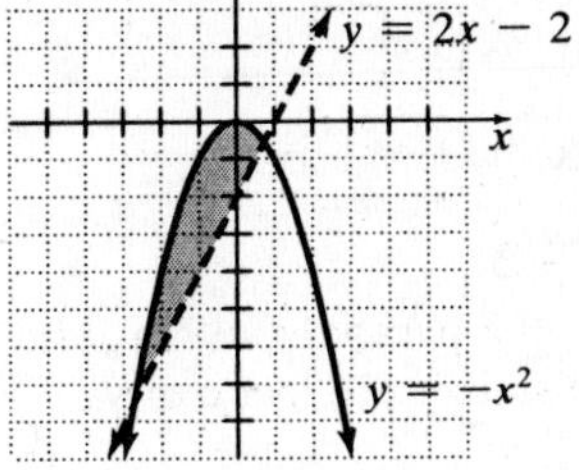

25. $\begin{cases} 4x + 3y \ge 12 \\ x^2 + y^2 < 16 \end{cases}$

First, graph the circle with a dashed line.

Does (0, 0) satisfy $x^2 + y^2 < 16$?

$0^2 + 0^2 < 16$

$0 < 16$ true

Shade the portion of the graph which contains (0, 0).

Next, graph the line with a solid line.

Does (0, 0) satisfy $4x + 3y \ge 12$?

$4(0) + 3(0) \ge 12$

$0 \ge 12$ false

Shade the portion of the graph which does not contain (0, 0).

The solution to the system is the overlapping region.

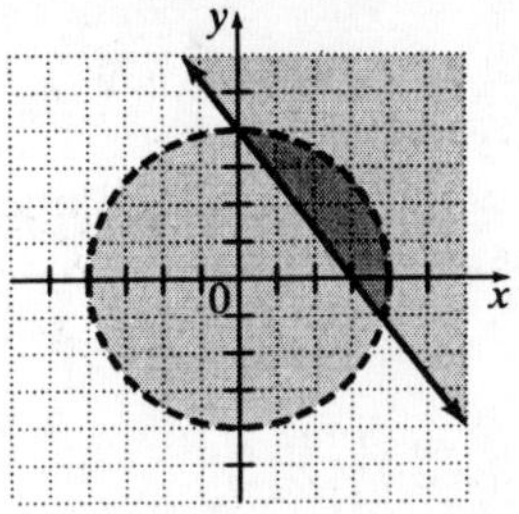

27. $\begin{cases} x^2 + y^2 \le 9 \\ x^2 + y^2 \ge 1 \end{cases}$

First, graph the circle with radius 1.

Does (0, 0) satisfy $x^2 + y^2 \ge 1$?

$0^2 + 0^2 \ge 1$

$0 \ge 1$ false

Shade the portion of the graph which does not contain (0, 0).

Next, graph the circle with radius 3.

Does (0, 0) satisfy $x^2 + y^2 \le 9$?

$0^2 + 0^2 \le 9$

$0 \le 9$ true

Shade the portion of the graph which contains (0, 0).

The solution to the system is the overlapping region.

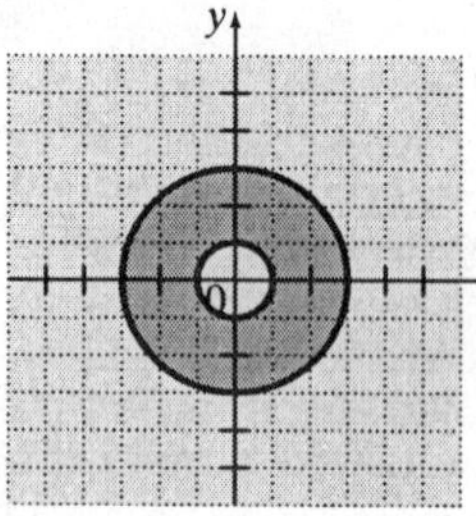

29. $\begin{cases} y > x^2 \\ y \geq 2x + 1 \end{cases}$

First, graph the parabola with a dashed line.

Does (0, 2) satisfy $y > x^2$?

$2 > 0^2$

$2 > 0$ true

Shade the portion of the graph which contains (0, 2).

Next, graph the line with a solid line.

Does (0, 0) satisfy $y \geq 2x + 1$?

$0 \geq 2(0) + 1$

$0 \geq 1$ false

Shade the portion of the graph which does not contain (0, 0).

The solution to the system is the overlapping region.

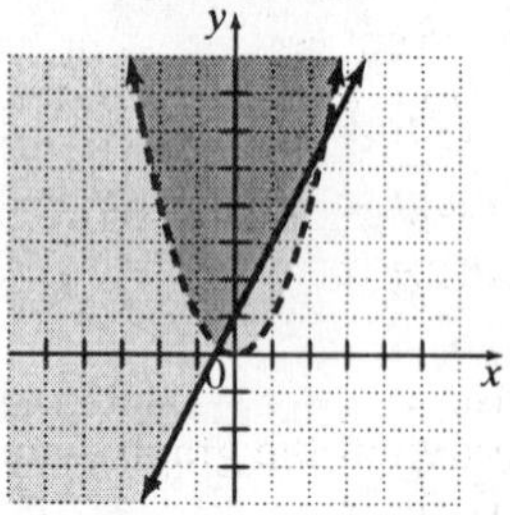

31. $\begin{cases} x > y^2 \\ y > 0 \end{cases}$

First graph the parabola with a dashed line.

Does (2, 0) satisfy $x > y^2$?

$2 > 0^2$

$2 > 0$ true

Shade the portion of the graph which contains (2, 0).

Next, graph the line with a dashed line.

Does (0, 1) satisfy $y > 0$?

$1 > 0$ true

Shade the portion of the graph which contains (0, 1).

The solution to the system is the overlapping region.

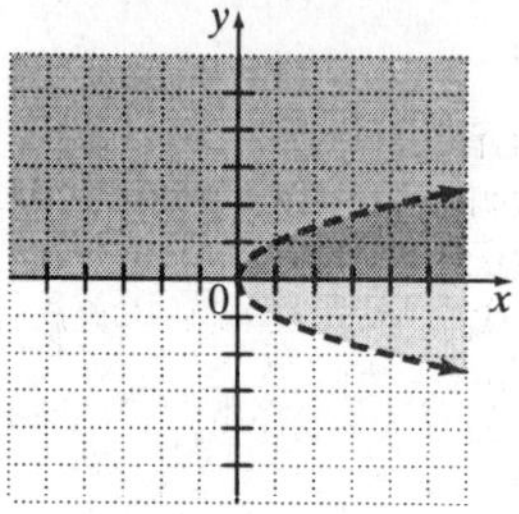

33. $\begin{cases} x^2 + y^2 > 9 \\ y > x^2 \end{cases}$

First, graph the circle with a dashed line.

Does (0, 0) satisfy $x^2 + y^2 > 9$?

$0^2 + 0^2 > 9$

$0 > 9$ false

Shade the portion of the graph which does not contain (0, 0).

Next, graph the parabola with a dashed line.

Does (0, 2) satisfy $y > x^2$?

$2 > 0^2$

$2 > 0$ true

Shade the portion of the graph which contains (0, 2).

The solution to the system is the overlapping region.

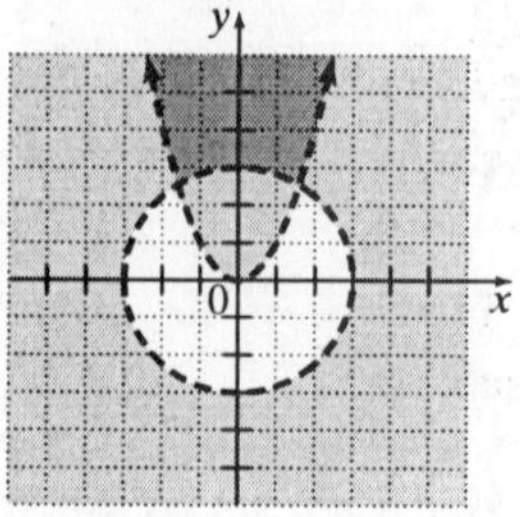

35. $\begin{cases} \dfrac{x^2}{4}+\dfrac{y^2}{9} \ge 1 \\ x^2+y^2 \ge 4 \end{cases}$

First, graph the ellipse. Does (0, 0) satisfy

$\dfrac{x^2}{4}+\dfrac{y^2}{9} \ge 1?$

$\dfrac{0^2}{4}+\dfrac{0^2}{9} \ge 1$

$0 \ge 1$ false

Shade the portion of the graph which does not contain (0, 0).

Next, graph the circle. Does (0, 0) satisfy

$x^2+y^2 \ge 4?$

$0^2+0^2 \ge 4$

$0 \ge 4$ false

Shade the portion of the graph which does not contain (0, 0).

The solution to the system is the overlapping region.

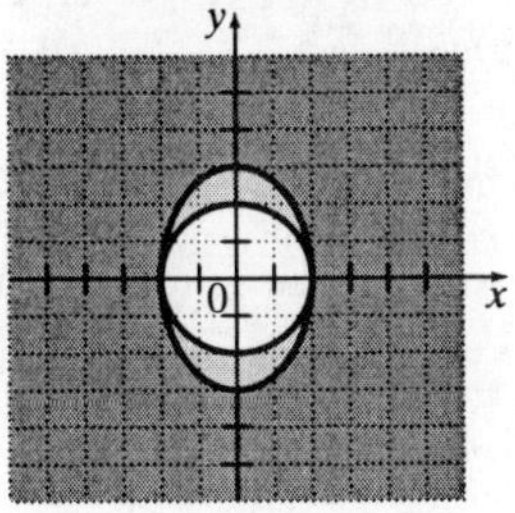

37. $\begin{cases} x^2-y^2 \ge 1 \\ y \ge 0 \end{cases}$

First, graph the hyperbola. Does (–2, 0) satisfy $x^2-y^2 \ge 1?$

$(-2)^2-0^2 \ge 1$

$4 \ge 1$ true

Does (0, 0) satisfy $x^2+y^2 \ge 1?$

$0^2+0^2 \ge 1$

$0 \ge 1$ false

Does (2, 0) satisfy $x^2+y^2 \ge 1?$

$2^2+0^2 \ge 1$

$4 \ge 1$ true

Shade the portion of the graph containing (–2, 0) and the portion of the graph containing (2, 0).

Next, graph the line. Does (0, 2) satisfy

$y \ge 0?$

$2 \ge 0$ true

Shade the portion of the graph containing (0, 2).

The solution to the system is the overlapping region.

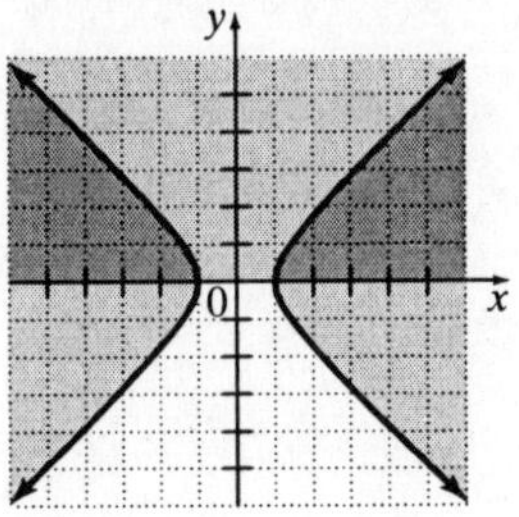

39. $\begin{cases} x+y \ge 1 \\ 2x+3y < 1 \\ x > -3 \end{cases}$

First, graph $x > -3$ with a dashed line.

Does (0, 0) satisfy $x > -3$?

$0 > -3$ true

Shade the portion of the graph containing (0, 0).

Next, graph $2x + 3y < 1$ with a dashed line.

Does (0, 0) satisfy $2x + 3y < 1$?

$2(0)+3(0) < 1$

$0 < 1$ true

Shade the portion of the graph containing (0, 0).

Next, graph $x + y \ge 1$ with a solid line.

Does (0, 0) satisfy $x + y \ge 1$?

$0+0 \ge 1$

$0 \ge 1$ false

Shade the portion of the graph which does not contain (0, 0).

The solution to the system is the overlapping region.

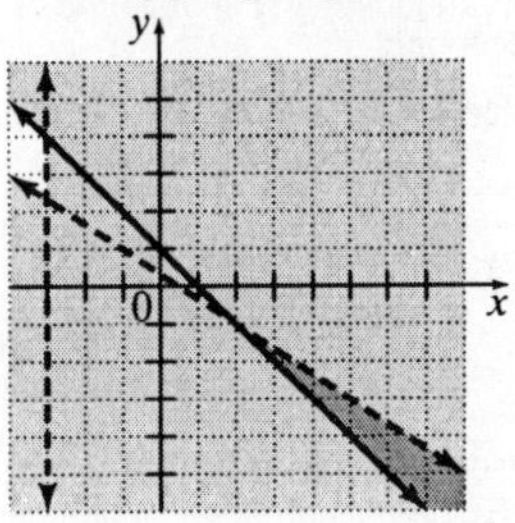

41. $\begin{cases} x^2 - y^2 < 1 \\ \frac{x^2}{16} + y^2 \le 1 \\ x \ge -2 \end{cases}$

First graph the hyperbola with a dashed line.

Does (–2, 0) satisfy $x^2 - y^2 < 1$?

$(-2)^2 - 0^2 < 1$

$4 < 1$ false

Does (0, 0) satisfy $x^2 - y^2 < 1$?

$0^2 - 0^2 < 1$

$0 < 1$ true

Does (2, 0) satisfy $x^2 - y^2 < 1$?

$2^2 - 0^2 < 1$

$4 < 1$ false

Shade the portion of the graph containing (0, 0).

Next, graph the ellipse with a solid line.

Does (0, 0) satisfy $\frac{x^2}{16} + y^2 \le 1$?

$\frac{0^2}{16} + 0^2 \le 1$

$0 \le 1$ true

Shade the portion of the graph containing (0, 0).

Next, graph the line.

Does (0, 0) satisfy $x \ge -2$?

$0 \ge -2$ true

Shade the portion of the graph containing (0, 0).

The solution to the system is the overlapping region.

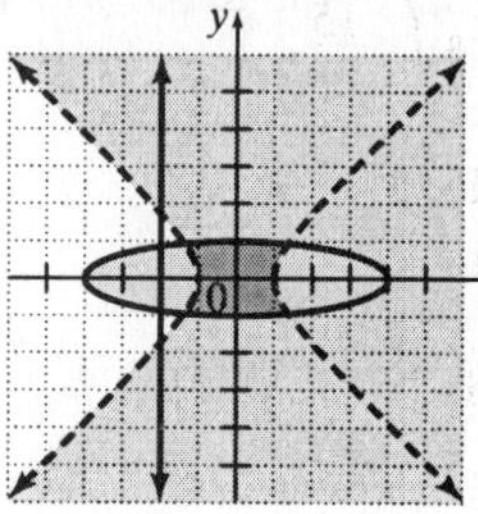

43. $\begin{cases} y \le x^2 \\ y \ge x + 2 \\ x \ge 0 \\ y \ge 0 \end{cases}$

First, graph the parabola with a solid line.

Does (0, 2) satisfy $y \le x^2$?

$2 \le 0^2$

$2 \le 0$ false

Shade the portion of the graph which does not contain (0, 2).

Next, graph $y \ge x + 2$ with a solid line.

Does (0, 0) satisfy $y \ge x + 2$?

$0 \ge 0 + 2$

$0 \ge 2$ false

Shade the portion of the graph which does not contain (0, 0).

Next, graph $x \ge 0$ with a solid line.

Does (1, 1) satisfy $x \ge 0$?

$1 \ge 0$ true

Shade the portion of the graph which contains (1, 1).

Next, graph $y \ge 0$ with a solid line.

Does (1, 1) satisfy $y \ge 0$?

$1 \ge 0$ true

Shade the portion of the graph which contains (1, 1).

The solution to the system is the overlapping region.

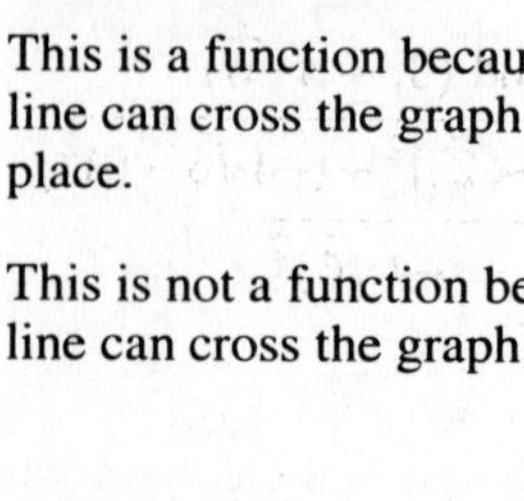

45. This is a function because a nonvertical line can cross the graph in more than one place.

47. This is not a function because a vertical line can cross the graph on two places.

49. $f(x) = 3x^2 - 2$

$f(-3) = 3(-3)^2 - 2$

$= 3(9) - 2$

$f(-3) = 2$

51. $f(x) = 3x^2 - 2$

$f(b) = 3b^2 - 2$

Chapter 9 Review

1. (–6, 3) and (8, 4)

$d = \sqrt{[8-(-6)]^2 + (4-3)^2}$

$d = \sqrt{(14)^2 + 1^2}$

$d = \sqrt{197}$

2. (3, 5) and (8, 9)

$d = \sqrt{(8-3)^2 + (9-5)^2}$

$d = \sqrt{5^2 + 4^2}$

$d = \sqrt{25+16}$

$d = \sqrt{41}$

3. (–4, –6) and (–1, 5)

$d = \sqrt{[-1-(-4)]^2 + [5-(-6)]^2}$

$d = \sqrt{3^2 + 11^2}$

$d = \sqrt{9+121} = \sqrt{130}$

4. (–1, 5) and (2, –3)

$d = \sqrt{[2-(-1)]^2 + (-3-5)^2}$

$d = \sqrt{3^2 + (-8)^2}$

$d = \sqrt{9+64} = \sqrt{73}$

5. $(-\sqrt{2},\ 0)$ and $(0,\ -4\sqrt{6})$

$d = \sqrt{[0-(-\sqrt{2})]^2 + (-4\sqrt{6}-0)^2}$

$d = \sqrt{(\sqrt{2})^2 + (-4\sqrt{6})^2}$

$d = \sqrt{2+96}$

$d = \sqrt{98}$

$d = 7\sqrt{2}$

6. $(-\sqrt{5},\ -\sqrt{11})$ and $(-\sqrt{5},\ -3\sqrt{11})$

$d = \sqrt{[-\sqrt{5}-(-\sqrt{5})]^2 + [-3\sqrt{11}-(-\sqrt{11})]^2}$

$d = \sqrt{0 + (-2\sqrt{11})^2}$

$d = \sqrt{44}$

$d = 2\sqrt{11}$

7. (7.4, –8.6) and (–1.2, 5.6)

$d = \sqrt{(-1.2-7.4)^2 + [5.6-(-8.6)]^2}$

$d = \sqrt{(-8.6)^2 + (14.2)^2}$

$d = \sqrt{275.6}$

$d \approx 16.60$ units

8. (2.3, 1.8) and (10.7, –9.2)

$d = \sqrt{(10.7-2.3)^2 + (-9.2-1.8)^2}$

$d = \sqrt{(8.4)^2 + (-11)^2}$

$d = \sqrt{191.56} \approx 13.84$ units

9. (2, 6) and (–12, 4)

$\left(\frac{2-12}{2}, \frac{6+4}{2}\right)$

The midpoint is (–5, 5).

10. (–3, 8) and (11, 24)

$\left(\frac{-3+11}{2}, \frac{8+24}{2}\right)$

The midpoint is (4, 16).

11. (–6, –5) and (–9, 7)

$\left(\frac{-6-9}{2}, \frac{-5+7}{2}\right)$

The midpoint is $\left(-\frac{15}{2},\ 1\right)$.

12. (4, –6) and (–15, 2)

$\left(\frac{4-15}{2}, \frac{-6+2}{2}\right)$

The midpoint is $\left(-\frac{11}{2},\ -2\right)$.

13. $\left(0, -\frac{3}{8}\right)$ and $\left(\frac{1}{10}, 0\right)$

$\left(\frac{0+\frac{1}{10}}{2}, \frac{-\frac{3}{8}+0}{2}\right)$

The midpoint is $\left(\frac{1}{20}, -\frac{3}{16}\right)$.

14. $\left(\frac{3}{4}, -\frac{1}{7}\right)$ and $\left(-\frac{1}{4}, -\frac{3}{7}\right)$

$\left(\frac{\frac{3}{4}-\frac{1}{4}}{2}, \frac{-\frac{1}{7}-\frac{3}{7}}{2}\right)$

The midpoint is $\left(\frac{1}{4}, -\frac{2}{7}\right)$.

15. $(\sqrt{3}, -2\sqrt{6})$ and $(\sqrt{3}, -4\sqrt{6})$

$\left(\frac{\sqrt{3}+\sqrt{3}}{2}, \frac{-2\sqrt{6}-4\sqrt{6}}{2}\right)$

The midpoint is $(\sqrt{3}, -3\sqrt{6})$.

16. $(-5\sqrt{3}, 2\sqrt{7})$ and $(-3\sqrt{3}, 10\sqrt{7})$

$\left(\frac{-5\sqrt{3}-3\sqrt{3}}{2}, \frac{2\sqrt{7}+10\sqrt{7}}{2}\right)$

The midpoint is $(-4\sqrt{3}, 6\sqrt{7})$.

17. center (–4, 4), radius 3

$(x-(-4))^2+(y-4)^2=3^2$

$(x+4)^2+(y-4)^2=9$

18. center (5, 0), radius 5

$(x-5)^2+(y-0)^2=5^2$

$(x-5)^2+y^2=25$

19. center (–7, –9), radius $\sqrt{11}$

$(x-(-7))^2+(y-(-9))^2=\left(\sqrt{11}\right)^2$

$(x+7)^2+(y+9)^2=11$

20. center (0, 0), radius $\frac{7}{2}$

$(x-0)^2+(y-0)^2=\left(\frac{7}{2}\right)^2$

$x^2+y^2=\frac{49}{4}$

21. $x^2+y^2=7$

or $(x-0)^2+(y-0)^2=\sqrt{7}^2$

The center is (0, 0).

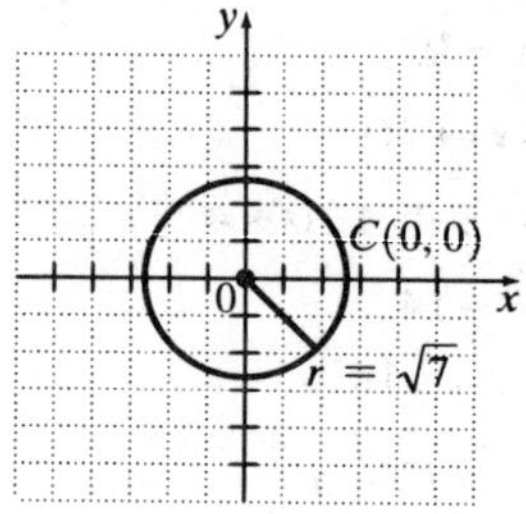

22. $x=2(y-5)^2+4$

The vertex is (4, 5).

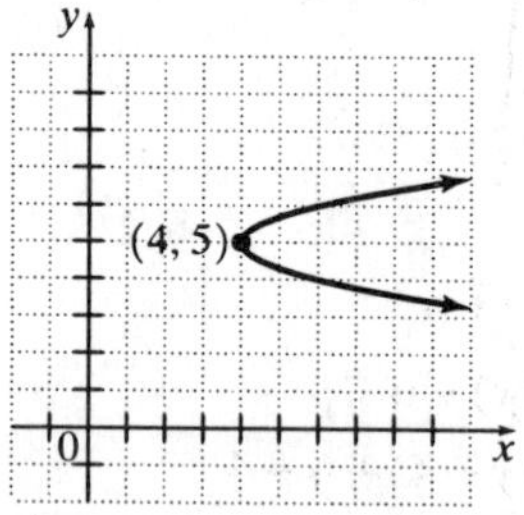

23. $x=-(y+2)^2+3$

The vertex is (3, –2).

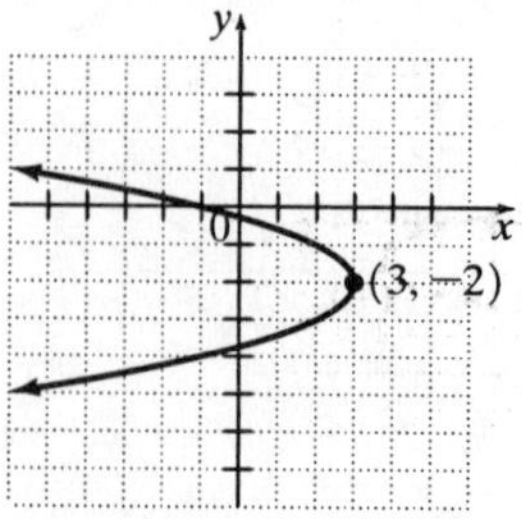

24. $(x-1)^2+(y-2)^2=4$

The center is (1, 2). The radius is 2.

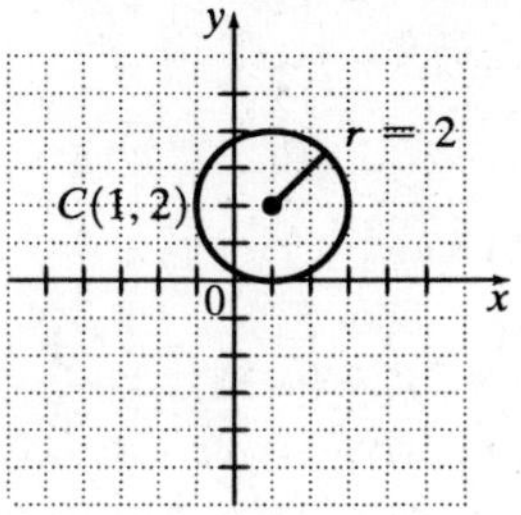

25. $y=-x^2+4x+10$

$y=-(x^2-4x)+10$

$y=-(x^2-4x+4)+10+4$

$y=-(x-2)^2+14$

The vertex is (2, 14).

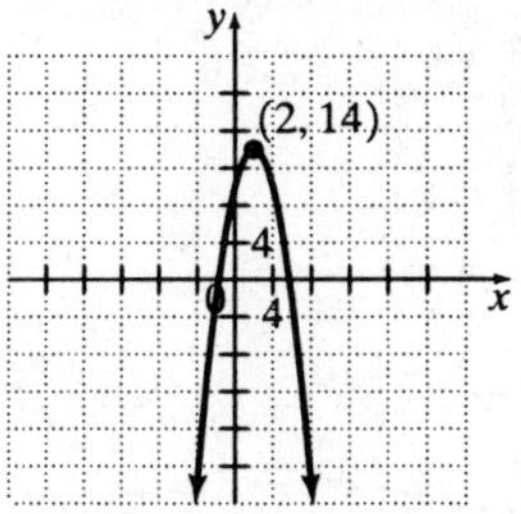

26. $x=-y^2-4y+6$

$x=-(y^2+4y)+6$

$x=-(y^2+4y+4)+6+4$

$x=-(y+2)^2+10$

The vertex is (10, –2).

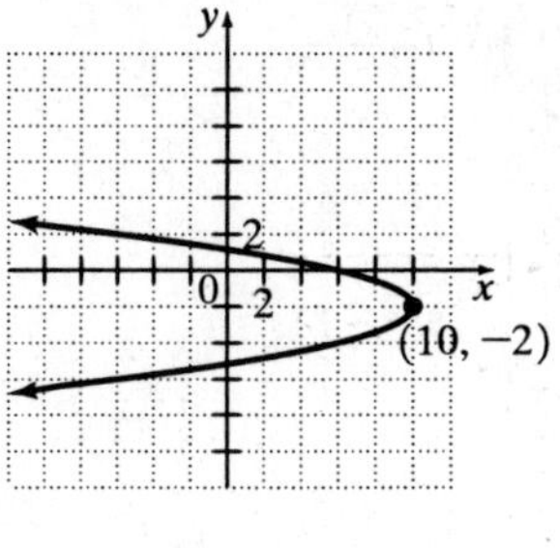

27. $x=\frac{1}{2}y^2+2y+1$

$x=\frac{1}{2}(y^2+4y)+1$

$x=\frac{1}{2}(y^2+4y+4)+1-2$

$x=\frac{1}{2}(y+2)^2-1$

The vertex is (–1, –2).

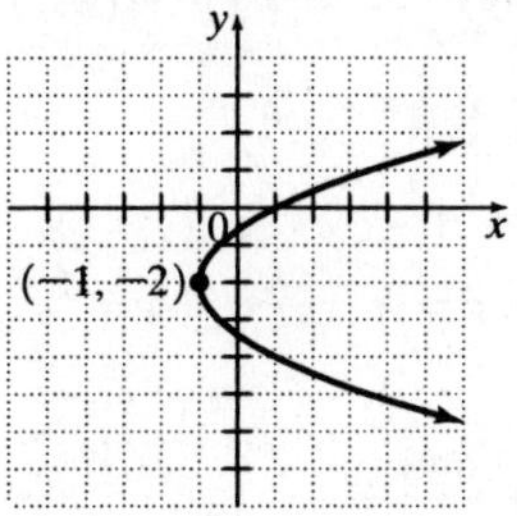

28. $y=-3x^2+\frac{1}{2}x+4$

$y=-3\left(x^2+\frac{1}{6}x\right)+4$

$y=-3\left(x^2+\frac{1}{6}+\frac{1}{144}\right)+4+\frac{3}{144}$

$y=-3\left(x+\frac{1}{12}\right)^2+\frac{193}{48}$

The vertex is $\left(\frac{1}{12}, \frac{193}{48}\right)$.

29.
$$x^2+y^2+2x+y=\frac{3}{4}$$
$$(x^2+2x+1)+\left(y^2+y+\frac{1}{4}\right)=\frac{3}{4}+1+\frac{1}{4}$$
$$(x+1)^2+\left(y+\frac{1}{2}\right)^2=2$$
$$(x+1)^2+\left(y+\frac{1}{2}\right)^2=\left(\sqrt{2}\right)^2$$

The center is $\left(-1,\ -\frac{1}{2}\right)$.

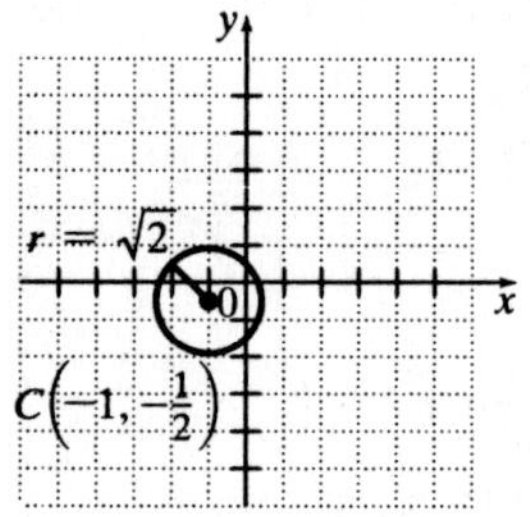

30.
$$x^2+y^2+3y=\frac{7}{4}$$
$$x^2+\left(y^2+3y+\frac{9}{4}\right)=\frac{7}{4}+\frac{9}{4}$$
$$x^2+\left(y+\frac{3}{2}\right)^2=4$$

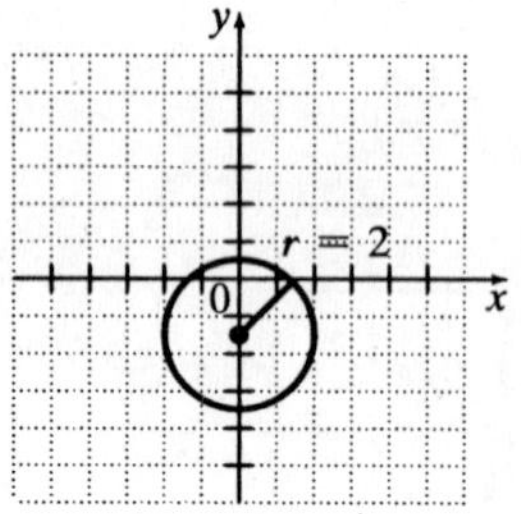

31.
$$4x^2+4y^2+16x+8y=1$$
$$x^2+y^2+4x+2y=\frac{1}{4}$$
$$x^2+4x+y^2+2y=\frac{1}{4}$$
$$(x^2+4x+4)+(y^2+2y+1)=\frac{1}{4}+4+1$$
$$(x+2)^2+(y+1)^2=\frac{21}{4}$$
$$(x+2)^2+(y+1)^2=\left(\frac{\sqrt{21}}{2}\right)^2$$

The center is (–2, –1).

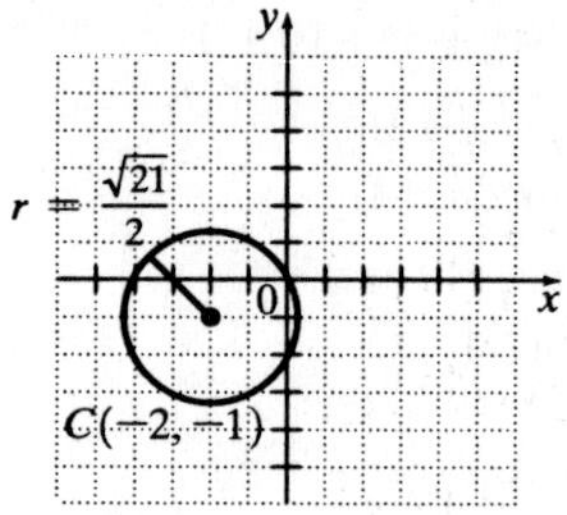

32.
$$3x^2+6x+3y^2=9$$
$$x^2+2x+y^2=3$$
$$(x^2+2x+1)+y^2=3+1$$
$$(x+1)^2+y^2=4$$

The center is (–1, 0).

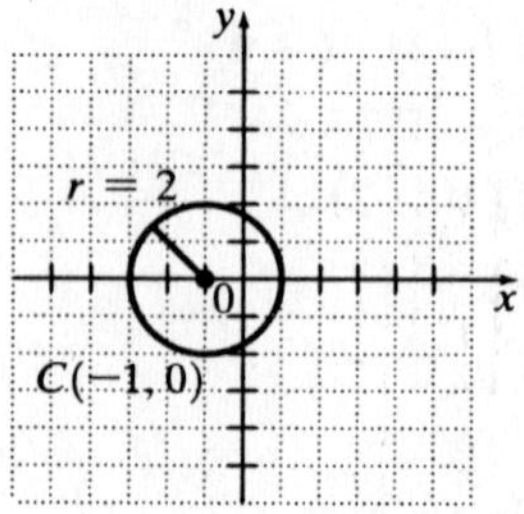

33. $y = x^2 + 6x + 9$

$y = (x+3)^2 + 0$

The vertex is (–3, 0).

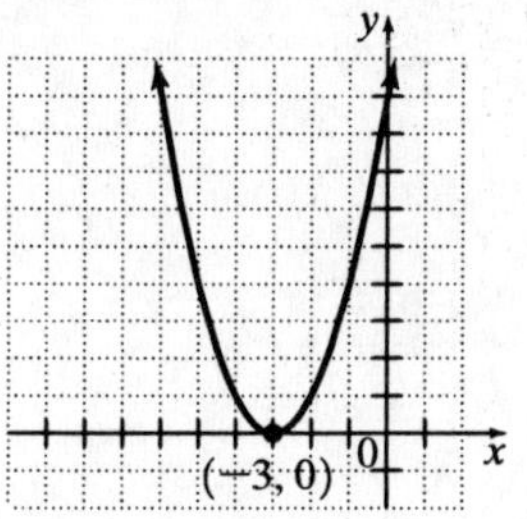

34. $x = y^2 + 6y + 9$

$x = (y+3)^2$

The vertex is (0, –3).

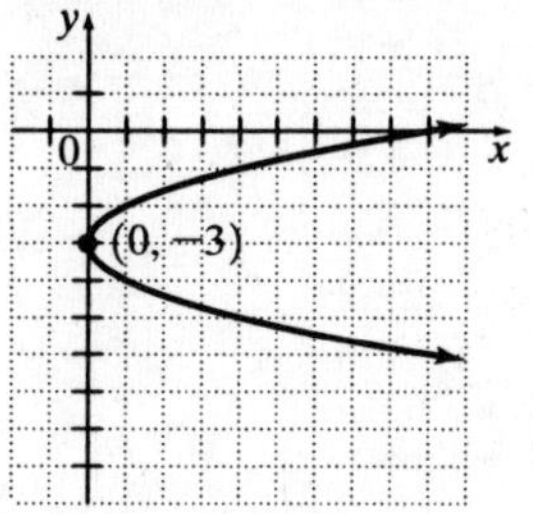

35. The center is (5.6, –2.4).

$\text{radius} = \frac{6.2}{2} = 3.1$

$(x-5.6)^2 + [y-(-2.4)]^2 = 3.1^2$

or $(x-5.6)^2 + (y+2.4)^2 = 9.61$

36. $x^2 + \frac{y^2}{4} = 1$

$a = 1$, $b = 2$, The center is (0, 0).

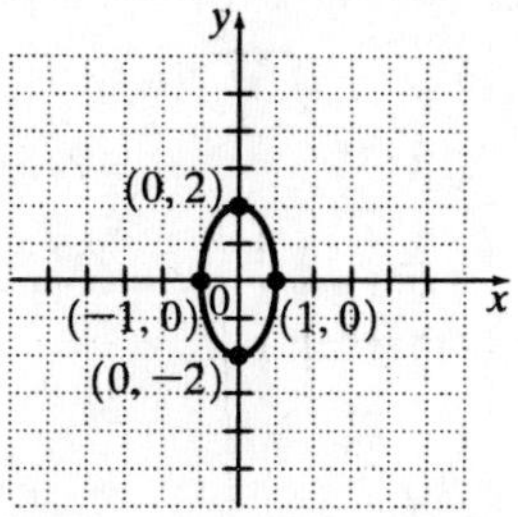

37. $x^2 - \frac{y^2}{4} = 1$

$a = 1$, $b = 2$

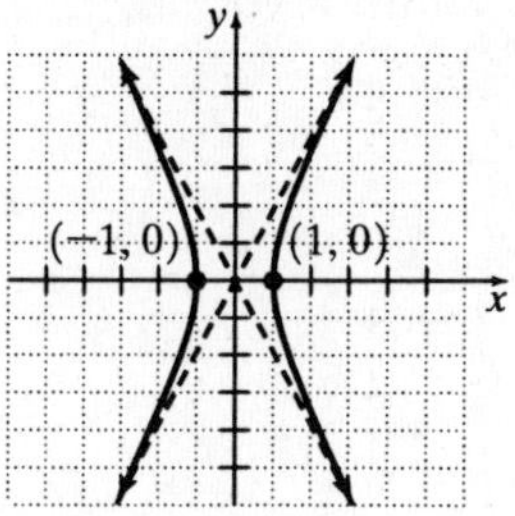

38. $\frac{y^2}{4} - \frac{x^2}{16} = 1$

$a = 4$, $b = 2$

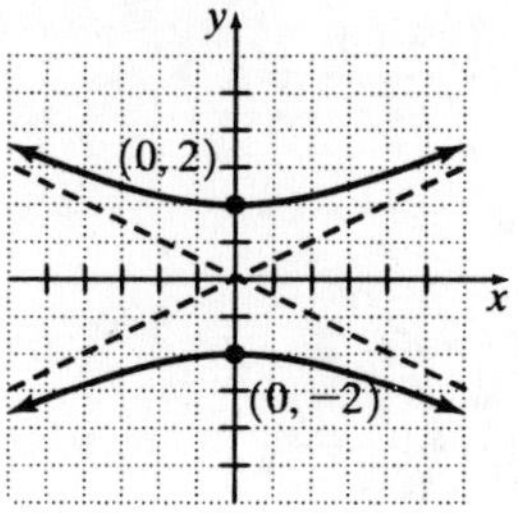

39. $\frac{y^2}{4} + \frac{x^2}{16} = 1$

$a = 4$, $b = 2$

40. $\frac{x^2}{5}+\frac{y^2}{5}=1$

Center (0, 0), radius = $\sqrt{5}$

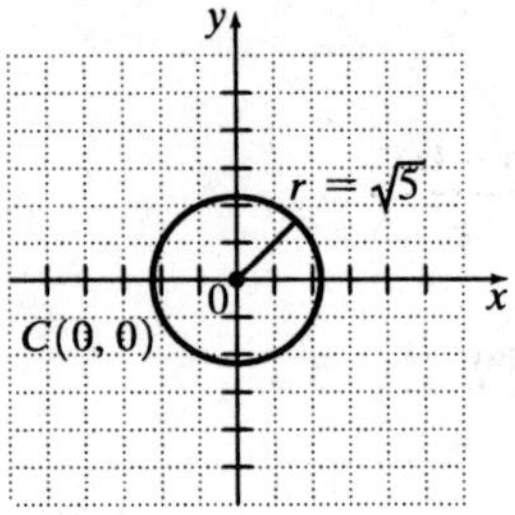

41. $\frac{x^2}{5}-\frac{y^2}{5}=1$

or $\frac{(x-0)^2}{\sqrt{5}^2}-\frac{(y-0)^2}{\sqrt{5}^2}=1$

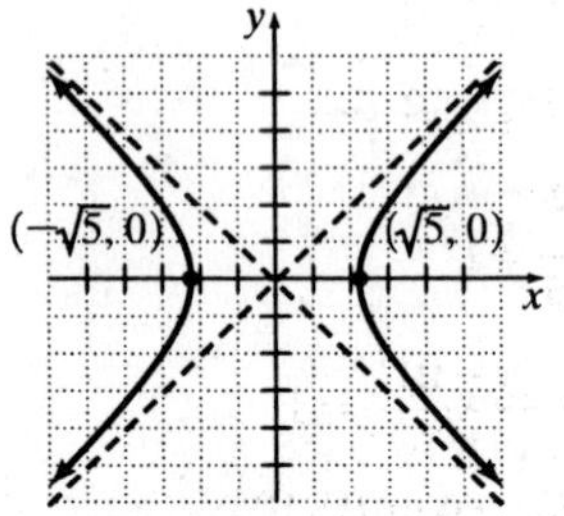

42. $-5x^2+25y^2=125$

$\frac{y^2}{5}-\frac{x^2}{25}=1$

$a=5,\ b=\sqrt{5}$

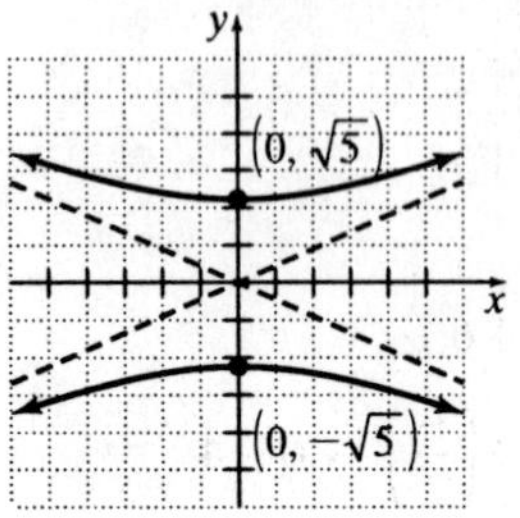

43. $4y^2+9x^2=36$

$\frac{y^2}{9}+\frac{x^2}{4}=1$

$a=2,\ b=3$

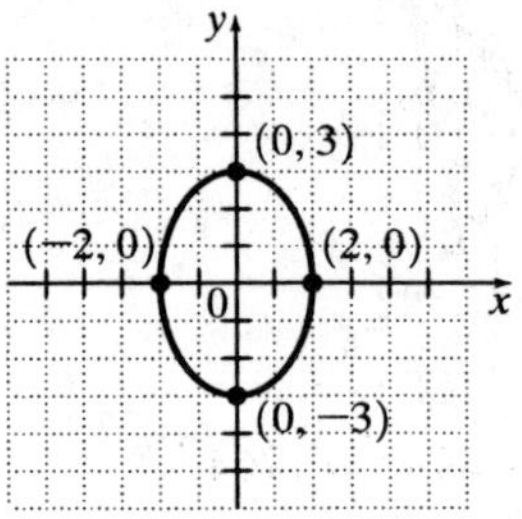

44. $\frac{(x-2)^2}{4}+(y-1)^2=1$

$\frac{(x-2)^2}{2^2}+\frac{(y-1)^2}{1^2}=1$

$a=2,\ b=1$, center (2, 1)

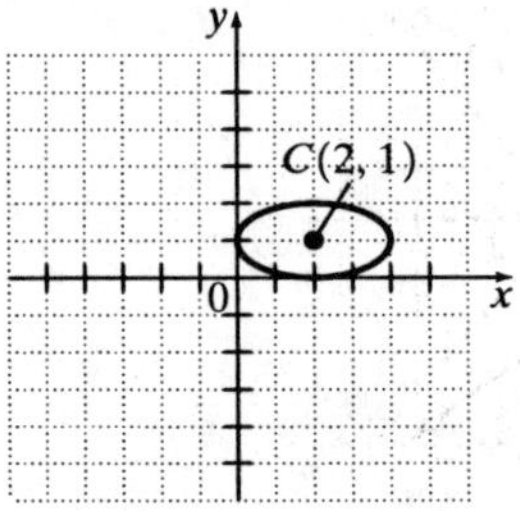

45. $\frac{(x+3)^2}{9}+\frac{(y-4)^2}{25}=1$

$\frac{(x+3)^2}{3^2}+\frac{(y-4)^2}{5^2}=1$

$a=3,\ b=5$, center (−3, 4)

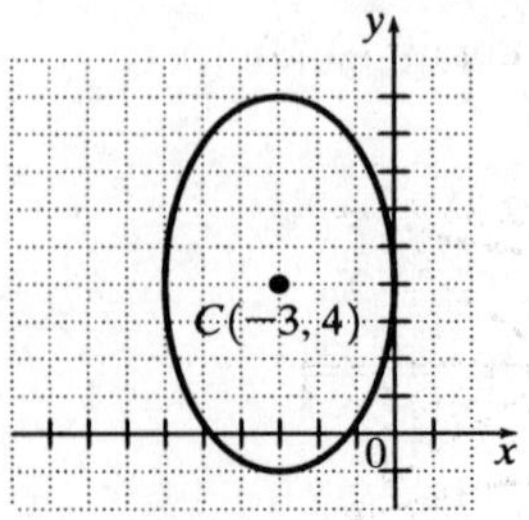

46. $x^2 - y^2 = 1$

$a = 1$, $b = 1$, center $(0, 0)$

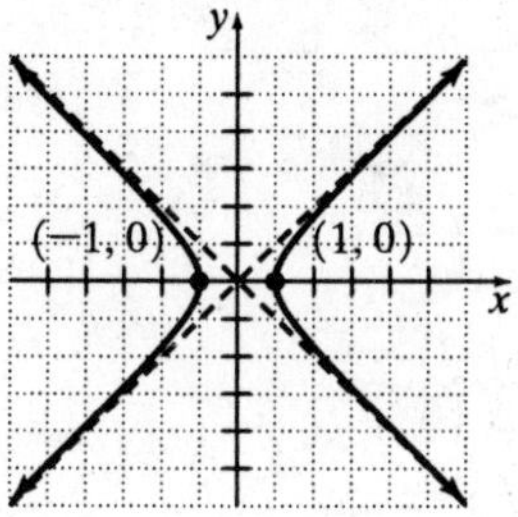

47.
$$36y^2 - 49x^2 = 1764$$
$$\frac{y^2}{49} - \frac{x^2}{36} = 1$$
$$\frac{(y-0)^2}{7^2} - \frac{(x-0)^2}{6^2} = 1$$

$a = 6$, $b = 7$, center $(0, 0)$

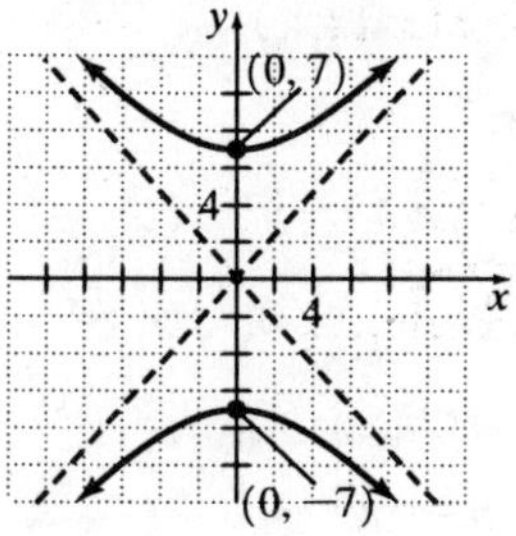

48. $y^2 = x^2 + 9$
$$\frac{y^2}{9} - \frac{x^2}{9} = 1$$
$$\frac{(y-0)^2}{3^2} - \frac{(x-0)^2}{3^2} = 1$$

$a = 3$, $b = 3$, center $(0, 0)$

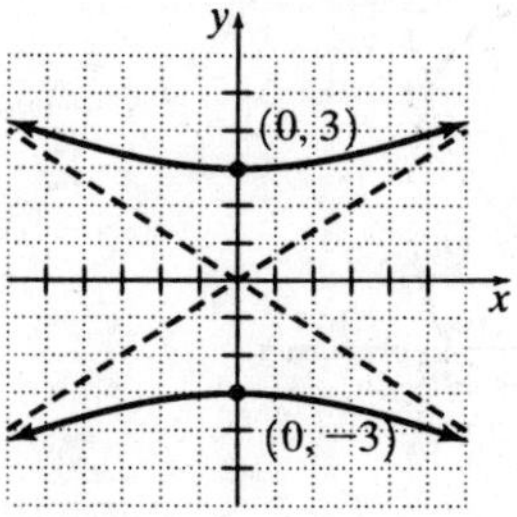

49. $x^2 = 4y^2 - 16$
$$16 = 4y^2 - x^2$$
$$1 = \frac{y^2}{4} - \frac{x^2}{16}$$
$$\frac{(y-0)^2}{2^2} - \frac{(x-0)^2}{4^2} = 1$$

$a = 4, b = 2$,
center $(0,0)$

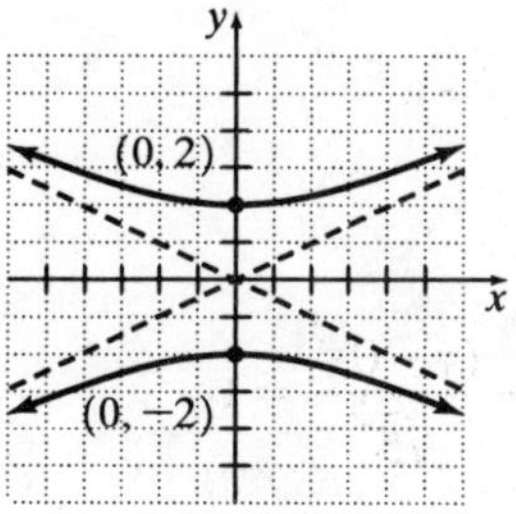

50. $100 - 25x^2 = 4y^2$
$$\frac{x^2}{4} + \frac{y^2}{25} = 1$$
$$\frac{(x-0)^2}{2^2} + \frac{(y-0)^2}{5^2} = 1$$

$a = 5$, $b = 2$, center $(0, 0)$

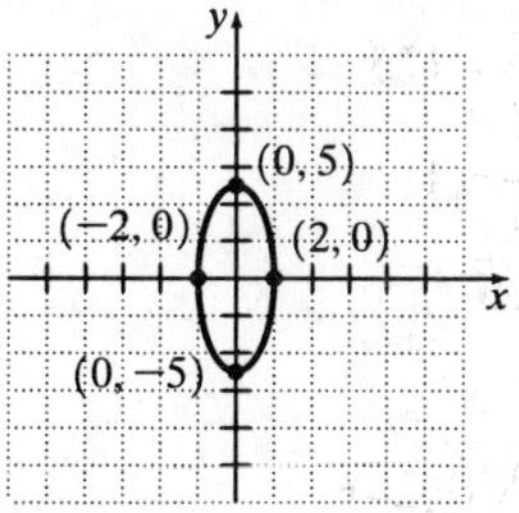

51. parabola
$$y = x^2 + 4x + 6$$
$$y = x^2 + 4x + \left(\frac{4}{2}\right)^2 + 6 - 4$$
$$y = (x+2)^2 + 2$$

vertex: (–2, 2)

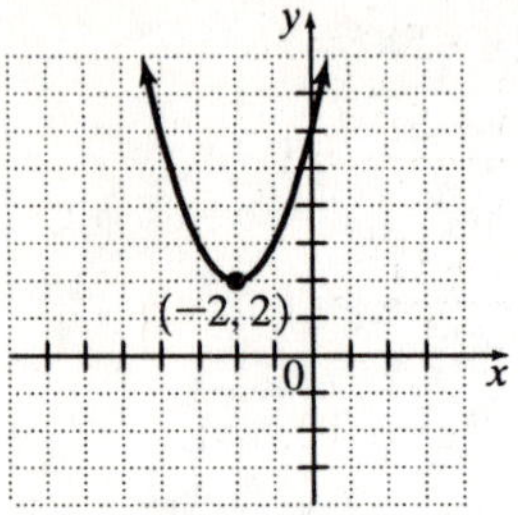

52. hyperbola

$$y^2 = x^2 + 6$$

$$\frac{y^2}{6} - \frac{x^2}{6} = 1$$

$$\frac{(y-0)^2}{(\sqrt{6})^2} - \frac{(x-0)^2}{(\sqrt{6})} = 1$$

$a = \sqrt{6},\ b = \sqrt{6}$, center (0, 0)

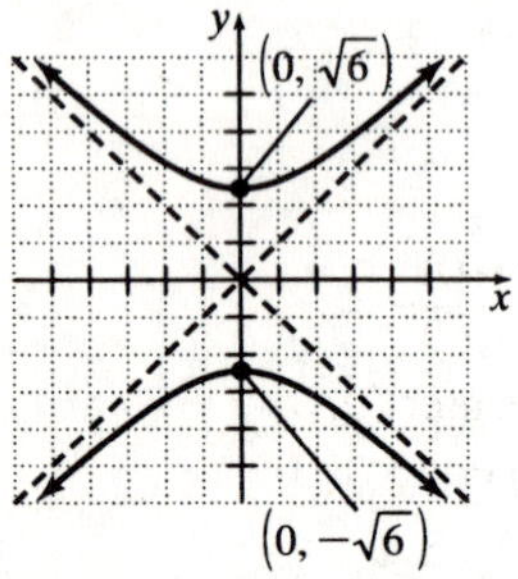

53. circle

$$y^2 + x^2 = 4x + 6$$

$$x^2 - 4x + y^2 = 6$$

$$x^2 - 4x + \left(\frac{4}{2}\right)^2 + y^2 = 6 + 4$$

$$(x-2)^2 + y^2 = 10$$

$$(x-2)^2 + (y-0)^2 = \sqrt{10}^2$$

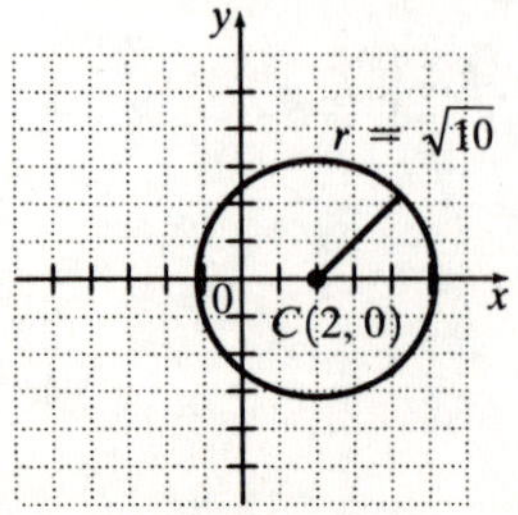

54. ellipse

$$y^2 + 2x^2 = 4x + 6$$

$$y^2 + 2(x^2 - 2x + 1) = 6 + 2$$

$$y^2 + 2(x-1)^2 = 8$$

$$\frac{y^2}{8} + \frac{(x-1)^2}{4} = 1$$

$$\frac{(y-0)^2}{(2\sqrt{2})^2} + \frac{(x-1)^2}{(2)^2} = 1$$

$a = 2,\ b = 2\sqrt{2}$, center (1, 0)

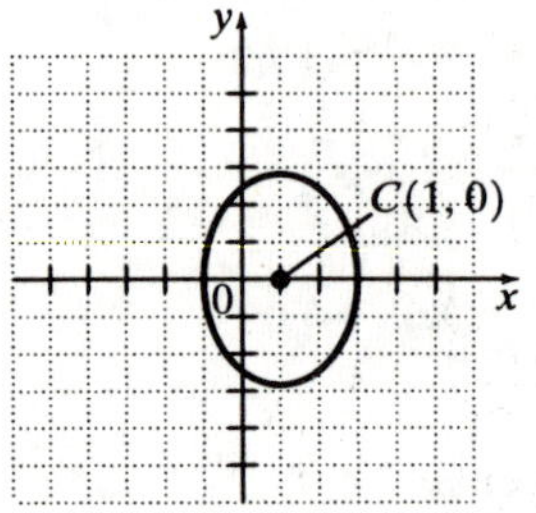

55. circle

$$x^2 + y^2 - 8y = 0$$

$$x^2 + y^2 - 8y + \left(\frac{8}{2}\right)^2 = 0 + 16$$

$$x^2 + (y-4)^2 = 16$$

$$(x-0)^2 + (y-4)^2 = 4^2$$

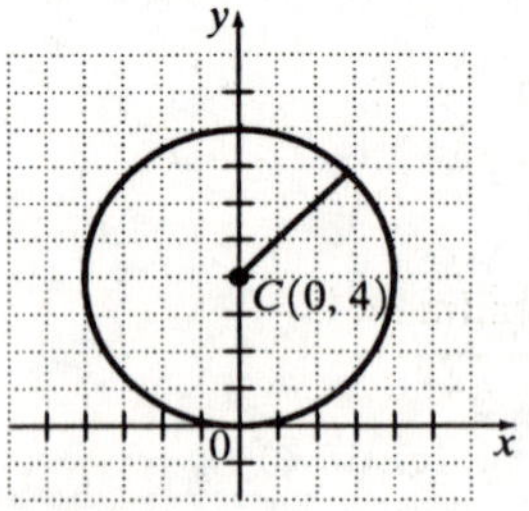

56. parabola

$$x - 4y = y^2$$

$$x = y^2 + 4y + 4 - 4$$

$$x = (y+2)^2 - 4$$

vertex: (–4, –2)

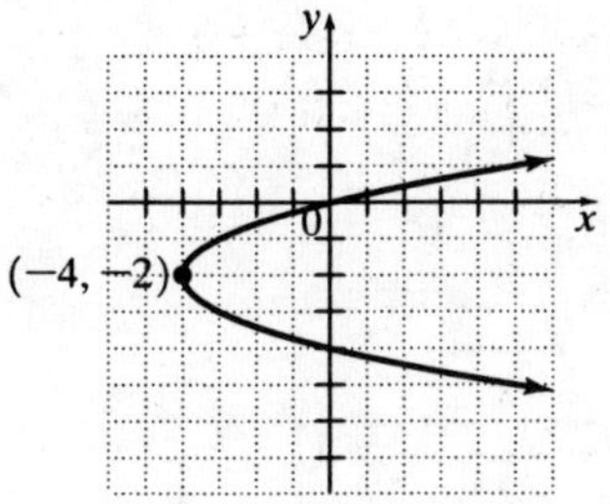

57. hyperbola

$$x^2 - 4 = y^2$$
$$x^2 - y^2 = 4$$
$$\frac{x^2}{4} - \frac{y^2}{4} = 1$$
$$\frac{(x-0)^2}{2^2} - \frac{(y-0)^2}{2^2} = 1$$

$a = 2$, $b = 2$, center (0, 0)

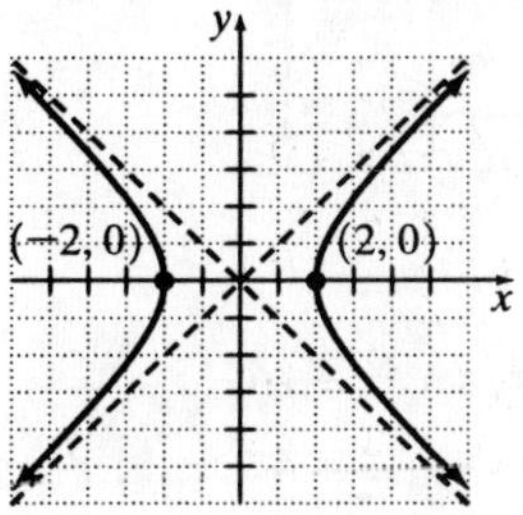

58. circle

$$x^2 = 4 - y^2$$
$$x^2 + y^2 = 4$$

center (0, 0), radius = 2

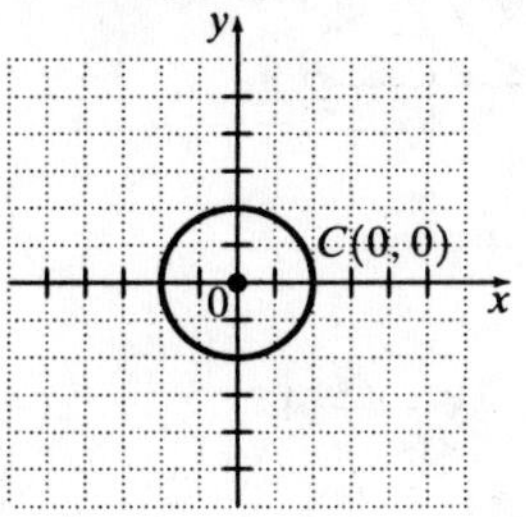

59. ellipse

$$6(x-2)^2 + 9(y+5)^2 = 36$$
$$\frac{(x-2)^2}{6} + \frac{(y+5)^2}{4} = 1$$
$$\frac{(x-2)^2}{\sqrt{6}^2} + \frac{(y+5)^2}{2^2} = 1$$

$a = \sqrt{6}$, $b = 2$, center (2, –5)

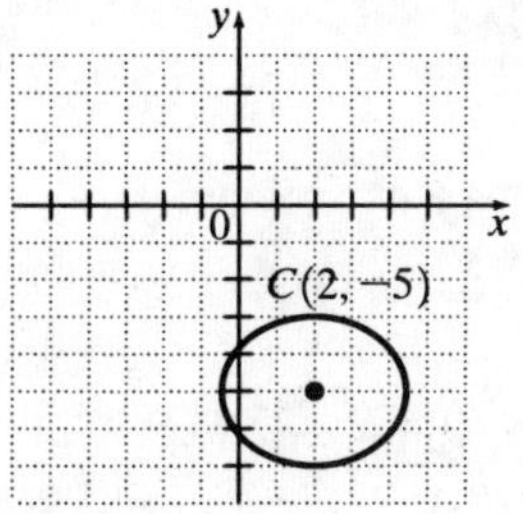

60. hyperbola

$$36y^2 = 576 + 16x^2$$
$$\frac{y^2}{16} - \frac{x^2}{36} = 1$$
$$\frac{(y-0)^2}{4^2} - \frac{(x-0)^2}{6^2} = 1$$

$a = 6$, $b = 4$, center (0, 0)

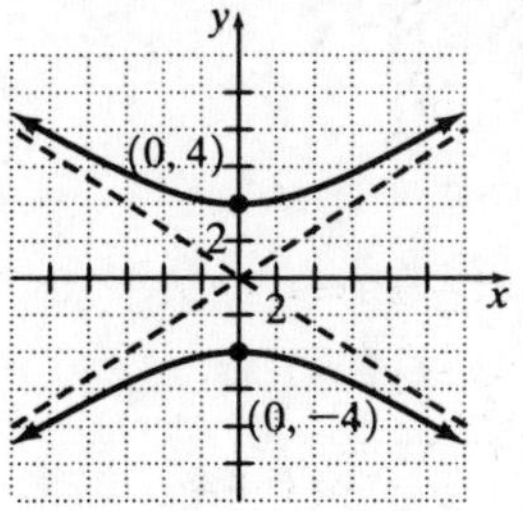

61. hyperbola

$$\frac{x^2}{16} - \frac{y^2}{25} = 1$$

or $\dfrac{(x-0)^2}{4^2} - \dfrac{(y-0)^2}{5^2} = 1$

$a = 4$, $b = 5$, center (0, 0)

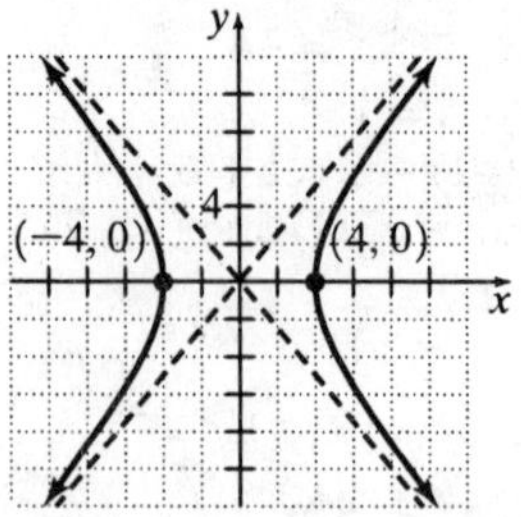

62. circle

$$3(x-7)^2 + 3(y+4)^2 = 1$$
$$(x-7)^2 + (y+4)^2 = \frac{1}{3}$$

center (7, −4), radius $= \frac{\sqrt{3}}{3}$

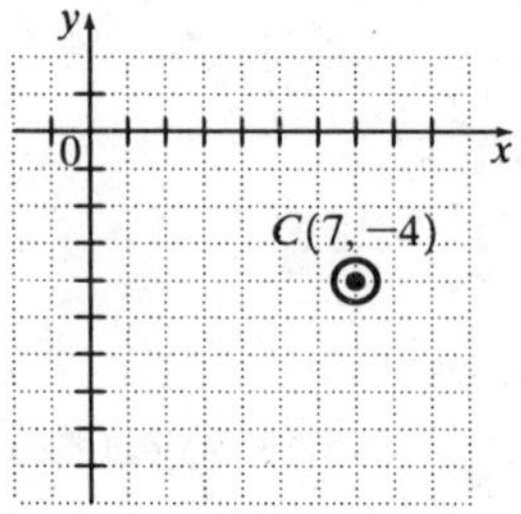

63. $\frac{y^2}{4} + \frac{x^2}{16} = 1$

$$4y^2 + x^2 = 16$$
$$4y^2 = 16 - x^2$$
$$y^2 = \frac{16 - x^2}{4}$$
$$y = \pm\sqrt{\frac{16 - x^2}{4}}$$

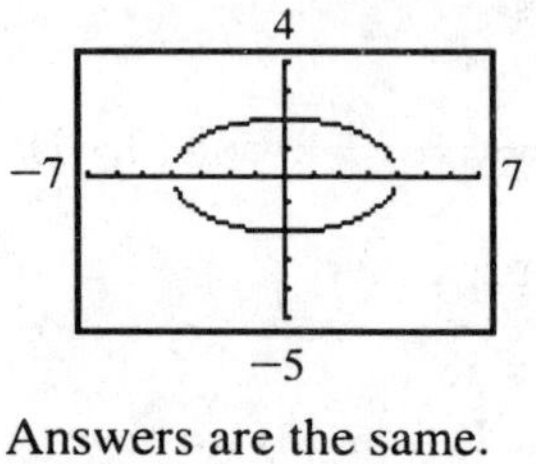

Answers are the same.

64. $\frac{x^2}{5} + \frac{y^2}{5} = 1$

$$x^2 + y^2 = 5$$
$$y^2 = 5 - x^2$$
$$y = \pm\sqrt{5 - x^2}$$

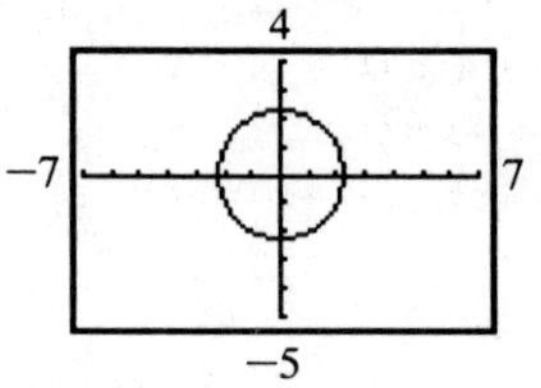

Answers are the same.

65. $y = x^2 + 4x + 6$

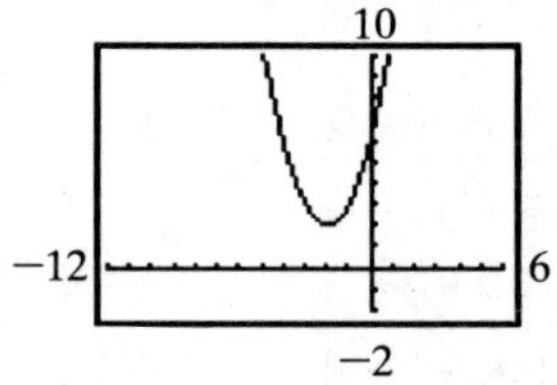

Answers are the same.

66. $x^2 = 4 - y^2$

$$y^2 = 4 - x^2$$
$$y = \pm\sqrt{4 - x^2}$$

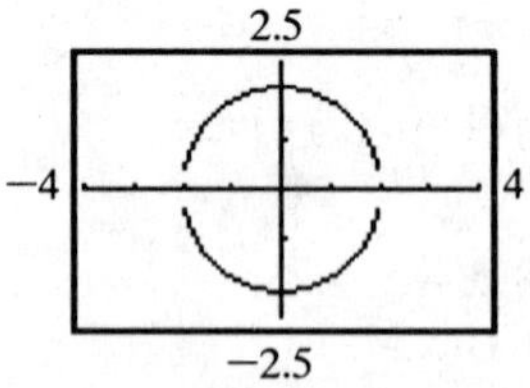

Answers are the same.

67. $\begin{cases} y = 2x - 4 \\ y^2 = 4x \end{cases}$

Substituting (1) in (2) gives

$$(2x-4)^2 = 4x$$
$$4x^2 - 16x + 16 = 4x$$
$$4x^2 - 20x + 16 = 0$$
$$x^2 - 5x + 4 = 0$$
$$(x-1)(x-4) = 0$$
$$x - 1 = 0 \quad \text{or} \quad x - 4 = 0$$
$$x = 1 \quad \text{or} \quad x = 4$$
$$y = -2 \quad \text{or} \quad y = 4$$

The solutions are (1, –2) and (4, 4).

68. $\begin{cases} x^2 + y^2 = 4 \\ x - y = 4 \text{ or } x = y + 4 \end{cases}$

Substitute.

$$(y+4)^2 + y^2 = 4$$
$$y^2 + 8y + 16 + y^2 = 4$$
$$2y^2 + 8y + 16 = 4$$
$$2y^2 + 8y + 12 = 0$$
$$y^2 + 4y + 6 = 0$$
$$b^2 - 4ac = 4^2 - 4 \cdot 1 \cdot 6 = 16 - 24 = -8$$

Therefore, no real solutions exist.
The solution is ∅.

69. $\begin{cases} y = x + 2 \\ y = x^2 \end{cases}$

Substituting (1) in (2) gives

$$x + 2 = x^2$$
$$0 = x^2 - x - 2$$
$$0 = (x+1)(x-2)$$
$$x + 1 = 0 \quad \text{or} \quad x - 2 = 0$$
$$x = -1 \quad \text{or} \quad x = 2$$
$$y = 1 \quad \text{or} \quad y = 4$$

The solutions are (–1, 1) and (2, 4).

70. $\begin{cases} y = x^2 - 5x + 1 \\ y = -x + 6 \end{cases}$

Substitute.

$$-x + 6 = x^2 - 5x + 1$$
$$0 = x^2 - 4x - 5$$
$$0 = (x-5)(x+1)$$
$$x - 5 = 0 \quad \text{or} \quad x + 1 = 0$$
$$x = 5 \quad \text{or} \quad x = -1$$
$$x = 5: y = -5 + 6 = 1$$
$$x = -1: y = -(-1) + 6 = 1 + 6 = 7$$

The solutions are (5, 1) and (–1, 7).

71. $\begin{cases} 4x - y^2 = 0 \\ 2x^2 + y^2 = 16 \end{cases}$

From (1) we have $y^2 = 4x$. Substituting in (2) gives

$$2x^2 + 4x = 16$$
$$2x^2 + 4x - 16 = 0$$
$$x^2 + 2x - 8 = 0$$
$$(x-2)(x+4) = 0$$
$$x - 2 = 0 \quad \text{or} \quad x + 4 = 0$$
$$x = 2 \quad \text{or} \quad x = -4$$
$$x = 2:\ 4(2) - y^2 = 0$$
$$y^2 = 8$$
$$y = \pm 2\sqrt{2}$$
$$x = -4:\ 4(-4) - y^2 = 0$$
$$y^2 = -16 \quad \text{reject}$$

The solutions are $(2, 2\sqrt{2})$ and $(2, -2\sqrt{2})$.

72. $\begin{cases} x^2 + 4y^2 = 16 \\ x^2 + y^2 = 4 \end{cases}$

Subtract.

$$3y^2 = 12$$
$$y^2 = 4$$
$$y = \pm 2$$

Substitute back.

$$x^2 + 4 = 4$$
$$x^2 = 0$$
$$x = 0$$

The solutions are (0, 2) and (0, –2).

73. $\begin{cases} x^2 + y^2 = 10 \\ 9x^2 + y^2 = 18 \end{cases}$

Subtracting (1) from (2) we have

$$8x^2 = 8$$
$$x^2 = 1$$
$$x = \pm 1$$

Substitution gives

$1+y^2=10$

$y^2=9$

$y=\pm 3$

The solutions are (–1, 3), (–1, –3), (1, 3), and (1, –3).

74. $\begin{cases} x^2+2y=9 \\ 5x-2y=5 \end{cases}$

Add.

$x^2+5x=14$

$x^2+5x-14=0$

$(x+7)(x-2)=0$

$x+7=0$ or $x-2=0$

$x=-7$ or $x=2$

$x=-7$: $(-7)^2+2y=9$

$49+2y=9$

$2y=-40$

$y=-20$

$x=2$: $2^2+2y=9$

$4+2y=9$

$2y=5$

$y=\frac{5}{2}$

The solutions are (–7, –20) and $\left(2, \frac{5}{2}\right)$.

75. $\begin{cases} y=3x^2+5x-4 \\ y=3x^2-x+2 \end{cases}$

Subtracting (2) from (1) gives

$0=6x-6$

$6x=6$

$x=1$

Substitution gives

$y=3(1)^2-1+2=4$.

The solution is (1, 4).

76. $\begin{cases} x^2-3y^2=1 \text{ or } x^2=3y^2+1 \\ 4x^2+5y^2=21 \end{cases}$

Substitute.

$4(3y^2+1)+5y^2=21$

$12y^2+4+5y^2=21$

$17y^2+4=21$

$17y^2=17$

$y^2=1$

$y=\pm 1$

Substitute back.

$x^2-3(1)=1$

$x^2-3=1$

$x^2=4$

$x=\pm 2$

The solutions are (2, 1), (2, –1), (–2, 1), and (–2, –1).

77. $\begin{cases} xy=150 \\ 2x+2y=50 \end{cases}$

$x=\frac{150}{y}$

$2\left(\frac{150}{y}\right)+2y=50$

$\frac{150}{y}+y=25$

$150+y^2=25y$

$y^2-25y+150=0$

$(y-15)(y-10)=0$

$y=15$ or $y=10$

$2x+2(15)=50$ $\quad$ $2x+2(10)=50$

$2x=20$ $\quad$ $2x=30$

$x=10$ $\quad$ $x=15$

The room is 15 feet by 10 feet.

78. Four real solutions

79. $y\le -x^2+3$

Graph $y\le -x^2+3$. Does (0, 0) satisfy $y\le x^2+3$?

$0\le -0^2+3$

$0\le 3$ true

Shade the portion of the graph which contains (0, 0).

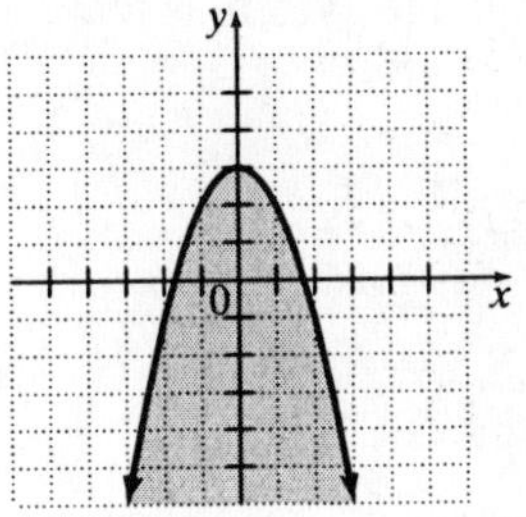

80. $x^2 + y^2 < 9$

Graph $x^2 + y^2 < 9$ with a dashed line.

Does (0, 0) satisfy $x^2 + y^2 < 9$?

$0 + 0 < 9$

$0 < 9$ true

Shade the portion of the graph which contains (0, 0).

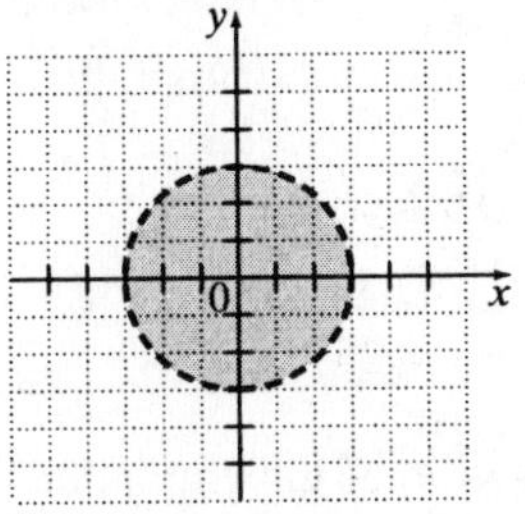

81. $x^2 - y^2 < 1$

Graph $x^2 - y^2 < 1$ with a dashed line.

Does (0, 0) satisfy the inequality?

$0 - 0 < 1$

$0 < 1$ true

Does (2, 0) satisfy the inequality?

$2^2 - 0^2 < 1$

$4 < 1$ false

Does $(-2, 0)$ satisfy the inequality?

$(-2)^2 - 0^2 < 1$

$4 < 1$ false

Shade the portion of the graph containing (0, 0).

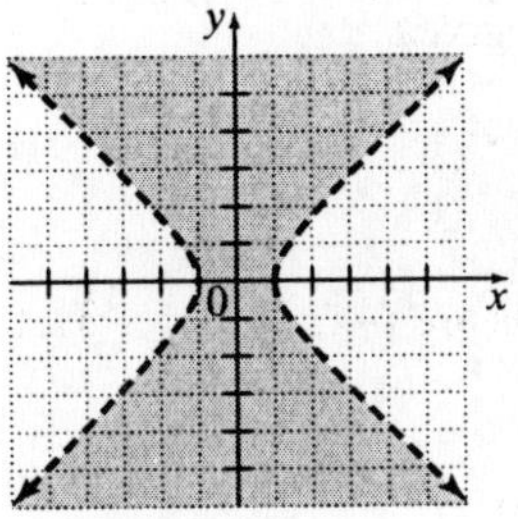

82. $\frac{x^2}{4} + \frac{y^2}{9} \geq 1$

Graph $\frac{x^2}{4} + \frac{y^2}{9} \geq 1$.

Does (0, 0) satisfy the inequality?

$\frac{0^2}{4} + \frac{0^2}{9} \geq 1$

$0 \geq 1$ false

Shade the portion of the graph that does not contain (0, 0).

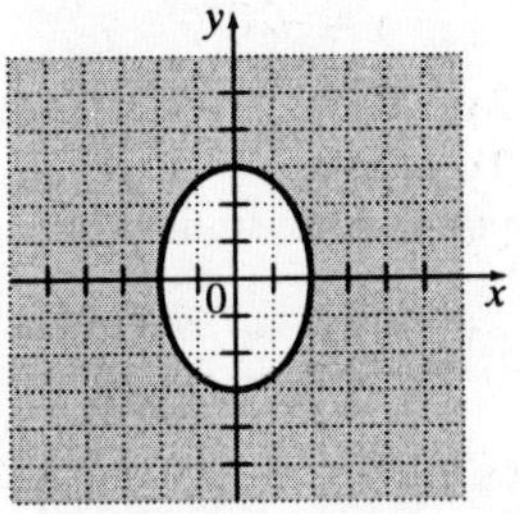

83. $\begin{cases} 2x \leq 4 \\ x + y \geq 1 \end{cases}$

First, graph $2x \leq 4$.

Does (0, 0) satisfy the inequality?

$2 \cdot 0 \leq 4$

$0 \leq 4$ true

Shade the portion of the graph that contains (0, 0).

Next, graph $x + y \geq 1$.

Does (0, 0) satisfy the inequality?

$0 + 0 \geq 1$

$0 \geq 1$ false

Shade the portion of the graph that does not contain (0, 0).

The solution to the system is the overlapping region.

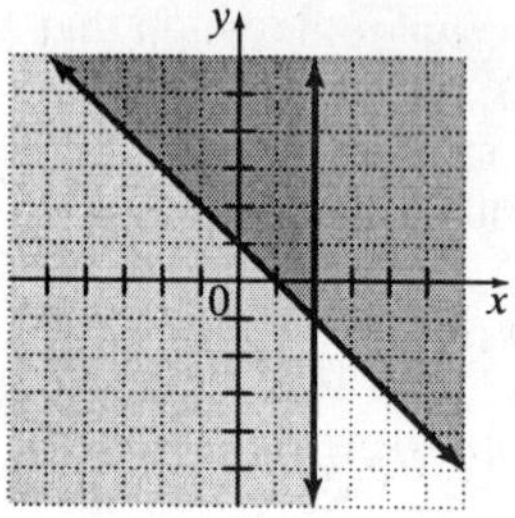

84. $\begin{cases} 3x+4y \le 12 \\ x-2y > 6 \end{cases}$

First, graph $3x+4y \le 12$

Does (0, 0) satisfy the inequality?

$3(0)+4(0) \le 12$

$0 \le 12$ true

Shade the portion of the graph containing (0, 0).

Next, graph $x-2y>6$ with a dashed line.

Does (0, 0) satisfy the inequality?

$0-2(0)>6$

$0>6$ false

Shade the portion of the graph that does not contain (0, 0).

The solution to the system is the overlapping region.

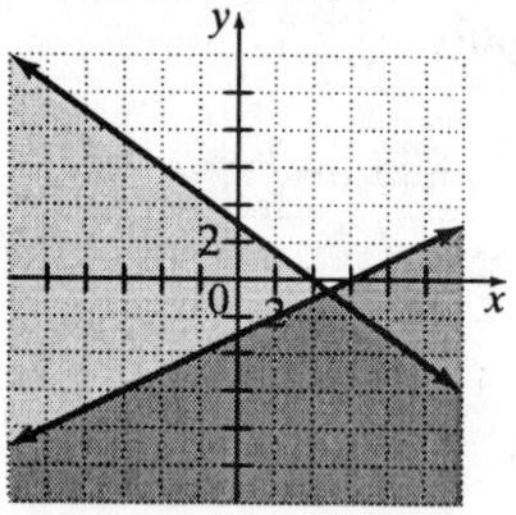

85. $\begin{cases} y > x^2 \\ x+y \ge 3 \end{cases}$

First, graph $y > x^2$ with a dashed line.

Does (0, 1) satisfy the inequality?

$0 > 1^2$

$0 > 1$ false

Shade the portion of the graph that does not contain (0, 0).

Next, graph $x+y \ge 3$.

Does (0, 0) satisfy the inequality?

$0+0 \ge 3$

$0 \ge 3$ false

Shade the portion of the graph that does not contain (0, 0).

The solution to the system is the overlapping region.

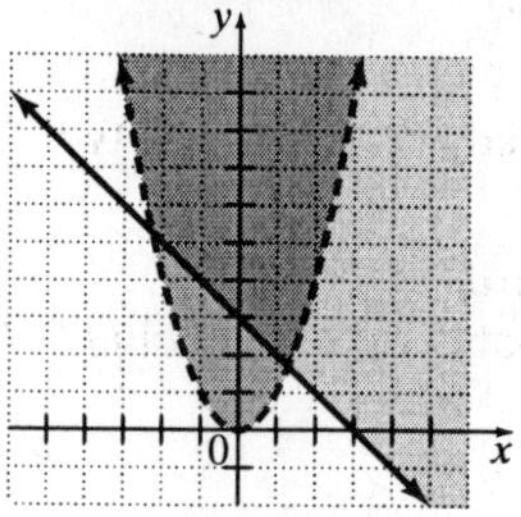

86. $\begin{cases} x^2+y^2 \le 16 \\ x^2+y^2 \ge 4 \end{cases}$

First, graph $x^2+y^2 \le 16$.

Does (0, 0) satisfy the inequality?

$0^2+0^2 \le 16$

$0 \le 16$ true

Shade the portion of the graph that contains (0, 0).

Next, graph $x^2+y^2 \ge 4$.

Does (0, 0) satisfy the inequality?

$0^2+0^2 \ge 4$

$0 \ge 4$ false

Shade the portion of the graph that does not contain (0, 0).

The solution to the system is the overlapping region.

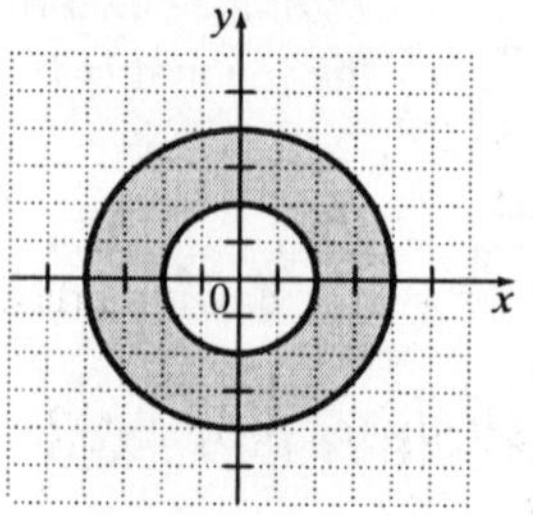

87. $\begin{cases} x^2+y^2 < 4 \\ x^2-y^2 \le 1 \end{cases}$

First, graph $x^2+y^2<4$ with a dashed line.

Does (0, 0) satisfy the inequality?

$0^2 + 0^2 < 4$
$0 < 4$ true
Shade the portion of the graph that contains (0, 0).
Next, graph $x^2 - y^2 \le 1$.
Does (0, 0) satisfy the inequality?
$0^2 - 0^2 \le 1$
$0 \le 1$ true
Does (2, 0) satisfy the inequality?
$2^2 - 0^2 \le 1$
$4 - 0 \le 1$
$4 \le 1$ false
Does (−2, 0) satisfy the inequality?
$(-2)^2 - 0^2 \le 1$
$4 - 0 \le 1$
$4 \le 1$ false
Shade the portion of the graph that contains (0, 0).
The solution to the system is the overlapping region.

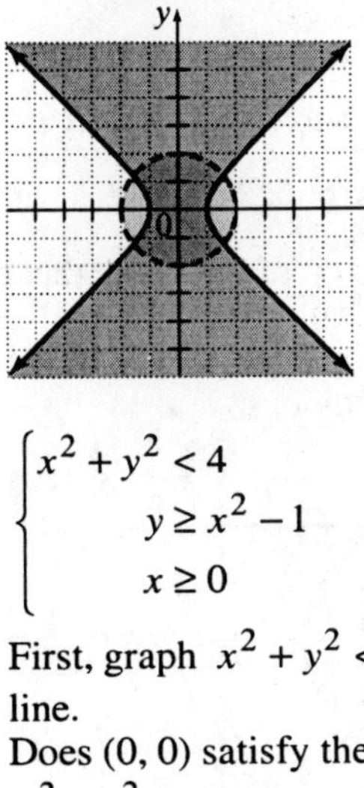

88. $\begin{cases} x^2 + y^2 < 4 \\ y \ge x^2 - 1 \\ x \ge 0 \end{cases}$

First, graph $x^2 + y^2 < 4$ with a dashed line.
Does (0, 0) satisfy the inequality?
$0^2 + 0^2 < 4$
$0 < 4$ true
Shade the portion of the graph that contains (0, 0).
Next, graph $y \ge x^2 - 1$.
Does (0, 0) satisfy the inequality?
$0 \ge 0^2 - 1$
$0 \ge -1$ true
Shade the portion of the graph that contains (0, 0).
Next, graph $x \ge 0$.
Does (1, 0) satisfy the inequality?
$1 \ge 0$ true
Shade the portion of the graph that contains (1, 0). The solution to the system is the overlapping region.

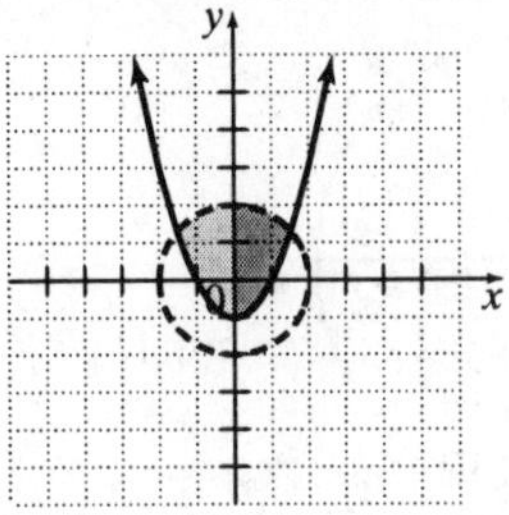

Chapter 9 Test

1. (−6, 3) and (−8, −7)
$d = \sqrt{(-8+6)^2 + (-7-3)^2}$
$d = \sqrt{(-2)^2 + (-10)^2}$
$d = \sqrt{4 + 100}$
$d = \sqrt{104}$
$d = 2\sqrt{26}$ units

2. $(-2\sqrt{5}, \sqrt{10})$ and $(-\sqrt{5}, 4\sqrt{10})$
$d = \sqrt{(-\sqrt{5} + 2\sqrt{5})^2 + (4\sqrt{10} - \sqrt{10})^2}$
$d = \sqrt{(\sqrt{5})^2 + (3\sqrt{10})^2}$
$d = \sqrt{5 + 90}$
$d = \sqrt{95}$ units

3. (−2, −5) and (−6, 12)
$\left(\frac{-2-6}{2}, \frac{-5+12}{2}\right)$
The midpoint of the line segment is $\left(-4, \frac{7}{2}\right)$.

4. $\left(-\frac{2}{3}, -\frac{1}{5}\right)$ and $\left(-\frac{1}{3}, \frac{4}{5}\right)$
$\left(\frac{-\frac{2}{3} - \frac{1}{3}}{2}, \frac{-\frac{1}{5} + \frac{4}{5}}{2}\right)$

The midpoint of the line segment is $\left(-\frac{1}{2}, \frac{3}{10}\right)$.

5. $x^2 + y^2 = 36$

$(x-0)^2 + (y-0)^2 = 6^2$

Circle: center (0, 0), radius = 6

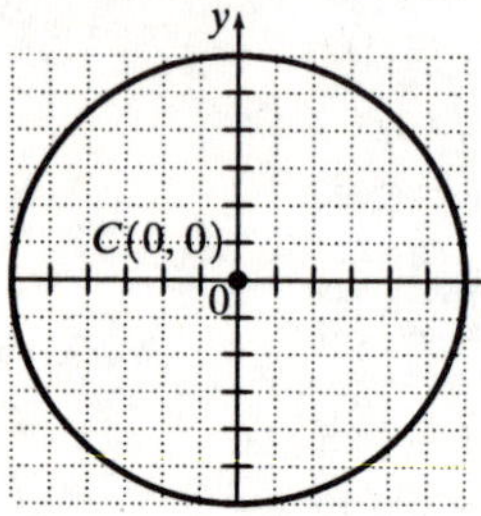

6. $x^2 - y^2 = 36$

$$\frac{x^2}{36} - \frac{y^2}{36} = 1$$

$$\frac{(x-0)^2}{6^2} - \frac{(y-0)^2}{6^2} = 1$$

Hyperbola:
$a = 6$, $b = 6$, center(0, 0)

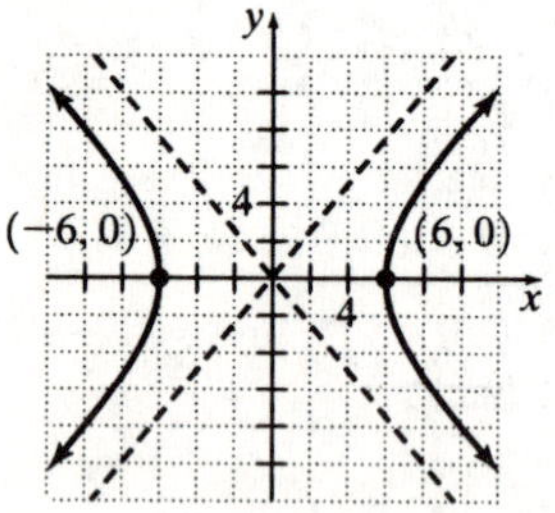

7. $16x^2 + 9y^2 = 144$

$$\frac{x^2}{9} + \frac{y^2}{16} = 1$$

$$\frac{(x-0)^2}{3^2} + \frac{(y-0)^2}{4^2} = 1$$

Ellipse: $a = 3$, $b = 4$, center (0, 0)

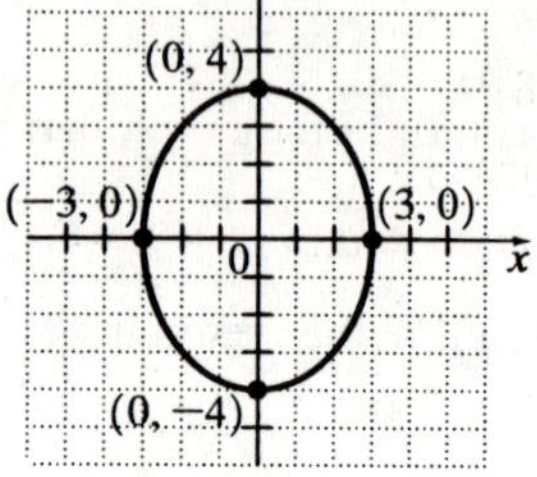

8. $y = x^2 - 8x + 16$

$y = (x-4)^2 + 0$

Parabola: The vertex is (4, 0).

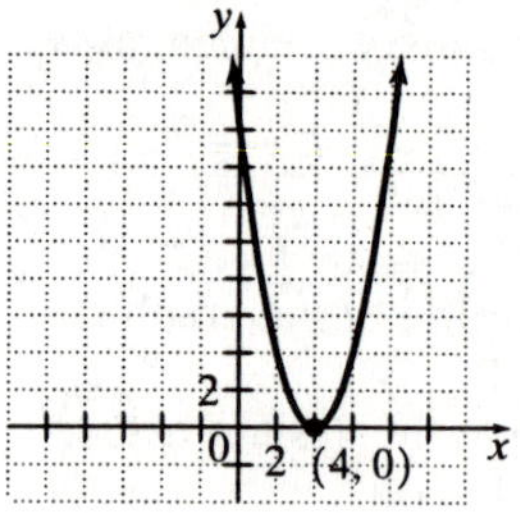

9. $x^2 + y^2 + 6x = 16$

$x^2 + 6x + y^2 = 16$

$(x^2 + 6x + 9) + y^2 = 16 + 9$

$(x+3)^2 + y^2 = 5^2$

Circle: center(−3, 0), radius 5

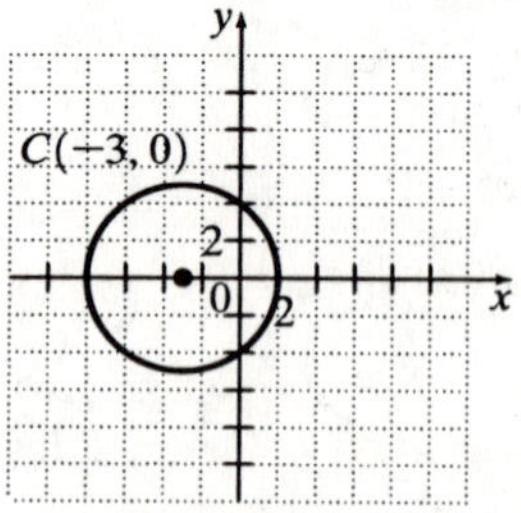

10. $x = y^2 + 8y - 3$

$x = (y^2 + 8y + 16) - 3 - 16$

$x = (y + 4)^2 - 19$

Parabola: The vertex is (–19, –4).

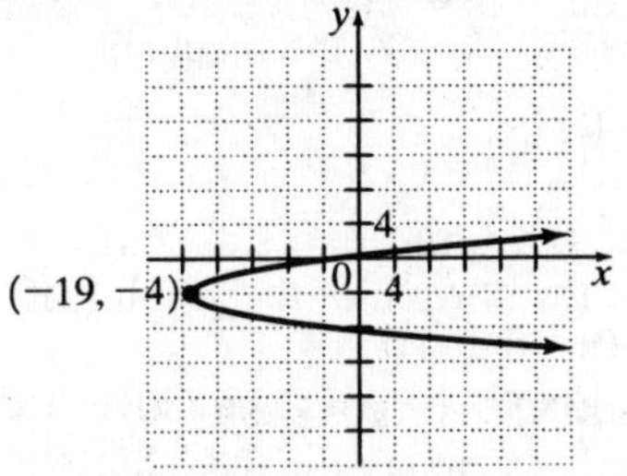

11. $\dfrac{(x-4)^2}{16} + \dfrac{(y-3)^2}{9} = 1$

$\dfrac{(x-4)^2}{4^2} + \dfrac{(y-3)^2}{3^2} = 1$

Ellipse: $a = 4$, $b = 3$, center (4, 3)

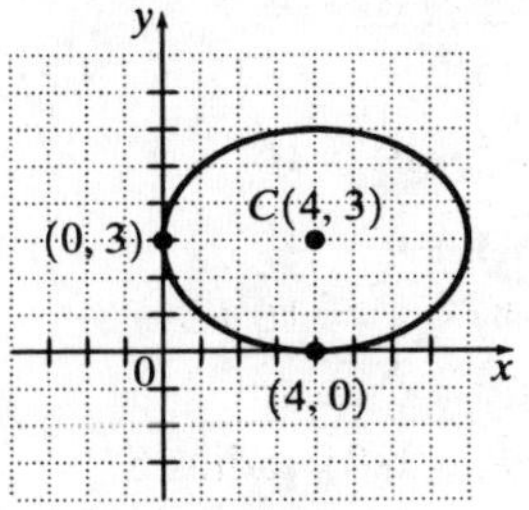

12. $y^2 - x^2 = 1$

$y^2 = x^2 + 1$

$y = \pm\sqrt{x^2 + 1}$

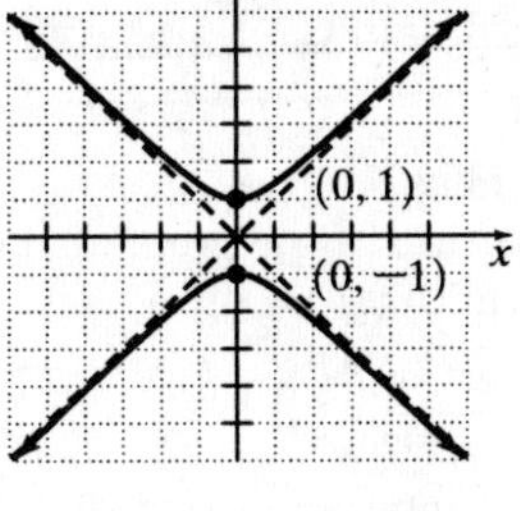

13. $\begin{cases} x^2 + y^2 = 169 \\ 5x + 12y = 0 \end{cases}$

$12y = -5x$

$y = -\dfrac{5x}{12}$

Substitute.

$x^2 + \left(-\dfrac{5x}{12}\right)^2 = 169$

$x^2 + \dfrac{25x^2}{144} = 169$

$\dfrac{169x^2}{144} = 169$

$x^2 = 144$

$x = \pm 12$

Substitute back.

$x = 12$: $y = -\dfrac{5}{12}(12) = -5$

$x = -12$: $y = -\dfrac{5}{12}(-12) = 5$

The solutions are (12, –5), (–12, 5).

14. $\begin{cases} x^2 + y^2 = 26 \\ x^2 - y^2 = 24 \end{cases}$

Add.

$2x^2 = 50$

$x^2 = 25$

$x = \pm 5$

Substitute back.

$25 + y^2 = 26$

$y^2 = 1$

$y = \pm 1$

The solutions are (5, 1), (5, –1), (–5, 1), and (–5, –1).

15. $\begin{cases} y = x^2 - 5x + 6 \\ y = 2x \end{cases}$

Substitute.

$2x = x^2 - 5x + 6$

$0 = x^2 - 7x + 6$

$0 = (x-1)(x-6)$

$x - 1 = 0$ or $x - 6 = 0$

$x = 1$ or $x = 6$

$x = 1: y = 2(1) = 2$
$x = 6: y = 2(6) = 12$
The solutions are (1, 2) and (6, 12).

16. $\begin{cases} x^2 + 4y^2 = 5 \\ y = x \end{cases}$

Substitute.

$x^2 + 4x^2 = 5$
$5x^2 = 5$
$x^2 = 1$
$x = \pm 1$

$x = 1: y = 1$
$x = -1: y = -1$
The solutions are (1, 1) and (–1, –1).

17. $\begin{cases} 2x + 5y \geq 10 \\ y \geq x^2 + 1 \end{cases}$

First, graph $2x + 5y \geq 10$.
Does (0, 0) satisfy the inequality?
$2(0) + 5(0) \geq 10$
$0 \geq 10$ false
Shade the portion of the graph that does not contain (0, 0).
Next, graph $y \geq x^2 - 1$.
Does (0, 0) satisfy the inequality?
$0 \geq 0^2 + 1$
$0 \geq 1$ false
Shade the portion of the graph that does not contain (0, 0).
The solution to the system is the overlapping region.

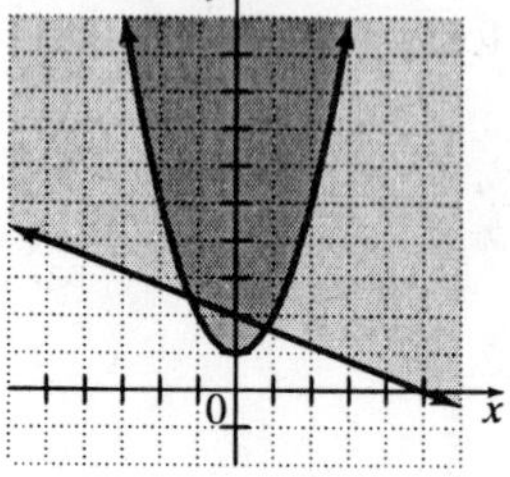

18. $\begin{cases} \frac{x^2}{4} + y^2 \leq 1 \\ x + y > 1 \end{cases}$

First, graph $\frac{x^2}{4} + y^2 \leq 1$.
Does (0, 0) satisfy the inequality?
$\frac{0^2}{4} + 0^2 \leq 1$
$0 \leq 1$ true
Shade the portion of the graph that contains (0, 0).
Next, graph $x + y > 1$ with a dashed line.
Does (0, 0) satisfy the inequality?
$0 + 0 > 1$
$0 > 1$ false
Shade the portion of the graph that does not contain (0, 0).
The solution to the system is the overlapping region.

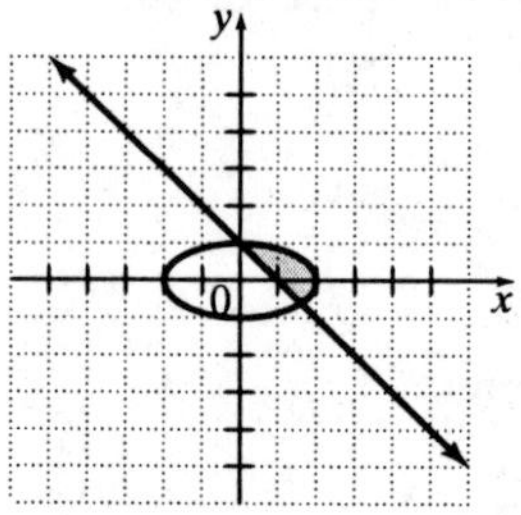

19. $\begin{cases} x^2 + y^2 > 1 \\ \frac{x^2}{4} - y^2 \geq 1 \end{cases}$

First, graph $x^2 + y^2 > 1$ with a dashed line.
Does (0, 0) satisfy the inequality?
$0^2 + 0^2 > 0$
$0 > 1$ false
Shade the portion of the graph that does not contain (0, 0).
Next, graph $\frac{x^2}{4} - y^2 \geq 1.$.
Does (0, 0) satisfy the inequality?
$\frac{0^2}{4} - 0^2 \geq 1$
$0 \geq 1$ false

Does (3, 0) satisfy the inequality?

$\frac{3^2}{4} - 0^2 \ge 1$

$\frac{9}{4} \ge 1$ true

Does $(-3, 0)$ satisfy the inequality?

$\frac{(-3)^2}{4} - 0^2 \ge 1$

$\frac{9}{4} \ge 1$ true

Shade the portions of the graph that contain $(3, 0)$ and $(-3, 0)$.
The solution to the system is the overlapping region.

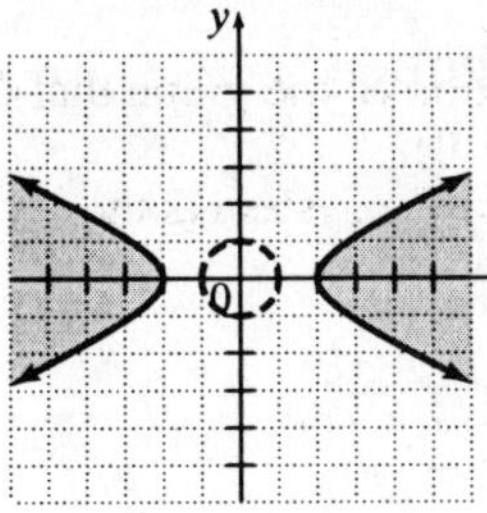

20. $\begin{cases} x^2 + y^2 \ge 4 \\ x^2 + y^2 < 16 \\ y \ge 0 \end{cases}$

First, graph $x^2 + y^2 \ge 4$.
Does (0, 0) satisfy the inequality?

$0^2 + 0^2 \ge 4$

$0 \ge 4$ false

Shade the portion of the graph that does not contain (0, 0).

Next, graph $x^2 + y^2 < 16$ with a dashed line.
Does (0, 0) satisfy the inequality?

$0^2 + 0^2 < 16$

$0 < 16$ true

Shade the portion of the graph that contains (0, 0).
Next, graph $y \ge 0$.
Does (0, 2) satisfy the inequality?

$2 \ge 0$ true

Shade the portion of the graph that contains (0, 2).
The solution to the system is the overlapping region.

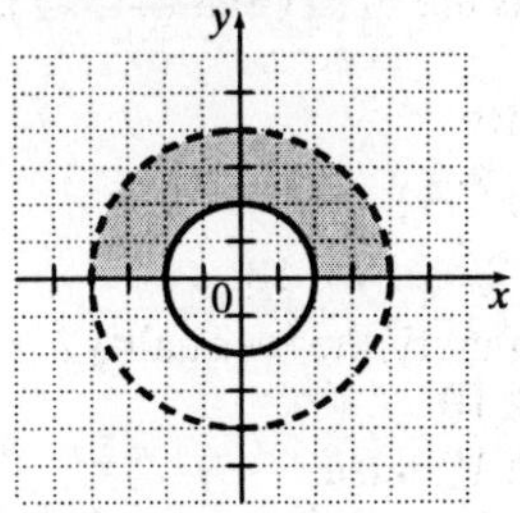

21. Graph B; vertex in second quadrant, opens to the right.

22. $100x^2 + 225y^2 = 22,500$

$\frac{x^2}{225} + \frac{y^2}{100} = 0$

$a = \sqrt{225} = 15$

$b = \sqrt{100} = 10$

Width = 15 + 15 = 30 feet
Height = 10 feet

Chapter 10

Exercise Set 10.1

1. a. $(f+g)(x) = x - 7 + 2x + 1 = 3x - 6$

b. $$\begin{aligned}(f-g)(x) &= x - 7 - (2x+1)\\ &= x - 7 - 2x - 1\\ &= -x - 8\end{aligned}$$

c. $$\begin{aligned}(f \cdot g)(x) &= (x-7)(2x+1)\\ &= 2x^2 - 13x - 7\end{aligned}$$

d. $\left(\dfrac{f}{g}\right)(x) = \dfrac{x-7}{2x+1}$, where $x \neq -\dfrac{1}{2}$

3. a. $(f+g)(x) = x^2 + 1 + 5x = x^2 + 5x + 1$

b. $(f-g)(x) = x^2 + 1 - 5x = x^2 - 5x + 1$

c. $(f \cdot g)(x) = (x^2+1)(5x) = 5x^3 + 5x$

d. $\left(\dfrac{f}{g}\right)(x) = \dfrac{x^2+1}{5x}$, where $x \neq 0$

5. a. $(f+g)(x) = \sqrt{x} + x + 5$

b. $(f-g)(x) = \sqrt{x} - x - 5$

c. $(f \cdot g)(x) = \sqrt{x}(x+5) = x\sqrt{x} + 5\sqrt{x}$

d. $\left(\dfrac{f}{g}\right)(x) = \dfrac{\sqrt{x}}{x+5}$, where $x \neq -5$

7. a. $(f+g)(x) = -3x + 5x^2 = 5x^2 - 3x$

b. $(f-g)(x) = -3x - 5x^2 = -5x^2 - 3x$

c. $(f \cdot g)(x) = -3x(5x^2) = -15x^3$

d. $\left(\dfrac{f}{g}\right)(x) = \dfrac{-3x}{5x^2} = -\dfrac{3}{5x}$, where $x \neq 0$.

9. $$\begin{aligned}(f \circ g)(2) &= f(g(2))\\ &= f(-4)\\ &= (-4)^2 - 6(-4) + 2\\ &= 16 + 24 + 2\\ &= 42\end{aligned}$$

11. $$\begin{aligned}(g \circ f)(-1) &= g(f(-1))\\ &= g(9)\\ &= -2(9)\\ &= -18\end{aligned}$$

13. $$\begin{aligned}(g \circ h)(0) &= g(h(0))\\ &= g(0)\\ &= -2(0)\\ &= 0\end{aligned}$$

15. $$\begin{aligned}(f \circ g)(x) &= f(g(x))\\ &= f(5x)\\ &= (5x)^2 + 1\\ &= 25x^2 + 1\end{aligned}$$

$$\begin{aligned}(g \circ f)(x) &= g(f(x))\\ &= g(x^2+1)\\ &= 5(x^2+1)\\ &= 5x^2 + 5\end{aligned}$$

17. $$\begin{aligned}(f \circ g)(x) &= f(g(x))\\ &= f(x+7)\\ &= 2(x+7) - 3\\ &= 2x + 14 - 3\\ &= 2x + 11\end{aligned}$$

$$\begin{aligned}(g \circ f)(x) &= g(f(x))\\ &= g(2x-3)\\ &= (2x-3) + 7\\ &= 2x + 4\end{aligned}$$

19. $(f \circ g)(x) = f(g(x))$
$= f(-2x)$
$= (-2x)^3 + (-2x) - 2$
$= -8x^3 - 2x - 2$

$(g \circ f)(x) = g(f(x))$
$= g(x^3 + x - 2)$
$= -2(x^3 + x - 2)$
$= -2x^3 - 2x + 4$

21. $(f \circ g)(x) = f(g(x))$
$= f(-5x + 2)$
$= \sqrt{-5x + 2}$

$(g \circ f)(x) = g(f(x)) = g\left(\sqrt{x}\right) = -5\sqrt{x} + 2$

23. $H(x) = (g \circ h)(x)$
$= g(h(x))$
$= g(x^2 + 2)$
$= \sqrt{x^2 + 2}$

25. $F(x) = (h \circ f)(x)$
$= h(f(x))$
$= h(3x)$
$= (3x)^2 + 2$
$= 9x^2 + 2$

27. $G(x) = (f \circ g)(x)$
$= f(g(x))$
$= f\left(\sqrt{x}\right)$
$= 3\sqrt{x}$

29. Answers may vary. For example,
$g(x) = x + 2$ and $f(x) = x^2$.

31. Answers may vary. For example,
$g(x) = x + 5$ and $f(x) = \sqrt{x} + 2$.

33. Answers may vary. For example,
$g(x) = 2x - 3$ and $f(x) = \frac{1}{x}$.

35. $(f + g)(2) = 7 + (-1) = 6$

37. $(f \circ g)(2) = f(g(2)) = f(-1) = 4$

39. $(f \cdot g)(7) = 1 \cdot 4 = 4$

41. $\left(\frac{f}{g}\right)(-1) = \frac{4}{-4} = -1$

43. $P(x) = R(x) - C(x)$

45. $x = y + 2$
$x - 2 = y + 2 - 2$
$x - 2 = y$
$y = x - 2$

47. $x = 3y$
$\frac{x}{3} = \frac{3y}{3}$
$\frac{x}{3} = y$
$y = \frac{x}{3}$

49. $x = -2y - 7$
$x + 7 = -2y$
$\frac{x+7}{-2} = y$
$y = -\frac{x+7}{2}$

Exercise Set 10.2

1. $f = \{(-1, -1), (1, 1), (0, 2), (2, 0)\}$ is a one-to-one function.
$f^{-1} = \{(-1, -1), (1, 1), (2, 0), (0, 2)\}$

3. $h = \{(10, 10)\}$ is a one-to-one function.
$h^{-1} = \{(10, 10)\}$

5. $f = \{(11, 12), (4, 3), (3, 4), (6, 6)\}$ is a one-to-one function.
$f^{-1} = \{(12, 11), (3, 4), (4, 3), (6, 6)\}$

7. This function is not one-to-one because there are two months with the same output: (January, 282) and (May, 282).

9. This function is one-to-one.

Rank in population (Input)	1	49	12	2	45
State (Output)	California	Vermont	Virginia	Texas	South Dakota

11. $f(x) = x^3 + 2$

a. $f(1) = 1^3 + 2 = 3$

b. $f^{-1}(3) = 1$

13. $f(x) = x^3 + 2$

a. $f(-1) = (-1)^3 + 2 = 1$

b. $f^{-1}(1) = -1$

15. The graph represents a one-to-one function because it passes the horizontal line test.

17. The graph does not represent a one-to-one function because it does not pass the horizontal line test.

19. The graph represents a one-to-one function because it passes the horizontal line test.

21. The graph does not represent a one-to-one function because it does not pass the horizontal line test.

23. $f(x) = x + 4$

$$y = x + 4$$
$$x = y + 4$$
$$y = x - 4$$
$$f^{-1}(x) = x - 4$$

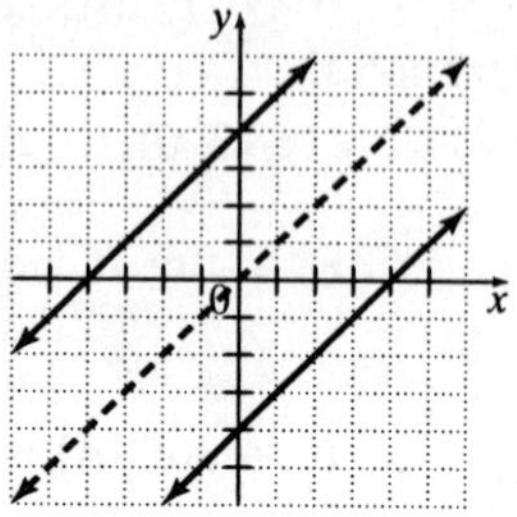

25. $f(x) = 2x - 3$

$y = 2x - 3$

$x = 2y - 3$

$2y = x + 3$

$y = \frac{x+3}{2}$

$f^{-1}(x) = \frac{x+3}{2}$

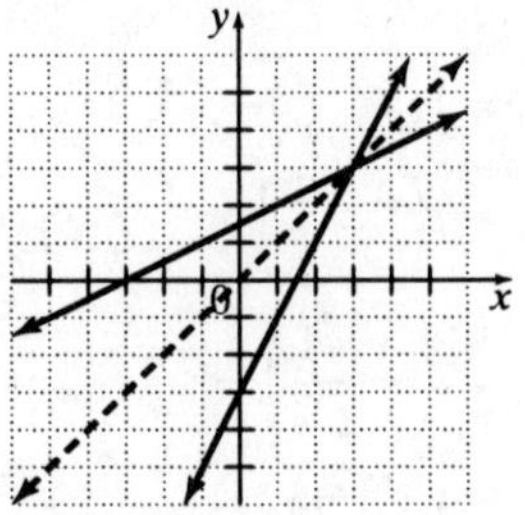

27. $f(x) = \frac{1}{2}x - 1$

$y = \frac{1}{2}x - 1$

$x = \frac{1}{2}y - 1$

$\frac{1}{2}y = x + 1$

$y = 2x + 2$

$f^{-1}(x) = 2x + 2$

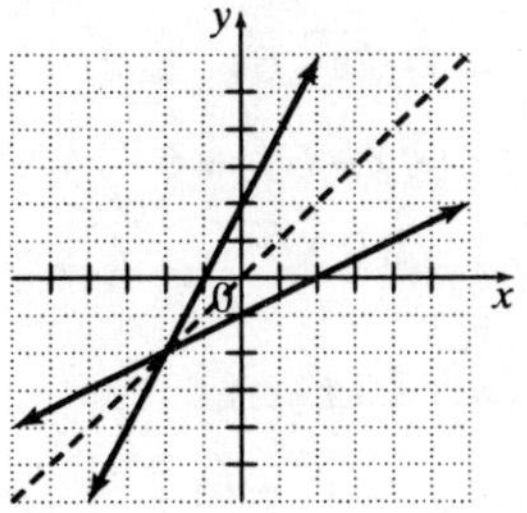

29. $f(x) = x^3$

$y = x^3$

$x = y^3$

$y = \sqrt[3]{x}$

$f^{-1}(x) = \sqrt[3]{x}$

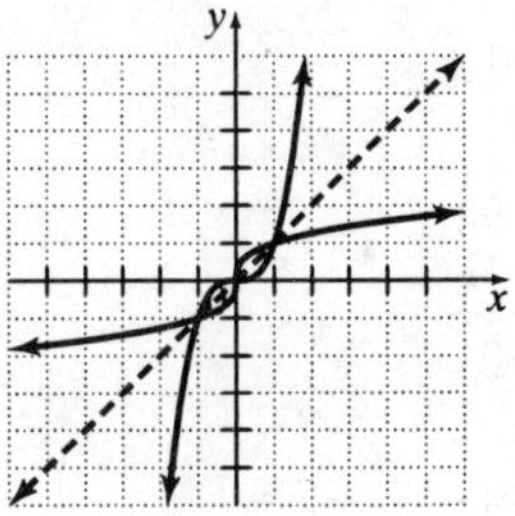

31. $f(x) = 5x + 2$

$y = 5x + 2$

$x = 5y + 2$

$5y = x - 2$

$y = \frac{x-2}{5}$

$f^{-1}(x) = \frac{x-2}{5}$

33. $f(x) = \frac{x-2}{5}$

$y = \frac{x-2}{5}$

$x = \frac{y-2}{5}$

$5x = y - 2$

$y = 5x + 2$

$f^{-1}(x) = 5x + 2$

35. $f(x) = \sqrt[3]{x}$

$y = \sqrt[3]{x}$

$x = \sqrt[3]{y}$

$x^3 = y$

$f^{-1}(x) = x^3$

37. $f(x)=\frac{5}{3x+1}$

$y=\frac{5}{3x+1}$

$x=\frac{5}{3y+1}$

$3y+1=\frac{5}{x}$

$3y=\frac{5}{x}-1$

$3y=\frac{5-x}{x}$

$y=\frac{5-x}{3x}$

$f^{-1}(x)=\frac{5-x}{3x}$

39. $f(x)=(x+2)^3$

$y=(x+2)^3$

$x=(y+2)^3$

$\sqrt[3]{x}=y+2$

$\sqrt[3]{x}-2=y$

$f^{-1}(x)=\sqrt[3]{x}-2$

41.

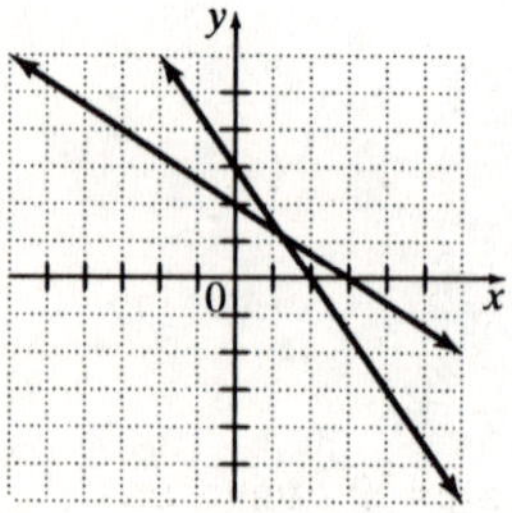

43.

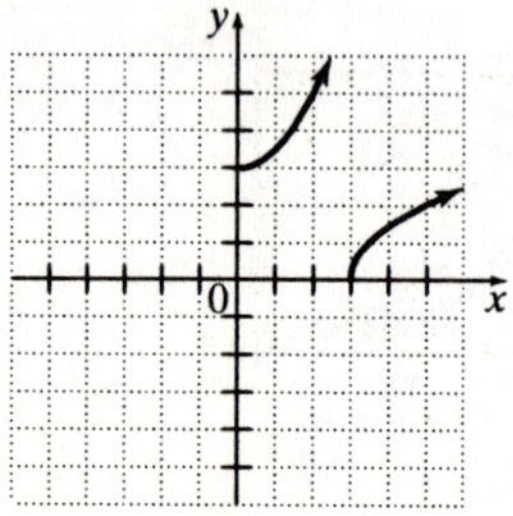

45.

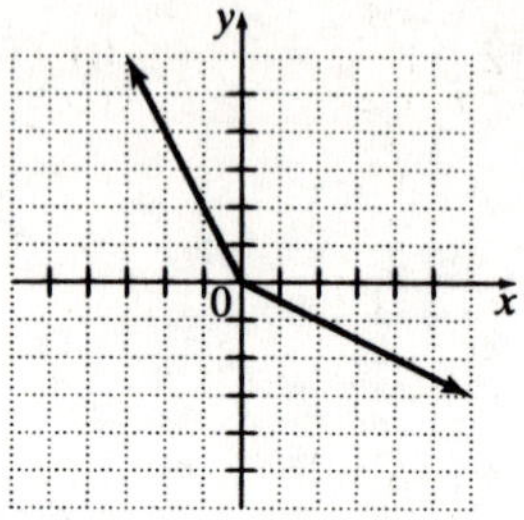

47. $(f\circ f^{-1})(x)=f(f^{-1}(x))$

$=f\left(\frac{x-1}{2}\right)$

$=2\left(\frac{x-1}{2}\right)+1$

$=x-1+1$

$=x$

$(f^{-1}\circ f)(x)=f^{-1}(f(x))$

$=f^{-1}(2x+1)$

$=\frac{(2x+1)-1}{2}$

$=\frac{2x}{2}$

$=x$

49. $(f\circ f^{-1})(x)=f(f^{-1}(x))$

$=f(\sqrt[3]{x-6})$

$=(\sqrt[3]{x-6})^3+6$

$=x-6+6$

$=x$

$(f^{-1}\circ f)(x)=f^{-1}(f(x))$

$=f^{-1}(x^3+6)$

$=\sqrt[3]{(x^3+6)-6}$

$=\sqrt[3]{x^3}$

$=x$

51. a. $\left(-2,\ \frac{1}{4}\right)$, $\left(-1,\ \frac{1}{2}\right)$, (0, 1), (1, 2), (2, 5)

b. $\left(\frac{1}{4}, -2\right), \left(\frac{1}{2}, -1\right)$, (1, 0), (2, 1), (5, 2)

c.

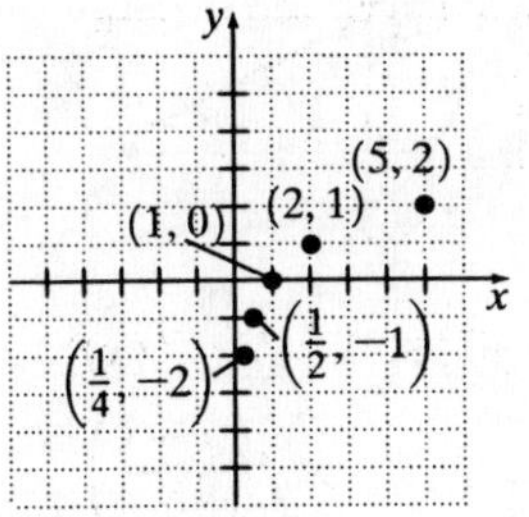

d.

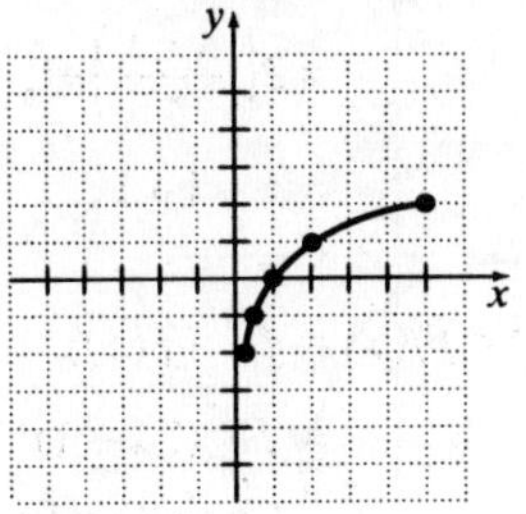

53. $f(x) = 3x + 1$ or $y = 3x + 1$
Find inverse

$$x = 3y + 1$$
$$3y = x - 1$$
$$y = \frac{x-1}{3}$$
$$f^{-1}(x) = \frac{x-1}{3}$$

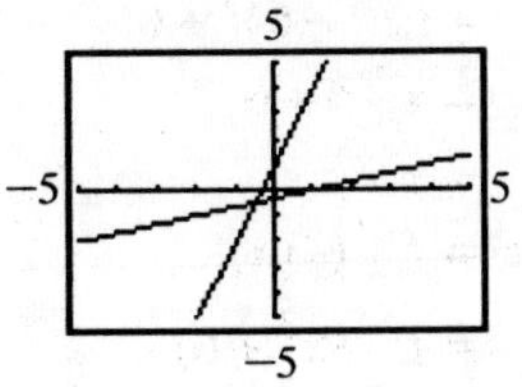

55. $f(x) = \sqrt[3]{x+1}$ or $y = \sqrt[3]{x+1}$
Find inverse

$$x = \sqrt[3]{y+1}$$
$$x^3 = y + 1$$
$$y = x^3 - 1$$
$$f^{-1}(x) = x^3 - 1$$

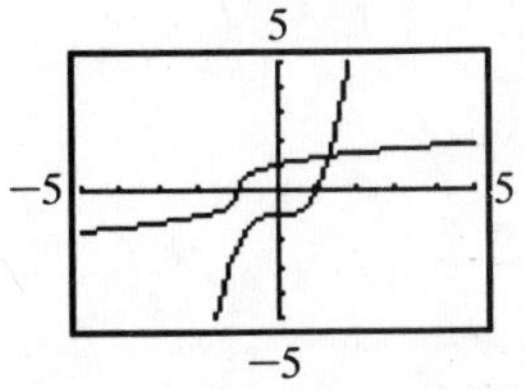

57. $25^{1/2} = \sqrt{25} = 5$

59. $16^{3/4} = (16^3)^{1/4} = (4096)^{1/4} = 8$

61. $9^{-3/2} = \frac{1}{9^{3/2}} = \frac{1}{\sqrt{9^3}} = \frac{1}{\sqrt{729}} = \frac{1}{27}$

63. $f(x) = 3^x$

$f(2) = 3^2 = 9$

65. $f(x) = 3^x$

$f\left(\frac{1}{2}\right) = 3^{1/2} = \sqrt{3} \approx 1.73$

Section 10.3

Graphing Calculator Explorations

1.

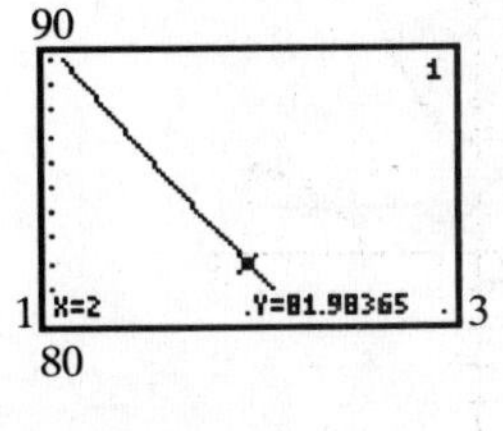

81.98%

3.

30

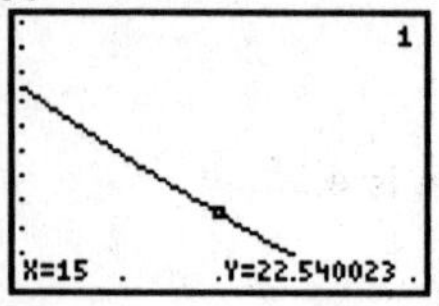

13 17

20

22.54%

7. $y=\left(\frac{1}{2}\right)^x-2$

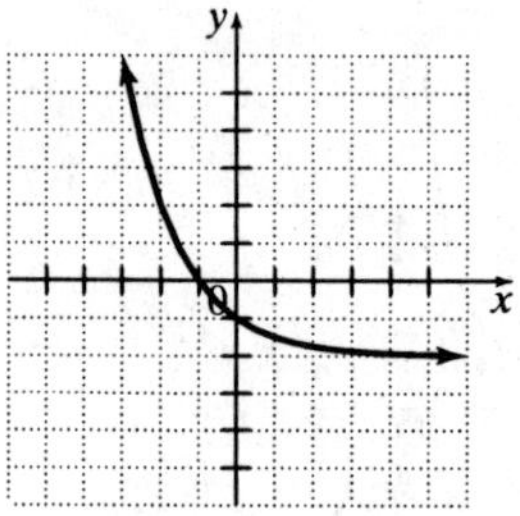

Exercise Set 10.3

1. $y=4^x$

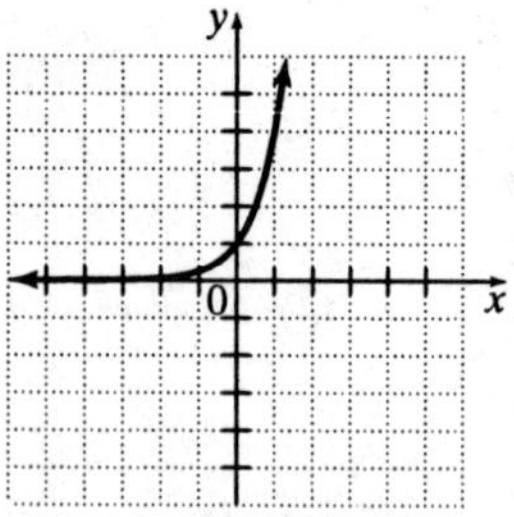

9. $y=-2^x$

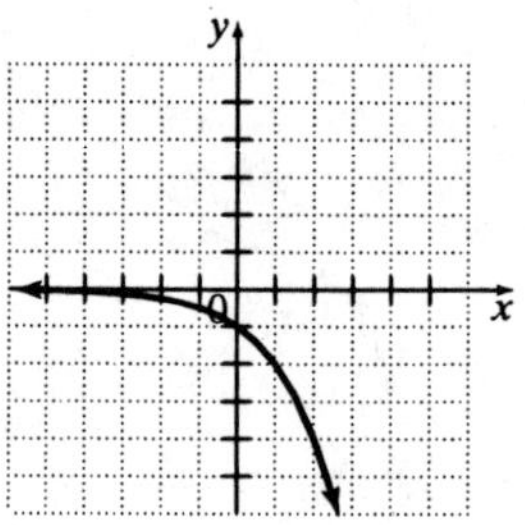

3. $y=1+2^x$ or $y=2^x+1$

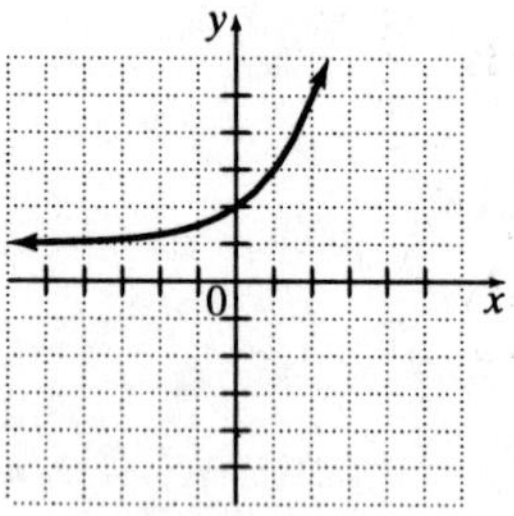

11. $y=-\left(\frac{1}{4}\right)^x$

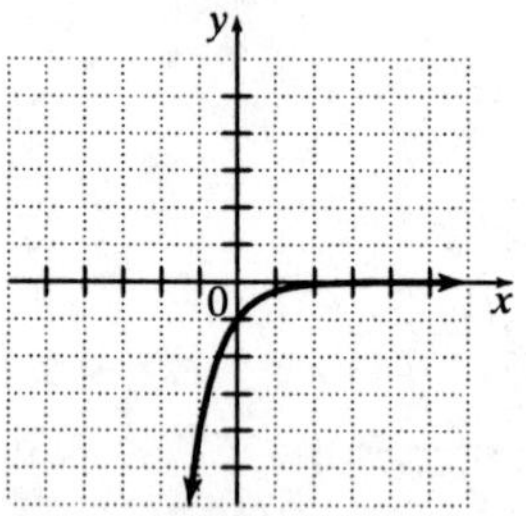

5. $y=\left(\frac{1}{4}\right)^x$

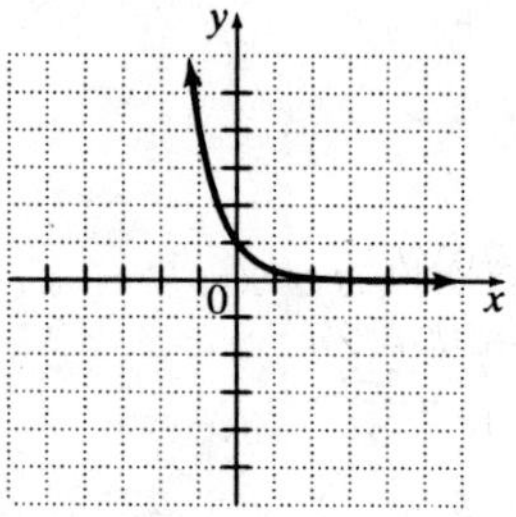

13. $f(x)=2^{x+1}$

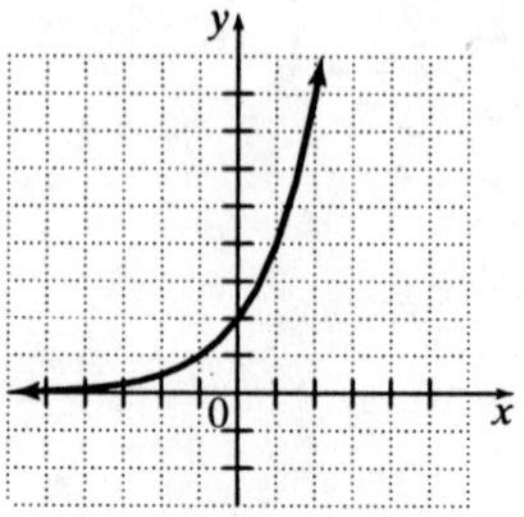

15. $f(x) = 4^{x-2}$

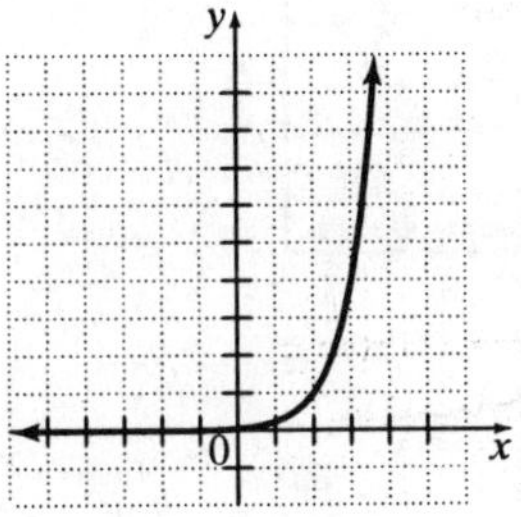

17. Answers may vary.

19. $3^x = 27$

$3^x = 3^3$

$x = 3$

The solution is 3.

21. $16^x = 8$

$(2^4)^x = 2^3$

$2^{4x} = 2^3$

$4x = 3$

$x = \frac{3}{4}$

The solution is $\frac{3}{4}$.

23. $32^{2x-3} = 2$

$(2^5)^{2x-3} = 2^1$

$10x - 15 = 1$

$10x = 16$

$x = \frac{8}{5}$

The solution is $\frac{8}{5}$.

25. $\frac{1}{4} = 2^{3x}$

$2^{-2} = 2^{3x}$

$3x = -2$

$x = -\frac{2}{3}$

The solution is $-\frac{2}{3}$.

27. $5^x = 625$

$5^x = 5^4$

$x = 4$

The solution is 4.

29. $4^x = 8$

$(2^2)^x = 2^3$

$2^{2x} = 2^3$

$2x = 3$

$x = \frac{3}{2}$

The solution is $\frac{3}{2}$.

31. $27^{x+1} = 9$

$(3^3)^{x+1} = 3^2$

$3^{3x+3} = 3^2$

$3x + 3 = 2$

$3x = -1$

$x = -\frac{1}{3}$

The solution is $-\frac{1}{3}$.

33. $81^{x-1} = 27^{2x}$

$(3^4)^{x-1} = (3^3)^{2x}$

$3^{4x-4} = 3^{6x}$

$4x - 4 = 6x$

$-4 = 2x$

$x = -2$

The solution is -2.

35. C

37. D

39. $y = 30(2.7)^{-0.004t}$, $t = 50$

$= 30(2.7)^{-(0.004)(50)}$

$= 30(2.7)^{-0.2}$

≈ 24.6

Therefore, approximately 24.6 pounds of uranium will remain after 50 days.

41. $y = 200(2.7)^{0.08t}, \ t = 12$
$= 200(2.7)^{0.08(12)}$
$= 200(2.7)^{0.96}$
≈ 519
There should be approximately 519 rats by next January.

43. $y = 5(2.7)^{-0.15t}, \ t = 10$
$= 5(2.7)^{-0.15(10)}$
$= 5(2.7)^{-1.5}$
≈ 1.1 grams

45. $y = 15{,}525{,}000(2.7)^{0.007t}, \ t = 6$
$y = 15{,}525{,}000(2.7)^{0.007(6)}$
$y = 16{,}186{,}349$
Approximately 16,190,000 residents.

47. $A = P\left(1 + \frac{r}{n}\right)^{nt}$
$t = 3$, $P = 6000$, $r = 0.08$, and $n = 12$.
Then
$A = 6000\left(1 + \frac{0.08}{12}\right)^{12(3)}$
$= 6000(1.006)^{36}$
≈ 7621.42
Erica would owe \$7621.42 after 3 years.

49. $A = P\left(1 + \frac{r}{n}\right)^{nt}$, where $P = 2000$,
$r = 0.06$, $n = 2$, and $t = 12$.
Then: $A = 2000\left(1 + \frac{0.06}{2}\right)^{2(12)}$
$= 2000(1.03)^{24}$
≈ 4065.59
Janina has approximately \$4065.59 in her savings account.

51.

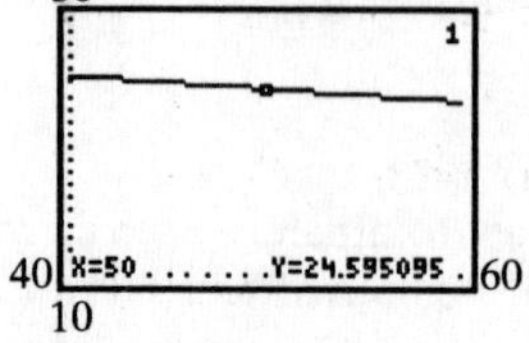

24.60 pounds

53.

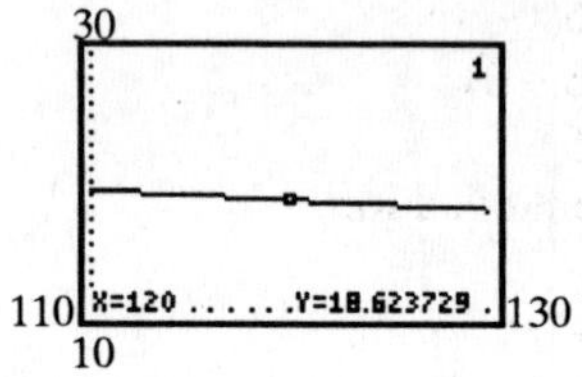

18.62 pounds

55.

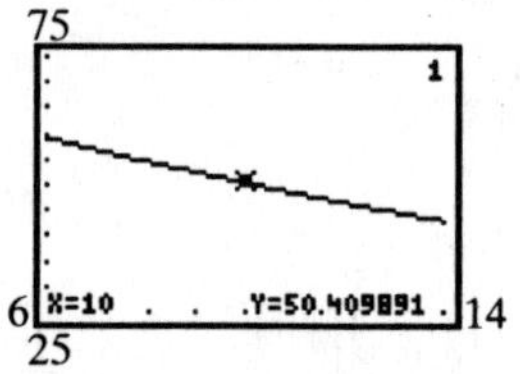

50.41 grams

57. $y = 5{,}926{,}466{,}814(2.7)^{0.0132t}, t = 7$
$y = 5{,}926{,}466{,}814(2.7)^{0.0132(7)}$
$y \approx 6{,}496{,}117{,}455$ people
Approximately 6,496,000 people.

59. $y = 358(1.004)^t$

a. $t = 2004 - 1994 = 10$
$y = 358(1.004)^{10}$
≈ 372.6 parts per million by volume

b. $t = 2025 - 1994 = 31$
$y = 358(1.004)^{31}$
≈ 405.2 parts per million by volume

61. $y = 6.052(1.378)^x$
$t = 2008 - 1991 = 17$
$y = 6.052(1.378)^{17} \approx 1410$
There will be approximately 1410 million cellular phone users in 2008.

63. $3x - 7 = 11$
$3x = 18$
$x = 6$

65. $2-6x=6(1-x)$
$2-6x=6-6x$
$2=6$
Inconsistent equation

67. $18=11x-x^2$
$x^2-11x+18=0$
$(x-2)(x-9)=0$
$x-2=0$ or $x-9=0$
$x=2$ or $x=9$

69. 2

71. 0

Exercise Set 10.4

1. $\log_6 36=2$
$6^2=36$

3. $\log_3 \frac{1}{27}=-3$
$3^{-3}=\frac{1}{27}$

5. $\log_{10} 1000=3$
$10^3=1000$

7. $\log_e x=4$
$e^4=x$

9. $\log_e \frac{1}{e^2}=-2$
$e^{-2}=\frac{1}{e^2}$

11. $\log_7 \sqrt{7}=\frac{1}{2}$
$7^{1/2}=\sqrt{7}$

13. $2^4=16$
$\log_2 16=4$

15. $10^2=100$
$\log_{10} 100=2$

17. $e^3=x$
$\log_e x=3$

19. $10^{-1}=\frac{1}{10}$
$\log_{10} \frac{1}{10}=-1$

21. $4^{-2}=\frac{1}{16}$
$\log_4 \frac{1}{16}=-2$

23. $5^{1/2}=\sqrt{5}$
$\log_5 \sqrt{5}=\frac{1}{2}$

25. $\log_2 8=3$ since $2^3=8$.

27. $\log_3 \frac{1}{9}=-2$ since $3^{-2}=\frac{1}{9}$.

29. $\log_{25} 5=\frac{1}{2}$ since $25^{1/2}=5$.

31. $\log_{1/2} 2=-1$ since $\left(\frac{1}{2}\right)^{-1}=2$.

33. $\log_7 1=0$

35. $\log_2 2^4=4$

37. $\log_{10} 100=2$

39. $3^{\log_3 5}=5$

41. $\log_3 81=4$

43. $\log_4\left(\frac{1}{64}\right)=-3$

45. Answers may vary.

47. $\log_3 9=x$
$2=x$

49. $\log_3 x = 4$

$x = 3^4 = 81$

51. $\log_x 49 = 2$

$x^2 = 49$

$x = \pm 7$

We discard the negative base.
$x = 7$

53. $\log_2 \frac{1}{8} = x$

$2^x = \frac{1}{8}$

$2^x = 2^{-3}$

$x = -3$

55. $\log_3\left(\frac{1}{27}\right) = x$

$\frac{1}{27} = 3^x$

$3^{-3} = 3^x$

$-3 = x$

57. $\log_8 x = \frac{1}{3}$

$x = 8^{1/3} = 2$

59. $\log_4 16 = x$

$4^x = 4^2$

$x = 2$

61. $\log_{3/4} x = 3$

$\left(\frac{3}{4}\right)^3 = x$

$x = \frac{3^3}{4^3} = \frac{27}{64}$

63. $\log_x 100 = 2$ or $x^2 = 100$
$x = \pm 10$ and we discard the negative base.
$x = 10$

65. $\log_5 5^3 = 3$

67. $2^{\log_2 3} = 3$

69. $\log_9 9 = 1$

71. $y = \log_3 x$
$y = 0$:
$\log_3 x = 0$

$x = 3^0 = 1$ is the only x - intercept.
$x = 0$: $y = \log 0$ which is not defined. No y-intercept exists.

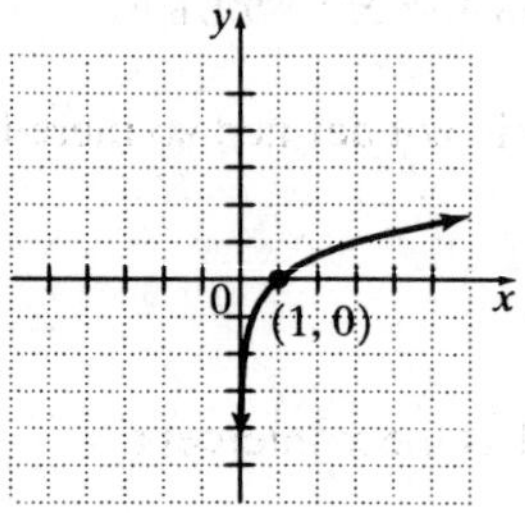

73. $f(x) = \log_{1/4} x$ or $y = \log_{1/4} x$
$y = 0$:
$0 = \log_{1/4} x$

$x = \left(\frac{1}{4}\right)^0 = 1$ is the x - intercept.

$x = 0$: $y = \log_{1/4} 0$ which is not defined.
No y-intercept exists.

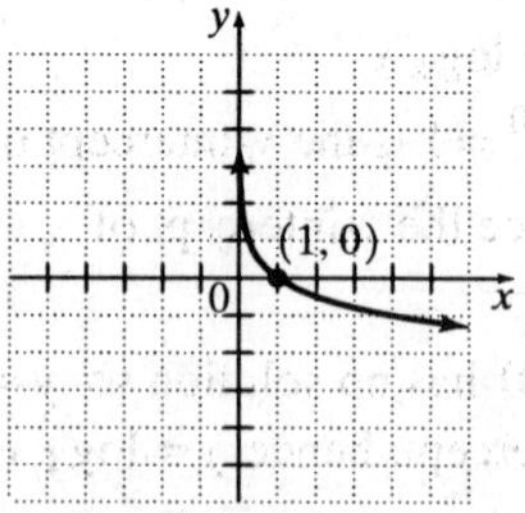

75. $f(x) = \log_5 x$ or $y = \log_5 x$
$x = 0$:
$y = \log_5 0$ is not defined so there is no y-intercept.
$y = 0$:
$0 = \log_5 x$

$x = 5^0 = 1$ is the x-intercept.

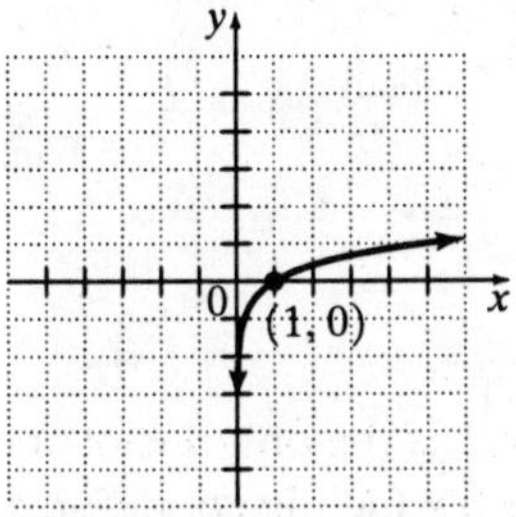

77. $f(x) = \log_{1/6} x$ or $y = \log_{1/6} x$

$x = 0$:

$y = \log_{1/6} 0$ is not defined so there is no y-intercept.

$y = 0$:

$0 = \log_{1/6} x$

$x = \left(\frac{1}{6}\right)^0 = 1$ is the x-intercept.

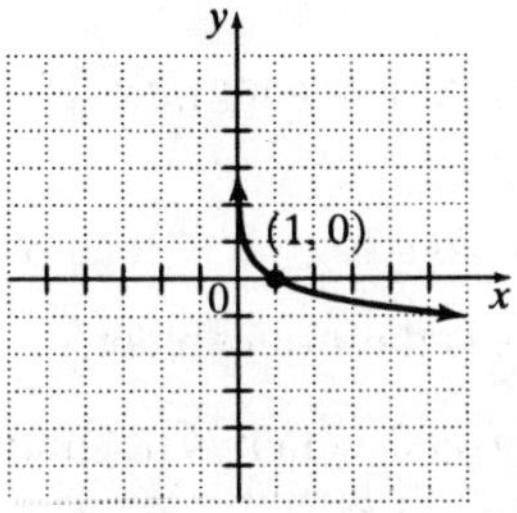

79. $y = 4^x$; $y = \log_4 x$

$x = 0$: $y = 4^0 = 1$ is the y-intercept of $y = 4^x$; hence the x-intercept of $y = \log_4 x$.

$y = 0$: $4^x = 0$ has no solution so $y = 4^x$ has no x-intercept; hence $y = \log_4 x$ has no y-intercept.

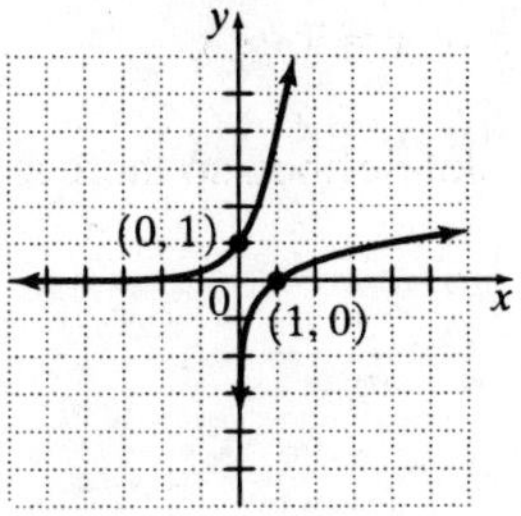

81. $y = \left(\frac{1}{3}\right)^x$; $y = \log_{1/3} x$

$x = 0$: $y = \left(\frac{1}{3}\right)^0 = 1$ is the y-intercept of $y = \left(\frac{1}{3}\right)^x$, and hence the x-intercept of $y = \log_{1/3} x$.

$y = 0$: $0 = \left(\frac{1}{3}\right)^x$ has no solution so $y = \left(\frac{1}{3}\right)^x$ has no x-intercept; hence $y = \log_{1/3} x$ has no y-intercept.

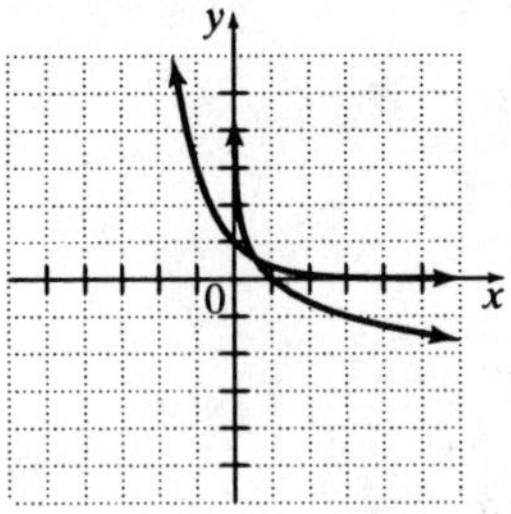

83. $\log_{10}(1-k) = \frac{-0.3}{H}$, $H = 8$

$$\log_{10}(1-k) = \frac{-0.3}{8} = -0.0375$$

$$1 - k = 10^{-0.0375}$$

$$1 - 10^{-0.0375} = k$$

$$k \approx 0.0827$$

85. $\log_3 10$ is between 2 and 3 because $3^2 = 9$ and $3^3 = 27$

87. $\frac{x-5}{5-x} = \frac{x-5}{-(x-5)} = \frac{1}{-1} = -1$

89. $\frac{x^2 - 3x - 10}{2 + x} = \frac{(x+2)(x-5)}{x+2} = x - 5$

91. $\frac{3x}{x+3} + \frac{9}{x+3} = \frac{3x+9}{x+3} = \frac{3(x+3)}{x+3} = 3$

93. $\frac{5}{y+1}-\frac{4}{y-1}=\frac{5(y-1)-4(y+1)}{(y+1)(y-1)}$
$=\frac{5y-5-4y-4}{y^2-1}$
$=\frac{y-9}{y^2-1}$

Exercise Set 10.5

1. $\log_5 2+\log_5 7=\log_5(2\cdot 7)=\log_5 14$

3. $\log_4 9+\log_4 x=\log_4 9x$

5. $\log_{10}5+\log_{10}2+\log_{10}(x^2+2)$
$=\log_{10}[5\cdot 2(x^2+2)]$
$=\log_{10}(10x^2+20)$

7. $\log_5 12-\log_5 4=\log_5\left(\frac{12}{4}\right)=\log_5 3$

9. $\log_2 x-\log_2 y=\log_2\frac{x}{y}$

11. $\log_4 2+\log_4 10-\log_4 5$
$=\log_4 2\cdot 10-\log_4 5$
$=\log_4\left(\frac{20}{5}\right)$
$=\log_4 4=1$

13. $\log_3 x^2=2\log_3 x$

15. $\log_4 5^{-1}=-\log_4 5$

17. $\log_5\sqrt{y}=\log_5 y^{1/2}=\frac{1}{2}\log_5 y$

19. $2\log_2 5=\log_2 5^2=\log_2 25$

21. $3\log_5 x+6\log_5 z$
$=\log_5 x^3+\log_5 z^6$
$=\log_5 x^3z^6$

23. $\log_{10}x-\log_{10}(x+1)+\log_{10}(x^2-2)$
$=\log_{10}\frac{x}{x+1}+\log_{10}(x^2-2)$
$=\log_{10}\frac{x(x^2-2)}{x+1}$
$=\log_{10}\frac{x^3-2x}{x+1}$

25. $\log_4 5+\log_4 7=\log_4(5\cdot 7)=\log_4 35$

27. $\log_3 8-\log_3 2=\log_3\left(\frac{8}{2}\right)=\log_3 4$

29. $\log_7 6+\log_7 3-\log_7 4$
$=\log_7(6\cdot 3)-\log_7 4$
$=\log_7\left(\frac{18}{4}\right)$
$=\log_7\frac{9}{2}$

31. $3\log_4 2+\log_4 6=\log_4 2^3+\log_4 6$
$=\log_4 8+\log_4 6$
$=\log_4(8\cdot 6)$
$=\log_4 48$

33. $3\log_2 x+\frac{1}{2}\log_2 x-2\log_2(x+1)$
$=\log_2 x^3+\log_2 x^{1/2}-\log_2(x+1)^2$
$=\log_2(x^3\cdot x^{1/2})-\log_2(x+1)^2$
$=\log_2 x^{7/2}-\log_2(x+1)^2$
$=\log_2\frac{x^{7/2}}{(x+1)^2}$

35. $2\log_8 x-\frac{2}{3}\log_8 x+4\log_8 x$
$=\left(2-\frac{2}{3}+4\right)\log_8 x$
$=\frac{16}{3}\log_8 x$
$=\log_8 x^{16/3}$

37. $\log_2 \frac{7\cdot 11}{3} = \log_2(7\cdot 11) - \log_2 3$
$= \log_2 7 + \log_2 11 - \log_2 3$

39. $\log_3\left(\frac{4y}{5}\right) = \log_3 4y - \log_3 5$
$= \log_3 4 + \log_3 y - \log_3 5$

41. $\log_2\left(\frac{x^3}{y}\right) = \log_2 x^3 - \log_2 y$
$= 3\log_2 x - \log_2 y$

43. $\log_b \sqrt{7x} = \log_b (7x)^{1/2}$
$= \frac{1}{2}\log_b(7x)$
$= \frac{1}{2}[\log_b 7 + \log_b x]$
$= \frac{1}{2}\log_b 7 + \frac{1}{2}\log_b x$

45. $\log_7\left(\frac{5x}{4}\right) = \log_7 5x - \log_7 4$
$= \log_7 5 + \log_7 x - \log_7 4$

47. $\log_5 x^3(x+1) = \log_5 x^3 + \log_5(x+1)$
$= 3\log_5 x + \log_5(x+1)$

49. $\log_6 \frac{x^2}{x+3} = \log_6 x^2 - \log_6(x+3)$
$= 2\log_6 x - \log_6(x+3)$

51. $\log_b\left(\frac{5}{3}\right) = \log_b 5 - \log_b 3$
$= 0.7 - 0.5 = 0.2$

53. $\log_b 15 = \log_b(5\cdot 3)$
$= \log_b 5 + \log_b 3$
$= 0.7 + 0.5 = 1.2$

55. $\log_b \sqrt[3]{5} = \log_b 5^{1/3}$
$= \frac{1}{3}\log_b 5$
$= \frac{1}{3}(0.7)$
≈ 0.233

57. $\log_2 x^3 = 3\log_2 x$
True

59. $\frac{\log_7 10}{\log_7 5} = \log_7 2$
False

61. $\frac{\log_7 x}{\log_7 y} = (\log_7 x) - (\log_7 y)$
False

63. $\log_b 8 = \log_b 2^3$
$= 3\log_b 2$
$= 3(0.43)$
$= 1.29$

65. $\log_b\left(\frac{3}{9}\right) = \log_b\left(\frac{1}{3}\right)$
$= \log_b 3^{-1}$
$= (-1)\log_b 3$
$= -(0.68)$
$= -0.68$

67. $\log_b \sqrt{\frac{2}{3}} = \log_b\left(\frac{2}{3}\right)^{1/2}$
$= \frac{1}{2}\log_b \frac{2}{3}$
$= \frac{1}{2}(\log_b 2 - \log_b 3)$
$= \frac{1}{2}(0.43 - 0.68)$
$= \frac{1}{2}(-0.25)$
$= -0.125$

69. $y = 10^x$ and $y = \log_{10} x$

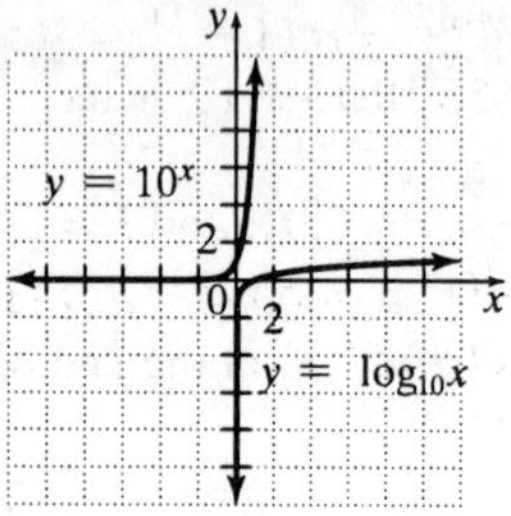

71. $\log_{10} \frac{1}{10} = x$

$10^x = \frac{1}{10}$

$10^{-1} = \frac{1}{10}$

$x = -1$

73. $\log_7 \sqrt{7} = x$

$7^x = \sqrt{7}$

$7^x = 7^{1/2}$

$x = \frac{1}{2}$

Exercise Set 10.6

1. $\log 8 \approx 0.9031$

3. $\log 2.31 \approx 0.3636$

5. $\ln 2 \approx 0.6931$

7. $\ln 0.0716 \approx -2.6367$

9. $\log 12.6 \approx 1.1004$

11. $\ln 5 \approx 1.6094$

13. $\log 41.5 \approx 1.6180$

15. Answers may vary.

17. $\log 100 = \log 10^2 = 2$

19. $\log \frac{1}{1000} = \log 10^{-3} = -3$

21. $\ln e^2 = 2$

23. $\ln \sqrt[4]{e} = \ln e^{1/4} = \frac{1}{4}$

25. $\log 10^3 = 3$

27. $\ln e^2 = 2$

29. $\log 0.0001 = \log 10^{-4} = -4$

31. $\ln \sqrt{e} = \ln e^{1/2} = \frac{1}{2}$

33. ln 50 is larger
Answers may vary.

35. $\log x = 1.3$

$x = 10^{1.3} \approx 19.9526$

37. $\log 2x = 1.1$

$2x = 10^{1.1}$

$x = \frac{10^{1.1}}{2} \approx 6.2946$

39. $\ln x = 1.4$

$x = e^{1.4} \approx 4.0552$

41. $\ln(3x - 4) = 2.3$

$3x - 4 = e^{2.3}$

$3x = 4 + e^{2.3}$

$x = \frac{4 + e^{2.3}}{3} \approx 4.6581$

43. $\log x = 2.3$

$x = 10^{2.3} \approx 199.5262$

45. $\ln x = -2.3$

$x = e^{-2.3} \approx 0.1003$

47. $\log(2x+1) = -0.5$
$$2x+1 = 10^{-0.5}$$
$$2x = 10^{-0.5} - 1$$
$$x = \frac{10^{-0.5}-1}{2} \approx 0.3419$$

49. $\ln 4x = 0.18$
$$4x = e^{0.18}$$
$$x = \frac{e^{0.18}}{4} \approx 0.2993$$

51. $\log_2 3 = \frac{\ln 3}{\ln 2} \approx 1.5850$

53. $\log_{1/2} 5 = \frac{\ln 5}{\ln\left(\frac{1}{2}\right)} \approx -2.3219$

55. $\log_4 9 = \frac{\ln 9}{\ln 4} \approx 1.5850$

57. $\log_3\left(\frac{1}{6}\right) = \log_3 6^{-1}$
$$= (-1)\log_3 6$$
$$= -\frac{\ln 6}{\ln 3} \approx -1.6309$$

59. $\log_8 6 = \frac{\ln 6}{\ln 8} \approx 0.8617$

61. $R = \log\left(\frac{a}{T}\right) + B$, $a = 200$, $T = 1.6$, and $B = 2.1$
$$R = \log\left(\frac{200}{1.6}\right) + 2.1 \approx 4.2$$
The earthquake measures 4.2 on the Richter scale.

63. $R = \log\left(\frac{a}{T}\right) + B$, $a = 400$, $T = 2.6$, and $B = 3.1$
$$R = \log\left(\frac{400}{2.6}\right) + 3.1 \approx 5.3$$
The earthquake measures 5.3 on the Richter scale.

65. $A = Pe^{rt}$, $t = 12$, $P = 1400$, and $r = 0.08$
$$A = 1400e^{(0.08)12} = 1400e^{0.96} \approx 3656.38$$
Dana has \$3656.38 after 12 years.

67. $A = Pe^{rt}$, $t = 4$, $r = 0.06$, and $P = 2000$
$$A = 2000e^{(0.06)4} = 2000e^{0.24} \approx 2542.50$$
Barbara owes \$2542.50 at the end of 4 years.

69. $f(x) = e^x$ or $y = e^x$

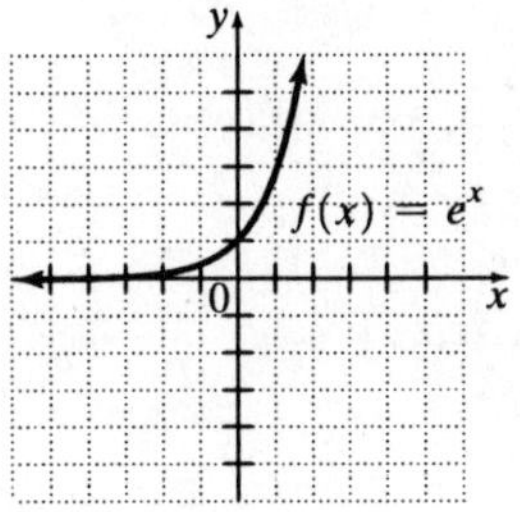

71. $f(x) = e^{-3x}$ or $y = e^{-3x}$

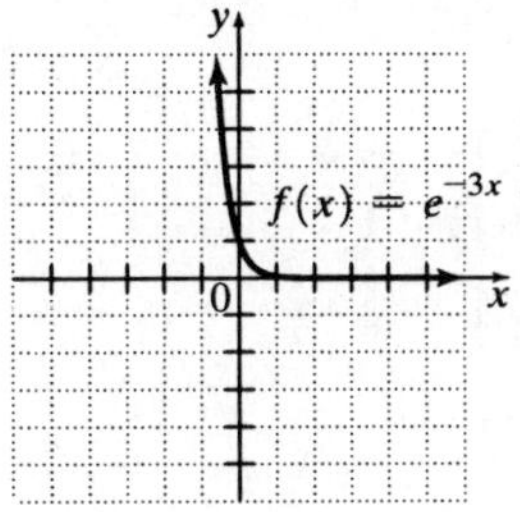

73. $f(x) = e^x + 2$ or $y = e^x + 2$

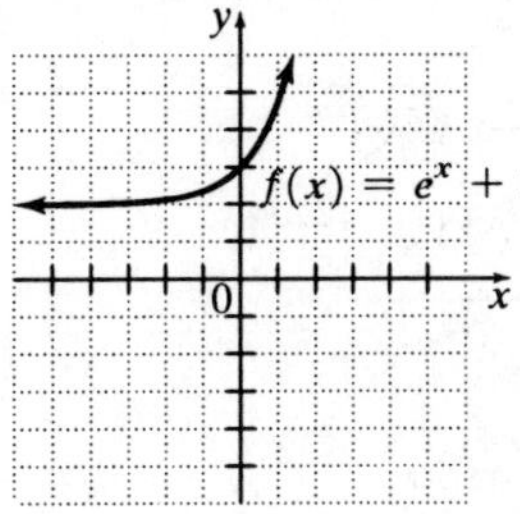

75. $f(x) = e^{x-1}$ or $y = e^{x-1}$

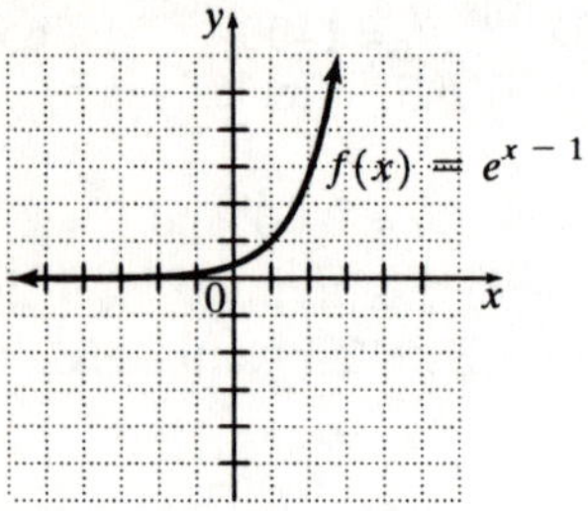

77. $f(x) = 3e^x$ or $y = 3e^x$

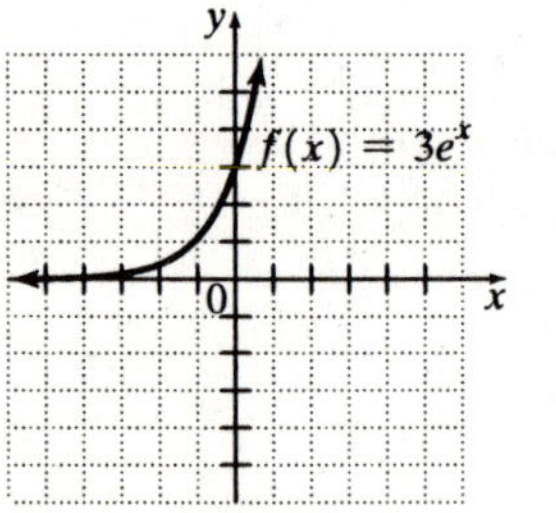

79. $f(x) = \ln x$ or $y = \ln x$

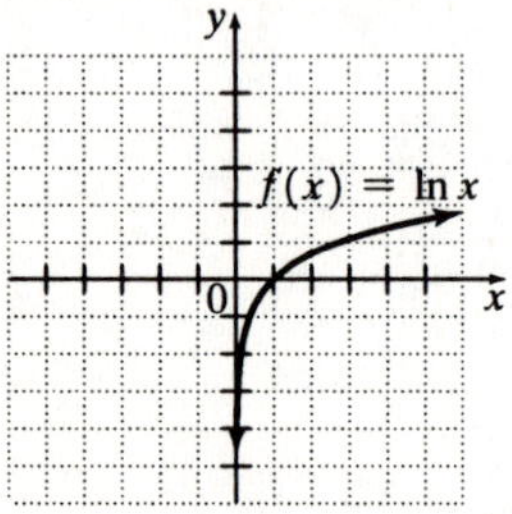

81. $f(x) = -2\log x$ or $y = -2\log x$

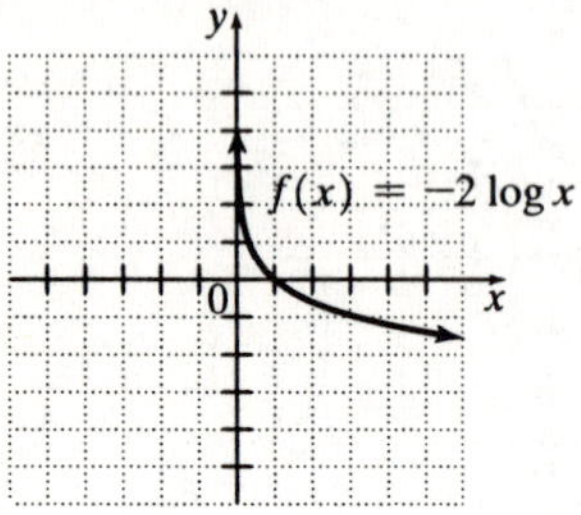

83. $f(x) = \log(x+2)$ or $y = \log(x+2)$

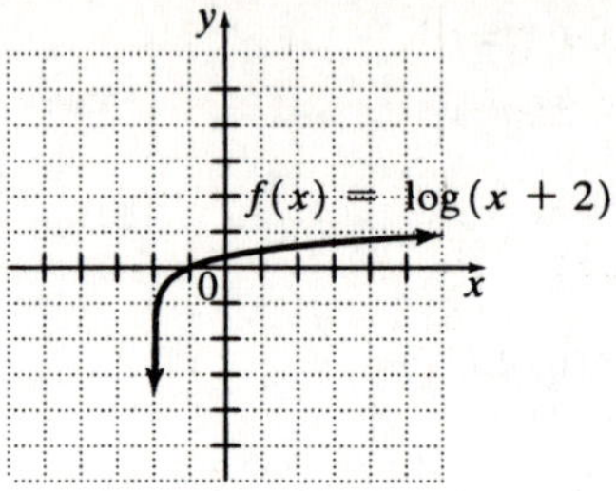

85. $f(x) = \ln x - 3$ or $y = \ln x - 3$

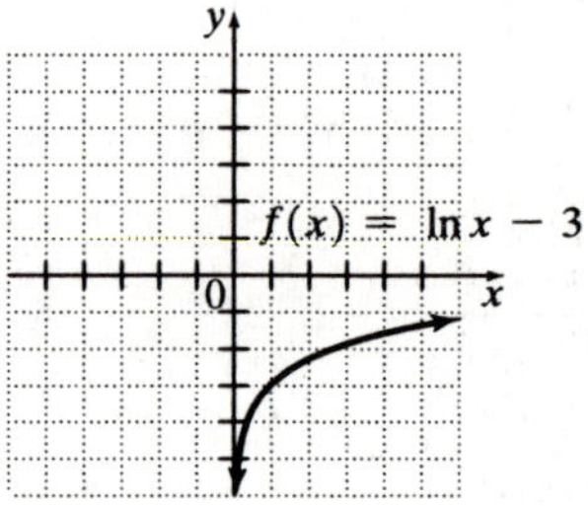

87. $f(x) = e^x$

$f(x) = e^x + 2$

$f(x) = e^x - 3$

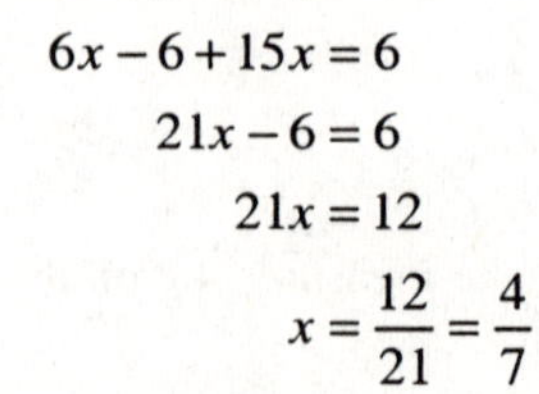

89. $6x - 3(2 - 5x) = 6$

$$6x - 6 + 15x = 6$$
$$21x - 6 = 6$$
$$21x = 12$$
$$x = \frac{12}{21} = \frac{4}{7}$$

91. $2x + 3y = 6x$

$$3y = 4x$$
$$x = \frac{3y}{4}$$

93. $x^2+7x=-6$

$x^2+7x+6=0$

$(x+6)(x+1)=0$

$x+6=0$ or $x+1=0$

$x=-6$ or $x=-1$

95. $\begin{cases} x+2y=-4 \\ 3x-y=9 \end{cases} \rightarrow \begin{cases} x+2y=-4 \\ 6x-2y=18 \end{cases}$

Add.

$7x=14$

$x=2$

Substitute back.

$2+2y=-4$

$2y=-6$

$y=-3$

$(2, -3)$

Section 10.7

Graphing Calculator Explorations

1. $Y_1=5000\left(1+\frac{0.05}{4}\right)^{4x}$

$Y_2=6000$

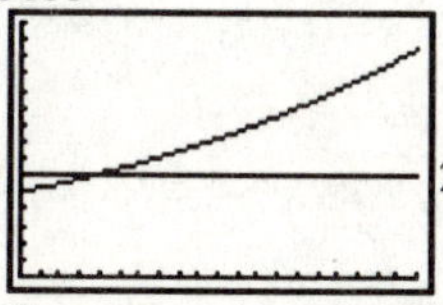

It takes 3.67 years, or 3 years and 8 months.

3. $Y_1=10000\left(1+\frac{0.06}{12}\right)^{12x}$

$Y_2=40000$

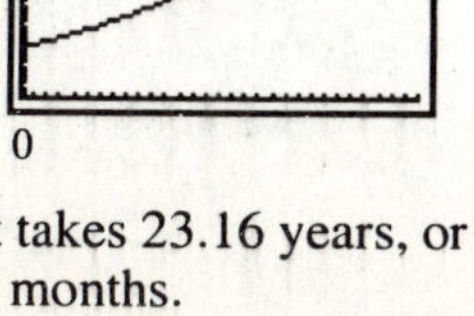

It takes 23.16 years, or 23 years and 2 months.

Exercise Set 10.7

1. $3^x=6$

$x=\log_3 6=\frac{\log 6}{\log 3}\approx 1.6309$

3. $3^{2x}=3.8$

$2x=\log_3 3.8=\frac{\log 3.8}{\log 3}$ so

$x=\frac{\log 3.8}{2\log 3}\approx 0.6076$

5. $2^{x-3}=5$

$x-3=\log_2 5$

$x=3+\log_2 5=3+\frac{\log 5}{\log 2}\approx 5.3219$

7. $9^x=5$

$x=\log_9 5=\frac{\log 5}{\log 9}\approx 0.7325$

9. $4^{x+7}=3$

$x+7=\log_4 3$

$x=-7+\log_4 3$

$=-7+\frac{\log 3}{\log 4}$

≈ -6.2075

11. $7^{3x-4}=11$

$3x-4=\log_7 11$

$3x=4+\log_7 11$

$x=\frac{1}{3}\left(4+\frac{\log 11}{\log 7}\right)\approx 1.7441$

13. $e^{6x}=5$

$6x=\ln 5$

$x=\frac{\ln 5}{6}\approx 0.2682$

15. $\log_2(x+5)=4$

$x+5=2^4$

$x+5=16$

$x=11$

17. $\log_3 x^2 = 4$
$x^2 = 3^4$
$x^2 = 81$
$x = \pm 9$

19. $\log_4 2 + \log_4 x = 0$
$\log_4(2x) = 0$
$2x = 4^0$
$2x = 1$
$x = \frac{1}{2}$

21. $\log_2 6 - \log_2 x = 3$
$\log_2\left(\frac{6}{x}\right) = 3$
$\frac{6}{x} = 2^3$
$\frac{6}{x} = 8$
$8x = 6$
$x = \frac{6}{8} = \frac{3}{4}$

23. $\log_4 x + \log_4(x+6) = 2$
$\log_4 x(x+6) = 2$
$x(x+6) = 4^2$
$x^2 + 6x = 16$
$x^2 + 6x - 16 = 0$
$(x+8)(x-2) = 0$
$x + 8 = 0$ or $x - 2 = 0$
$x = -8$ or $x = 2$
We discard –8 as extraneous.

25. $\log_5(x+3) - \log_5 x = 2$
$\log_5\left(\frac{x+3}{x}\right) = 2$
$\frac{x+3}{x} = 5^2$
$\frac{x+3}{x} = 25$
$x + 3 = 25x$
$3 = 24x$
$x = \frac{3}{24} = \frac{1}{8}$

27. $\log_3(x-2) = 2$
$x - 2 = 3^2$
$x - 2 = 9$
$x = 11$

29. $\log_4(x^2 - 3x) = 1$
$x^2 - 3x = 4^1$
$x^2 - 3x = 4$
$x^2 - 3x - 4 = 0$
$(x-4)(x+1) = 0$
$x - 4 = 0$ or $x + 1 = 0$
$x = 4$ or $x = -1$

31. $\ln 5 + \ln x = 0$
$\ln(5x) = 0$
$e^0 = 5x$
$1 = 5x$
$\frac{1}{5} = x$

33. $3\log x - \log x^2 = 2$
$3\log x - 2\log x = 2$
$\log x = 2$
$x = 10^2$
$x = 100$

35. $\log_2 x + \log_2(x+5) = 1$

$\log_2 x(x+5) = 1$

$x(x+5) = 2^1$

$x^2 + 5x - 2 = 0$

$a = 1$, $b = 5$, and $c = -2$

$$x = \frac{-5 \pm \sqrt{5^2 - 4(1)(-2)}}{2(1)} = \frac{-5 \pm \sqrt{25+8}}{2} = \frac{-5 \pm \sqrt{33}}{2}$$

Discard $\frac{-5 - \sqrt{33}}{2}$

37. $\log_4 x - \log_4(2x-3) = 3$

$\log_4\left(\frac{x}{2x-3}\right) = 3$

$\frac{x}{2x-3} = 4^3$

$\frac{x}{2x-3} = 64$

$x = 64(2x-3)$

$x = 128x - 192$

$192 = 127x$

$x = \frac{192}{127}$

39. $\log_2 x + \log_2(3x+1) = 1$

$\log_2 x(3x+1) = 1$

$x(3x+1) = 2^1$

$3x^2 + x = 2$

$3x^2 + x - 2 = 0$

$(3x-2)(x+1) = 0$

$3x - 2 = 0$ or $x + 1 = 0$

$3x = 2$ or $x = -1$

$x = \frac{2}{3}$

We discard −1 as extraneous.

41. $y = y_0 e^{0.043t}$, $y_0 = 83$ and $t = 5$

$y = 83e^{0.043(5)} = 83e^{0.215} \approx 103$

There should be 103 wolves in 5 years.

43. $y = y_0 e^{0.026t}$, $y_0 = 10{,}052{,}000$ and $t = 6$

$y = 10{,}052{,}000 e^{0.026(6)} \approx 11{,}750{,}000$

There will be approximately 11,750,000 inhabitants in 2005.

45. $y = y_0 e^{-0.005t}$,

$y_0 = 146{,}394$, and $y = 120{,}000$

$120{,}000 = 146{,}394 e^{-0.005t}$

$\frac{120{,}000}{146{,}394} = e^{-0.005t}$

$$t = \frac{\ln\left(\frac{120{,}000}{146{,}394}\right)}{-0.005} \approx 39.8$$

It will take approximately 39.8 years to reach 120,000 thousand.

47. $A = P\left(1 + \frac{r}{n}\right)^{nt}$, $P = 600$,

$A = 2(600) = 1200$, $r = 0.07$, and $n = 12$

$1200 = 600\left(1 + \frac{0.07}{12}\right)^{12t}$

$2 = (1.0058\overline{3})^{12t}$

$12t = \log_{1.0058\overline{3}}(2)$

$$t = \frac{1}{12}\log_{1.0058\overline{3}}(2) = \frac{1}{12}\frac{\ln 2}{\ln(1.0058\overline{3})} \approx 10$$

It would take approximately 10 years for the \$600 to double.

49. $A = P\left(1 + \frac{r}{n}\right)^{nt}$, $P = 1200$,

$A = P + I = 1200 + 200 = 1400$,

$r = 0.09$, and $n = 4$

$$1400 = 1200\left(1+\frac{0.09}{4}\right)^{4t}$$
$$\frac{7}{6} = (1.0225)^{4t}$$
$$4t = \log_{1.0225}\left(\frac{7}{6}\right)$$
$$t = \frac{1}{4}\log_{1.0225}\left(\frac{7}{6}\right)$$
$$= \frac{1}{4}\frac{\ln\left(\frac{7}{6}\right)}{\ln 1.0225} \approx 1.73$$

It would take the investment approximately 1.7 years to earn $200.

51. $A = P\left(1+\frac{r}{n}\right)^{nt}$, $P = 1000$, $A = 2(1000) = 2000$, $r = 0.08$, and $n = 2$.
$$2000 = 1000\left(1+\frac{0.08}{2}\right)^{2t}$$
$$2 = (1.04)^{2t}$$
$$2t = \log_{1.04} 2 = \frac{\ln 2}{\ln 1.04}$$
$$t = \frac{1}{2}\frac{\ln 2}{\ln 1.04} \approx 8.8$$

It would take approximately 8.8 years for the $1000 to double.

53. $w = 0.00185h^{2.67}$, and $h = 35$
$$w = 0.00185(35)^{2.67} \approx 24.5$$

The expected weight of a boy 35 inches tall is 24.5 pounds.

55. $w = 0.00185h^{2.67}$, $w = 85$
$$85 = 0.00185h^{2.67}$$
$$\frac{85}{0.00185} = h^{2.67}$$
$$h = \left(\frac{85}{0.00185}\right)^{1/2.67} \approx 55.7$$

The expected height of the boy is approximately 55.7 inches.

57. $P = 14.7e^{-0.21x}$, $x = 1$
$$= 14.7e^{-0.21(1)}$$
$$= 14.7e^{-0.21}$$
$$\approx 11.9$$

The average atmospheric pressure of Denver is approximately 11.9 pounds per square inch.

59. $P = 14.7e^{-0.21x}$, $P = 7.5$
$$7.5 = 14.7e^{-0.21x}$$
$$\frac{7.5}{14.7} = e^{-0.21x}$$
$$-0.21x = \ln\left(\frac{7.5}{14.7}\right)$$
$$x = -\frac{1}{0.21}\ln\left(\frac{7.5}{14.7}\right) \approx 3.2$$

The elevation of the jet is approximately 3.2 miles.

61. $t = \frac{1}{c}\ln\left(\frac{A}{A-N}\right)$
$$t = \frac{1}{0.09}\ln\left(\frac{75}{75-50}\right)$$
$$t = \frac{1}{0.09}\ln(3)$$
$$t \approx 12.21$$

It will take 12 weeks.

63. $t = \frac{1}{c}\ln\left(\frac{A}{A-N}\right)$
$$t = \frac{1}{0.07}\ln\left(\frac{210}{210-150}\right)$$
$$t = \frac{1}{0.07}\ln(3.5)$$
$$t \approx 17.9$$

It will take 18 weeks.

65. $Y_1 = e^{0.3x}$ and $Y_2 = 8$

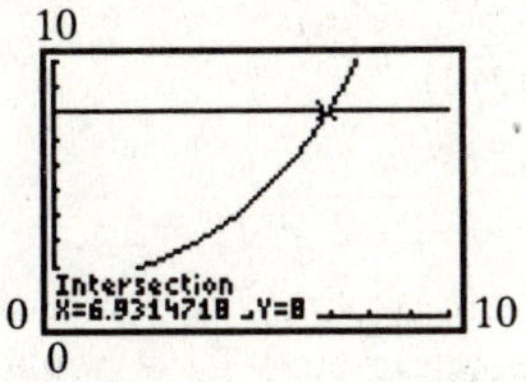

$x \approx 6.93$

67. $Y_1 = 2\log(-5.6x + 1.3)$
$Y_2 = -x - 1$

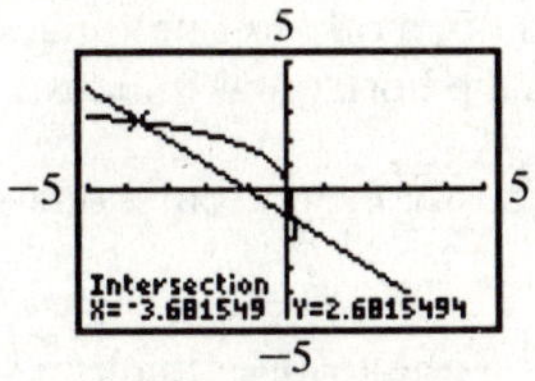

$x \approx -3.68$ or $x \approx 0.19$

69. $Y_1 = 7^{3x-4}$
$Y_2 = 11$

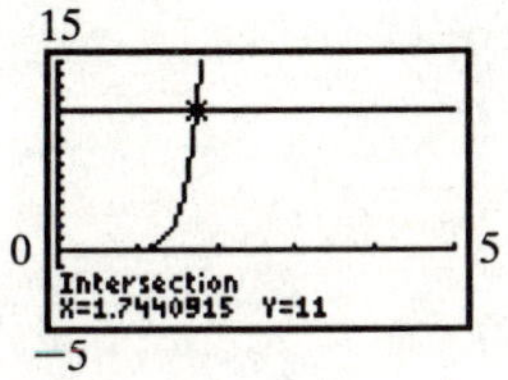

$x \approx 1.74$

71. $Y_1 = \ln 5 + \ln x$
$Y_2 = 0$

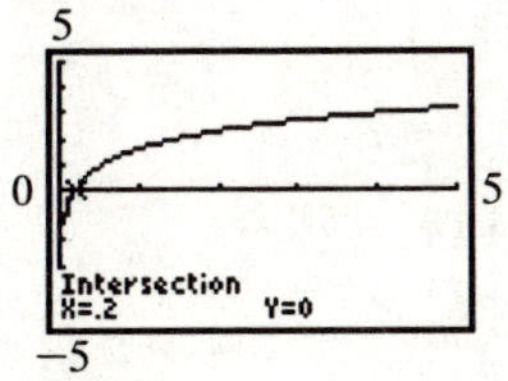

$x = 0.2$

73. $\frac{x^2 - y + 2z}{3x} = \frac{(-2)^2 - (0) + 2(3)}{3(-2)}$
$= \frac{4+6}{-6}$
$= \frac{-10}{6}$
$= -\frac{5}{3}$

75. $\frac{3z - 4x + y}{x + 2z} = \frac{3(3) - 4(-2) + 0}{-2 + 2(3)}$
$= \frac{9+8}{-2+6}$
$= \frac{17}{4}$

77. $f(x) = 5x + 2$ so $f^{-1}(x) = \frac{x-2}{5}$

Chapter 10 Review

1. $(f+g)(x) = f(x) + g(x)$
$= (x-5) + (2x+1)$
$= x - 5 + 2x + 1$
$= 3x - 4$

2. $(f-g)(x) = f(x) - g(x)$
$= (x-5) - (2x+1)$
$= x - 5 - 2x - 1$
$= -x - 6$

3. $(f \cdot g)(x) = f(x) \cdot g(x)$
$= (x-5)(2x+1)$
$= 2x^2 + x - 10x - 5$
$= 2x^2 - 9x - 5$

4. $\left(\frac{g}{f}\right)(x) = \frac{g(x)}{f(x)} = \frac{2x+1}{x-5},\ x \neq 5$

5. $(f \circ g)(x) = f(g(x))$
$= f(x+1)$
$= (x+1)^2 - 2$
$= x^2 + 2x + 1 - 2$
$= x^2 + 2x - 1$

6. $(g \circ f)(x) = g(f(x))$
$= g(x^2 - 2)$
$= x^2 - 2 + 1$
$= x^2 - 1$

7. $(h \circ g)(2) = h(g(2))$
$= h(3)$
$= 3^3 - 3^2$
$= 18$

8. $(f \circ f)(x) = f(f(x))$
$= f(x^2 - 2)$
$= (x^2 - 2)^2 - 2$
$= x^4 - 4x^2 + 4 - 2$
$= x^4 - 4x^2 + 2$

9. $(f \circ g)(-1) = f(g(-1))$
$= f(0)$
$= 0^2 - 2$
$= -2$

10. $(h \circ h)(2) = h(h(2))$
$= h(4)$
$= 4^3 - 4^2$
$= 48$

11. The function is one-to-one.
h^{-1}
$= \{(14, -9), (8, 6), (12, -11), (15, 15)\}$

12. The function is not one-to-one.

13. The function is one-to-one.

Rank in Automobile Thefts (Input)	2	4	1	3
U.S. Region (Output)	West	Midwest	South	Northeast

14. The function is not one-to-one.

15. $f(x) = \sqrt{x+2}$

a. $f(7) = \sqrt{7+2} = \sqrt{9} = 3$

b. $f^{-1}(3) = 7$

16. $f(x) = \sqrt{x+2}$

a. $f(-1) = \sqrt{-1+2} = \sqrt{1} = 1$

b. $f^{-1}(1) = -1$

17. The graph does not represent a one-to-one function.

18. The graph is not a one-to-one function since it fails the horizontal line test.

19. The graph does not represent a one-to-one function.

20. The graph is a one-to-one function since it passes the horizontal line test.

21. $f(x) = x - 9$
$y = x - 9$
$x = y - 9$
$y = x + 9$
$f^{-1}(x) = x + 9$

22. $f(x) = x + 8$
$y = x + 8$
$x = y + 8$
$y = x - 8$
$f^{-1}(x) = x - 8$

23. $f(x) = 6x + 11$
$y = 6x + 11$
$x = 6y + 11$
$6y = x - 11$
$y = \frac{x - 11}{6}$
$f^{-1}(x) = \frac{x - 11}{6}$

24.
$$f(x) = 12x$$
$$y = 12x$$
$$x = 12y$$
$$y = \frac{x}{12}$$
$$f^{-1}(x) = \frac{x}{12}$$

25.
$$f(x) = x^3 - 5$$
$$y = x^3 - 5$$
$$x = y^3 - 5$$
$$y^3 = x + 5$$
$$y = \sqrt[3]{x+5}$$
$$f^{-1}(x) = \sqrt[3]{x+5}$$

26.
$$f(x) = \sqrt[3]{x+2}$$
$$y = \sqrt[3]{x+2}$$
$$x = \sqrt[3]{y+2}$$
$$x^3 = y + 2$$
$$y = x^3 - 2$$
$$f^{-1}(x) = x^3 - 2$$

27.
$$g(x) = \frac{12x - 7}{6}$$
$$y = \frac{12x - 7}{6}$$
$$x = \frac{12y - 7}{6}$$
$$6x = 12y - 7$$
$$6x + 7 = 12y$$
$$\frac{6x + 7}{12} = y$$
$$g^{-1}(x) = \frac{6x + 7}{12}$$

28.
$$r(x) = \frac{13}{2}x - 4$$
$$y = \frac{13}{2}x - 4$$
$$x = \frac{13}{2}y - 4$$
$$x + 4 = \frac{13}{2}y$$
$$2(x + 4) = 13y$$
$$\frac{2(x + 4)}{13} = y$$
$$r^{-1}(x) = \frac{2(x + 4)}{13}$$

29.
$$y = g(x) = \sqrt{x}$$
$$x = \sqrt{y}$$
$$x^2 = y = g^{-1}(x), \quad x \geq 0$$

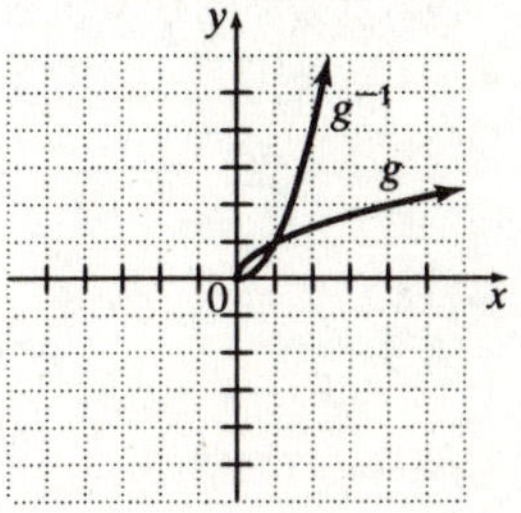

30.
$$h(x) = 5x - 5$$
$$y = 5x - 5$$
$$x = 5y - 5$$
$$5y = x + 5$$
$$y = \frac{x + 5}{5}$$
$$h^{-1}(x) = \frac{x + 5}{5}$$

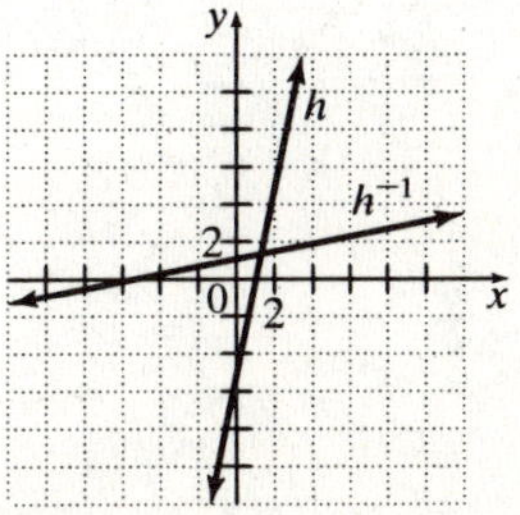

31. $f(x) = 2x - 3$ or $y = 2x - 3$

Find inverse

$x = 2y - 3$

$y = \frac{x+3}{2}$

$f^{-1}(x) = \frac{x+3}{2}$

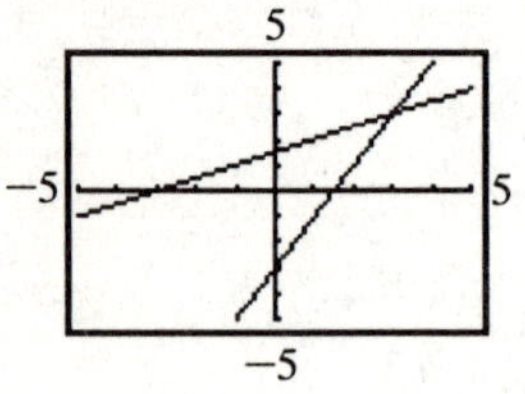

32. $4^x = 64$

$4^x = 4^3$

$x = 3$

33. $3^x = \frac{1}{9}$

$3^x = 3^{-2}$

$x = -2$

34. $2^{3x} = \frac{1}{16}$

$2^{3x} = 2^{-4}$

$3x = -4$

$x = -\frac{4}{3}$

35. $5^{2x} = 125$

$5^{2x} = 5^3$

$2x = 3$

$x = \frac{3}{2}$

36. $9^{x+1} = 243$

$(3^2)^{x+1} = 3^5$

$3^{2x+2} = 3^5$

$2x + 2 = 5$

$2x = 3$

$x = \frac{3}{2}$

37. $8^{3x-2} = 4$

$(2^3)^{3x-2} = 2^2$

$2^{9x-6} = 2^2$

$9x - 6 = 2$

$9x = 8$

$x = \frac{8}{9}$

38.

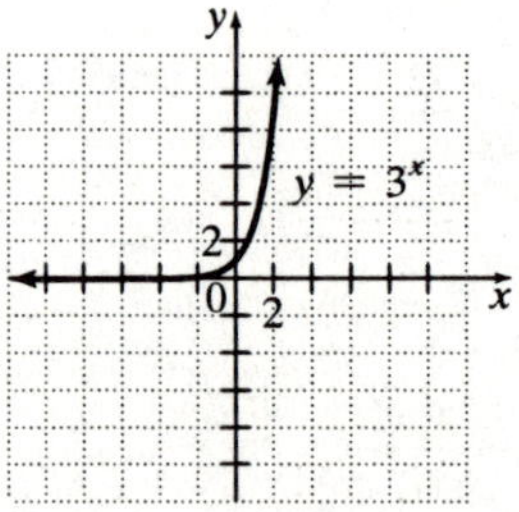

39.

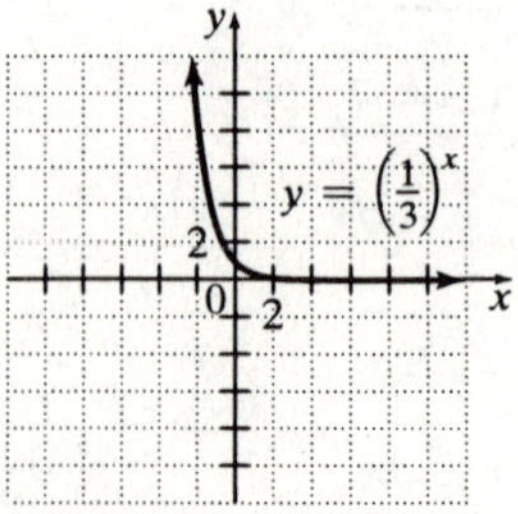

40.

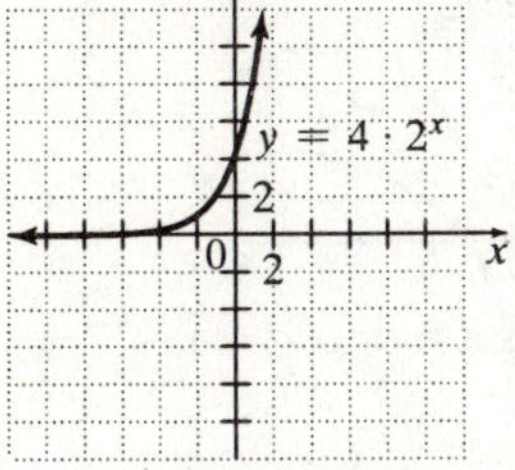

41.

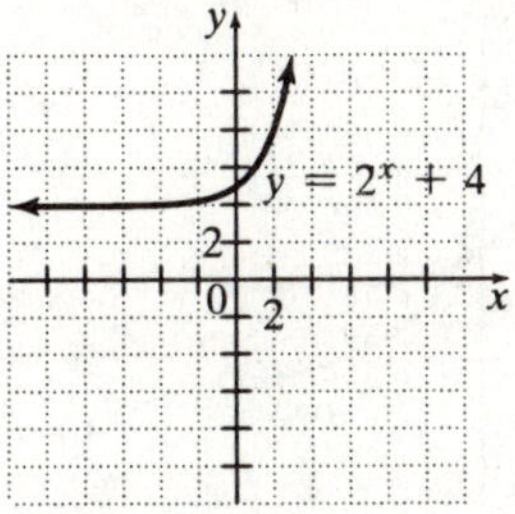

42. $A = P\left(1+\frac{r}{n}\right)^{nt}$

$A = 1600\left(1+\frac{0.09}{2}\right)^{(2)(7)}$

$A = \$2963.11$

43. $A = P\left(1+\frac{r}{n}\right)^{nt}$

$A = 800\left(1+\frac{0.07}{4}\right)^{4(5)}$

$= 800(1.0175)^{20}$

≈ 1131.82

The certificate is worth \$1131.82 at the end of 5 years.

44.

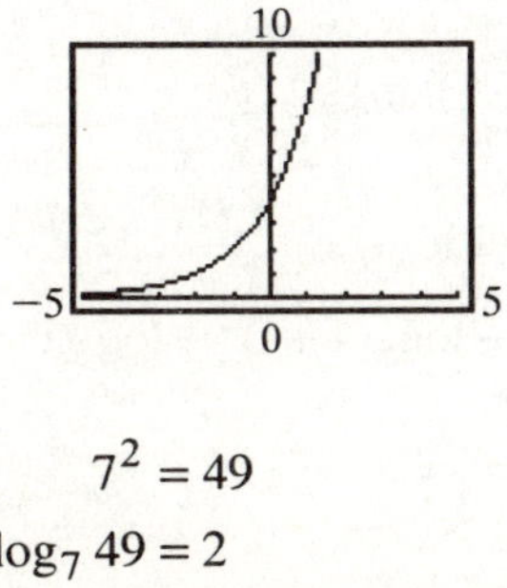

45. $7^2 = 49$

$\log_7 49 = 2$

46. $2^{-4} = \frac{1}{16}$

$\log_2 \frac{1}{16} = -4$

47. $\log_{1/2} 16 = -4$

$\left(\frac{1}{2}\right)^{-4} = 16$

48. $\log_{0.4} 0.064 = 3$

$0.4^3 = 0.064$

49. $\log_4 x = -3$

$x = 4^{-3} = \frac{1}{64}$

50. $\log_3 x = 2$

$x = 3^2 = 9$

51. $\log_3 1 = x$

$3^x = 1$

$3^x = 3^0$

$x = 0$

52. $\log_4 64 = x$

$4^x = 64$

$4^x = 4^3$

$x = 3$

53. $\log_x 64 = 2$

$x^2 = 64$

$x = \pm\sqrt{64} = \pm 8$

$x = 8$ since base > 0

54. $\log_x 81 = 4$

$x^4 = 81$

$x = \pm 3$

We discard the negative base −3.

55. $\log_4 4^5 = x$

$x = 5$

56. $\log_7 7^{-2} = x$

$x = -2$

57. $5^{\log_5 4} = x$

$x = 4$

58. $2^{\log_2 9} = x$

$9 = x$

59. $\log_2(3x-1) = 4$

$3x - 1 = 2^4 = 16$

$3x = 17$

$x = \dfrac{17}{3}$

60. $\log_3(2x+5) = 2$

$2x + 5 = 3^2$

$2x + 5 = 9$

$2x = 4$

$x = 2$

61. $\log_4(x^2 - 3x) = 1$

$x^2 - 3x = 4^1 = 4$

$x^2 - 3x - 4 = 0$

$(x+1)(x-4) = 0$

$x + 1 = 0$ or $x - 4 = 0$

$x = -1$ or $x = 4$

62. $\log_8(x^2 + 7x) = 1$

$x^2 + 7x = 8^1$

$x^2 + 7x = 8$

$x^2 + 7x - 8 = 0$

$(x+8)(x-1) = 0$

$x + 8 = 0$ or $x - 1 = 0$

$x = -8$ or $x = 1$

63.

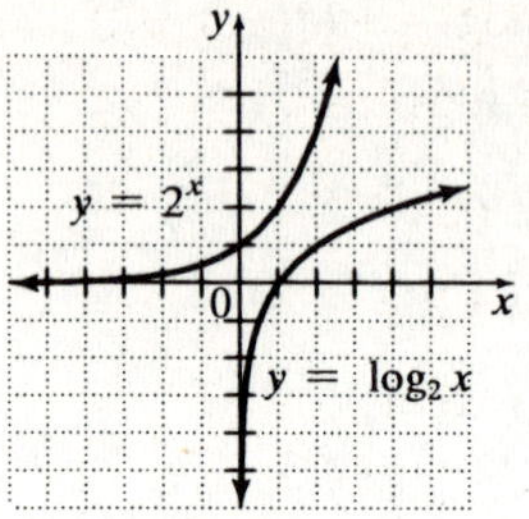

64.

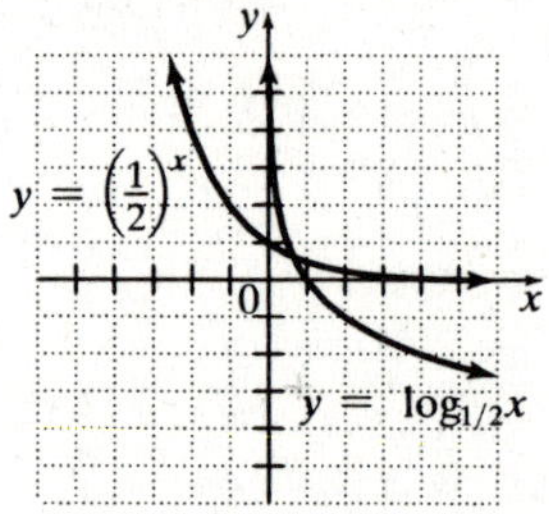

65. $\log_3 8 + \log_3 4 = \log_3(8)(4) = \log_3 32$

66. $\log_2 6 + \log_2 3 = \log_2(6 \cdot 3) = \log_2 18$

67. $\log_7 15 - \log_7 20 = \log_7 \dfrac{15}{20} = \log_7 \dfrac{3}{4}$

68. $\log 18 - \log 12 = \log \dfrac{18}{12} = \log \dfrac{3}{2}$

69. $\log_{11} 8 + \log_{11} 3 - \log_{11} 6 = \log_{11} \dfrac{(8)(3)}{6}$

$= \log_{11} 4$

70. $\log_5 14 + \log_5 3 - \log_5 21$

$= \log_5(14 \cdot 3) - \log_5 21$

$= \log_5\left(\dfrac{42}{21}\right) = \log_5 2$

71. $2\log_5 x - 2\log_5(x+1) + \log_5 x$

$= \log_5 x^2 - \log_5(x+1)^2 + \log_5 x$

$= \log_5 \dfrac{(x^2)(x)}{(x+1)^2}$

$= \log_5 \dfrac{x^3}{(x+1)^2}$

72. $4\log_3 x - \log_3 x + \log_3(x+2)$
$= 3\log_3 x + \log_3(x+2)$
$= \log_3 x^3 + \log_3(x+2)$
$= \log_3[x^3(x+2)]$
$= \log_3(x^4 + 2x^3)$

73. $\log_3 \frac{x^3}{x+2} = \log_3 x^3 - \log_3(x+2)$
$= 3\log_3 x - \log_3(x+2)$

74. $\log_4 \frac{x+5}{x^2} = \log_4(x+5) - \log_4 x^2$
$= \log_4(x+5) - 2\log_4 x$

75. $\log_2 \frac{3x^2y}{z}$
$= \log_2 3 + \log_2 x^2 + \log_2 y - \log_2 z$
$= \log_2 3 + 2\log_2 x + \log_2 y - \log_2 z$

76. $\log_7 \frac{yz^3}{x} = \log_7(yz^3) - \log_7 x$
$= \log_7 y + \log_y z^3 - \log_7 x$
$= \log_7 y + 3\log_7 z - \log_7 x$

77. $\log_6 50 = \log_6(5)(5)(2)$
$= \log_6(5) + \log_6(5) + \log_6(2)$
$= 0.83 + 0.83 + 0.36$
$= 2.02$

78. $\log_b \frac{4}{5} = \log_b 4 - \log_b 5$
$= \log_b 2^2 - \log_b 5$
$= 2\log_b 2 - \log_b 5$
$= 2(0.36) - 0.83$
$= 0.72 - 0.83$
$= -0.11$

79. $\log 3.6 \approx 0.5563$

80. $\log 0.15 \approx -0.8239$

81. $\ln 1.25 \approx 0.2231$

82. $\ln 4.63 \approx 1.5326$

83. $\log 1000 = 3$

84. $\log \frac{1}{10} = \log 10^{-1} = -1$

85. $\ln \frac{1}{e} = \ln 1 - \ln e = 0 - 1 = -1$

86. $\ln(e^4) = 4$

87. $\ln(2x) = 2$
$2x = e^2$
$x = \frac{e^2}{2}$

88. $\ln(3x) = 1.6$
$3x = e^{1.6}$
$x = \frac{e^{1.6}}{3}$

89. $\ln(2x-3) = -1$
$2x - 3 = e^{-1}$
$x = \frac{e^{-1}+3}{2}$

90. $\ln(3x+1) = 2$
$3x + 1 = e^2$
$3x = e^2 - 1$
$x = \frac{e^2-1}{3}$

91. $\ln \frac{I}{I_0} = -kx$
$\ln \frac{0.03I_0}{I_0} = -2.1x$
$\ln 0.03 = -2.1x$
$\frac{\ln 0.03}{-2.1} = x$
$x \approx 1.67$ mm

92. $\ln\left(\frac{I}{I_0}\right) = -kx$

$\ln\frac{0.02I_0}{I_0} = -3.2x$

$\ln 0.02 = -3.2x$

$x = \frac{\ln 0.02}{-3.2} \approx 1.22$

2% of the original radioactivity will penetrate at a depth of approximately 1.22 millimeters.

93. $\log_5 1.6 = \frac{\log 1.6}{\log 5} = 0.2920$

94. $\log_3 4 = \frac{\log 4}{\log 3} \approx 1.2619$

95. $A = Pe^{rt}$

$A = 1450e^{(0.06)(5)}$

$A = \$1957.30$

96. $A = Pe^{rt}$

$A = 940e^{0.11(3)} = 940e^{0.33} \approx 1307.51$

The $940 investment grows to $1307.51 in 3 years.

97. $3^{2x} = 7$

$2x\log 3 = \log 7$

$x = \frac{\log 7}{2\log 3} \approx 0.8856$

98. $6^{3x} = 5$

$3x = \log_6 5$

$x = \frac{1}{3}\log_6 5 = \frac{\log 5}{3\log 6} \approx 0.2994$

99. $3^{2x+1} = 6$

$(2x+1)\log 3 = \log 6$

$2x = \frac{\log 6}{\log 3} - 1$

$x = \frac{1}{2}\left(\frac{\log 6}{\log 3} - 1\right) \approx 0.3155$

100. $4^{3x+2} = 9$

$3x + 2 = \log_4 9$

$3x = \log_4 9 - 2$

$x = \frac{1}{3}\left(\frac{\log 9}{\log 4} - 2\right) \approx -0.1383$

101. $5^{3x-5} = 4$

$(3x-5)\log 5 = \log 4$

$3x = \frac{\log 4}{\log 5} + 5$

$x = \frac{1}{3}\left(\frac{\log 4}{\log 5} + 5\right)$

$x \approx 1.9538$

102. $8^{4x-2} = 3$

$4x - 2 = \log_8 3$

$4x = \log_8 3 + 2$

$4x = \frac{\log 3}{\log 8} + 2$

$x = \frac{1}{4}\left(\frac{\log 3}{\log 8} + 2\right)$

$x \approx 0.6321$

103. $2 \cdot 5^{x-1} = 1$

$\log 2 + (x-1)\log 5 = \log 1$

$(x-1)\log 5 = -\log 2$

$x = -\frac{\log 2}{\log 5} + 1$

$x \approx 0.5693$

104. $3 \cdot 4^{x+5} = 2$

$4^{x+5} = \frac{2}{3}$

$x + 5 = \log_4\left(\frac{2}{3}\right)$

$x = \log_4\left(\frac{2}{3}\right) - 5$

$x = \frac{\log\frac{2}{3}}{\log 4} - 5$

$x \approx -5.2925$

105. $\log_5 2 + \log_5 x = 2$

$$\log_5 2x = 2$$
$$2x = 5^2 = 25$$
$$x = \frac{25}{2}$$

106. $\log_3 x + \log_3 10 = 2$

$$\log_3(10x) = 2$$
$$10x = 3^2$$
$$10x = 9$$
$$x = \frac{9}{10}$$

107. $\log(5x) - \log(x+1) = 4$

$$\log\frac{5x}{x+1} = 4$$
$$\frac{5x}{x+1} = 10^4 = 10{,}000$$
$$5x = 10{,}000x + 10{,}000$$
$$-10{,}000 = 9995x$$
$$-1.0005 = x$$

no solution

108. $\ln(3x) - \ln(x-3) = 2$

$$\ln\left(\frac{3x}{x-3}\right) = 2$$
$$\frac{3x}{x-3} = e^2$$
$$3x = e^2(x-3)$$
$$3x = e^2x - 3e^2$$
$$3x - e^2x = -3e^2$$
$$(3-e^2)x = -3e^2$$
$$x = \frac{-3e^2}{3-e^2} = \frac{3e^2}{e^2-3}$$

109. $\log_2 x + \log_2 2x - 3 = 1$

$$\log_2(x)(2x) = 4$$
$$2x^2 = 2^4 = 16$$
$$x^2 = 8$$
$$x = \pm\sqrt{8} = \pm 2\sqrt{2}$$
$$x = 2\sqrt{2}$$

110. $-\log_6(4x+7) + \log_6 x = 1$

$$\log_6\frac{x}{4x+7} = 1$$
$$\frac{x}{4x+7} = 6^1$$
$$\frac{x}{4x+7} = 6$$
$$x = 6(4x+7)$$
$$x = 24x + 42$$
$$-42 = 23x$$
$$x = -\frac{42}{23}$$

We discard $x = -\frac{42}{23}$ as extraneous leaving no solution.

111. $y = y_0e^{kt}$

$$y = 155{,}000e^{(0.06)(4)}$$
$$y = 197{,}044 \text{ ducks}$$

112. $y = y_0e^{kt}$

$$y = 212{,}942{,}000e^{0.015(8)}$$
$$= 212{,}942{,}000e^{0.12}$$
$$\approx 240{,}091{,}435$$

The expected population of Indonesia by the year 2006 is approximately 240,091,435.

113. $y = y_0e^{kt}$

$$140{,}000{,}000 = 125{,}932{,}000e^{0.002t}$$
$$t = \frac{\ln\frac{140{,}000{,}000}{125{,}932{,}000}}{0.002}$$
$$t \approx 53$$

It will take approximately 53 years.

114.
$$y = y_0 e^{kt}$$
$$2(30,675,000) = 30,675,000e^{0.011t}$$
$$2 = e^{0.011t}$$
$$t = \frac{\ln 2}{0.011}$$
$$t \approx 63$$
It will take approximately 63 years.

115. $2(66,050,000) = 66,050,000e^{(0.019)t}$
$$2 = e^{0.019t}$$
$$t = \frac{\ln 2}{0.019} \approx 36$$
It will take approximately 36 years.

116.
$$A = P\left(1+\frac{r}{n}\right)^{nt}$$
$$10,000 = 5000\left(1+\frac{0.08}{4}\right)^{4t}$$
$$2 = (1.02)^{4t}$$
$$\log 2 = 4t \log 1.02$$
$$t = \frac{\log 2}{4 \log 1.02} \approx 8.8 \text{ years}$$

117. $A = P\left(1+\frac{r}{n}\right)^{nt}$
$$10,000 = 6,000\left(1+\frac{0.06}{12}\right)^{12t}$$
$$\frac{5}{3} = (1.005)^{12t}$$
$$12t = \log_{1.005}\left(\frac{5}{3}\right)$$
$$t = \frac{1}{12}\log_{1.005}\left(\frac{5}{3}\right)$$
$$= \frac{1}{12}\frac{\log\left(\frac{5}{3}\right)}{\log(1.005)}$$
$$\approx 8.5$$
It was invested for approximately 8.5 years.

118.

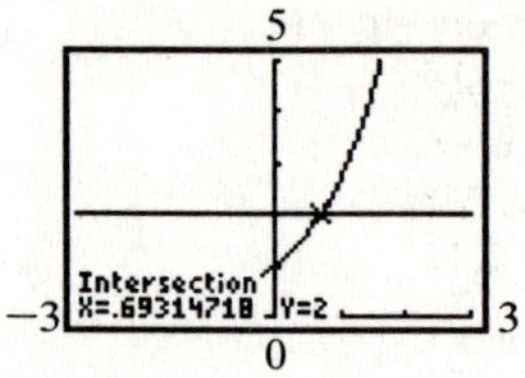

$x \approx 0.69$

119.

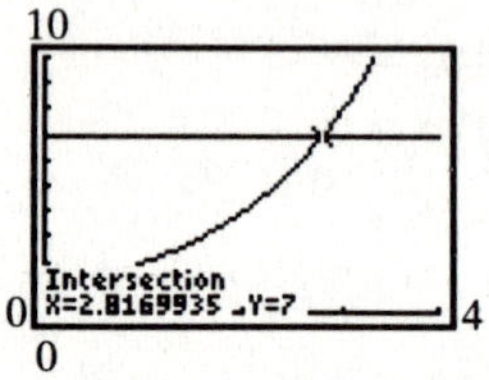

$x \approx 2.82$

Chapter 10 Test

1. $(f \circ h)(0) = f(h(0)) = f(5) = 5$

2. $(g \circ f)(x) = g(f(x)) = g(x) = x - 7$

3. $(g \circ h)(x) = g(h(x))$
$$= g(x^2 - 6x + 5)$$
$$= x^2 - 6x + 5 - 7$$
$$= x^2 - 6x - 2$$

4. $f(x) = 7x - 14$ so $f^{-1}(x) = \frac{x+14}{7}$

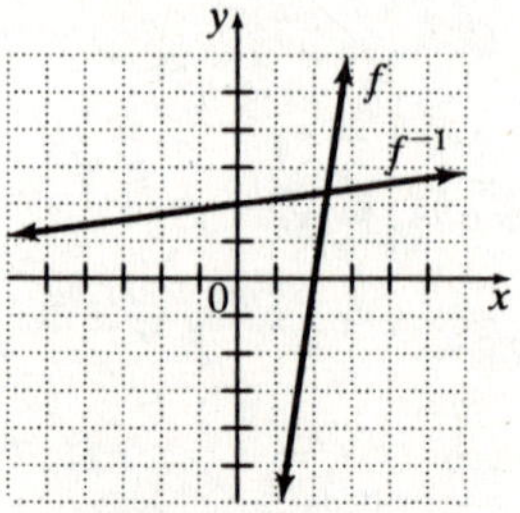

5. The graph does represent a one-to-one function.

6. The graph does not represent a function, hence does not represent a one-to-one function.

7. $y = 6 - 2x$ is one-to-one.

$$x = 6 - 2y$$
$$2y = -x + 6$$
$$y = \frac{-x+6}{2}$$
$$f^{-1}(x) = \frac{-x+6}{2}$$

8. $f = \{(0, 0), (2, 3), (-1, 5)\}$ is one-to-one.

$$f^{-1} = \{(0, 0), (3, 2), (5, -1)\}$$

9. This function is not one-to-one.

10. $\log_3 6 + \log_3 4 = \log_3(6 \cdot 4) = \log_3 24$

11.
$$\log_5 x + 3\log_5 x - \log_5(x+1)$$
$$= 4\log_5 x - \log_5(x+1)$$
$$= \log_5 x^4 - \log_5(x+1)$$
$$= \log_5 \frac{x^4}{x+1}$$

12.
$$\log_6 \frac{2x}{y^3} = \log_6 2x - \log_6 y^3$$
$$= \log_6 2 + \log_6 x - 3\log_6 y$$

13.
$$\log_b\left(\frac{3}{25}\right) = \log_b 3 - \log_b 25$$
$$= \log_b 3 - \log_b 5^2$$
$$= \log_b 3 - 2\log_b 5$$
$$= 0.79 - 2(1.16)$$
$$= -1.53$$

14. $\log_7 8 = \dfrac{\ln 8}{\ln 7} \approx 1.0686$

15.
$$8^{x-1} = \frac{1}{64}$$
$$8^{x-1} = 8^{-2}$$
$$x - 1 = -2$$
$$x = -1$$

16.
$$3^{2x+5} = 4$$
$$2x + 5 = \log_3 4$$
$$2x + 5 = \frac{\log 4}{\log 3}$$
$$2x = \frac{\log 4}{\log 3} - 5$$
$$x = \frac{1}{2}\left(\frac{\log 4}{\log 3} - 5\right)$$
$$x \approx -1.8691$$

17.
$$\log_3 x = -2$$
$$x = 3^{-2}$$
$$x = \frac{1}{9}$$

18.
$$\ln\sqrt{e} = x$$
$$\ln e^{1/2} = x$$
$$\frac{1}{2} = x$$

19.
$$\log_8(3x-2) = 2$$
$$3x - 2 = 8^2$$
$$3x - 2 = 64$$
$$3x = 66$$
$$x = \frac{66}{3} = 22$$

20.
$$\log_5 x + \log_5 3 = 2$$
$$\log_5(3x) = 2$$
$$3x = 5^2$$
$$3x = 25$$
$$x = \frac{25}{3}$$

21. $\log_4(x+1) - \log_4(x-2) = 3$

$$\log_4 \frac{x+1}{x-2} = 3$$
$$\frac{x+1}{x-2} = 4^3$$
$$x+1 = 64(x-2)$$
$$x+1 = 64x - 128$$
$$129 = 63x$$
$$x = \frac{43}{21}$$

22. $\ln(3x+7) = 1.31$

$$3x+7 = e^{1.31}$$
$$3x = e^{1.31} - 7$$
$$x = \frac{e^{1.31} - 7}{3} \approx -1.0979$$

23. $y = \left(\frac{1}{2}\right)^x + 1$

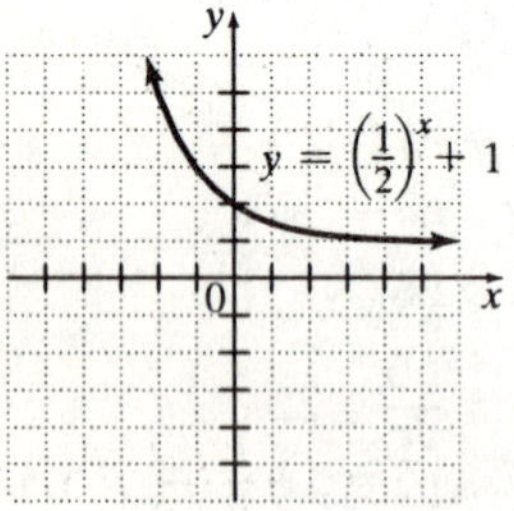

24. $y = 3^x$ and $y = \log_3 x$

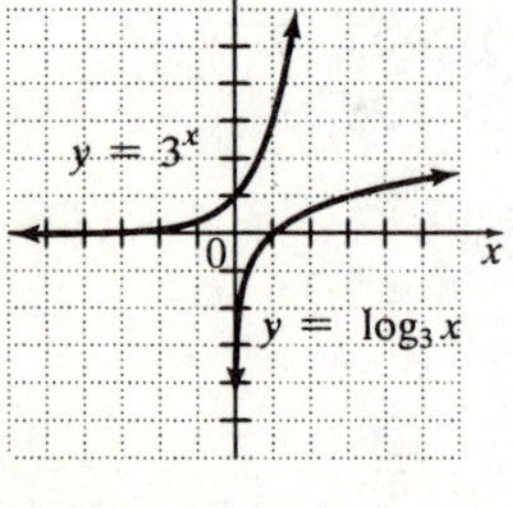

25. $A = P\left(1 + \frac{r}{n}\right)^{nt}$, $P = 4000$, $t = 3$, $r = 0.09$, and $n = 12$

$$A = 4000\left(1 + \frac{0.09}{12}\right)^{12(3)}$$
$$= 4000\left(1 + \frac{0.09}{12}\right)^{12(3)}$$
$$= 4000(1.0075)^{36}$$
$$\approx 5234.58$$

\$5234.58 will be in the account.

26. $A = P\left(1 + \frac{r}{n}\right)^{nt}$, $P = 2000$, $A = 3000$, $r = 0.07$, and $n = 2$

$$3000 = 2000\left(1 + \frac{0.07}{2}\right)^{2t}$$
$$1.5 = (1.035)^{2t}$$
$$2t = \log_{1.035} 1.5$$
$$t = \frac{1}{2}\log_{1.035} 1.5$$
$$= \frac{1}{2} \frac{\ln 1.5}{\ln 1.035}$$
$$\approx 5.9$$

It would take 6 years for the investment to reach \$3000.

27. $y = y_0 e^{kt}$, $y_0 = 57{,}000$, $k = 0.026$, and $t = 5$

$$y = 57{,}000e^{0.026(5)}$$
$$= 57{,}000e^{0.13}$$
$$\approx 64{,}913$$

There will be approximately 64,913 prairie dogs 5 years from now.

28. $y = y_0 e^{kt}$, $y_0 = 400$, $y = 1000$, and $k = 0.062$

$$1000 = 400e^{0.062t}$$
$$2.5 = e^{0.062t}$$
$$0.062t = \ln 2.5$$
$$t = \frac{\ln 2.5}{0.062}$$
$$\approx 14.8$$

It will take the naturalists approximately 15 years to reach their goal.

29. $\log(1+k) = \dfrac{0.3}{D},\ D = 56$

$\log(1+k) = \dfrac{0.3}{56}$

$1+k = 10^{0.3/56}$

$k = -1 + 10^{0.3/56}$

$k \approx .012$

The rate of population increase is approximately 1.2%.

30. $Y_1 = e^{0.2x}$

$Y_2 = e^{-0.4x} + 2$

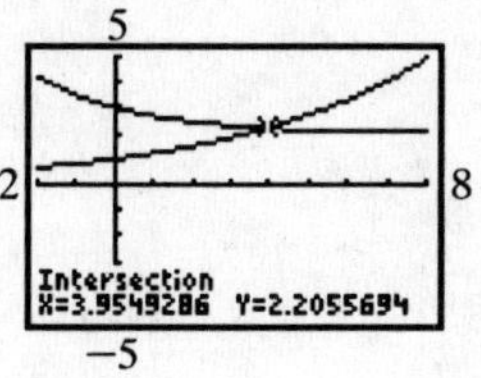

$x \approx 3.95$

Chapter 11

Exercise Set 11.1

1. $a_n = n+4$

$a_1 = 1+4 = 5$

$a_2 = 2+4 = 6$

$a_3 = 3+4 = 7$

$a_4 = 4+4 = 8$

$a_5 = 5+4 = 9$

Thus, the first five terms of the sequence $a_n = n+4$ are or 5, 6, 7, 8, 9.

3. $a_n = (-1)^n$

$a_1 = (-1)^1 = -1$

$a_2 = (-1)^2 = 1$

$a_3 = (-1)^3 = -1$

$a_4 = (-1)^4 = 1$

$a_5 = (-1)^5 = -1$

Thus, the first five terms of the sequence $a_n = (-1)^n$ are –1, 1, –1, 1, –1.

5. $a_n = \frac{1}{n+3}$

$a_1 = \frac{1}{1+3} = \frac{1}{4}$

$a_2 = \frac{1}{2+3} = \frac{1}{5}$

$a_3 = \frac{1}{3+3} = \frac{1}{6}$

$a_4 = \frac{1}{4+3} = \frac{1}{7}$

$a_5 = \frac{1}{5+3} = \frac{1}{8}$

Thus, the first five terms of the sequence $a_n = \frac{1}{n+3}$ are $\frac{1}{4}, \frac{1}{5}, \frac{1}{6}, \frac{1}{7}, \frac{1}{8}$.

7. $a_n = 2n$

$a_1 = 2(1) = 2$

$a_2 = 2(2) = 4$

$a_3 = 2(3) = 6$

$a_4 = 2(4) = 8$

$a_5 = 2(5) = 10$

or 2, 4, 6, 8, 10

9. $a_n = -n^2$

$a_1 = -1^2 = -1$

$a_2 = -2^2 = -4$

$a_3 = -3^2 = -9$

$a_4 = -4_2 = -16$

$a_5 = -5^2 = -25$

Thus, the first five terms of the sequence $a_n = -n^2$ are –1, –4, –8, –16, –25.

11. $a_n = 2^n$

$a_1 = 2^1 = 2$

$a_2 = 2^2 = 4$

$a_3 = 2^3 = 8$

$a_4 = 2^4 = 16$

$a_5 = 2^5 = 32$

Thus, the first five terms of the sequence $a_n = 2^n$ are 2, 4, 8, 16, 32.

13. $a_n = 2n+5$

$a_1 = 2(1)+5 = 2+5 = 7$

$a_2 = 2(2)+5 = 4+5 = 9$

$a_3 = 2(3)+5 = 6+5 = 11$

$a_4 = 2(4)+5 = 8+5 = 13$

$a_5 = 2(5)+5 = 10+5 = 15$

Thus, the first five terms of the sequence $a_n = 2n+5$ are 7, 9, 11, 13, 15.

15. $a_n = (-1)^n n^2$

$a_1 = (-1)^1(1^2) = -1(1) = -1$

$a_2 = (-1)^2(2)^2 = 1(4) = 4$

$a_3 = (-1)^3(3)^2 = -1(9) = -9$

$a_4 = (-1)^4(4)^2 = 1(16) = 16$

$a_5 = (-1)^5(5)^2 = -1(25) = -25$

Thus, the first five terms of the sequence $a_n = (-1)^n n^2$ are –1, 4, –9, 16, –25.

17. $a_n = 3n^2$

$a_5 = 3(5)^2 = 3(25) = 75$

19. $a_n = 6n-2$

$a_{20} = 6(20)-2 = 120-2 = 118$

21. $a_n = \dfrac{n+3}{n}$

$a_{15} = \dfrac{15+3}{15} = \dfrac{18}{15} = \dfrac{6}{5}$

23. $a_n = (-3)^n$

$a_6 = (-3)^6 = 729$

25. $a_n = \dfrac{n-2}{n+1}$

$a_6 = \dfrac{6-2}{6+1} = \dfrac{4}{7}$

27. $a_n = \dfrac{(-1)^n}{n}$

$a_8 = \dfrac{(-1)^8}{8} = \dfrac{1}{8}$

29. $a_n = -n^2+5$

$a_{10} = -10^2+5 = -100+5 = -95$

31. $a_n = \dfrac{(-1)^n}{n+6}$

$a_{19} = \dfrac{(-1)^{19}}{19+6} = -\dfrac{1}{25}$

33. 3, 7, 11, 15 or
4(1) – 1, 4(2) – 1, 4(3) – 1, 4(4) – 1
In general, $a_n = 4n-1$

35. –2, –4, –8, –16, or $-2, -2^2, -2^3, -2^4$
In general, $a_n = -2^n$

37. $\dfrac{1}{3}, \dfrac{1}{9}, \dfrac{1}{27}, \dfrac{1}{81}$, or

$\dfrac{1}{3}, \dfrac{1}{3^2}, \dfrac{1}{3^3}, \dfrac{1}{3^4}$

In general, $a_n = \dfrac{1}{3^n}$

39. $a_n = 32n-16$

$a_2 = 32(2)-16 = 64-16 = 48$ ft

$a_3 = 32(3)-16 = 96-16 = 80$ ft

$a_4 = 32(4)-16 = 128-16 = 112$ ft

41. 0.10, 0.20, 0.40 or
$0.10, 0.10(2), 0.10(2)^2$
In general, $a_n = 0.10(2)^{n-1}$

$a_{14} = 0.10(2)^{13} \approx \$819.20$

43. $a_n = 75(2)^{n-1}$

$a_6 = 75(2)^5 = 75(32) = 2400$ cases

$a_1 = 75(2)^0 = 75(1) = 75$ cases

45. $a_n = \frac{1}{2}a_{n-1}$ for $n > 1$, $a_1 = 800$

In 2000, $n = 1$ and $a_1 = 800$.

In 2001, $n = 2$ and $a_2 = \frac{1}{2}(800) = 400$.

In 2002, $n = 3$ and $a_3 = \frac{1}{2}(400) = 200$.

In 2003, $n = 4$ and $a_4 = \frac{1}{2}(200) = 100$.

In 2004, $n = 5$ and $a_5 = \frac{1}{2}(100) = 50$.

The population estimate for 2004 is 50 sparrows.

Continuing the sequence:

in 2005, $n = 6$ and $a_6 = \frac{1}{2}(50) = 25$;

in 2006, $n = 7$ and $a_7 = \frac{1}{2}(25) = 12$;

in 2007, $n = 8$ and $a_8 = \frac{1}{2}(12) = 6$;

in 2008, $n = 9$ and $a_9 = \frac{1}{2}(6) = 3$;

in 2009, $n = 10$ and $a_{10} = \frac{1}{2}(3) \approx 1$;

in 2010, $n = 11$ and $a_{11} = \frac{1}{2}(1) \approx 0$.

The population is estimated to become extinct in 2010.

47.
$$a_n = \frac{1}{\sqrt{n}}$$
$$a_1 = \frac{1}{\sqrt{1}} = \frac{1}{1} = 1$$
$$a_2 = \frac{1}{\sqrt{2}} \approx 0.7071$$
$$a_3 = \frac{1}{\sqrt{3}} \approx 0.5774$$
$$a_4 = \frac{1}{\sqrt{4}} = \frac{1}{2} = 0.5$$
$$a_5 = \frac{1}{\sqrt{5}} \approx 0.4472$$

Thus, the first five terms of the sequence $a_n = \frac{1}{\sqrt{n}}$ are $1, 0.7071, 0.5774, 0.5, 0.4472$.

49.
$$a_n = \left(1+\frac{1}{n}\right)^n$$
$$a_1 = \left(1+\frac{1}{1}\right)^1 = (2)^1 = 2$$
$$a_2 = \left(1+\frac{1}{2}\right)^2 = \left(\frac{3}{2}\right)^2 = 2.25$$
$$a_3 = \left(1+\frac{1}{3}\right)^3 = \left(\frac{4}{3}\right)^3 \approx 2.3704$$
$$a_4 = \left(1+\frac{1}{4}\right)^4 = \left(\frac{5}{4}\right)^4 \approx 2.4414$$
$$a_5 = \left(1+\frac{1}{5}\right)^5 = \left(\frac{6}{5}\right)^5 \approx 2.4883$$

Thus, the first five terms of the sequence $a_n = \left(1+\frac{1}{n}\right)^n$ are 2, 2.25, 2.3704, 2.4414, 2.4883.

51. $f(x) = (x-1)^2 + 3$

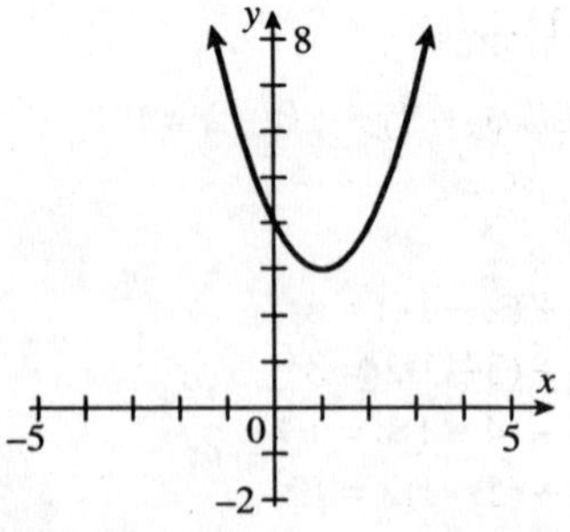

53. $f(x) = 2(x+4)^2 + 2$

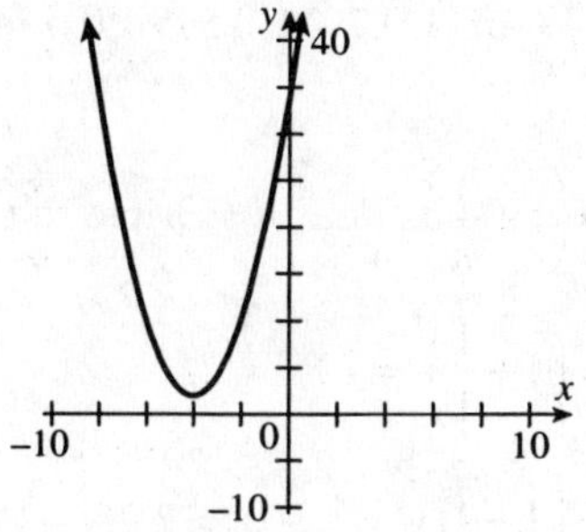

55. $(-4,-1)$ and $(-7,-3)$

$d = \sqrt{[-7-(-4)]^2 + [-3-(-1)]^2}$

$d = \sqrt{(-7+4)^2 + (-3+1)^2}$

$d = \sqrt{(-3)^2 + (-2)^2}$

$d = \sqrt{9+4} = \sqrt{13}$ units

57. $(2,-7)$ and $(-3,-3)$

$d = \sqrt{(-3-2)^2 + [-3-(-7)]^2}$

$d = \sqrt{(-5)^2 + (-3+7)^2}$

$d = \sqrt{(-5)^2 + (4)^2}$

$d = \sqrt{25+16} = \sqrt{41}$ units

Exercise Set 11.2

1. $a_n = a_1 + (n-1)d$

$a_1 = 4;\ d = 2$

$a_1 = 4$

$a_2 = 4 + (2-1)2 = 6$

$a_3 = 4 + (3-1)2 = 8$

$a_4 = 4 + (4-1)2 = 10$

$a_5 = 4 + (5-1)2 = 12$

The first five terms are 4, 6, 8, 10, 12.

3. $a_n = a_1 + (n-1)d$

$a_1 = 6,\ d = -2$

$a_1 = 6$

$a_2 = 6 + (2-1)(-2) = 4$

$a_3 = 6 + (3-1)(-2) = 2$

$a_4 = 6 + (4-1)(-2) = 0$

$a_5 = 6 + (5-1)(-2) = -2$

The first five terms are 6, 4, 2, 0, –2.

5. $a_n = a_1 r^{n-1}$

$a_1 = 1,\ r = 3$

$a_1 = 1(3)^{1-1} = 1$

$a_2 = 1(3)^{2-1} = 3$

$a_3 = 1(3)^{3-1} = 9$

$a_4 = 1(3)^{4-1} = 27$

$a_5 = 1(3)^{5-1} = 81$

The first five terms are 1, 3, 9, 27, 81.

7. $a_n = a_1 r^{n-1}$

$a_1 = 48,\ r = \frac{1}{2}$

$a_1 = 48\left(\frac{1}{2}\right)^{1-1} = 48$

$a_2 = 48\left(\frac{1}{2}\right)^{2-1} = 24$

$a_3 = 48\left(\frac{1}{2}\right)^{3-1} = 12$

$a_4 = 48\left(\frac{1}{2}\right)^{4-1} = 6$

$a_5 = 48\left(\frac{1}{2}\right)^{5-1} = 3$

The first five terms are 48, 24, 12, 6, 3.

9. $a_n = a_1 + (n-1)d$

$a_1 = 12,\ d = 3$

$a_n = 12 + (n-1)3$

$a_8 = 12 + 7(3) = 12 + 21 = 33$

11. $a_n = a_1 r^{n-1}$

$a_1 = 7,\ r = -5$

$a_n = a_1 r^{n-1}$

$a_4 = 7(-5)^3 = 7(-125) = -875$

13. $a_n = a_1 + (n-1)d$

$a_1 = -4,\ d = -4$

$a_n = -4 + (n-1)(-4)$

$a_{15} = -4 + 14(-4) = -4 - 56 = -60$

15. 0, 12, 24

$a_1 = 0$ and $d = 12$

$a_n = 0 + (n-1)12$

$a_9 = 8(12) = 96$

17. 20, 18, 16

$a_1 = 20$ and $d = -2$

$a_n = 20 + (n-1)(-2)$

$a_{25} = 20 + 24(-2) = 20 - 48 = -28$

19. 2, −10, 50

$a_1 = 2$ and $r = -5$

$a_n = 2(-5)^{n-1}$

$a_5 = 2(-5)^4 = 2(625) = 1250$

21. $a_4 = 19,\ a_{15} = 52$

$\begin{cases} a_4 = a_1 + (4-1)d \\ a_{15} = a_1 + (15-1)d \end{cases}$ or $\begin{cases} 19 = a_1 + 3d \\ 52 = a_1 + 14d \end{cases}$

Multiply both sides of the second equation by −1 in order to solve for *d*.

$$\begin{aligned} 19 &= a_1 + 3d \\ -52 &= -a_1 - 14d \\ \hline -33 &= \quad -11d \end{aligned}$$

$3 = d$

To find a_1, substitute $d = 3$ into the first equation.

$19 = a_1 + 3(3)$

$19 = a_1 + 9$

$10 = a_1$

Thus, $a_1 = 10$ and $d = 3$, so

$$\begin{aligned} a_n &= 10 + (n-1)3 \\ &= 10 + 3n - 3 \\ &= 7 + 3n \end{aligned}$$

and $a_8 = 7 + 3(8)$

$= 7 + 24$

$a_8 = 31$

23. $a_2 = -1$ and $a_4 = 5$

$\begin{cases} a_2 = a_1 + (2-1)d \\ a_4 = a_1 + (4-1)d \end{cases}$ or $\begin{cases} -1 = a_1 + d \\ 5 = a_1 + 3d \end{cases}$

Multiply both sides of the second equation by −1 in order to solve for *d*.

$$\begin{aligned} -1 &= a_1 + d \\ -5 &= -a_1 - 3d \\ \hline -6 &= \quad -2d \end{aligned}$$

$3 = d$

To find a_1, substitute $d = 3$ into the first equation.

$-1 = a_1 + 3$

$-4 = a_1$

Thus, $a_1 = -4$ and $d = 3$, so

$$\begin{aligned} a_n &= -4(n-1)3 \\ &= -4 + 3n - 3 \\ &= -7 + 3n \end{aligned}$$

and $a_9 = -7 + 3(9)$

$= -7 + 27$

$a_9 = 20$

25. $a_2 = -\frac{4}{3}$ and $a_3 = \frac{8}{3}$

Notice that $\frac{8}{3} \div \frac{-4}{3} = \frac{8}{3} \cdot -\frac{3}{4} = -2$,

so $r = -2$. Then

$a_2 = a_1(-2)^{2-1}$

$-\frac{4}{3} = a_1(-2)$

$\frac{2}{3} = a,$

The first term is $\frac{2}{3}$ and the common ratio is −2.

27. Answers may vary.

29. 2, 4, 6 is an arithmetic sequence.

$a_1 = 2$ and $d = 2$

31. 5, 10, 20 is a geometric sequence.

$a_1 = 5$ and $r = 2$

33. $\frac{1}{2}, \frac{1}{10}, \frac{1}{50}$ is a geometric sequence.

$a_1 = \frac{1}{2}$ and $r = \frac{1}{5}$

35. *x*, 5*x*, 25*x* is a geometric sequence.

$a_1 = x$ and $r = 5$

37. $p, p+4, p+8$ is an arithmetic sequence.
$a_1 = p$ and $d = 4$

39. $a_1 = 14$ and $d = \frac{1}{4}$
$a_n = 14 + (n-1)\frac{1}{4}$
$a_{21} = 14 + 20\left(\frac{1}{4}\right) = 14 + 5 = 19$

41. $a_1 = 3$ and $r = -\frac{2}{3}$
$a_n = 3\left(-\frac{2}{3}\right)^{n-1}$
$a_4 = 3\left(-\frac{2}{3}\right)^3 = 3\left(-\frac{8}{27}\right) = -\frac{8}{9}$

43. $\frac{3}{2}, 2, \frac{5}{2}, \ldots$
$a_1 = \frac{3}{2}$ and $d = \frac{1}{2}$
$a_n = \frac{3}{2} + (n-1)\frac{1}{2}$
$a_{15} = \frac{3}{2} + 14\left(\frac{1}{2}\right) = \frac{17}{2}$

45. $24, 8, \frac{8}{3}, \ldots$
$a_1 = 24$ and $r = \frac{1}{3}$
$a_n = 24\left(\frac{1}{3}\right)^{n-1}$
$a_6 = 24\left(\frac{1}{3}\right)^5 = 24\left(\frac{1}{243}\right) = \frac{8}{81}$

47. $a_3 = 2$ and $a_{17} = -40$

$$\begin{cases} a_3 = a_1 + (3-1)d \\ a_{17} = a_1 + (17-1)d \end{cases} \text{ or } \begin{array}{l} 2 = a_1 + 2d \\ -40 = a_1 + 16d \end{array}$$

Multiply both sides of the second equation by -1 in order to solve for d.

$$\begin{array}{rl} 2 = & a_1 + 2d \\ 40 = & -a_1 - 16d \\ \hline 42 = & -14d \end{array}$$

$-3 = d$

To find a_1, substitute $d = -3$ into the first equation.

$2 = a_1 + 2(-3)$
$2 = a_1 - 6$
$8 = a_1$

Thus, $a_1 = 8$ and $d = -3$, so
$a_n = 8 + (n-1)(-3)$
$= 8 - 3n + 3$
$= 11 - 3n$
and $a_{10} = 11 - 3(10)$
$= 11 - 30$
$a_{10} = -19$

49. 54, 58, 62
$a_1 = 54$ and $d = 4$
$a_n = 54 + (n-1)4$
$a_{20} = 54 + 19(4) = 54 + 76 = 130$
The general term of the sequence is $a_n = 4n + 50$.
There are 130 seats in the twentieth row.

51. $a_1 = 6$ and $r = 3$
$a_n = 6(3)^{n-1} = 2 \cdot 3 \cdot 3^{n-1} = 2(3)^n$
The general term of the sequence is $a_n = 6(3)^{n-1}$ or $a_n = 2(3)^n$.

53. $a_1 = 486$ and $r = \frac{1}{3}$

Initial Height $= a_1 = 486\left(\frac{1}{3}\right)^{1-1} = 486$

Rebound 1 $= a_2 = 486\left(\frac{1}{3}\right)^{2-1} = 162$

Rebound 2 $= a_3 = 486\left(\frac{1}{3}\right)^{3-1} = 54$

Rebound 3 $= a_4 = 486\left(\frac{1}{3}\right)^{4-1} = 18$

Rebound 4 $= a_5 = 486\left(\frac{1}{3}\right)^{5-1} = 6$

The first five terms of the sequence are 486, 162, 54, 18, 6.

$a_n = 486\left(\frac{1}{3}\right)^{n-1}$

Solve $486\left(\frac{1}{3}\right)^{n-1} = 1$

$$\left(\frac{1}{3}\right)^{n-1} = \frac{1}{486}$$
$$\frac{1}{3^{n-1}} = \frac{1}{486}$$
$$3^{n-1} = 486$$
$$3^n = 1458$$
$$n = \log_3 1458$$
$$n = \frac{\ln 1458}{\ln 3}$$
$$\approx 6.6$$

Since a_7 is less than a foot and a_7 corresponds to the 6th rebound, the ball will rebound less than a foot on the 6th bounce.

55. $a_1 = 4000$ and $d = 125$

$a_n = 4000 + (n-1)125$ or

$a_n = 3875 + 125_n$

$a_{12} = 4000 + 11(125)$

$a_{12} = 5375$

His salary for his last month of training is \$5375.

57. $a_1 = 400$ and $r = \frac{1}{2}$

12 hrs = 4(3 hrs), so we seek the fourth term after a_1, namely a_5.

$a_n = a_1 r^{n-1}$

$a_5 = 400\left(\frac{1}{2}\right)^4 = \frac{400}{16} = 25$

25 grams of the radioactive material remains after 12 hours.

59. $a_1 = \$11,782.40$

$r = 0.5$

$a_2 = (11,782.40)(0.5) = \$5891.20$

$a_3 = (5891.20)(0.5) = \$2945.60$

$a_4 = (2945.60)(0.5) = \$1472.80$

The first four terms of the sequence are \$11,782.40, \$5891.20, \$2945.60, \$1472.80.

61. $a_1 = 19.652$ and $d = -0.034$

$a_2 = 19.652 - 0.034 = 19.618$

$a_3 = 19.618 - 0.034 = 19.584$

$a_4 = 19.584 - 0.034 = 19.550$

The first four terms of the sequence are 19.652, 19.618, 19.584, 19.550.

63. Answers may vary.

65. $\frac{1}{3(1)} + \frac{1}{3(2)} + \frac{1}{3(3)}$

$= \frac{1}{3}\left(\frac{1}{1} + \frac{1}{2} + \frac{1}{3}\right)$

$= \frac{1}{3}\left(\frac{6}{6} + \frac{3}{6} + \frac{2}{6}\right)$

$= \frac{1}{3}\left(\frac{11}{6}\right) = \frac{11}{18}$

67. $3^0 + 3^1 + 3^2 + 3^3 = 1 + 3 + 9 + 27 = 40$

69. $\frac{8-1}{8+1} + \frac{8-2}{8+2} + \frac{8-3}{8+3}$

$= \frac{7}{9} + \frac{6}{10} + \frac{5}{11}$

$= \frac{770}{990} + \frac{594}{990} + \frac{450}{990}$

$= \frac{1814}{990} = \frac{907}{495}$

Exercise Set 11.3

1. $\sum_{i=1}^{4}(i-3)$
$= (1-3)+(2-3)+(3-3)+(4-3)$
$= -2+(-1)+0+1 = -2$

3. $\sum_{i=4}^{7}(2i+4)$
$= [2(4)+4]+[2(5)+4]+[2(6)+4] + [2(7)+4]$
$= 12+14+16+18 = 60$

5. $\sum_{i=2}^{4}(i^2-3)$
$= (2^2-3)+(3^2-3)+(4^2-3)$
$= 1+6+13 = 20$

7. $\sum_{i=1}^{3}\frac{1}{i+5}$
$= \frac{1}{1+5}+\frac{1}{2+5}+\frac{1}{3+5}$
$= \frac{1}{6}+\frac{1}{7}+\frac{1}{8} = \frac{28}{168}+\frac{24}{168}+\frac{21}{168} = \frac{73}{168}$

9. $\sum_{i=1}^{3}\frac{1}{6i} = \frac{1}{6(1)}+\frac{1}{6(2)}+\frac{1}{6(3)}$
$= \frac{1}{6}+\frac{1}{12}+\frac{1}{18} = \frac{6+3+2}{36} = \frac{11}{36}$

11. $\sum_{i=2}^{6}3i$
$= 3(2)+3(3)+3(4)+3(5)+3(6)$
$= 6+9+12+15+18 = 60$

13. $\sum_{i=3}^{5}i(i+2)$
$= 3(3+2)+4(4+2)+5(5+2)$
$= 15+24+35 = 74$

15. $\sum_{i=1}^{5}2^i = 2^1+2^2+2^3+2^4+2^5$
$= 2+4+8+16+32 = 62$

17. $\sum_{i=1}^{4}\frac{4i}{i+3} = \frac{4(1)}{1+3}+\frac{4(2)}{2+3}+\frac{4(3)}{3+3}+\frac{4(4)}{4+3}$
$= 1+\frac{8}{5}+2+\frac{16}{7} = \frac{105}{35}+\frac{56}{35}+\frac{80}{35} = \frac{241}{35}$

19. $1+3+5+7+9$
$= [(2)-1]+[2(2)-1]+[2(3)-1] + [2(4)-1]+[2(5)-1]$
$= \sum_{i=1}^{5}(2i-1)$

21. $4+12+36+108$
$= 4+4(3)+4(3^2)+4(3^3)$
$= \sum_{i=1}^{4}4(3)^{i-1}$

23. $12+9+6+3+0+(-3)$
$= [-3(1)+15]+[-3(2)+15] + [-3(3)+15]+[-3(4)+15] + [-3(5)+15]+[-3(6)+15]$
$= \sum_{i=1}^{6}(-3i+15)$

25. $12+4+\frac{4}{3}+\frac{4}{9} = \frac{4}{3^{-1}}+\frac{4}{3^0}+\frac{4}{3^1}+\frac{4}{3^2}$
$= \sum_{i=1}^{4}\frac{4}{(3)^{i-2}}$

27. $1+4+9+16+25+36+49$
$= 1^2+2^2+3^2+4^2+5^2+6^2+7^2$
$= \sum_{i=1}^{7}i^2$

29. $a_n = (n+2)(n-5)$
$a_1 = (1+2)(1-5) = 3(-4) = -12$
$a_2 = (2+2)(2-5) = 4(-3) = -12$
$a_1+a_2 = -12+(-12) = -24$

31. $a_n = n(n-6)$
$= a_1 + a_2 = 1(1-6) + 2(2-6)$
$= 1(-5) + 2(-4) = -13$

33. $a_n = (n+3)(n+1)$
$a_1 = (1+3)(1+1) = 4(2) = 8$
$a_2 = (2+3)(2+1) = 5(3) = 15$
$a_3 = (3+3)(3+1) = 6(4) = 24$
$a_4 = (4+3)(4+1) = 7(5) = 35$
$\sum_{i=1}^{4} a_i = 8 + 15 + 24 + 35 = 82$

35. $a_n = -2n$
$\sum_{i=1}^{4} (-2i)$
$= -2(1) + (-2)(2) + (-2)(3) + (-2)(4)$
$= -2 - 4 - 6 - 8 = -20$

37. $a_n = -\frac{n}{3}$
$a_1 + a_2 + a_3 = -\frac{1}{3} - \frac{2}{3} - \frac{3}{3} = -2$

39. 1, 2, 3, …, 10
$a_n = n$
$\sum_{i=1}^{10} i = 1 + 2 + 3 + \cdots + 10$
$= \frac{10(11)}{2} = 55$
A total of 55 trees were planted.

41. $a_1 = 6$ and $r = 2$
$a_n = 6 \cdot 2^{n-1}$
$a_5 = 6 \cdot 2^4 = 6 \cdot 16 = 96$
There will be 96 fungus units at the beginning of the 5th day.

43. $a_1 = 50$ and $r = 2$
Since 48 = 4(12), we seek the fourth term after a_1, namely a_5.
The general term of the sequence is $a_n = 50(2)^{n-1}$, where *n* represents the number of 12-hr periods.
$a_5 = 50(2)^4 = 50(16) = 800$
There are 800 bacteria after 48 hours.

45. $a_n = (n+1)(n+2)$
$a_4 = (4+1)(4+2) = 5(6) = 30$ opossums
$a_1 = (1+1)(1+2) = 2(3) = 6$
$a_2 = (2+1)(2+2) = 3(4) = 12$
$a_3 = (3+1)(3+2) = 4(5) = 20$
$\sum_{i=1}^{4} a_i = 6 + 12 + 20 + 30 = 68$ opossums

47. $a_n = 100(0.5)^n$
$a_4 = 100(0.5)^4 = 6.25$ lbs of decay.
$a_1 = 100(0.5)^1 = 50$
$a_2 = 100(0.5)^2 = 25$
$a_3 = 100(0.5)^3 = 12.5$
$\sum_{i=1}^{4} a_i = 50 + 25 + 12.5 + 6.25$
$= 93.75$ lbs of decay

49. $a_1 = 40$ and $r = \frac{4}{5}$
$a_5 = 40\left(\frac{4}{5}\right)^4 = 16.384$ or 16.4 in.
$a_2 = 40\left(\frac{4}{5}\right)^1 = 32$
$a_3 = 40\left(\frac{4}{5}\right)^2 = 25.6$
$a_4 = 40\left(\frac{4}{5}\right)^3 = 20.48$
$\sum_{i=1}^{5} a_i = 40 + 32 + 25.6 + 20.48 + 16.384$
$= 134.464$ or 134.5 in.

51. a. $\sum_{i=1}^{7} i + i^2$
$= (1+1^2) + (2+2^2) + (3+3^2)$
$+ (4+4^2) + (5+5^2) + (6+6^2)$
$+ (7+7^2)$
$= 2 + 6 + 12 + 20 + 30 + 42 + 56$

b. $\sum_{i=1}^{7} i + \sum_{i=1}^{7} i^2$
$= 1 + 2 + 3 + 4 + 5 + 6 + 7 + 1 + 4$
$+ 9 + 16 + 25 + 36 + 49$

c. They are equal; 168

d. True

53. $\frac{5}{1-\frac{1}{2}} = \frac{5}{\frac{1}{2}} = 5 \cdot \frac{2}{1} = 10$

55. $\frac{\frac{1}{3}}{1-\frac{1}{10}} = \frac{\frac{1}{3}}{\frac{9}{10}} = \frac{1}{3} \cdot \frac{10}{9} = \frac{10}{27}$

57. $\frac{3(1-2^4)}{1-2} = \frac{3(1-16)}{-1}$
$= \frac{3(-15)}{-1} = \frac{-45}{-1} = 45$

59. $\frac{10}{2}(3+15) = \frac{10}{2}(18) = \frac{180}{2} = 90$

Exercise Set 11.4

1. 1, 3, 5, 7, ...
The first term of this arithmetic sequence is $a_1 = 1$ and the sixth term is $a_6 = 11$.

$$S_6 = \frac{6}{2}(a_1 + a_6)$$
$$= 3(1+11)$$
$$= 3(12)$$
$$S_6 = 36$$

3. 4, 12, 36, ...
$a_1 = 4,\ r = 3, n = 5$

$$S_5 = \frac{4(1-3^5)}{1-3} = -2(1-243)$$
$$S_5 = -2(-242) = 484$$

5. 3, 6, 9, ...
The first term of this arithmetic sequence is $a_1 = 3$ and the sixth term is $a_6 = 18$.

$$S_6 = \frac{6}{2}(a_1 + a_6)$$
$$= 3(3+18)$$
$$= 3(21)$$
$$S_6 = 63$$

7. 2, $\frac{2}{5}$, $\frac{2}{25}$, ...
$a_1 = 2,\ r = \frac{1}{5},\ n = 4$

$$S_4 = \frac{2\left[1-\left(\frac{1}{5}\right)^4\right]}{1-\frac{1}{5}} = \frac{5}{2}\left(1-\frac{1}{625}\right)$$
$$S_4 = \frac{5}{2}\left(\frac{624}{625}\right) = \frac{312}{125} = 2.496$$

9. 1, 2, 3, ..., 10
The first term of this arithmetic sequence is $a_1 = 1$ and the tenth term is $a_{10} = 10$.

$$S_{10} = \frac{10}{2}(a_1 + a_{10})$$
$$= 5(1+10)$$
$$= 5(11)$$
$$S_{10} = 55$$

11. 1, 3, 5, 7
The first term of this arithmetic sequence is $a_1 = 1$ and the fourth term is $a_4 = 7$.

$$S_4 = \frac{4}{2}(a_1 + a_4)$$
$$= 2(1+7)$$
$$= 2(8)$$
$$S_4 = 16$$

13. 12, 6, 3, ...
$a_1 = 12$ and $r = \frac{1}{2}$

$$S_\infty = \frac{12}{1-\frac{1}{2}} = \frac{12}{\frac{1}{2}} = 24$$

15. $\frac{1}{10}, \frac{1}{100}, \frac{1}{1000}, \ldots$

$a_1 = \frac{1}{10}$ and $r = \frac{1}{10}$

$S_\infty = \frac{\frac{1}{10}}{1-\frac{1}{10}} = \frac{\frac{1}{10}}{\frac{9}{10}} = \frac{1}{9}$

17. $-10, -5, -\frac{5}{2}, \ldots$

$a_1 = -10$ and $r = \frac{1}{2}$

$S_\infty = \frac{-10}{1-\frac{1}{2}} = \frac{-10}{\frac{1}{2}} = -20$

19. $2, -\frac{1}{4}, \frac{1}{32}, \ldots$

$a_1 = 2$ and $r = -\frac{1}{8}$

$S_\infty = \frac{2}{1-\left(-\frac{1}{8}\right)} = \frac{2}{\frac{9}{8}} = \frac{16}{9}$

21. $\frac{2}{3}, -\frac{1}{3}, \frac{1}{6}, \ldots$

$a_1 = \frac{2}{3}$ and $r = -\frac{1}{2}$

$S_\infty = \frac{\frac{2}{3}}{1-\left(-\frac{1}{2}\right)} = \frac{\frac{2}{3}}{\frac{3}{2}} = \frac{4}{9}$

23. $-4, 1, 6, \ldots, 41$

The first term of this arithmetic sequence is $a_1 = -4$ and the tenth term is $a_{10} = 41$.

$S_{10} = \frac{10}{2}(a_1 + a_{10})$

$= 5(-4+41)$

$= 5(37)$

$S_{10} = 185$

25. $3, \frac{3}{2}, \frac{3}{4}, \ldots$

$a_1 = 3,\ r = \frac{1}{2},\ n = 7$

$S_7 = \frac{3\left[1-\left(\frac{1}{2}\right)^7\right]}{1-\frac{1}{2}} = 6\left(1-\frac{1}{128}\right)$

$S_7 = 6\left(\frac{127}{128}\right) = \frac{381}{64}$

27. $-12, 6, -3, \ldots$

$a_1 = -12,\ r = -\frac{1}{2},\ n = 5$

$S_5 = \frac{-12\left[1-\left(-\frac{1}{2}\right)^5\right]}{1-\left(-\frac{1}{2}\right)}$

$S_5 = -8\left(1+\frac{1}{32}\right) = -8\left(\frac{33}{32}\right)$

$S_5 = -\frac{33}{4}$ or -8.25

29. $\frac{1}{2}, \frac{1}{4}, 0, \ldots, -\frac{17}{4}$

The first term of this arithmetic sequence is $a_1 = \frac{1}{2}$ and the twentieth term is $a_{20} = -\frac{17}{4}$.

$S_{20} = \frac{20}{2}(a_1 + a_{20})$

$= 10\left(\frac{1}{2} - \frac{17}{4}\right)$

$= 10\left(-\frac{15}{4}\right)$

$S_{20} = -\frac{75}{2}$

31. $a_1 = 8,\ r = -\frac{2}{3},\ n = 3$

$S_3 = \frac{8\left[1-\left(-\frac{2}{3}\right)^3\right]}{1-\left(-\frac{2}{3}\right)} = \frac{24}{5}\left(1+\frac{8}{27}\right)$

$S_3 = \frac{24}{5}\left(\frac{35}{27}\right) = \frac{8}{1}\left(\frac{7}{9}\right) = \frac{56}{9}$

33. The first five terms of the sequence are 4000, 3950, 3900, 3850, 3800

$a_1 = 4000,\ d = -50,\ n = 12$

$a_{12} = 4000 + 11(-50)$

$a_{12} = 4000 - 550$

$a_{12} = 3450$ cars sold in month 12.

$S_{12} = \frac{12}{2}[4000 + 3450]$

$S_{12} = 6(7450)$

$S_{12} = 44,700$ cars sold in the first 12 months.

35. Firm A:
The first term of this arithmetic sequence is $a_1 = 22,000$ and the tenth term is $a_{10} = 31,000$.

$$S_{10} = \frac{10}{2}(a_1 + a_{10})$$
$$= 5(22,000 + 31,000)$$
$$= 5(53,000)$$
$$S_{10} = \$265,000$$

Firm B:
The first term of this arithmetic sequence is $a_1 = 20,000$ and the tenth term is $a_{10} = 30,800$.

$$S_{10} = \frac{10}{2}(a_1 + a_{10})$$
$$= \frac{10}{2}(20,000 + 30,800)$$
$$= 5(50,800)$$
$$S_{10} = \$254,000$$

Thus, Firm A is making the more profitable offer.

37. $a_1 = 30,000,\ r = 1.10, n = 4$

$a_4 = 30,000(1.10)^{4-1}$

$a_4 = \$39,930$ made during her fourth year of business.

$$S_4 = \frac{30,000(1-1.10^4)}{1-1.10}$$

$S_4 = \$139,230$ made during the first four years.

39. $a_1 = 30,\ r = 0.9, n = 5$

$a_5 = 30(0.9)^{5-1} = 19.683$ or approximately 20 minutes to assemble the first computer.

$$S_5 = \frac{30(1-0.9^5)}{1-0.9} = 122.853, \text{ or}$$

approximately 123 minutes to assemble the first 5 computers.

41. $a_1 = 20$ and $r = \frac{4}{5}$

$$S_\infty = \frac{20}{1-\frac{4}{5}} = \frac{20}{\frac{1}{5}} = 100$$

We double the number (to account for the flight up as well as down) and subtract 20 (since the first bounce was preceded by only a downward flight). Thus, the ball traveled a total distance of $2(100) - 20 = 180$ feet.

43. Player A:
The first term of this arithmetic sequence is $a_1 = 1$ and the ninth term is $a_9 = 9$.

$$S_9 = \frac{9}{2}(a_1 + a_9)$$
$$S_9 = \frac{2}{9}(1+9)$$
$$= \frac{9}{2}(10)$$
$$S_{10} = 45 \text{ points}$$

Player B:
The first term of this arithmetic sequence is $a_1 = 10$ and the sixth term is $a_6 = 15$.

$$S_6 = \frac{6}{2}(a_1 + a_6)$$
$$= 3(10+15)$$
$$= 3(25)$$
$$S_6 = 75 \text{ points}$$

45. The first term of this arithmetic sequence is $a_1 = 200$ and the twentieth term is $a_{20} = 105$.

$$S_{20} = \frac{20}{2}(a_1 + a_{20})$$
$$S_9 = 10(200 + 105)$$
$$= 10(305)$$
$$S_{20} = 3050$$

Thus, \$3050 rent is paid for 20 days during the holiday rush.

47. $a_1 = 0.01,\ r = 2, n = 30$

$$S_{30} = \frac{0.01(1-2^{30})}{1-2} = 10,737,418.23$$

He would pay \$10,737,418.23 in room and board for the 30 days.

49. $0.88\overline{8} = 0.8 + 0.08 + 0.008 + \cdots$
$= \frac{8}{10} + \frac{8}{100} + \frac{8}{1000} + \cdots$

This is a geometric series with $a_1 = \frac{8}{10}$ and $r = \frac{1}{10}$

$$S_\infty = \frac{\frac{8}{10}}{1 - \frac{1}{10}} = \frac{\frac{8}{10}}{\frac{9}{10}} = \frac{8}{10} \cdot \frac{10}{9} = \frac{8}{9}$$

51. Answers may vary.

53. $6 \cdot 5 \cdot 4 \cdot 3 \cdot 2 \cdot 1 = (30)(12)(2) = 720$

55. $\frac{3 \cdot 2 \cdot 1}{2 \cdot 1} = 3$

57. $(x+5)^2 = x^2 + 2(5x) + 25$
$= x^2 + 10x + 25$

59. $(2x-1)^3$
$= (2x-1)(2x-1)^2$
$= (2x-1)(4x^2 - 4x + 1)$
$= 8x^3 - 8x^2 + 2x - 4x^2 + 4x - 1$
$= 8x^3 - 12x^2 + 6x - 1$

Exercise Set 11.5

1. $(m+n)^3$
$= 1 \cdot m^3 + 3 \cdot m^2 n + 3 \cdot mn^2 + 1 \cdot n^3$
$= m^3 + 3m^2 n + 3mn^2 + n^3$

3. $(c+d)^5$
$= 1 \cdot c^5 + 5 \cdot c^4 d + 10 \cdot c^3 d^2 + 10 \cdot c^2 d^3 + 5 \cdot cd^4 + 1 \cdot d^5$
$= c^5 + 5c^4 d + 10c^3 d^2 + 10c^2 d^3 + 5cd^4 + d^5$

5. $(y-x)^5 = [y + (-x)]^5$
$= 1y^5 + 5y^4(-x) + 10y^3(-x)^2 + 10y^2(-x)^3 + 5y(-x)^4 + (-x)^5$
$= y^5 - 5y^4 x + 10y^3 x^2 - 10y^2 x^3 + 5yx^4 - x^5$

7. Answers may vary.

9. $\frac{8!}{7!} = \frac{8 \cdot 7!}{7!} = 8$

11. $\frac{7!}{5!} = \frac{7 \cdot 6 \cdot 5!}{5!} = 7 \cdot 6 = 42$

13. $\frac{10!}{7!2!} = \frac{10\cdot 9\cdot 8\cdot 7!}{7!2!} = \frac{10\cdot 9\cdot 8}{2\cdot 1} = \frac{720}{2} = 360$

15. $\frac{8!}{6!0!} = \frac{8\cdot 7\cdot 6!}{6!\cdot 1} = 8\cdot 7 = 56$

17. $(a+b)^7$

$= a^7 + 7a^6b + \frac{7\cdot 6}{2!}a^5b^2 + \frac{7\cdot 6\cdot 5}{3!}a^4b^3 + \frac{7\cdot 6\cdot 5\cdot 4}{4!}a^3b^4 + \frac{7\cdot 6\cdot 5\cdot 4\cdot 3}{5!}a^2b^5$

$+ \frac{7\cdot 6\cdot 5\cdot 4\cdot 3\cdot 2}{6!}ab^6 + b^7$

$a^7 + 7a^6b + 21a^5b^2 + 35a^4b^3 + 35a^3b^4 + 21a^2b^5 + 7ab^6 + b^7$

19. $(a+2b)^5$

$= a^5 + 5a^4(2b) + \frac{5\cdot 4}{2!}a^3(2b)^2 + \frac{5\cdot 4\cdot 3}{3!}a^2(2b)^3 + \frac{5\cdot 4\cdot 3\cdot 2}{4!}a(2b)^4 + (2b)^5$

$= a^5 + 10a^4b + 40a^3b^2 + 80a^2b^3 + 80ab^4 + 32b^5$

21. $(q+r)^9$

$= q^9 + 9q^8r + \frac{9\cdot 8}{2!}q^7r^2 + \frac{9\cdot 8\cdot 7}{3!}q^6r^3 + \frac{9\cdot 8\cdot 7\cdot 6}{4!}q^5r^4 + \frac{9\cdot 8\cdot 7\cdot 6\cdot 5}{5!}q^4r^5 + \frac{9\cdot 8\cdot 7\cdot 6\cdot 5\cdot 4}{6!}q^3r^6$

$+ \frac{9\cdot 8\cdot 7\cdot 6\cdot 5\cdot 4\cdot 3}{7!}q^2r^7 + \frac{9\cdot 8\cdot 7\cdot 6\cdot 5\cdot 4\cdot 3\cdot 2}{8!}qr^8 + r^9$

$= q^9 + 9q^8r + 36q^7r^2 + 84q^6r^3 + 126q^5r^4 + 126q^4r^5 + 84q^3r^6 + 36q^2r^7 + 9qr^8 + r^9$

23. $(4a+b)^5$

$= (4a)^5 + 5(4a)^4b + \frac{5\cdot 4}{2!}(4a)^3b^2 + \frac{5\cdot 4\cdot 3}{3!}(4a)^2b^3 + \frac{5\cdot 4\cdot 3\cdot 2}{4!}(4a)b^4 + b^5$

$= 1024a^5 + 1280a^4b + 640a^3b^2 + 160a^2b^3 + 20ab^4 + b^5$

25. $(5a-2b)^4$

$= (5a)^4 + 4(5a)^3(-2b) + \frac{4\cdot 3}{2!}(5a)^2(-2b)^2 + \frac{4\cdot 3\cdot 2}{3!}(5a)(-2b)^3 + (-2b)^4$

$= 625a^4 - 1000a^3b + 600a^2b^2 - 160ab^3 + 16b^4$

27. $(2a+3b)^3$

$= (2a)^3 + 3(2a)^2(3b) + \frac{3\cdot 2}{2!}(2a)(3b)^2 + (3b)^3$

$= 8a^3 + 36a^2b + 54ab^2 + 27b^3$

29. $(x+2)^5$

$= x^5 + 5x^4(2) + \frac{5\cdot 4}{2!}x^3(2^2) + \frac{5\cdot 4\cdot 3}{3!}x^2(2^3) + \frac{5\cdot 4\cdot 3\cdot 2}{4!}x(2^4) + 2^5$

$= x^5 + 10x^4 + 40x^3 + 80x^2 + 80x + 32$

31. 5th term of $(c-d)^5$ corresponds to $r = 4$:

$\frac{5!}{4!(5-4)!}c^{5-4}(-d)^4 = 5cd^4$

33. 8th term of $(2c+d)^7$ corresponds to $r = 7$:

$$\frac{7!}{7!(7-7)!}(2c)^{7-7}d^7 = d^7$$

35. 4th term of $(2r-s)^5$ corresponds to $r = 3$:

$$\frac{5!}{3!(5-3)!}(2r)^{5-3}(-s)^3 = -40r^2s^3$$

37. 3rd term of $(x+y)^4$ corresponds to $r = 2$:

$$\frac{4!}{2!(4-2)!}x^{4-2}y^2 = 6x^2y^2$$

39. 2nd term of $(a+3b)^{10}$ corresponds to $r = 1$:

$$\frac{10!}{1!(10-1)!}a^{10-1}(3b)^1 = 30a^9b$$

41. $f(x) = |x|$

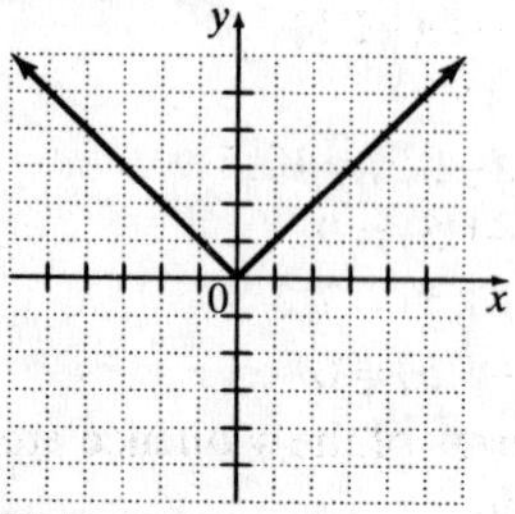

Not one-to-one

43. $H(x) = 2x + 3$

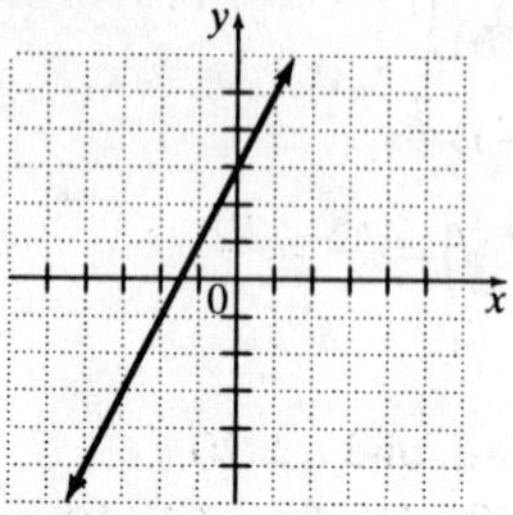

One-to-one

45. $f(x) = x^2 + 3$

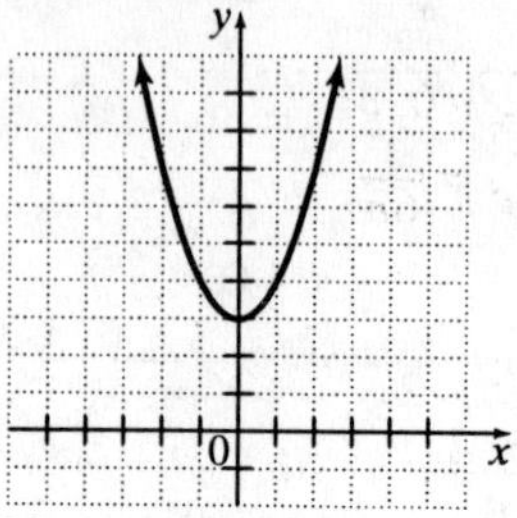

Not one-to-one

Chapter 11 Review

1. $a_n = -3n^2$

$a_1 = -3(1)^2 = -3$

$a_2 = -3(2)^2 = -12$

$a_3 = -3(3)^2 = -27$

$a_4 = -3(4)^2 = -48$

$a_5 = -3(5)^2 = -75$

The first five terms of the sequence are −3, −12, −27, −48, −75.

2. $a_n = n^2 + 2n$

$a_1 = 1^2 + 2(1) = 1 + 2 = 3$

$a_2 = 2^2 + 2(2) = 4 + 4 = 8$

$a_3 = 3^2 + 2(3) = 9 + 6 = 15$

$a_4 = 4^2 + 2(4) = 16 + 8 = 24$

$a_5 = 5^2 + 2(5) = 25 + 10 = 35$

The first five terms of the sequence are 3, 8, 15, 24, 35.

3. $a_n = \frac{(-1)^n}{100}$

$a_{100} = \frac{(-1)^{100}}{100} = \frac{1}{100}$

4. $a_n = \frac{2n}{(-1)^2}$

$a_{50} = \frac{2(50)}{(-1)^2} = 100$

5. $\frac{1}{6}, \frac{1}{12}, \frac{1}{18}, \ldots$
or $\frac{1}{6\cdot 1}, \frac{1}{6\cdot 2}, \frac{1}{6\cdot 3}, \ldots$
In general, $a_n = \frac{1}{6n}$

6. $-1, 4, -9, 16, \ldots$
$a_n = (-1)^n n^2$

7. $a_n = 32n - 16$
$a_5 = 32(5) - 16 = 160 - 16 = 144$ ft
$a_6 = 32(6) - 16 = 192 - 16 = 176$ ft
$a_7 = 32(7) - 16 = 224 - 16 = 208$ ft

8. $a_n = 100(2)^{n-1}$
$10{,}000 = 100(2)^{n-1}$
$100 = 2^{n-1}$
$\log 100 = (n-1)\log 2$
$n = \frac{\log 100}{\log 2} + 1 \approx 7.6$
Eighth day culture will be at least 10,000.
Originally, $a_1 = 100(2)^0 = 100$.

9. $a_1 = 450$
$a_2 = 3(450) = 1350$
$a_3 = 3(1350) = 4050$
$a_4 = 3(4050) = 12{,}150$
$a_5 = 3(12{,}150) = 36{,}450$
This predicts that in 2003 (when $n = 5$) the number of infected people should be 36,450.

10. $a_n = 50 + (n-1)8$
$a_1 = 50$
$a_2 = 50 + 8 = 58$
$a_3 = 50 + 2(8) = 66$
$a_4 = 50 + 3(8) = 74$
$a_5 = 50 + 4(8) = 82$
$a_6 = 50 + 5(8) = 90$
$a_7 = 50 + 6(8) = 98$
$a_8 = 50 + 7(8) = 106$
$a_9 = 50 + 8(8) = 114$
$a_{10} = 50 + 9(8) = 122$
There are 122 seats in the tenth row.

11. $a_1 = -2$ and $r = \frac{2}{3}$
The first 5 terms of the sequence are
$-2, -\frac{4}{3}, -\frac{8}{9}, -\frac{16}{27}, -\frac{32}{81}$.

12. $a_n = 12 + (n-1)(-1.5)$
$a_1 = 12$
$a_2 = 12 + (1)(-1.5) = 10.5$
$a_3 = 12 + 2(-1.5) = 9$
$a_4 = 12 + 3(-1.5) = 7.5$
$a_5 = 12 + 4(-1.5) = 6$
The first 5 terms of the sequence are 12, 10.5, 9, 7.5, 6.

13. $a_1 = -5$, $d = 4$, and $n = 30$
$a_{30} = 5 + (30-1)4 = -5 + 116 = 111$

14. $a_n = 2 + (n-1)\frac{3}{4}$
$a_{11} = 2 + 10\left(\frac{3}{4}\right) = \frac{38}{4} = \frac{19}{2}$

15. $12, 7, 2, \ldots$
$a_1 = 12$, $d = -5$, and $n = 20$
$a_{20} = 12 + (20-1)(-5) = 12 - 95 = -83$

16. $a_n = a_1 r^{n-1}$
$a_n = 4\left(\frac{3}{2}\right)^{n-1}$
$a_6 = 4\left(\frac{3}{2}\right)^{6-1} = 4\left(\frac{3}{2}\right)^5 = \frac{4(243)}{32} = \frac{243}{8}$

17. $a_4 = 18$ and $a_{20} = 98$
Use the relationship:
$a_4 + 16d = a_{20}$
$18 + 16d = 98$
$16d = 80$
$d = 5$
Now use the relationship:
$a_4 = a_1 + 3d$
$18 = a_1 + 3(5)$
$18 = a_1 + 15$
$a_1 = 3$

18. $-48 = a_3 = a_1 r^{3-1}$
$192 = a_4 = a_1 r^{4-1}$
$-48 = a_1 r^2$
$192 = a_1 r^3$
Dividing we have $r = -4$.
Substitution gives $-48 = 16a_1$ or $a_1 = -3$.

19. $\frac{3}{10}, \frac{3}{100}, \frac{3}{1000}, \ldots$ or
$\frac{3}{10}, \frac{3}{10^2}, \frac{3}{10^3}, \ldots$
In general, $a_n = \frac{3}{10^n}$

20. 50, 58, 66, ...
$a_n = 50 + (n-1)8$ or $a_n = 42 + 8n$

21. $\frac{8}{3}$, 4, 6, ...
Geometric, since $\frac{4}{\frac{8}{3}} = \frac{6}{4} = \frac{3}{2}$
$a_1 = \frac{8}{3}$ and $r = \frac{3}{2}$

22. arithmetic; $a_1 = -10.5$, $d = 4.4$

23. $7x, -14x, 28x$
Geometric, since
$\frac{-14x}{7x} = \frac{28x}{-14x} = -2$
$a_1 = 7x$ and $r = -2$

24. neither

25. $a_1 = 8$ and $r = 0.75$
8, 6, 4.5, 3.4, 2.5, 1.9
Yes, a ball that rebounds to a height of 2.5 feet after the fifth bounce is good, since $2.5 \geq 1.9$.

26. $a_n = 25 + (n-1)(-4)$
$a_n = 25 + 6(-4) = 1$
Continuing the progression as far as possible leaves 1 can in the top row.

27. $a_1 = 1$ and $r = 2$
$a_n = 1 \cdot 2^{n-1}$ or
$a_n = 2^{n-1}$
$a_{10} = 2^{10-1} = 2^9 = \$512$
$a_{30} = 2^{30-1} = 2^{29} = \$536,870,912$

28. $a_n = a_1 r^{n-1}$
$a_5 = 30(.7)^4 = 7.203$ in.

29. $a_1 = 900$ and $d = 150$
$a_n = 900 + (n-1)150$
$a_6 = 900 + (6-1)150$
$= 900 + 750 = \$1650$/month

30. $\frac{1}{512}, \frac{1}{256}, \frac{1}{128}, \ldots$
first fold $a_1 = \frac{1}{256}$, $r = 2$
$a_n = a_1 r^{n-1}$
$a_{15} = \frac{1}{256}(2)^{15-1} = 64$ inches

31. $\sum_{i=1}^{5}(2i-1)$
$= [2(1) - 1] + [2(2) - 1] + [2(3) - 1] + [2(4) - 1] + [2(5) - 1]$
$= 1 + 3 + 5 + 7 + 9 = 25$

32. $\sum_{i=1}^{5} i(i+2)$
$= 1(1 + 2) + 2(2 + 2) + 3(3 + 2) + 4(4 + 2) + 5(5 + 2)$
$= 3 + 8 + 15 + 24 + 35 = 85$

33. $\sum_{i=2}^{4} \frac{(-1)^i}{2i}$

$= \frac{(-1)^2}{2(2)} + \frac{(-1)^3}{2(3)} + \frac{(-1)^4}{2(4)}$

$= \frac{1}{4} - \frac{1}{6} + \frac{1}{8}$

$= \frac{6-4+3}{24} = \frac{5}{24}$

34. $\sum_{i=3}^{5} 5(-1)^{i-1}$

$= 5(-1)^{3-1} + 5(-1)^{4-1} + 5(-1)^{5-1}$

$= 5(1) + 5(-1) + 5(1) = 5$

35. $a_n = (n-3)(n+2)$

$S_4 = (1-3)(1+2) + (2-3)(2+2)$
$\quad + (3-3)(3+2) + (4-3)(4+2)$

$= -6 - 4 + 0 + 6 = -4$

36. $a_n = n^2$

$S_6 = (1)^2 + (2)^2 + (3)^2 + (4)^2 + (5)^2 + (6)^2$

$= 1 + 4 + 9 + 16 + 25 + 36 = 91$

37. $a_n = -8 + (n-1)3$

$a_1 = -8 + (1-1)3 = -8 + 0 = -8$

$a_2 = -8 + (2-1)3 = -8 + 3 = -5$

$a_3 = -8 + (3-1)3 = -8 + 6 = -2$

$a_4 = -8 + (4-1)3 = -8 + 9 = 1$

$a_5 = -8 + (5-1)3 = -8 + 12 = 4$

So $S_5 = -8 + (-5) + (-2) + 1 + 4 = -10$.

38. $a_n = 5(4)^{n-1}$

$S_3 = 5(4)^0 + 5(4)^1 + 5(4)^2$

$= 5 + 20 + 80 = 105$

39. $1 + 3 + 9 + 27 + 81 + 243$

$= 3^0 + 3^1 + 3^2 + 3^3 + 3^4 + 3^5$

$= \sum_{i=1}^{6} 3^{i-1}$

40. $6 + 2 + (-2) + (-6) + (-10) + (-14)$
$\quad + (-18)$

$a_n = 6 + (n-1)(-4)$

$\sum_{i=1}^{7} 6 + (i-1)(-4)$

41. $\frac{1}{4} + \frac{1}{16} + \frac{1}{64} + \frac{1}{256} = \frac{1}{4^1} + \frac{1}{4^2} + \frac{1}{4^3} + \frac{1}{4^4}$

$\sum_{i=1}^{4} \frac{1}{4^i}$

42. $1 + \left(-\frac{3}{2}\right) + \frac{9}{4}$

$a_n = 1\left(-\frac{3}{2}\right)^{n-1}$

$\sum_{i=1}^{3} \left(-\frac{3}{2}\right)^{i-1}$

43. $a_1 = 20$ and $r = 2$

$a_n = 20(2)^n$ represents the number of yeast, where n represents the number of 8-hr periods. Since $48 = 6(8)$ here, $n = 6$.

$a_6 = 20(2)^6 = 1280$ yeast

44. $a_n = n^2 + 2n - 1$

$a_4 = (4)^2 + 2(4) - 1 = 23$ cranes

$\sum_{i=1}^{4} i^2 + 2i - 1$

$= (1 + 2 - 1) + (4 + 4 - 1)$
$\quad + (9 + 6 - 1) + (16 + 8 - 1)$

$= 2 + 7 + 14 + 23 = 46$ cranes

45. For Job A: $a_1 = 39,500$ and $d = 2200$;

$a_5 = 39,500 + (5-1)2200 = \$48,300$

For Job B: $a_1 = 41,000$ and $d = 1400$;

$a_5 = 41,000 + (5-1)1400 = \$46,600$

For the fifth year, Job A has a higher salary.

46. $a_n = 200(0.5)^n$

$a_3 = 200(0.5)^3 = 25$ kg

$\sum_{i=1}^{3} 200(0.5)^i$

$= 200(0.5) + 200(0.5)^2 + 200(0.5)^3$

$= 100 + 50 + 25 = 175$ kg

47. 15, 19, 23, ...

$a_1 = 15$ and $d = 4$

$S_6 = \frac{6}{2}[2(15) + (6-1)4]$

$= 3[30 + 20] = 3(50) = 150$

48. 5, −10, 20, ⋯

$a_1 = 5,\ r = -2$

$S_n = \frac{a_1(1-r^n)}{1-r}$

$S_9 = \frac{5(1-(-2)^9)}{1-(-2)} = 855$

49. $a_1 = 1$, $d = 2$, and $n = 30$

$S_{30} = \frac{30}{2}[2(1) + (30-1)2]$

$= 15[2 + 58] = 15(60) = 900$

50. 7, 14, 21, 28, ...

$a_n = 7 + (n-1)7$

$a_{20} = 7 + (20-1)7 = 140$

$S_n = \frac{n}{2}(a_1 + a_n)$

$S_{20} = \frac{20}{2}(7 + 140) = 1470$

51. 8, 5, 2, ...

$a_1 = 8$, $d = -3$, and $n = 20$

$S_{20} = \frac{20}{2}[2(8) + (20-1)(-3)]$

$= 10[16 - 57] = 10(-41) = -410$

52. $\frac{3}{4}, \frac{9}{4}, \frac{27}{4}, \ldots$

$a_1 = \frac{3}{4},\ r = 3$

$S_n = \frac{a_1(1-r^n)}{1-r}$

$S_8 = \frac{\frac{3}{4}(1-3^8)}{1-3} = 2460$

53. $a_1 = 6$ and $r = 5$

$S_4 = \frac{6(1-5^4)}{1-5} = \frac{-3}{2}(1-625)$

$= \frac{3}{2}(624) = 936$

54. $a_1 = -3,\ d = -6$

$a_n = -3 + (n-1)(-6)$

$a_{100} = -3 + (100-1)(-6) = -597$

$S_n = \frac{n}{2}(a_1 + a_n)$

$S_{100} = \frac{100}{2}(-3 + (-597))$

$S_{100} = -30,000$

55. $5, \frac{5}{2}, \frac{5}{4}, \cdots$

$a_1 = 5$ and $r = \frac{1}{2}$

$S_\infty = \frac{5}{1-\frac{1}{2}} = \frac{5}{\frac{1}{2}} = 10$

56. $18, -2, \frac{2}{9}, \cdots$

$a_1 = 18,\ r = -\frac{1}{9}$

$S_\infty = \frac{a_1}{1-r} = \frac{18}{1+\frac{1}{9}} = \frac{81}{5}$

57. $-20, -4, -\frac{4}{5}, \cdots$

$a_1 = -20$ and $r = \frac{1}{5}$

$S_\infty = \frac{-20}{1-\frac{1}{5}} = \frac{-20}{\frac{4}{5}} = -25$

58. $0.2,\ 0.02,\ 0.002,\ \cdots$

$a_1 = 0.2,\ r = \frac{1}{10}$

$S_\infty = \frac{a_1}{1-r} = \frac{.2}{1-\frac{1}{10}} = \frac{2}{9}$

59. $a_1 = 20{,}000$, $r = 1.15$, and $n = 4$.

$a_4 = 20{,}000(1.15)^{4-1} = \$30{,}418$

earned in his fourth year.

$S_4 = \frac{20{,}000(1-1.15^4)}{1-1.15}$ = \$99,868 earned in his first four years.

60. $a_n = 40(.8)^{n-1}$

$a_4 = 40(.8)^{4-1} = 20.48$ min

$S_n = \frac{a_1(1-r^n)}{1-r} = \frac{40(1-.8^4)}{1-.8} = 118$ min

61. $a_1 = 100$, $d = -7$, and $n = 7$

$a_7 = 100 + (7-1)(-7) = 100 - 42$

= \$58 rent paid for the seventh day.

$S_7 = \frac{7}{2}[2(100) + (7-1)(-7)]$

$= \frac{7}{2}[200 - 42] = \frac{7}{2}(158)$

= \$553 rent paid for the first seven days.

62. $a_1 = 15,\ r = .8$

$S_\infty = \frac{a_1}{1-r} = \frac{15}{1-.8} = 75$ feet downward

$a_1 = 12,\ r = .8$

$S_\infty = \frac{a_1}{1-r} = \frac{12}{1-.8} = 60$ feet upward

The total is 135 feet.

63. 1800, 600, 200, ...

$a_1 = 1800$, $r = \frac{1}{3}$, and $n = 6$

$$S_6 = 1800\frac{\left(1-\left(\frac{1}{3}\right)^6\right)}{1-\frac{1}{3}} = 2700\left(1-\frac{1}{729}\right) = 2700\left(\frac{728}{729}\right) = \frac{72800}{27}$$

≈ 2696 mosquitoes killed during the first six days after the spraying

64. 1800, 600, 200, ...
For which n is $a_n > 1$?
$a_n = a_1 r^{n-1} = 1800\left(\frac{1}{3}\right)^{n-1} > 1$
$(n-1)\log\left(\frac{1}{3}\right) > \log\frac{1}{1800}$
$(n-1)(-0.4771213) > (-3.2552725)$
$n - 1 < 6.8$
$n < 7.8$
No longer effective on the 8th day. About 2700 mosquitos were killed.

65. $0.55\overline{5} = .5 + .05 + .005 + \cdots$
$a_1 = .5$ and $r = .1$
$S_\infty = \frac{.5}{1-.1} = \frac{.5}{.9} = \frac{5}{9}$
$0.55\overline{5} = \frac{5}{9}$

66. 27, 30, 33, ⋯
$a_n = 27 + (n-1)(3)$
$a_{20} = 27 + (20-1)(3) = 84$
$S_n = \frac{n}{2}(a_1 + a_n)$
$S_{20} = \frac{20}{2}(27 + 84) = 1110$
1110 seats

67. $(x+z)^5 = 1\cdot x^5 + 5\cdot x^4 z + 10\cdot x^3 z^2 + 10\cdot x^2 z^3 + 5\cdot xz^4 + 1\cdot z^5$
$= x^5 + 5x^4 z + 10x^3 z^2 + 10x^2 z^3 + 5xz^4 + z^5$

68. $(y-r)^6 = y^6 - 6y^5 r + 15y^4 r^2 - 20y^3 r^3 + 15y^2 r^4 - 6yr^5 + r^6$

69. $(2x+y)^4 = 1\cdot(2x)^4 + 4\cdot(2x)^3 y + 6\cdot(2x)^2 y^2 + 4\cdot(2x)y^3 + 1\cdot y^4$
$= 16x^4 + 32x^3 y + 24x^2 y^2 + 8xy^3 + y^4$

70. $(3y-z)^4 = (3y)^4 + 4(3y)^3(-z) + 6(3y)^2(-z)^2 + 4(3y)(-z)^3 + (-z)^4$
$= 81y^4 - 108y^3 z + 54y^2 z^2 - 12yz^3 + z^4$

71. $(b+c)^8$
$= b^8 + 8b^7 c + \frac{8\cdot 7}{2!}b^6 c^2 + \frac{8\cdot 7\cdot 6}{3!}b^5 c^3 + \frac{8\cdot 7\cdot 6\cdot 5}{4!}b^4 c^4 + \frac{8\cdot 7\cdot 6\cdot 5\cdot 4}{5!}b^3 c^5 + \frac{8\cdot 7\cdot 6\cdot 5\cdot 4\cdot 3}{6!}b^2 c^6$
$+ \frac{8\cdot 7\cdot 6\cdot 5\cdot 4\cdot 3\cdot 2}{7!}bc^7 + c^8$
$= b^8 + 8b^7 c + 28b^6 c^2 + 56b^5 c^3 + 70b^4 c^4 + 56b^3 c^5 + 28b^2 c^6 + 8bc^7 + c^8$

72. $(x-w)^7 = x^7 - 7x^6 w + 21x^5 w^2 - 35x^4 w^3 + 35x^3 w^4 - 21x^2 w^5 + 7xw^6 - w^7$

73. $(4m-n)^4 = (4m+(-n))^4$

$= (4m)^4 + 4(4m)^3(-n) + \frac{4\cdot 3}{2!}(4m)^2(-n)^2 + \frac{4\cdot 3\cdot 2}{3!}(4m)(-n)^3 + (-n)^4$

$= 256m^4 - 256m^3n + 96m^2n^2 - 16mn^3 + n^4$

74. $(p-2r)^5 = p^5 + 5p^4(-2r) + 10p^3(-2r)^2 + 10p^2(-2r)^3 + 5p(-2r)^4 + (-2r)^5$

$= p^5 - 10p^4r + 40p^3r^2 - 80p^2r^3 + 80pr^4 - 32r^5$

75. 4th term of $(a+b)^7$ corresponds to $r = 3$.

$\frac{7!}{3!(7-3)!}a^{7-3}b^3 = 35a^4b^3$

76. The 11th term is $\frac{n!}{r!(n-r)!}a^{n-r}b^r$ where $n = 10$, $r = 10$, $a = y$, $b = 2z$.

$\frac{10!}{10!0!}y^{10-10}(2z)^{10} = 1024z^{10}$

Chapter 11 Test

1. $a_n = \frac{(-1)^n}{n+4}$

$a_1 = \frac{(-1)^1}{1+4} = -\frac{1}{5}$

$a_2 = \frac{(-1)^2}{2+4} = \frac{1}{6}$

$a_3 = \frac{(-1)^3}{3+4} = -\frac{1}{7}$

$a_4 = \frac{(-1)^4}{4+4} = \frac{1}{8}$

$a_5 = \frac{(-1)^5}{5+4} = -\frac{1}{9}$

or $-\frac{1}{5}, \frac{1}{6}, -\frac{1}{7}, \frac{1}{8}, -\frac{1}{9}$

2. $a_n = \dfrac{3}{(-1)^n}$

$a_1 = \dfrac{3}{(-1)^1} = -3$

$a_2 = \dfrac{3}{(-1)^2} = 3$

$a_3 = \dfrac{3}{(-1)^3} = -3$

$a_4 = \dfrac{3}{(-1)^4} = 3$

$a_5 = \dfrac{3}{(-1)^5} = -3$

or $-3, 3, -3, 3, -3$

3. $a_n = 10 + 3(n-1)$

$a_{80} = 10 + 3(80-1) = 10 + 237 = 247$

4. $a_n = (n+1)(n-1)(-1)^n$

$a_{200} = (200+1)(200-1)(-1)^{200}$

$= 200^2 - 1^2 = 40000 - 1 = 39{,}999$

5. $\frac{2}{5}, \frac{2}{25}, \frac{2}{125}, \cdots$ or $\frac{2}{5}, \frac{2}{5^2}, \frac{2}{5^3}, \cdots$

In general, $a_n = \dfrac{2}{5^n}$

6. $-9, 18, -27, 36, \cdots$

or $(-1)^1 9 \cdot 1, (-1)^2 9 \cdot 2,$

$(-1)^3 9 \cdot 3, (-1)^4 9 \cdot 4, \cdots$

In general, $a_n = (-1)^n 9n$

7. $a_n = 5(2)^{n-1}$

Geometric sequence with $a_1 = 5$ and $r = 2$.

$S_5 = \dfrac{5(1-2^5)}{1-2} = -5(1-32) = 5(31) = 155$

8. $a_n = 18 + (n-1)(-2)$

Arithmetic sequence with $a_1 = 18$ and $d = -2$.

$S_{30} = \frac{30}{2}[2(18) + (30-1)(-2)]$

$= 15[36 - 58] = 15(-22) = -330$

9. $a_1 = 24$ and $r = \frac{1}{6}$

$S_\infty = \dfrac{24}{1-\frac{1}{6}} = \dfrac{24}{\frac{5}{6}} = \dfrac{144}{5}$

10. $\frac{3}{2}, -\frac{3}{4}, \frac{3}{8}, \cdots$

$a_1 = \frac{3}{2}$ and $r = -\frac{1}{2}$

$S_\infty = \dfrac{\frac{3}{2}}{1-\left(-\frac{1}{2}\right)} = \dfrac{\frac{3}{2}}{\frac{3}{2}} = 1$

11. $\displaystyle\sum_{i=1}^{4} i(i-2)$

$= 1(1-2) + 2(2-2) + 3(3-2) + 4(4-2)$

$= -1 + 0 + 3 + 8 = 10$

12. $\sum_{i=2}^{4} 5(2)^i(-1)^{i-1}$

$= 5(2)^2(-1)^{2-1} + 5(2)^3(-1)^{3-1} + 5(2)^4(-1)^{4-1} = -20 + 40 - 80 = -60$

13. $(a-b)^6 = (a+(-b))^6$

$= 1\cdot a^6 + 6\cdot a^5(-b) + 15\cdot a^4(-b)^2 + 20a^3(-b)^3 + 15\cdot a^2(-b)^4 + 6\cdot a(-b)^5 + 1\cdot(-b)^6$

$= a^6 - 6a^5b + 15a^4b^2 - 20a^3b^3 + 15a^2b^4 - 6ab^5 + b^6$

14. $(2x+y)^5 = 1\cdot(2x)^5 + 5\cdot(2x)^4y + 10\cdot(2x)^3y^2 + 10\cdot(2x)^2y^3 + 5\cdot(2x)y^4 + y^5$

$= 32x^5 + 80x^4y + 80x^3y^2 + 40x^2y^3 + 10xy^4 + y^5$

15. $(y+z)^8 = y^8 + 8y^7z + \frac{8\cdot 7}{2!}y^6z^2 + \frac{8\cdot 7\cdot 6}{3!}y^5z^3 + \frac{8\cdot 7\cdot 6\cdot 5}{4!}y^4z^4 + \frac{8\cdot 7\cdot 6\cdot 5\cdot 4}{5!}y^3z^5$

$+ \frac{8\cdot 7\cdot 6\cdot 5\cdot 4\cdot 3}{6!}y^2z^6 + \frac{8\cdot 7\cdot 6\cdot 5\cdot 4\cdot 3\cdot 2}{7!}yz^7 + z^8$

$= y^8 + 8y^7z + 28y^6z^2 + 56y^5z^3 + 70y^4z^4 + 56y^3z^5 + 28y^2z^6 + 8yz^7 + z^8$

16. $(2p+r)^7 = (2p)^7 + 7(2p)^6r + \frac{7\cdot 6}{2!}(2p)^5r^2 + \frac{7\cdot 6\cdot 5}{3!}(2p)^4r^3 + \frac{7\cdot 6\cdot 5\cdot 4}{4!}(2p)^3r^4$

$+ \frac{7\cdot 6\cdot 5\cdot 4\cdot 3}{5!}(2p)^2r^5 + \frac{7\cdot 6\cdot 5\cdot 4\cdot 3\cdot 2}{6!}(2p)r^6 + r^7$

$= 128p^7 + 448p^6r + 672p^5r^2 + 560p^4r^3 + 280p^3r^4 + 84p^2r^5 + 14pr^6 + r^7$

17. $a_n = 250 + 75(n-1)$

$a_{10} = 250 + 75(10-1) = 250 + 675 = 925$

There were 925 people in the town at the beginning of the tenth year.

$a_1 = 250 + 75(1-1) = 250$

There were 250 people in the town at the beginning of the first year.

18. 1, 3, 5, ...

$a_1 = 1,\ d = 2,$ and $n = 8$

$a_8 = 1 + (8-1)2 = 1 + 14 = 15$

We want 1+ 3 + 5 + ... + 15

$S_8 = \frac{8}{2}[1+15] = 4(16) = 64$

There were 64 shrubs planted in the 8 rows.

19. $a_1 = 80,\ r = \frac{3}{4},$ and $n = 4$

$a_4 = 80\left(\frac{3}{4}\right)^{4-1} = 80\left(\frac{27}{64}\right) = \frac{135}{4}$ or 33.75

The arc length is 33.75 cm on the 4th swing.

$S_4 = \frac{80\left(1-\left(\frac{3}{4}\right)^4\right)}{1-\frac{3}{4}}$

$= 320\left(1 - \frac{81}{256}\right)$

$= 320\left(\frac{175}{256}\right)$

$= \frac{875}{4}$

$= 218.75$

The total of the arc lengths is 218.75 cm for the first 4 swings.

20. $a_1 = 80$ and $r = \frac{3}{4}$

$S_\infty = \frac{80}{1-\frac{3}{4}} = \frac{80}{\frac{1}{4}} = 320$

The total of the arc lengths is 320 cm before the pendulum comes to rest.

21. 16, 48, 80, ...

$a_1 = 16$, $d = 32$, and $n = 10$

$a_{10} = 16 + (10-1)32 = 16 + 288 = 304$

He falls 304 feet during the 10th second.

$S_{10} = \frac{10}{2}[16 + 304] = 5(320) = 1600$

He falls 1600 feet during the first 10 seconds.

22. $0.42\overline{42} = 0.42 + 0.0042 + 0.000042 + \cdots$

$a_1 = 0.42$ and $r = 0.01$

$S_\infty = \frac{0.42}{1-0.01} = \frac{0.42}{0.99} = \frac{42}{99} = \frac{14}{33}$

Thus, $0.42\overline{42} = \frac{14}{33}$.

Appendices

Appendix A Exercise Set

1. $90° - 19° = 71°$

3. $90° - 70.8° = 19.2°$

5. $90° - 11\frac{1}{4}° = 78\frac{3}{4}°$

7. $180° - 150° = 30°$

9. $180° - 30.2° = 149.8°$

11. $180° - 79\frac{1}{2}° = 100\frac{1}{2}°$

13. $m\angle 1 = 100°$, since it and the given angle are vertical angles.
$m\angle 2 = m\angle 3 = 70°$, since they are supplementary to $\angle 1$ and the given angle.
$m\angle 4 = 70°$ and $m\angle 5 = 110°$, since they are alternate interior angles with $\angle 3$ and the given angle, respectively.
$m\angle 6 = 70°$, since angles 4 and 6 are vertical angles.
$m\angle 7 = 110°$, since angles 5 and 7 are vertical angles.

15. $180° - 11° - 79° = 90°$

17. $180° - 25° - 65° = 90°$

19. $180° - 30° - 60° = 90°$

21. $90° - 45° = 45°$, so the other two angles are 45° and 90°.

23. $90° - 17° = 73°$, so the other two angles are 73° and 90°.

25. $90° - 39\frac{3}{4}° = 50\frac{1}{4}°$, so the other two angles are $50\frac{1}{4}°$ and $90°$.

27.
$$\frac{12}{4} = \frac{18}{x}$$
$$4x\left(\frac{12}{4}\right) = 4x\left(\frac{18}{x}\right)$$
$$12x = 72$$
$$x = 6$$

29.
$$\frac{6}{9} = \frac{3}{x}$$
$$9x\left(\frac{6}{9}\right) = 9x\left(\frac{3}{x}\right)$$
$$6x = 27$$
$$x = 4.5$$

31.
$$6^2 + 8^2 = c^2$$
$$36 + 64 = c^2$$
$$100 = c^2$$
$$c = 10$$

33.
$$5^2 + b^2 = 13^2$$
$$25 + b^2 = 169$$
$$b^2 = 144$$
$$b = 12$$

Appendix C Exercise Set

1.
$$\begin{aligned} V &= lwh \\ &= (6 \text{ in.})(4 \text{ in.})(3 \text{ in.}) = 72 \text{ cubic inches} \\ SA &= 2lh + 2wh + 2lw \\ &= 2(6 \text{ in.})(3 \text{ in.}) + 2(4 \text{ in.})(3 \text{ in.}) \\ &\quad + 2(6 \text{ in.})(4 \text{ in.}) \\ &= 36 \text{ sq. in.} + 24 \text{ sq. in.} + 48 \text{ sq. in.} \\ &= 108 \text{ sq. in.} \end{aligned}$$

3.
$$V = s^3 = (8 \text{ cm})^3 = 512 \text{ cu. cm}$$
$$SA = 6s^2 = 6(8 \text{ cm})^2 = 384 \text{ sq. cm}$$

5. $V = \frac{1}{3}\pi r^2 h$
$= \frac{1}{3}\pi(2 \text{ yd})^2(3 \text{ yd}) = 4\pi \text{ cu. yd}$
$\approx 4\left(\frac{22}{7}\right) \text{ cu. yd}$
$= 12\frac{4}{7} \text{ cu. yd}$
$SA = \pi r\sqrt{r^2 + h^2} + \pi r^2$
$= \pi(2 \text{ yd})\left(\sqrt{(2 \text{ yd})^2 + (3 \text{ yd})^2}\right) + \pi(2 \text{ yd})^2$
$= \pi(2 \text{ yd})(\sqrt{13} \text{ yd}) + \pi(4 \text{ sq. yd})$
$= \pi(2\sqrt{13} + 4) \text{ sq. yd}$
$\approx 35.20 \text{ sq. yd}$

7. $V = \frac{4}{3}\pi r^3$
$= \frac{4}{3}\pi(5 \text{ in.})^3$
$= \frac{500}{3}\pi \text{ cu. in.}$
$\approx \frac{500}{3}\left(\frac{22}{7}\right) \text{ cu. in.}$
$= 523\frac{17}{21} \text{ cu. in.}$
$SA = 4\pi r^2$
$= 4\pi(5 \text{ in.})^2$
$= 100\pi \text{ sq. in.}$
$\approx 100\left(\frac{22}{7}\right) \text{ sq. in.}$
$= 314\frac{2}{7} \text{ sq. in.}$

9. $V = \frac{1}{3}s^2 h$
$= \frac{1}{3}(6 \text{ cm})^2(4 \text{ cm})$
$= 48 \text{ cu. cm}$
$SA = B + \frac{1}{2}pl$
$= (6 \text{ cm})^2 + \frac{1}{2}(24 \text{ cm})(5 \text{ cm})$
$= 36 \text{ sq. cm} + 60 \text{ sq. cm}$
$= 96 \text{ sq. cm}$

11. $V = s^3$
$= \left(1\frac{1}{3} \text{ in.}\right)^3$
$= \frac{64}{27} \text{ cu. in.}$
$= 2\frac{10}{27} \text{ cu. in.}$

13. $SA = 2(2 \text{ ft})(1.4 \text{ ft}) + 2(2 \text{ ft})(3 \text{ ft})$
$+ 2(1.4 \text{ ft})(3 \text{ ft})$
$= 5.6 \text{ sq. ft} + 12 \text{ sq. ft} + 8.4 \text{ sq. ft}$
$= 26 \text{ sq. ft}$

15. $V = \frac{1}{3}s^2 h$
$= \frac{1}{3}(5 \text{ in.})^2(1.3 \text{ in.})$
$= 10\frac{5}{6} \text{ cu. in.}$

17. $V = \frac{1}{3}s^2 h$
$= \frac{1}{3}(12 \text{ cm})^2(20 \text{ cm})$
$= 960 \text{ cu. cm}$

19. $SA = 4\pi r^2 = 4\pi(7 \text{ in.})^2 = 196\pi \text{ sq. in.}$

21. $V = lwh = (2 \text{ ft})\left(2\frac{1}{2} \text{ ft}\right)\left(1\frac{1}{2} \text{ ft}\right) = 7\frac{1}{2} \text{ cu. ft}$

23. $V = \frac{1}{3}\pi r^2 h$

$\approx \frac{1}{3}\left(\frac{22}{7}\right)(2 \text{ cm})^2(3 \text{ cm})$

$= 12\frac{4}{7}$ cu. cm

Appendix D Exercise Set

1. Yes, since every coordinate is between –10 and 10.

3. No, since –11 is less than –10.

5. Any values such that Xmin < –90, Ymin < –80, Xmax > 55, and Ymax > 80.

7. Any values such that Xmin < –11, Ymin < –5, Xmax > 7, and Ymax > 2.

9. Any values such that Xmin < 50, Ymin < –50, Xmax > 200, and Ymax > 200.

11. Xmin = –12 Ymin = –12
Xmax = 12 Ymax = 12
Xscl = 3 Yscl = 3

13. Xmin = –9 Ymin = –12
Xmax = 9 Ymax = 12
Xscl = 1 Yscl = 2

15. Xmin = –10 Ymin = –25
Xmax = 10 Ymax = 25
Xscl = 2 Yscl = 5

17. Xmin = –10 Ymin = –30
Xmax = 10 Ymax = 30
Xscl = 1 Yscl = 3

19. Xmin = –20 Ymin = –30
Xmax = 30 Ymax = 50
Xscl = 5 Yscl = 10

1. Setting A:

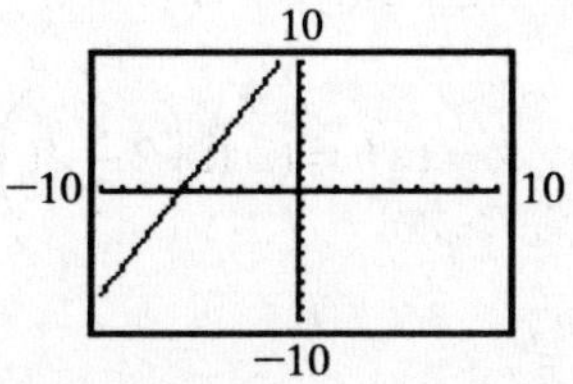

Setting B:

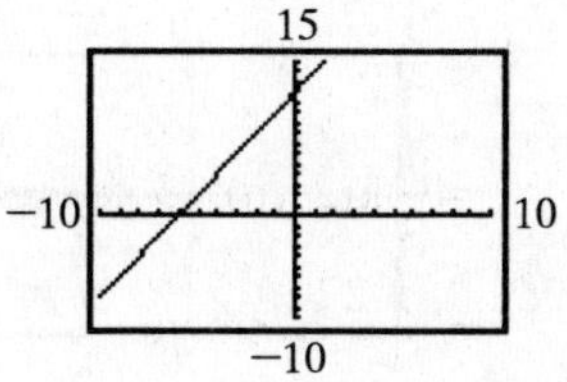

Setting B shows all the intercepts.

3. Setting A:

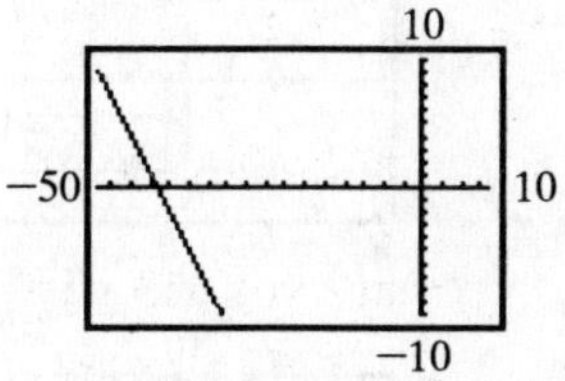

Setting B:

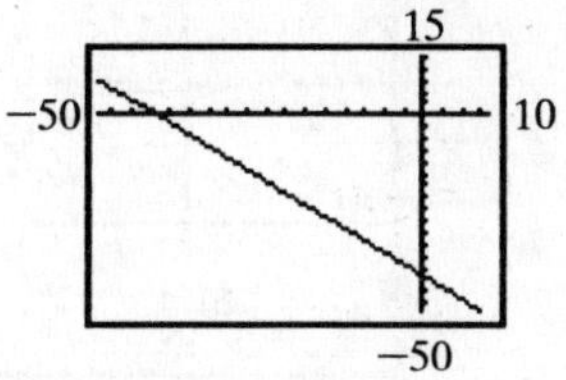

Setting B shows all the intercepts.

5. Setting A:

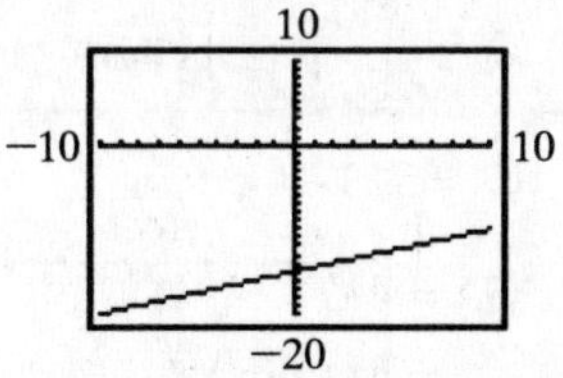

Setting B:

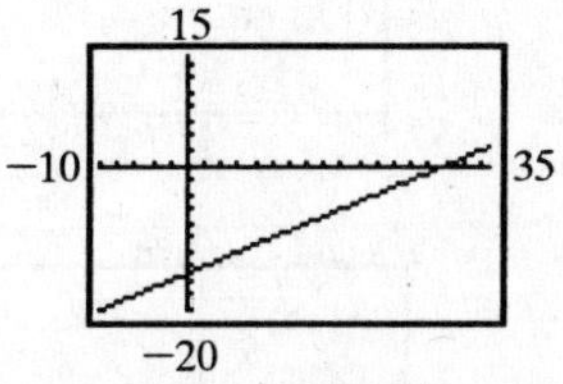

Setting B shows all the intercepts.

7.

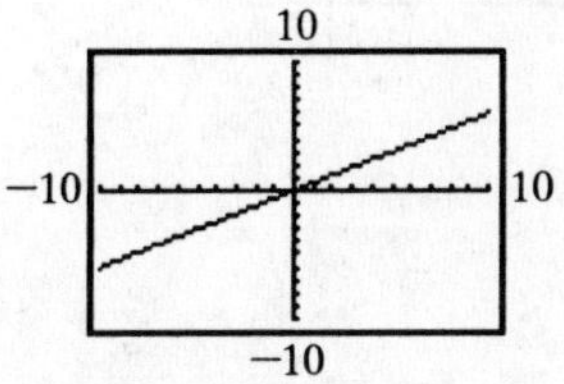

9.

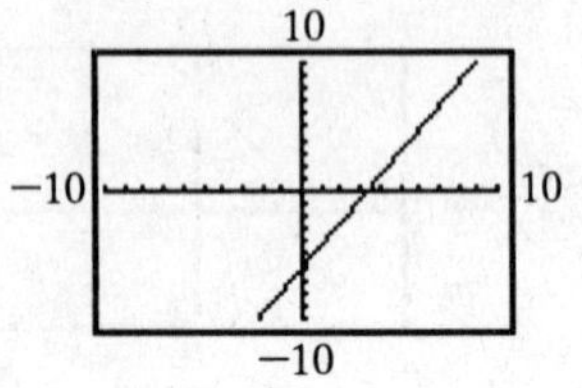

11.

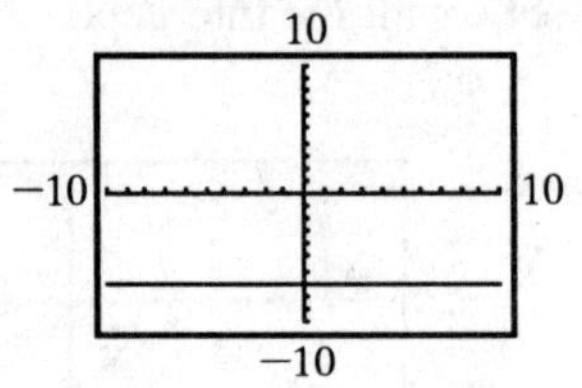

13.

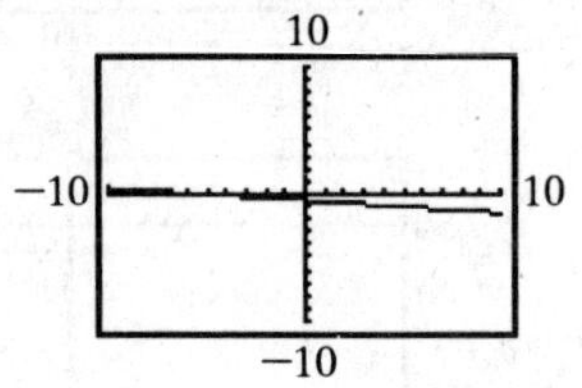

15.

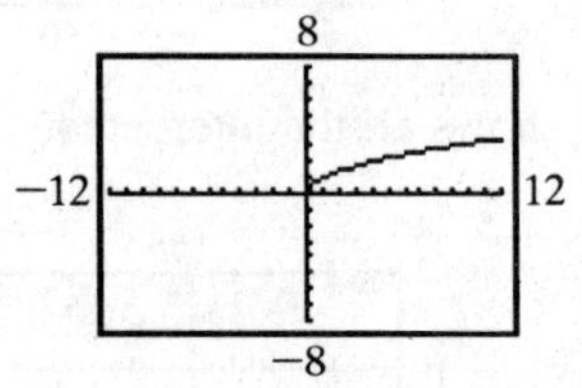

17.

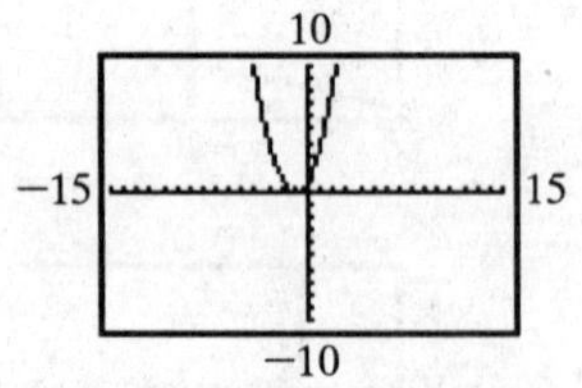

19.

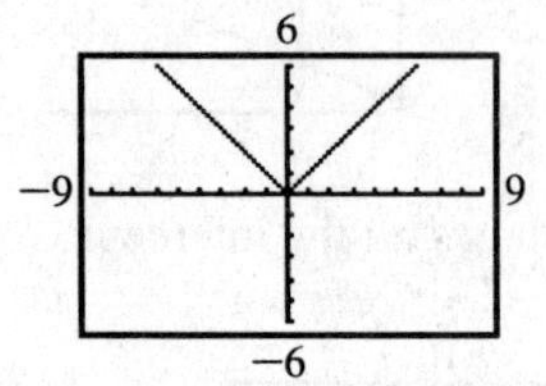

21. Standard window:

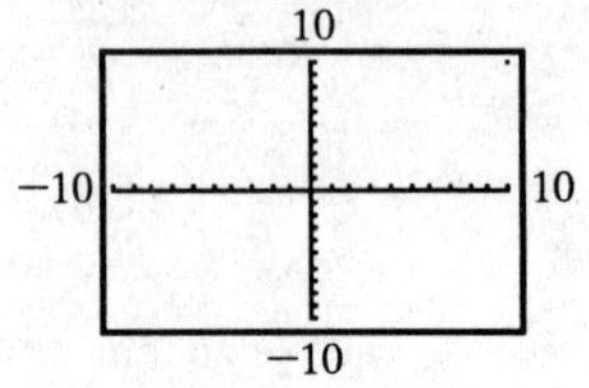

Adjusted window:

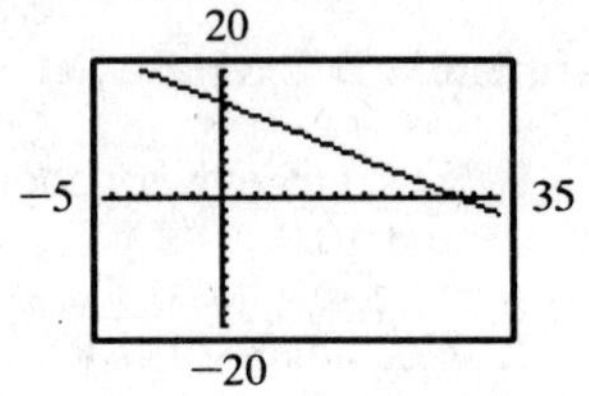